The Plant Family Brassicaceae

Mirza Hasanuzzaman
Editor

The Plant Family Brassicaceae

Biology and Physiological Responses to Environmental Stresses

Springer

Editor
Mirza Hasanuzzaman
Department of Agronomy
Sher-e-Bangla Agricultural University
Dhaka, Bangladesh

ISBN 978-981-15-6344-7 ISBN 978-981-15-6345-4 (eBook)
https://doi.org/10.1007/978-981-15-6345-4

This Springer imprint is published by the registered company Springer Nature Singapore Pte Ltd.
The registered company address is: 152 Beach Road, #21-01/04 Gateway East, Singapore 189721, Singapore

This book is dedicated to

My Sons

Mirza Arian Zaman

and

Mirza Ahyan Zaman

Preface

Brassicaceae (formerly, Cruciferae) is among the largest angiosperm families belonging to the order Brassicales. It consists of 12–15 tribes with 338–360 genera and about 3709 species distributed all around the globe in all continents, excluding Antarctica. Among the plant families, Brassicaceae has particular agri-horticultural importance, and diverse uses apart from the basic needs. The family consists of various essential genera that have different economic as well as agronomic uses in exploring the world of knowledge using them as model plants. Thus it implies and covers all possible sources by which a plant can be of a bliss/benefit to humankind. The Brassicaceae family comprises many different edible species such as fodder, oilseed, vegetables, and condiments. Rapeseed and mustard are the most crucial oil crops of this family, which is ranked as the third most vital oil source and contains about 14% of the world's edible vegetable oil. Members of this family also uptake heavy metals and serve as hyper-accumulators in soil polluted with heavy metals. Some species can hyper-accumulate toxic metals/metalloids.

Considering the economic importance and the unique adaptive mechanisms, further research is still needed to understand the response of these plants toward abiotic stresses. This information needs to be translated into improved elite lines that can contribute to achieving food security. The knowledge of the physiological and molecular mechanisms acting on these plants needs to be further extended. In the last decades, a growing body of research has reported an interesting finding on the physiology and stress responses in both Brassicaceae plants. Many research works have also been done to understand their tolerance mechanisms when facing abiotic stresses. Recent advances and developments in molecular and biotechnological tools have contributed to easing and widening this mission. However, most of these results and reports are published sporadically, and there are no comprehensive books dealing with this vital plant family, and their research progresses related to abiotic stress physiology.

The Plant Family Brassicaceae—Biology and Physiological Responses to Environmental Stresses combines a group of twenty-three chapters written by expert researchers that organizes the most recent information with up-to-date citations, which will provide comprehensive literature of recent advances on

Brassicaceae plant physiology. This book includes several chapters addressing general and unique aspects and questions of Brassicaceae biology and economic importance, as well as several chapters devoted to the Brassicaceae responses and adaptation to environmental stresses as well as their potential to phytoremediation. This book will be an important source of information both for students and researchers working on biology, physiology, environmental interactions, and biotechnology of Fabaceae and Brassicaceae plants.

I would like to give special thanks to the authors for their outstanding and timely work in producing such excellent chapters. I am highly thankful to Dr. Mei Hann Lee (Senior Editor, Life Science) Springer, Japan, for her prompt responses during the acquisition. I am also grateful to Arulmurugan Venkatasalam, Project Coordinator of this book, and all other editorial staff for their precious help in formatting and incorporating editorial changes in the manuscripts. Special thanks to Dr. Sarvajeet Singh Gill (MD University, India) for his critical review of the initial contents of the book and Dr. Md. Mahabub Alam (Sher-e-Bangla Agricultural University, Bangladesh) for his generous help in formatting the manuscripts. I believe that this book is useful for undergraduate and graduate students, teachers, and researchers, particularly from the field of botany, agriculture, plant physiology, agronomy, environmental sciences, plant breeding, biotechnology, and food sciences.

Dhaka, Bangladesh Mirza Hasanuzzaman

Contents

Editors and Contributors

About the Editor

Dr. Mirza Hasanuzzaman is a Professor of Agronomy at Sher-e-Bangla Agricultural University, Dhaka, Bangladesh. He received his PhD on 'Plant Stress Physiology and Antioxidant Metabolism' from the United Graduate School of Agricultural Sciences, Ehime University, Japan, with Japanese Government (MEXT) Scholarship. Later, he completed his postdoctoral research in Center of Molecular Biosciences (COMB), University of the Ryukyus, Okinawa, Japan, with 'Japan Society for the Promotion of Science (JSPS)' postdoctoral fellowship. Subsequently, he joined as Adjunct Senior Researcher at the University of Tasmania with Australian Government's Endeavour Research Fellowship. Professor Hasanuzzaman has been devoting himself to research in the field of Crop Science, especially focused on Environmental Stress Physiology since 2004. Professor Hasanuzzaman published over 100 articles in peer-reviewed journals and books. He has edited 15 books and written 35 book chapters on important aspects of plant physiology, plant stress responses, and environmental problems in relation to plant species. These books were published by internationally renowned publishers. Professor Hasanuzzaman is a research supervisor of undergraduate and graduate students and supervised 20 MS students so far. He is Editor and Reviewer of more than 50 peer-reviewed international journals, and recipient of 'Publons Global Peer Review Award 2017, 2018, and 2019'. Professor

Hasanuzzaman is an active member of about 40 professional societies and acting as Publication Secretary of Bangladesh Society of Agronomy. He has been honored by different authorities due to his outstanding performance in different fields like research and education. He received the World Academy of Science (TWAS) Young Scientist Award 2014. He attended and presented 25 papers and posters in national and international conferences in different countries (USA, UK, Germany, Australia, Japan, Austria, Sweden, Russia, etc.).

Contributors

Sharafat Ali Institute of Crop Science and Zhejiang Key Laboratory of Crop Germplasm, Zhejiang University, Hangzhou, China

Mohammad Israil Ansari Department of Botany, University of Lucknow, Lucknow, India

M. H. M. Borhannuddin Bhuyan Citrus Research Station, Bangladesh Agricultural Research Institute, Jaintapur, Sylhet, Bangladesh

Koushik Chakraborty Division of Crop Physiology & Biochemistry, ICAR National Rice Research Institute, Cuttack, India

Sidra Charagh Centre of Agricultural Biochemistry and Biotechnology (CABB), University of Agriculture, Faisalabad, Pakistan

Heba T. Ebeed Botany and Microbiology Department, Faculty of Science, Damietta University, Damietta, Egypt

Muhammad Bilal Gill International Centre for Environmental Membrane Biology, Foshan University, Foshan, China;
Tasmanian Institute of Agriculture, College of Science and Engineering, University of Tasmania, Hobart, TAS, Australia

Muhammad Bilal Hafeez College of Agronomy, Northwest A&F University, Yangling, China

Mirza Hasanuzzaman Department of Agronomy, Faculty of Agriculture, Sher-e-Bangla Agricultural University, Dhaka, Bangladesh

Nusrat Jabeen Biosaline Laboratory, Department of Botany, University of Karachi, Karachi, Pakistan

Syed Uzma Jalil Amity Institute of Biotechnology, Amity University Uttar Pradesh, Lucknow, India

Rida Javed Centre of Agricultural Biochemistry and Biotechnology (CABB), University of Agriculture, Faisalabad, Pakistan

Wanmei Jin Key Laboratory of Biology and Genetic Improvement of Horticultural Crops (North China), Beijing Academy of Forestry and Pomology Sciences, Beijing, China

K. S. Karthika ICAR-National Bureau of Soil Survey and Land Use Planning, Regional Centre, Bangalore, India

Rao Sohail Ahmad Khan Centre of Agricultural Biochemistry and Biotechnology (CABB), University of Agriculture, Faisalabad, Pakistan

Mojtaba Kordrostami Nuclear Agriculture Research School, Nuclear Science and Technology Research Institute (NSTRI), Karaj, Iran

Pankaj Kumar Science and Engineering Research Board, Department of Science & Technology, Government of India, New Delhi, India;
CSIR-IHBT, Palampur, India

Mohammad Mafakheri Department of Horticultural Sciences, Faculty of Agricultural Sciences, University of Guilan, Rasht, Iran

Jubayer Al Mahmud Department of Agroforestry and Environmental Science, Sher-e-Bangla Agricultural University, Sher-e-Bangla Nagar, Dhaka, Bangladesh

Sayed Mohammad Mohsin Laboratory of Plant Stress Responses, Department of Applied Biological Sciences, Faculty of Agriculture, Kagawa University, Miki-Cho, Kita-Gun, Kagawa, Japan;
Department of Plant Pathology, Sher-e-Bangla Agricultural University, Sher-e-Bangla Nagar, Dhaka, Bangladesh

Sanchita Mondal Department of Agronomy, Bidhan Chandra Krishi Viswavidyalaya, Mohanpur, Nadia, West Bengal, India;
Division of Crop Physiology & Biochemistry, ICAR National Rice Research Institute, Cuttack, India

Subhankar Mondal Division of Crop Physiology & Biochemistry, ICAR National Rice Research Institute, Cuttack, India

Muhammad Mudassir Nazir Institute of Crop Science and Zhejiang Key Laboratory of Crop Germplasm, Zhejiang University, Hangzhou, China

S. Neenu ICAR-Central Plantation Crops Research Institute, Kasaragod, Kerala, India

Larysa V. Nishchenko Institute of Food Biotechnology and Genomics, National Academy of Sciences of Ukraine, Kiev, Ukraine

Prabha Susan Philip ICAR-National Bureau of Soil Survey and Land Use Planning, Regional Centre, Delhi, India

Mrinalini Prasad Faculty of Science, Department of Botany, Dayalbagh Educational Institute (Deemed University), Dayalbagh, Agra-5, India

Rajiv Ranjan Faculty of Science, Department of Botany, Dayalbagh Educational Institute (Deemed University), Dayalbagh, Agra-5, India

Ali Raza Oil Crops Research Institute, Chinese Academy of Agricultural Sciences (CAAS), Wuhan, China

Ali Razzaq Centre of Agricultural Biochemistry and Biotechnology (CABB), University of Agriculture, Faisalabad, Pakistan

Shyamashree Roy Regional Research Station, Old Alluvial Zone, Uttar Banga Krishi Viswavidyalaya, Majhian, Dakshin Dinajpur, West Bengal, India

Nida Sadaqat Institute of Molecular Biology and Biotechnology (IMBB), University of Lahore, Lahore, Pakistan

Pushp Sharma Department of Plant Breeding and Genetics, Punjab Agricultural University, Ludhiana, India

Kanval Shaukat Department of Botany, University of Balochistan, Quetta, Pakistan

Dinesh Kumar Srivastava Department of Biotechnology, Dr. Y. S. Parmar University of Horticulture and Forestry, Solan, Himachal Pradesh, India

Neerja Srivastava Department of Biochemistry, IBSBT, CSJM University, Kanpur, UP, India

Javaria Tabassum China National Rice Research Institute, Chinese Academy of Agricultural Science (CAAS), Hangzhou, Zhejiang, China

Zaid Ulhassan Institute of Crop Science and Zhejiang Key Laboratory of Crop Germplasm, Zhejiang University, Hangzhou, China

Shaheena Umbreen Department of Botany, University of Agriculture, Faisalabad, Pakistan

Noreen Zahra Department of Botany, University of Agriculture, Faisalabad, Pakistan

Muhammad Zeeshan Institute of Crop Science and Zhejiang Key Laboratory of Crop Germplasm, Zhejiang University, Hangzhou, China

The Plant Family Brassicaceae: Introduction, Biology, And Importance

Ali Raza, Muhammad Bilal Hafeez, Noreen Zahra, Kanval Shaukat, Shaheena Umbreen, Javaria Tabassum, Sidra Charagh, Rao Sohail Ahmad Khan, and Mirza Hasanuzzaman

Abstract The model plant family Brassicaceae, also known as Cruciferae, is among the largest angiospermic family belonging to the order Brassicales. The family consists of annuals, biennials as well as herbaceous perennials. The Brassicaceae family comprises many different edible species such as fodder, oilseed, vegetables, and condiments. This family is crucial for vitamins A, B1–2, B6, C, E, K, and minerals such as magnesium, iron, and calcium. Members of the family Brassicaceae produce secondary metabolites that are not only family-specific but also species- and genus-specific. The family consists of various important genera that have diverse economic as well as agronomic use in exploring the world of knowledge using them as model plants. This family has precisely documented advances in the understanding of phylogeny, polyploidy, and genomics in the members of the Brassicaceae family in a very brief and concise review. There are numerous plants having great economic and agronomic importance in family Brassicaceae to scientific

A. Raza (✉)
Oil Crops Research Institute, Chinese Academy of Agricultural Sciences (CAAS), Wuhan 430062, China
e-mail: alirazamughal143@gmail.com

M. B. Hafeez
College of Agronomy, Northwest A&F University, Yangling 712100, China

N. Zahra · S. Umbreen
Department of Botany, University of Agriculture, Faisalabad 38040, Pakistan

K. Shaukat
Department of Botany, University of Balochistan, Quetta, Pakistan

J. Tabassum
China National Rice Research Institute, Chinese Academy of Agricultural Science (CAAS), Hangzhou, Zhejiang, China

S. Charagh · R. S. A. Khan
Centre of Agricultural Biochemistry and Biotechnology (CABB), University of Agriculture, Faisalabad 38040, Pakistan

M. Hasanuzzaman
Department of Agronomy, Faculty of Agriculture, Sher-e Bangla Agricultural University, Dhaka 1207, Bangladesh

M. Hasanuzzaman (ed.), *The Plant Family Brassicaceae*,
https://doi.org/10.1007/978-981-15-6345-4_1

and medicinal significance. Moreover, various species of Brassicaceae are explored for their pharmacological potential.

Keywords *Arabidopsis* · *Brassicas* · Brassicaceae Evolution · Genome duplication · Polyploidy

1 Introduction, Biology, and Economic Importance of Brassicaceae Family

1.1 Introduction

Family Brassicaceae (Cruciferae) is among the largest angiospermic families belonging to the order Brassicales. It consists of 12–15 tribes with 338–360 genera and about 3709 species distributed all around the globe in all continents excluding Antarctica (Al-Shehbaz 1973; Al-Shehbaz and Warwick 2006; Appel and Al-Shehbaz 2003). Besides different geographical distribution of the family from the northern to the southern hemisphere, its distribution has been reported to be limited on mountainous and alpine regions in the tropics. However, *Arabis alpina* has demonstrated its occurrence in both hemispheres from mountains to alpines and Arctic habitats (Koch and Kiefer 2006). According to Price et al. (1994), Central Asia, Mediterranean, and Southwestern regions are enormously dominated by the plants of family Brassicaceae. *Draba* is the largest genera of the family with about 365 species followed by *Erysimum, Lepidium, Cardamine,* and *Alyssum* with 225, 230, 200, and 195 species, respectively (Anjum et al. 2012).

The vast literature has been reported regarding the taxonomic and systematic essay of Brassicaceae plants. von Hayek (1911) was the pioneer to publish about the family; later, various other taxonomists contributed to the ocean of family Brassicaceae. Vaughan and MacLeod (1976) were amid the writers who successfully published a book entitled "The Biology and Chemistry of the Cruciferae" discussing and compiling the vast knowledge of family Brassicaceae. The family is well identified and differentiated from other families by its distinctive flower architecture. Numbers of taxonomists/researchers like (Al-Shehbaz 1984; von Hayek 1911; Schulz 1936) have made an effort to give a natural system of classification to divide the family into different tribes. von Hayek (1911) and Schulz (1936) have divided the family into 10 and 19 tribes, respectively, based on fruit shape, an arrangement of the seed, and embryo. However, Janchen (1942) divided Brassicaceae into 15 tribes. Schranz et al. (2005) precisely documented advances in the understanding of phylogeny, polyploidy, and genomics in the members of the Brassicaceae family in a very brief and concise review.

The plants which revolutionized modern biology are *Arabidopsis* and *Brassica*. Therefore, it is considered that there would be no life for model plants, or we would lack many more crop plants without this particular Brassicaceae family (Franzke et al. 2011). Besides being as a model plant to explore the genomic and molecular arrays, various species of Brassicaceae are used to explore the flowering time and floral architecture (C. *bursa-pastoris*), fruit and flower morphology (*Iberis* spp.), seed physiology, and morphology of fruit, (*Lepidium* spp.) adaptive modification of water-usage gradient (*Cardamine hirsute*), trace metal tolerance and hyper-accumulation (*Arabidopsis lyrata* and *A. suecica*), salt stress (*Eutrema-Thellungiella* spp.) and contact between plant and pathogen (*Arabis alpine*), etc. (Franzke et al. 2011). There are numerous plants having great economic and agronomic importance in family Brassicaceae (Bailey et al. 2007) to scientific (Hall et al. 2002a; Koch and Mummenhoff 2006) and medicinal significance. Moreover, various species of Brassicaceae are explored for their pharmacological potential.

1.2 Biology of Family Brassicaceae

The family consists of annuals, biennials as well as herbaceous perennials. Mostly the plants are herbs with the least percentage (5%) of typically woody like *Faresetia somalensis, Heliophila scandens, Cramboxylon* spp., *Zilla spinosa, Vella* spp., *Parolinia* spp., etc. (Franzke et al. 2011). The main root system of *Brassicaceae* plants is tap root which sometimes may be swollen (when it is the storage organ of the plant) or may be conical shaped, e.g., radish and napiform in a turnip. Leaves of the Brassicaceae show alternate arrangements rarely opposite. In shrubby species of Mediterranean, the leaves are arranged in a terminal rosette. Leaves are exstipulate and often pinnately incised. However, the pungent smell from the crushed leaves of Brassicaceae plants is among the distinctive characteristic attributes of the family. Moreover, leaves are usually covered with hair-like structures called trichomes. The flower morphology is strictly uniform in all members of the family Brassicaceae (exceptions in color do exist). Hence, the family name Cruciferae comes from the appearance of the flora with 4 petals arranged in the form of a cross, usually, hermaphrodite and actinomorphic, rarely zygomorphic. Other floral characters of the family are as follows: (1) sepals—four—arranged in 2 whorls; (2) stamens—six—with the exception in *Lepidium* spp. Having 2–4 and *Megacarpaea polyandra* with 8–24 stamens; (3) ovary—superior; (4) nectary—varies from species to species, both in position and dominant nectar compound (Davis et al. 1998); (5) arrangements of flowers—bracteates racemose—frequently corymb-like at apex; (6) pollination—entomogamy; (7) fruit—vary but typically dehiscent bivalve capsule (silique)—maybe indehiscent or achene, rarely nutlet, schizocarp, or drupe; (8) seed size—vary from minute (*Mancoa mexicana*) to large (*Megacarpaea gigantea*) seeds (Al-Shehbaz 1984, 1986); (9)—mode of propagation—mostly by seeds—very few species show vegetative propagation, e.g., *A. rusticana* (horseradish) and *Neobeckia lacustris* (American watercress).

Mode of photosynthesis in Brassicaceae is typically C3 in the majority of the members (Uprety et al. 1995); however, some spp. of the genus *Moricandia* like *M. arvensis, M. sinaica, M. suffruticosa, M. spinosa, and M. nitens* are C3–C4 intermediates (Apel et al. 1997; Razmjoo et al. 1996; Rylott et al. 1998).

Members of the family Brassicaceae produce secondary metabolites that are not only family-specific but also species- and genus-specific. Secondary metabolites play an essential role in the defense of plants against pathogens, herbivory, and weed eradication (Angelini et al. 1998; Bednarek 2012; Clauss et al. 2006). Secondary metabolites produced by the Brassicaceae include glucosinolates, oils, and seed fatty acids. Fahey et al. (2001) reported that greater than 96 glucosinolates are present in family Brassicaceae, of which many are specific to genera and spp.

1.3 Economic Importance of Family Brassicaceae

The family is enormously blessed with plants that have a wide range of economic importance from food to fodder, medicines to use, as research model plants, to high yielding crops to ornamental plants. Thus it implies and covers all possible sources that a plant can be of a bliss/benefit to humankind. The Brassicaceae family comprises many different edible species such as fodder, oilseed, vegetables, and condiments. This family is crucial for vitamins, A, B1–2, B6, C, E, K, and minerals such as magnesium, iron, and calcium (Wink and Van Wyk 2008). The most common vegetable crops are kohlrabi, broccoli, cabbage, turnip, and cauliflower enriched with dietary fibers (Shankar et al. 2019; Table 1). The family also has ornamental value, and various species are cultivated for its aesthetic value like *Iberis, Lobularia, Cheiranthus*, *Erysimum*, *Hesperis,* etc. It has many plants that are used as dyes, including *Isatis tinctoria,* which is a medicinal plant and also being used as a dye for color indigo (Hamburger 2002). *Amoracia rusticana*, *Cheiranthes cheiri*, kale and collards, *Matthiola incana*, *Raphanus raphanistrum* (Austin 2003), and *Lepidium ruderale* (Buchanan 1995) are also used for various dyes.

Furthermore, members of this family also uptake the heavy metals and serve as hyper-accumulators in soil polluted with heavy metals. Some species can hyper-accumulate cadmium, zinc, and nickel. About 25% of the Brassicaceae members are known to be hyper-accumulators. *Brassica juncea* and *Brassica oleracea* can aggregate and tolerate heavy metals (Devi 2017). Moreover, some members can uptake excess nitrates (Kingsbury 1964). *Erysmium cheiri* seeds are enriched with cardiac glycosides (Wink and Van Wyk 2008). Additionally, Canola and *Brassica napus* are the most important oil crops of this family (Alagoz and Toorchi 2018). After soybean and palm oil, Brassica seed oil is ranked as the third most crucial oil source (Friedt et al. 2018; Raza et al. 2019a) and contains about 14% of the world's edible vegetable oil (Shankar et al. 2019).

Moreover, it also includes some protein and biodiesel crops such as *Brassica carinata, Crambe abyssinica, Eruca vesicaria,* and *Camelina sativa* (Gugel and Falk 2006; Warwick et al. 2007; Warwick and Gugel 2003). E*rysimum* spp., *Sinapis alba, Armoracia rusticana,* and *Brassica juncea* of *Brassicaceae* are used as spices (Gugel and Falk 2006; Warwick et al. 2007). According to Agusdinata et al. (2011) and Shonnard et al. (2010), *Camelina* oil can cut GHG (greenhouse gas emissions) up to 75 percent than that of petroleum-based jet fuel. *Arabidopsis* and *Capsella*

Table 1 A table summarizing some of the most vital members of family Brassicaceae used as vegetables in the human diet Adapted from Šamec and Salopek-Sondi (2019)

Genus, species	Cultivar (group)	Common name	Edible part
B. oleracea	var. *capitata*	Cabbage (white, red, cone, etc.)	Leaves
	var. *acephala*	Kale	Leaves
	var. *viridis*	Collard greens	Leaves
	var. *alboglabra*	Chinese broccoli, kai-lan	Leaves
	var. *gemmifera*	Brussels sprouts	Buds
	var. *gongylodes*	Kohlrabi	Stem
	var. *botrytis*	Cauliflower	Inflorescence
	var. *italic*	Broccoli	Inflorescence
	var. *botrytis*	Romanesco broccoli	Inflorescence
	var. *italica* × *alboglabra*	Broccollini, broccoflower	Inflorescence
B. rapa	ssp. *Rapa*	Turnip	Root
	ssp. *Pekinensis*	Chinese cabbage, napa cabbage	Leaves
	ssp. *narinosa* (or *rosularis*)	Asian greens	Leaves
	ssp. *Chinensis*	Bok choy, pak choy	Leaves
	ssp. *Pervidis*	Komatsuna, Japanese mustard spinach	Leaves
	ssp. *Nipposinica*	Mizuna	Leaves
	ssp. *Parachinensis*	Rapini, broccoli rabe	Leaves, stem, flower buds
B. napus	var. *napobrassica*	Rutabaga (swede)	Root
	var. *pabularia*	Siberian kale	Leaves
	var. *oleifera*	Rapeseed	Seeds
B. juncea	var. *rugosa* (or *integrifolia*)	Mustard greens	Leaves
		Brown Indian mustard	Seeds
B. nigra		Black mustard	Seeds
B. carinata		Ethiopian mustard	Leaves, seeds
Brassica hirta		White mustard	Seeds
Amoracia rusticana		Horseradish	Root
Barbarea verna		Land cress	Leaves

(continued)

Table 1 (continued)

Genus, species	Cultivar (group)	Common name	Edible part
Eruca vesicaria		Arugula (rocket)	Leaves, stems
Lepidium sativum		Garden cress	Leaves, stems
Nasturtium officinale		Watercress	Leaves, stems
Raphanus sativus		Radish	Root
	var. *longipinnatus*	Daikon	Root
Wasabia japonica		Wasabi	Root

are well-known model plants (Soengas Fernández et al. 2011). For bio-fumigation processes, *Brassica* is extensively used as a biocontrol agent (Ahuja et al. 2011).

A wide range of the plant species of the family is blessed to have the medicinal potential, and the local people of the area use it as medicine to treat various diseases (Table 2). Brassicaceae members like cauliflower, Brussels sprout, kale, green mustard, cabbage, and broccoli decrease the risk of several types of cancer (Wang et al. 2004). The anticarcinogenic potential of these vegetables is attributed to glucosinolates, iso-thiocyanate, indole antioxidants, and other phytoalexins (Zukalová and Vasak 2002). An important component of Brassicaceae vegetables known as indole-3-carbinol acts as an anticarcinogenic compound through different hormonal and metabolic changes (Hanf and Gonder 2005), and also reduces the tumor especially in reproductive organs (Staub et al. 2002). The extracts from *Brassica* species proved helpful to control the hepatoma cells (Steinkellner et al. 2001). *Brassica* plants contain amino acids such as L-glutamine, L-histidine, L-alanine, L-aspartic acid, L-valine, L-tryptophan, L-threonine, L-phenylalanine, and L-methionine; moreover, different phytochemicals such as indole phytoalexins (N-Methoxyspirobrassinol methyl ether, 1-methoxyspirobrassinol, 1-methoxyspirobrassinin, camalexin, brassilexin, spirobrassinin, brassinin, glucosinolates (neoglucobrassicin, gluconapin, glucobrassicin, gluconasturtiin, glucobrassicanapin, glucoalyssin, glucoraphanin, and glucoiberin), and phenolics (kamepferol, anthocyanins, *p*-coumaric, quercetin, sinapic acid, ferulic, caffeic, chlorogenic, neochlorogenic, hydroxybenzoic, isoferuloylcholine, and feruloyl) have antioxidant, cardiovascular protective, and anticarcinogenic properties (Jahangir et al. 2009).

2 General Description of the Brassicaceae Family

The family consists of various important genera have diverse economic as well as agronomic use in exploring the world of knowledge using them as model plants.

Table 2 The summary of the members of family Brassicaceae used for their pharmacological potential

S. No	Botanical name	Common name	Part used	Pharmacological activity	References
1	*Brassica rupestris* L.	Brown mustard	Whole plant	Anticancer and antioxidant activity	Amri (2014)
2	*Brassica tournefortii Gouan*	Asian mustard	Whole plant	Anticancer and antioxidant activity	Amri (2014)
3	*Brassica napus* L.	Rapeseed	Whole plant	Anticancer, anti-goat, antioxidant, analgesic, diuretic, and anti-catarrhal activity, anti-scurvy and anti-inflammatory of bladder	Amri (2014); Rahman et al. (2018); Saeidnia and Gohari (2012)
4	*Brassica* L. *var. perviridis*	Mustard spinach	Whole plant	Anticancer and antioxidant activity	Amri (2014)
5	*Brassica rapa* L. *var. rapifera*	Turnips	Whole plant	Anticancer and antioxidant activity	Amri (2014)
6	*Brassica rapa* L. *var. chinensis*	Bok choy	Whole plant	Anticancer and antioxidant activity	Amri (2014)
7	*Brassica rapa* L. *var.pekinensis*	Chinese cabbage	Whole plant	Anticancer and antioxidant activity	Amri (2014)
8	*Brassica oleracea*	Cauliflower	Leaves	Antibacterial activity	Prasad (2014)
9	*Brassica carinata* A. Braun.	Ethiopian or Abyssinian mustard	Whole plant	Utilized in bio-fumigant, to repress pathogens and soil-borne pests Potential as new edible oil/protein crops	Rahman et al. (2018); Warwick (2011)
10	*Malcolmia africana* (L.) *R.Br.*	African mustard	Spices	Antioxidant activity and phenol content	Owis (2015)
11	*Brassica oleracea* L. *var. capitata*	Cabbage	Raw and processed cabbage	Antioxidant, anti-inflammatory, and antibacterial properties	Prasad (2014); Rokayya et al. (2013)

(continued)

Table 2 (continued)

S. No	Botanical name	Common name	Part used	Pharmacological activity	References
12	*Brassica rapa* L.	Broccoli raab	Vegetables	Anticancer, diuretic, analgesic, anti-gout potential, aphrodisiac activity, anti-inflammatory and anthelmintic activity Improves insulin resistance in type 2 diabetic patients	Amri (2014); Gul et al. (2013); Ravikumar (2015)
13	*Brassica oleracea var. capitata f. rubra*	Red cabbage	Leaves	It is competent in stomach diseases, diabetic, anti-cancer activity, antioxidant, hypolipidemic, anti-hyperglycemic, and cardio-protective	Chauhan et al. (2016)
14	*Brassica juncea* L.	Mustard	Seed Leaves Dried leaf and flower Total plant	Anticancer, anti-diabetic, diuretic, analgesic, emetic activity, and rubefacient Antihyperglycemic, antioxidant, antiatherogenic, antifungal activity, allergenicity and antitumor activity The antiatherogenic effect,antioxidant, and fungicidal activity. Used to treat dengue fever, splenic disorders, and dyspepsia	Amri (2014); Kumar et al. (2011); Rahman et al. (2018)
15	*Brassica campestris* Linn.	Sarson	Seed oil	Used to remove dandruff from hair, used to make an ointment for curing itching and skin infections, hair tonic and laxative	Bhatnagar et al. (2016)

(continued)

Table 2 (continued)

S. No	Botanical name	Common name	Part used	Pharmacological activity	References
16	*Raphanus sativus*	Radish	Leaves and seeds Underground parts	Antimicrobial activity Treatment of intestinal parasites, asthma and chest pain	Parvaiz et al. (2013); Prasad (2014); Singh and Singh (2013)
17	*Lepidium sativum* L.	Garden cress	Seeds	Used in treating dysentery and bone fracture, healing in human and migraine; used as a saag and anthelmintic, anti-arthritic activity valuable in the cure of cough asthma, with expectoration, poultices for sprains, leucorrhoea, leprosy, skin infections, diarrhea, dysentery, dyspepsia, lumbago, splenomegaly scurvy, and seminal weakness	Dutta et al. (2014); El Sayed and Aly (2014); Khan et al. (2013); Raval (2016)
18	*Nasturtium Officinale R.Br.*	Water cress	Vegetative shoot	It has many uses in the form of salad, pot herb, and its saag have been used as appetizers which help as a diuretic, antibiotic, and also relieves from problems of the chest	Khan et al. (2013)

(continued)

Besides, the family is enormously blessed with plants that possess pharmaceutical properties and are used as medicine to treat various diseases. Some of the important genera of the family are briefly discussed as follows.

Table 2 (continued)

S. No	Botanical name	Common name	Part used	Pharmacological activity	References
19	*Sisymbrium Irio* L.	London rocket	Leaves and seeds Whole plant	Antipyretic, anti-vomiting, diarrhea, and cough. A tonic herb with a mustard-like aroma. It has laxative, diuretic, and expectorant effects, and benefits the digestion, internally used for bronchitis, coughs, laryngitis, and bronchial catarrh.	Gulshan et al. (2012); Khan et al. (2013)
20	*Brassica nigra* L.	Black mustard	Seeds	Anticancer, anti-diabetic, diuretic, activity in cold and influenza, stimulant activity, emetic, antibacterial, anti-catarrhal activity, and laxative. Anti-spasmodic, aphrodisiac activity, appetizing, digestive and aperitif activity Used against alopecia, Anti-dandruff activity, Used in neuralgia Used for common cold and arthritis	Amri (2014); Obi et al. (2009); Rahman et al. (2018); Tomar and Shrivastava (2014)
21	*Armoracia rusticana*	Horseradish	Roots and leaves	Anti-lipase and antioxidant activity	Calabrone et al. (2015)
22	*Calepina irregularis*	White ball mustard	Mustard extracts	Analgesic activity	Rahman et al. (2018)
23	*Lepidium meyenii*	Maca	Leaves	Restores the levels of testosterone in the males Hypoglycaemic and anti-obesity effect	Pachiappan et al. (2017)

(continued)

Table 2 (continued)

S. No	Botanical name	Common name	Part used	Pharmacological activity	References
24	*Capsella bursa-pastoris Moench*	Bambaisa	Whole plant Seeds	Used in fertility regulation Astringent	Sher et al. (2011)
25	*Cheirantus cheiri* L.	Wallflower	Flower and seed	Diuretic, aphrodisiac, jaundice, tumors	Erum et al. (2017)
26	*Aethionema oppositifolium*	Opposite-leaf candytuft	Spices	Antioxidant activity	Owis (2015)
27	*Cardamine Hirsuta* Linn	Hairy bittercress	Whole plant	Used for indigestion	Nag (2013)
28	*Rorippa indica* (Linn.) Hiern	Indian yellow cress	Whole plant	Used for a toothache, sore throat, rheumatic arthritis, hepatitis, abdominal and blood disorders	Nag (2013)
29	*Descurainia sophia* (L.) Webb	Skhabootay	Flowers and leaves Seeds	Antiscorbutic. Used as Cardiotonic, demulcent, diuretic, expectorant, febrifuge, and laxative	Haq (2012); Sher et al. (2011)
30	*Nasturtium officinale* R.Br.	Talmeera	Shoot Leaves	Purgative, emetic Effective in cough	Nag (2013); Shah et al. (2015)
31	*Alliaria petiolata (M.Bieb.)*	Garlic mustard	Leaves	Antimicrobial activities It acts as an antiseptic to ulcers and also plays a role in wound healing, especially complications caused in bronchi, could also be used disinfectant.	Cipollini and Cipollini (2016); Rahman et al. (2018)
32	*Raphanus sativus var. longipinnatus*	White radius	Leaves	Antimicrobial activities	Singh et al. (2018)

(continued)

Table 2 (continued)

S. No	Botanical name	Common name	Part used	Pharmacological activity	References
33	*Brassica alba Boiss.*	White or yellow mustard	Seedling leaves Seeds	It is helpful for blood purification. It has strong sanitizer properties and is utilized to preserve food, used for the cure of cold, cough, and sore throats	Rahman et al. (2018)
34	*Sisymbrium officinale L. Scop.*	English watercress	Whole plant	It is used to treat sore throat and as an expectorant to cure common asthma and cold	Rahman et al. (2018)
35	*Neslia paniculata*	Ball mustard	Whole plant	It is a forage crop used as food for cattle, skin disorders.	Rahman et al. (2018)
36	*Sisymbrium erysimoides*	Smooth mustard	Whole plant	Used to treat bronchitis and has anti-inflammatory activity	Rahman et al. (2018)
37	*Sisymbrium orientale*	Asian hedge mustard	Whole plant	Used to treat bronchitis	Rahman et al. (2018)
38	*Sisymbrium officinale*	Hedge mustard	Whole plant	Used to treat bronchitis and snake-bite antidote Anti-asthmatic, anti-spasmodic, and anti- addiction activity	Rahman et al. (2018)
39	*Camelina sativa*	Camelina	Whole plant	Likely to be used as food, fodder, and biofuel. Also used in paints, cosmetics, and dyes. Likely to be used as edible oil and protein crops	Warwick (2011)
40	*Crambe abyssinica*	Crambe	Whole plant	Use as erucamide. Likely to be used as edible oil and protein crops	Warwick (2011)

(continued)

Table 2 (continued)

S. No	Botanical name	Common name	Part used	Pharmacological activity	References
41	*E. vesicaria*	Rocket	Seed oil	It is utilized as a lubricator, hair oil, and for massage and relaxation. It used as a blistering agent and pickling. Likely to be used as edible oil and protein crops	Warwick (2011)
42	*Aethionema grandiflorum*	Persian stonecress	Whole plant	Used to treat meningitis, bacterial infections, and typhoid	Parvaiz et al. (2013)
43	*Erysimum kotschyan*	Wallflower	Spices	Antioxidant activity	Owis (2015)
44	*Sterigmostemum incanum*	–	Spices	Antioxidant activity	Owis (2015)
45	*Aethionema dumanii*	–	Spices	Antioxidant activity	Owis (2015)
46	*Brassica hirta*	White mustard	Extracts	Anti-microbial activity	Rahman et al. (2018)
47	*Eruca sativa*	Rocket salad	Leaves and whole plant	It posses astringent properties, digestive, emollient, depurative, laxative, rubefacient, tonic properties, stomachic, anti-inflammatory, antibacterial activity, diuretic. hair tonic, antidandruff, and antioxidant activity antidiabetic activity	Barlas et al. (2011); Koubaa et al. (2015); Salma et al. (2018); Sher et al. (2011)

2.1 *Arabidopsis*

Arabidopsis is one of the most important genera of the family Brassicaceae on the basis of its morphology and molecular phylogeny various aspects of its taxonomical history has been explored and reported (Al-Shehbaz and O'Kane 2002; Al-Shehbaz et al. 1999). Genus Arabidopsis has about 9 species and 8 subspecies (Koch et al. 2008). Arabidopsis has no well-documented economic importance, but its small genome and short life cycle paved the way to better study, understand, and explore the development, physiology, and gene function of plants through genetic approaches, thus making it a highly valuable and cherished model plant.

2.1.1 *Arabidopsis thaliana* (L.) Heynh

It is the key species of the genus *Arabidopsis.* It is a small herbaceous annual plant about 10-40 cm in height, cosmopolitan in its distribution being native to Europe and Middle Asian mountain ranges. More than 750 accessions of the species have been collected around the globe. It has a short life cycle of about 5–7 weeks with a flowering period from April to early June and produces about 10,000 seeds/plants. Moreover, the species is self-fertile and self-compatible. It has a small genome that has been completely sequenced (Initiative 2000). Thus, it is the 1st plant and the 3rd multicellular organism after *Caenorhabditis elegans* (Blaxter 1998), and *Drosophila melanogaster* (Adams et al. 2000) to have their genome completely sequenced. Through various available mutants, *A. thaliana* can be transformed through vector *Agrobacterium tumefaciens* (http://www.geochembio.com/biology/organisms/arabidopsis/) (Al-Shehbaz and O'Kane 2002; Koornneef et al. 2004). Hence, *Arabidopsis thaliana* helped to explore every aspect of plant biology and provide insight to increase crop production through various molecular and biotechnological approaches (Buell and Last 2010). Moreover, it that has been explored to understand plant responses under different stress conditions (Dinneny et al. 2008; Kreps et al. 2002; Nakashima et al. 2009; Roosens et al. 2008) and variation in adaptive attributes (Alonso-Blanco and Koornneef 2000; Koornneef et al. 2004; Vaughn et al. 2007; Weigel 2012). Besides other species of *Arabidopsis* like *A. lyrata, A. halleri, A. arenosa,* etc. are explored for the variation in complex traits like self-incompatibility, heavy metal tolerance and hybridization-polyploidization (Clauss and Koch 2006; Roosens et al. 2008).

2.1.2 *Arabidopsis Halleri*

It is previously named as *Cardaminopsis halleri* (L.) Hayek. However, later, on the basis of molecular evidence, it was placed in genus *Arabidopsis.* It is reported to occur on heavy metal contaminated soils and is among the important hyperaccumulators of zinc (Zn) and cadmium (Cd) (Bert et al. 2000; Küpper et al. 2000). Interestingly it is closely related to *Arabidopsis lyrata,* which is infertile and a non-accumulator of heavy metals (Macnair et al. 1999).

2.1.3 *Arabidopsis Suecica*

It is an allotetraploid species of genus that is formed by the hybridization of *Arabidopsis thaliana* with *Arabidopsis arenosa* (Koornneef et al. 2004; Pontes et al. 2004).

2.2 *Brassicas*

Brassica is the essential genera of the family, consisting of 37 species (Gómez-Campo 1980). Most of the crop plants are included in this genus which has edible buds, flowers, leaves, roots, seeds, and stems. Most of the wild species are used for hybrid seed development. These hybrid seeds and modified species are therefore of prime importance in developing resistant varieties against various pests (OECD 2016; Rakow 2004).

2.2.1 *Brassica Oleracea* L

Brassica oleracea shows a very distinguish phenotype with 9 chromosomes. It is widely distributed in Asia, Turkey, and Southern Greece. It has fleshy leaves. The plant has a woody appearance and is perennial (OECD 2016; Rakow 2004). There are six groups of the *Brassica oleracea,* e.g., var. *acephala,* var. *botrytis,* var. *fruticosa,* var. *gongylodes, and* var. *italic.* These are used as vegetables, oil, and as a fodder crop (Rakow 2004).

2.2.2 *Brassica Rapa* L

Brassica rapa possesses 10 chromosomes and sometimes it is called *B. campestris. B. rapa* has been reported to be of Mediterranean origin excluding the coastal areas. These areas are mountainous and have a very cold temperature; therefore, the species show preferred rapid growth to low-temperature conditions. It is spread throughout Germany, Eastern Europe, while introduced in the Western Asian countries. In south Asian countries, such as India, also it is cultivated (OECD 2016). It is cultivated as an oilseed crop in China, Canada, and Finland. Seven different varietal groups of *B. rapa* are now being reported to be separate/independent species which was var. *campestris,* var. *chinensis,* var. *japonica,* var. *narinosa,* var. *parachinensis,* var. *pekinensis,* and var. *rapa.* The need for the separation was due to diversity and variability in the said species due to their isolated evolution. For example, var. *pekinensis* (Chinese cabbage) is adapted to a more relaxed environment and grown as an oilseed crop. The var. *chinensis* and var. *japonica* both are leaf vegetables in China and Japan, respectively. The most primitive leaf vegetable is var. *campestris*. Var. *rapa* has been reported to be cultivated all over the globe and utilized enormously as a fodder crop, var. *narinosa* shows high tolerance and resistance to chilling stress (Rakow 2004).

2.2.3 *Brassica Nigra* (L.) Koch

B. nigra is species with 8 chromosomes. It is a noxious weed found in the Mediterranean region, Ethiopia, and Turkey. The plant tends to have a height of almost 2 m and does not need vernalization treatment for flowering (OECD 2016; Rakow 2004).

2.2.4 Crossed and Cultivated Species

Brassica carinata, Brassica juncea, and Brassica napus are the species which are cultivated and formed after interspecific crosses between the wild species of *Brassica* to acquire the desirable character which may enhance its economic value.

Brassica Carinata A. Braun

It is the species which comprises 17 chromosomes and is formed by crossing *B. nigra* which is a species of 8 chromosomes with *B. oleracea* with 9 chromosomes (OECD 2016; Rakow 2004). Cultivation of this species is restricted to Ethiopia. *Brassica carinata* is a slow-growing plant which might be an inherited character from *B. oleracea*. Its seeds are rich in oil content (Cardone et al. 2003; Rakow 2004).

Brassica Juncea (L.) Czern and Coss

It is derived from an interspecific cross of *Brassica nigra* (n = 9) and *Brassica rapa* ($n = 10$). Ukraine, India, Italy, and China are the main regions of its diverse population. The difference between Chinese and Indian spp. is made in terms of their seed color as Chinese seeds are yellow while the Indian ones are brown and larger (OECD 2016; Rakow 2004). Mostly it is grown for oilseeds and is commonly called Indian mustard. Other than the oils, their leaf and root (turnip) are also of great economic importance. The species are grown widely in China, but it cannot be considered native to the region because none of its wild relatives exist in the area. In some Western countries such as Canada, *Brassica juncea* is cultivated as an important condiment production crop (Rakow 2004).

Brassica Napus L

It is an amphidiploid species and consists of 19 chromosomes, which has been obtained as a result of an interspecific cross between *B. rapa* (n = 9) and *B. oleracea* (n = 9) (Parkin et al. 1995; Rakow 2004). Wild species other than the cultivated forms of *B. napus* could be found on beaches, coastal areas of Britain, Gotland, Netherlands, and Sweden (Rakow 2004). In many countries, *Brassica napus* are the most productive cultivated species grown as oilseeds. The potential of yield is pretty much high as the photosynthetic activity is high concerning per unit of the leaf which is further positive and proportional to several chloroplast present per unit area of a leaf (Liu et al. 2009).

2.3 Etymology

The name Brassicaceae is derived from new Latin from *Brassica* meaning the type genus (Webster 1889) and -aceae, standard name representing family name used in the modern taxonomy system. *Brassica* (genus name) is derived from Classical Latin meaning cabbage and other crucifers. Brassicaceae's old name was Cruciferae referring to "cross-bearing", representing four petals of mustard flowers that resemble the Cross symbol. Cruciferae is among the eight plant families whose names does not derive from any genus nor have the suffix -aceae, but it is an authorized alternative name (ICBN: International Code of Botanical Nomenclature).

3 Taxonomy and the Phylogenetic Relationship Between the Brassicaceae Family

3.1 Taxonomy

Family Brassicaceae is among the important angiospermic families with diverse economic and scientific importance, e.g., *Arabidopsis thaliana and Brassica,* etc. It has been delineated into 321 genera, 51 tribes, and 3660 species (Al-Shehbaz 2012; Al-Shehbaz 2014; German and Friesen 2014). It is a remarkable model family to explore the phylogenetic relationship and evolution through the course of time due to its diverse morphological, ecological, and genetic diversity. In its early evolutionary history, Brassicaceae experienced whole-genome duplications (Edger et al. 2015; Kagale et al. 2014), providing excellent opportunities to explore these critical evolutionary processes (Kagale et al. 2014). In short, the Brassicaceae family has many benefits to be a model family for relative and evolutionary studies (Couvreur et al. 2009; Franzke et al. 2011; Mitchell-Olds 2001).

Carl Linnaeus was the pioneer taxonomist who classified the Brassicaceae family in 1753, naming them "Klass Tetradynamia". After that, Rendle positioned this family in the order Rhoedales; additionally, Bentham and Hooker (1862-1883) placed it in cohort Parietales, now known as Violates. Consequently, Hutchinson (1948-1964) explored a close relationship of Brassicaceae with family Papaveraceae (Hutchinson 1934), while Judd et al. (1994) suggested to position Capparaceae in Brassicaceae. Early DNA linkage showed Capparaceae a paraphyletic genera closet to Brassicaceae and was assigned into Cleomaceae (Hall et al. 2002b). According to Lysak and Koch (2011) and Schmidt and Bancroft (2011), Brassicaceae and Cleomaceae diverged from each other over 41 million years ago. Hall et al. (2002b) placed all these three families in one order Brassicales or Capparales. Angiosperm Phylogeny Group II system (APG II) combined Brassicaceae and Cleomaceae into one group. However, other classification systems still identify the Capparaceae family with some limitations, by placing Cleome and their closer lineages in the Brassicaceae family or segregate it into the Cleomaceae family. As the consensus arises, the APG II system might be changed with the passage of time.

The taxonomy of the family Brassicaceae has been overwhelmed by a morphological feature which is used to define tribes and genera and their phylogenetic relation and evolution through the course of time. Some of the important taxonomical attributes of family Brassicaceae are discussed in relation to their taxonomic and phylogenetic relationships.

3.1.1 Leaf

In the Brassicaceae family, leaf shapes vary from simple to compound (Piazza et al. 2010), *Arabidopsis thaliana* shows small serration around the leaf margin. However, *Arabidopsis lyrata* develops lobes, and the pattern also varies along the leaf margin. In *A. lyrata*, lyrate leaves are present in which lobes are present in concentrated form at the basal end with an enlarged lobe at the terminal position. In *Capsella rubella,* pinnatifid lobes are present in natural form with deep dissection near the margins (Sicard et al. 2014). Moreover, leaf taxonomy has been delineated about phylogeny by Streubel et al. (2018), and Nikolov et al. (2019); Fig. 1). For more details about the leaf taxonomy, see the article published by Nikolov et al. (2019); Streubel et al. (2018).

3.1.2 Flower

Brassicaceae represents ideal taxon and provides a detailed view of how monosymmetric flowers evolved from polysymmetric ancestors. A characteristic trait specific to this family is the formation of polysymmetric corolla, which have four petals equal in size and shape (Franzke et al. 2011). Among the 338 genera, only six genera have monosymmetric corolla with unequal petals. Monosymmetric is most apparent in genus *Iberis*. Monosymmetric corolla is also present in *Streptanthus* (Rollins 1993),

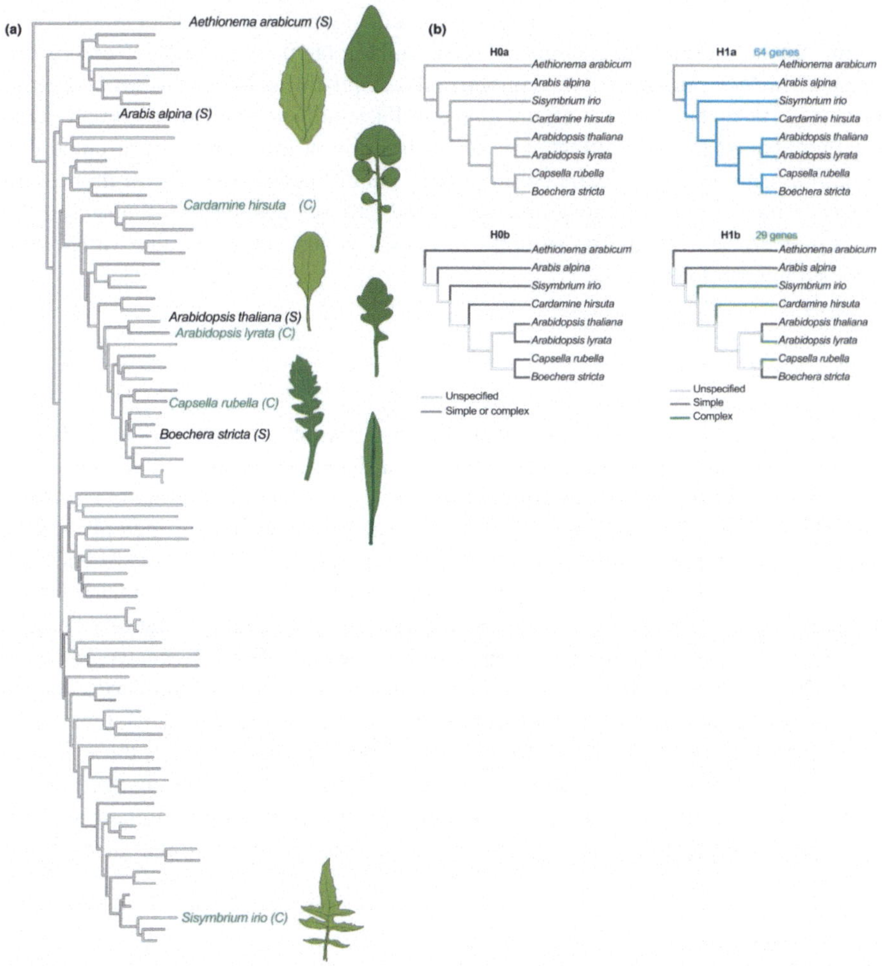

Fig. 1 Leaf-shaped diversity in Brassicaceae. **a** Species included in the transcriptome analyses and their respective leaf morphologies, simple (black) or complex (including compound; green). **b** Models compared to identify genes exhibiting a significant directional change in expression from neutral expectations in different taxa. Models H0a and H1a are compared to test for shifts in gene expression in the core Brassicaceae (64 genes, blue). Models H0b and H1b are compared to test for shifts in gene expression in species with complex leaves (29 genes, green) compared to species with simple leaves (dark gray); the character state along the internal branches is unspecified (light gray) Taken from Nikolov et al. (2019)

Notoceras, and *Ionopsidium* (Schulz, 1936). *Calepina irregularis* consists of only one species that is monotypic (Al-Shehbaz et al. 2006); however, genus *Teesdalia* has polysymmetric *T. coronopifolia* and monosymmetric *T. nudicaulis* species (Tutin 1993).

Recently, Busch et al. (2011) analyzed that monosymmetric character is controlled by the transcriptional factor, and it is strongly related to the CYC2 clade. For instance, in *Iberis amara* monosymmetric development is due to the expression of the CYC2 clade gene; it develops unequal corolla, smaller adaxial petals as compared to larger abaxial petals. Phylogenetic reconstruction of this family represented that *Teesdalia, Calepina,* and *Iberis* belong to one major lineage, specific for monosymmetric. In these genera, monosymmetric is more pronounced in *Iberis* than *Teesdalia* and *Calepina,* as well as the strength of CYC2 gene expression also varies with monosymmetric formation.

3.1.3 Fruit

Organization and composition of fruit architectural tissues are highly conserved in the Brassicaceae family; great diversity in their fruit shape exists that includes spherical, cylindrical, or heart-shaped structures (Eldridge et al. 2016; Łangowski et al. 2016). *Capsella* and *Arabidopsis* fruits have lateral valves attached with medial replum, and the fruit is topped with stigmatic tissues and style (Eldridge 2014; Roeder and Yanofsky 2006). Although some species like *Neslia paniculata* have spherical fruit (Francis and Warwick 2003), *Lepidium campestre* has a medial orientation (*Alyssum maritime*) and lateerally flattened round fruit (Bowman 2006). Recently, Eldridge et al. (2016) evaluated the growth pattern conserved in two different fruit-shaped species, near cylindrical (*Arabidopsis*) to heart-shaped (*Capsella*). Different clonal and growth analysis showed anisotropic shape growth in approximately three phases. They observed tissue deformation as it grows. Anisotropic shape change activity is provided by a fruitful gene (valve identity gene). Comparative mutant analysis revealed that this gene regulates shape change at post-fertilization in both species.

3.1.4 Trichomes

Dennert (1884) was the first author who paid attention to different trichomes present in the family Brassicaceae. He categorized these hairs as simple, stellate, and dichotomous. To Dennert's (1884) work, Prantl and Engler (1887) added to the knowledge regarding the family, they used trichomes for the classification of Brassicaceae and divided it into four tribes. Inamdar and Rao (1983) explored the significance of trichomes and identified differences in 35 species of Brassicaceae family. Metcalfe and Chalk (1950) reported that its trichomes are always unicellular rarely glandular. They recorded glandular trichomes in *Sterigma, Parrya, Matthiola, Hesperis, Bunias, Chorispora,* and *Anchonium* species. Theobald et al. (1979) noted 2–5 armed, scaled, stellate, dendritic, and unicellular trichomes in this family. Rollins and Banerjee (1979) investigated trichome differences in *Physaria* and *Lesquerella.* Abdel (2005) documented that multicellular trichomes of Brassicaceae family are glandular in shape, while unicellular trichomes of this family could be stellate, hooked, clavate simple glandular, Y-shaped (3-4 fids), vasculate, or appressed medifixed trefixed.

3.1.5 Seed

Many taxonomists and systematists believed that data related to seed structure is essential for angiosperms classification. In the Brassicaceae family, seed shape separates closely associated genera's such as *Diceratella* (suborbicular to elliptical), *Matthiola* (oblong seeds), and *Moretti* (ovoid seeds) (Jonsell 1979; Khalik and Van der Maesen 2002). Moreover, variation and changes in the seed coat were reported by Khalik and Van der Maesen (2002) in various Brassicaceae species. For more detail about the variation in seed morphology, size, and structure, see the article published by Kasem et al. (2011); Khalik and Van der Maesen (2002); Tantawy et al. (2004). According to Kasem et al. (2011), kidney-shaped seeds are present in *Farsetia aegyptiaca*, circular in *Diplotaxis erucoides*, *Cardamine hirsute*, *Descurainia*, and Coronopus; globose seeds are present in three species of *Sinapis* and *Thlaspi perfoliatum* and in six species of genus of *Brassica*, while circular in *Farsetia aegyptiaca*, and obliquely ovate in *Raphanus sativus*. Most of the seeds have no wings, but in *Farsetia aegyptia* seed wings are present at the margin or apical position. Wings' presence or absence may be used as a diagnostic character for classification (Khalik and Van der Maesen 2002). Largest seed size was noted in *Erysimum cheiri* and *Farsetia aegyptia*, and relatively large size was measured in 5 taxa, i.e., *Eruca sativa, Schouwia thebaica, Sinapis alba, Matthiola longipetala,* and *Brassica oleracea,* while *the* smallest seed taxa are *Moricandia sinaica* and *Diplotaxis tenifolia* (Tantawy et al. 2004). Furthermore, seed color varies from shiny brown, orange-brown, pale brown-black, dark brown, and brown (Barthlott 1984).

3.1.6 Pollen

Chiguriaeva (1973) was the first to examine and document pollen morphology about taxonomy as a specific character of family Brassicaceae. Moreover, Jonsell (1986) used the monographic study to examine pollen variations in this family. Brassicaceae family is stenophyllous, and pollen grains vary from prolate to spherical and rarely spheroidal and oblate. Pollens are usually 3 colpate, but 4–8 colpate was also reported in *Erysimum stocksiana* (Erdtman 1952). However, Appel and Al-Shehbaz (2003) also examined tricolpate pollen in this family. They also reported 10 colpate pollen in a few genera of family Brassicaceae and examined punctate tectum and sparsely spinulose pollen in *Heliophila* genus. Based on exine thickness, Erdtam et al. (1963) divided the Brassicaceae family into two groups. Khalik (2002) classified this family on the basis of lumina size and Webb and Webb (1983) divided the family on the basis of reticulate pollen. Keshavarzi et al. (2012) and Karaismailoğlu and Erol (2019) documented variations in *Clypeola* and *Thlaspi* species, respectively. For more knowledge about the pollen documentation in Brassicaceae, read the following articles published by Fuentes-Soriano and Al-Shehbaz (2013); Karaismailoğlu and Erol (2019); Khan (2004); Maciejewska-Rutkowska et al. (2007); Perveen et al. (2004).

3.2 Phylogenetic Relationship Within the Family

Old classification was based on morphological differences due to convergent evaluation, while the phylogenetic relationship was inconsistent. However, many efforts have been made even through molecular phylogenetic techniques, but the controversy among the Brassicaceae family is not settled yet. From previous studies, it was quite clear that Aethionema is the sister taxon of the remainders of family Brassicaceae (Al-Shehbaz 2012). Based on 2014 study analysis, the relationship among 39 tribes is represented in the phylogenetic tree (Edger et al. 2014).

The phylogenetic study of *Brassica* species has not been well understood because of the traditional use of traits like fruit length, orientation of radical, fruit compression and dehiscence, a number of rows of seeds in locule, trichome type, and features of nectarines to define different tribes of the family Brassicaceae (Warwick et al. 2009). Different workers attempted to organize a genus of the family Brassicaceae into tribes. Subsequently, based on later research reports by Al-Shehbaz et al. (2006), each tribe has been defined morphologically and cytologically. Altogether, depending on Internal transcribed spacer of nuclear ribosomal DNA (ITS)-based phylogenetic studies, and the other recent techniques, 33 tribes are recognized in the family on the basis of molecular phylogeny and re-evaluation of morphological features (Warwick et al. 2009).

Additionally, molecular phylogenetic analysis at the tribal- and family level was reported by Franzke et al. (2011). Although many molecular and morphological conflicts are solved by the discovery of tribes and genera, the progress on generic sampling and the number of genes is still lacking. To assist the use of Brassicaceae as an evolutionary model family, a phylogeny with well-organized and supported relationships is essential. Many previous phylogenetic studies provide the current perceptive of Brassicaceae phylogeny (Al-Shehbaz et al. 2006; Hohmann et al. 2015). Notably, Al-Shehbaz et al. (2006) conducted an experiment on the chloroplast *ndh*F gene from 113 species of 101 genera and discovered 3 major lineages (i.e., Lineages I – III and LI – III) with low to moderate supports. The group of additional genera associated with LII with low support was referred to as Expanded Lineage II (EII) by (Franzke et al. 2011). In another study conducted by Busch et al. (2011), it was analyzed that monosymmetric character is controlled by a transcriptional factor, and it is strongly related to CYC2 clades through various molecular and biotech approaches, e.g., RT-PCR and in situ hybridization, and detected variations among different crucifer lineages.

Recently, Huang et al. (2016) conducted phylogenetic analyses of family Brassicaceae and documented three major lineages (I, II, and III) (Figs. 2 and 3). They represented phylogeny with 6 clades using nuclear markers from newly sequenced transcriptomes of 32 species and additional taxa of about 55 Brassicaceae species covering 29 tribes of 51 reported tribes. Many phylogenetic markers have been tested as effective markers such as low or single-copy nuclear genes (Yang et al. 2015; Zeng et al. 2014; Zhang et al. 2012; Zimmer and Wen 2013; Zimmer and Wen 2015). However, Brassicaceae still requires strong support and the origin of novel traits

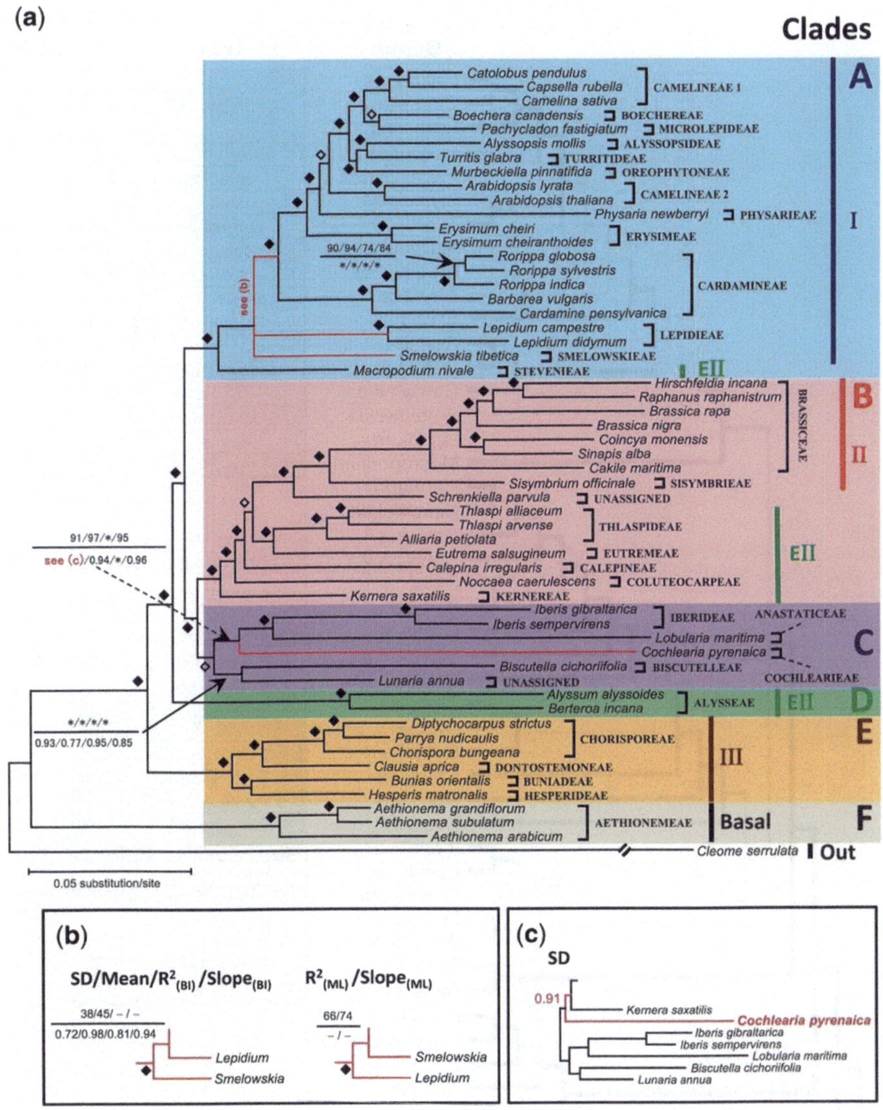

Fig. 2 Phylogenetic tree from ML and BI analyses of concatenations of gene sets after sequence bias tests. Solid and open diamonds indicate maximum (BS = 100 or PP = 1) and support of _90 (BS) or 0.9 (PP) in all tests, respectively. Values correspond to the statistical support of the bootstrap value of ML analysis (BS; above) and Bayesian posterior probabilities (PP; below) in the sequence of the tests. Stars (*) indicate maximum support in at least one of the analyses. A complete phylogeny consistent in all tests is in **a** relationships between Smelowskia and Lepidium and their supports in each test are shown in **b** position of Cochlearia and its support in BI analysis of data set screened by the standard deviation of LB score (SD) are shown in **c** I, II, III, and EII indicate Lineage I, II, III, and Expanded Lineage II, respectively *Adapted* from Huang et al. (2016)

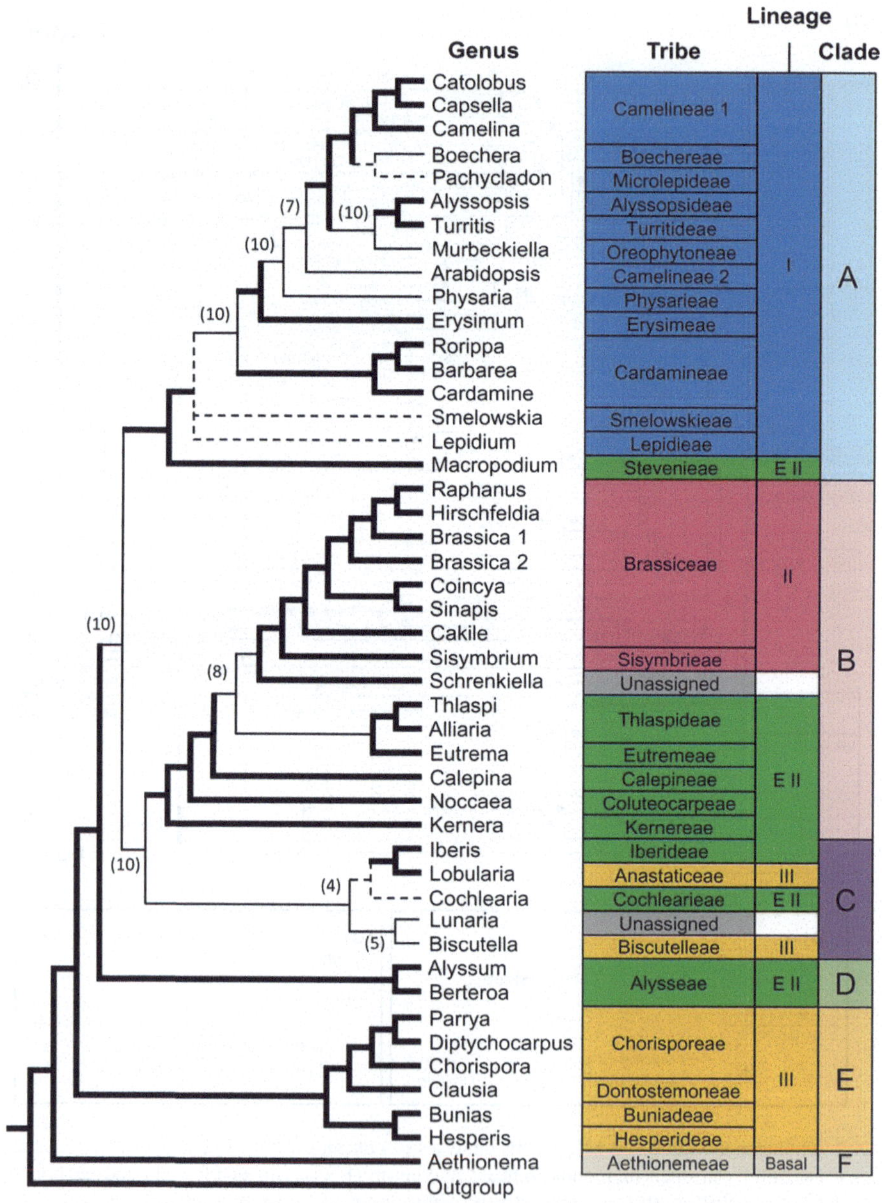

Fig. 3 A summarized Brassicaceae phylogeny. This model is proposed according to their results from concatenation (using 113 genes as markers), and coalescent methods. Structures consistent in all reference trees are drawn in thick (all showing maximum support) or thin solid lines (with values showing the number of results with maximum support). Dashed lines indicate uncertainties *Adapted* from Huang et al. (2016)

(Edger et al. 2015). Even at the family level, multigenes to resolve a consistent and highly confident phylogenetic backbone are rare (Karl and Koch 2013).

4 The Model Plant Family Brassicaceae for Studying Genome Evolution Followed by Polyploidy

Polyploidy or whole-genome duplication plays a vital role in innovation, and it is widely distributed across the plant's life, although its long-term evolutionary importance is still arguable. While polyploidy expanded much slower than diploidy, polyploidy has fast extinction and biological evolution rate. It is also evidenced that polyploidy has played a vital part in the shaping of the current-day differences among species' richness within the Brassicaceae family, along with the future intimations in explaining the pattern of genetic diversity beyond the plant's life (Román-Palacios et al. 2019; Van de Peer et al. 2017). The primary task of genome duplication in plant evolutions is still arguable. However, there are some successful proofs about the long-term operation of polyploidy which may be linked with environmental changes, various harsh conditions, and time to extinction (Yao et al. 2019).

All of the six species under the genus *Brassica* basically have A, B, and C genomes which are further split into different sub-genomes (Ar/An/Aj, Cn/Co/Cc, and Bj/Bc/Bn), which possess plenty of variations at both genome and sub-genome level (Chalhoub et al. 2014; Cheng et al. 2016; Liu et al. 2014; Navabi et al. 2013; Parkin et al. 2014; Pires et al. 2004; Yang et al. 2016; Zou et al. 2016). The *Brassica* species plays an eminent role in interpreting the evolutionary history of genomes followed by polyploidy, particularly at α whole-genome duplication events. The term "diploidization," (genome has stability among genome formation and genetic material), a basic phase of polyploidy in *Brassica* species, proves helpful to determine their effects for a series of time spans. The core Brassicales, including the Brassicaceae, are well known for tetraploidy, (the α and β whole-genome duplications) (Hall et al. 2004). The whole-genome duplication by causes a delay in Cleomaceae–Brassicaceae split (Schranz and Mitchell-Olds 2006). The duplication of the genome happened to be very supportive in conserving specific nucleotide identity for cross-hybridization to work on physical mapping possible in the *Brassica* species through gene duplication (O'neill and Bancroft 2000). Earlier than event, and events also have some influence on genome analyses, so there is a need to have comparative studies among all events on the nucleotide level. However, the young genome of an allopolyploid provides its quick genome evolution contrived by interspecific and intraspecific crosses (Hu et al. 2019).

The two diploid sub-genomes, *B. oleracea* (C sub-genome) and *B. rapa* (A sub-genome) known for historical polyploids, still have an ancestry of triplicated genome structure, combined to form *B. napus* (oilseed rape). However, there is no evidence found for wild oilseed rape species; its genomic evolution is merely processed artificially by human needs and agricultural systems (Chalhoub et al. 2014). The former series of genome doubling of *Brassica* are basics of early cytogenetic work. One of the significant features of the Brassicaceae family under ploidy levels is Mesopolyploidy, an ancestral hexaploid structure that is entirely identified by molecular genetics (Lysak et al. 2007).

The genome of *B. napus* endures to change for generations, causing homoeologous non-reciprocal transpositions, ultimately altering the qualitative gene expression (Gaeta et al. 2007). In *Brassica*, ploidy manipulation concurrent with the modern molecular cytogenetics techniques is helpful to expose the stages of meiosis and genetic interactions (Grandont et al. 2014; Rousseau-Gueutin et al. 2017; Suay et al. 2014; Szadkowski et al. 2010; Xiong et al. 2011). Undoubtedly, it seems impossible without next-generation sequencing technologies to resequence the genome of *B. napus* on a large scale, with alterations in germplasm pool generating polyploid evolution and breeding up till the present scenario (Chalhoub et al. 2014; Rousseau-Gueutin et al. 2017; Schmutzer et al. 2015). Genes responsible for meiosis are found as a single copy after polyploidization (Lloyd et al. 2014); however, this gene copy number is also persuaded by chromosomal aberrations among different genomes of *B. napus* (Schiessl et al. 2014). Genomic selection by humans also had an impact on evolution (Chalhoub et al. 2014).

Arabidopsis thaliana, a model plant in Kingdom Plantae, along with its corresponding species *Arabidopsis suecica* and *Arabidopsis arenosa* made great intuitions during evolution accompanied by meiosis (Henry et al. 2014; Yant et al. 2013). Using Arabidopsis as a model plant, the genomic evolution of the genus *Brassica* (particularly *B. napus*) has been genome-wide explored via polyploidy and interspecific hybridization along with expressed gene copies comparative to *Arabidopsis* genes through human selection (Parkin et al. 2010). A comparative study of *Arabidopsis* and *Brassicas* was highly appreciable for some important traits, i.e., flowering stage and susceptibility/resistance against stress (Lagercrantz et al. 1996; Mun et al. 2009; Osborn et al. 1997; Robert et al. 1998; Sillito et al. 2000; Tadege et al. 2001) to identify candidate genes by sequencing data considering *Arabidopsis* as a reference genome (Initiative 2000). The positive and close genomic relationships between *Brassica* and *Arabidopsis* anticipates comparing similarities or differences by their sequences (Parkin et al. 2005), and genetic interpretation (Wang et al. 2011).

Genome sequence of *B. rapa* (Wang et al. 2011), followed by two *B. oleracea* and *B. napus* genomes, was identified as a prestigious genome (Chalhoub et al. 2014; Liu et al. 2014; Parkin et al. 2014). Most chromosomal aberrations such as translocations, deletions, duplications, and inversions are now frequently found than before in a young neopolyploid, *B. napus* (Osborn et al. 2003). Higher similarity among subgenomes of *Brassica* and non-homologous chromosome pairing are major reasons for these aberrations (Grandont et al. 2014). Fragments of a genome from ancestral paleopolyploidy events was entirely mapped by considering *B. rapa* as a reference genome among all genomes of *Brassica* (Cheng et al. 2013). Study of homoeologous non-reciprocal translocations in resequenced *B. napus* genomes opens their ways to identifying variation responsible for a de novo polyploid, evolution, and breeding under agricultural selection (Chalhoub et al. 2014; Snowdon et al. 2015).

Homologous and duplicated sequences result in a complex polyploid genome of crops, i.e., *B. napus*, demanding a hard trial of genome assembly (Edwards et al. 2013; Feuillet et al. 2011). The crop phenotype, as well as genome evolution, are critically influenced by transposable elements (Fu et al. 2016). Evolution of newly arising genomes, their organization, and polyploidy cannot be studied deeply due to these transposable elements; however, third-generation sequencing will overcome this problem (Fu et al. 2016; Yao et al. 2019). Alterations in similar genetic constitution onto an ordered pan-transcriptome can be determined in diploid variants and progenies by transcriptome sequencing, via mapping of *B. napus* transcripts under all genomes (He et al. 2015). Modifications in the early stages of allotetraploid *B. napus* by making the sub-genomic variation among inter- and intraspecies through the firm selection and recombination, deliver an understanding to swift genome evolution due to whole genome reorganization (Hu et al. 2019).

5 The Model Plant Family Brassicaceae for Studying Morphological and Environmental Factors

5.1 Abiotic Stresses

Different environmental stress factors affect the growth and productivity of plant crops (Hasanuzzaman et al. 2012; Hasanuzzaman and Fujita 2012; Raza et al. 2019b, 2019c, 2019d). Drought is among the major environmental stress factors affecting the yield of *Brassica* crops. Phytohormones play a vital part in the development of drought tolerance in plants. Researchers have found that drought resistance in Brassicaceae mainly depends on the increased endogenous volume of auxin, indole-3-acetic acid (IAA), cytokinins (CKs), abscisic acid (ABA), as well as salicylic acid (SA), whereas it depends on decreased active brassinosteroids (BRs) levels (Pavlović et al. 2018; Raza et al. 2019c, 2020). It has been studied that insects help to alleviate plant drought stress without increasing susceptibility of plants to herbivores (above ground). Dung beetles increased the retention of soil water by almost 10% and improved plant growth around 280% that are affected by drought, removing the water stress in plants. Plants were grown in drought conditions with dung beetles and without dung beetles. Dung beetles increased the nitrogen content 2.7% fold and carbon content greater than 3 fold. Dung beetles increased the root carbon concentrations in both ambients as well as raised precipitation levels. Dung beetles improved the growth and nitrogen content of plants suffering from drought (Johnson et al. 2016). Another soil-borne plant pathogen (fungal type) Verticillium longisporum triggered the VASCULAR-RELATED NAC DOMAIN7-Dependent de Novo Xylem Formation and increased the drought resistance in *Brassica* crops (Reusche et al. 2012).

Soil salinity is another important abiotic stress factor that lowered the productivity of cultivated crops. In plants, salinity resulted in oxidative stress, which enhanced

ROS production (Hasanuzzaman et al. 2013a, b). Ascorbic acid behaves as an effective antioxidant which is important for the development of salinity tolerance. In plants, antioxidant responses to salinity differ significantly in *B. rapa* plants. Scientists have studied the Ascorbic acid effects on *B. rapa* plants grown in a saline environment. Ascorbic acid was applied in different levels (50 mM, 100 mM), salinity considerably lowered the *B. rapa* plants' growth. 100 mM was found effective in lowering the salinity stress, which determined that salinity tolerance improved by Ascorbic acid (Mittal et al. 2018). One member of transcription factor family APETALA2/ERF is ethylene response factor (ERFs) have a major role in growth, development, and different abiotic stress responses in plants. A transcription factor ERF named as *BoERF1* from broccoli and its expression were induced through NaCl and *S. sclerotiorum.* Transgenic plants overexpressed with *BoERF1* showed better seed germination as well as low chlorophyll loss in salt-stressed situations than wild type plants, and in transgenic lines, chlorophyll content increased nearly two-fold. Transgenic broccoli overexpressed with *BoERF1* increased the activities of peroxidase (POD), catalase (CAT), superoxide dismutase (SOD), and free proline while it decreased the malondialdehyde (MDA), hydrogen peroxide (H_2O_2), and relative electrical conductivity (ERC) (Jiang et al. 2019). Species-specific phenolic acids are found in Brassicaceae plants and some are involved in the formation of salt-stress tolerance. Varieties that are salt-tolerant contained phenolic acid in higher levels and metabolic stress has minimum effects in salt-stressed conditions (Linić et al. 2019).

Temperature is a major abiotic stress factor affecting growth as well as plant development. Plant as sessile organisms needs to sense momentary variations plus seasonal temperature changes and reply to these alterations through adjusting plant's biology to fit in resultant temperature environments (Kaur et al. 2018; Raza et al. 2020). Brassica crops, *B. oleracea* having biennial life cycle, face seasonal alterations in temperature. Increased temperatures (heat and cold) affect yield as well as photosynthetic activity of kale and cabbage. Under extreme temperatures, the plant increases the antioxidant defense system, and reactive oxygen species increased. In chloroplast, ROS accumulation decreased the chlorophyll content and stimulated photoinhibition which resulted in low CO2 fixation and low dry weight. Low temperature raised the antioxidant defense as well as lowered chlorophyll content than heat situations (Soengas et al. 2018). Currently, it has been found that 28-Homobrassinolide (28-homoBL) has potential as an effective stress-protector particularly in *B. juncea* plants from oxidative damage in temperature stress (Sirhindi et al. 2017).

Waterlogging, as well as extreme temperature, reduced the production of *Brassica oleracea* (broccoli), mainly leaf chlorophyll and water content decreased while hydrogen peroxide levels increased in cells (Pucciariello et al. 2012). Waterlogging and high temperature provoke the expression of different heat shock factors plus proteins and protect plants from environmental stresses. Different gene sets activating under abiotic stresses in plants have been identified. Under waterlogging and high temperature, 31 differentially expressed proteins (DEPs) are identified from TSS-AVRDC and B-75 broccoli cultivars; heat tolerant and heat susceptible through MALDI-TOF MS and PF2D (two-dimensional liquid phase fractionation). Further genes of stress-responsive Rubisco plus their transcripts levels are identified under stressed conditions. In TSS-AVRDC-2, increased Rubisco protein levels resulted in improved efficiency of carbon fixation and provide maximum energy for stress tolerance in waterlogging at 40°C (Lin et al. 2015). Brassica crops (*B. rapa*) have a variety of aquaporin proteins (AQP) that play a role as transporters for nutrients and water molecules via membranes. Aquaporin proteins are involved in water uptake contributing leaves' water-homeostasis by roots (Kayum et al. 2017). A form of hemoglobin present in Vitreoscilla (gram-negative bacteria) called Vitreoscilla hemoglobin (VHb) is overexpressed in *Arabidopsis* plants and improved resistance to waterlogging stress (Du et al. 2016). In *B. napus* plants, ethanol content is simultaneous with waterlogging resistance and is controlled by citrate synthase and lactate dehydrogenase (Xu et al. 2016). An ethylene-responsive gene (*BnERF2.4*) isolated from *B. napus* improves submergence resistance in *Arabidopsis* plants by alleviating oxidative damage (Lv et al. 2016).

Drought and salinity cause osmotic stress in plants which lower crop productivity throughout the world. Osmotic and drought-stress tolerance improvement without affecting productivity has been an excellent task in brassica crop improvement. A protein of HD-Zip family class IV, AtEDT1/HDG11 (*Arabidopsis ENHANCED DROUGHT TOLERANCE1/HOMEODOMAIN GLABROUS11*), has been identified to enhanced *Arabidopsis* drought resistance. *AtEDT1/HDG11* improve osmotic and drought tolerance in Chinese kale. Proline content and activities of ROS-scavenging enzymes increased in transgenic kale upon osmotic and drought stress than wild-type kale. Plants overexpressed with *AtEDT1/HDG11* result in ABA hypersensitivity and ABA, auxin, and stress-associated gene expressions also changed in both normal and stressed environments. It was concluded that *AtEDT1/HDG11* in Chinese kale enhanced abiotic stress tolerance via ABA and auxin-mediated signaling response (Zhu et al. 2016). In *Arabidopsis thaliana*, grapevine *VvABF2* gene overexpression could improve osmotic-stress resistance but is completely needed for abscisic acid signaling (Liu et al. 2019). The heterologous expression of *Arabidopsis rty* enhances drought tolerance in transgenic strawberry by modulating the ABA content, and by increasing the antioxidant enzyme activites (Li et al. 2020).

Heavy metal stress considered as one of the critical environmental stress limits the growth and development of oilseed crops. Antioxidant defense system increases plant resistance toward toxicity induced by heavy metals via up-regulating diverse antioxidant enzymes plus non-enzymatic antioxidants which are tangled in free radicals-detoxification in stressed plants (Farooq et al. 2016; Nahar et al. 2016, 2017). Numerous species of the Brassicaceae family are resistant toward heavy metals and have the aptitude to hyper-accumulate Ni, Se, Zn, Cd, and Sr metals (Warwick 2011). Low molecular-weight Metallothionein (MT) is rich in cysteine amino acids; metal-binding proteins have a vital role in defending plants from heavy-metal toxicity. To date, 60 Metallothionein proteins have been identified from Arabidopsis while five from Brassica species participating in detoxification of heavy metals. In *Brassica napus*, *BnaMT3C* have a major role As^{3+} stress response (Pan et al. 2018). In *Arabidopsis*, *bHLH104* is found to be tangled in positive regulation of Cd tolerance. bHLH104 identified as a positive regulator for Cd tolerance recommends that this may be used in new approaches to engineer crops for Cd-stress adaptation (Yao et al. 2018).

5.2 *Biotic Stresses*

Plants constantly defend themselves against a variety of different pathogenic organisms. Plants have two kinds of immune systems, used to protect themselves from various kinds of pathogens. A pattern recognition receptor (PRR) on the plant cell surface is used to recognize the attack of pathogens first. PRR generates signaling that transfers to the nucleus activating defense-related genes. Plant cell produces callose, ROS, and tyloses in a considerable amount that deposits inside the cell wall to stop pathogen invasion further. Second, plant cells secrete ROS in high amounts which also activates pathogenesis-related proteins that control pathogen (Kayum et al. 2016a).

In plants, biotic stresses result in alterations in metabolite biosynthesis (Singh et al. 2015). Cruciferous plants have developed a precise and elaborate defense system like glucosinolates (GSLs). These glucosinolates, as well as their biologically active hydrolysis products, are used in plant defense systems in various economically important Brassicaceae members such as mustard seed, broccoli, and cabbage (Evivie et al. 2019). A wide variety of pathogens and pests affects carrot production. Bacterial pathogens caused almost five carrot diseases, 36 from oomycete and fungal pathogens, 2 and 13 diseases by phytoplasmas and viruses, respectively. New strategies have developed to determine resistant material and speed up the traditional breeding through molecular breeding tools like SSR markers (simple sequence repeat) as well as carrot genome deep-coverage libraries. These approaches will increase efforts to determine and breed carrot resistance to pathogens and pests (du Toit et al. 2019).

To date, various genes have been documented expressed differentially to biotic stresses in the Brassicaceae family. Six *BrWRKY* genes (*BrWRKY4, 65, 72, 97, 133,* and *141*) in *Brassica rapa* displayed considerably high expression about 8-, 6-, 6-, 3-, and 5-fold upon infection with *F. oxysporum.* After infection with P. carotovorum, *BrWRKY141* displayed an enhanced expression of 180-fold (Kayum et al. 2015a). *BrAL2, 3, 7, 9, 13, 14,* plus *15* have shown expression to biotic as well as abiotic stresses, and *BrAL2, 3, 4, 7, 9, 10, 13, 14,* then *15* displayed improved expression of several-fold upon infection with *F. oxysporum* (Kayum et al. 2015b). Brassica chitinase-like proteins (*BrCLP1, BrCLP2*, and *BrCLP3*) have been identified from *Brassica rapa* and showed high expression in Chinese cabbage when infected with *Pectobacterium carotovorum*, suggesting their role in resistance development in plants to biotic stresses (Ahmed et al. 2012). In *Brassica oleracea,* two genes (*BoAL8* and *BoAL12*); Alfin-like (AL-transcription factor) have exhibited significant expression upon treatment with *P. carotovorum,* and these *BoAL* genes could be used in the improvement of stress-tolerant cultivars Brassica (Kayum et al.

2016b). Thaumatin-like proteins (TLPs) have a central role in resistance enhancement to stresses. From *B. oleracea*, 12 TLPs have been identified, then analyzed, and compared to their sequences with published TLPs (pathogenesis-related). Three out of 12 TLPs have differentially expressed in cabbage plants when inoculated with *Pectobacterium carotovorum* (Ahmed et al. 2013). Gene family *17 TIR-NBS-LRR* have determined from RNA-sequencing data (80-upregulated). 8 genes have displayed major responses toward biotic as well as abiotic stresses. Three genes in *B. oleracea* (*Bol007132, Bol016084,* and *Bol030522*) have shown resistance to *Fusarium oxysporum* than other lines (Kim et al. 2015).

5.3 Morphological Parameters

Seeds and fruits (diaspores) have evolved special germination as well as dispersal abilities to support plants' angiosperm life cycle in a changing environment (Willis et al. 2014). Climatic change triggers the shifts in species range then elimination, a worldwide plant diversity threat (Walck et al. 2011). Dispersed fruit or seed in different plant species is the single way in plants life cycle when it travels and has potential to move a complete plant, population as well as the species (Kesseler and Stuppy 2006). In plants seeds and fruits, inherent biomechanical, morphological shapes plus structures are required for suitable germination time and effective dispersal (Arshad et al. 2019).

Brassicaceae has four main fruit morphological traits parallel to dispersion ability: pericarp, joints, indehiscence, and abscission zones features (Hall et al. 2011). Indehiscence means dehiscence zones absence, which inhibits the seeds to be released after fruit ripening (Hall et al. 2006). Joints isolate the fruit to the distal and proximal segment, while pericarp states thin and smooth Brassicaceous pericarp modification like barbs, wings, and hooks (Willis et al. 2014). Pericarp features enhance dispersal ability. In *Brassica villosa* and *B. incana*, trichomes (hairs) on juvenile leaves as well as on cotyledons play a role in hindering insects. In *Brassica oleracea*, increased waxiness, as well as thickness provide insect and drought tolerance. Also, pod-shattering resistance has been described in *B. tournefortii, B. hilarionis, B. juncea, B. macrocarpa*, and *C. orientalis* (Warwick 2011).

Family Brassicaceae have highly conserved floral structure; four yellow petals are organized in shape like a cross; however, ther color exceptions like purple, white, and pink. Species (*C. orientalis*) with small or no petals are of major agronomic interest; stamens are in six numbers while in *Lepidium,* 2-4 stamens, and in *Megacarpaea polyandra* 8-24 stamens. Also, they have varied nectary kinds but a few species have a median, lateral, sucrose, and others are glucose dominant (Warwick 2011). The Brassicaceae family is stenopalynous, with tricolpate pollen grains having reticulate exine. Many authors have studied the Brassicaceae pollen morphology. Eight *Alyssum L.* species were studied by Bolurian and they found three different pollen kinds: subplate, prolate-spheroidal, and prolate having reticulate exine embellishment (Keshavarzi et al. 2012).

Primarily, the family Brassicaceae is herbaceous while the woody species are just 5% and have limited growth form alterations. These have large shrubs such as *Parolinia* (Canary Islands), *Foleyola* (N, Africa), small trees like *Farsetia somalensis* (NE Africa), and woody plus vine climbers like *Cremolobus peruvianus* (Peru), *Lepidium scandens* (Australia) *Heliopila scandens* (South Africa), and Subshrubs (*Vella* spp.). Fruits are above the ground, but in some species fruits are buried inside the ground and independently evolve in peanut-like manner including *Geococcus pusillus* (Australia), *Morisia monanthos* (Corsica and Sardinia), and *Cardamine chenopodiifolia* (South America) (Warwick 2011).

Acknowledgements The authors are grateful to all the researchers whose excellent work has been cited and has helped us to prepare this chapter.

References

Abdel KK (2005) Morphological studies on trichomes of Brassicaceae in Egypt and taxonomic significance. Acta Bot Croat 64:57–73

Adams MD, Celniker SE, Holt RA, Evans CA, Gocayne JD, Amanatides PG, Scherer SE, Li PW, Hoskins RA, Galle RF, George RA (2000) The genome sequence of Drosophila melanogaster. Science 287:2185–2195

Agusdinata DB, Zhao F, Ileleji K, DeLaurentis D (2011) Life cycle assessment of potential biojet fuel production in the United States. Environ Sci Technol 45:9133–9143

Ahmed NU, Park JI, Jung HJ, Chung MY, Cho YG, Nou IS (2013) Characterization of thaumatin-like gene family and identification of Pectobacterium carotovorum subsp. *Carotovorum* inducible genes in *Brassica oleracea*. Plant Breed Biotechnol 1:111–121

Ahmed NU, Park JI, Seo M-S, Kumar TS, Lee IH, Park BS, Nou IS (2012) Identification and expression analysis of chitinase genes related to biotic stress resistance in *Brassica*. Mol Biol Rep 39:3649–3657

Ahuja I, Rohloff J, Bones AM (2011) Defence mechanisms of Brassicaceae: implications for plant-insect interactions and potential for integrated pest management. In: Sustainable agriculture vol 2. Springer, pp 623–670

Al-Shehbaz I, Beilstein MA, Kellogg E (2006) Systematics and phylogeny of the Brassicaceae (Cruciferae): an overview. Plant Syst Evol 259:89–120

Al-Shehbaz IA (1973) The biosystematics of the genus *Thelypodium* (Cruciferae). Contributions from the Gray Herbarium of Harvard University 3–148

Al-Shehbaz IA (1984) The tribes of Cruciferae (Brassicaceae) in the southeastern United States. J Arnold Arbor 65:343–373

Al-Shehbaz IA (1986) The genera of *Lepidieae* (cruciferae; Brassicaceae) in the southeastern United States. J Arnold Arbor 67:265–311

Al-Shehbaz IA (2012) A generic and tribal synopsis of the Brassicaceae (Cruciferae). Taxon 61:931–954

Al-Shehbaz IA (2014) A synopsis of the genus *Noccaea* (Coluteocarpeae, Brassicaceae). Harv Pap Bot 19:25–52

Al-Shehbaz IA, O'Kane SL Jr (2002) Taxonomy and phylogeny of Arabidopsis (Brassicaceae). Arabidopsis Book/Am Soc Plant Biologists 1:e0001

Al-Shehbaz IA, O'Kane SL Jr, Price RA (1999) Generic placement of species excluded from Arabidopsis (Brassicaceae). Novon 9:296–307

Al-Shehbaz IA, Warwick SI (2006) A synopsis of Smelowskia (Brassicaceae). Harv Pap Bot 11:91–100

Alagoz SM, TOORCHI M (2018) An investigation of some key morpho-physiological attributes and leaf proteome profile in canola (*Brassica napus* L.) under salinity stress. Pak J Bot 50:847–852

Alonso-Blanco C, Koornneef M (2000) Naturally occurring variation in *Arabidopsis*: an underexploited resource for plant genetics. Trend Plant Sci 5:22–29

Amri E (2014) The role of selected plant families with dietary Ethnomedicinal species used as anticancer. J Medicinal Plant 2:28–39

Angelini LG, Lazzeri L, Galletti S, Al Cozzani, Macchia M, Palmieri S (1998) Antigerminative activity of three glucosinolate-derived products generated by mirosinase hydrolysis. SST 26:771–780

Anjum NA, Gill SS, Ahmad I, Duarte AC, Umar S, Khan NA, Pereira E (2012) Metals and metalloids accumulation variability in Brassica species–a review. Phytotechnologies: remediation of environmental contaminants CRC Press/Taylor and Francis Group, Boca Raton, USA

Apel P, Horstmann C, Pfeffer M (1997) The Moricandia syndrome in species of the Brassicaceae-evolutionary aspects. Photosynthetica 33:205–215

Appel O, Al-Shehbaz I (2003) Cruciferae. In: Flowering Plants Dicotyledons. Springer, pp 75–174

Arshad W, Sperber K, Steinbrecher T, Nichols B, Jansen VA, Leubner-Metzger G, Mummenhoff K (2019) Dispersal biophysics and adaptive significance of dimorphic diaspores in the annual Aethionema arabicum (Brassicaceae). New Phytol 221:1434

Austin DF (2003) Dye Plants and Dyeing. Econ Bot 57:288–289

Bailey CD, Al-Shehbaz IA, Rajanikanth G (2007) Generic limits in tribe Halimolobeae and description of the new genus Exhalimolobos (Brassicaceae). Syst Bot 32:140–156

Barlas NT, Irget ME, Tepecik M (2011) Mineral content of the rocket plant (*Eruca sativa*). Afr J Biotechnol 10:14080–14082

Barthlott W (1984) Microstructural features of seed surfaces. In: Heywood VH, Moree DM (eds) Current Concepts in Plant Taxonomy. Academic Press, London, pp 95–105

Bednarek P (2012) Sulfur-containing secondary metabolites from *Arabidopsis thaliana* and other Brassicaceae with function in plant immunity. ChemBioChem 13:1846–1859

Bert V, Macnair M, De Laguerie P, Saumitou-Laprade P, Petit D (2000) Zinc tolerance and accumulation in metallicolous and nonmetallicolous populations of *Arabidopsis halleri* (Brassicaceae). New Phytol 146:225–233

Bhatnagar A, Bhushan B, Malik M (2016) Traditional medicinal plants of district Bijnor, UP, India. Int J Recent Sci Res 7:10670–10677

Blaxter M (1998) *Caenorhabditis elegans* is a nematode. Science 282:2041–2046

Bowman J (2006) Molecules and morphology: comparative developmental genetics of the Brassicaceae. Plant Syst Evol 259:199–215

Buchanan R (1995) A dyer's garden: from plant to pot: growing dyes for natural fibers. Interweave Press, Loveland, CO

Buell CR, Last RL (2010) Twenty-first century plant biology: impacts of the *Arabidopsis* genome on plant biology and agriculture. Plant Physiol 154:497–500

Busch A, Horn S, Mühlhausen A, Mummenhoff K, Zachgo S (2011) Corolla monosymmetry: evolution of a morphological novelty in the Brassicaceae family. Mol Biol Evol 29:1241–1254

Calabrone L, Larocca M, Marzocco S, Martelli G, Rossano R (2015) Total phenols and flavonoids content, antioxidant capacity and lipase inhibition of root and leaf horseradish (*Armoracia rusticana*) extracts. Food Nutr Sci 6:64

Cardone M, Mazzoncini M, Menini S, Rocco V, Senatore A, Seggiani M, Vitolo S (2003) *Brassica carinata* as an alternative oil crop for the production of biodiesel in Italy: agronomic evaluation, fuel production by transesterification and characterization. Biomass Bioenerg 25:623–636

Chalhoub B, Denoeud F, Liu S, Parkin IA, Tang H, Wang X, Chiquet J, Belcram H, Tong C, Samans B, Corréa M (2014) Early allopolyploid evolution in the post-Neolithic *Brassica napus* oilseed genome. Science 345:950–953

Chauhan ES, Tiwari A, Singh A (2016) Phytochemical screening of red cabbage (*Brassica oleracea*) powder and juice-A comparative study. J Med Plants Stud 4:196–199
Cheng F, Mandáková T, Wu J, Xie Q, Lysak MA, Wang X (2013) Deciphering the diploid ancestral genome of the mesohexaploid *Brassica rapa*. Plant Cell 25:1541–1554
Cheng F, Sun R, Hou X, Zheng H, Zhang F, Zhang Y, Liu B, Liang J, Zhuang M, Liu Y, Liu D (2016) Subgenome parallel selection is associated with morphotype diversification and convergent crop domestication in *Brassica rapa* and *Brassica oleracea*. Nat Genet 48:1218
Chiguriaeva A (1973) Pollen morphology of Cruciferae in pollen and spore morphology of recent plants (in Russian). In: Pora-3rd Int. Palynology Conf. Acad. Sci. USSR, pp 93–98
Cipollini D, Cipollini K (2016) A review of garlic mustard (*Alliaria petiolata, Brassicaceae*) as an allelopathic plant. J Torrey Bot Soc 143:339–348
Clauss MJ, Dietel S, Schubert G, Mitchell-Olds T (2006) Glucosinolate and trichome defenses in a natural *Arabidopsis lyrata* population. J Chem Ecol 32:2351–2373
Clauss MJ, Koch MA (2006) Poorly known relatives of *Arabidopsis thaliana*. Trends Plant Sci 11:449–459
Couvreur TL, Franzke A, Al-Shehbaz IA, Bakker FT, Koch MA, Mummenhoff K (2009) Molecular phylogenetics, temporal diversification, and principles of evolution in the mustard family (Brassicaceae). Mol Biol Evol 27:55–71
Davis AR, Pylatuik JD, Paradis JC, Low NH (1998) Nectar-carbohydrate production and composition vary in relation to nectary anatomy and location within individual flowers of several species of Brassicaceae. Planta 205:305–318
Dennert E (1884) Beiträge zur vergleichenden Anatomie des Laubstengels der Cruciferen. Doctoral dissertation, Universität Marburg, Marburg, Germany
Devi M (2017) Biological properties of soil and nutrient uptake in cauliflower (*Brassica oleracea var botrytis L.*) as influenced by integrated nutrient management. J Pharmacogn Phytochem 6:325–328
Dinneny JR, Long TA, Wang JY, Jung JW, Mace D, Pointer S, Barron C, Brady SM, Schiefelbein J, Benfey PN (2008) Cell identity mediates the response of *Arabidopsis* roots to abiotic stress. Science 320:942–945
Du H, Shen X, Huang Y, Huang M, Zhang Z (2016) Overexpression of *Vitreoscilla hemoglobin* increases waterlogging tolerance in *Arabidopsis* and maize. BMC Plant Biol 16:35
du Toit LJ, Le Clerc V, Briard M (2019) Genetics and genomics of carrot Biotic stress. In: The carrot genome. Springer, pp 317–362
Dutta KN, Chetia P, Lahkar S, Das S (2014) Herbal plants used as diuretics: a comprehensive review. J Pharm Chem Biol Sci 2:27–32
Edger PP, Heidel-Fischer HM, Bekaert M, Rota J, Glöckner G, Platts AE, Heckel DG, Der JP, Wafula EK, Tang M, Hofberger JA (2015) The butterfly plant arms-race escalated by gene and genome duplications. Proc Nati Acad Sci 112:8362–8366
Edger PP, Tang M, Bird KA, Mayfield DR, Conant G, Mummenhoff K, Koch MA, Pires JC (2014) Secondary structure analyses of the nuclear rRNA internal transcribed spacers and assessment of its phylogenetic utility across the Brassicaceae (Mustards). PLoS One 9:e101341
Edwards D, Batley J, Snowdon RJ (2013) Accessing complex crop genomes with next-generation sequencing. Theor Appl Genet 126:1–11
El Sayed HESA, Aly MM (2014) Antibacterial activities of six medicinal plants used traditionally by Saudi people to treat common diseases. Br Biotechnol J 4:499
Eldridge T (2014) An integrative analysis of fruit shape in *Capsella* rubella and *Arabidopsis thaliana*. Doctoral dissertation, University of East Anglia, Norwich, UK
Eldridge T, Łangowski Ł, Stacey N, Jantzen F, Moubayidin L, Sicard A, Southam P, Kennaway R, Lenhard M, Coen ES, Østergaard L (2016) Fruit shape diversity in the Brassicaceae is generated by varying patterns of anisotropy. Development 143:3394–3406
Erdtam G, Praglowski J, Nilsson S (1963) An Introduction to a Scandinavian Pollen Flora. II. Almqvist & Wicksell, Stockholm
Erdtman G (1952) Pollen morphology and plant taxonomy. New York:6–24

Erum A, Aslam M, Jafri M, Ahmed MA, Yousuf AW (2017) Phytochemical and ethnopharmacological review of "Tudri surkh"(*Cheiranthus cheiri*). World J Pharm Res 6:352–359
Evivie ER, Ogwu MC, Cang W, Xu R, Li J (2019) Progress and prospects of glucosinolate pathogen resistance in some brassica plants. J Appl Nat Sci 11:556–567
Fahey JW, Zalcmann AT, Talalay P (2001) The chemical diversity and distribution of glucosinolates and isothiocyanates among plants. Phytochemistry 56:5–51
Farooq MA, Ali B, Gill RA, Islam F, Cui P, Zhou W (2016) Breeding oil crops for sustainable production: heavy metal tolerance. In: Breeding oilseed crops for sustainable production. Elsevier, pp 19–31
Feuillet C, Leach JE, Rogers J, Schnable PS, Eversole K (2011) Crop genome sequencing: lessons and rationales. Trends Plant Sci 16:77–88
Francis A, Warwick S (2003) The biology of Canadian weeds. 120. Neslia paniculata (L.) Desv. Can J Plant Sci 83:441–451
Franzke A, Lysak MA, Al-Shehbaz IA, Koch MA, Mummenhoff K (2011) Cabbage family affairs: the evolutionary history of Brassicaceae. Trends Plant Sci 16:108–116
Friedt W, Tu J, Fu T (2018) Academic and economic importance of *Brassica napus* rapeseed. In: The *Brassica napus* genome. Springer, pp 1–20
Fu D, Mason AS, Xiao M, Yan H (2016) Effects of genome structure variation, homeologous genes and repetitive DNA on polyploid crop research in the age of genomics. Plant Sci 242:37–46
Fuentes-Soriano S, Al-Shehbaz I (2013) Phylogenetic relationships of mustards with multiaperturate pollen (Physarieae, Brassicaceae) based on the plastid ndhF gene: implications for morphological diversification. Syst Bot 38:178–191
Gaeta RT, Pires JC, Iniguez-Luy F, Leon E, Osborn TC (2007) Genomic changes in resynthesized *Brassica napus* and their effect on gene expression and phenotype. Plant Cell 19:3403–3417
German D, Friesen N (2014) Shehbazia (Shehbazieae, Cruciferae), a new monotypic genus and tribe of hybrid origin from Tibet. Turczaninowia 17:17–23
Gómez-Campo C (1980) Morphology and morpho-taxonomy of the tribe Brassiceae. Morphology and morpho-taxonomy of the tribe Brassiceae 3–31
Grandont L, Cuñado N, Coriton O, Huteau V, Eber F, Chèvre AM, Grelon M, Chelysheva L, Jenczewski E (2014) Homoeologous chromosome sorting and progression of meiotic recombination in *Brassica napus*: ploidy does matter! Plant Cell 26:1448–1463
Gugel R, Falk K (2006) Agronomic and seed quality evaluation of *Camelina sativa* in western Canada. Can J Plant Sci 86:1047–1058
Gul S, Ahmed S, Gul H, Shad KF, Zia-Ul-Haq M, Badiu D (2013) The antioxidant potential of *Brassica rapa* L. on glutathione peroxidase, superoxide dismutase enzymes and total antioxidant status. Rev Rom Med Lab 21:161–169
Gulshan AB, Dasti AA, Hussain S, Atta MI, Amin-ud-Din M (2012) Indigenous uses of medicinal plants in rural areas of Dera Ghazi Khan, Punjab, Pakistan. ARPN J Agric Biol Sci 7:750–762
Hall AE, Fiebig A, Preuss D (2002a) Beyond the *Arabidopsis* genome: opportunities for comparative genomics. Plant Physiol 129:1439–1447
Hall JC, Iltis HH, Sytsma KJ (2004) Molecular phylogenetics of core Brassicales, placement of orphan genera *Emblingia*, *Forchhammeria*, *Tirania*, and character evolution. Syst Bot 29:654–669
Hall JC, Sytsma KJ, Iltis HH (2002b) Phylogeny of Capparaceae and Brassicaceae based on chloroplast sequence data. Am J Bot 89:1826–1842
Hall JC, Tisdale TE, Donohue K, Kramer EM (2006) Developmental basis of an anatomical novelty: heteroarthrocarpy in Cakile lanceolata and *Erucaria erucarioides* (Brassicaceae). Int J Plant Sci 167:771–789
Hall JC, Tisdale TE, Donohue K, Wheeler A, Al-Yahya MA, Kramer EM (2011) Convergent evolution of a complex fruit structure in the tribe Brassiceae (Brassicaceae). Am J Bot 98:1989–2003
Hamburger M (2002) Isatis *tinctoria*—from the rediscovery of an ancient medicinal plant towards a novel anti-inflammatory phytopharmaceutical. Phytochem Rev 1:333

Hanf V, Gonder U (2005) Nutrition and primary prevention of breast cancer: foods, nutrients and breast cancer risk. Eur J Obstet Gynecol Reprod Biol 123:139–149
Haq F (2012) The ethno botanical uses of medicinal plants of Allai Valley, Western Himalaya Pakistan. Int J Plant Res 2:21–34
Hasanuzzaman M, Fujita M (2012) Heavy metals in the environment: Current status, toxic effects on plants and possible phytoremediation. In: Anjum NA, Pereira MA, Ahmad I, Duarte AC, Umar S, Khan NA (eds) Phytotechnologies: remediation of environmental contaminants. CRC Press, Boca Raton, pp 7–73
Hasanuzzaman M, Hossain MA, Silva JAT, Fujita M (2012) Plant responses and tolerance to abiotic oxidative stress: antioxidant defense is a key factor. In: Bandi V, Shanker AK, Shanker C, Mandapaka M (eds) Crop Stress and its management: perspectives and strategies. Springer, New York, pp 261–316
Hasanuzzaman M, Nahar K, Fujita M, Ahmad P, Chandna R, Prasad MNV, Ozturk M (2013a) Enhancing plant productivity under salt stress—relevance of poly-omics. In: Ahmad P, Azooz MM, Prasad MNV (eds) Salt stress in plants: omics, signaling and responses. Springer, Berlin, pp 113–156
Hasanuzzaman M, Nahar K, Fujita M (2013b) Plant response to salt stress and role of exogenous protectants to mitigate salt-induced damages. In: Ahmad P, Azooz MM, Prasad MNV (eds) Ecophysiology and responses of plants under salt stress. Springer NY, USA, pp 25–87
He Z, Cheng F, Li Y, Wang X, Parkin IA, Chalhoub B, Liu S, Bancroft I (2015) Construction of *Brassica* A and C genome-based ordered pan-transcriptomes for use in rapeseed genomic research. Data Brief 4:357–362
Henry IM, Dilkes BP, Tyagi A, Gao J, Christensen B, Comai L (2014) The boy named SUE quantitative trait locus confers increased meiotic stability to an adapted natural allopolyploid of *Arabidopsis*. Plant Cell 26:181–194
Hohmann N, Wolf EM, Lysak MA, Koch MA (2015) A time-calibrated road map of Brassicaceae species radiation and evolutionary history. Plant Cell 27:2770–2784
Hu D, Zhang W, Zhang Y, Chang S, Chen L, Chen Y, Shi Y, Shen J, Meng J, Zou J (2019) Reconstituting the genome of a young allopolyploid crop, *Brassica napus*, with its related species. Plant Biotechnol J 17:1106–1118
Huang CH, Sun R, Hu Y, Zeng L, Zhang N, Cai L, Zhang Q, Koch MA, Al-Shehbaz I, Edger PP, Pires JC (2016) Resolution of Brassicaceae phylogeny using nuclear genes uncovers nested radiations and supports convergent morphological evolution. Mol Biol Evol 33:394–412
Hutchinson J (1934) The families of flowering plants: arranged according to a new system based on their probable phylogeny. Clarendon Press
Inamdar J, Rao N (1983) Light and scanning electron microscopic studies on trichomes of some Brassicaceae. Feddes Repertorium
Initiative AG (2000) Analysis of the genome sequence of the flowering plant *Arabidopsis thaliana*. Nature 408:796
Jahangir M, Kim HK, Choi YH, Verpoorte R (2009) Health-affecting compounds in Brassicaceae. Compr Rev Food Sci F 8:31–43
Janchen E (1942) Das system der Cruciferen. Plant Syst Evol 91:1–28
Jiang M, Ye Z-h, Zhang H-j, Miao L-x (2019) Broccoli plants over-expressing an ERF transcription factor gene BoERF1 facilitates both salt stress and *Sclerotinia* stem rot resistance. J Plant Growth Regul 38:1–13
Johnson SN, Lopaticki G, Barnett K, Facey SL, Powell JR, Hartley SE (2016) An insect ecosystem engineer alleviates drought stress in plants without increasing plant susceptibility to an above-ground herbivore. Funct Ecol 30:894–902
Jonsell B (1979) New taxa of Cruciferae from East tropical Africa and Madagascar. Bot Not 132:521–535
Jonsell B (1986) A monograph of *Farsetia* (Cruciferae). Syst Bot Ups 25:1–106
Judd WS, SANDERS RW, DONOGHUE MJ (1994) Angiosperm family pairs: preliminary phylogenetic analyses. Harv Pap Bot 1:1–51

Kagale S, Robinson SJ, Nixon J, Xiao R, Huebert T, Condie J, Kessler D, Clarke WE, Edger PP, Links MG, Sharpe AG (2014) Polyploid evolution of the Brassicaceae during the *Cenozoic* era. Plant Cell 26:2777–2791

Karaismailoğlu MC, Erol O (2019) Pollen morphology of some taxa of *Thlaspi* L. sensu lato (Brassicaceae) from Turkey, and its taxonomical importance. Palynology 43:244–254

Karl R, Koch MA (2013) A world-wide perspective on crucifer speciation and evolution: phylogenetics, biogeography and trait evolution in tribe *Arabideae*. Ann Bot 112:983–1001

Kasem W, Ghareeb A, Marwa E (2011) Seed morphology and seed coat sculpturing of 32 taxa of family Brassicaceae. Am J Sci 7:166–178

Kaur H, Sirhindi G, Bhardwaj R, Alyemeni M, Siddique KH, Ahmad P (2018) 28-homobrassinolide regulates antioxidant enzyme activities and gene expression in response to salt-and temperature-induced oxidative stress in *Brassica juncea*. Sci Rep 8:8735

Kayum MA, Jung H-J, Park J-I, Ahmed NU, Saha G, Yang T-J, Nou I-S (2015a) Identification and expression analysis of WRKY family genes under biotic and abiotic stresses in *Brassica rapa*. Mol Genet Gen 290:79–95

Kayum MA, Kim H-T, Nath UK, Park J-I, Kho KH, Cho Y-G, Nou I-S (2016a) Research on biotic and abiotic stress related genes exploration and prediction in *Brassica rapa* and *B. oleracea*: a review. Plant Breed Biotechnol 4:135–144

Kayum MA, Park J-I, Ahmed NU, Jung H-J, Saha G, Kang J-G, Nou I-S (2015b) Characterization and stress-induced expression analysis of Alfin-like transcription factors in *Brassica rapa*. Mol Genet Gen 290:1299–1311

Kayum MA, Park J-I, Ahmed NU, Saha G, Chung M-Y, Kang J-G, Nou I-S (2016b) Alfin-like transcription factor family: characterization and expression profiling against stresses in *Brassica oleracea*. Acta Physiol Plant 38:127

Kayum MA, Park J-I, Nath UK, Biswas MK, Kim H-T, Nou I-S (2017) Genome-wide expression profiling of aquaporin genes confer responses to abiotic and biotic stresses in *Brassica rapa*. BMC Plant Biol 17:23

Keshavarzi M, Abassian S, Sheidai M (2012) Pollen morphology of the genus *Clypeola* (Brassicaceae) in Iran. Phytol Balcan 18:17–24

Kesseler R, Stuppy W (2006) Seeds: time capsules of life. Papadakis Publisher, London, UK, p 264

Khalik KA (2002) Biosystematic studies on Brassicaceae (Cruciferae) in Egypt. Dissertation, Wageningen University, Netherlands

Khalik KA, Van der Maesen L (2002) Seed morphology of some tribes of Brassicaceae (implications for taxonomy and species identification for the flora of Egypt). Blumea 47:363–383

Khan R (2004) Studies on the pollen morphology of the genus *Arabidopsis* (Brassicaceae) from Pakistan. Pak J Bot 36:229–234

Khan R, Mehmood S, Khan S, Khan A, Shah I, Bokhari T (2013) Medicinal value of indigenous flora in the vicinity of district Bannu, Khyber Pakhtunkhwa, Pakistan. Adv Pharma Ethnomed 1:7–14

Kim Y-W, Jung H-J, Park J-I, Hur Y, Nou I-S (2015) Response of NBS encoding resistance genes linked to both heat and fungal stress in *Brassica oleracea*. Plant Physiol Biochem 86:130–136

Kingsbury JM (1964) Poisonous plants of the United States and Canada. Soil Sci 98:349

Koch M, Kiefer C (2006) Molecules and migration: biogeographical studies in cruciferous plants. Plant Syst Evol 259:121–142

Koch MA, Mummenhoff K (2006) Evolution and phylogeny of the Brassicaceae. Plant Syst Evol 259:81–83

Koch MA, Wernisch M, Schmickl R (2008) *Arabidopsis thaliana's* wild relatives: an updated overview on systematics, taxonomy and evolution. Taxon 57:933–933

Koornneef M, Alonso-Blanco C, Vreugdenhil D (2004) Naturally occurring genetic variation in *Arabidopsis thaliana*. Annu Rev Plant Biol 55:141–172

Koubaa M, Driss D, Bouaziz F, Ghorbel RE, Chaabouni SE (2015) Antioxidant and antimicrobial activities of solvent extract obtained from rocket (*Eruca sativa* L.) flowers. Free Rad Antiox 5:29–34

Kreps JA, Wu Y, Chang H-S, Zhu T, Wang X, Harper JF (2002) Transcriptome changes for *Arabidopsis* in response to salt, osmotic, and cold stress. Plant Physiol 130:2129–2141

Kumar V, Thakur AK, Barothia ND, Chatterjee SS (2011) Therapeutic potentials of *Brassica juncea*: an overview. Tang [Humanitas Medicine] 1:2.1–2.16

Küpper H, Lombi E, Zhao F-J, McGrath SP (2000) Cellular compartmentation of cadmium and zinc in relation to other elements in the hyperaccumulator *Arabidopsis halleri*. Planta 212:75–84

Lagercrantz U, Putterill J, Coupland G, Lydiate D (1996) Comparative mapping in *Arabidopsis* and Brassica, fine scale genome collinearity and congruence of genes controlling flowering time. Plant J 9:13–20

Łangowski Ł, Stacey N, Østergaard L (2016) Diversification of fruit shape in the Brassicaceae family. Plant Reprod 29:149–163

Li M, Yang Y, Raza A, Yin S, Wang H, Zhang Y, Dong J, Wang G, Zhong C, Zhang H, Liu J (2020) Heterologous Expression of *Arabidopsis* rty Enhances Drought Tolerance in Strawberry (*Fragaria*× *ananassa* Duch.). https://doi.org/10.21203/rs.3.rs-34565/v1

Lin H-H, Lin K-H, Chen S-C, Shen Y-H, Lo H-F (2015) Proteomic analysis of broccoli (*Brassica oleracea*) under high temperature and waterlogging stresses. Bot Stud 56:18

Linić I, Šamec D, Grúz J, Vujčić Bok V, Strnad M, Salopek-Sondi B (2019) Involvement of phenolic acids in short-term adaptation to salinity stress is species-specific among Brassicaceae. Plants 8:155

Liu J, Chu J, Ma C, Jiang Y, Ma Y, Xiong J, Cheng Z-M (2019) Overexpression of an ABA-dependent grapevine bZIP transcription factor, VvABF2, enhances osmotic stress in *Arabidopsis*. Plant Cell Rep 38:587–596

Liu S, Liu Y, Yang X, Tong C, Edwards D, Parkin IA, Zhao M, Ma J, Yu J, Huang S, Wang X (2014) The *Brassica oleracea* genome reveals the asymmetrical evolution of polyploid genomes. Nat Commun 5:3930

Liu T, Zhang C, Yang G, Wu J, Xie G, Zeng H, Yin C, Liu T (2009) Central composite design-based analysis of specific leaf area and related agronomic factors in cultivars of rapeseed (*Brassica napus* L.). Field Crops Res 111:92–96

Lloyd AH, Ranoux M, Vautrin S, Glover N, Fourment J, Charif D, Choulet F, Lassalle G, Marande W, Tran J, Granier F (2014) Meiotic gene evolution: can you teach a new dog new tricks? Mol Biol Evol 31:1724–1727

Lv Y, Fu S, Chen S, Zhang W, Qi C (2016) Ethylene response factor BnERF2-like (ERF2. 4) from *Brassica napus* L. ehances submergence tolerance and alleviates oxidative damage caused by submergence in *Arabidopsis thaliana*. Crop J 4:199–211

Lysak MA, Cheung K, Kitschke M, Bureš P (2007) Ancestral chromosomal blocks are triplicated in Brassiceae species with varying chromosome number and genome size. Plant Physiol 145:402–410

Lysak MA, Koch MA (2011) Phylogeny, genome, and karyotype evolution of crucifers (Brassicaceae). In: Genetics and genomics of the Brassicaceae. Springer, pp 1–31

Maciejewska-Rutkowska I, Bednorz L, Fujiki T (2007) SEM observations of pollen grains, fruits and seeds of the Pieniny Mountains [South Poland] endemic species *Erysimum pieninicum* [Zapal.] Pawl.[Brassicaceae]. Acta Soc Bot Pol 76

Macnair MR, Bert V, Huitson SB, Saumitou-Laprade P, Petit D (1999) Zinc tolerance and hyperaccumulation are genetically independent characters. Proc R Soc Lond B Biol Sci 266:2175–2179

Metcalfe C, Chalk L (1950) Anatomy of dicotyledons, vol 1. Clarendon Press, Oxford, UK, pp xiv + 1–724

Mitchell-Olds T (2001) *Arabidopsis thaliana* and its wild relatives: a model system for ecology and evolution. Trends Ecol Evol 16:693–700

Mittal N, Thakur S, Verma H, Kaur A (2018) Interactive effect of salinity and ascorbic acid on *Brassica rapa* L. plants. GJBB 7:27–29

Mun J-H, Yu H-J, Park S, Park B-S (2009) Genome-wide identification of NBS-encoding resistance genes in *Brassica rapa*. Mol Genet Gen 282:617

Nag K (2013) Study of some rare medicinal wild herbs from gardens of Bhopal city, Madhya Pradesh (India). Int J Pharma Life Sci 4:2437–2439

Nahar K, Hasanuzzaman M, Alam MM, Rahman A, Suzuki T, Fujita M (2016) Polyamine and nitric oxide crosstalk: antagonistic effects on cadmium toxicity in mung bean plants through upregulating the metal detoxification, antioxidant defense, and methylglyoxal detoxification systems. Ecotoxicol Environ Saf 126:245–255

Nahar K, Hasanuzzaman M, Suzuki T, Fujita M (2017) Polyamine-induced aluminium tolerance in mung bean: a study on antioxidant defense and methylglyoxal detoxification systems. Ecotoxicology 26(1):58–73

Nakashima K, Ito Y, Yamaguchi-Shinozaki K (2009) Transcriptional regulatory networks in response to abiotic stresses in *Arabidopsis* and grasses. Plant Physiol 149:88–95

Navabi Z-K, Huebert T, Sharpe AG, O'Neill CM, Bancroft I, Parkin IA (2013) Conserved microstructure of the *Brassica* B genome of *Brassica nigra* in relation to homologous regions of *Arabidopsis thaliana, B. rapa* and *B. oleracea.* BMC Genomics 14:250

Nikolov LA, Shushkov P, Nevado B, Gan X, Al-Shehbaz IA, Filatov D, Bailey CD, Tsiantis M (2019) Resolving the backbone of the Brassicaceae phylogeny for investigating trait diversity. New Phytol 222:1638–1651

O'neill CM, Bancroft I (2000) Comparative physical mapping of segments of the genome of *Brassica oleracea* var. Alboglabra that are homoeologous to sequenced regions of chromosomes 4 and 5 of *Arabidopsis thaliana.* Plant J 23:233–243

Obi R, Nwanebu F, Ndubuisi U, Orji N (2009) Antibacterial qualities and phytochemical screening of the oils of *Curcubita pepo* and *Brassica nigra.* J. Med. Plants Res 3:429–432

OECD (2016) Safety assessment of transgenic organisms in the environment 5. https://doi.org/10.1787/9789264253018-en

Osborn T, Kole C, Parkin I, Sharpe A, Kuiper M, Lydiate D, Trick M (1997) Comparison of flowering time genes in *Brassica rapa,* B. napus and *Arabidopsis thaliana.* Genetics 146:1123–1129

Osborn TC, Butrulle DV, Sharpe AG, Pickering KJ, Parkin IA, Parker JS, Lydiate DJ (2003) Detection and effects of a homeologous reciprocal transposition in *Brassica napus.* Genetics 165:1569–1577

Owis A (2015) Broccoli; the green beauty: a review. J Pharma Sci Res 7:696

Pachiappan S, Matheswaran S, Pushkalai P (2017) Medicinal plants for polycystic ovary syndrome: a review of phytomedicine research. Int J Herb Med 5:78–80

Pan Y, Zhu M, Wang S, Ma G, Huang X, Qiao C, Wang R, Xu X, Liang Y, Lu K, Li J (2018) Genome-Wide characterization and analysis of Metallothionein family genes that function in metal stress tolerance in *Brassica napus* L. Int J Mol Sci 19:2181

Parkin I, Sharpe A, Keith D, Lydiate D (1995) Identification of the A and C genomes of amphidiploid *Brassica napus* (oilseed rape). Genome 38:1122–1131

Parkin IA, Clarke WE, Sidebottom C, Zhang W, Robinson SJ, Links MG, Karcz S, Higgins EE, Fobert P, Sharpe AG (2010) Towards unambiguous transcript mapping in the allotetraploid *Brassica napus.* Genome 53:929–938

Parkin IA, Gulden SM, Sharpe AG, Lukens L, Trick M, Osborn TC, Lydiate DJ (2005) Segmental structure of the *Brassica napus* genome based on comparative analysis with *Arabidopsis thaliana.* Genetics 171:765–781

Parkin IA, Koh C, Tang H, Robinson SJ, Kagale S, Clarke WE, Town CD, Nixon J, Krishnakumar V, Bidwell SL, Denoeud F (2014) Transcriptome and methylome profiling reveals relics of genome dominance in the mesopolyploid *Brassica oleracea.* Genome Biol 15:R77

Parvaiz M, Bhatti K, Nawaz K, Hussain Z, Khan R, Hussain A (2013) Ethno-botanical studies of medicinal plants of Dinga, district Gujrat, Punjab, Pakistan. World Appl Sci J 26:826–833

Pavlović I, Petřík I, Tarkowská D, Lepeduš H, Vujčić Bok V, Radić Brkanac S, Novák O, Salopek-Sondi B (2018) Correlations between phytohormones and drought tolerance in selected Brassica crops: Chinese cabbage, white cabbage and kale. Int J Mol Sci 19:2866

Perveen A, Qaiser M, Khan R (2004) Pollen flora of Pakistan–XLII. Brassicaceae Pak J Bot 36:683–700

Piazza P, Bailey CD, Cartolano M, Krieger J, Cao J, Ossowski S, Schneeberger K, He F, de Meaux J, Hall N, MacLeod N (2010) *Arabidopsis thaliana* leaf form evolved via loss of KNOX expression in leaves in association with a selective sweep. Curr Biol 20:2223–2228

Pires JC, Zhao J, Schranz ME, Leon EJ, Quijada PA, Lukens LN, Osborn TC (2004) Flowering time divergence and genomic rearrangements in resynthesized *Brassica* polyploids (Brassicaceae). Biol J Linnean Soc 82:675–688

Pontes O, Neves N, Silva M, Lewis MS, Madlung A, Comai L, Viegas W, Pikaard CS (2004) Chromosomal locus rearrangements are a rapid response to formation of the allotetraploid *Arabidopsis suecica* genome. Proc Natl Acad Sci 101:18240–18245

Prantl KAE, Engler A (1887) Cruciferae. W Engelmann

Prasad M (2014) Antimicrobial potential of Brassicaceae family against clinical isolates. Int J Pure Appl Biosci 2:158–162

Price RA, Palmer JD, Al-Shehbaz IA (1994) Systematic relationships of *Arabidopsis*: a molecular and morphological perspective. Cold Spring Harb Monogr Arch 27:7–19

Pucciariello C, Banti V, Perata P (2012) ROS signaling as common element in low oxygen and heat stresses. Plant Physiol Biochem 59:3–10

Rahman M, Khatun A, Liu L, Barkla B (2018) Brassicaceae mustards: traditional and agronomic uses in Australia and New Zealand. Molecules 23:231

Rakow G (2004) Species origin and economic importance of Brassica. In: Brassica. Springer, pp 3–11

Raval N (2016) A comprehensive review of *Lepidium sativum* Linn, A traditional medicinal plant. World J Pharm Pharma Sci 5:593–1601

Ravikumar C (2015) Therapeutic potential of *Brassica oleracea* (broccoli)–a review. Int J Drug Dev Res 7:009–010

Raza A, Ashraf F, Zou X, Zhang X, Tosif H (2020) Plant Adaptation and Tolerance to Environmental Stresses: Mechanisms and Perspectives. In: Hasanuzzaman M (eds) Plant Ecophysiology and Adaptation under Climate Change: Mechanisms and Perspectives I. Springer, Singapore, pp 117–145

Raza A, Mehmood SS, Ashraf F, Khan RSA (2019a) Genetic diversity analysis of Brassica species using PCR-based SSR markers. Gesunde Pflanz 71:1–7

Raza A, Mehmood SS, Shah T, Zou X, Yan L, Zhang X, Khan RS (2019d) Applications of molecular markers to develop resistance against abiotic stresses in wheat. In: Hasanuzzaman M, Nahar K, Hossain M (eds) Wheat production in changing environments. Springer, Singapore, pp 393–420

Raza A, Mehmood SS, Tabassum J, Batool R (2019c) Targeting plant hormones to develop abiotic stress resistance in wheat. In: Hasanuzzaman M, Nahar K, Hossain M (eds) Wheat production in changing environments. Springer, Singapore, pp 557–577

Raza A, Razzaq A, Mehmood SS, Zou X, Zhang X, Lv Y, Xu J (2019b) Impact of climate change on crops adaptation and strategies to tackle its outcome: a review. Plants 8:34

Razmjoo K, Toriyama K, Ishii R, Hinata K (1996) Photosynthetic properties of hybrids between *Diplotaxis* muralis DC, a C3 species, and *Moricandia arvensis* (L.) DC, a C3–C4 intermediate species in Brassicaceae. Genes Genet Syst 71:189–192

Reusche M, Thole K, Janz D, Truskina J, Rindfleisch S, Drübert C, Polle A, Lipka V, Teichmann T (2012) Verticillium infection triggers vascular-related NAC domain 7–dependent de novo xylem formation and enhances drought tolerance in *Arabidopsis*. Plant Cell 24:3823–3837

Robert LS, Robson F, Sharpe A, Lydiate D, Coupland G (1998) Conserved structure and function of the *Arabidopsis* flowering time gene constans in *Brassica napus*. Plant Mol Biol 37:763–772

Roeder AH, Yanofsky MF (2006) Fruit development in *Arabidopsis*. Arabidopsis Book/Am Soc Plant Biologists 4:e0075

Rokayya S, Li C-J, Zhao Y, Li Y, Sun C-H (2013) Cabbage (*Brassica oleracea* L. var. capitata) phytochemicals with antioxidant and anti-inflammatory potential. Asian Pac J Cancer Prev 14:6657–6662

Rollins R, Banerjee U (1979) Trichome patterns in physaria. Publication of the Bussey Institute, Harvard University 1979:65–77

Rollins RC (1993) The Cruciferae of continental North America: systematics of the mustard family from the arctic to panama. Stanford University Press

Román-Palacios C, Molina-Henao YF, Barker MS (2019) Polyploidy increases overall diversity despite higher turnover than diploids in the Brassicaceae. bioRxiv 717306

Roosens NH, Willems G, Saumitou-Laprade P (2008) Using *Arabidopsis* to explore zinc tolerance and hyperaccumulation. Trends Plant Sci 13:208–215

Rousseau-Gueutin M, Morice J, Coriton O, Huteau V, Trotoux G, Nègre S, Falentin C, Deniot G, Gilet M, Eber F, Pelé A (2017) The impact of open pollination on the structural evolutionary dynamics, meiotic behavior, and fertility of Resynthesized Allotetraploid *Brassica napus* L. G3: genes gen genet 7:705–717

Rylott EL, Metzlaff K, Rawsthorne S (1998) Developmental and environmental effects on the expression of the C3–C4 Intermediate phenotype in *Moricandia arvensis*. Plant Physiol 118:1277–1284

Saeidnia S, Gohari AR (2012) Importance of *Brassica napus* as a medicinal food plant. J Med Plants Res 6:2700–2703

Shankar S, Segaran G, Sundar RD, Settu S, Sathiavelu M (2019) Brassicaceae—a classical review on its pharmacological Activities. Int J Pharm Sci Rev Res 55:107–113

Salma U, Khan T, Shah AJ (2018) Antihypertensive effect of the methanolic extract from Eruca sativa Mill., (Brassicaceae) in rats: muscarinic receptor-linked vasorelaxant and cardiotonic effects. J Ethnopharmacol 224:409–420

Schulz OE (1936) Cruciferae. In: Engler A, Harms H (eds) The natural plant families. Wilhelm Engelmann, Leipzig, pp 227–658

Šamec D, Salopek-Sondi B (2019) Cruciferous (Brassicaceae) vegetables. In: Nonvitamin and nonmineral nutritional supplements. Elsevier, pp 195–202

Schiessl S, Samans B, Hüttel B, Reinhard R, Snowdon RJ (2014) Capturing sequence variation among flowering-time regulatory gene homologs in the allopolyploid crop species *Brassica napus*. Front. Plant Sci 5:404

Schmidt R, Bancroft I (2011) Genetics and Genomics of the Brassicaceae. Springer, New York

Schmutzer T, Samans B, Dyrszka E, Ulpinnis C, Weise S, Stengel D, Colmsee C, Lespinasse D, Micic Z, Abel S, Duchscherer P (2015) Species-wide genome sequence and nucleotide polymorphisms from the model allopolyploid plant *Brassica napus*. Sci Data 2:150072

Schranz ME, Dobeš C, Koch MA, Mitchell-Olds T (2005) Sexual reproduction, hybridization, apomixis, and polyploidization in the genus *Boechera* (Brassicaceae). Am J Bot 92:1797–1810

Schranz ME, Mitchell-Olds T (2006) Independent ancient polyploidy events in the sister families Brassicaceae and Cleomaceae. Plant Cell 18:1152–1165

Shah GM, Hussain M, Abbasi AM (2015) Medicinal plants used to treat respiratory tract illness in Kaghan Valley. Himalayan Region-Pakistan, SMGE book, p 5

Sher Z, Khan Z, Hussain F (2011) Ethnobotanical studies of some plants of Chagharzai valley, district Buner, Pakistan. Pak J Bot 43:1445–1452

Shonnard DR, Williams L, Kalnes TN (2010) Camelina-derived jet fuel and diesel: Sustainable advanced biofuels. Environ Prog Sustain Energy 29:382–392

Sicard A, Thamm A, Marona C, Lee YW, Wahl V, Stinchcombe JR, Wright SI, Kappel C, Lenhard M (2014) Repeated evolutionary changes of leaf morphology caused by mutations to a homeobox gene. Curr Biol 24:1880–1886

Sillito D, Parkin IA, Mayerhofer R, Lydiate D, Good A (2000) *Arabidopsis thaliana*: a source of candidate disease-resistance genes for *Brassica napus*. Genome 43:452–460

Singh A, Guest D, Copeland L (2015) Associations between glucosinolates, white rust, and plant defense activators in Brassica plants: a review. Int J Veg Sci 21:297–313

Singh A, Sharma B, Deswal R (2018) Green silver nanoparticles from novel Brassicaceae cultivars with enhanced antimicrobial potential than earlier reported Brassicaceae members. J Trace Elem Med Bio 47:1–11

Singh P, Singh J (2013) Medicinal and therapeutic utilities of *Raphanus sativus*. Int J Plant Anim Environ Sci 3:103–105

Sirhindi G, Kaur H, Bhardwaj R, Sharma P, Mushtaq R (2017) 28-Homobrassinolide potential for oxidative interface in *Brassica juncea* under temperature stress. Acta Physiol Plant 39:228

Snowdon RJ, Abbadi A, Kox T, Schmutzer T, Leckband G (2015) Heterotic haplotype capture: precision breeding for hybrid performance. Trends Plant Sci 20:410–413

Soengas Fernández MdP, Sotelo Pérez T, Velasco Pazos P, Cartea González ME (2011) Antioxidant properties of *Brassica* vegetables. Funt Plant Sci Biotechnol 5:43–55

Soengas P, Rodríguez VM, Velasco P, Cartea ME (2018) Effect of temperature stress on Antioxidant defenses in *Brassica oleracea*. ACS Omega 3:5237–5243

Staub RE, Feng C, Onisko B, Bailey GS, Firestone GL, Bjeldanes LF (2002) Fate of indole-3-carbinol in cultured human breast tumor cells. Chem Res Toxicol 15:101–109

Steinkellner H, Rabot S, Freywald C, Nobis E, Scharf G, Chabicovsky M, Knasmüller S, Kassie F (2001) Effects of cruciferous vegetables and their constituents on drug metabolizing enzymes involved in the bioactivation of DNA-reactive dietary carcinogens. Mutat Res-Fund Mol M 480:285–297

Streubel S, Fritz MA, Teltow M, Kappel C, Sicard A (2018) Successive duplication-divergence mechanisms at the RCO locus contributed to leaf shape diversity in the Brassicaceae. Development 145:164301

Suay L, Zhang D, Eber F, Jouy H, Lodé M, Huteau V, Coriton O, Szadkowski E, Leflon M, Martin OC, Falque M (2014) Crossover rate between homologous chromosomes and interference are regulated by the addition of specific unpaired chromosomes in *B rassica*. New Phytol 201:645–656

Szadkowski E, Eber F, Huteau V, Lode M, Huneau C, Belcram H, Coriton O, Manzanares-Dauleux MJ, Delourme R, King GJ, Chalhoub B (2010) The first meiosis of resynthesized *Brassica napus*, a genome blender. New Phytol 186:102–112

Tadege M, Sheldon CC, Helliwell CA, Stoutjesdijk P, Dennis ES, Peacock WJ (2001) Control of flowering time by FLC orthologues in *Brassica napus*. Plant J 28:545–553

Tantawy M, Khalifa S, Hassan S, Al-Rabiai G (2004) Seed exomorphic characters of some Brassicaceae (LM and SEM study). Int J Agric Biol 6:821–830

Theobald W, Krahulik J, Rollins R (1979) Trichome description and classification. In: Metcalfe CR, Chalk L (eds) Anatomy of the dicotyledons. Clarendon Press, Oxford, pp 40–53

Tomar RS, Shrivastava V (2014) Efficacy evaluation of ethanolic extract of *Brassica nigra* as potential antimicrobial agent against selected microorganisms. IJPHC 3:117–123

Tutin T, Tutin TG, Burges NA, Chater AO, Edmondson JR, Heywood VH, Moore DM, Valentine DH, Walters SM, Webb DA (1993) Flora Europaea I, 2nd ed, Cambridge University Press, Cambridge

Uprety D, Prakash S, Abrol Y (1995) Variability for photosynthesis in Brassica and allied genera. Indian J Plant Physiol 38:207–213

Van de Peer Y, Mizrachi E, Marchal K (2017) The evolutionary significance of polyploidy. Nat Rev Genet 18:411

Vaughan JG, MacLeod AJ (1976) Jones BMG Biology and chemistry of the cruciferae. In: Conference on the biology of the Cruciferae (1974: London, England), Academic Press

Vaughn MW, Tanurdžić M, Lippman Z, Jiang H, Carrasquillo R, Rabinowicz PD, Dedhia N, McCombie WR, Agier N, Bulski A, Colot V (2007) Epigenetic natural variation in *Arabidopsis thaliana*. PLoS Biol 5:e174

von Hayek A (1911) Entwurf eines Cruciferen-systems auf phylogenetischer Grundlage. Beih Bot Centralbl 27:127–335

Walck JL, Hidayati SN, Dixon KW, Thompson K, Poschlod P (2011) Climate change and plant regeneration from seed. Global Change Biol 17:2145–2161

Wang LI, Giovannucci EL, Hunter D, Neuberg D, Su L, Christiani DC (2004) Dietary intake of cruciferous vegetables, glutathione S-transferase (GST) polymorphisms and lung cancer risk in a Caucasian population. Cancer Causes Control 15:977–985

Wang X, Wang H, Wang J, Sun R, Wu J, Liu S, Bai Y, Mun JH, Bancroft I, Cheng F, Huang S (2011) The genome of the mesopolyploid crop species *Brassica rapa*. Nat Genet 43:1035

Warwick S, Gugel R, Gomez-Campo C, James T (2007) Genetic variation in *Eruca vesicaria* (L.) Cav. Plant Genet Res 5:142–153
Warwick SI (2011) Brassicaceae in agriculture. In: Genetics and genomics of the Brassicaceae. Springer, pp 33–65
Warwick SI, Gugel RK (2003) Genetic variation in the *Crambe abyssinica*-C. *hispanica*-C. *glabrata* complex. Genet Resour Crop Ev 50:291–305
Warwick SI, Sauder CA, Mayer MS, Al-Shehbaz IA (2009) Phylogenetic relationships in the tribes Schizopetaleae and Thelypodieae (Brassicaceae) based on nuclear ribosomal ITS region and plastid ndh F DNA sequences. Bot 87:961–985
Webb J, Webb J (1983) An illustrated guide to pollen analysis. Hodder and Stoughton, London
Webster N (1889) Webster's unabridged dictionary. CC Merriam & Company
Weigel D (2012) Natural variation in *Arabidopsis*: from molecular genetics to ecological genomics. Plant Physiol 158:2–22
Willis C, Hall J, Rubio de Casas R, Wang T, Donohue K (2014) Diversification and the evolution of dispersal ability in the tribe Brassiceae (Brassicaceae). Ann Bot 114:1675–1686
Wink M, Van Wyk BE (2008) Mind-altering and poisonous plants of the world. Timber Press
Xiong Z, Gaeta RT, Pires JC (2011) Homoeologous shuffling and chromosome compensation maintain genome balance in resynthesized allopolyploid *Brassica napus*. Proc Natl Acad Sci 108:7908–7913
Xu B, Cheng Y, Zou X, Zhang X (2016) Ethanol content in plants of *Brassica napus* L. correlated with waterlogging tolerance index and regulated by lactate dehydrogenase and citrate synthase. Acta Physiol Plant 38:81
Yang J, Liu D, Wang X, Ji C, Cheng F, Liu B, Hu Z, Chen S, Pental D, Ju Y, Yao P (2016) The genome sequence of allopolyploid *Brassica juncea* and analysis of differential homoeolog gene expression influencing selection. Nat Genet 48:1225
Yang Y, Moore MJ, Brockington SF, Soltis DE, Wong GK, Carpenter EJ, Zhang Y, Chen L, Yan Z, Xie Y, Sage RF (2015) Dissecting molecular evolution in the highly diverse plant clade Caryophyllales using transcriptome sequencing. Mol Biol Evol 32:2001–2014
Yant L, Hollister JD, Wright KM, Arnold BJ, Higgins JD, Franklin FCH, Bomblies K (2013) Meiotic adaptation to genome duplication in *Arabidopsis arenosa*. Curr Biol 23:2151–2156
Yao X, Cai Y, Yu D, Liang G (2018) bHLH104 confers tolerance to cadmium stress in *Arabidopsis thaliana*. J Integr Plant Biol 60:691–702
Yao Y, Carretero-Paulet L, Van de Peer Y (2019) Using digital organisms to investigate the effect of whole genome duplication in (artificial) evolution. bioRxiv 521112
Zeng L, Zhang Q, Sun R, Kong H, Zhang N, Ma H (2014) Resolution of deep angiosperm phylogeny using conserved nuclear genes and estimates of early divergence times. Nat Commun 5:4956
Zhang N, Zeng L, Shan H, Ma H (2012) Highly conserved low-copy nuclear genes as effective markers for phylogenetic analyses in angiosperms. New Phytol 195:923–937
Zhu Z, Sun B, Xu X, Chen H, Zou L, Chen G, Cao B, Chen C, Lei J (2016) Overexpression of AtEDT1/HDG11 in Chinese kale (*Brassica oleracea* var. alboglabra) enhances drought and osmotic stress tolerance. Front Plant Sci 7:1285
Zimmer EA, Wen J (2013) Reprint of: using nuclear gene data for plant phylogenetics: progress and prospects. Mol Phylogenet Evol 66:539–550
Zimmer EA, Wen J (2015) Using nuclear gene data for plant phylogenetics: progress and prospects II. Next-gen approaches J Syst Evol 53:371–379
Zou J, Hu D, Liu P, Raman H, Liu Z, Liu X, Parkin IA, Chalhoub B, Meng J (2016) Co-linearity and divergence of the a subgenome of *Brassica juncea* compared with other *Brassica* species carrying different a subgenomes. BMC Genom 17:18
Zukalová H, Vasak J (2002) The role and effects of glucosinolates of Brassica species-a review. Rost Vyroba 48:175–180

Agricultural, Economic and Societal Importance of Brassicaceae Plants

Nusrat Jabeen

Abstract This chapter reviews the importance of Brassicaceae (Cruciferae) or mustard family and summarizes the role of some representative plant species of Brassicaceae in agriculture, economy and society. The family Brassicaceae is one of the largest dicot families with more than 360 genera and 4000 species. It is grown and highly diversified almost all over the world for its edible roots, leaves, stems, buds, flowers and oilseed. The wild germplasm of this family could be used to develop cytoplasmic male sterility for the production of hybrid seeds and some weedy member provides an experimental platform for the progression of modern biology. The morphology, biogeography and ecology of current crops of Brassicaceae are reviewed. Some physiological traits like tolerance to biotic and abiotic stresses and resistance to different diseases or pests by providing nuclear genes are also discussed with the oil content and fatty acid profile. The purpose of this review is to highlight the potential values of Brassicaceae plants under any circumstances and their uses in agriculture.

Keywords *Brassica* · Crop production · Economic botany · Oilseed crops · Mustard

1 Introduction

The plants in family Brassicaceae are herbs or shrubs, with a sharply strong taste or smell due to the presence of glucosinolates, may be annual, biennial or perennial. The Brassicaceae can easily be recognized by having distinctive flowers with four petals, alternate with sepals, shaped like a cross, i.e. cruciform; therefore, the earlier name of the family is Cruciferae. The plants are normally with simple alternatively arranged leaves, tetradynamous stamens and a carpel that matures into capsular fruits called siliques.

N. Jabeen (✉)
Biosaline Laboratory, Department of Botany, University of Karachi, Karachi, Pakistan
e-mail: jabeennusrat@yahoo.com

M. Hasanuzzaman (ed.), *The Plant Family Brassicaceae*,
https://doi.org/10.1007/978-981-15-6345-4_2

This is a well-defined group of almost 360 genera and 4000 species found in temperate and alpine areas of all continents except Antarctica. The highest diversity is observed in the central and western parts of Asia, Mediterranean Europe and western North America. According to Bennett (2011), some of the genera like *Draba, Cardamine, Erysimum, Lepidium, Alyssum, Arabis* and *Physaria* contain more than 100 species. Brassicaceae contain such important crop species which contribute to raise in the agriculture and its economy. *Brassica* plants are the oldest cultivated plants known to humans as the source of medicines and vegetables (Shankar et al. 2019). *Brassica napus* (canola) is the most important edible oilseed crop. *Brassica juncea* (Indian mustard), *Armoracia rusticana* (horseradish), *Sinapis alba* (white mustard) and *Eutrena japonica* (wasabi) are condiment crops. Cole crops are the source of dietary fibre, carotenoids, phytochemical constituents with strong antioxidant potential (Nawaz et al. 2018a, b), vitamins A and C, and some macro- and micronutrients like calcium, magnesium and iron. It includes varieties of the species *Brassica oleracea* such as collards, broccoli, kale, cauliflower, mustard, cabbage, turnips, and radishes (Guerena 2006). *Brassica carinata, Crambe abyssinica, Camelina sativa, Eruca vesicaria* are found to have potential which provide a platform for bioproducts and molecular farming. These crop varieties have also been proved as a source of new edible oil, protein and biodiesel fuel (Gugel and Falk 2006; Warwick et al. 2006a, b, 2007). According to Warwick (2010), this family also contains more than 120 agriculturally important weedy species like *Sinapis arvensis, Raphanus sativus* and *Thlaspi arvense.* Several have the potential to increase weediness through the exchange of genes with crops naturally or by genetic engineering techniques (Warwick et al. 2003, 2008). *Arabidopsis thaliana, Capsella, Arabis* and *Brassica* species are well-known model organisms for genome studies (Parkin et al. 2005; Koch and Kiefer 2005; Schranz et al. 2006).

According to an estimation of FAOSTAT (2013) about 34 million hectares of the world's agricultural lands are being occupied annually by *Brassica* oilseed crops and is considered to be the third most important oilseed crop that yields about 15% of vegetable oil, after palm and soybean, in the world (Table 1) (Bennett 2011; OECD 2016).

Taxonomic databases are available to study Brassicaceae, which help plant breeders to choose correct taxa for phylogenetic study and to screen genetic potential, agronomic and economic traits. In the databases, Warwick et al. (2006a) identified 338 genera and 3709 species; 9000 chromosome number have been reported by Warwick and Al-Shehbaz (2006) from 232 genera and 1558 species and a complete guide of morphology, cytology, cytodeme status, life cycle, biogeographical distribution of all species in the tribe Brassicaceae is maintained by Warwick et al. (2009).

A comprehensive phylogenetic framework is essential to study and identify closely related species within the Brassicaceae that could serve as a source to estimate evolutionary distances, times variance between groups and various agronomic and economic characters. Phylogenetic studies on broad scales provide a comparative framework of the broad diversity of nature.

Table 1 World production of major edible vegetable oils (million metric tons)

Crops production	2015/16	2016/17	2017/18	2018/19	2019/20 (July)	2019/20 (August)
Oil, Palm	58.90	65.27	70.61	74.08	76.01	76.01
Oil, Soybean	51.56	53.81	55.18	56.04	57.52	57.41
Oil, *Rape/mustard*	27.34	27.54	28.08	27.58	28.02	27.56
Oil, Sunflower	15.39	18.16	18.48	19.71	19.72	20.28
Oil, Palm Kernel	7.01	7.64	8.34	8.59	8.81	8.81
Oil, Groundnut	5.42	5.77	5.96	5.69	5.87	5.90
Oil, Cottonseed	4.30	4.43	5.18	5.17	5.36	5.34
Oil, Coconut	3.31	3.39	3.66	3.67	3.58	3.58
Oil, Olive	3.13	2.48	3.26	3.09	3.36	3.36
Total	176.35	188.48	198.76	203.62	208.25	208.25

Source USDA Foreign Agricultural Service (2019)

According to a recent phylogenetic analysis, Brassicaceae has allotted most of the species to 52 monophyletic tribes (Al-Shehbaz 2012) in three major lineages (I, II and III) (Beilstein et al. 2006, 2008) with an unclear relationship. Huang et al. (2015) presented a comprehensive phylogeny with six clades (A–F) by using nuclear marker derived from transcriptomes of 32 Brassicaceae species representing 29 of the 51 tribes.

Nikolov et al. (2019) reformed the evolutionary history of Brassicaceae and reached to a conclusion that Brassicaceae falls within six major lineages with a novel relationship among them. They used a dataset of 1827 nuclear genes, spread throughout the genome, of 79 species of 50 of the 52 tribes in the family. One-third of the sampling was from herbarium material for a broad taxonomic coverage of the family. The authors also uncovered previously unidentified clades on a genetic basis and resolved their position.

2 Morphological Features of Brassicaceae

Most of the morphological traits in the family are effective from an agricultural point of view. Most obvious is variation in growth, limited to herbaceous woody climbers, subshrubs, large shrubs and small trees. Roots are conical, fusiform or napiform and store food. Waxy thick leaves provide resistance against drought and insects. The presence of trichomes/hairs on cotyledons and young leaves hampers insect herbivory. The floral structure includes four yellow petals, in a cross shape, but in *Streptanthus* the flower is bilaterally symmetrical. Petals are small and absent in *Cruicifer orientalis*. Usually, stamens are six in number but 2–4 are reported in *Lepidium* and 8–24 in *Megacarpaea polyandra*. Nectar may be lateral, median or of both types, have glucose or sucrose exudates (Davis et al. 1998). Fruits are usually borne above the soil but geopcarpy is also observed in some species like *Cardamine chenopodiifolia* and *Geococcus pusillus*, in which plants bend their stems and buried their fruit underground in a peanut-like manner (Warwick 2010). Fruit types vary with species mostly bear silique or silicule, which are dehiscent bivalvate capsules, some bear indehiscent lomentum, nutlet, samara, schizocarp or a drupe. Dehiscence may or may not be explosive while spines present on fruits help for animal dispersal. Seeds size and shape show a wide range of variations. The largest seed size is 2.8–3.5 × 0.7–1.1 mm in oval and globose seeds of *Neslia paniculata* and *Rapistrum rugosum*, medium in *S. alba* (2.0–2.8 mm), *S. arvensis* (1.9–2.9 mm) or small in *Diplotaxis virgate* (0.3–0.5 mm) (Hani et al. 2017). The capacity of some *Brassica* species for vegetative propagation led to the use of them in phytoremediation efforts, e.g. *A. rusticana* (Palmer et al. 2001) and *B. juncea* (Singh and Fulekar 2012).

3 Relationship of Some *Brassica* Crop Species and Genera

An old genus *Brassica* of Brassicaceae belongs to the tribe Brassiceae with the characteristic feature of conduplicate cotyledon and transversely segmented fruits. Both the segments of the fruits contain seeds or undeveloped ovules with simple trichome (if present) (OECD 2016). Three of the nine subtribes of tribe Brassicaceae are close relatives of the *Brassica* crops, namely Brassicinae, Moricandiinae and Raphaninae. The closely related genera within these subtribes include *Brassica, Diplotaxis, Eruca, Erucastrum, Guiraoa, Hirschfeldia, Muricaria, Otocarpus, Rapistrum, Raphanus, Sinapis* and *Trachystoma* are of primary interest (OECD 2016; Warwick 2010).

Brassica plants have always been the subject of primary scientific interest for their agricultural importance. Most are annual herbs, some are perennial and small shrubs. Triangle of U (1935) theory established a genomic relationship of six cultivated *Brassica* species. Three diploid species are *Brassica nigra* ($2n = 16$, genome BB), *B. oleracea* ($2n = 18$, genome CC), and *Brassica rapa* ($2n = 20$, genome AA) and three amphidiploid, derived by hybridization and polyploidization of three earlier species are *B. carinata* ($2n = 34$, BBCC), *B. juncea* ($2n = 36$, AABB), and *B. napus* ($2n = 38$, AACC). This genetic relationship (Fig. 1) is established by chloroplast and nuclear DNA analysis and suggests to assign *B. rapa* and *B. oleracea* (including wild CC genome species) to one group with *Diplotaxis erucoides* ($n = 7$) or a close relative as the ancient ancestor, and *B. nigra* assigned to a second group with *S. arvensis* ($n = 9$) or a close relative as the ancestor species (Warwick and Sauder 2005; Warwick 2010).

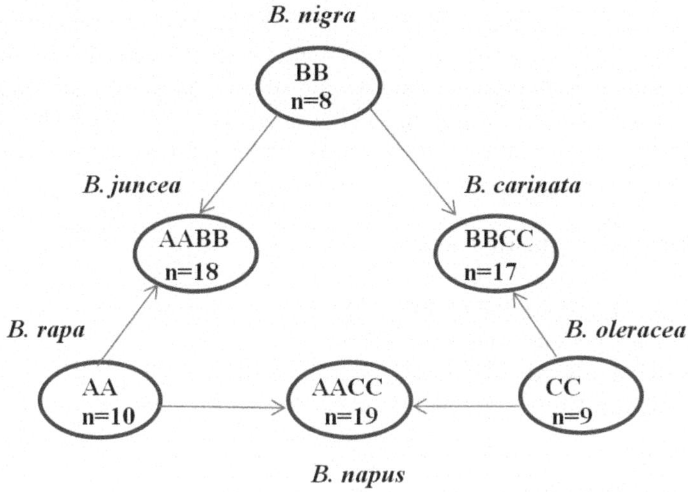

Fig. 1 Triangle of U showing the genetic relationship among six cultivated *Brassica* species (adapted from UN 1935)

4 Economically Important Crucifer Crops in Agriculture

The *Brassica* and other crucifer crops demonstrate tremendous morphological diversity and great genetic variability and their relationship has only been clarified with the help of advanced genetic techniques and chemical investigations. The crops are not only used as ornamentals but also a big source of edible roots, stems, leaves, buds, flowers, seeds, forage and good quality oil.

4.1 B. nigra *L.* (n = *8 B) (Black Mustard)*

4.1.1 Origin and Geographic Distribution

- Originated in the Asia Minor-Iran area
- Presently wildly distributed in the Mediterranean region, throughout central Europe, in the Middle East, Ethiopian highlands, Australasia, Americas and the Indian subcontinent (USDA-ARS 2013; OECD 2016).

4.1.2 Description

This widespread annual herb ranges in height from 0.6 to 2.4 m with a strong taproot. The stem is erect and hairy with large lower leaves (up to 16 cm × 5 cm) and reduced upper leaves. Fruit silique is erect, 2.5 cm long, appressed to the inflorescence axis with 4-flat sides, 0.6 cm long beak at the apex and contains 4–10 small (1 mm in diameter) brown to black seeds which are minutely pitted. Seeds dehisce when ripe and tend to germinate throughout the growing season. Its fruits shattering make it unsuitable for mechanized cultivation and it has largely been replaced by the higher yielding brown or Indian mustard *B. juncea* L., suitable for mechanical harvesting.

4.1.3 Tolerance to Abiotic Factors

- Tolerates an annual rainfall of 300–1700 mm
- Tolerates a temperature range of 6–27 °C
- Tolerates a pH of sandy loams ranging from 4.9 to 8.2 (Center for New Crops and Plant Products 2013).

4.1.4 Disease Resistance

Brassicaceae plants contain different ranges of amino acid derivative, glucosinolates. In *B. nigra*, sinigrin represents 90–99% of the total glucosinolate concentration. Glucosinolate hydrolyzes in the presence of an enzyme myrosinase into toxic byproducts like isothiocyanates, which build resistance in plant against herbivory and diseases (Lankau 2007) like Black rot (*X. campestris*) (Marthe et al. 2004); White rust (*Albugo candida*) (Gulati et al. 1991; Kolte et al. 1991) and black leaf spot (*Alternaria brassicae*) (Westman et al. 1999; Warwick 2010). Crops when grown can also be affected by insect pests like seed-pod weevils (*Ceutorhynchus* spp.), flea beetles (*Phyllotreta* spp.) and aphids (*Brevicoryne brassicae*) (Cal-IPC 2004).

4.1.5 Allelopathy

There have been numerous reports of allelopathic effects of *B. nigra*, it reduces soil biodiversity and inhibits the establishment of other species (Orrock et al. 2008; Lankau et al. 2011) and support growth of nearly monotypic mustard plants. Extracts of leaves, flowers, stems and roots of the *B. nigra* is reported to impede *Avena fatua* (wild oat) germination and growth (Turk and Tawaha 2003).

4.1.6 Hybridization

Hybridization has been reported between *B. nigra* and 27 species (FitzJohn et al. 2007). Hybrids are successfully produced at a high rate between the cross of *B. nigra* and *B. carinata* of which *B. nigra* acts as an ancestor (Mizushima 1950). Successful interspecific and intergeneric hybridization have been reported between *B. nigra* and *B. fruticulosa, Diplotaxis tenuifolia, Hirschfeldia incana* (Salisbury 1991), *B. maurorum* (Chrungu et al. 1999), *B. tournefortii* (Narain and Prakash

1972), *Orychophragmus violaceus* (Li and Heneen 1999), *S. alba* (Choudhary and Joshi 2000) and *S. arvensis* (Banga and Labana 1991). Crosses between *B. nigra* and 14 other species have been reported to fail (FitzJohn et al. 2007).

4.1.7 Fatty Acid Composition

There are seven major fatty acids palmitic (C16:0), stearic (C18:0), oleic (C18:1), linoleic (C18:2), linolenic (C18:3), eicosanoic (C22:0) and erucic (C22:1) acids. Presence of significant amount of long-chain monounsaturated fatty acids, eicosenoic (C20:1) and erucic acid (C22:1) in *Brassica* plant seeds make it unsuitable for human consumption but for industrial use. Through breeding and selection, the eicosenoic (C20:1) and erucic (C22:1) biosynthesis can be blocked to increase the percentage of oleic and linoleic acids. The oil with high oleic and linoleic acid are considered to be the good quality oil and suitable for human consumption (Sharafi et al. 2015; OECD 2016).

Black mustard oil (*B. nigra*) is pungent flavoured due to the presence of allyl isothiocyanate. Seeds of two accessions contain on an average 21% oil with 4% palmitic acid (C16:0), 1% stearic acid (C18:0), 13% oleic acid (C18:1), 17% linoleic acid (C18:2), 19% linolenic acid (C18:3) and 35% erucic acid (C22:1) (Sharafi et al. 2015).

4.2 B. napus *L. (*n = *19 AC) (Canola/Rapeseeds)*

4.2.1 Origin and Geographic Distribution

- Originated in the coastal Mediterranean region or southern Europe.
- Multiple centres of origin have also been considered (Rakow 2004; OECD 2012).
- A wide range of *B. napus* is cultivated in Asia, Africa, Australia, Canada, Europe, Mexico, New Zealand, Russia, United States and Central and South America (USDA-ARS 2017).

4.2.2 Description

B. napus is an amphidiploid resulted from the interspecific hybridization between plants of *B. oleracea* and *B. rapa* (OECD 2012). It is 1.5 m tall annual or biennial species with erect and branched stem, partially clasped by waxy leaves. Flowers are typical cruciform. Fruits are linear cylindrical silique, enclose a single row of 15 or more seeds, with the dehiscent valves present in the lower segments (4–10 cm). The upper segments of slique are narrow, thick (3.5–5.0 mm) and seedless (Callihan et al. 2000; Gulden et al. 2008; OECD 2016). Green full-sized seeds (1.8–2.7 mm in diameter) turn to black or brown when mature.

The rutabaga or swede (spp. *rapifera*); a biennial vegetable and an annual oilseed or fodder crop (spp. *napus* or spp. *oleifera*) is a common sub-species of *B. napus*.

4.2.3 Tolerance to Abiotic/Biotic Factors

- Tolerates annual rainfall of 325–700 mm
- Tolerates mean summer temperatures range 13.0–16.0 °C, mean winter temperature range −14.5 to −8.0 °C
- Tolerates pH of soil range 5.5–8.3
  (USDA-ERS 2014; USDA-ARS 2017; OGTR 2008)
- Tolerates glufosinate-ammonium (phosphinothricin) based herbicides for weed control. Genetically engineered herbicide tolerant canola variety has been developed for four, glyphosate, glufosinate-ammonium, bromoxynil and imidazoline ingredients.

4.2.4 Disease Resistance

Fungal diseases like Blackleg (*Leptosphaeria maculans*), Sclerotinia stem rot (Sclerotinia spp.), Clubroot (*Plasmodiophora brassicae*), downy mildew (*Peronospora parasitica*) and Alternaria leaf spot (*A. brassicae*) can reduce crop yield (Oilseeds WA 2006). *B. napus* varieties with resistance to disease are commercially used. Strelkov et al. (2016) reported resistance in clubroot pathogen against the varieties with a single resistance trait gene. Pests like cutworms (*Agrotis infusa*), cabbage moths (*Plutella xylostella*), Heliothis caterpillars (*Helicoverpa punctigera*), some aphids and bugs can cause serious damage to canola crop. 10–15% crop loss by viral diseases has also been reported in Europe. These viral diseases, Beet Western Yellow Virus and Cauliflower Mosaic Virus, spread by aphids (OGTR 2008).

4.2.5 Allelopathy

The allelopathic effects of *B. napus* also results in the inhibition of germination and development of plants but comparatively lesser toxic than other crucifer crops. Waddington and Bowren (1978) found that *B. napus* residue does not inhibit barley, bromegrass development. Infact Almond et al. (1986) provided evidence of good productivity of some cereal crops following oilseed rape than another cereal (OECD 2016). A 90% reduction was found in yellow nutsedge on sweet potatoes by the allelopathic effect of rapeseed (Anonymous 1993). Crop residues on the soil surface can reduce 75–90% weed emergence (Sullivan 2003).

4.2.6 Hybridization

Spontaneous hybridization between *B. rapa* and *B. oleracea* could several times result in *B. napus* (Olsson 1960; OECD 2016). The study on chloroplast and mitochondrial DNA proposed that *B. napus* evolved from multiple hybridizations between *B. oleracea* and the closely related $n = 9$ species, *Brassica montana* and *B. rapa* (Palmer et al. 1983).

Breeders are interested in the development of hybrid vigour from crosses of genetically distinct parents by ovule, ovary and embryo culture and protoplast fusion techniques to enhance crop yield, agronomic performance and increased resistance to any type of disease. To some extent, they are successful to overcome the natural barriers to interspecific/generic hybridization.

4.2.7 Intraspecific Hybridization

Intraspecific hybridization, i.e. the transfer of gene between the individuals of *B. napus*, has been summarized by Beckie et al. (2003) and Hüsken and Dietz-Pfeilstetter (2007). Cleistogamy, the self-fertilization in permanently closed flower can decrease intraspecific hybridization in *B. napus* (Gruber et al. 2012). The average outcrossing rate in *B. napus* is 30% and varies according to variety, local geographical and ecological conditions and availability of pollinators (Beckie et al. 2003; Halfhill et al. 2004; Damgaar and Kjellsson 2005; Yoshimura et al. 2006; Hüsken and Dietz-Pfeilstetter 2007). Hoyle et al. (2007) manipulated from pollen dispersal modelling that wind is the main vector of long-distance gene flow in *B. napus*.

Herbicide tolerance trait was developed in *B. napus* from intraspecific hybridization (Hall et al. 2000; Knispel et al. 2008). Recombinant DNA technology is used to develop tolerance in *B. napus* for glyphosate, glufosinate-ammonium and bromoxynil and mutagenesis for imidazoline ingredients (Simard et al. 2002; Johnson et al. 2004).

4.2.8 Interspecific/Generic Hybridization

Interspecific/generic hybridization in *B. napus* has been reviewed by many authors (Chèvre et al. 2004; FitzJohn et al. 2007; Warwick et al. 2009; Katche et al. 2019). The reports of interspecific and intergeneric sexual crossing between related species and *B. napus* are summarized in Table 2.

In vitro *B. napus* hybrids development through ovary, ovule and embryo culture and through protoplast fusion is summarized in Table 3.

4.2.9 Cytoplasmic Male Sterility

Cytoplasmic male sterility (CMS)-restorer system is a practical and efficient system for hybrid seed production worldwide. China developed the first *B. napus* commercial

Table 2 Inter-species and inter-genera hybridization (sexual crosses) of *Brassica napus* with related species

Interspecific cross	Potential	Report	References
Female × Male	Natural cross introgression		
B. carinata *B. napus*	Low Low	Successful	Fernandez-Escobar et al. (1988)
		Successful 26 seeds/cross	Getinet et al. (1997)
		Successful with 1 F1 seed/unreported number of crosses	La Mura et al. (2010)
		Yield 1.56 seeds/pollination	Niemann et al. (2014)
B. napus *B. carinata*	Low Low	Successful	Fernandez-Escobar et al. (1988), Chen and Heneen (1992)
		Successful yield 8.2 F1 seeds/pollination	Rashid et al. (1994)
		Yield 90 siliques/110 number of pollinated flowers	Gupta (1997)
		0–0.6 seeds/pollination. Pollen is not observed in F1	Chang et al. (2007)
		Yield 4 seeds in F1/unreported number of crosses	La Mura et al. (2010)
		Yield 3.44 seeds/pollination	Niemann et al. (2014)
		Successful in 1 or 2 *B. carinata* genotypes. Demonstrate 15–20% male sterility in F1	Sheikh et al. (2014)
B. juncea *B. napus*	High High	Successfully inherited resistance to blackleg	Roy (1978)
		Successfully inherited resistance to white rust	Subudhi and Raut (1994)
		Successful	Anand et al. (1985)
		Yield successfully 4.6 seeds in F1/pollination	Frello et al. (1995)
B. napus *B. juncea*	High High	Successfully inherited male sterility	Fan et al. (1986)

(continued)

Table 2 (continued)

Interspecific cross	Potential	Report	References
Female × Male	Natural cross introgression		
		Successfully produced vigorous hybrid plants in F1 with 29% pollen fertility. 1–2 seeds/silique	Prakash and Chopra (1990)
		Successfully obtained 2.1 seeds/pollinated bud	Rashid et al. (1994)
B. maurorum *B. napus*	Extremely low Extremely low	Two hybrid seedlings were established with sterile pollen. Number of seeds per cross was 7	Chrungu et al. (1999)
B. nigra *B. napus*	Very low Low	Cross attempted but failed	Diederichsen and Sacristan (1988), Kerlan et al. (1992), Daniels et al. (2005), Wei and Darmency (2008)
		Successful	Struss et al. (1991)
B. napus *B. nigra*	Low Low	Successful	Zhu and Struss (1991)
B. oleracea *B. napus*	Very low Very low	Successfully produced 4–19 seeds/silique	Chiang et al. (1977)
B. napus *B. oleracea*	Very low Very low	Successful	Chiang et al. (1977)
		Did not get the expected result	Ford et al. (2006)
B. rapa *B. napus*	High High	Successfully yielded 278 seeds and 122 hybrids	Palmer (1962)
		Successfully produced 35 seeds out of which 8 seeds grown into plants	Kamala (1976)
		Successfully produced 52 viable seeds	Beversdorf et al. (1980)
		Successfully produced average 7.6/silique	Jenkins et al. (2005)
		Numerous hybrid seeds produced successfully	Brown and Brown (1996), Hauser et al. (1997), Hauser et al. (1998)

(continued)

Table 2 (continued)

Interspecific cross	Potential	Report	References
Female × Male	Natural cross introgression		
		Successful	Anand and Downey 1981), Hansen et al. (2001), Zhu et al. 2004
		Successfully produced clubroot-resistant hybrid	Gowers (1982)
		Successfully produced numerous F1 hybrids from the cross of *B. rapa* and 147 *B. napus* cultivars and lines	Pellan-Delourme and Renard (1987)
		Successfully yielded 3 hybrids from 230 pollinated flowers	Metz et al. (1997)
		Successfully produced 19–100% F1 hybrids	Halfhill et al. (2001)
		Obtained 16.9–0.7% hybridization frequency	Halfhill et al. (2002)
B. napus *B. rapa*	High High	Successfully produced 226 seeds and 160 hybrids from 165 plants	Palmer (1962)
		Successfully produced many seeds	McNaughton (1973), Mackay (1977), Brown and Brown (1996), Hauser et al. (1998)
		Successfully transferred resistance to clubroot race 3 hybrids	Johnston (1974)
		Successfully produced 87 seeds out of which 10 seeds grown into plants	Kamala (1976)
		Successful	Nwankiti (1970), Beversdorf et al. (1980), Anand and Downey (1981), Hauser et al. (1997), Hu et al. (1997), Verma et al. (2000), Liu et al. (2002)
		Successfully produced an average of 4 seeds/silique	Cheng et al. (1994)

(continued)

Table 2 (continued)

Interspecific cross Female × Male	Potential Natural cross introgression	Report	References
		Successfully set an average of 9.8 seeds/pollination	Mikkelsen et al. (1996)
		Successfully introgressed long pod character from a *B. napus* into *B. rapa*	Lewis et al. (2001)
D. erucoides *B. napus*		Successfully harvested 74 viable seeds from 78 pods. Produced 3 hybrids	Ringdahl et al. (1987)
		Unsuccessful	Vyas et al. (1995)
B. napus *D. erucoides*	Very low Very low	Sexual and filed cross not reported	
B. napus *B. tournefortii*	Low Low	Successful with average 0.69 seeds/cross	Gupta (1997), Salisbury (2002a)
B. tournefortii *B. napus*	Very low Very low	Sexual cross attempt failed	Salisbury (2002a)
B. napus *D. muralis*	Low Very low	Successful	Bijral and Sharma (1996)
D. muralis *B. napus*	Not reported	Successful	Fan et al. (1985)
		Successfully harvested 607 viable seeds from 157 pods. On germination formed 31 hybrids	Ringdahl et al. (1987)
B. napus *Eruca sativa*	Low Very low		Bijral and Sharma (1996)
Eruca sativa *B. napus*	Not reported	Not reported	
B. napus *E. gallicum*	Very low Very low	Successfully obtained numerous viable seeds. 1 hybrid plant produced/100 pollinated flowers	Lefol et al. (1997)
E. gallicum *B. napus*	Very low Very low	Unsuccessful	Lefol et al. (1997)
B. napus *H. incana*	High Low	Successful with 2 seeds/cross	Chadoeuf et al. (1998)
H. incana *B. napus*	High Low	Successfully produced 2 × 10^{-5} seeds/cross	Chèvre et al. (1996), Wei and Darmency (2008)

(continued)

Table 2 (continued)

Interspecific cross	Potential	Report	References
Female × Male	Natural cross introgression		
B. napus *R. raphanistrum*	High Low	Successfully obtained 2 viable seeds/100 harvested seeds	Lefol et al. (1997)
		Successfully produced an average of 0.12 seeds/100 flowers and 0.78 seeds/plant	Chèvre et al. (1998)
R. raphanistrum *B. napus*	Very low Very low	Unsuccessful	Kerlan et al. (1992), Lefol et al. (1997), Warwick et al. (2003)
B. napus *R. sativus*	Very low Extremely low	Successfully produced intergeneric hybrids	Huang et al. (2002)
R. sativus *B. napus*	Very low Extremely low	Unsuccessful	Lelivelt et al. (1993a), Metz et al. (1995)
B. napus *S. alba*	Very low Extremely low	Unsuccessful	Lelivelt et al. (1993b), Sridevi and Sarla (1996)
S. alba *B. napus*	Extremely low Extremely low	Unsuccessful	Lelivelt et al. (1993b), Sridevi and Sarla (1996)
B. napus *S. arvensis*	Low Very low	Successful	Inomata (1988), Moyes et al. (1999), Moyes et al. (2002)
		Unsuccessful	Kerlan et al. (1992), Bing et al. (1995)
S. arvensis *B. napus*	Extremely low Extremely low	Successfully obtained 1 hybrid	Moyes et al. (2002)
		Unsuccessful	Kerlan et al. (1992), Bing et al. (1995), Moyes et al. (1999)

CMS system, known as the Polima system (Fu et al. 1995a, b; OECD 2016). In Ogura CMS system alien cytoplasm was obtained from crossing *B. napus* to *R. sativus* (Ogura 1968). The Ogura CMS system was widely used in *B. napus, B. juncea* and *B. oleracea* (Yamagishi and Bhat 2014). Protoplast fusion is used to transfer Ogu cytoplasmic male sterility gene from *B. napus* to *B. juncea* to improve male-sterile lines in hybrid breeding systems (Kirti et al. 1995).

Table 3 In vitro *B. napus* hybrids development through ovary, ovule and embryo culture and through protoplast fusion

Parental combination ♀ × ♂	Cross-type	References
B. napus + *A. thaliana*	**Protoplast fusion** Callus formed. From each callus arise 29 shoots. 25 were hybrids, 14 hybrids were female fertile and set seeds when backcrossed to *B. napus*	Forsberg et al. (1994)
	Protoplast fusion 1520 calli formed which produced 155 shoots. When roots formed 109 were transplanted to soil. 68 plants established in the soil. Seeds are produced in 56 plants after self-pollination. The leftover plants remain flowerless or seedless	Yamagishi et al. (2002)
B. napus + *Barbarea vulgaris*	**Protoplast fusion** 5 experiments resulted in 1414 calli. 6 were reported hybrids from 9 sprouts of 102 established shoots. The development of mature plants failed outside in vitro conditions	Fahleson et al. (1994a)
B. napus × *B. bourgeaui*	**Ovary culture** Ovary culture yielded interspecific F_1 hybrids with fertile pollen. Many F_2 and BC_1 seeds were harvested from F1 hybrids after self- or open pollination	Inomata (2002)
B. napus + *B. carinata*	**Protoplast fusion** From calli, 13 plants were regenerated	Klíma et al. (2009)
	Protoplast fusion From 800 calli, 13 were differentiated into shoots and 5 were successfully transferred outside in vitro conditions. From 1500 calli, 38 produced shoots and 20 hybrid plants	Sundberg and Glimelius (1991)
B. napus × *B. carinata*	**Ovary, ovule culture** Yielded 2 seeds from 44 pollinated flowers in F1. Male-sterile hybrids were reported in F1	Sabharwal and Doležel (1993)

(continued)

Table 3 (continued)

Parental combination ♀ × ♂	Cross-type	References
B. napus × *B. juncea*	**Ovary/Ovule/Embryo culture** Embryos were formed from controlled pollination and cultured in basal media. 37.8–80% of seeds were produced from ovaries. Hybrids were transferred in the field to maturity	Bajaj et al. (1986)
	Ovary/Ovule culture 2346 ovules obtained from controlled crosses were cultured and 249 hybrids were produced	Sacristan and Gerdemann (1986)
	Ovary/Ovule culture Ovules from 95 controlled crosses were developed and placed in the culture medium, for seedlings establishment then transferred to greenhouse for development. 14 seeds were produced from 44 pods in F1	Sabharwal and Dolezel (1993)
	Embryo Culture On average, 36–50% calli were produced from hybrid cultured embryos, resulting from controlled crosses. 18.62% of cultured embryo-derived plants were obtained	Zhang et al. (2003)
B. napus × *B. nigra*	**Embryo culture** Embryo-derived hybrid plants were obtained from a mutual cross in a control environment	Busso et al. (1987)
B. napus + *B. nigra*	**Protoplast fusion** 332 hybrid calli were produced from X-ray irradiated protoplasts fusion from which 30 developed shoots and transferred into the soil for further growth in growth chambers	Gerdemann-Knorck et al. (1994, 1995)

(continued)

Table 3 (continued)

Parental combination ♀ × ♂	Cross-type	References
B. napus × *B. oleracea*	**Ovary culture** After controlled pollination, ovary was cultured in two different mediums. Then seedlings were transferred to pots in greenhouse. Embryos obtained from the two varieties of *B. oleracea* with *B. napus* as a female parent were 21.5 and 13.2%, with *B. napus* as a male parent it was 1.8 and 1.2%. Seedlings obtained with *B. napus* as a female parent were 13.7 and 5.7%, with *B. napus* as a male parent it was 0.7 and 0.2%	Kerlan et al. (1992)
	Ovules culture Developing ovules from controlled crosses were cultured in medium. The transformation of embryos to plants was reported to be from 0 to 22%	Ripley and Beversdorf (2003)
	Embryo culture Hybrid embryos were saved by ovule culture in culture medium added with 300 mg l-1 casein hydrolysate, 200 mg l-1 glutamine and 13% sucrose	Bennett et al. (2008)
B. napus + *B. oleracea*	**Protoplast fusion** 8.9% shoots were regenerated from 1128 calli. 93% of hybrid plants were obtained	Sundberg and Glimelius (1991)
B. napus × *B. rapa*	**Ovule culture** Ovule culture and colchicine treatment were used to develop lines of *B. napus* through mutual interspecific crosses between *B. oleracea* and *B. rapa*	Zhang et al. (2002)
B. napus × *C. bursa-pastoris*	**Embryos culture** Embryos from 9248 controlled pollination were cultured in medium. 169 plants were obtained in F1	Chen et al. (2007)

(continued)

Table 3 (continued)

Parental combination ♀ × ♂	Cross-type	References
B. napus + *Descurainia sophia*	**Protoplast fusion** 19 somatic hybrid plants were obtained from protoplast fusion. 8 did not flower	Rongzhan et al. (2007a)
B. napus × *Diplotaxis harra*	**Ovary culture** Ovaries from controlled pollinations were cultured in basal medium. 75 cultured ovaries yielded 3 seeds. 2 died at the seedling stage and the remaining 1 matured as F1 hybrid plant	Inomata (2005)
B. napus + *Eruca sativa*	**Protoplast fusion** Protoplast fusion was cultured in a modified medium then transferred to a basal medium for rooting. Average 5.4% calli were differentiated into shoots and 23 hybrids plants were transferred to the greenhouse	Fahleson et al. (1988)
B. napus + *Microthlaspi perfoliatum*	**Protoplast fusion** 27 hybrid or partially hybrid calli yielded 19 plants	Fahleson et al. (1994b)
B. napus × *R. raphanistrum*	**Ovary Culture** Developing ovaries obtained from controlled crosses were cultured in the medium for seedling development. Transferred to pots in greenhouse. 243 embryos yielded 109 interspecific hybrid plants	Kerlan et al. (1992)
B. napus × *R. sativus*	**Ovule Culture** 14 cross combinations yield 5 combinations of hybrids by ovule culture	Takeshita et al. (1980)
	Ovary Culture 14 hybrid plants were obtained from 58 cultured ovaries	Lou et al. (2000)
B. napus +*R. sativus*	**Protoplast fusion** 5% of 364 calli regenerated shoots and obtained 100% hybrid plants	Sundberg and Glimelius (1991)
	Protoplast fusion PEG-mediated protoplast fusion produced 12 somatic hybrids	Wang et al. (2006)

(continued)

Table 3 (continued)

Parental combination ♀ × ♂	Cross-type	References
B. napus + *Rorippa indica*	**Protoplast fusion** Protoplast fusion in cultured medium produced somatic hybrids. Then 890 F_2 plants were produced by 3-way intergeneric hybridization and embryo rescue	Rongzhan et al. (2007b)
B. napus × *S. alba*	**Ovary culture** Ovaries from mutual crosses were cultured in medium and produced 2.2% and 1.9% of interspecific hybrids when *S. alba* was used as the female and male parent, respectively	Chevre et al. (1994)
B. napus + *S. alba*	**Protoplast fusion** Electrically induced protoplast fusion produced 7 somatic hybrids. Hybrids at maturity set seeds by self-pollination or backcross with *B. napus*	Wang et al. (2005)
B. napus + *S. arvensis*	**Protoplast fusion** Protoplast fusion produced 54 symmetric and 4 asymmetric somatic hybrids with 1.4% plant regeneration efficiency	Hu et al. (2002)
B. napus × *S. arvensis*	**Ovules culture** 30 embryos and 3 seedlings from the 808 ovaries cultured with *B. napus* as the female parent contrasted with no embryos and no seedlings from 732 ovaries cultured with *B. napus* as the male parent	Kerlan et al. (1992)
	Ovary culture Produced 1.2% *B. napus* and 0.1% *S. arvensis* hybrids/number of flowers pollinated	Lefol et al. (1996)
B. napus + *B. tournefortii*	**Protoplast fusion** Male-sterile cybrids were produced by the fusion of protoplasts	Liu Clarke et al. (1999)

4.2.10 Fatty Acid Composition

Seeds of three accessions of *B. napus* contain average 47% oil with 5% palmitic acid (C16:0), 2% stearic acid (C18:0), 62% oleic acid (C18:1), 17% linoleic acid (C18:2), 7% linolenic acid (C18:3) and 1% erucic acid (C22:1). Whereas two cultivars of *B. napus* Okapi and Opera, originated from France and Germany, respectively, are reported as free-erucic acid genotypes (Sharafi et al. 2015). The genotypes with high oleic acid and low or zero erucic acid are ranked in a higher level of nutrition.

In one of the research experiment conducted on 26 genotypes of *Brassica* species, Rai et al. (2018) found 37.82–40.56% oil content in *B. napus* with the Palmitic acid, 3.70–5.15%; oleic acid, 16.15–37.98%; Linoleic acid, 18.57–26.93%; linolenic acid, 9.99–17.23%; Erucic acid, 10.04–34.96%.

4.3 B. rapa *L.* (n = *10 A) (Turnip Rape)*

4.3.1 Origin and Geographic Distribution

- Originated in the mountainous region close to the Mediterranean Sea. Afghanistan–Pakistan region is also considered as the centre of origin.
- Widely distributed throughout Europe, Africa, Northwest China, Asia and the Indian subcontinent (OECD 2016).

4.3.2 Description

Metzger (1833) combined the taxa, *B. rapa* and *Brassica campestris* under the name *B. rapa* because of the similar ranks (Toxopeus et al. 1984).

B. rapa includes both annual and biennial forms. The above portion of tap roots expands to develop a bulbous storage organ and a lower narrow portion extends below the storage organ. Leaves are glabrous, upper ones are sessile, partially clasp stem and lower ones are pinnatifid and petioled. Stem is branched which originates in the axils of highest leaves and ends in an elongated raceme with typical crucifer flower and ¾ to 4 inches long siliques. Mature siliques split at the base to release the seeds. Reddish-grey to black seeds are about 1/16 inch wide almost round in shape (Turner and Gustafson 2006; DiTomaso and Healy 2007; Warwick 2010; Young-Mathews 2012)

Three major groups of *B. rapa* can be differentiated on the basis of morphology:

1. The oleiferous or oil-type rape.

Canola is a specific form with low erucic acid in oil and low glucosinolate in protein meal.

2. The leafy-type include

   the Chinensis group (Pak Choi, celery mustard).
   the Pekinensis group (Chinese cabbage).
   the Perviridis group (tender green).

3. The rapiferous-type include

   the rapifera group (turnip, rapini),
   the ruvo group (turnip broccoli, Italian turnip).

4.3.3 Tolerance to Abiotic/Biotic Factors

- Tolerates annual rainfall of 3.5–41.0 dm,
- Tolerates temperature range of 10–27.5 °C (Duke 1983),
- Tolerates pH of soil range 4.8–8.5 (Hannaway and Larson 2004),
- Tolerates drought conditions, moderate temperature and low fertile soil (Clark 2007),
- Triazine-resistant biotypes have been reported in *B. rapa* (Warwick 2010).

4.3.4 Disease Resistance

Fungal diseases like Clubroot (*P. brassicae*) and Black rot (*Xanthomonas campestris*), White rust (*Albugo candids*), Black rust (*A. brassicae*), Gray mold (*Botrytis cinerea*) can cause crop yield reduction. Clubroot and Black rot resistance has been reported in *B. rapa* (Ignatov et al. 1999; Zhang et al. 2015). Some parasite like *Orobanche cernua* and bacteria like *Agrobacterium tumefaciens* and *Bacterium aroideae* can attack *rapa* crop. Bacterial soft-rot-resistant variety is produced through protoplast culture between *B. rapa* and *B. oleracea*. Turnip mosaic viruses, beet mild yellowing, cauliflower mosaic viruses are also isolated. Some aphids, root maggot and flea beetles can cause serious damage to turnip *rapa* crop (CFIA 2014).

4.3.5 Allelopathy

Barnum and Franks (2013) suggested that compounds in the extract-treated seeds of *B. rapa* appeared to be allelopathic and inhibit germination. The reduction in germination, growth and yield of crop from the residues of *B. rapa* are reported to be less toxic than other members of Brassicaceae (OECD 2016).

4.3.6 Hybridization

Interspecific and intergeneric sexual crossing attempts lead to the production of a new improved hybrid with useful agronomic traits in agriculture. In interspecific or generic hybridization, a trait has to be incorporated into a species genome to produce F1 hybrids and repeated backcross of the plant of that species to F1 results in a viable and fertile progeny which is necessary. Some hybrids are resulted from field outcrossing hybridizations such as *B. rapa* × *B. napus*, *B. napus* × *B. rapa* and *B. rapa* × *B. nigra* (Bing et al. 1991, 1996a, b) and some are achieved through hand pollination. Interspecific crosses may result in partial fertile or sterile progeny. The number of maternal and paternal chromosomes improves the chance for a successful cross. Flowering periods of *B. rapa* are critical for hybridization events to occur. Hybridization potential can be evaluated by calculating the flowering period of cultivated plant and related species, their physical distance, occurrence of the pollinators, its frequency, activity and behaviour.

According to the findings of Bing et al. (1991), the hybrids produced by the crosses between *B. napus*, *B. juncea* and *B. rapa* have the potential to produce viable seed with fertile progeny and these species are widely grown for the production of both canola and mustard. Some interspecific/intergeneric sexual attempt which produced hybrids in F1 progeny, include *B. rapa* × *B. oleracea* (Mattsson 1988), *B. oleracea* × *B. rapa* (Wojciechowski 1985), *B. rapa* × *B. carinata* (Falk 1991), *B. carinata* × *B. rapa* (Mizushima 1950), *B. juncea* × *B. rapa* (Katiyar and Chamola 1995), *R. sativus* × *B. rapa* (Lou et al. 2000; Matsuzawa et al. 2000), *B. rapa* × *S. arvensis* (Mizushima 1950), *D. erucoides* × *B. rapa* (Vyas et al. 1995); *D. Muralis* × *B. rapa* (Salisbury 1991; Lefol et al. 1997); *D. tenuifolia* × *B. rapa* (Salisbury 1991); *B. rapa* × *Erucastrum gallicum* (Lefol et al. 1997). Hybridization is found unsuccessful for 18 other intergeneric combinations (FitzJohn et al. 2007). Protoplast fusion technology does not only allow intra-generic hybridization, but also intergeneric hybrids and cybrids production that are sexually incompatible. The fusion of *B. rapa* and *B. oleracea* protoplasts result in bacterial soft-rot-resistant hybrids (Ren et al. 2000; Park et al. 2012).

4.3.7 Fatty Acid Composition

The double low commercial genotypes of *B. rapa* is dominating in oilseed *Brassica* production area in developed countries. It is characterized by having a low content of erucic acid (<1%) and glucosinolates (<18 μmoles/g seed at 8.5% moisture) (CFIA 2014). Sharafi et al. (2015) reported considerable variations in oil content and fatty acid profile in seeds of 5 accessions of *B. rapa*. Average 40% oil content was with 3% palmitic acid (C16:0), 1.4% stearic acid (C18:0), 19% oleic acid (C18:1), 14% linoleic acid (C18:2), 16% linolenic acid (C18:3) and 40% erucic acid (C22:1). In one of the experiments, Cartea et al. (2019) evaluated the potential of genotypes to assess oilseed crop suitability for edible vegetable oil or industry. The seed fatty acid profile showed 12% oleic acid, 13% linoleic acid, 8–9% linolenic acid, 8–9% eicosenoic acid and 50–51% erucic acid with an average 47.3% oil content. The presence of high erucic acid content in genotypes is suitable for industrial purposes only.

4.4 B. carinata *A. Braun.* (n = *17 BC) (Abyssinian Mustard)*

4.4.1 Origin and Geographic Distribution

- Originated in the highland Ethiopian plateau of northeast Africa and the Mediterranean coast.
- Modern cultivation is limited to southern Europe, Australia and India and commercial cultivation is to Ethiopia and neighbouring countries, Yemen, Eritrea, Kenya, Uganda and Tanzania (Warwick et al. 2009; Prakash et al. 2012; USDA and ARS 2014; Marillia et al. 2014).
- Recently, it has been introduced in India as an oil crop and in Canada and Spain for commercial cultivation.

4.4.2 Description

B. carinata is an amphidiploid species derived from cross between wild-growing *B. nigra* as the female and cultivated kale-like forms of *B. oleracea* as the male (Palmer et al. 1983; Warwick 2010). *B. carinata* is a herbaceous annual 1.4 m (average) tall plant with a well-developed taproot and highly branched shoots (Barro and Martín 1999; Zanetti et al. 2013). Stems are waxy and 2 cm in diameter. Leaves with short petiole and simple trichomes are alternate, slightly hairy and waxy (Mnzava and Schippers 2007; Al-Shehbaz 2012). Lower leaf-blades are large up to 20 cm long and 10 cm wide and upper leaves are small and narrow. Inflorescences are compound racemes with typical cruciform flowers (Mnzava and Schippers 2007). 5 cm long siliques with a 2–7 mm straight or curved conical beak are non-dehiscent due to their thick and highly lignified valve margins (Banga et al. 2011). Siliques contain yellow to brown up to 20 globose, 1–1.5 mm in diameter and finely reticulated seeds (Mnzava and Schippers 2004; Rahman and Tahir 2010).

4.4.3 Tolerance to Abiotic/Biotic Factors

- Tolerates annual rainfall of 800–1700 mm but preferable is in the range of 600–1000 mm.
- Tolerates extreme temperatures range of 5–35 °C but plants do well in the cool range of 14–18 °C.
- Tolerates pH of soil range 5–8.5. Preferable pH is in the range of 5.5–8. (CFIA 2017; http://ecocrop.fao.org).

4.4.4 Disease Resistance

B. carinata is resistant to blackleg (*Verticillium longisporum*; Navabi et al. 2010) sclerotinia stem rot (*Sclerotinia sclerotiorum*; Mustard 2013), black rot (*X. campestris;*

Sharma et al. 2017a, b), powdery mildew (Erysiphe polygoni; Tonguç and Griffiths 2004) white rust (*A. candida*; Naresh 2014), and alternaria (*A. brassicae*; Chavan and Kamble 2014). It is susceptible to clubroot (*P. brassicae*) and clubroot-resistant variety of *B. carinata* might be expected through hybridization with *B. rapa* (Peng et al. 2013). It is reported to be less susceptible to flea beetles (*Phyllotreta* spp.), diamondback moth (*P. xylostella* L.) plant bugs and aphids (Ulmer et al. 2001, 2002; Cárcamo et al. 2007).

4.4.5 Hybridization

Interspecific hybridization in *B. carinata* was not studied extensively. Hybridization between *B. carinata* with *Brassica maurorum*, *Brassica tournefortii*, *Erucastrum abyssinicum*, *R. sativus* and *S. alba* have failed when *B. carinata* is the pollen donor (FitzJohn et al. 2007; Cheung et al. 2015; CFIA 2017). Successful hybridization is reported between the *B. carinata* with other members of Brassicaceae like *B. napus, B. nigra, B. oleracea, B. rapa, B. juncea, R. sativus* and *S. arvensis.* The results of such hybridization are summarized in Table 4. The results of in vitro hybridization are summarized in Table 5.

4.4.6 Fatty Acid Composition

Considerable variations for oil content in *B. carinata* are reported in different accessions and it depends on the cultivar type and growth conditions. The seeds contain on an average 25–47% oil with 35–44% erucic acid, 15–22% linoleic acid, 16–20% linolenic acid, 10–12% oleic acid, 7–9% eicosenoic acid, 2–4% palmitic acids (Getinet et al. 1997; Cardone et al. 2003; Mnzava and Schippers 2007; Taylor et al. 2010; CFIA 2017). The seed protein and glucosinolate contents are 25–45% and 150 mmol g^{-1}, respectively. Sharafi et al. (2015) reported in three accessions of *B. carinata,* the average amount of oil is 26.86% with 3.16% palmitic acid, 0.93% stearic acid, 9.8% oleic acid, 17.3% linoleic, 18.31% linolenic acid, 41.25% erucic acid. *B. carinata* oil is not suitable for humans due to the presence of a high amount of erucic acid in it. It is extracted for the conversion of it to fuel (Kasim et al. 2017).

4.5 B. oleracea *L. (n = 9 C) (Cole Crops)*

4.5.1 Origin and Geographic Distribution

- Originated in the Mediterranean region.
- The wild forms grow along the coast of the Mediterranean Sea and the Atlantic Ocean from Greece to Britain.
- Cultivated forms are introduced and grown worldwide except tropical areas. (Warwick 2010; OECD 2016)

Table 4 Inter-species and inter-genera hybridization (sexual crosses) of *B. carinata* with related species

Interspecific cross Female × Male	Reports	References
B. carinata *B. juncea*	375 pollinations yield 11 seeds in F1. Average of 0.02 seeds per pollination	Getinet et al. (1997)
	Yield 6 seeds in F1 but the number of crosses is not reported	La Mura et al. (2010)
B. carinata *B. napus*	Yield average is 0.08 seeds/pollination	Getinet et al. (1997)
	Yield 1 seed in F1 but the number of crosses is not known	La Mura et al. (2010)
	Yield average is 1.56 seeds/pollination in F1	Niemann et al. (2014)
B. carinata *B. nigra*	Successful	Chang et al. (2011)
B. carinata *B. oleracea var. alboglabra*	Yield average is 7.8 seeds/pollination in F1 Yield average is 7.9 fertilized ovules/silique	Rahman (2001, 2004)
B. carinata *B. oleracea*	Yield 2 seeds in F1 from 27 flowers; Pollen produced from plants of F1 was not viable	Chang et al. (2011)
B. carinata *B. rapa*	Successful; Yield 6 hybrids in F1 hybrids but the number of crosses is not reported	Struss et al. (1991, 1992)
	Successful; Yield average is 3.3 seeds/pollination in F1; Yield average is 11.1–11.3 fertilized ovules/silique	Rahman (2001, 2002, 2004)
	Yield average is 0.23 seeds/pollination in F1	Choudhary et al. (2000)
	Yield average is 0.05–0.16 hybrids/pollination in F1 with 4.4–7.6% pollen viability	Li et al. (2005)
	Yield more than 80 seeds in F1/100 buds from crosses with 107 *B. carinata* accessions	Jiang et al. (2007)
	Yield 642 seeds in F1 from an unknown number of crosses	Liu et al. (2009)
	Yield 2 seeds in F1 seeds from an unknown number of crosses	La Mura et al. (2010)

(continued)

Table 4 (continued)

Interspecific cross Female × Male	Reports	References
B. carinata *B. tournefortii*	Yield average is 0.21 seeds/pollination with the recovery of 1 F1 hybrid and 2.3% pollen viability	Choudhary and Joshi (2012)
B. carinata *R. sativus*	Yield 1 seed in F1 from a number of unreported crosses	La Mura et al. (2010)
B. carinata *S. alba*	Yield average is 0.175 seeds/pollination. 26 hybrids germinate in F1	Sridevi and Sarla (2005)
B. carinata *S. arvensis*	Yield 6 seeds in F1 from the number of crosses not reported	La Mura et al. (2010)
	Total 731 hybrids were produced from 997 crosses at a rate of 6.4%	Cheung et al. (2015)
B. carinata *O. violaceus*	Yield average is 0.67–1.56 hybrids in F1/100 pollinations; Yield 8 hybrids in F1	Li et al. (1998, 2003)
B. fruticulosa *B. carinata*	Multivalent associations were found high in hybrids	Bijral et al. (1994)
B. juncea *B. carinata*	Yield 108 seeds in F1 from an unknown number of pollinations. 5 hybrids in F1 were male-sterile. backcross was not successful	Getinet et al. (1994)
	Yield 48 siliques/60 pollinated flowers	Gupta (1997)
	Yield average is 0.75 viable seeds/pollination in F1	GhoshDastidar and Varma (1999)
	Yield average is 0.7 seeds/pod in F1 22.2% pollen viability	Chang et al. (2007)
	Yield 3 seeds in F1 from an unreported number of crosses	La Mura et al. (2010)
	Fertile hybrids produced from 2 out of 9 crosses resulted in A6 generation with natural *Brassica napus* genome (AACC, $2n = 38$)	Chatterjee et al. (2016)
B. napus *B. carinata*	Yield 90 siliques from 110 pollinated flowers	Gupta (1997)

(continued)

Table 4 (continued)

Interspecific cross Female × Male	Reports	References
	Yield average is 0–0.6 seeds/pollination with sterile male in F1	Chang et al. (2007)
	Yield 4 seeds in F1from an unreported number of crosses	La Mura et al. (2010)
	Yield average is 3.44 seeds/pollination	Niemann et al. (2014)
B. oleracea *B. carinata*	Yield average is 7.2–8.2 fertilized ovules/silique	Rahman (2004)
B. rapa *B. carinata*	Yield 2 hybrids in F1 from an undocumented number of crosses	Struss et al. (1991)
	Yield average is 1.17 seeds/pollination	Meng et al. (1998)
	Successful; Yield average is 5.2 seeds/pollination	Rahman (2001, 2002)
	Yield 1 seed in F1 from an unreported number of crosses	La Mura et al. (2010)
S. arvensis *B. carinata*	A single hybrid was produced from 1109 crosses and it produced less than 1% of the parent's pollen	Cheung et al. (2015)

4.5.2 Description

B. oleracea is a group of vegetables called the cole crops, which means mainly stem plants. Various cultivated forms have been developed from the wild *B. oleracea*. Plants may be biennial or perennial with a thin taproot and attain 60–100 cm height. All forms in common have succulent, waxy, hairless leaves and the surface of which look grey-green or blue-green. The structure and arrangement of plants parts vary in *B. oleracea* cultivars which are as follows:

Leaves: loose, branched and open foliage (kale and collards; *B. oleracea*, var. *acephala* var. *costata*, var. *medullosa*) folded into compact tightly packed heads—large terminal heads (cabbage; *B. oleracea*, var. *capitate*, var. *sabauda*) small axillary tightly packed heads (Brussels sprouts; *B. oleracea*, var. *gemmifera*);

Stem: much expanded to a bulbous structure (kohlrabi; *B. oleracea*, var. *gongylodes*);

Inflorescences: Raceme, flowers tiny, not modified (broccoli; *B. oleracea*, var. *italic*) and flowers much thick and modified (cauliflower; *B. oleracea*, var. *botrytis*).

Table 5 In vitro *B. carinata* hybrids development through ovary, ovule and embryo culture and through protoplast fusion

Parental combination ♀ × ♂	Cross-type	References
B. carinata + *B. napus*	**Protoplast fusion** 13 plants were produced from protoplast fusion in polyethylene glycol media	Klíma et al. (2009)
	Ovary, ovule culture Yield 2 seeds in F1 from 44 pollinated flowers, with male-sterile F1 hybrids	Sabharwal and Doležel (1993)
B. carinata × *B. nigra*	**Embryo culture** Successful	Attia et al. (1987)
B. carinata × *B. oleracea*	**Embryo culture** Average 6.0–7.2 embryos/pollination were rescued @ 57–96% survival rate	Rahman (2004)
	Embryo culture Five embryos sprouted from 45 pollinations, four were true hybrids and male-sterile	Tonguç and Griffiths (2004)
B. carinata var. *botrytis* + *B. oleracea*	**Protoplast fusion** 31 hybrids are reported by asymmetric protoplast fusion of 374 regenerated plants	Scholze et al. (2010)
B. carinata var. *capitata* + *B. oleracea*	**Protoplast fusion** 05 hybrids are reported by symmetric protoplast fusion of 21 regenerated plants	Scholze et al. (2010)
B. carinata × *B. rapa*	**Embryo culture** Successful	Quiros et al. (1985), Meng et al. (1998)
	Embryo culture Average of 6.6–8.0 embryos/pollination were rescued @ 73–96% survival rate	Rahman (2004)

(continued)

Table 5 (continued)

Parental combination ♀ × ♂	Cross-type	References
B. carinata + *B. rapa*	**Protoplast fusion** Protoplast fusion in media produced 58 calluses and 14 shoots, 60% of seedlings were hybrids	Beránek et al. (2007)
B. carinata × *B. fruticulosa*	**Embryo culture** Successful with male-sterile hybrids	Harberd and McArthur (1980) Chen et al. (2012)
B. carinata × *B. maurorum*	**Embryo culture** 09 hybrid seedlings were produced from 642 pollinated flowers in F1 with 25% pollen viability	Yao et al. (2012)
B. fruticulosa × *B. carinata*	**Embryo culture** Successful with male-sterile hybrids	Chen et al. (2012)
B. juncea × *B. carinata*	**Embryo culture** 13–17 bivalents are successfully reported in 27 cells	Harberd and McArthur (1980)
	Ovary culture 226 ovaries were cultured in media. 91 poorly developed and shrivelled seeds formed in F1	Sharma and Singh (1992)
B. maurorum × *B. carinata*	**Embryo culture** Seven hybrids sprouted from 368 pollinated flowers with 1.90% crossability	Yao et al. (2012)
B. napus × *B. carinata*	**Embryo culture** 09 bivalents are successfully observed in 50 cells	Harberd and McArthur (1980)
	Ovule culture Yield average is 17.0–64.1% hybrid with 0–30% pollen viability in F1 plants	Sacristan and Gerdemann (1986)

(continued)

Table 5 (continued)

Parental combination ♀ × ♂	Cross-type	References
B. oleracea var. *alboglabra* × *B. carinata*	**Embryo culture** 12 hybrids were produced from 249 cross-pollinations with 5.8% pollen viability in F1 plants	Rahman (2001)
B. oleracea × *B. carinata*	**Embryo culture** Average of 0.02–0.35 embryos/pollination were rescued at 16.7% of survival rate	Rahman (2004)
B. rapa × *B. carinata*	**Embryo culture** Successful	Busso et al. (1987)
B. carinata + *C. sativa*	**Protoplast fusion** Protoplast fusion produced 227 calluses and only 03 shoots. Plants did not reach maturity	Narasimhulu et al. (1994)
B. carinata × *D. assurgens*	**Embryo culture** 3–10 bivalents are successfully observed in 55 cells	Harberd and McArthur (1980)
B. carinata × *D. tenuisiliqua*	**Embryo culture** 1–10 bivalents are successfully observed in 83 cells	Harberd and McArthur (1980)
B. carinata × *D. virgata*	**Embryo culture** 4–11 bivalents are successfully observed in 36 cells	Harberd and McArthur (1980)
B. carinata × *E. gallicum*	**Embryo culture** 5–12 bivalents are successfully observed in 77 cells	Harberd and McArthur (1980)
B. carinata × *R. sativus*	**Embryo culture** 0–4 bivalents are successfully observed in 142 cells	Harberd and McArthur (1980)
B. carinata × *S. alba*	**Ovule culture** 08 ovules were cultured from 45 pollinations but no hybrid is produced	Momotaz et al. (1998)

(continued)

Table 5 (continued)

Parental combination ♀ × ♂	Cross-type	References
	Ovary and ovule culture 27 ovules cultured from 249 ovaries, 2 plants were formed from 4 germinated ovules and were confirmed as matromorphs	Sridevi and Sarla (2005)
B. carinata × *S. arvensis*	**Embryo culture** 0–9 bivalents are successfully seen in 72 cells	Harberd and McArthur (1980)
	Ovule culture 269 ovules were cultured from 96 pollinations, 29 hybrids were produced with unfertile pollen	Momotaz et al. (1998)
B. carinata × *Sinapis turgida*	**Ovule culture** 166 ovules were cultured from 41 pollinations, 08 hybrids were produced with unfertile pollen	Momotaz et al. (1998)
E. lyratus × *B. carinata*	**Ovule culture** 1 hybrid produced from 54 pollinated ovaries with 2% pollen fertility in F1	Gundimeda et al. (1992)
E. abyssinicum × *B. carinata*	**Ovary culture** Successful with pollen sterile F1 hybrids	Rao et al. (1996)
S. alba × *B. carinata*	**Ovule culture** 06 ovules were cultured from 45 pollinations, no ovule and hybrid in F1 reported	Momotaz et al. (1998)
	Ovary and ovule culture 11 ovules were cultured from 153 ovaries, resulted in F1 hybrid	Sridevi and Sarla (2005)
S. arvensis × *B. carinata*	**Ovule culture** 32 ovules were cultured from 33 pollinations, no ovule and hybrid in F1 reported	Momotaz et al. (1998)

(continued)

Table 5 (continued)

Parental combination ♀ × ♂	Cross-type	References
Sinapis turgida × *B. carinata*	**Ovule culture** 11 ovules were cultured from 21 pollinations, no ovule and hybrid in F1 reported	Momotaz et al. (1998)

Chinese kale is cultivated in China as a white-flowering crop and considered it a separate species as *Brassica alboglabra*, imported from the Mediterranean region.

The heads of horticultural varieties may be pointed, globular or flat in shape, soft to hard in structure, look green, grey-green and magenta or red in colour and less than 1 kg to more than 3 kg in weight. If flowers left unharvested it produces seed pods, a silique that opens through dehiscence when matured. Seeds are brown or black, small and circular in shape about 1 mm in diameter (Warwick 2010; Rochfort and Jones 2011; https://www.britannica.com/plant/cabbage).

4.5.3 Tolerance to Abiotic/Biotic Factors

- Tolerates frost of −6 °C and some tolerate hard freezing of −10 °C at some of the growth stages. Hot weather is harmful to its growth and quality. Temperature range of 10–24 °C is suitable for growth.
- Tolerates pH of soil range between 6.0 and 6.8.
- Water requirement is from 380 to 500 mm but depends on weather and growing period (Bradley et al. 2009; FAO 2019).

4.5.4 Disease Resistance

B. oleracea is resistant to black rot (*X. campestris* pv. *campestris*; Afrin et al. 2018). Partial resistance has been found to Turnip mosaic virus (TuMV), Cauliflower mosaic virus (CaMV), beet western yellow virus (BWYV) and to turnip yellow mosaic virus in *B. oleracea* cultivars (Pitrat 2012).

Cabbage aphid damage is susceptible cultivar of *B. oleracea* var. capitata by 85% but do not harm resistant cultivars (Munthali 2009). Most local varieties of *Brassica olearceae* were reported to be highly susceptible to *Mamestra brassicae* and commercial hybrids were found to be resistant in terms of head foliage consumption and a number of larvae per plant (Cartea et al. 2010). Root-knot nematodes and cabbage maggots are very common and result in stunted growth of plants with yellow leaves.

Resistance against pmildew has been transferred in *B. oleracea* from *B. carinata* through embryo rescue followed by backcrossing to *B. oleracea* (Tonguç and Griffiths 2004; Katche et al. 2019). The resistance against black spot (*A. brassicae* and *A. brassicicola*) disease is transferred in *B. oleracea* vegetables from *C. sativa,* highly resistant to Alternaria spp. through protoplast culture.

4.5.5 Allelopathy

B. oleracea var. *medullosa* (Marrow stem kale) inhibits germination, root growth, plant height, dry weight, tiller number and seed yield through its allelopathic effects (OECD 2016).

4.5.6 Hybridization

Crosses have been made between *B. oleracea* with *B. carinata*, of which *B. oleracea* is a progenitor, at a very low rate of hybridization (Morinaga 1933; Mizushima 1950; Barcikowska et al. 1983; Rahman 2001). High rates of hybrid production (0.63 hybrids/pollination are reported in the cross between *B. tournefortii* and *B. oleracea* (Ljungberg et al. 1993), whereas a much lower rate (0.003 hybrids/pollination) is reported by Narain and Prakash (1972). Warwick et al. (2000) reported successful hybridization between *B. oleracea* and *B. oleracea* cytodeme: *B. bourgeaui, B. cretica, B. incana, B. insularis, B. macrocarpa, B. montana, B. rupestris,* and *B. villosa*. Successful hybridization is reported between *B. oleracea* and *B. maurorum* (Chrungu et al. 1999). Crosses with *B. balearica* (Snogerup and Persson 1983) and *B. nigra* (Prakash et al. 1984; Sarla et al. 1987; Sarla and Raut 1988, 1991) were reported unsuccessful. Successful intergeneric hybridization has been reported between *B. oleracea* and *E. vesicaria* (Eenink 1974), *E. abyssinicum* (Sarmah and Sarla 1998), *H. incana* (Quiros et al. 1988), *Moricandia arvensis* (Apel et al. 1984), *R. sativus* (Ellerstrom 1978; McCollum 1979; Sarashima and Matsuzawa 1979; Namai 1980), *S. alba* (U et al. 1937) and *S. arvensis* (Mizushima 1950). Evidence for one-sided incompatibility was reported successful in the cross between *B. oleracea* and *R. sativus*, where *B. oleracea* was the female parent (Opena and Lo 1978), and 13 results where *B. oleracea* was the male parent were reported successful (FitzJohn et al. 2007). *B. bourgeaui, B. cretica, B. incana, B. insularis, B. juncea, B. macrocarpa, B. maurorum, B. montana, B. nigra, B. rupestris, B. tournefortii, B. villosa, E. vesicaria, H. incana or S. arvensis*, are the species that formed F1 hybrids with *B. oleracea* with unreported backcross attempts. F2 hybrid productions are not documented with *B. bourgeaui, B. juncea, B. maurorum, B. tournefortii, E. abyssinicum or S. arvensis* (FitzJohn et al. 2007).

4.5.7 Cytoplasmic Male Sterility

Ogura cytoplasmic male sterility (CMS) is widely used in *B. oleracea* to facilitate hybrid production. Hybrids produced by Ogura CMS system in cabbage are reported to be lower in quality (Haser-Krause 1989; Leung and Williams 1983, Yoon et al. 1988) whereas Wei and Li (1989) produced Ogura-based Chinese cabbage hybrids with no maternal effects on plant weight, head diameter, number of leaves, core length and heading index. de Melo and de Giordano (1994) reported that Ogura-based hybrid showed a reduction in plant and head weight, core length, head length and width. However, grading index, shape and core indexes and head compactness in cabbage hybrids remained unaffected. It can be concluded that the selection of parental lineages is important for the production of good quality hybrids. Transfer of cytoplasmic male sterility from *B. juncea* to *B. oleracea* and resistance against turnip mosaic virus is possible by somatic hybridization (Lian et al. 2011; Chen et al. 2005).

4.5.8 Fatty Acid Composition

Ahuja et al. (1987) reported higher oil content in turnip and radish as compared to cauliflower. Levels of erucic acid were in the order of radish < turnip < cauliflower. Radish seeds contained comparatively higher amounts of palmitic, oleic and linolenic + eicosenoic acids and lower amounts of linoleic and erucic acids than cauliflower and turnip seeds. Sharafi et al. (2015) reported in three accessions of *B. oleracea,* the average amount of oil is 28% with 3.3% palmitic acid, 0.83% stearic acid, 15.5% oleic acid, 12.4% linoleic, 15.1% linolenic acid, 46.3% erucic acid.

4.6 B. juncea *(L.) Czern.* (n = *18 AB) (Brown and Oriental Mustard)*

4.6.1 Origin and Geographic Distribution

- Primary centre of origin is China from where divergent forms evolved.
- Secondary centre of origin is thought to be Afghanistan and adjoining regions from where it spread to the Indian subcontinent and became a major oilseed crop.
- Native to temperate Asia including China, Kazakhstan, Kyrgyzstan, Mongolia, Eastern And Western Siberia, Tajikistan and Turkmenistan.
- Introduced as a condiment crop to Europe, America, Australia and New Zealand.
- As a weed, it occurs in the south of the European part of Russia, the Caucasus, Central Asia and Southern Siberia and as a casual or feral plant in south and Southeast Asia, Africa and America.
  (USDA-ARS 2011; CFIA 2012; OECD 2016)

4.6.2 Description

B. juncea, a group of plants grown for their seed oil or condiment production, *is* believed to have derived from interspecific hybridization between plants of *B. nigra* and *B. rapa.* Due to ecogeographical variation some morphologically different forms, oleiferous, semi-oleiferous, rapiferous and leafy types are available. On the basis of the differences in morphology, quality and uses *B. juncea* can be divided into four sub-species (Spect and Diederichsen 2001).

1. sub-species *integrifolia*; leafy vegetable;
2. sub-species *juncea*; cultivated mostly for its seed, occasionally as fodder;
3. sub-species *napiformis*; root-tuber vegetable;
4. sub-species *taisai;* stalks and leafs vegetables.

Spring-sown annual crops with tuberous root grow up to 1.2 m and winter crop attain 2.1 m height. Stem is sparsely branched. The lower leaves are thin, elliptic to obovate and lobed, the upper is small, narrow and do not clasp the stem. The inflorescence is an elongated raceme. The siliques are 2.5–7 cm in length with 0.5–1 cm long beak containing 20 seeds. Seeds are round, yellow or brown, weighing 2.5–3.0 g/1000 seeds.

4.6.3 Tolerance to Abiotic/Biotic Factors

- Tolerates annual precipitation range from 500 to 4200 mm.
- Tolerant to heat and drought stress and tolerate an annual temperature of 6–27 °C.
- Tolerates a pH range from 4.3 to 8.3 (Duke 1983).

4.6.4 Disease Resistance

B. juncea is susceptible to some insect pests and disease. *A. candida* (white rust), *Alternaria* spp. (blackspot), *P. brassicae* (clubroot), *L. maculans* (blackleg) and some prevalent virus like rape mosaic, turnip yellow mosaic, yellow virus, radish mosaic cause diseases and damage crop. Resistance against diseases like blackleg, white rust, Alternaria blackspot and Fusarium wilt have to be built through breeding new tolerant cultivars. Resistance to *Brassica* leaf blight caused by *A. brassicae* transferred from *B. hirta* to *B. juncea* (Mohapatra and Bajaj 1987). Some of the cultivars of *B. juncea* are reported to be of Alternaria blight tolerant (Raghav et al. 2014). *B. juncea* type was found to be completely resistant to blackleg (*L. maculans*) and was transferred from *B. juncea* to blackleg susceptible *B. napus* through *the* interspecific cross (Roy 1984). Seed treatments are usually used to control flea beetles and seedling diseases.

4.6.5 Allelopathy

Mason-Sedun et al. (1986) found the effect of water extracts from dry residues of *B. juncea* on the coleoptile growth of *Triticum aestivum*. They observed a significant reduction in plant height, grain yield, plant dry weight and tiller production with the greatest inhibition level. The inhibiting compounds from dead or decaying stems and leaves are leached by water, reside in the upper layer of soil for a short period and then dissolve.

4.6.6 Hybridization

All varieties and species of *B. juncea* intercross freely. Some successful interspecific and intergeneric sexual crosses between *B. juncea* with its relatives and in vitro hybrid development through ovary, ovule, Emyro culture and protoplast fusion are listed in Table 6.

Table 6 Inter-species/genera hybridization (sexual crosses) of *B. juncea* with related species (extracted from Warwick et al. 2000) and in vitro hybrid development through ovary, ovule and embryo culture and through protoplast fusion

Parental combination ♀ × ♂	Cross-type		References
	Sexual cross	In vitro hybrid development	
B. juncea × *B. carinata*	The reciprocal cross has been successful		Alam et al. (1992), Ghosh Dastidar and Varma (1999)
B. juncea × *B. carinata*	Hybrid was obtained sexually		Anand et al. (1985), Sheikh et al. (2009a)
B. juncea × *Brassica maurorum*	Hybrid was obtained sexually		Bijral et al. (1995)
B. juncea × *Brassica napus*	Hybrid was obtained sexually		Anand et al. (1985)
B. juncea × *Brassica napus*		OVARY/OVULE/EMBR culture the reciprocal cross has been successful	Bajaj et al. (1986)
B. juncea × *Brassica napus*		OVARY culture the reciprocal cross has been successful	Bajaj (1990)
B. juncea × *Brassica napus*	The reciprocal cross has been successful		Bing et al. (1991), Alam et al. (1992), Zhao et al. (2003)
B. juncea × *Brassica napus*	Hybrid was obtained sexually		Sharma and Singh (1992), Sandhu and Gupta (2000), Choudhary and Joshi (2001), Wen et al. (2008)
B. juncea × *Brassica napus*		EMBRYO culture the reciprocal cross has been successful	Zhang et al. (2003)
B. juncea × *B. nigra*		EMBRYO culture the reciprocal cross has been successful	Anand et al. (1985)
B. juncea × *B. nigra*	The reciprocal cross has been successful		Bing et al. (1991)
B. juncea × *B. nigra*	Hybrid was obtained sexually		Rao and Shivanna (1997)

(continued)

Table 6 (continued)

Parental combination ♀ × ♂	Cross-type		References
	Sexual cross	In vitro hybrid development	
B. juncea × *B. oleracea*	Hybrid was obtained sexually		Gupta (1997)
B. juncea + *B. oleracea*		PROTOPLAST fusion	Chen et al. (2005)
B. juncea × *Brassica rapa*		OVULE culture the reciprocal cross has been successful	Takeshita et al. (1980)
B. juncea × *Brassica rapa*		OVARY culture the reciprocal cross has been successful	Bajaj (1990)
B. juncea × *Brassica rapa*	Hybrid was obtained sexually		Choudhary et al. (2002), Choudhary and Joshi (2001)
B. juncea × *Brassica rapa*	The reciprocal cross has been successful		Choudhary and Joshi (1999)
B. juncea × *Brassica tournefortii*		OVULE/OVARY culture the reciprocal cross has been successful	Lokanadha and Sarla (1994)
B. juncea × *Brassica tournefortii*		EMBRYO culture	Goyal et al. (1997)
B. juncea × *Diplotaxis erucoides*		OVARY culture	Inomata (1998)
B. juncea × *Diplotaxis muralis*	Hybrid was obtained sexually		Bijral and Sharma (1995)
B. juncea × *Diplotaxis virgata*		OVARY culture, F1 & BC	Inomata (2003)
B. juncea × *Eruca sativa*	Hybrid was obtained sexually		Goswami and Devi (2002)
B. juncea × *Erucastrum virgatum*		OVARY culture	Inomata (2001)
B. juncea × *R. raphanistrum*	Hybrid was obtained sexually		Kamala (1983)

(continued)

Table 6 (continued)

Parental combination ♀ × ♂	Cross-type		References
	Sexual cross	In vitro hybrid development	
B. juncea × *R. sativus*	The reciprocal cross has been successful		Rhee et al. (1997)
B. juncea × *S. alba*		OVARY/OVULE culture; the reciprocal cross has been successful	Sridevi and Sarla (1996)
B. juncea × *S. alba*	Hybrid was obtained sexually		Sharma and Singh (1992)
B. juncea × *S. arvensis*		EMBRYO culture; the reciprocal cross has been tried and not been successful	Harberd and McArthur (1980)
B. juncea × *S. arvensis*	The reciprocal cross has been tried and not been successful		Bing et al. (1991)
B. carinata × *B. juncea*	The reciprocal cross has been successful		Alam et al. (1992), Ghosh Dastidar and Varma (1999)
B. maurorum × *B. juncea*		EMBRYO culture	Chrungu et al. (1999)
B. napus × *B. juncea*		OVARY/OVUL/EMBRYO culture; the reciprocal cross has been successful	Bajaj et al. (1986)
B. napus × *B. juncea*		OVARY/OVULE culture; the reciprocal cross has been successful	Sabharwal and Doležel (1993)
B. napus × *B. juncea*	The reciprocal cross has been successful		Bing et al. (1991), Alam et al. (1992), Choudhary and Joshi (1999)
R. sativus × *B. juncea*	The reciprocal cross has been successful		Rhee et al. (1997)

(continued)

Table 6 (continued)

Parental combination ♀ × ♂	Cross-type		References
	Sexual cross	In vitro hybrid development	
S. alba × *B. juncea*		OVARY/OVULE, culture; the reciprocal cross has been successful	Sridevi and Sarla (1996)
[*B. carinata* × *B. juncea* F1] × *B. carinata*	Hybrid was obtained sexually with F1 & BC		Getinet et al. (1994, 1997)
[*B. juncea* × *B. carinata* F1] × *B. carinata*	Hybrid was obtained sexually with F1 & BC		Sheikh et al. (2009a)
[*B. juncea* × *B. napus* F1] × *B. juncea*	Hybrid was obtained sexually with F1 & BC		Schelfhout et al. (2008)
[*B. napus* × *B. juncea* F1] × *B. juncea*	Hybrid was obtained sexually with F1 & BC		Sheikh et al. (200b)
[*B. oleracea* × *B. juncea* F1]	Hybrid was obtained sexually with F1		Struss et al. (1991)
[*B. tournefortii* × *B. nigra* F1] × *B. juncea*	Hybrid was obtained sexually with BC		Pradhan et al. (1991)
[*B. rapa* × *B. nigra* F1] × *B. juncea*	Hybrid was obtained sexually with BC		Lukens et al. (2004)
[*Diplotaxis erucoides* × *B. juncea* F1] × *B. juncea*		OVARY/OVULE culture with F1 and BC	Vyas et al. (1995)
[*D. erucoides* × *B. rapa* F1] × *B. juncea*	Hybrid was obtained sexually with BC		Malik et al. (1999)

(continued)

Table 6 (continued)

Parental combination ♀ × ♂	Cross-type		References
	Sexual cross	In vitro hybrid development	
[*E. abyssinicum* × *B. juncea* F1] × *B. juncea*		OVARY/OVULE culture with BC	Sarmah and Sarla (1997)
R. sativus × *B. oleracea* F1] × *B. juncea*	The reciprocal cross has been successful		Chen and Wu (2008)
Cytoplasmic male sterility			
[*M. arvensis* + *B. juncea* F1] × *B. juncea*		Successfully obtained both stable cytoplasmic male-sterile *B. juncea* and an introgression line carrying the restorer gene through somatic hybridization between *M. arvensis* and *B. juncea* followed by backcrossing with *B. juncea*	Prakash et al. (1998)
[*B. juncea* + *B. fruticulosa* F1, F2]		Gene for fertility restoration could be successfully introgressed from cytoplasm donor species, *B. juncea* in cytoplasmic background of recipient *B. fruticulosa*, results in cytoplasmic male sterility	Atri et al. (2016)

4.6.7 Fatty Acid Composition

Sharafi et al. (2015) reported in four accessions of *B. juncea* the average amount of oil is 25.7% with 3.6% palmitic acid, 1.3% stearic acid, 14% oleic acid, 19.5% linoleic, 20% linolenic acid and 32% erucic acid. Rai et al. (2018) observed significant differences among the 14 genotypes of *B. juncea* for oil and fatty acids content and its quality. Oil content was in the range of 32.67–39.47%, the saturated fatty acid (Palmitic acid) was in the range of 3.08–3.85%, oleic acid 0.80–48.70%, linoleic acid 11.00–45.30%, Linolenic acid 11.10–26.72%, eicosenoic acid 0.00–34.90% and erucic acid was in the range of 0.80–49.40. The genotypes containing low erucic acid are fit for commercial cultivation and for introgression of low erucic acid traits in higher yielding cultivars to develop qualitative genotypes.

4.7 C. sativa *(L.) Crantz (*n = *20) (Gold-of-Pleasure, False Flax)*

4.7.1 Origin and Geographic Distribution

- Originated in Southeastern Europe and Southwestern Asia
- Introduced and naturalized in Africa, Australasia, Canada, South and North America and the United States.

4.7.2 Description

C. sativa is an annual herbaceous plant with a taproot, attaining 30–90 cm height. Leaves are lance-shaped, sessile, clasping, 2–8 cm long and 2–10 mm wide, hairy and withered at flowering. Stems are single, branched and become woody with maturity. Inflorescences are racemes with terminal ebracteate flowers in clusters. Silique is pear-shaped smooth and leathery, 7–9 mm long with small yellow to brown 2–3 mm long deeply ridged surface seeds (Francis and Warwick 2009). Seeds contain 38–43% oil and 27–32% protein (Gugel and Falk 2006).

4.7.3 Tolerance to Abiotic/Biotic Factors

- Tolerates to some pre-emergence and post-emergence herbicide.
- Tolerates temperatures as low as −11 °C without harm.
- Few varieties are salt tolerant.
- Grows well in acidic soil.

  (Plessers et al. 1962; Johnson et al. 2007, 2008, 2011; Ehrensing and Guy 2008; Walsh et al. 2013).

4.7.4 Disease Resistance

C. sativa is resistant to black spot (*A. brassicae*), blackleg (*L. maculans*) (Li et al. 2005; Séguin-Swartz et al. 2009), stem rot (*S. sclerotiorum*) and downy mildew (*P. parasitica*) which suggest that breeding could result in the development of resistant cultivars. Resistance against *A. brassicae* is due to the presence of camalexin and methoxycamalexin (Conn et al. 1988; Browne et al. 1991). It is not found to be resistant to clubroot (*P. brassicae*), aster yellows and white rust (*A. candida*).

4.7.5 Allelopathy

On the basis of a few evidences, *Camelina* was not found to have effective allelopathic traits. Saucke and Ackermann (2006) found a similar result in the case of *Camelina* and *Pisum sativum* L. grown together. Lovett and Jackson (1980) have suggested that *Camelina* has allelopathic traits in the case of radicle length of germinating flax (*Linum usitatissimum*) and wheat (*T. aestivum* L.) in the presence of certain bacteria.

4.7.6 Hybridization

Table 7 summarizes reports of interspecific/generic and somatic hybridization between *C. sativa* and its relatives.

4.7.7 Fatty Acid Composition

Genotype and environmental conditions affect *Camelina* oil content and fatty acid composition. Presence of higher content of unsaturated fatty acid, *Camelina* oil has high potential to be used in different industrial productions, e.g. biodiesel and renewable jet fuels (Moser 2010; Shonnard et al. 2010; Soriano and Narani 2012), adhesives, coatings, resins and gums (Zaleckas et al. 2012; Kim et al. 2015; Zhu et al. 2017). *Camelina* genetically can be modified by transformation (floral dip transformation) to create unique oil compositions useful for fuel and industrial applications (Bansal and Durrett 2016).

Seed of *Camelina* contains an average of 30–48% oil content (Vollmann et al. 2007; Pavlista et al. 2012; Jiang et al. 2014; Sintim et al. 2016). The oil contains approx. 60% polyunsaturated fatty acids, mainly linoleic acid (approx. 15%) and a-linolenic acid (approx. 35–45%), 30% monounsaturated fatty acids and 6–10% saturated fatty acids (Zubr and Matthaus 2002; Kirkhus et al. 2013; Jiang et al. 2014).

Fatty acid composition of *Camelina* oil from several literature are reported as average palmitic (16:0) 5.7%; stearic (18:0) 2.5%; oleic (18:1) 15%; linoleic (18:2) 16%; a-linolenic (18:3) 36.3%; 11-eicosenoic (20:1)15%; behenic (22:0) 0.25%; erucic (22:1) 2.7% (Obour et al. 2017).

Table 7 Inter-species/genera and somatic hybridization between *C. sativa* and its relatives

Parental combination ♀ × ♂	Cross-type		References
	Sexual	Somatic	
C. sativa × *C. alyssum*	Yield 1273 seeds/578 cross Yield 2.2 seeds/cross and no fertile hybrids		Séguin-Swartz et al. (2010, 2013)
C. alyssum × *C. sativa*	Yield 918 seeds/639 cross Yield 1.4 seeds/cross and fertile hybrids with 95.8% pollen viability		Séguin-Swartz et al. (2010, 2013)
C. sativa × *C. microcarpa*	Yield 708 seeds/590 cross		Séguin-Swartz et al. (2010)
C. microcarpa × *C. sativa*	Yield 1233 seeds/608 cross Yield F1 fertile hybrids from one accession of *C. microcarpa*		Séguin-Swartz et al. (2010, 2013)
C. sativa × *C. rumelica*	Yield 309 shrivelled and non-viable seeds/245 cross Yield fertile hybrids with 1.2–4.6% pollen viability		Séguin-Swartz et al. (2010, 2013)
C. rumelica × *C. sativa*	Yield 32 shrivelled and non-viable seeds/180 cross		Séguin-Swartz et al. (2010, 2013)
C. bursa-pastoris × *C. sativa*	Yield 586 seeds/186 cross and 7 non-fertile hybrids with 1.9% pollen viability		Martin et al. (2015)
C. hirsuta × *C. sativa*	Yield 8 seeds/64 cross		Julié-Galau et al. (2014)
C. sativa + *B. carinata*		**Protoplast fusion** 227 calli are produced from 3 fusions. Among that 3 regenerated shoots but failed to produce roots	Narasimhulu et al. (1994)
C. sativa + *B. napus*		**Protoplast electrofusion** No shoot is formed from the callus of electrofused protoplast	Müller and Sonntag (2000)

(continued)

Table 7 (continued)

Parental combination ♀ × ♂	Cross-type		References
	Sexual	Somatic	
		Protoplast electrofusion 25 shoots are formed from 385 calli. 03 were confirmed to be hybrids in F1. Seed set is reported to be very low in all the generation	Jiang et al. (2009)
C. sativa + *B. oleracea*		**Protoplast fusion** 2903 calli are produced from 4 protoplast fusion. 14 regenerated shoots but failed for development	Hansen (1998)
		5158 calli are formed from 3 fusions, 19 form shoots with undeveloped roots. Sterile flowers are reported	Sigareva and Earle (1999)

4.8 A. thaliana *L. Heynh* (n = *5) (Mouse-Ear Cress, Thale Cress)*

A. thaliana is a member of Brassicaceae and a model organism in genetics and molecular biology research with no major agronomic significance.

4.8.1 Origin and Geographic Distribution

- Native to Europe, Asia, tropical alpine ecosystems in Africa and South Africa.
- Distributed from the Mediterranean to Scandinavia and Spain to Greece.
- Introduced now worldwide.
  (Hoffmann 2002; Sharbel et al. 2000; Fulgione and Hancock 2018).

4.8.2 Description

A. thaliana is a 20–25 cm tall annual plant. The single primary root grows vertically downward with lateral small roots. These roots build interactions with rhizosphere bacteria. The leaves are with serrated margin and trichrome, basal leaves are subsessile and 1.5–5 cm long and 2–10 mm broad and upper leaves are small and sessile. Inflorescence corymb with cruciform flowers 3 mm in diameter. The silique is 5–20 mm long containing 20–30 seeds (López-Bucio et al. 2007).

4.8.3 Use of *A. thaliana* as a Research Organism

The name *thaliana* is on name Johannes Thal, who discovered it in the sixteenth century in the Harz Mountain. Alexander Braun in 1873 reported the first mutant in *A. thaliana.* Friedrich Laibach proposed in 1943 *A. thaliana* as a model organism though he did his research in 1907 on its chromosome number. In 1947, Erna Reinholz described the first collection of the induced mutant by X-ray mutagenesis in *A. thaliana.*

In the 1950s, John Langridge and in the 1960s, George Rédei recognized *A. thaliana* as a useful organism for biological laboratory experiments and the first *Arabidopsis* Information Service (AIS) was established in 1964 with the first International Conference on *Arabidopsis* in the following year in Germany. T-DNA-mediated transformation and the first cloned *A. thaliana* gene were described in 1986 (Lloyd et al. 1986; Chang and Meyerowitz 1986). A large group of natural accession (750) of *A. thaliana* is reported from around the world.

For over 40 years, *A. thaliana* has been the focus of biological study because of its numerous traits; it has a very fast life cycle which requires only light, air, water and a few minerals for its completion. It produces self-progeny and can easily be grown in an indoor growth chamber and produce many seeds. Because of its small genome with about 135 mega base pairs and 5 chromosomes, it can easily be used in genetic engineering for the improvement in crop species. Arabidopsis Information Resource maintained its up-to-date version and they assigned functions to its 27,000 genes and encode 35,000 proteins. The genome of plastid is a 154,478 base pair long DNA molecule with 136 genes whereas the mitochondrial genome is a complex branched, linear molecule with 367,808 base pairs and 57 genes. The largest repeated regions recombine to isomerize the genome (Klein et al. 1994; Gualberto et al. 2014). Post-genomic research on the set of metabolites present within species explained how environmental factors affect metabolic processes (Bundy et al. 2009; Lake et al. 2009).

Genetic transformation takes place by *A. tumefaciens* to transfer DNA into the plant genome or by floral dip method of Agrobacterium-mediated transformation (Clough and Bent 1998; Zhang et al. 2006). Characterized accessions and mutant lines of *A. thaliana* are used in laboratory experiments.

A. thaliana is extensively studied as a model for leaf and flower development. The developing flower has four organs and homeotic mutations result in the change of one organ to another, for example, stamens change to petals and carpels become a flower, resulting in a repeated sepal–petal–petal pattern. The development of leaf shape is divided into three stages, the initiation of the leaf primordium, the establishment of dorsiventrality and the development of a marginal meristem. Leaf primordium initiates by the suppression of KNOX genes which result in increases of gibberellic acid level to initiate leaf primordium growth.

There are five genes in *A. thaliana* identified that respond to touch, on the basis of this response it was genetically engineered with luciferase gene to glow in the dark when touched and the effect can only be observed by using ultra-sensitive cameras (Bioresearch Online 2000).

Plants are affected by multiple pathogens throughout their lifespan. In response to the presence of pathogens, plants have evolved receptors on the cell surface to detect and respond to pathogens (Bent et al. 1994). *A. thaliana* is also used for the determination of defense mechanisms of plant–pathogen resistance. These plants detect pathogen by special receptors FLS2 (bacterial flagellin receptor) and EF-Tu (bacterial EF-Tu protein) present on their cell surfaces which inhibit pathogen growth (Zipfel et al. 2004). Plants with FLS2 and EF-Tu receptors have shown to have increased fitness in the population (Delaney et al. 1994). *A. thaliana* has also been used to study systemic acquired resistance (SAR). In which Benzothiadiazol induces transcription factors, mRNA, of SAR genes. The accumulation of transcription factors ultimately inhibits pathogen-related genes (Lawton et al. 1996).

Somatic hybridization between *A. thaliana* and other Brassicaceae crop results in desirable traits for introgression. Hybrids obtained through protoplast fusion of *A. thaliana* and *B. nigra are* resistant to flea beetles, cold tolerant and have short life cycles (Siemens and Sacristan 1995). *A. thaliana* + *B. oleracea* are for plastome transformation (Nitovskaya and Shakhovskyi 1998; Yamagishi and Nakagawa 2004; Nitovskaya et al.2006a). Likewise, *A. thaliana* + *B. rapa* is for experimental demonstration (Gleba and Hoffmann 1980); *A. thaliana* + *B. juncea* is to build resistance to Phosphinothricin (Ovcharenko et al. 2004); *A. thaliana* + *B. napus* for herbicide and blackleg resistance (Bauer-Weston et al. 1993; Yamagishi et al. 2002).

4.9 **S. arvensis** *L. (***n** *= 9 Sar) (Wild Mustard, Charlock)*

4.9.1 Origin and Geographic Distribution

- A Native to North Africa, Europe and temperate Asia and naturalized throughout North and South America, South Africa, Japan, Australia and New Zealand (USDA-ARS 2016).
- Widespread in the United States (USDA-NRCS 2016).

4.9.2 Description

S. arvensis is an annual plant on an average 7.9–31.5 in. tall, can attain more than 1 m height in an ideal condition. The stems are erect, branched and hairy at the base. Petiolated leave is 0.39–1.57 in. in length. Basal leaves are oval, 1.6–7.1 in. long and 0.79–1.97 in. broad. The upper leaves are much reduced with short petiole to sessile. The inflorescence is a raceme. Silique is 3–5 cm long with 1–2 cm long beak. The seeds are globose, compressed at the hilum, usually black can be reddish to brown, 1–2 mm in diameter. The seeds and leaves of *S. arvensis* contain glucosinolates including sinalbin that causes severe illness in livestock (Warwick et al. 2000). *S. arvensis* often grow in high densities in cereal and raw crops, compete for soil nutrients and reduces the yield of crop. *Sinapis arevensis* can tolerate temperatures of −20 °C in dormant condition and grow in nutrient-rich heavy alkaline soils in the wild. Oil extracted from the seed is used for lubricating machinery.

There are no extensive reports about hybridization in *S. arvensis.* Daniels et al. (2005) did not get any resistant plants in a test of 60 768 progeny of 818 *S. arvensis* plants, growing in 23 glufosinate resistant *B. napus* fields. Warwick et al. (2003) found the same result from 43 828 *S. arvensis* plants growing in Herbicide-Resistant *B. napus* fields. A similar finding is also reported by Bing et al. (1996b) and Moyes et al. (2002). However, when male-sterile *B. napus* plants were crossed with *S. arvensis* pollen, a small number of hybrids were produced. Chèvre et al. (1996) reported 0.18 hybrids/100 pollinations while Lefol et al. (1996) found 6 hybrids in 50,000 seeds analysed. Bang et al. (1996) reported successful reciprocal cross between *S. arvensis* and *R. sativus, a* hybrid was obtained with the ovary and embryo culture. Moyes et al. (2002) reported one completely sterile hybrid in a hand cross-test with *S. arvensis* females. Hybrids were produced but weak and could not be able to survive in nature (Moyes et al. 2002). From all the results, it is concluded that gene flow between *B. napus* and *S. arvensis* is extremely low to non-existent.

4.10 Raphanus raphanistrum *L.* (n = *9, R) (Wild Radish)*

4.10.1 Origin and Geographic Distribution

- Native to North Africa, Europe and West Asia and the Caucasus.
- Introduced in North and South America except the central region, Africa, Asia, Australia and New Zealand.

4.10.2 Description

R. raphanistrum is annual or biennial with taproot, unequally divided leaves and 2 feet tall inflorescence. Siliques, born below the flower head, are narrow oval, jointed, 4.0–5.5 mm long, 2.5–3.0 mm wide; contain 4–10 reddish-brown seeds.

Wild radish seeds may be dispersed as a contaminant of commercial cereals and remain viable in the soil for 15–20 years, grow well in acidic soil (Warwick and Francis 2005; CABI 2016). *R. raphanistrum* is thought to be the ancestor of the domestic radish (*R. sativus*), and they can readily hybridize. Hand crosses between *B. napus* as female and *R. raphanistrum* have produced a high number of reciprocal hybrids (Kerlan et al. 1992: Chèvre et al. 1996). In the same situation, only three hybrids of tens of thousands of seeds, have been reported (Chèvre et al. 2000; Darmency and Fleury 2000). Chèvre et al. (2000) estimated the hybridization rate to be 10^{-7} to 10^{-5} while other studies reported 4×10^{-8} (Rieger et al. 2001) and 3×10^{-5} (Warwick et al. 2003). The hybridization rate can vary depending on the parental variety (that may be *B. napus)* and the population of *R. raphanistrum.* Guéritaine et al. (2003) reported the lower and slower emergence of hybrids in F1and survives less than that of both parents. No intergeneric cross is reported in the UK monitoring programme of natural populations of *R. raphanistrum* growing near fields of herbicide-resistant *B. napus* (Eastham and Sweet 2002).

5 Economical and Societal Importance of Brassicaceae

Brassicaceae contributes a lot to the economy and society as a source of nutritious vegetables, oil, medicines, condiments, fodder, ornament, etc. With these values, the abilities and activities of Brassicaceae members are highlighted in Table 8.

Table 8 Brassicaceae family members with their activities and abilities

Crop/Species	Phytochemical contents, biological activity and abilities	References
B. nigra L. Koch	Source of seed oil, spices and culinary herbs. Seed oil contains totally phenolics 142.86 μg/ml and total flavonoid content 23.43 μg CE/ml. Mustard meal with water was used for plaster preparation and to treat skin diseases and arthritis. The seeds possess antibacterial, anticancer, diaphoretic, diuretic, antidiabetic, antidandruff, emetic, expectorant, irritant, antioxidant, antiradical and stimulant activities. Crushed seeds with hot water cure sore throat, headaches and cold. Mustard oil stimulates hair growth. Form its nectar yield mild-flavoured honey. Allyl isothiocyanate is used in cat and dog repellents	Cal-IPC (2004), Obi et al. (2009), Amri (2014), Tomar and Shrivastava (2014), Danlami et al. (2016), OECD (2016), Olgun et al. (2017)
Brassica napus L.	Source of high-quality edible oil and high-protein meal to feed cattle, poultry, swine and fish. Some varieties are used as a biofuel feedstock. The whole plant is anticancer, antioxidant, analgesic, diuretic, anti-scurvy, anti-inflammatory of bladder and anti-goat. Root and leaves of napobrassica possess antioxidant activity, inhibit lipid peroxidation and increase the superoxide dismutase (SOD) and Glutathione peroxidase activity (GPx). According to an estimation, canola oil is used in 80% of the salad oil market, 56% of the shortening market and 42% of the margarine market. Oil is used in cooking, sandwich spreads, coffee whiteners, creamers, industrial lubricants, cosmetics, printing inks, oiled fabrics, plasticizers, plastic wraps and pesticides. Oil is used to remove hair dandruff and as a hair tonic, as skin ointment and laxative	Saeidnia and Gohari (2012), Amri (2014), Canola Council of Canada (2014d), Bhatnagar et al. (2016), Jovičić et al. (2017), CFIA (2017)

(continued)

Table 8 (continued)

Crop/Species	Phytochemical contents, biological activity and abilities	References
B. rapa L. var. *rapifera* or *B. campestris* var. *chinensis* var. *pekinensis*	The whole plant parts contain phenolics and organic acids. Total phenolic content (TPC) of root shoot and leaves is 30–78 mg/100 g fw, total flavonoid content (TFC) 4.1–8.5 mg RE/g fw and ascorbic acid content (AAC) 0.13–0.25 mg/g. Stems and leaves have anticancer, roots have cytotoxic, antioxidant activity and superoxide anion (SOA) radical scavenging ability. An ointment derived from its flowers helps in skin cancer Leaves contain phenolics 42.32–42.92 mg/g extract, flavonoids 49–133 mg QE/g extract and anthocyanin. Leaves contain anti-carcinogens and antioxidants Leaves contain phenolics 150–347 mg/100 g, flavonoids 61.9–328.70 mg RE/g fw, and have anti-cancer and antioxidant activity. Able to absorb Fe reducing, oxygen radical	Amri (2014), Fernandes et al. (2007), Ryu et al. (2012), Beltagy(2014), Iqbal et al. (2013), Unal et al. (2014), Seong et al. (2016)
B. carinata A. Braun.	The mean value of total phenolic content of leaves in different *B. carinata* accessions is reported as 8.57 mg/g dry matter and can vary from 2.20 to 12.70 mg/g DM. *B. carinata* is able to reduce soil-borne plant pathogens as it is used as biofumigants, biopesticides and bionematicides. *B. carinata* has also been used to make biofuel for jet aircraft. The derivative of *B. carinata* oil is used in protein meal for animal fodder	Wang and Zhu (2006), Fang et al. (2008), Warwick (2010), Pratt (2017), Rahman et al. (2018), Martínez-Valdivieso et al. (2019); http://www.ncbi.nlm.nih.gov; http://plants.usda.gov; http://www.asdnews.com

(continued)

Table 8 (continued)

Crop/Species	Phytochemical contents, biological activity and abilities	References
B. oleracea L. var. *capitata*	Main edible part is the leaf. Leaves contain 3.64 μM/g dw and flower buds contain 4.14 mM/g dw phenolic content. Leaves are antibacterial and antioxidant and possess antioxidant enzymes that inhibit DNA methylation, damage, protect from cancer and cardiovascular diseases	Sotelo et al. (2014), Ahmed et al. (2012), Ogbede et al. (2014), Grønbæk (2014), Thaipratum (2014), Prasad (2014), Nawaz et al. (2018a, b), Jaiswal et al. (2012), Simlai et al. (2014), Cartea et al. (2010), Sosnowska et al. (2006), Sharma et al. (2015), Nicoletto et al. (2016), Sharma and Kapoor (2015), Miraj (2016), Singh et al. (2011), Wu et al. (2015), Shah et al. (2016), Ligen et al. (2017), Chauhan et al. (2016), Hedges and Lister (2006), Sikora and Bodziarczyk (2012)
var. *capitate f. alba*	Leaves are a rich source of phytochemicals including total phenolic, 14.78–18.7 mg/g extract and flavonoids 4.12–8.80 mg QE/g extract, and vitamin C with antioxidant activity	
var. *capitata f. rubra*	Leaves contain total phenolic content 134–171 mg/100 g fw, carotenoids, glucosinolates and rich vitamin C with free radical scavenging abilities	
var. *italica*	The edible part is the usually immature flower stalk. Florets and stem are a rich source of phytochemicals. Florets contains total phenolic 17.9–23.6 mg/g extract and total flavonoids 12.5–17.5 mg CE/100 g. Rich source of vitamin C and possesses antioxidant enzymes which inhibit DNA methylation, damage and protect from cancer and cardiovascular diseases. It is also anti-proliferative (inhibit tumour cell formation), neuroprotective, antidiabetic, and anti-genotoxic. Seeds are also reported as antioxidants	

(continued)

Table 8 (continued)

Crop/Species	Phytochemical contents, biological activity and abilities	References
var. *gemmifera*	Axillary buds are the main edible parts. Leaves and sprouts are rich sources of phytochemicals. Sprouts contain total phenolic 18.12–20.4 mg/g extract, and total flavonoids 12.1–15.4 mg CE/100 g. Good antioxidant with enzymes to inhibit DNA methylation and its damage. Protect from cancer and cardiovascular diseases	
var. *acephala* L.	Edible leaves with total phenolic content 574.9 mg/100 g fw. Good source of Vitamin C and carotenoids with antioxidant activity	
B. juncea (L.) Czern.	Leaves stem and seeds are the edible parts. Leaves contain rich amount of phytochemical with reducing sugars and vitamin C. Total phenolic content of leaf and stem is 3.01–3.85 mg/100 g sample. Leaves and seeds oil possess antioxidant and cytotoxic activity. Seeds have sinigrin, quercetin, catechin, and vitamin E. Use for Phytoremediation	Cartea et al. (2010), Chaudhary et al. (2016), Parikh and Khanna (2014), Sharma et al. (2017), Kumari et al. (2016), Singh and Malik(2017)
B. tournefortii Gouan	The volatiles content of the seeds possesses cytotoxic, anticancer and antioxidant activity	Amri (2014), Shabana et al. (2013)
B. rupestris L.	Antioxidant and protect from cancer	Amri (2014)

(continued)

Table 8 (continued)

Crop/Species	Phytochemical contents, biological activity and abilities	References
Eruca sativa	Rich in vitamin C and potassium. The whole plant is edible and possesses antioxidant, antibacterial, antidandruff, antidiabetic activity. Used as a hair tonic. Leaves are used as an astringent, diuretic, digestive, emollient, depurative, laxative, rubefacient, tonic, and stomachic anti-inflammatory. Also used in cosmetics industries because of its effective phytochemical content. Taramira oil is extracted from seeds, used in pickling and cooking. The seed cake is animal feed	Barlas et al. (2011), Koubaa et al. (2015), Zudkafh (2011), Salma et al. 2018), Grubben and Denton (2004), Srinabas et al. (2004)
C. sativa (L.) Crantz	Source of food and animal feed. Plant stems yield fibre. Cold-pressed, unrefined *Camelina* oil is used in food because of the presence of omega-3 fatty acid α-linolenic acid. Oil is used in lamp fuel and in industries for the production of renewable biofuels such as biodiesel and jet fuels. *Camelina* oil has been proposed to use in paints, inks, soaps, varnishes, lubricants, cosmetics and as a plastic additive. *Camelina* meal is used as animal feed. As a medicine, it is used in burns, wounds, eye inflammations, to cure stomach ulcers and as a tonic	Sturtevant (1919), Rode (2002), Francis and Warwick (2009), Abramovič and Abram (2005), McVay and Lamb (2007), El Bassam (2010), Ní Eidhin and O'Beirne (2010)
A. thaliana L. Heynh	*A. thaliana* as a model organism is widely used to study and understand plant science, genetics and evolution. Its sequenced genome is a tool understanding the molecular biology of many plant characters, like flower development and light sensing	Saeidfirozeh et al. (2018), Hasegawa et al. (2016)

(continued)

Table 8 (continued)

Crop/Species	Phytochemical contents, biological activity and abilities	References
S. arvensis L. *S. alba* L.	Leaves and seeds are toxic to livestock. Juvenile raw or cooked leaves are edible. Plants cause a decline in yield of field crops. The seeds contain up to 30% fatty oil with glucosinolate and sinalbin, and Gibberellic acid which break the dormancy of the seeds. Oil is used in food, industries to lubricate machinery, in making soap, and for lighting. Oilcake is used for mustard preparation Leaf and young flower stalk are edible parts. Good source of seeds (used as condiments), fodder crop or green manure. Suppress weeds growth, resistant to white beet-cyst nematode and protect soil from erosion. 6.96% sinapin/extract in seeds is reported. Possess anti-proliferative, proapoptotic, antioxidant and anti-microbial activities. Have potential in chemoprevention and food preservation	wildfooduk.com; plantlife.org.uk; Edwards (1976), Warwick et al. (2000), Popova and Morra (2014) Rahman et al. (2018), Guerena (2006), Boscaro et al. (2018)

(continued)

Table 8 (continued)

Crop/Species	Phytochemical contents, biological activity and abilities	References
R. raphanistrum L. *R. sativus* var. *longipinnatus*	Used as famine food, in biological control programme; poisonous to a mammal, reduce the yield of field crops, antirheumatic. Seed contains up to 48% oil and is a potential source of biofuel The whole plant is edible. Leaf radish is used as green manure and forage for animals. Beneficial phytochemicals protect from cancer and cardiovascular diseases. Sulphoraphane of seed oil improves intestinal function and solve digestive problems. Oil is potential source for food, medicine and cosmetic industries The daikon (*R. longipinnatus*) use in industries and for soil improvement. The whole plant is edible. It grows as a cover crop to increase soil fertility as it scavenge nutirents, suppress weeds, help in the alleviation of soil compaction and prevent soil erosion in winter. Seed oil is a natural emollient used in cosmetics	Holm et al. (1997), Cheam and Code (1995), Triska (1975); https://www.cabi.org/iscdatasheet/46796/; Kuang et al. (2013), Zhang et al. (2016), Wada et al. (2007), Cavigelli et al. (2014), Gruber et al. (2012)

(continued)

Table 8 (continued)

Crop/Species	Phytochemical contents, biological activity and abilities	References
Lepidium sativum L.	Edible herb can be used as a functional food in raw or cooked form. Its seeds oil and powder are full of protein, fat, minerals, fibres and beneficial phytochemicals, to be incorporated in beverages and foods. It is the best antioxidant and contains vitamin A, E and eugenol. A balanced amount of polyunsaturated fatty acids (46.8%) and monounsaturated fatty acids (37.6%) is present in seeds. High linoleic (32.18%) and oleic acid content (30.5%) reduce the risk of coronary heart diseases. Benzyl isothiocyanate is present in seeds which induce carcinogen-detoxifying enzymes. The seeds are anti-anaemic, galactagogue, anti-inflammatory, analgesic, coagulant, antipyretic, thermogenic, depurative, diuretic, rubefacient, aphrodisiac, ophthalmic, antiscorbutic, antihistaminic, and act as a tonic. Various diseases such as asthma, coughs with expectoration, diarrhea, dysentery, poultices for sprains, leprosy, skin disease, splenomegaly, dyspepsia, lumbago, leucorrhoea, scurvy and seminal weakness can be treated by seed. The diet of lactating women can be supplemented with it to increase milk secretion. The leaves are used against syphilis	Sahsrabudde (1962), Vohora and Khan (1977), Gokavi et al. (2004), Williams et al. (2009), Mali et al. (2007), Raghavendra and Akhilender (2011), Singh and Paswan (2017)

(continued)

Table 8 (continued)

Crop/Species	Phytochemical contents, biological activity and abilities	References
Lepidium meyenii	Root is an edible part to be used commercially with different colours, red, yellow, brown, purple, black, etc. Highly nutritive, aphrodisiac, energizer and fertility booster in men and animals. It has a hypoglycaemic and anti-obesity effect. It contains antioxidant compounds like phenols, glucosinolates, alkamides and polysaccharides. The black variety is reported to have more total phenols than red and yellow. The total average amount of phenols 5.5–7.6 mg GAE/g, is reported	Korkmaz (2018), Campos et al. (2013), Melnikovova et al. (2015), Zhang et al. (2019)
Sisymbrium irio L.	Whole plant is edible and a rich source of flavonoids and glucosinates. The plant is laxative antipyretic, diuretic, expectorant, restorative, anti-vomiting, diarrhea and cough. Seeds are stimulants used to treat asthma, coughs and chest congestion and also used as stimulating poultice. It is also used to relieve rheumatism, to detoxify the liver and spleen, and to reduce swelling and clean wounds. The Bedouin used the whole plant as food and leaves as a tobacco substitute	Gulshan et al. (2012), Khan et al. (2013), Efraima (2003), Bailey and Danin (1981), Barton (1977), Al-Qudah and Abu Zarga (2010)

(continued)

Table 8 (continued)

Crop/Species	Phytochemical contents, biological activity and abilities	References
A. rusticana	A root vegetable used as spices and prepared as condiments. Roots and leaves, used for seasoning or as preservatives, possess anti-lipase, antioxidant, antimicrobial and anti-carcinogenic activity. Presence of allyl isothiocyanate repell animal and prevent from herbivory. It may be poisonous to mammals. The enzyme horseradish peroxidase is used in molecular biology and biochemistry to amplify a weak signal and increase the detectability of a target molecule. It is used to visualize under microscopy and measure non-quantitatively the permeability of brain capillaries. Horseradish paste is used to cure body pains, sciatica pain, arthritis, relieve rheumatism, colic, increased urination, kill worms in children, treat scurvy, influenza, used as a poultice, and syrup is used as a respiratory aid to cure common cold to tuberculosis	Agneta et al. (2013), Calabrone et al. (2015), Lossinsky and Shivers (2004), Shehata et al. (2009), McCann (2004), Pieroni et al. (2004), Pieroni and Quave (2005), Sarli et al. (2012), Talalay and Fahey (2001), Bladh and Olsson (2011)
C. bursa-pastoris Moench	*C. bursa-pastoris* is commercially grown as food crop used, to supplement animal feed, in cosmetics and in medicine. It possesses stimulating, astringent, expectorant, and diuretic activities. The whole plant is used in fertility regulation. The extract has the property of controlling chronic hemorrhages, chronic diarrhea, dysentery and constant desire to urinate with phosphate deposits. The fresh herb is effective in *ecchymosis* and *rheumatic pains*	Zudkafh (2011), David (2014), Felter and Lloyd (1898); http://www.henriettes-herb.com

(continued)

Table 8 (continued)

Crop/Species	Phytochemical contents, biological activity and abilities	References
Nasturtium officinale R. Br.	*N. officinale* is a source of vegetables and salads. It is grown as a healthy food because it contains anticancer compound. It is a mild stimulant, diuretic, an expectorant, digestive aid, used for the treatment of colds, coughs, eczema and as an abortifacient. It improves blood circulation to treat renal colic and liver diseases	https://www.cabi.org/isc/datasheet/35646; Duke (2013), Matsushima et al. (2006), Guarrera et al. (2005)
Alliaria petiolata (M. Bieb.)	*A. petiolata* flower, leaf and young fruit are used raw as salad, or cooked both as a vegetable and to flavour cooked food. The leaves contain a higher amount of vitamin C than oranges and more vitamin A than spinach. The leaves are commonly used to flavour beverages. Leaves possess antimicrobial activities, used as a sudorific when taken internally, and externally used as an antiseptic to treat infection, cuts, and ulcers, heal wounds and bronchial complications. Leaf juices alone or with honey are used to treat oedema	https://www.cabi.org/isc/datasheet/3941; Simonetti (1990), Cipollini and Cipollini (2016), Debby (2015), Zennie and Ogzewalla, (1977), Ravindran (2017)

(continued)

Table 8 (continued)

Crop/Species	Phytochemical contents, biological activity and abilities	References
C. abyssinica	*C. abyssinica* is grown commercially for industrial purposes. Oil contains 55–60% erucic acid and the compounds derived from it are used for the production of many important industrial products like pharmaceuticals, waxes, lubricants, plasticisers, base for paints and coatings, cosmetics and many other products. Plant is used as erucamide. The crambe meal contains 46% proteins and is used as a protein supplement in different ratios for animal feed. The use as forage is very limited because of the presence of toxic compounds glucosinolates, tannins and inositol phosphate. Crambe seed oil has the potential to be utilized as biofuels	Warwick (2010), Zhu (2016), Lalas et al. (2012), IENICA (2012), Falasca et al. (2010), Hartwig et al. (2005), Eduardo et al. (2017)
Aethionema grandiflorum	Whole plant is used to treat meningitis, bacterial infections and typhoid	Parvaiz et al. (2013)

References

Abramovič H, Abram V (2005) Physico-chemical properties, composition and oxidative stability of *Camelina sativa* oil. Food Technol Biotechnol 43:63–70

Afrin KS, Rahim MA, Jong-In P, Natarajan S, Rubel M, Kim H, Nou, Ill-Sup (2018) Screening of cabbage (*Brassica oleracea* L.) germplasm for resistance to black rot. Plant Breed Biotech 6(1):30–43

Agneta R, Möllers C, Rivelli AR (2013) Horseradish (Armoracia rusticana), a neglected medical and condiment species with a relevant glucosinolate profile: a review. Genet Resour Crop Evol 60(7)

Ahmed MF, Rao AS, Ahemad SR, Ibrahim M (2012) Phytochemical studies and antioxidant activities of *Brassica oleracea* L. Var. Capitata. Int J Phar Pharmaceut Sci 4:374–378

Ahuja, KL, Singh H, Raheja RK, Labana KS (1987) The oil content and fatty acid composition of various genotypes of cauliflower, turnip and radish. Plant Food Hum Nutr 37:33–40

Alam M, Ahmad H, Quazi MH, Khawaja HIT (1992) Cross compatibility studies within the genus *Brassica* 1. Amphidiplpoid combinations. Sci Khyber 5:89–92

Almond JA, Dawkins TCK, Askew MF (1986) Aspects of crop husbandry. In: Scarisbrick DH, RW Daniels (eds) Oilseed rape. Collins, London, pp 127–175

Al-Qudah MA, Abu Zarga MH (2010) Chemical constituents of Sisymbrium irio L. from Jordan. Natu Prod Res 24(5):448–456

Al-Shehbaz IA (2012) A generic and tribal synopsis of the Brassicaceae (Cruciferae). Taxon 61(5):931–954

Amri E (2014) The role of selected plant families with dietary Ethnomedicinal species used as anticancer. J Med Plants Stud 2(1):28–39

Anand IJ, Downey RK (1981) A study of erucic acid alleles in digenomic rapeseed (*Brassica napus* L.). Can J Plant Sci 61:199–203

Anand IJ, Mishra PK, Rawat DS (1985) Mechanism of male sterility in *Brassica juncea*. I. Manifestation of sterility and fertility restoration. Cruciferae Newslett 10:44–46

Anonymous (1993) Sweet potato plants vs. weeds. HortIdeas. January p 8

Apel P, Bauwe H, Ohle H (1984) Hybrids between Brassica alboglabra and Moricandia arvensis and their photosynthetic properties. Biochem Physiol Pflanz 179:793–797

Atri C, Kaur B, Sharma S, Gandhi N, Verma H, Goyal A et al (2016) Substituting nuclear genome of *Brassica juncea* (L.) Czern & Coss. In cytoplasmic background of *Brassica fruticulosa* results in cytoplasmic male sterility. Euphytica 209:31–40

Attia T, Busso C, Röbbelen G (1987) Digenomic triploids for an assessment of chromosome relationships in the cultivated diploid Brassica species. Genome 29:326–330

Bajaj YPS (1990) Wide hybridization in legumes and oilseed crops through embryo, ovule and ovary culture. In: Bajaj YPS (ed) Biotechnology in agriculture and forestry 10. Legumes and oilseed crops I. Springer, Berlin, pp 3–37

Bajaj YPS, Mahajan SK, Labana KS (1986) Interspecific hybridization of *Brassica napus* and *B. juncea* through ovary, ovule and embryo culture. Euphytica 35(103–109)

Bang SW, Kaneko Y, Matsuzawa Y (1996) Production of intergeneric hybrids between Raphanus and Sinapis and the cytogenetics of their progenies. Breed Sci 46:45–51

Banga SS, Labana KS (1991) Cytoplasmic-genetic relationship between *Brassica nigra* and *Sinapis allioni*. Cruciferae Newslett 14(15):12–13

Banga S, Kaur G, Grewal N, Salisbury PA, Banga SS (2011) Transfer of resistance to seed shattering from *Brassica carinata* to *Brassica napus*. In: Proceedings of the 13th international rapeseed congress, Prague, Czech Republic, pp 863–866

Bansal S, Durrett TP (2016) *Camelina sativa*: an ideal platform for the metabolic engineering and field production of industrial lipids. Biochimie 120:9–16

Barcikowska B, Balicka M, Zwierzykowska E (1983) On the way to yellow seeded *Brassica napus*. I. Crossings between *Brassica oleracea* and *B. carinata*. Cruciferae Newslett 18:20

Barlas NT, Irget ME, Tepecik M (2011) Mineral content of the rocket plant (*Eruca sativa).* African J Biotechnol 10(64):14080–14082

Barnum K, Franks SJ (2013) Seed extracts impede germination in Brassica rapa plants. Int J Plant Biol 4(1):e2

Barro F, Martín A (1999) Response of different genotypes of *Brassica carinata* to microspore culture. Plant Breed 118(1):79–81

Barton WH (1977) Wildflowers of the Davis Mountains and the Marathon Basin, Texas, p 111

Bauer-Weston B, Keller W, Webb J, Gleddie S (1993) Production and characterization of asymmetric somatic hybrids between *Arabidopsis thaliana* and *Brassica napu.* Theo Appl Genet 86(2–3):150–158

Beckie HJ, Warwick SI, Nair H, Seguin-Swartz G (2003) Gene flow in commercial fields of herbicide-resistant canola (*Brassica napus* L.). Ecol Appl 13:1276–1294

Beilstein MA, Al-Shehbaz IA, Kellogg EA (2006) Brassicaceae phylogeny and trichome evolution. Am J Bot 93:607–619

Beilstein MA, Al-Shehbaz IA, Mathews S, Kellogg EA (2008) Brassicaceae phylogeny inferred from phytochrome A and ndhF sequence data: tribes and trichomes revisited. Am J Bot 95:1307–1327

Beltagy AM (2014) Investigation of new antimicrobial and antioxidant activities of *Brassica rapa* L. Int J Pharm Pharmaceut Sci 6:84–88

Bennett BC (2011) Twenty-five economically important plant families. Encyclopedia of life support systems. http://www.eolss.net/Sample-Chapters/C09/E6–118-03.pdf

Bennett RA, Thiagarajah MR, King JR, Rahman MH (2008) Interspecific cross of *Brassica oleracea* var. alboglabra and *B. napus*: effects of growth condition and silique age on the efficiency of hybrid production and inheritance of erucic acid in the self-pollinated backcross generation. Euphytica 164:593–601

Bent AF, Kunkel BN, Dahlbeck D, Brown KL, Schmidt R, Giraudat J, Leung J, Staskawicz BJ (1994) RPS2 of Arabidopsis thaliana: a leucine-rich repeat class of plant disease resistance genes. Science 265(5180):1856–1860

Beránek M, Bechyne M, Klíma M (2007) Protoplast isolation and fusion between *Brassica carinata* A. Braun. and *Brassica rapa* L. Agricultura Tropica et Subtropica 40(1):1–6

Beversdorf WD, Weiss-Lerman J, Erickson LR, Machado VS (1980) Transfer of cytoplasmically-inherited triazine resistance from bird's rape to cultivated oilseed rape (*Brassica campestris* and *B. napus*). Can J Genet Cytol 22:167–172

Bhatnagar A, Bhushan B, Malik M (2016) Traditional medicinal plants of district Bijnor, U.P., India. Int J Recent Sci Res 7(4):10670–10677

Bijral JS, Sharma TR (1995) A sexual hybrid between *Brassica juncea* and *Diplotaxis muralis.* Indian J Genet Plant Breed 55:170–172

Bijral JS, Sharma TR (1996) Intergeneric hybridization between *Brassica napus* and *Diplotaxis muralis.* Cruciferae Newslett 18:10–11

Bijral JS, Kanwal KS, Sharma TR (1994) *Brassica cossoneana* x *Brassica carinata* hybrids. Cruciferae Newslett 16(22)

Bing DJ, Downey RK, Rakow GFW (1991) Potential of gene transfer among oilseed *Brassica* and their weedy relatives. In: GCIRC 1991 congress, pp 1022–1027

Bing DJ, Downey RK, Rakow GFW (1995) An evaluation of the potential of intergeneric gene transfer between *Brassica napus* and *Sinapis arvensis.* Plant Breed 114:481–484

Bing DJ, Downey RK, Rakow GFW (1996a) Assessment of transgene escape from *Brassica rapa* (*B. campestris*) into *B. nigra* or *Sinapis arvensis.* Plant Breed 115:1–4

Bing DJ, Downey RK, Rakow GFW (1996b) Hybridizations among Brassica napus, B. rapa and B. juncea and their two weedy relatives B. nigra and Sinapis arvensis under open pollination conditions in the field. Plant Breed 115:470–473

Bioresearch Online (2000) Plants that glow in the dark. Bioresearch Online

Bladh KW, Olsson KM (2011) Introduction and use of horseradish (*Armoracia rusticana*) as food and medicine from antiquity to the present: emphasis on the Nordic countries. J Herbs Spices Med Plants 17(3):197–213

Boscaro V, Boffa L, Binello A, Amisano G, Fornasero S, Cravotto G, Gallicchio M (2018) Antiproliferative, proapoptotic, antioxidant and antimicrobial effects of *Sinapis nigra* L. and *Sinapis alba* L. extracts. Molecules *23*(11):3004

Bradley FM, Ellis B W, Martin DL (eds) (2009) The organic gardener's handbook of natural pest and disease control. Rodale, Inc. ISBN 978-1-60529-677-7

Brown J, Brown AP (1996) Gene transfer between canola (*Brassica napus* L. and *B. campestris* L.) and related weed species. Ann Appl Biol 129:513–522

Browne LM, Conn KL, Ayer WA, Tewari JP (1991) The camalexins: new phytoalexins produced in the leaves of *Camelina sativa* (Cruciferae). Tetrahedron 47:3909–3914

Bundy JG, Davey MP, Viant MR (2009) Environmental metabolomics: a critical review and future perspectives. Metabolomics 5(3–21):3–21

Busso C, Attia T, Röbbelen G (1987) Trigenomic combinations for the analysis of meiotic control in the cultivated *Brassica* species. Genome 29:331–333

CABI (2016) Centre for Agriculture and Bioscience International. Invasive species compendium. www.cabi.org/isc

Calabrone L, Larocca M, Marzocco S, Martelli G, Rossano R (2015) Total phenols and flavonoids content, antioxidant capacity and lipase inhibition of root and leaf horseradish (*Armoracia rusticana)* extracts. Food Nutr Sci 6(2015):64–74

Cal-IPC (2004) Cal-IPC plant assessment form for *Brassica nigra*. California Invasive Plant Council, Berkeley, California, USA. http://www.cal-ipc.org/paf/site/paf/500

Callihan B, Brennan J, Miller T, Brown J, Moore M (2000) Mustards in mustards: guide to identification of canola, mustard, rapeseed and related weeds. University of Idaho

Campos D, Chirinos R, Barreto O, Noratto G, Pedreschi R (2013) Optimized methodology for the simultaneous extraction ofglucosinolates, phenolic compounds and antioxidant capacity from maca (*Lepidium meyenii*). Indus Crops Prod 49:747–754

Canola Council of Canada (2014d) What is Canola? http://www.canolacouncil.org/oil-and-meal/what-is-canola/

Cárcamo H, Olfert O, Dosdall L, Herle C, Beres B, Soroka J (2007) Resistance to cabbage seedpod weevil among selected Brassicaceae germplasm. The Can Entomol 139:658–669

Cardone M, Mazzoncini M, Menini S, Rocco V, Senatore A, Seggiani M, Vitolo S (2003) Brassica carinata as an alternative oil crop for the production of biodiesel in Italy: agronomic evaluation, fuel production by transesterification and characterization. Biom Bioener 25(6):623–636

Cartea M, Francisco M, Lema M, Soengas P, Velasco P (2010) Resistance of Cabbage (Brassica oleracea capitata Group) Crops to Mamestra brassicae. J Eco Entomol 103:1866–1874

Cartea E, De Haro-Bailón A, Padilla G, Obregón-Cano S, Del Rio-Celestino M, Ordás A (2019) Seed Oil Quality of *Brassica napus* and *Brassica rapa* Germplasm from Northwestern Spain. Foods 8(8):292

Cavigelli MA, Martin TE, Mutch DR (2014) Oilseed radish. Michigan State University

Center for New Crops & Plant Products (2013) CropINDEX. West Lafayette, Indiana, USA: Center for New Crops & Plant Products, Purdue University. http://www.hort.purdue.edu/newcrop/Indices/index_ab.html

CFIA (2012) Canadian food inspection Agency.https://www.inspection.gc.ca/plants/plants-with-novel-traits/applicants/directive-94-08/biology-documents/brassica-juncea/eng/1330727837568/1330727899677

CFIA (2014) Canadian food inspection agency. https://inspection.gc.ca/plants/plants-with-novel-traits/applicants/directive-94-08/biology-documents/brassica-rapa-l-/eng/1330965093062/1330987674945

CFIA (2017) Canadian food inspection agency. https://inspection.gc.ca/plants/plants-with-novel-traits/applicants/directive-94-08/biology-documents/brassica-carinata/eng/1501087371874/1501087468251

Chadoeuf R, Darmency H, Maillet J (1998) Survival of buried seeds of interspecific hybrids between oilseed rape, hoary mustard and wild radish. Field Crops Res 58(3):197–204

Chang C, Meyerowitz EM (1986) Molecular cloning and DNA sequence of the *Arabidopsi thaliana* alcohol dehydrogenase gene. Proc Natl Acad Sci USA 83(5):1408–1412

Chang CT, Uesugi R, Hondo K, Kakihara F, Kato M (2007) The effect of the cytoplasms of *Brassica napus* and *B. juncea* on some characteristics of *B. carinata*, including flower morphology. Euphytica 158(1–2):261–270

Chang C, Kakihara F, Hondo K, Kato M (2011) The cytoplasm effect comparison between *Brassica napus* and *Brassica carinata* on floral characteristics of *Brassica oleracea*. Plant Breed 120(1):73–79

Chatterjee D, Banga S, Gupta M, Bharti S, Salisbury PA, Banga SS (2016) Resynthesis of *Brassica napus* through hybridization between *B. juncea* and *B. carinata*. Theor Appl Genet 1–14

Chaudhary A, Choudhary S, Sharma U, Vig AP, Arora S (2016) In vitro evaluation of *Brassica* sprouts for its antioxidant and Antiproliferative potential. Ind J Pharmaceut Sci 78:615–623

Chauhan ES, Tiwari A, Singh A (2016) Phytochemical screening of red cabbage (*Brassica oleracea*) powder and juice-a comparative study. J Med Plants 4:196–199

Chavan V, Kamble A (2014) Induction of total phenolics and defence-related enzymes during beta-aminobutyric acid-induced resistance in *Brassica carinata* against *Alternaria* blight. Archives Phytopathol Plant Protect 47:2200–2212

Cheam AH, Code GR (1995) The biology of Australian weeds. 24. Raphanus raphanistrum L. Plant Prot Quart 10(1):2–13

Chen BY, Heneen WK (1992) Inheritance of seed colour in *Brassica campestris* L. and breeding for yellow-seeded *B. napus* L. Euphytica 59:157–163

Chen HG, Wu JS (2008) Characterization of fertile amphidiploid between *Raphanus sativus* and *Brassica alboglabra* and the crossability with *Brassica* species. Genet Resour Crop Evol 55:143–150

Chen LP, Zhang MF, Li CS, Hirata Y (2005) Production of interspecific somatic hybrids between tuber mustard (*Brassica juncea*) and red cabbage (*Brassica oleracea*). Plant Cell Tissue Organ Cult 80:305–311

Chen H-F, Wang H, Li Z-Y (2007) Production and genetic analysis of partial hybrids in intertribal crosses between *Brassica* species (*B. rapa*, *B. napus*) and *Capsella bursa-pastoris*. Plant Cell Rep 26:1791–1800

Chen JP, Ge XH, Yao XC, Li ZY (2012) Genome affinity and meiotic behavior in trigenomic hybrids and their doubled allohexaploids between three cultivated *Brassica* allotetraploids and *Brassica fruticulosa*. Genome 55:164–171

Cheng BF, Chen BY, Heneen WK (1994) Addition of *Brassica* alboglabra Bailey chromosomes to *B. campestris* L. with special emphasis on seed colour. Heredity 73:185–189

Cheung KW, Razeq FM, Sauder CA, James T, Martin SL (2015) Bidirectional but asymmetrical sexual hybridization between *Brassica carinata* and *Sinapis arvensis* (Brassicaceae). J Plant Res 128(2):1–12

Chevre AM, Eber F, Marbale E, Kerlan MC, Primad C, Vedel F, Delseny M, Pelletier G (1994) Comparison of somatic and sexual *Brassica napus - Sinapis alba* hybrids and their progeny by cytogenetic studies and molecular characterization. Genome 37:367–374

Chevre AM, Eber F, Baranger A, Kerlan MC, Barret P, Festoc G, Vallee P, Renard M (1996) Interspecific gene flow as a component of risk assessment for transgenic Brassicas. Acta Hort 407:169–179

Chèvre AM, Eber F, Baranger A, Hureau G, Barret P, Picault H, Renard M (1998) Characterization of backcross generations obtained under field conditions from oilseed rape-wild radish F1 interspecific hybrids: an assessment of transgene dispersal. Theor Appl Genet 97:90–98

Chèvre AM, Eber F, Darmency H, Fleury A, Picault H, Letanneur JC, Renard M (2000) Assessment of interspecific hybridization between transgenic oilseed rape and wild radish under normal agronomic conditions. Theor App Genet 100(8):1233–1239

Chèvre AM, Ammitzbøll H, Breckling B, Dietz-Pfeilstetter A, Eber F et al (2004) A review on interspecific gene flow from oilseed rape to wild relatives. In: den Nijs HCM, Bartsch D, Sweet J (eds) Introgression from genetically modified plants into wild relatives. CABI, Wallingford, pp 235–251

Chiang MS, Chiang BY, Grant WF (1977) Transfer of resistance to race 2 of *Plasmodiophora brassicae* from *Brassica napus* to cabbage (*B. oleracea* var. *capitata*). I. Interspecific hybridization between *B. napus* and *B. oleracea* var. *capitate*. Euphytica 26(2):319–336

Choudhary BR, Joshi P (1999) Interspecific hybridization in *Brassica*. In: Proceedings of the 10th international rapeseed congress, Australia. Contribution No 510

Choudhary BR, Joshi P (2000) Cytomorphology of intergeneric hybrid *Sinapis alba* · *Brassica nigra*. J Genet Breed 54:157–160

Choudhary BR, Joshi P (2001) Genetic diversity in advanced derivatives of *Brassica* interspecific hybrids. Euphytica 121:1–7

Choudhary BR, Joshi P (2012) Crossability of *Brassica carinata* and *B. tournefortii*, and cytomorphology of their F1 hybrid. Cytologia 77(4):453–458

Choudhary BR, Joshi P, Ramarao S (2000) Interspecific hybridization between *Brassica carinata* and *Brassica rapa*. Plant Breed 119:417–420

Choudhary BR, Joshi P, Rama Rao S (2002) Cytogenetics of *Brassica juncea* X *Brassica rapa* hybrids and patterns of variation in the hybrid derivatives. Plant Breed 121:292–296

Chrungu B, Verma N, Mohanty A, Pradhan A, Shivanna KR (1999) Production and characterization of interspecific hybrids between *Brassica maurorum* and crop brassicas. Theor Appl Genet 98:608–613

Cipollini D, Cipollini K (2016) A review of garlic mustard (*Alliaria petiolata*, Brassicaceae) as an allelopathic plant. J Torrey Bot Soci 143(4):339–348

Clark A (2007) Managing cover crops profitably, 3rd ed. National SARE Outreach Handbook Series Book 9. Nat Agric Lab Beltsville MD

Clough SJ, Bent AF (1998) Floral dip: a simplified method for Agrobacterium-mediated transformation of Arabidopsis thaliana. The Plant J 16(6):735–743

Conn KL, Tewari JP, Dahiya JS (1988) Resistance to *Alternaria brassicae* and phytoalexin-elicitation in rapeseed and other crucifers. Plant Sci 56:21–25

Damgaar C, Kjellsson G (2005) Gene flow of oilseed rape (*Brassica napus*) according to isolation distance and buffer zone. Agri Eco Environ 108:291–301

Daniels R, Boffey C, Mogg R, Bond J, Clarke R (2005) The potential for dispersal of herbicide tolerance genes from genetically-modified, herbicide-tolerant oilseed rape crops to wild relatives. Report to the Department for the Environment, Food and Rural Affairs (DEFRA) No. EPG 1/5/151. http://www.defra.gov.uk/environment/gm/research/pdf/epg_1–5–151.pdf

Danlami U, Orishadipe AT, Lawal DR (2016) Phytochemical, nutritional and antimicrobial evaluations of the aqueous extract of *Brassica Nigra* (Brassicaceae) seeds. Am J Appl Chem 4:161–163

Darmency H, Fleury A (2000) Mating system in *Hirschfeldia incana* and hybridization to oilseed rape. Weed Res 40(2):231–238

David M (2014) Nature's restaurant: fields, forests & wetlands foods of Eastern North America—a complete wild food guide

Davis AR, Pylatuik JD, Paradis JC et al (1998) Nectar-carbohydrate production and composition vary in relation to nectary anatomy and location within individual flowers of several species of Brassicaceae. Planta 205:305–318

de Melo PE, de Giordano LB (1994) Effect of Ogura male-sterile cytoplasm on theperformance of cabbage hybrid variety. II. Commercial characteristics. Euphytica 78:149–154

Debby K (2015) Introduced species summary project garlic mustard (*Alliaria petiolata*)

Delaney TP, Uknes S, Vernooij B, Friedrich L, Weymann K, Negrotto D, Gaffney T, Gut-Rella M, Kessmann H, Ward E, Ryals J (1994) A central role of salicylic acid in plant disease resistance. Science 266(5188):1247–1250

Diederichsen E, Sacristan MD (1988) Interspecific hybridizations in the genus *Brassica* followed by in ovule embryo culture. Cruciferae Newslett 13:20–21

DiTomaso JM, Healy EA (2007) Weeds of California and other western states. UC ANR Publ 3488. Univ California Div of Agric. and Nat. Resources, Davis

Duke JA (1983) *Brassica rapa* L. In: Handbook of energy crops. Center for New Crops & Plant Products, Purdue Univ. www.hort.purdue.edu/ newcrop/duke_energy/Brassica_rapa.html

Duke JA (2013) Nasturtium officinale (Brassicaceae). Dr. Duke's phytochemical and ethnobotanical databases: ethnobotanical uses

Eastham K, Sweet J (2002) Genetically modified organisms (GMOs): the significance of gene flow through pollen transfer. Environ Issue Report, No. 28, Europe Environ Agency Copenhagen

Eduardo VS, Lorrane SV, Carlos FSC, Luciano ML, Francisco FGN, Pedro SMO (2017) Chromatographic characterization of the crambe (Crambe abyssinica Hochst) oil and modeling of some parameters for its conversion in biodiesel. Indus Crops Prod 97:545–551

Edwards M (1976) Dormancy in seeds of Charlock (*Sinapis arvensis* L.). Plant Physiol 58(5):626–630

Eenink AH (1974) Matromorphy in Brassica oleracea L. III. The influence of temperature, delayed prickle pollination and growth regulators on the number of matromorphic seeds formed. Euphytica 23:711–718

Efraima L (2003) "*Sisymbrium irio*" Medicinal substances in Jerusalem from early times to the present day. Archaeopress, Oxford, UK, p 62. 978-1-84171-490-5

Ehrensing DT, Guy SO (2008) *Camelina*. EM 8953-E. Oregon State University Extension Service, Corvallis, OR, USA. http://extension.oregonstate.edu/catalog/pdf/em/em8953-e.pdf

El Bassam N (2010) Handbook of bioenergy crops. A complete reference to species, development and applications. Earthscan LLC, Washington, DC, USA

Ellerstrom s (1978) Species crosses and sterility in Brassica and Raphanus. Cruciferae Newslett 3:16–17

Fahleson J, Rahlen L, Glimelius K (1988) Analysis of plants regenerated from protoplast fusions between *Brassica napus* and *Eruca sativa*. Theor Appl Genet 76:507–512

Fahleson J, Eriksson I, Glimelius K (1994a) Intertribal somatic hybrids between *Brassica napus* and *Barbarea vulgaris*—production of in vitro plantlets. Plant Cell Rep 13:411–416

Fahleson J, Eriksson I, Landgren M, Stymne S, Giimelius K (1994b) Intertribal somatic hybrids between *Brassica napus* and *Thlaspi perfoliatum* with high content of the *T. perfoliatum*-specific nervonic acid. Theor Appl Genet 87:795–804

Falasca SL, Flores N, Lamas MC, Carballo SM, Anschau A (2010) Crambe abyssinica: an almost unknown crop with a promissory future to produce biodiesel in Argentina. Int J Hydrogen Ener 35(11):5808–5812

Falk KC (1991) Heterosis in summer turnip rape (*Brassica campestris* L.) and cytoplasmic substitution in the genus Brassica. PhD thesis, Dept. of Crop Science and Plant Ecology, University of Saskatchewan, Saskatoon, Saskatchewan, Canada

Fan Z, Tai W, Stefansson BR (1985) Male sterility in *Brassica napus* L. associated with an extra chromosome. Can J Genet Cytol 27:467–471

Fan Z, Stefansson BR, Sernyk JL (1986) Maintainers and restorers for three male sterility-inducing cytoplasms in rape (*Brassica napus* L.). Can J Plant Sci C66:229–234

Fang ZX, Hu YX, Liu DH, Chen JC, Ye XQ (2008) Changes of phenolic acids and antioxidant activities during potherb mustard (Brassica juncea, Coss.) pickling. Food Chem 108:811–817

FAO (2019) The state of the world's biodiversity for food and agriculture. In: Bélanger J, Pilling D (eds) FAO commission on genetic resources for food and agriculture assessments, Rome, 572 pp. http://www.fao.org/3/CA3129EN/CA3129EN.pdf

FAOSTAT (2013) FAO Statistics online database, Production/Crops–rapeseed. http://faostat3.fao.org/home/E

Felter HW, Lloyd JU (1898) King's American Dispensatory. http://www.henriettes-herb.com

Fernandes F, Valentão P, Sousa C, Pereira JA, Seabra RM, Andrade PB (2007) Chemical and antioxidative assessment of dietary turnip (*Brassica rapa* var. rapa L.). Food Chem 105:1003–1010
Fernandez-Escobar J, Dominguez J, Martin A, Fernandez-Martinez JM (1988) Genetics of the erucic acid content in interspecific hybrids of Ethiopian mustard (*Brassica carinata* Braun) and rapeseed (*B. napus* L.). Plant Breed 100:310–315
FitzJohn RG, Armstrong TT, Newstrom-Lloyd LE, Wilton DA, Cochrane M (2007) Hybridisation within Brassica and allied genera: evaluation of potential for transgene Escape. Euphyt 158:209–230
Forsberg J, Landgren M, Glimelius K (1994) Fertile somatic hybrids between *Brassica napus* and *Arabidopsis thaliana*. Plant Sci 95:213–223
Francis A, Warwick SI (2009) The Biology of Canadian Weeds. 142. Camelina alyssum (Mill.) Thell; C. microcarpa Andrz. ex DC; *C. sativa* (L.) Crantz. Can J Plant Sci 89:791–810
Frello S, Hansen KR, Jensen J, Jørgensen RB (1995) Inheritance of rapeseed (*Brassica napus*)-specific RAPD markers and a transgene in the cross *B. juncea* × (*B. juncea* × *B. napus*). Theor Appl Genet 91:236–241
Fu TD, Yang GS, Yang XN, Ma CZ (1995a) Discovery, study and utilization of polima cytoplasmic male sterility in *Brassica napus* L. Prog Natl Sci 5:169–177
Fu TD, Yang GS, Yang XN, Ma CZ (1995b) Discovery, study and utilization of polima cytoplasmic male sterility in *Brassica napus* L. Prog Nat Sci 5:169–177
Fulgione A, Hancock AM (2018) Archaic lineages broaden our view on the history of Arabidopsis thaliana. The New Phytol 219(4):1194–1198
Gerdemann-Knorck M, Sacristan MD, Braatz C, Schieder O (1994) Utilization of asymmetric somatic hybridization for the transfer of disease resistance from *Brassica nigra* to *Brassica napus*. Plant Breed 113:106–113
Gerdemann-Knorck M, Nielen S, Tzscheetzsch C, Iglisch J, Schieder O (1995) Transfer of disease resistance within the genus *Brassica* through asymmetric somatic hybridization. Euphytica 85:247–253
Getinet A, Rakow G, Raney JP, Downey RK (1994) Development of zero erucic acid Ethiopian mustard through an interspecific cross with zero erucic acid Oriental mustard. Can J Plant Sci 74(4):793–795
Getinet A, Rakow G, Raney JP, Downey RK (1997) Glucosinolate content in interspecific crosses of *Brassica carinata* with *B. juncea* and *B. napus*. Plant Breed 116:39–46
Ghosh Dastidar N, Varma NS (1999) A study on intercrossing between transgenic *B. juncea* and other related species. In: Proceedigs of the 10th international rapeseed congress, Australia. Contribution No. 244
Gleba YY, Hoffmann F (1980) Arabidobrassica: a novel plant obtained by protoplast fusion. Planta 149:112–117
Gokavi SS, Malleshi NG, Guo M (2004) Chemical composition of garden cress (*Lepidium sativum*) seeds and its fractions and use of bran as a functional ingredient. Plant Foods for Human Nutr 59(3):105–111
Goswami R, Devi J (2002) Intergeneric hybridization of Indian mustard (*Brassica juncea*) with taramira (*Eruca sativa*). Indian J Agric Sci 72:436–438
Gowers S (1982) The transfer of characters from *Brassica campestris* L. to *Brassica napus* L.: production of clubroot-resistant oil-seed rape (*B. napus* ssp. *oleifera*). Euphytica 31:971–976
Goyal RK, Chowdhury JB, Jain RK (1997) Development of fertile *Brassica juncea* x *B. tournefortii* hybrids through embryo rescue. Cruciferae Newslett Eucarpia 19:19–20
Grønbæk M (2014) Effects of cultivation strategies on phytochemicals and sensory properties of cabbage (*Brassica oleracea L. var. capitata L.*) and curly kale (*Brassica oleracea* L. var. sabellica L.). Aarhus Uni Dep Food Sci
Grubben GJH, Denton OA (ed) (2004) Vegetables. Plant Resour Trop Africa 2:295

Gruber S, Husken A, Dietz-Pfeilstetter A, Mollers C, Weber EA, Stockmann F, Thole H, Gruver J, Weil R, White C, Lawley Y (2012) Radishes—a new cover crop for organic farming systems. e-Organic. Michigan State University

Gualberto JM, Mileshina D, Wallet C, Niazi AK, Weber-Lotfi F, Dietrich A (2014) The plant mitochondrial genome: dynamics and maintenance. Biochimie 100:107–120

Guarrera PM, Salerna G, Caneva G (2005) Folk phytotherapeutical plans from Maratea area (Basilicata, Italy). J Ethnopharmacol 99(3):367–378

Guerena (2006) Cole crops and other Brassicas: Organic production. NCAT Agriculture Specialist. www.attra.ncat.org

Guéritaine G, Bazot S, Darmency H (2003) Emergence and growth of hybrids between *Brassica napus* and *Raphanus raphanistrum*. New Phytol 158(3):561–567

Gugel RK, Falk KC (2006) Agronomic and seed quality evaluation of *Camelina sativa* in eastern Canada. Can J Plant Sci 86:1047–1058

Gulati SC, Varma NS, Mani N et al (1991) Resistance to white rust (*Albugo candida*) in Indian mustard. In: Proceedings of the GCIRC 8th international rapeseed congress, pp 256–261

Gulden RH, Warwick SI, Thomas AG (2008) The biology of Canadian weeds. 137. *Brassica napus* L. and *B. rapa* L. Can J Plant Sci 88:951–996

Gulshan AB, Dasti AA, Sabir H, Atta MI, Aminud-din M (2012) Indigenous uses of medicinal plants in rural areas of Dera Ghazi Khan, Punjab, Pakistan. ARPN J Agric Biol Sci 7(9):750–762

Gundimeda HR, Prakash S, Shivanna KR (1992) Intergeneric hybrids between *Enarthrocarpus lyratus*, a wild species, and crop brassicas. Theor Appl Genet 83:655–662

Gupta SK (1997) Production of interpsecific and intergenric hybrids in *Brassica* and *Raphanus*. Cruciferae Newslett 19:21–22

Halfhill MD, Richards HA, Mabon SA, Stewart CN (2001) Expression of GFP and Bt transgenes in *Brassica napus* and hybridization with *Brassica rapa*. Theor Appl Genet 103:659–667

Halfhill MD, Millwood RJ, Raymer PL, Stewart CN Jr (2002) Bt-transgenic oilseed rape hybridization with its weedy relative, *Brassica rapa*. Environ Biosaf 1:19–28

Halfhill MD, Zhu B, Warwick SI, Raymer PL, Millwood RJ, Weissinger AK, Stewart CN (2004) Hybridization and backcrossing between transgenic oilseed rape and two weed species under field conditions. Environ Biosaf Res 3:73–81

Hall L, Topinka K, Huffman J, Davis L, Good A (2000) Pollen flow between herbicide-resistant *Brassica napus* is the cause of multiple-resistant *B. napus* volunteers. Weed Sci 48(6):688–694

Hani M, Lebazda R, Fenni M (2017) Studies of morphological characteristics and production of seeds weeds of species of family Brassicaceae (Cruciferous) in Setifian High Plateau, Algeria. ARRB 12(5):1–9

Hannaway DB, Larson C (2004) Forage fact sheet: field mustard (*Brassica rapa* L. var. *rapa*). Oregon State Univ., Corvallis. http://forages.oregonstate.edu/php/fact_sheet_print_ffor.php?SpecID=152&use=Forage

Hansen LN (1998) Intertribal somatic hybridization between rapid cycling *Brassica oleracea* L. and *Camelina sativa* (L.) Crantz. Euphytica 104:173–179

Hansen LB, Siegismund HR, Jørgensen RB (2001) Introgression between oilseed rape (*Brassica napus* L.) and its weedy relative *B. rapa* L. in a natural population. Gen Resour Crop Evol 48:621–627

Harberd DJ, McArthur ED (1980) Meiotic analysis of some species and genus hybrids in the Brassiceae. In: Tsunoda S, Hinata K, Gómez-Campo C (eds) Brassica crops and wild allies. Jap Sci Soc Press, Tokyo, pp 65–87

Hartwig B, Detlef K, Peter L, Gerhard F (2005) Feeding value of crambe press cake and extracted meal as well as production responses of growing-finishing pigs and dairy cows fed these by-products. Arch Animal Nutr 59(2):111–122

Hasegawa J, Sakamoto Y, Nakagami S, Aida M, Sawa S, Matsunaga S (2016) Three dimensional imaging of plant organs using a simple and rapid transparency technique. Plant Cell Phyiol 57(3):462–472

Haser-Krause J (1989) Comparison of the suitability of sources of cytoplasmic male sterility (CMS) and self-incompatibility for breeding F1 hybrids of cauliflower. Biuletyn Warzywniczy 29–34
Hauser TP, Jørgensen RB, Østergård H (1997) Preferential exclusion of hybrids in mixedpollinations between oilseed rape (*Brassica napus*) and weedy *B. campestris* (*Brassicaceae*). Am J Bot 84:756–762
Hauser TP, Shaw RG, Østergård H (1998) Fitness of F1 hybrids between weedy *Brassica rapa* and oilseed rape (*B. napus*). Heredity 81:429–435
Hedges LJ, Lister CE (2006) Nutritional attributes of Brassica vegetables. Crop Food Res Confid Rep
Hoffmann MH (2002) Biogeography of Arabidopsis thaliana (L.) Heynh. (Brassicaceae). J Biogeol 29:125–134
Holm L, Doll J, Holm E, Pancho J, Herberger J (1997) World weeds. Natural histories and distribution. Wiley, NY USA
Hoyle M, Hayter K, Cresswell JE (2007) Effect of pollinator abundance on self-fertilization and gene flow: application to GM canola. Ecol Appl 17(7):2123–2135
Hu B, Chen F, Li Q (1997) Sterility and variation resulting from the transfer of polimacytoplasmic male sterility from *Brassica napus* into *Brassica* chinensis. J Agric Sci 128:299–301
Hu Q, Andersen SB, Dixelius C, Hansen LN (2002) Production of fertile intergeneric somatic hybrids between *Brassica napus* and *Sinapis arvensis* for the enrichment of the rapeseed gene pool. Plant Cell Rep 21:147–152
Huang B, Liu Y, Wu W, Xue X (2002) Production and cytogenetics of intergeneric hybrids between Ogura CMS *Brassica napus* and *Raphanus sativus*. Cruciferae Newslett 24:25–27
Huang CH, Sun R, Hu Y, Zeng L, Zhang N, Cai L, Zhang Q, Koch MA, Al-Shehbaz IA, Edger PP, Pires JC, Tan DY, Zhong Y, Ma H (2015) Resolution of Brassicaceae phylogeny using nuclear genes uncovers nested radiations and supports convergent morphological evolution. Mol Biol Evol 33(2):394–412
Hüsken A, Dietz-Pfeilstetter A (2007) Pollen-mediated intraspecific gene flow from herbicide resistant oilseed rape (*Brassica napus* L.). Transgenic Res 16:557–569
IENICA (2012) Interactive European Network for Industrial Crops and their Applications (IENICA). http://www.ienica.net/crops/crambe.pdf
Ignatov A, Kuginuki Y, Hida K (1999) Disease reaction to *Xanthomonas campestris* pv. *Campestris* races 1, 4 and 5 in weedy and cultivated *Brassica rapa* L. Cruciferae Newsl Eucarpia 21:123–124
Inomata N (1988) Intergeneric hybridization between *Brassica napus* and *Sinapis arvensis* and their crossability. Cruciferae Newslett 13:22–23
Inomata N (2001) Intergeneric hybridization between *Brassica juncea* and *Erucastrum virgatum* and the meiotic behavior of F_1 hybrids. Cruciferae Newslett Eucarpia 23:17–18
Inomata N (2002) A cytogenetic study of the progenies of hybrids between *Brassica napus* and *Brassica oleracea*, *Brassica bourgeaui*, *Brassica cretica* and *Brassica montana*. Plant Breed 121:174–176
Inomata N (2003) Production of intergeneric hybrids between *Brassica juncea* and *Diplotaxis virgata* through ovary culture, and the cytology and crossability of their progenies. Euphytica 133:57–64
Inomata N (2005) Intergeneric hybrid between *Brassica napus* and *Diplotaxis harra* through ovary culture and the cytogenetic analysis of their progenies. Euphytica 145:87–93
Iqbal S, Younas U, Chan KW, Saeed Z, Shaheen MA, Akhtar N, Majeed A (2013) Growth and antioxidant response of *Brassica rapa* var. rapa L. (turnip) irrigated with different compositions of paper and board mill (PBM) effluent. Chemosphere 91:1196–1202
Jaiswal AK, Abu-Ghannam N, Gupta S (2012) A comparative study on the polyphenolic content, antibacterial activity and antioxidant capacity of different solvent extracts of *Brassica oleracea* vegetables. Int J Food Sci Technol 47:223–231
Jenkins TE, Frampton CM, Conner AJ (2005) Population variability in wild turnip (*Brassica rapa* var. *oleifera*) for interspecific hybridisation with herbicideresistant rape (*Brassica napus*) pollen. New Zealand J Crop Hort Sci 33:9–16

Jiang Y, Tian E, Li R, Chen L, Meng J (2007) Genetic diversity of *Brassica carinata* with emphasis on the interspecific crossability with *B. rapa*. Plant Breed 126(5):487–491

Jiang JJ, Zhao XX, Tian W, Li TB, Wang YP (2009) Intertribal somatic hybrids between *Brassica napus* and *Camelina sativa* with high linolenic acid content. Plant Cell Tiss Organ Cult 99:91–95

Jiang Y, Caldwell CD, Falk KC (2014) Camelina seed quality in response to applied nitrogen, genotype and environment. Can J Plant Sci 94:971–980

Johnson EN, Beckie HJ, Warwick S, Shirtliffe SJ, Gulden RH, Séguin-Swartz G, Légère A, Simard MJ, Harker KN, Thomas G (2004) Ecology and management of volunteer canola. http://ws373847.websoon.com/uploads/managing_vol_canola.pdf

Johnson E, Vera C, Klein-Gebbinck H, Gan Y, Falk K, Bauche C (2007) Agronomy of *Camelina sativa*. Western Applied Research Corporation

Johnson EN, Falk K, Klein-Gebbinck H, Lewis L, Vera C, Malhi S, Shirtliffe S, Gan Y, Hall L, Topinka K, Nybo B, Sluth D, Bauche C, Phelps S (2008) Agronomy of *Camelina sativa* and *Brassica carinata*. Annual report. Western Applied Research Corporation

Johnson EN, Falk K, Klein-Gebbinck H, Lewis L, Malhi S, Leach D, Shirtliffe S, Holm F A, Sapsford K, Hall L, Topinka K, May W, Nybo B, Sluth D, Gan Y, Phelps S (2011) Agronomy of *Camelina sativa* and *Brassica carinata*. Final Report

Johnston TD (1974) Transfer of disease resistance from *Brassica campestris* L. to rape (*B. napus* L.). Euphytica 23:681–683

Jovičić D, Jovica V, Zorica N, Gordana P, Gordana T, Maja I, Dragana M (2017) Antioxidant capacity of oilseed rape (Brassica napus) in different soil types. Turk J Agric 41:463–471

Julié-Galau S, Bellec Y, Faure J-D, Tepfer M (2014) Evaluation of the potential for interspecific hybridization between *Camelina sativa* and related wild *Brassicaceae* in anticipation of field trials of GM Camelina. Transgenic Res 23:67–74

Kamala T (1976) Interspecific hybrids in *Brassica*. Cytologia 41:407–415

Kamala T (1983) A study on the cytogenetic homeologies between *Raphanus* and *Brassica* genomes. Ind J Bot 6:131–140

Kasim A, Kumar S, Dhaliwal HS (2017) Biodiesel production from Abyssinian Mustard (*Brassica carinata*). Int J Eng Tech Manag Appl Sci 5(6):710–722

Katche E, Quezada-Martinez D, Ihien Katche E, Vasquez-Teuber P, Mason AS (2019) Interspecific hybridization for *Brassica* crop improvement. Crop Breed Genet Genom 1:e190007. https://doi.org/10.20900/cbgg20190007

Katiyar RK, Chamola R (1995) Useful end products from *Brassica juncea* x *B. carinata* and *Brassica juncea* x *B. campestris* crosses. Cruciferae Newslett 17:20–21

Kerlan MC, Chèvre AM, Eber F, Baranger A, Renard M (1992) Risk assessment of outcrossing of transgenic rapeseed to related species: I. Interspecific hybrid production under optimal conditions with emphasis on pollination and fertilization. Euphytica 62:145–153

Khan RU, Mehmood S, Khan SU, Khan A, Shah IA, Bokhari TZ (2013) Medicinal value of indigenous flora in the vicinity of district bannu, Khyber Pakhtunkhwa, Pakistan. Adv Pharm Ethnomed 1 (1):7–14

Kim N, Li Y, Sun S (2015) Epoxidation of *Camelina sativa* oil and peel adhesion properties. Ind Crops Products 64

Kirkhus B, Lundon AR, Hauge n JE, Vogt G, Borge GI, Henriksen BI (2013) Effects of environmental factors on edible oil quality of organically grown *Camelina sativa*. J. Ag ic Food Chem. 61:3179–3185

Kirti P, Banga S, Prakash S, Chopra V (1995) Transfer of Ogu cytoplasmic male sterility to *Brassica juncea* and improvement of the male sterile line through somatic cell fusion. Theor Appl Genet 91(3):517–521

Klein M, Eckert-Ossenkopp U, Schmiedeberg I, Brandt P, Unseld M, Brennicke A, Schuster W (1994) Physical mapping of the mitochondrial genome of *Arabidopsis thaliana* by cosmid and YAC clones. Plant J 6(3):447–455

Klíma M, Abraha E, Vyvadilová M, Bechyn M (2009) Protoplast culture and fusion between *brassica carinata* and *brassica napus*. Agric Tropica et Subtropica 42(1)

Knispel AL, McLachlan M, Van Acker RC, Friesen LF (2008) Gene flow and multiple herbicide resistance in escaped canola populations. Weed Sci 56:72–80

Koch MA, Kiefer M (2005) Genome evolution among cruciferous plants: a lecture from the comparison of the genetic maps of three diploid species—*Capsella rubella*, *Arabidopsis lyrata* subsp. *petraea*, and *A. thaliana*. Am J Bot 92:761–767

Kolte SJ, Bordoloi DK, Awasthi RP (1991) The search for resistance to major diseases of rapeseed and mustard in India. In: Proceedings of the GCIRC 8th International Rapeseed Congress, pp 219–225

Korkmaz S (2018) Antioxidants in Maca (Lepidium meyenii) as a supplement in nutrition. Antioxidants Foods Appl. https://doi.org/10.5772/intechopen.75582

Koubaa M, Driss D, Bouaziz F, Ghorbel RE, Chaabouni SE (2015) Antioxidant and antimicrobial activities of solvent extract obtained from rocket (*Eruca sativa* L.) flowers. Free Radicals Antioxidants 5(1):20–22

Kuang P, Song D, Yuan Q, Yi R, Lv X, Liang H (2013) Separation and purification of sulforaphene from radish seeds using macroporous resin and preparative high performance liquid chromatography. Food Chem 136:342–347

Kumari N, Avtar R, Thakral BSN (2016) Antioxidant potential in seed meal of different Indian mustard genotypes. J Oilseed Brassica 1:63–67

La Mura M, Norris C, Sporle S, Jayaweera D, Greenland A, Lee D (2010) Development of genome-specific 5S rDNA markers in *Brassica* and related species for hybrid testing. Genome 53(8):643–649

Lake JA, Field KJ, Davey MP, Beerling DJ, Lomax BH (2009) Metabolomic and physiological responses reveal multi-phasic acclimation of Arabidopsis thaliana to chronic UV radiation. Plant Cell Environ 32(32):1377–1389

Lalas S, Gortzi O, Athanasiadis V, Dourtoglou T, Dourtoglou V (2012) Full characterisation of Crambe abyssinica Hochst. Seed Oil. J Am Oil Chem Soc 89(12): 2253–2258

Lankau RA (2007) Specialist and generalist herbivores exert opposing selection on a chemical defense. New Phytol 175:176–184

Lankau RA, Wheeler E, Bennett AE, Strauss SY (2011) Plant-soil feedbacks contribute to an intransitive competitive network that promotes both genetic and species diversity. J Ecol (Oxford) 99(1):176–185

Lawton K, Friedrich L, Hunt M (1996) Benzothiadizaole induces disease resistance by a citation of the systemic acquired resistance signal transduction pathway. The Plant J 10:71–82

Lefol E, Danielou V, Darmency H (1996) Predicting hybridization between transgenic oilseed rape and wild mustard. Field Crops Res 45:153–161

Lefol E, Séguin-Swartz G, Downey RK (1997) Sexual hybridization in crosses of cultivated *Brassica* species with the crucifers *Erucastrum gallicum* and *Raphanus raphanistrum*: potential for gene introduction. Euphytica 95:127–139

Lelivelt CLC, Lange W, Dolstra O (1993a) Intergeneric crosses for the transfer of resistance to the beet cyst nematode from *Raphanus sativus* to *Brassica napus*. Euphytica 68:111–120

Lelivelt CLC, Leunissen EHM, Frederiks HJ, Helsper JPFG, Krens FA (1993b) Transfer of resistance to the beet cyst nematode (*Heterodera schachtii* Schm.) from *Sinapis alba* L. (white mustard) to the *Brassica napus* L. gene pool by means of sexual and somatic hybridization. Theor Appl Genet 85:688–696

Leung H, Williams PH (1983) Cytoplasmic sterile Brassica campestris breeding lines with resistance to club root, turnip mosaic and downy mildew. Hort Sci 18:774–775

Lewis LJ, Woods DL, Cheng BF (2001) Introgression of long pod genotype from spring rape (*Brassica napus* L.) into summer turnip rape (*Brassica rapa* L.). Can J Plant Sci 81:59–60

Li Z, Heneen WK (1999) Production and cytogenetics of intergeneric hybrids between the three cultivated Brassica diploids and Orychophragmus violaceus. Theor Appl Genet 99:694–704

Li Z, Wu JG, Liu Y, Liu HL, Heneen WK (1998) Production and cytgenetics of intergenric hybrids *Brassica juncea* x *Orychopharagmus violaceus* and *B. carinata* x *O. violaceus*. Theor Appl Genet 96:251–265

Li Z, Ceccarelli M, Minelli S, Contento A, Liu Y, Cionini PG (2003) High efficiency production and genomic *in situ* hybridization analysis of *Brassica* aneuploids and homozygous plants. Sci China Ser C: Life Sci 46(1):104–112

Li H, Barbetti MJ, Sivasithamparam K (2005) Hazard from reliance on cruciferous hosts as sources of major gene-based resistance for managing blackleg (*Leptosphaeria maculans*) disease. Field Crops Res 91:185–198

Lian Y, Lin G, Zhao X, Lim H (2011) Production and genetic characterization of somatic hybrids between leaf mustard (*Brassica juncea*) and broccoli (*Brassica oleracea*). Vitro Cell Dev Plant 47:289–296

Ligen Z, Yuanfeng W, Yuke S, Lei Z, Mupunga J, Jianwei M, Shiwang L (2017) Broccoli seed extracts but not sulforaphane have strong free radical scavenging activities. Int J Food Sci Technol 52:2374–2381

Liu Clarke JH, Chèvre AM, Landgren M, Glimelius K (1999) Characterization of sexual progenies of male-sterile somatic cybrids between Brassica napus and Brassica tournefortii. Theor Appl Genet 99:605–610

Liu R, Qian W, Meng J (2002) Association of RFLP markers and biomass heterosis in trigenomic hybrids of oilseed rape (*Brassica napus* × *B. campestris*). Theor Appl Genet 105:1050–1057

Liu J, Wang H, Yu L, Li D, Li M (2009) Morphology and cytology of flower chimeras in hybrids of *Brassica carinata* x *Brassica rapa*. Afric J Biotech 8(5):801–806

Ljungberg A, Cheng B, Heneen WK (1993) Investigation of hybrids between Brassica tournefortii Gouan and B. alboglabra Bailey. Sver Utsa¨desfo¨ren Tidsk 103:191–197

Lloyd AM, Barnason AR, Rogers SG, Byrne MC, Fraley RT, Horsch RB (1986) Transformation of *Arabidopsis thaliana* with *Agrobacterium tumefaciens*. Science 234(4775):464–66

Lokanadha RD, Sarla N (1994) Hybridization of *Brassica tournefortii* and cultivated Brassicas. Cruciferae Newslett Eucarpia 16:32–33

López-Bucio J, Campos-Cuevas JC, Hernández-Calderón E, Velásquez-Becerra C, Farías-Rodríguez R, Macías-Rodríguez LI, Valencia-Cantero E (2007) Bacillus megaterium rhizobacteria promote growth and alter root-system architecture through an auxin- and ethylene-independent signaling mechanism in Arabidopsis thaliana. Mol Plant-Microbe Inter 20(2):207–217

Lossinsky AS, Shivers RR (2004) Structural pathways for macromolecular and cellular transport across the blood-brain barrier during inflammatory conditions. Rev Histol Histopathol 19(2):535–564

Lou P, Lan ZQ, Deng J, Wang ZQ (2000) Application of in vitro organ culture in wide–cross breeding of rapeseed. Euphytica 114:217–221

Lovett JV, Jackson HF (1980) Allelopathic activity of *Camelina sativa* (L.) Crantz in relation to its phyllosphere bacteria. New Phytol 86:273–277

Lukens LN, Quijada PA, Udall J, Pires JC, Schranz ME, Osborn TC (2004) Genome redundancy and plasticity within ancient and recent *Brassica* crop species. Biol J Linn Soc 82:654–665

Mackay GR (1977) The introgression of S alleles into forage rape, *Brassica napus* L. from turnip, *Brassica campestris* L. ssp. *rapifera*. Euphytica 26:511–519

Mali RG, Mahajan SG, Mehta AA (2007) *Lepidium sativum* (garden cress) a review of contemporary literature and medicinal properties. Oriental Pharm Exp Med 7(4):331–335

Malik M, Vyas PO, Rangaswamy NS, Shivanna KR (1999) Development of two new cytoplasmic male-sterile lines in *Brassica juncea* through wide hybridization. Plant Breed 118:75–78

Marillia EF, Francis T, Falk KC, Smith M, Taylor DC (2014) Palliser's promise: *Brassica carinata*, an emerging western Canadian crop for delivery of new bio-industrial oil feedstocks. Biocatal Agric Biotech 3(1):65–74

Marthe F, Richter K, Schrader O (2004) Cabbage (Brassica oleracea) with new resistance to black rot (*Xanthomonas campestris pv. campestris*) from black mustard (Brassica nigra). Brassica 2004, 4th ISHS Symposium Brassicas/Proc 14th Crucifer Genet Workshop, p 175

Martin SL, Sauder CA, James T, Cheung KW, Razeq FM, Kron P, Hall L (2015) Sexual hybridization between *Capsella bursa-pastoris* (L.) Medik and *Camelina sativa* (L.) Crantz (*Brassicaceae*). Plant Breed 134:212–220

Martínez-Valdivieso D, Font R, Del Río-Celestino M (2019) Prediction of agro-morphological and nutritional traits in Ethiopian Mustard Leaves (Brassica Carinata A. Braun) by visible-near-infrared spectroscopy. Foods 8:6

Mason-Sedun W, Jesspo RS, Lovett JV (1986) Differential phytotoxicity among species and cultivars of the genus *Brassica* to wheat. I. Laboratory and field screening of species. Plant Soil 93(1):3–16

Matsushima K, Nemoto K, Nakashima N, Dema D, Thapa L, Watanabe A, Maegawa F, Baba T, Matsushita G (2006) Report of investigation for wild edible plants and their traditional knowledge in Bhutan. J Facul Agri Shinshu Uni 42(1/2):37–46

Matsuzawa Y, Funayama T, Kamibayashi M, Konnai M, Bang SW, Kaneko Y (2000) Synthetic *Brassica rapa-Raphanus sativus* amphidiploid lines developed by reciprocal hybridization. Plant Breed 119:357–359

Mattsson B (1988) Interspecific crosses within the genus *Brassica* and some related genera. Sveriges Utsadesforenings Tidskrift 98:187–212

McCann J (2004) The horseradish plant. www.globalgourmet.com/food/egg/egg1296/horsplnt.html

McCollum GD (1979) Sterility in successive backcrosses of Raphanobrassica ($2n = 4x = 36$) with recurrent *Brassica oleracea* ($2n = 2x = 18$). Can J Genet Cytol 21:479–485

McNaughton IH (1973) *Brassica napocampestris* L. ($2n = 58$). 1. Synthesis, cytology, fertility and general considerations. Euphytica 22:301–309

McVay KA, Lamb PF (2007) Camelina production in Montana. Field Crops No. D-16. Montana State Uni Ext, Bozeman, MT, USA

Melnikovova I, Fait T, Kolarova M, Fernandez EC, Milella L (2015) Effect of *Lepidium meyenii* Walp. On semen parameters and serum hormone levels in healthy adult men: A doubleblind, randomized, placebocontrolled pilot study. Evidence-based Complementary and Alternative Medicine. https://doi.org/10.1155/2015/324369

Meng J, Shi S, Gan L, Li Z, Qu X (1998) The production of yellow-seeded *Brassica napus* (AACC) through crossing interspecific hybrids of *B. campestris* (AA) and *B. carinata* (BBCC) with *B. napus*. Euphytica 103(3):329–333

Metz PLJ, Nap JP, Stiekema WJ (1995) Hybridization of radish (*Raphanus sativus* L.) and oilseed rape (*Brassica napus* L.) through a flower-culture method. Euphytica 83:159–168

Metz PLJ, Jacobsen E, Nap JP, Pereira A, Stiekema WJ (1997) The impact on biosafety of the phosphinothricin-tolerance transgene in inter-specific *B. rapa* × *B. napus* hybrids and their successive backcrosses. Theor Appl Genet 95:442–450

Metzger J (1833) Systematische Beschreibung der Kultivirten Kohlarten. Heidelberg 2:1–68

Mikkelsen TR, Jensen J, Jørgensen RB (1996) Inheritance of oilseed rape (*Brassica napus*) RAPD markers in a backcross progeny with *Brassica campestris*. Theor Appl Genet 92:492–497

Miraj S (2016) Broccoli (*Brassica oleracea* var. Italica): potential candidate in the health management. Der Pharmacia Lettre 8:61–65

Mizushima U (1950) Karyogenetic studies of species and genus hybrids in the tribe Brassiceae of Cruciferae. Tohoku J Agric Res 1:1–14

Mnzava NA, Schippers RR (2004) *Brassica carinata* A. Braun. In: Grubben GJH, Denton OA (eds) Plant resources of Tropical Africa 2: vegetables. PROTA Foundation, Wageningen, Netherlands/Backhuys Publishers, Leiden, Netherlands/CTA, Wageningen, Netherlands, pp 119–123

Mnzava, NA, Schippers, RR (2007) *Brassica carinata* A. Braun. In: van der Vossen HAM, Mkamilo GS (eds) Plant resources of Tropical Africa 14: Vegetable oils/Oléagineux. PROTA, Wageningen, Netherlands. http://database.prota.org/PROTAhtml/Brassica%20carinata_En.htm

Mohapatra D, Bajaj YPS (1987) Interspecific hybridization in *Brassica juncea—Brassica hirta* using embryo rescue. Euphytic 36(1):321–326

Momotaz A, Kato M, Kakihara F (1998) Production of intergeneric hybrids between *Brassica* and *Sinapis* species by means of embryo rescue techniques. Euphytica 103(1):123–130

Morinaga T (1933) Interspecific hybridization in *Brassica*. V. The cytology of F1 hybrids of *B. carinata* and *B. alboglabra*. Jap J Bot 6:467–475

Moser BR (2010) Camelina (Camelina sativa L.) oil as a biofuels feedstock: Golden opportunity or false hope. Lipid Tech 22(120):270–273

Moyes CL, Cole SG, Casais CA, Dale PJ (1999) Sexual compatibility between oilseed rape and *Sinapis arvensis*. In: Proceedings of the 10th international rapeseed congress, Canberra, Australia, Contrib No. 529

Moyes CL, Lilley JM, Casais CA, Cole SG, Haeger PD, Dale PJ (2002) Barriers to gene flow from oilseed rape (*Brassica napus*) into populations of Sinapis arvensis. Mol Ecol 11:103–112

Müller J, Sonntag K (2000) Electrofusion of protoplasts in Brassicaceae. Eucarpia Cruciferae Newslett 22:25–26

Munthali DC (2009) Evaluation of cabbage varieties for resistance to the cabbage aphid. Afr Entomol 17(1):1–7

Mustard S (2013) Saskatchewan Mustard Development Commission. Carinata production: a guide to best management practices. http://www.saskmustard.ca/grower/growing/pdfs/Carinata_Production_Manual_080213.pdf

Namai H (1980) Effect of flower bud age at bud pollination on cross fertility in reciprocal intergeneric crosses, *Raphanus sativus* · *Brassica oleracea* and *B. campestris*. Cruciferae Newslett 5:26–27

Narain A, Prakash S (1972) Investigations on the artificial synthesis of amphidiploids of *Brassica tournefortii* Gouan with the other elementary species of Brassica. I. Genomic relationships. Genetica 43:90–97

Narasimhulu SB, Kirti PB, Bhatt SR, Prakash S, Chopra VL (1994) Intergeneric protoplast fusion between *Brassica carinata* and *Camelina sativa*. Plant Cell Rep 13(11):657–660

Naresh M (2014) Epidemiology and forecasting for the management of rapeseed-mustard diseases. J Mycol Plant Pathol 44:131–147

Navabi ZK, Strelkov SE, Good AG, Thiagarajah MR, Rahman MH (2010) Brassica B-genome resistance to stem rot (Sclerotinia sclerotiorum) in a doubled haploid population of Brassica napus × Brassica carinata. Can J Plant Pathol 32(2):237–246

Nawaz H, Shad MA, Rauf A (2018a) Optimization of extraction yield and antioxidant properties of *Brassica oleracea* Convar Capitata Var L. leaf extracts. Food Chem 242:182–187

Nawaz H, Shad MA, Muzaffar S (2018b) Phytochemical composition and antioxidant potential of Brassica. https://doi.org/10.5772/intechopen.76120

Ní Eidhin D, O'Beirne D (2010) Oxidative stability and acceptability of Camelina oil blended with selected fish oils. Eur J Lipid Technol 112:878–886

Nicoletto C, Santagata S, Pino S, Sambo P (2016) Antioxidant characterization of different Italian broccoli landraces. Horti Brasileira 34:74–79

Niemann J, Kotlarski S, Wojciechowski A (2014) The evaluation of self-incompatibility and cross-ability in chosen *Brassica* species based on the observation of pollen tubes growth and seed set. Acta Scientiarum Polonorum Agric 13(1)

Nikolov LA, Shushkov P, Nevado B, Gan X, Al-Shehbaz IA, Filatov D, Bailey CD, Tsiantis M (2019) Resolving the backbone of the Brassicaceae phylogeny for investigating trait diversity. New Phytol 222:1638–1651

Nitovskaya IO, Shakhovskyi AM (1998) Obtaining of asymmetrical somatic hybrids between *Brassica oleracea* L. and *Arabidopsis thaliana* L. (in Russian, English abstr.). Tsitol Genet 32:72–81

Nitovskaya IO, Shakhovskyi AM, Komarnyts'kyi IK, Kuchuk MV (2006a) Production of *brassica oleracea* (+*Arabidopsis thaliana*) and *brassica napus* cell lines resistant to spectinomycin/streptomycin as a result of plastome genetic transformation. Tsitol Genet 40:3–10

Nwankiti O (1970) Cytogenetic and breeding studies with *Brassica* I. Cytogenetic experiments with Brassica napocampestris. Hereditas 66:109–126

Obi RK, Nwanebu FC, Ndubuisi UU, Orji NM (2009) Antibacterial qualities and phytochemical screening of the oils of *Curcubita pepo* and *Brassica nigra*. J Med Plants Res 3(5):429–432

Obour AK, Obeng E, Mohammed YA, Campitti I, Durrett TP, Aznar-Moreno JA, Chen C (2017) Camelina seed yield and fatty acids as influenced by genotype and environment. Agro J 109

OECD (2012) Organisation for Economic Co-operation and Development. Consensus document on the biology of the *brassica* crops (*Brassica* spp.). In: Series on Harmonisation of regulatory oversight of biotechnology, vol 54, p 142

OECD (2016) Brassica crops (Brassica species). In: Safety assessment of transgenic organisms in the environment, Volume 5: OECD consensus documents. OECD Publishing, Paris. https://doi.org/10.1787/9789264253018-6-en

Ogbede S, Saidu A, Kabiru A (2014) Phytochemical compositions, Antihyperlipidemic and Hepatoprotective effects of *Brassica oleracea* Var. Capitata L. leaf extracts on tritoninduced Hyperlipidemic rats. Int J Med Sci Clinic Inven 1:345–351

OGTR (2008) The biology of *brassica napus* L. Version 2. Department of Health and Ageing. Australian Government Office of the Gene Technology Regulator. http://www.ogtr.gov.au/internet/ogtr/publishing.nsf/content/canola3/$FILE/biologycanola08_2.pdf

Ogura H (1968) Studies on the new male-sterility in Japanese radish, with special reference to the utilization of this sterility towards the practical raising of hybrid seeds. Mem Fac Agric Kagoshina Univ 6:39–78

Oilseeds WA (2006) Growing western Canola: an overview of canola production in western Australia. In: Duff J, Sermon D, Walton G, Mangano P, Newman C, Walden K, Barbetti M, Addison B, Eksteen D, Pol E, Leach B (eds). Oil Seeds Industry Association of Western Australia

Olgun Ç, Özkan OE, Güney B, Pattabanoglu ES, Güney K, Gür M (2017) Chemical composition and antimicrobial activity in cold press oil of fennel, Anise, white and black mustard seeds. Ind J Pharm Edu Res 51:S200–S204

Olsson G (1960) Species crosses within the genus *Brassica*. II. Artificial Brassica juncea Coss. Hereditas 46:171–222

Opena RT, Lo SH (1978) Derivation of matroclinal diploids in Chinese cabbage and evaluation of their significance in breeding. J Am Soc Hort Sci 103:820–823

Orrock JL, Witter MS, Reichman OJ (2008) Apparent competition with an exotic plant reduces native plant establishment. Ecology 89(4):1168–1174

Ovcharenko OO, Komarnyts'kyi IK,Cherep MM, Hleba II, Kuchuk MV (2004) Obtaining of intertribal *brassica juncea* + Arabidopsis thaliana somatic hybrids and study of transgenic trait behavior. Tsitol Genet 38:3–8

Palmer TP (1962) Population structure, breeding system, inter-specific hybridisation and allopolyploidy. Heredity 17:278–293

Palmer JD, Shields CR, Cohen DB, Orton TJ (1983) Chloroplast DNA evolution and the origin of amphiploid *Brassica* species. Theor App Genet 65(3):181–189

Palmer CE, Warwick SI, Keller W (2001) Brassicaceae (Cruciferae) family, plant biotechnology and phytoremediation. Int J Phytoremediat 3:245–287

Parikh H, Khanna A (2014) Pharmacognosy and phytochemical analysis of *Brassica juncea* seeds. Pharmacog J 6

Park J, Ahmed NU, Kim H, Nou I (2012) Advances in in vitro culture of the Brassicaceae crop plants. J Plant Biotechnol 39:13–22

Parkin IA, Gulden SM, Sharpe AG Lukens L, Trick M, Osborn TC, Lydiate DJ (2005) Segmental structure of the *Brassica napus* genome based on comparative analysis with *Arabidopsis thaliana*. Genetics 171:765–781

Parvaiz M, Bhatti KH, Nawaz K, Hussain Z, Khan RMW, Hussian A (2013) Ethno-botanical studies of medicinal plants of Dinga. District Gujarat, Punjab, Pakistan 26(6):826–833

Pavlista AD, Baltensperger DD, IsbellTA Hergert GW (2012) Comparative growth of spring-planted canola, brown mustard and camelina. Ind Crops Prod 36:9–13

Pellan-Delourme R, Renard M (1987) Identification of maintainer genes in *Brassica napus* L. for the male-sterility-inducing cytoplasm of *Diplotaxis muralis* L. Plant Breed 99:89–97

Peng G, Falk KC, Gugel RK, Franke C, Yu F, James B, Strelkov SE, Hwang S-F, McGregor L (2013) Sources of resistance to *Plasmodiophora brassicae* (clubroot) pathotypes virulent on canola. Can J Plant Pathol 36:89–99
Pieroni A, Quave CL (2005) Traditional pharmacopoeias and medicines among Albanians and Italians in southern Italy: a comparison. J Ethnopharmacol 101:258–270
Pieroni A, Quavec CL, Santorod RF (2004) Folk pharmaceutical knowledge in the territory of the Dolomiti Lucane, inland southern Italy. J Ethnopharmacol 95:373–384
Pitrat M (2012) Vegetable crops in the Mediterranean Basin with an overview of virus resistance. Adv Virus Res 84:1–29
Plessers AG, McGregor WG, Carson RB, Nakoneshny W (1962) Species trials with oilseed plants. II. Camelina. Can J Plant Sci 42:452–459
Popova IE, Morra MJ (2014) Simultaneous quantification of sinigrin, sinalbin, and anionic glucosinolate hydrolysis products in Brassica juncea and Sinapis alba seed extracts using ion chromatography. J Agric Food Chem 62(44):10687–10693
Pradhan AK, Mukhopadhyay A, Pental D (1991) Identification of the putative cytoplasmic donor of a CMS system in *Brassica juncea*. Plant Breed 106:204–208
Prakash S, Chopra VL (1990) Reconstruction of allopolyploid *Brassicas* through nonhomologous recombination: introgression of resistance to pod shatter in *Brassica napus*. Gen Res 56:1–2
Prakash S, Gupta SK, Raut RN, Kalra AK (1984) Synthetic Brassica carinata – a preliminary report. Cruciferae Newslett 9:36
Prakash S, Kirti PB, Bhat SR, Gaikwad K, Kumar VD, Chopra VL (1998) A *Moricandia arvensis*-based cytoplasmic male sterility and fertility restoration system in *Brassica juncea*. Theor Appl Genet 97:488–492
Prakash S, Wu X-M, Bhat SR (2012) History, evolution and domestication of Brassica crops. In: Janick J (ed) Plant breeding reviews, vol 35, pp 19–84
Prasad MP (2014) Antimicrobial potential of Brassicaceae family against clinical isolates. Int J Pure Appl Biosci 2(2):158–162
Pratt S (2017) Firm eager to ride the 'canola train' with carinata. The Western Producer
Quiros CF, Kianian SF, Ochoa O, Douches D (1985) Genome evolution in Brassica: use of molecular markers and cytogenetic stocks. Cruciferae Newslett Eucarpia 10:21–23
Quiros CF, Ochoa O, Douches DS (1988) Exploring the role of $x = 7$ species in Brassica evolution: hybridization with B. nigra and B. oleracea. J Heredity 79:351–358
Raghav PP, Meena PD, Singh B, Meena HP, Ss Meena, Sharma P, Majumdar R, Singh D (2014) Development and evaluation of Alternaria blight tolerant lines in Indian mustard (*Brassica juncea*). J Oilseed Brassica 5:9
Raghavendra RH, Akhilender NK (2011) Eugenol and n-3 rich garden cress seed oil as modulators of platelet aggregation and eicosanoids in Wistar albino rats. The Open Nutraceuticals J 4:144–150
Rahman MH (2001) Production of yellow-seeded *Brassica napus* through interspecific crosses. Plant Breed 120:463–472
Rahman MH (2002) Fatty acid composition of resynthesized *Brassica napus* and trigenomic *Brassica* void of genes for erucic acid in their A genomes. Plant Breed 121:357–359
Rahman MH (2004) Optimum age of siliques for rescue of hybrid embryos from crosses between *Brassica oleracea*, *B. rapa* and *B. carinata*. Can J Plant Sci 84:965–969
Rahman M, Tahir M (2010) Inheritance of seed coat color of ethiopian mustard (*Brassica carinata* A. Braun). Can J Plant Sci 90(3):279–281
Rahman M, Khatun A, Liu L, Barkla BJ (2018) Brassicaceae mustards: traditional and agronomic uses in Australia and New Zealand Mahmudur. Molec 23:1–18
Rai GK, Bagati S, Rai PK, Rai SK, Singh M (2018) Fatty acid profiling in rapeseed mustard (*Brassica* species). Int J Curr Microbiol App Sci 7(05):148–157
Rakow G (2004) Species origin and economic importance of *Brassica*. In: Pua EC, Douglas CJ (eds) *Brassica* biotechnology in agriculture and forestry. Berlin (Germany), pp 3–11
Rao GU, Shivanna KR (1997) Alloplasmics of *B. juncea* as bridge-species for development of alloplasmics of other crop brassicas. Cruciferae Newslett Eucarpia 19:29–30

Rao GU, Lakshmikumaran M, Shivanna KR (1996) Production of hybrids, amphiploids and back-cross progenies between a cold-tolerant wild species, *Erucastrum abyssinicum* and crop brassicas. Theor Appl Genet 92:786–790

Rashid A, Rakow G, Downey RK (1994) Development of yellow seeded *Brassica napus* through interspecific crosses. Plant Breed 112:127–134

Ravindran PN (2017) The encyclopedia of herbs & spices. CAB International.xlv+, 1,2 Wallingford, UK, 1128 pp. http://www.cabi.org/cabebooks/ebook/20173378261

Ren JP, Dickson MH, Earle ED (2000) Improved resistance to bacterial soft rot by protoplast fusion between Brassica rapa and B. oleracea. Theo Appl Genet 100:810–819

Rhee WY, Cho YH, Paek KY (1997) Seed formation and phenotypic expression of intra- and inter-specific hybrids of *Brassica* and of intergeneric hybrids obtained by crossing with *Raphanus*. J Korean Soc Hort Sci 38:353–360

Rieger MA, Potter TD, Preston C, Powles SB (2001) Hybridization between *Brassica napus* L. and *Raphanus raphanistrum* L. under agronomic field conditions. Theor Appl Genet 103:555–560

Ringdahl EA, McVetty PBE, Sernyk JL (1987) Intergeneric hybridization of *Diplotaxus* spp. with *Brassica napus*: a source of new CMS systems? Can J Plant Sci 67:239–243

Ripley VL, Beversdorf WD (2003) Development of self-incompatible *Brassica napus*: (II) *Brassica oleracea* S-allele expression in *B. napus*. Plant Breed 122:6–11

Rochfort S, Jones R (2011) Glucosinolate phytochemicals from Broccoli (Brassica oleracea L. var. botrytis L.) Seeds and their potential health effects. Nuts Seeds Health Dis Prev 253–261. https://doi.org/10.1016/b978-0-12-375688-6.10030-1

Rode J (2002) Study of autochthon *Camelina sativa* (L.) Crantz in Slovenia. J Herb Spic Med Plants 9:313–318

Rongzhan G, Shuhui J, Ruying X, Hongsheng Z (2007a) Studies on rapeseed germplasm enhancement by use of cruciferous weed *Descurainia sophia*. Gen and Breeding, Gen Germplasm, pp 261–265

Rongzhan G, Shuhui J, Ruying X, Hongsheng Z (2007b) Studies on rapeseed germplasm enhancement by use of cruciferous weed *Rorippa indica*. Gen Breed Gen Germplasm 329–332

Roy NN (1978) A study on disease variation in the populations of an interspecific cross of *Brassica juncea* L. × *B. napus* L. Euphytica 27:145–149

Roy NN (1984) Interspecific transfer of *brassica juncea*-type high blackleg resistance to *brassica napus*. Euphytica 33:295–303

Ryu JP, Kim DC, In M-J, Chae HJ, Lee SD (2012) Antioxidant potential of ethanol extract of *Brassica rapa* L. root. J Med Plants Res 6:1581–1584

Sabharwal PS, Dolezel J (1993) Interspecific hybridization in *Brassica*: application of flow cytometry for analysis of pioidy and genome composition in hybrid plants. Biol Plant 35(2):169–177

Sacristan MD, Gerdemann M (1986) Different behavior of *Brassica juncea* and *B. carinata* as sources of *Phoma lingam* resistance in experiments of interspecific transfer to *B. napus*. Plant Breed 97:304–314

Saeidfirozeh H, Shafiekhani A, Cifra M, Masoudi AA (2018) Endogenous chemiluminescence from germinating *Arabidopsis thaliana* seeds. Sci Rep 8:16231

Saeidnia S, Gohari AR (2012) Importance of Brassica napus as a medicinal food plant. J Med Plants Res 6:2700–2703

Sahsrabudde MB (1962) The wealth of India, raw materials. Publication and information Directorate, CSIR, De NN, New Delhi. Curr Sci 12:23–24

Salisbury PA (1991) Genetic variability in Australian wild crucifers and its potential utilization in oilseed Brassica species. PhD thesis, La Trobe University, Victoria, Australia

Salisbury P (2002a) Gene flow between *Brassica napus* and other *Brassicaceae* species. Report PAS0201. Institute of Land and Food Resources. University of Melbourne, 45 pp

Salma U, Khan T, Shah AJ (2018) Antihypertensive effect of the methanolic extract from *Eruca sativa* Mill. (Brassicaceae) in rats: muscarinic receptor-linked vasorelaxant and cardiotonic effects. J Ethano pharmacol 1–44

Sandhu SK, Gupta VP (2000) Interspecific hybridization among digenomic species of *Brassica*. Crop Improvement 27:195–197

Sarashima M, Matsuzawa Y (1979) Possibility to breed cabbage lines with the cytoplasm of radish. Cruciferae Newslett 4:32

Sarla N, Raut RN (1988) Synthesis of *Brassica carinata* from *Brassica nigra Brassica oleracea* hybrids obtained by ovary culture. Theor Appl Genet 76:846–849

Sarla N, Raut RN (1991) Cytogenetical studies on *Brassica nigra B. oleracea* hybrids. Indian J Genet Plant Breed 51:408–413

Sarla N, Raut R, Shyam P (1987) Synthesis of *Brassica carinata* A. Br Curr Sci 56:779–781

Sarli G, De Lisi A, Agneta R, Grieco S, Ierardi G, Montemurro F, Negro D, Montesano V (2012) Collecting horseradish (Armoracia rusticana, Brassicaceae): local uses and morphological characterization in Basilicata (Southern Italy). Genet Resour Crop Evol 59(5):889–899

Sarmah BK, Sarla N (1997) In vitro flowering and back crossing of *Brassica* intergeneric hybrids. Curr Sci (Bangalore) 72:702–704

Sarmah BK, Sarla N (1998) Erucastrum abyssinicum · Brassica oleracea hybrids obtained by ovary and ovule culture. Euphytica 102:37–45

Saucke H, Ackermann K (2006) Weed suppression in mixed cropped grain peas and false flax (*Camelina sativa*). Weed Res 46:453–461

Schelfhout CJ, Wroth JM, Yan G, Cowling WA (2008) Enhancement of genetic diversity in canola-quality *Brassica napus* and *B. juncea* by interspecific hybridisation. Austral J Agric Res 59:918–925

Scholze P, Krämer R, Ryschka U, Klocke E, Schumann G (2010) Somatic hybrids of vegetable brassicas as source for new resistances to fungal and virus diseases. Euphytica 176:1–14

Schranz ME, Lysak MA, Mitchell-Olds T (2006) The ABC's of comparative genomics in the *Brassicaceae*: building blocks of crucifer genomes. Trends Plant Sci 11:535–542

Séguin-Swartz G, Eynck C, Gugel RK, Strelkov SE, Olivier CY, Li JL, Klein-Gebbinck H, Borhan H, Caldwell CD, Falk KC (2009) Diseases of *Camelina sativa* (false flax). Can J Plant Pathol 31:375–386

Séguin-Swartz G, Warwick SI, Gugel RK, Olivier CY, Soroka J, Strelkov SE, Klein-Gebbinck H, Falk KC (2010) Biology of *Camelina sativa* (L.) Crantz (false flax). Final report to the Can Food Inspec Agency

Séguin-Swartz G, Nettleton JA, Sauder C, Warwick SI, Gugel RK (2013) Hybridization between *Camelina sativa* (L.) Crantz (false flax) and North American Camelina species. Plant Breed 132:390–396

Seong G-U, Hwang I-W, Chung S-K (2016) Antioxidant capacities and polyphenolics of Chinese cabbage (*Brassica rapa* L. ssp. Pekinensis) leaves. Food Chem 199:612–618

Shabana MM, Fathy FI, Salama MM, Hashem MM (2013) Cytotoxic and antioxidant activities of the volatile constituents of *Brassica tournefortii* Gouan: growing in Egypt. Cancer Sci Res Open Access 1(1):4

Shah MA, Sarker MMR, Gousuddin M (2016) Antidiabetic potential of *Brassica oleracea* Var. Italica in type 2 diabetic Sprague dawley (sd) rats. Int J Pharm Phytochem Res 8:462–469

Shankar S, Segaran G, Sundar RDV, Settu S, Sathiavelu M (2019) Brassicaceae—a classical review on its pharmacological activities. Int J Pharm Sci Rev Res 20:107–113

Sharafi Y, Majidi MM, Goli SAH, Rashidi F (2015) Oil content and fatty acids composition in *Brassica* species. Int J Food Prop 18(10):2145–2154

Sharbel TF, Haubold B, Mitchell-Olds T (2000) Genetic isolation by distance in *Arabidopsis thaliana*: biogeography and postglacial colonization of Europe. Mol Ecol 9(12):2109–2118

Sharma P, Kapoor S (2015) Biopharmaceutical aspects of *Brassica* vegetables. J Pharm Phytochem 4

Sharma TR, Singh BM (1992) Transfer of resistance to *Alternaria brassicae* in *Brassica juncea* through interspecific hybridization among Brassicas. J Genet Breed 46:373–378

Sharma A, Kumar V, Kanwar MK, Thukral AK, Bhardwaj R (2017) Phytochemical profiling of the leaves of *Brassica* juncea L. using GC-MS. Int Food Res J:24

Sharma BB, Pritam K, Dinesh S, Tilak RS (2017b) Introgression of black rot resistance from *Brassica carinata* to cauliflower (*Brassica oleracea* botrytis Group) through embryo rescue. Front Plant Sci 8:1255

Shehata AM, Mulwa RMS, Babadoost M, Uchanski M, Norton MA, Skirvin R, Walters SA (2009) Horseradish: botany, horticulture, breeding, in horticultural reviews. In: Janick J (ed) vol 35. Wiley, Hoboken

Sheikh FA, Banga S, Banga SS, Najeeb S, Parray GA, Rather AG (2009a) Cytogenetic studies of F_1 and backcross generations of Ethiopian mustard (*Brassica carinata*) and Indian mustard (*Brassica juncea*). Cruciferae Newslett Eucarpia 28:10–11

Sheikh FA, Banga S, Banga SS, Najeeb S, Rather AG (2009b) Induction of genetic variability in Ethiopian mustard (*Brassica carinata*) for quality traits through interspecific hybridization. Cruciferae Newslett Eucarpia 28:12–13

Sheikh FA, Banga S, Banga SS (2014) Broadening the genetic base of Abyssinian mustard (*Brassica carinata* A. Braun) through introgession of genes from related allotetrapoloid species. Span J Agri Res 12(3):742–742

Shonnard DR, Williams L, Tom NK (2010) Camelina-derived jet fuel and diesel: sustainable advanced biofuels. Environ Prog Sustain Energy 29:382–392

Siemens J, Sacristan MD (1995) Production and characterization of somatic hybrids between *Arabidopsis thaliana* and *Brassica nigra*. Plant Sci 111:95–106

Sigareva MA, Earle ED (1999) Camalexin induction in intertribal somatic hybrids between *Camelina sativa* and rapid-cycling *Brassica oleracea*. Theor Appl Genet 98:164–170

Sikora E, Bodziarczyk I (2012) Composition and antioxidant activity of kale (*Brassica* oleracea L.var. acephala) raw and cooked. Acta Scientiarum Polonorum Technol Alimentaria 11:239–248

Simard MJ, Légère A, Pageau D, Lajeunesse J, Warwick S (2002) The frequency and persistence of volunteer canola (*Brassica napus*) in Quebec. Weed Technol 16:433–439

Simlai A, Chatterjee K, Roy A (2014) A comparative study on antioxidant potentials of some leafy vegetables consumed widely in India. J Food Biochem 38:365–373

Simonetti G (1990). In: Schuler S (ed) Simon & Schuster's guide to herbs and spices. Simon & Schuster, Inc. ISBN 0-671-73489-X

Singh A, Fulekar MH (2012) Phytoremediation of heavy metals by *Brassica juncea* in aquatic and terrestrial Environment. In: Anjum N, Ahmad I, Pereira M, Duarte A, Umar S, Khan N (eds) The plant family Brassicaceae. Environ Pollut 21. https://doi.org/10.1007/978-94-007-3913-0_6

Singh Y, Malik CP (2017) Phenols and their antioxidant activity in *Brassica juncea* seedlings growing under $HgCl_2$ stress. J Microbiol Biotech Res 1:124–130

Singh CS, Paswan VK (2017) The potential of garden cress (Lepidium sativum L.) seeds for development of functional foods. Adv Seed Biol

Singh B, Chaturvedi S, Walia S, Kaushik G, Thakur S (2011) Antioxidant potential of broccoli stalk: a preliminary investigation. Mediterr J Nutr Metabol 4:227–230

Sintim HY, Zheljazkov VD, Obour AK, Garcia AG, Foulke TK (2016) Evaluating agronomic responses of Camelina to seeding date under rain-fed conditions. Agron J 108:349–357

Snogerup S, Persson D (1983) Hybridization between Brassica insularis Moris and *B. balearica*. Pers Hereditas 99:187–190

Soriano NU, Narani A (2012) Evaluation of biodiesel derived from Camelina sativa oil. J Am Oil Chem Soc 89(5):917–923

Sosnowska D, Redzynia M, Anders B (2006) Antioxidant capacity and content of *Brassica oleracea* dietary antioxidants. Int J Food Sci Technol 41:49–58

Sotelo T, Cartea ME, Velasco P, Soengas P (2014) Identification of antioxidant capacity-related QTLs in *Brassica oleracea*. PLoS ONE 9:e107290

Spect CE, Diederichsen A (2001) Brassica. In: Hanelt P (ed) Mansfeld's encyclopedia of agricultural and horticultural crops, 6 vols. Springer, Berlin, Heidelberg, New York, pp 1453–1456

Sridevi O, Sarla N (1996) Reciprocal hybridization between *Sinapis alba* and *Brassica* species. Cruciferae Newslett 18:16

Sridevi O, Sarla N (2005) Production of intergeneric hybrids between *Sinapis alba* and *Brassica carinata*. Genet Res Crop Evol 52(7):839–845

Srinabas D, Tyagi K, Harjit Kaur H (2004) Evaluation of taramira oil-cake and reduction of its glucosinolate content by different treatments. Ind J Ani Sci 73(6):687–691

Strelkov SE, Hwang SF, Manolii VP, Cao T, Feindel D (2016) Emergence of new virulence phenotypes of *Plasmodiophora brassicae* on canola (*Brassica napus*) in Alberta, Canada. Eur J Plant Pathol 1–13

Struss D, Bellin U, Robbelen G (1991) Development of B-genome chromosome addition lines of *B. napus* using different interspecific *Brassica* hybrids. Plant Breed 106:209–214

Struss D, Quiros CF, Röbbelen G (1992) Mapping of molecular markers on *Brassica* B-genome chromosomes added to *Brassica napus*. Plant Breed (New York) 108:320–323

Sturtevant EL (1919) Sturtevant's notes on edible plants. In: Hedrick UP (ed). JB Lyon, Albany, NY, USA

Subudhi PK, Raut RN (1994) White rust resistance and its association with parental species type and leaf waxiness in *Brassica juncea* L Czern & Coss × *Brassica napus* L. crosses under the action of EDTA and gamma-ray. Euphytica 74:1–7

Sullivan P (2003) Principles of sustainable weed management for croplands. ATTRA Publication #IP039

Sundberg E, Glimelius K (1991) Effects of parental ploidy level and genetic divergence on chromosome elimination and chloroplast segregation in somatic hybrids within *Brassicaceae*. Theor Appl Genet 83:81–88

Takeshita M, Kato M, Tokumasu S (1980) Application of ovule culture to the production of intergeneric or interspecific hybrids in *Brassica* and *Raphanus*. Jap J Genet 55:373–387

Talalay P, Fahey JW (2001) Phytochemicals from cruciferous plants protect against cancer by modulating carcinogen metabolism. J Nutr 131(11):3027S–3033S

Taylor DC, Falk KC, Palmer CD, Hammerlindl J, Babic V, Mietkiewska E, Jadhav A, Marillia EF, Francis T, Hoffman T, Giblin EM, Katavic V, Keller WA (2010) *Brassica carinata*—a new molecular farming platform for delivering bio-industrial oil feedstocks: Case studies of genetic modifications to improve very long-chain fatty acid and oil content in seeds. Biofuels Bioprod Biorefin 4(5):538–561

Thaipratum R (2014) Evaluation of antioxidant activities of cabbage (*Brassica oleracea* L. var. capitata L.). World Acad Sci Eng Technol Int J Bio Biomol Agric Food Biotech Eng 8:591–593

Tomar RS, Shrivastava V (2014) Efficacy evaluation of ethanolic extract of Brassica nigra as potential antimicrobial agent against selected microorganisms. Int J Pharm Sci Health Care 3:117–123

Tonguç M, Griffiths PD (2004) Transfer of powdery mildew resistance from *Brassica carinata* to *Brassica oleracea* through embryo rescue. Plant Breed 123:587–589

Toxopeus H, Oost EH, Reuling G (1984) Current aspects of the taxonomy of cultivated Brassica species. The use of *B. rapa* L. versus *B. campestris* L. and a proposal for a new intraspecific classification of *B. rapa* L. Crucifer Newslett 9:55–57

Triska Dr (1975) Hamlyn encyclopaedia of plants. Hamlyn ISBN 0-600-33545-3

Turk MA, Tawaha AM (2003) Allelopathic effect of black mustard (*Brassica nigra* L.) on germination and growth of wild oat (*Avena fatua* L.). Crop Prot 22(4):673–677

Turner M, Gustafson P (2006) Wildflowers of the Pacific Northwest. Timber Press Inc, Portland, OR

Ulmer B, Gillott C, Erlandson M (2002) Oviposition preferences of bertha armyworm *Mamestra configurata* Walker (Lepidoptera: Noctuidae) on different crucifer cultivars and growth stages. Environ Entomol 31:1135–1141

U N, Nagamatu T, Midusima U (1937) A report on meiosis in the two hybrids, *Brassica alba* Rabh. (female) · *B. oleracea* L. (male) and *Eruca sativa* Lam. (female) · *B. oleracea* L. (male). Cytologia, Fujii Jubilee vol, pp 437–441

UN (1935) Genome analysis in *Brassica* with special reference to the experimental formation of *B. napus* and peculiar mode of fertilization. Jpn J Bot 7:389–452

Unal K, Susanti D, Taher M (2014) Polyphenol content and antioxidant capacity in organically and conventionally grown vegetables. J Coast Life Med 2:864–871

USDA ARS (2011) National genetic resources program. Germplasm resources information network (GRIN). National Germplasm Resources Laboratory, Beltsville, MD, USA. http://www.ars-grin.gov/cgi-bin/npgs/html/taxon.pl?

USDA, ARS (2014) United States Department of Agriculture, National Genetic Resources Program. Germplasm Resources Information Network (GRIN). National Germplasm Resources Laboratory, Beltsvilld, MD, USA. [Online] http://www.ars-grin.gov/cgi-bin/npgs/html/taxon.pl?7642

USDA Foreign Agricultural Service (2019) Oil seeds world markets and trades. United States Department of Agriculture Ulmer B, Gillott C, Erlandson M (2001) Feeding preferences, growth, and development of *Mamestra configurata* (Lepidoptera: Noctuidae) on Brassicaceae. The CanEntomol 133:509–519

USDA-ARS (2013) Germplasm resources information network (GRIN) online database. National Germplasm Resources Laboratory, Beltsville Maryland USA. https://npgsweb.ars-grin.gov/gringlobal/taxon/taxonomysearch.aspx

USDA-ARS (2016) Germplasm resources information network (GRIN). https://npgsweb.arsgrin.gov/gringlobal/taxon/taxonomysimple.aspx

USDA-ARS (2017) Germplasm resources information network (GRIN) taxonomy for plants. Taxon: *Brassica napus* L. United States Department of Agriculture. Agricultural Research Service, Beltsville Area. http://www.ars-grin.gov/cgi-bin/npgs/html/taxon.pl?7661

USDA-ERS (2014) Soybeans & oil crops—Canola. United States Department of Agriculture. Economic Research Service. http://www.ers.usda.gov/topics/crops/soybeans-oil-crops/canola.aspx

USDA-NRCS (2016) The PLANTS database. National Plant Data Center, Baton Rouge, LA, USA. https://plants.usda.gov

Verma JK, Sodhi YS, Mukhopadhyay A, Arumugam N, Gupta V, Pental D, Pradhan AK (2000) Identification of stable maintainer and fertility restorer lines for 'Polima' CMS in *Brassica campestris*. Plant Breed 119:90–92

Vohora SB, Khan MSY (1977) Pharmacological studies on *Lepidium sativum* Linn. Ind J Physiol Pharmacol 21:118–120

Vollmann J, Moritz T, Karg C, Baumgartner S, Wagentrist H (2007) Agronomic evaluation of Camelina genotypes selected for seed quality characteristics. Ind Crops Prod 26:270–277

Vyas P, Prakash S, Shivanna KR (1995) Production of wide hybrids and backcross progenies between *Diplotaxis erucoides* and crop *Brassicas*. Theor Appl Genet 90:549–553

Wada M, Kido H, Ohyama K, Ichibangase T, Kishikawa N, Ohba Y, Nakashima MN, Kuroda N, Nakashima K (2007) Chemiluminescent screening of quenching effects of natural colorants against reactive oxygen species: evaluation of grape seed, monascus, gardenia and red radish extracts as multi-functional food additives. Food Chem 101:980–986

Waddington J, Bowren KE (1978) Effect of crop residue on production of barley, bromegrass and alfalfa in the greenhouse and in the field. Can J Plant Sci 58:249–225

Walsh KD, Raatz LL, Topinka KC, Hall LM (2013) Transient seed bank of Camelina contributes to a low weedy propensity in western Canadian cropping systems. Crop Sci 53:2176–2185

Wang P, Zhu Z (2006) Effect of different harvest seasons on the flavonoids content and antioxidant activities of leaf mustard. Acta Horticult Sin 33:745–750

Wang YP, Sonntag K, Rudloff E, Chen JM (2005) Intergeneric somatic hybridization between *Brassica napus* L. and *Sinapis alba* L. J Integr Plant Biol 47(1):84–91

Wang YP, Sonntag K, Rudloff E, Groeneveld I, Gramenz J, Chu CC (2006) Production and characterization of somatic hybrids between *Brassica napus* and *Raphanus sativus*. Plant Cell Tiss Organ Cult 86:279–283

Warwick SI (2010) Brassicaceae in agriculture, in genetics and genomics of the Brassicaceae (eds Schmidt R, Bancroft I), pp 33–65. https://doi.org/10.1007/978-1-4419-7118-0_2

Warwick SI, Al-Shehbaz IA (2006) Brassicaceae: chromosome number index and database on CD-Rom. Pl Syst Evol 259:237–248

Warwick SI, Francis A (2005) The biology of Canadian weeds. 132. Raphanus raphanistrum. Can J Plant Sci 85:709–733

Warwick SI, Sauder C (2005) Phylogeny of tribe Brassiceae based on chloroplast restriction site polymorphisms and nuclear ribosomal internal transcribed spacer (ITS) and chloroplast *trn*L intron sequences. Can J Bot 83:467–483

Warwick SI, Francis A, La Fleche J (2000) Guide to the wild germplasm of brassica and allied crops (tribe Brassiceae, Brassicaceae), 2nd edn. Agriculture and Agri-Food Canada Eastern Cereal and Oilseeds Research Centre, Ottowa

Warwick SI, Simard MJ, Légère A, Beckie HJ, Braun L, Zhu B, Mason P, Seguin-Swartz G, Stewart CN (2003) Hybridization between transgenic *Brassica napus* L. and its wild relatives: *Brassica rapa* L., *Raphanus raphanistrum* L., *Sinapis arvensis* L. and *Erucastrum gallicum* (Willd.) O.E. Schulz. Theor Appl Genet 107:528–539

Warwick SI, Gugel RK, McDonald T, Falk KC (2006a) Genetic variation and agronomic potential of Ethiopian mustard (*Brassica carinata*) in western Canada. Genet Resour Crop Evol 53:297–312

Warwick SI, Francis A, Al-Shehbaz IA (2006b) Brassicaceae: species checklist and database on CD-Rom. Plant Syst Evol 259:249–258

Warwick SI, Gugel RK, Gómez-Campo C, James T (2007) Genetic variation in the *Eruca vesicaria* (L.) Cav. Plant Genet Resour Charact Util 5:142–153

Warwick SI, Légère A, Simard MJ, James T (2008) Do escaped transgenes persist in nature? The case of an herbicide resistance transgene in a weedy *Brassica rapa* population. Mol Ecol 17:1387–1395

Warwick SI, Francis A, Gugel RK (2009) Guide to wild germplasm. *Brassica* and allied crops (tribe *Brassiceae, Brassicaceae*), 3rd ed. Agriculture and Agri-Food Canada (AAFC), Eastern Cereal and Oilseeds Research Centre, Canada K1A0C6. http://www.brassica.info/info/publications/guide-wild-germplasm.php

Wei W, Darmency H (2008) Gene flow hampered by low seed size of hybrids between oilseed rape and five wild relatives. Seed Sci Res 18(2):115–123

Wei BQ, Li BJ (1989) Pairing Chinese cabbage hybrid combinations from cytoplasmic male-sterile line. J Shenyang Agric Uni 20:9–14

Wen J, Tu J-X, Li Z-Y, Fu T-D, Ma C-Z, Shen J-X (2008) Improving ovary and embryo culture techniques for efficient resynthesis of *Brassica napus* from reciprocal crosses between yellow-seeded diploids *B. rapa* and *B. oleracea*. Euphytica 162:81–89

Westman AL, Kresovich S, Dickson MH (1999) Regional variation in *Brassica nigra* and other weedy crucifers for disease reaction to *Alternaria brassicicola* and *Xanthomonas campestris* pv. *campestris*. Euphytica 106:253–259

Williams DJ, Critchley C, Pun S, Chaliha M, Timothy JO (2009) Differing mechanisms of simple nitrile formation on glucosinolate degradation in *Lepidium sativum* and *Nasturtium officinale seeds*. Phytochem 70:1401–1409

Wojciechowski A (1985) Interspecific hybrids between Brassica campestris and *B. oleracea* L. 1. Effectiveness of crossing, pollen tube growth, embryogenesis. Genetica Polonica 26:423–436

Wu H, Zhu J, Yang L, Wang R, Wang C (2015) Ultrasonic-assisted enzymatic extraction of phenolics from broccoli (*Brassica oleracea* L. var. italica) inflorescences and evaluation of antioxidant activity in vitro. Food Sci Technol Int 21:306–319

Yamagishi H, Bhat SR (2014) Cytoplasmic male sterility in Brassicaceae crops. Breed Sci 64(1):38–47. https://doi.org/10.1270/jsbbs.64.38

Yamagishi H, Nakagawa S (2004) Somatic hybrids between *Arabidopsis thaliana* and cabbage (*Brassica oleracea*). J Jap Soc Hort Sci 73:319–323

Yamagishi H, Landgren M, Forsberg J, Glimelius K (2002) Production of asymmetric hybrids between Arabidopsis thaliana and Brassica napus utilizing an efficient protoplast culture system. Theo Appl Genet 104:959–964

Yao X, Ge X, Li Z (2012) Different fertility and meiotic regularity in allohexaploids derived from trigenomic hybrids between three cultivated *Brassica* allotetraploids and *B. maurorum*. Plant Cell Rep 31:781–788
Yoon JY, Opena RT, Chang LC (1988) Evaluation of radish derived cytoplasmic male sterile for use in tropical heading Chinese cabbage. Cruciferae News 13:70–71
Yoshimura Y, Beckie HJ, Matsuo K (2006) Transgenic oilseed rape along transportation routes and port of Vancouver in western Canada. Environ Biosaf Res 5:67–75
Young-Mathews A (2012) Plant guide for field mustard (*Brassica rapa* var. *rapa)*. USDA Natural Resources Conservation Service, Corvallis Plant Materials Center, Corvallis, OR
Zaleckas E, Makarevičienė V, Sendžikienė W (2012) Possibilities of using *Camelina sativa* oil for producing biodiesel fuel. Transport 27:60–66
Zanetti F, Monti A, Berti MT (2013) Challenges and opportunities for new industrial oilseed crops in EU-27: a review. Indust Crops Prod 50:580–595
Zennie TM, Ogzewalla D (1977) Ascorbic acid and Vitamin A content of edible wild plants of Ohio and Kentucky. Eco Bot 31(1):76–79
Zhang B, Lu CM, Kakihara F, Kato M (2002) Effect of genome composition and cytoplasm on petal colour in resynthesized amphidiploids and sesquidiploids derived from crosses between *Brassica rapa* and *Brassica oleracea*. Plant Breed 121:297–300
Zhang GQ, Zhou WJ, Gu HH, Song WJ, Momoh EJJ (2003) Plant regeneration from the hybridization of *Brassica juncea* and *B. napus* through embryo culture. J Agron Crop Sci 189:347–350
Zhang X, Henriques R, Lin SS, Niu QW, Chua NH (2006) Agrobacterium-mediated transformation of Arabidopsis thaliana using the floral dip method. Nat Protocols 1(2):641–646
Zhang H, Feng J, Hwang SF, Strelkov SE, Falak I, Huang X, Sun R (2015) Mapping of clubroot (*Plasmodiophora brassicae*) resistance in canola (*Brassica napus*). Plant Pathol 65:435–440
Zhang J, Zhou X, Fu M (2016) Integrated utilization of red radish seeds for the efficient production of seed oil and sulforaphene. Food Chem 192:541–547
Zhang Y, Zhou F, Ge F (2019) Effects of combined extracts of *Lepidium meyenii* and *Allium tuberosum* Rottl. on erectile dysfunction. BMC Complement Altern Med 19:135
Zhao H-C, Du D-Z, Liu Q-Y, Zhan M-N, Yu Q-L (2003) Interspecific distant hybridization between *Brassica juncea* with multiloculus and *Brassica napus*. Xibei Zhiwu Xuebao 23:1587–1591
Zhu L-H (2016) Crambe (*Crambe abyssinica*). Ind Oil Crops 195–205
Zhu J, Struss D (1991) Transfer of *Phoma lingam* resistance from *B. nigra* into *B. napus*. Cruciferae Newslett 14/15:16–17
Zhu B, Lawrence J, Warwick SI, Mason P, Braun L, Halfhill MD, Stewart CN Jr (2004) Stable Bacillus thuringiensis (Bt) toxin content in interspecific F1 and backcross populations of wild *Brassica rapa* after Bt gene transfer. Mol Ecol 13:237–241
Zhu X, Wang D, Sun XS (2017) Carbodiimide stabilizes the ultrasound-pretreated Camelina protein structure with improved water resistance. Ind Crop Prod 97:196–200
Zipfel C, Robatzek S, Navarro L, Oakeley EJ, Jones JD, Felix G, Boller T (2004) Bacterial disease resistance in Arabidopsis through flagellin perception. Nature 428(6984):764–767
Zubr J, Matthaus B (2002) Effects of growth conditions on fatty acids and tocopherols in *Camelina sativa* oil. Ind Crops Prod 15:155–162
Zudkafh ZS (2011) Ethnobotanical studies of some plants of Chagharzai valley, district Buner, Pakistan. Pak J Bot 43(3):1445–1452

Arabidopsis thaliana: Model Plant for the Study of Abiotic Stress Responses

Ali Raza, Sidra Charagh, Nida Sadaqat and Wanmei Jin

Abstract In different regions around the world, abiotic stresses, including cold, drought, nutrient deficiency, toxicity, salinity, and flooding, minimize the rate of crop production. Abiotic stress problem has become an issue mainly in developing countries where they reason the large population food security, poverty, especially in rural areas. In environmental field conditions, plants experience various environmental stresses at once. According to research plant response to different stresses are varies from in comparison to individual stresses which produce non-additive effects. To understand abiotic stress on plants is a crucial topic in the field of plant research. Molecular biology and physiological analysis have been helped out to draw a map line to understand abiotic stresses in different plants and also *Arabidopsis thaliana* determination of its genome sequence had quite a great effect on plant research. *A. thaliana* research has been furnishing a useful understanding of all aspects of modern molecular biology. Complete genome sequences availability made it easy to access necessary information required for all genes, for example, including transcripts level, gene products, and their function, alternative patterns of splicing and putative *cis*-regulatory elements. Moreover, in natural stress response research in multiple plants instead of *A. thaliana* give rise to our knowledge related to the plant stress tolerance mechanisms. Based on this knowledge, progress in stress tolerance in crops has been strived by the meaning of marker-assisted breeding and gene transfer. In this chapter, we have summarized current progress in abiotic stress studies and have discussed new range perspectives in new directions for the future in a model plant.

A. Raza (✉)
Oil Crops Research Institute, Chinese Academy of Agricultural Sciences, Wuhan 430062, China
e-mail: alirazamughal143@gmail.com

S. Charagh
Centre of Agricultural Biochemistry and Biotechnology (CABB), University of Agriculture, Faisalabad 38040, Pakistan

N. Sadaqat
Institute of Molecular Biology and Biotechnology (IMBB), University of Lahore, Lahore, Pakistan

W. Jin
Key Laboratory of Biology and Genetic Improvement of Horticultural Crops (North China), Beijing Academy of Forestry and Pomology Sciences, Beijing 100093, China

M. Hasanuzzaman (ed.), *The Plant Family Brassicaceae*,
https://doi.org/10.1007/978-981-15-6345-4_3

Keywords Abiotic stress · *Arabidopsis thaliana* · Cold stress · Drought stress · Molecular mechanism · Salinity · Stress signaling · Transcription factors

1 Introduction

In plants, stress could be explained as which can originate change from the optimized developmental conditions and absolute growth (Coolen et al. 2016; Raza et al. 2020). Stresses could be differentiated as biotic and abiotic, where abiotic stresses are reasoned by various factors present in the surroundings including temperature, drought, salinity, osmotic stress, waterlogging, high and low intensifies, nutrients deficiency, and metal toxicity level (Coolen et al. 2016). Abiotic stress considered crucial in the case of sessile organisms such as plants because sessile organisms could not be able to thrive until they can bare the environmental changes (Pereira 2016; Raza et al. 2020). Sometimes plants are associated with facing unfavorable environmental abiotic factor conditions which cause a result of abiotic stresses, which usually are important factor keys to determine crop yield production (Hasanuzzaman et al. 2012, 2013a, b; Raza et al. 2019a, b, c). Moreover, it also includes differential distribution among the different plant species from various types of environments. The capability of plants to acclimatize to various environmental conditions is somehow associated with photosynthesis resilience indirectly or directly in emergence with other different procedures to understand the growth of a plant and its development, mainly plant reproduction (Verslues et al. 2006; Yamori 2016). An exceptional characteristic related to plant modification in response against abiotic stresses includes triggering of various reactions, which involve complicated gene interactions and interference among various molecular pathways (Lämke and Bäurle 2017). Plants are reliant on their surrounding environment because of their nature of inability of motion, due to that occurrence of abiotic stress cannot be prevented to harm plants (Rizhsky et al. 2004). Depletion in yield of food crops has been found due to abiotic stresses around the world are said to be the main factor that plays a role in the reduction of agriculture production (Hasanuzzaman et al. 2012; Hasanuzzaman and Fujita 2012; Raza et al. 2019a, b, c). According to the report by Intergovernmental Panel on Climate Change (IPCC) (http://www.ipcc.ch/), abiotic factors are predicted to get higher in rate due to climate change in the future (Pachauri et al. 2014).

A significant milestone in the development of this field is the identification and characterization of the *Arabidopsis thaliana* genome sequence in 2000 (Initiative 2000). By using the obtained genome information via the microarrays, expression files of all anticipated genes in one single time could be analyzed. Accessible genome data made the potential *cis*-regulatory elements and transcription factors (TFs) identification enabled (Hirayama and Shinozaki 2010). Based on large-scale mutant analysis and wide genome transcript profiles, the knowledge and data have been assembled associated with biotic or abiotic stress responses with model plant *A. thaliana*. Moreover, random mutagenesis in combination with particular cell and treatment in the expression of reporter gene has been characterized in various factors in stimulus to stress, but in particular, observations based on microarray resulted in transcriptional response on wide-scale genome level (Cheong et al. 2002; Rasmussen et al. 2013; Seki et al. 2002). Although most analysis studies have been limited to individual treatments and genes, different plant-specific database sources like The Arabidopsis Information Resource (TAIR) (https://www.arabidopsis.org/), GENEVESTIGATOR (https://www.genevestigator.ethz.ch), and the Nottingham Arabidopsis Stock Centre (NASC) (http://arabidopsis.info/BrowsePage), Arabidopsis thaliana Protein Interactome Database (AtPID) (http://www.megabionet.org/atpid) is an integrated database for protein interaction network and functional annotation of *Arabidopsis*, and Arabidopsis Gene Expression Database (http://www.arexdb.org/) which assembled the data and combine that particular data with features based on genome sequence (Craigon et al. 2004; Initiative 2000; Lv et al. 2016a; Rhee et al. 2003; Zimmermann et al. 2004), usually the data could be found in separated form. To combine these multiple wide-level datasets remains a great challenge to assemble a uniform database of plant responses against environmental effects. To single link genes, MapMan and Signal Transduction Knowledge Environment (STKE) tools have been developed, which further are linked to various pathways and circuits (Gough 2002; Usadel et al. 2005).

Complicated cellular responses are generated by abiotic stresses which have been triggered by development via exploration and comprehension of plant abiotic reactions at the extent of biochemical, physiological, molecular, and cellular levels of the whole plant (Grover et al. 2001). The main issue to recent agriculture sustainable growth is to gain understanding regarding the development of engineering plants which could allow seeking new desire agronomical traits (Le et al. 2007). The development associated with stress-tolerant crops via methods of conventional breeding or genetic engineering techniques covered up almost all the features of plant science and their application is in use of both private and public researchers (Atkinson and Urwin 2012).

Therefore, knowledge of abiotic stresses and plant stimulus responses to these environmental changes is a crucial topic regarding plant research. The main development in the field of research is due to molecular biology applications. Due to molecular biology procedure applications in plant sciences, field-inducible genes associated with abiotic stress isolation have been done, and in transgenic plants, their characterization has been done based on their function. The understanding of this knowledge enhanced the area to understand our view regarding the response of abiotic stress and plant tolerance. In this chapter, we have discussed the regulation and existence of abiotic stresses in *A. thaliana.*

2 Physiological and Biochemical Mechanisms Underlying Abiotic Stress Responses

Physics was the first discipline describing the notion of "Stress." Through the theory of elasticity, the concept of stress was described as this theory deals with the elastic solids stress and displacement exerted by forces (Kranner et al. 2010). Looking from a biological perspective, the term stress can generally be demarcated as an exterior influence having a detrimental effect on plants (Fleta-Soriano and Munné-Bosch 2016). On the other hand, substantial deviance from the peak circumstances of life could be termed as stress (Larcher 2003).

The three focal points involved in stress actions and reactions of the plant can be put as (1) the alarm point, (2) the resistance point, and (3) the enervation point (Larcher 2003). Afterward, the fourth point was added by Lichtenthaler (1988), named as the rejuvenation point, that takes place only when the stress is recovered before the onset of any drastic impairment. This, in return, allows the complete or fractional rejuvenation of the physical functionality of the plant. The first phase, i.e. the alarm point, is triggered via stress reaction that is categorized by functionality failures caused by stress factors, which are counterbalanced by a restoration of counter-responses leading to conversion in resistance point. The distinct and independent onset of stress factors on a plant infrequently happens, as their action occurs concurrently in combination. For instance, the repeated permutation of different conditions, i.e. parched summer stages, sunny and warm periods, heat, H_2O, and elevated sunlight stress (Lichtenthaler 1996).

Whenever certain environmental factors, i.e. abiotic or biotic factors, set off from their optimal levels, sensing is the first episode that a plant experiences (Nahar et al. 2016, 2017; Razzaq et al. 2019). The sensing of stress is an intricate matter because several shared sensing mechanisms are involved for different stresses. For instance, various stressors have an uninterrupted effect on subversive plant parts, i.e. drought stress, flood stress, etc. On the other hand, different stresses affect the plant parts that are above the ground, i.e. photo-inhibition. Therefore, the involvement of different mechanisms related to sensing is observed. The binding of a chemical ligand with that of an explicit receptor is the commonly exerted model to sense a stimulus of external nature (Takahashi and Shinozaki 2019). Nonetheless, the stated model is only appropriate for chemical-based stressors that are nutrient diminution stress and heavy-metal stress. This model is not for stresses that are related to physical conditions, i.e. temperature stress, freezing, or icy temperatures as there is not any involvement of chemical-based ligands. A similar case is with radiation-related stress; however, correspondence could be made amid ligand–receptor besides photon–receptor. The general ligand–receptor model identity is considered controversial despite the involvement of molecules, whereas, in concerns of a rooting system, it is not still clear whether the cells can sense the concentration of water in soil (Davila Olivas et al. 2017; Salim and Raza 2020). Contrastingly, investigational indications opined to the likelihood of detecting cell water-related homeostasis. The segregated trans-membraned hybrid histidine kinase from that of *A. thaliana* is the source of investigational indication regarding osmotic sensors among high plants (Urao et al. 1999). Moreover, other factors such as photosynthetically produced sugars and breakdown of carbon in tissues also played a paramount role in cell signaling and sensed pathways, growth modulation, progression, and responses of stress (Li and Sheen 2016; Rolland et al. 2006).

Some intricate monitoring responses as well as changing gene expression worldwide, alterations in proteins, and changed metabolic conformation; are all lead by stress (Apel and Hirt 2004; Gilmour et al. 2000). Some of the changes that are prompt via abiotic stress include; changes in morphology, physiology, and biochemical changes in structural and functionality of membranes, the amount of water in tissues, lipids, proteins, as well as the composition of secondary metabolites (Hirayama and Shinozaki 2010). Oxidative or osmosis-related secondary stress may occur when osmotic homeostasis of cells is disturbed by abiotic stressors (Vinocur and Altman 2005). Various downstream signal cassettes such as overly sensitive kinases (SOS), mitogen-activated protein kinases (MAPK), and calcium-dependent protein kinases (CDPKs) are activated via secondary messages, i.e. calcium ions, reactive oxidative species (ROS), phosphatidic acid (PtdOH), and osmotic sensors such as (*A. thaliana* histidine kinase-1) *AtHK1* (Qiu et al. 2002; Shi et al. 2003; Hossain et al. 2015; Choudhury et al. 2017). These signal cassettes prompt paramount response transcriptional factors which included heat shock factors/proteins (HSF/HSP), carbon-repeated binding factor or dehydratase response element-binding protein (CBF/DREB), induced CBF expression 1 (ICE1), ABA response element-binding factor/ABA-responsive element (ABF/ABRE) families, basic-leucine-zipper TFs (bZIP) along with MYB/MYC (Jaglo-Ottosen et al. 1998; Chinnusamy et al. 2003; Zhang et al. 2004a; Kim 2006; Nakashima and Yamaguchi-Shinozaki 2013; Zhao et al. 2016b).

The genes that have direct or indirect involvement in exerting abiotic stress in a plant are regulated by the inducible TFs regulating the function and regulatory proteins (Hirayama and Shinozaki 2010; Raza et al. 2020). For instance, the chaperones, i.e. HSP and late embryogenesis abundant (LEA); moreover, during the exposure of the plant to stress the structure of proteins are protected (Wang et al. 2004; Hundertmark and Hincha 2008). Superoxide dismutase (SOD) and peroxidase (PX) have a detoxifying function for O_2 radicals that intensify in stress and or else would cause mutilation of the cell (Savvides et al. 2016). Similarly, the osmosis integrity of a cell is guarded by proteins that take part in H_2O and ion transport, just like aquaporin (Vera-Estrella et al. 2004). Lastly, the synthesis of proteins essential for producing stress linked metabolites will occur for the functionality of defense against abiotic stress (Bowne et al. 2018). These comprise of the metabolites that can be utilized as osmotic protectants or else well-suited solutes for the protein–membrane maintenance besides turgor stress preservation; for instance, amino acids as well as their byproducts, derivatives and sugar, phosphor-di-esters, polyols along with quarter ammonium compounds (Kang et al. 2002; Shahjee et al. 2002). In correspondence to the function of PX and SOD for reactive oxygen species (ROS) detoxification, these metabolites also function as anti-oxidants. This anti-oxidative role also implies for proteins oxidation–reduction reactions and protection of cellular constituents which also take account of alcoholic sugars, simple sugars along glutathione (GSH) (Inupakutika et al. 2016; Nishizawa et al. 2008).

3 Recent Advances to Study Abiotic Stresses in a Model Plant

The plant yield has been under the effect of various environmental stresses, including biotic and abiotic. The abiotic stresses include salinity, extreme temperature, water-logging, drought, osmotic stress, and heavy metal stress (Razzaq et al. 2019; Raza et al. 2019a, b, c; Raza et al. 2020). However, Fig. 1 showing the general abiotic stress signaling pathways. Under the era of various climatic conditions, these stresses often interrelate. These stresses could be individual or in combination with each other. There are some major TFs (DREB/CBF, AP2/ERF, bZIP, WRKY, NAC, etc.) that have been extensively studied in a model plant for multiple abiotic stresses. In this section, we have discussed the recent advances to study above-mentioned abiotic stress in *A. thaliana*. However, a detailed list of abiotic stress-responsive genes in *A. thaliana* have been disclosed in Table 1, and a list of transgenic *Arabidopsis* plants have been described in Table 2.

3.1 Drought

One of the crucial environmental stress factors is drought, which alters the growth of plants as well as their development. Stress like drought results in different biochemical changes, morphological plus physiological that aids to maintain plant growth (Raza et al. 2019a, b; Raza et al. 2020). In climatic areas, soil water scarcity and drought might be chronic through low water availability; unpredictable and random as a result of weather conditions fluctuations during the plant growth period. With the increase in water scarcity and climate changes, the drought effects also increased (Harb et al. 2010). Although plants response have been studied extensively when exposed to severe drought, knowledge regarding how plants adapt themselves in response to drought stress situation is very little (Clauw et al. 2015). Abscisic acid (ABA) controls numerous genes functioning during drought response through ABA-dependent and -independent pathways. In *A. thaliana*, transcriptome sequencing, as well as network analysis, have been used for sightseeing the interplays among ABA acid-dependent then ABA independent pathways replying drought. Under drought 211 ABA-dependent and 1,118 ABA-independent differentially expressed genes (DEGs) have been identified. Functional analysis disclosed that, under ABA stimulus and water deprivation, ABA-dependent DEGs expression increased, while, under gibberellin (GA), jasmonic acid (JA), and salicylic acid (SA), ABA-independent DEGs

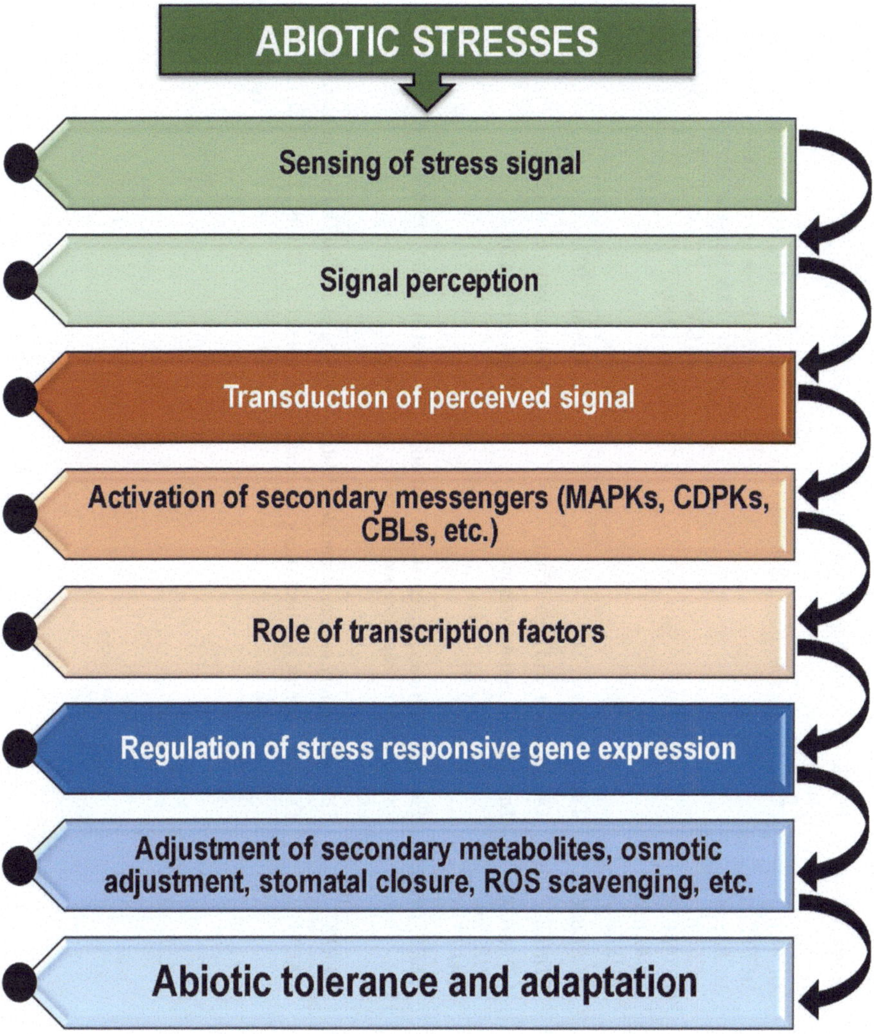

Fig. 1 Step-wise presentation of general abiotic stress signalling pathways

preferentially increased. Ninety-four genes have been found acting as core interacting components among ABA-dependent plus independent pathways by combing network analysis (Liu et al. 2018b).

In *Arabidopsis*, the drought-inducible genes products recognized by the microarray analysis and are classified into two sets. Proteins that mostly function in abiotic stress tolerance placed into the first group, for instance, osmotin, chaperones, LEA proteins, water channel proteins, proline transporters, as well as different proteases. Protein factors take part in stress-responsive gene expression and further signal transduction regulation placed into the second group named as regulatory proteins.

Table 1 *Arabidopsis thaliana* genes used for abiotic stress tolerance and their physiological and biochemical functions on plant growth

Transcription factor types/subfamily	Genes	Physiological and biochemical functions on plant growth	References
DREB/CBF	*AtCBF1, AtCBF2, AtCBF3, AtCBF4, AtDREB1A, AtDREB2A, AtDREB1B*	Cold, drought tolerance and dehydration	Behnam et al. (2007); Chinnusamy et al. (2003); Fursova et al. (2009); Gilmour et al. (2000); Haake et al. (2002); Miura et al. (2007); Park et al. (2015); Sakuma et al. (2006a)
	AtDREB2C	Thermotolerance	Lim et al. (2007)
	AtDREB19	Drought, heat, and high salinity	Krishnaswamy et al. (2011)
	AtDREB2A, AtHB7 and AtABF3	Drought and salinity tolerance	Pruthvi et al. (2014)
AP2/ERF	*AtERF71/HRE2*	Flooding tolerance, a regulator of osmotic stress, hypoxia response, and root cell expansion	Lee et al. (2015); Park et al. (2011)
	AtERF53	Drought, salinity tolerance, and dehydration	Cheng et al. (2012); Hsieh et al. (2013)
	AtERF1	Drought, salinity, and heat stress tolerance	Cheng et al. (2013b); Yang et al. (2009b)

(continued)

Table 1 (continued)

Transcription factor types/subfamily	Genes	Physiological and biochemical functions on plant growth	References
	AtERF6	Positive regulator of antioxidants (ROS) signaling during plant growth and responses to abiotic stresses	Sewelam et al. (2013)
	AtHARDY	Drought and high salinity tolerance	Abogadallah et al. (2011); Karaba et al. (2007)
	RAP2.2	Submergence tolerance and hypoxia response	Hinz et al. (2010)
	RAP2.4 and RAP2.4A	Drought stress tolerance	Lin et al. (2008a); Zhu et al. (2014)
	RAP2.6	Positively regulate abiotic stresses (salinity, freezing and osmotic stress tolerance) and ABA response	Zhu et al. (2010)
	RAP2.11	Positive regulator of signaling pathway in low-potassium conditions including ethylene, Ca^{+} signaling, and ROS production	Kim et al. (2012)

(continued)

Table 1 (continued)

Transcription factor types/subfamily	Genes	Physiological and biochemical functions on plant growth	References
	AtERF109	Salinity tolerance and responses	Bahieldin et al. (2018)
MYB	*AtMYB2*	Osmotic stress tolerance	Abe et al. (2003)
	AtMYB15	Drought, cold, and salinity tolerance	Agarwal et al. (2006); Ding et al. (2009)
	AtMYB44	Stomatal closure to confer abiotic stresses	Jung et al. (2008)
	AtMYB96	Drought stress tolerance	Seo et al. (2009)
	AtMYB41	Osmotic stress tolerance	Lippold et al. (2009)
	AtMYB44	Drought and salinity tolerance	Seo et al. (2012)
	AtMYB12	Drought and salinity tolerance	Wang et al. (2016a)
	AtMYB15	Freezing tolerance	Kim et al. (2017b)
	AtMYB111	Salinity tolerance	Li et al. (2019a)
WRKY	*AtWRKY25*, *AtWRKY33*	Heat and salinity tolerance	Jiang and Deyholos (2009); Li et al. (2011); Li et al. (2009)
	AtWRKY39	Heat tolerance	Li et al. (2010)
	AtWRKY18, *AtWRKY40*, *AtWRKY60*	Salinity and osmotic stress tolerance	Chen et al. (2010)
	AtWRKY26	Heat tolerance	Li et al. (2011)

(continued)

Table 1 (continued)

Transcription factor types/subfamily	Genes	Physiological and biochemical functions on plant growth	References
	AtWRKY28	Salinity tolerance	Babitha et al. (2013)
NAC	*ANAC055*	Dehydration and drought stress tolerance	Tran et al. (2004); Tran (2004)
	ANAC072/RD26	Dehydration	Fujita et al. (2004); Tran (2004)
	ANAC019	Cold tolerance	Jensen et al. (2010)
	ANAC096	Dehydration and osmotic stress responses	Xu et al. (2013)
bZIP	*AtABP9*	Drought and heat stress tolerance	Zhang et al. (2008)
	AtbZIP17, *AtbZIP28*	Drought, salt, and cold stress tolerance	Liu et al. (2008)
	AtbZIP19, *AtbZIP23*, *AtbZIP24*	Drought, salt, and freezing tolerance	Assunção et al. (2010); Yang et al. (2009a)
	AtbZIP9	Drought tolerance	Silveira et al. (2007)
	AtbZIP63	Cold and heat stress tolerance	Matiolli et al. (2011)
	AtbZIP1	Drought, salinity, and osmotic stress tolerance	Sun et al. (2012)
	AtTGA4	Drought tolerance	Zhong et al. (2015)
AREB/ABF	*AREB1*, *AREB2/ABF4* and *ABF3*	Drought and dehydration stress tolerance	Fujita et al. (2005); Yoshida et al. (2015)

(continued)

Table 1 (continued)

Transcription factor types/subfamily	Genes	Physiological and biochemical functions on plant growth	References
	ABF2	Osmotic stress tolerance	Kim et al. (2004)
NF-X1	*AtNFXL1*	Salt stress tolerance	Lisso et al. (2006)
NF-Y (HAP)	*AtNF-YB1*	Drought stress tolerance	Nelson et al. (2007)
SOS	*AtSOS1*, *AtSOS2/AtSOS3*	Salinity tolerance	Sako et al. (2015)
PDC	*TSPO*	Drought stress tolerance	Rasheed et al. 2018)
HDAC	*HDA6*, *HDA19*	Salinity, cold, and ABA-responsive	Chen and Wu (2010); To et al. (2011)
RNP	*AtRNP1*	Drought and salinity tolerance	Wang et al. (2016c)
PME	*AtPME41*	Chilling stress tolerance	Qu et al. (2011)
SAP	*AtSAP13*	Drought, salinity, and toxic metal tolerance	Dixit et al. (2018)
PDF	*AtPDF2.5*	Cadmium tolerance and accumulation	Luo et al. (2019)
HKT	*AtHKT1;1*	Salinity responsive	Tada (2019)

Table 2 A list of transgenic *A. thaliana* and genes used for abiotic stress tolerance

Transcription factor type/subfamily	Gene	Donor plant	Accepter plant	Stress type	References
AP2/DREB/CBF	*SlCBF1*	*Solanum lycopersicum*	*A. thaliana*	Freezing	Zhang et al. (2004b)
	HvCBF3	*Hordeum vulgare*	*A. thaliana*	Freezing	Skinner et al. (2005)
	TaICE141, *TaICE187*	*Triticum aestivum*	*A. thaliana*	Freezing	Badawi et al. (2008)
	LpCBF3	*Lolium perenne*	*A. thaliana*	Freezing	Zhao and Bughrara (2008)
	BpCBF1	*Betula pendula*	*A. thaliana*	Freezing	Welling and Palva (2008)
	GmDREB3	*Glycine max*	*A. thaliana*	Drought, salinity and freezing	Chen et al. (2008)
	DgDREB1A	*Chrysanthemum*	*A. thaliana*	Drought and freezing	Tong et al. (2009)
	HvDREB1	*H. vulgare*	*A. thaliana*	Drought	Xu et al. (2009)
	OsDREB2B	*Oryza sativa*	*A. thaliana*	Drought and heat	Matsukura et al. (2010)
	MbDREB1	*Malus baccata*	*A. thaliana*	Drought, salinity, and cold	Yang et al. (2011)
	VrCBF1, *VrCBF4*	*Vitis riparia*	*A. thaliana*	Drought and freezing	Siddiqua and Nassuth (2011)
	LcDREB3a	*Leymus chinensis*	*A. thaliana*	Drought and salinity	Peng et al. (2013)

(continued)

Table 2 (continued)

Transcription factor type/subfamily	Gene	Donor plant	Accepter plant	Stress type	References
	LcDREB2	*L. chinensis*	*A. thaliana*	Salinity	Peng et al. (2013)
	VrDREB2A	*Vigna radiata*	*A. thaliana*	Drought and salinity	Chen et al. (2016)
	AmDREB2C	*Ammopiptanthus mongolicus*	*A. thaliana*	Drought, heat, and freezing	Yin et al. (2018)
	StCBF1, *ScCBF1*	*Solanum tuberosum*, *Solanum commersonii*	*A. thaliana*	Drought and freezing	Li et al. (2018)
	ScDREB10	*Syntrichia caninervis*	*A. thaliana*	Salinity and osmotic stress	Li et al. (2019c)
AP2/ERF	*HvRAF*	*H. vulgare*	*A. thaliana*	High salinity	Jung et al. (2007)
	MsERF11	*Medicago sativa*	*A. thaliana*	Salinity	Chen et al. (2012)
	OsAP21	*O. sativa*	*A. thaliana*	Drought and salinity	Jin et al. (2013)
	BnaERF-B3-hy15-mu3	*Brassica napus*	*A. thaliana*	Cold	Xiong et al. (2013)
	LcERF054	*Lotus corniculatus*	*A. thaliana*	Salinity	Sun et al. (2014)
	BnERF2.4	*B. napus*	*A. thaliana*	Submergence	Lv et al. (2016b)
	AdPDC1	*Actinidia deliciosa*	*A. thaliana*	Waterlogging	Zhang et al. (2016)
	PsnERF75	*Populus simonii* × *P. nigra*	*A. thaliana*	Salinity	Wang et al. (2018b)

(continued)

Table 2 (continued)

Transcription factor type/subfamily	Gene	Donor plant	Accepter plant	Stress type	References
	LoERF017	*Larix olgensis*	*A. thaliana*	Salinity and osmotic stress	Hu et al. (2018)
	MaRAP2-4	*Mentha arvensis*	*A. thaliana*	Waterlogging	Phukan et al. (2018)
	IbRAP2-12	*Ipomoea batatas*	*A. thaliana*	Drought and salinity	Li et al. (2019d)
	LeERF-1	*Lithospermum erythrorhizon*	*A. thaliana*	Drought	Fang et al. (2019)
	VaERF080, *VaERF087*	*Vitis amurensis*	*A. thaliana*	Cold	Sun et al. (2019)
	LkERF-B2	*Larix kaempferi*	*A. thaliana*	Salinity	Cao et al. (2019)
MYB	*TaMYB33*	*T. aestivum*	*A. thaliana*	Drought and salinity	Qin et al. (2012)
	LcMYB1	*L. chinensis*	*A. thaliana*	Salinity	Cheng et al. (2013a)
	TaMYB19	*T. aestivum*	*A. thaliana*	Salinity	Zhang et al. (2014)
	GmMYBJ1	*G. max*	*A. thaliana*	Drought and cold	Su et al. (2014)
	TaMYB3R1	*T. aestivum*	*A. thaliana*	Drought and salinity	Cai et al. (2015)
	FtMYB9	*Fagopyrum tataricum*	*A. thaliana*	Drought and salinity	Gao et al. (2017)
	GhMYB73	*Gossypium hirsutum*	*A. thaliana*	Salinity	Zhao et al. (2019)

(continued)

Table 2 (continued)

Transcription factor type/subfamily	Gene	Donor plant	Accepter plant	Stress type	References
	MsMYB2L	*M. sativa*	*A. thaliana*	Drought and salinity	Song et al. (2019)
	TaSIM	*T. aestivum*	*A. thaliana*	Drought	Yu et al. (2019)
WRKY	*GmWRKY13*, *GmWRKY21*, *GmWRKY54*	*G. max*	*A. thaliana*	Drought, cold and salinity	Zhou et al. (2008)
	OsWRKY45	*O. sativa*	*A. thaliana*	Drought and salinity	Qiu and Yu (2009)
	OsWRKY08	*O. sativa*	*A. thaliana*	Drought and salinity	Song et al. (2009)
	VvWRKY11	*Vitis vinifera*	*A. thaliana*	Drought	Liu et al. (2011)
	GsWRKY20	*Glycine soja*	*A. thaliana*	Drought	Luo et al. (2013)
	ZmWRKY33	*Zea mays*	*A. thaliana*	Salinity	Li et al. (2013)
	TaWRKY79	*T. aestivum*	*A. thaliana*	Drought and salinity	Qin et al. (2013)
	TaWRKY93	*T. aestivum*	*A. thaliana*	Drought, salinity, and cold	Qin et al. (2015)
	GhWRKY34	*G. hirsutum*	*A. thaliana*	Salinity	Zhou et al. (2015)
NAC	*ONAC063*	*O. sativa*	*A. thaliana*	Salinity and osmotic stress	Yokotani et al. (2009)
	GmNAC20	*G. max*	*A. thaliana*	Salinity and low temperature	Hao et al. (2011)

(continued)

Table 2 (continued)

Transcription factor type/subfamily	Gene	Donor plant	Accepter plant	Stress type	References
	ZmSNAC1	*Z. mays*	*A. thaliana*	Salinity, drought and cold	Lu et al. (2012)
	TaNAC2	*T. aestivum*	*A. thaliana*	Drought, salinity and freezing	Mao et al. (2012)
	BnNAC5	*B. napus*	*A. thaliana*	Salinity	Zhong et al. (2012)
	TaNAC67	*T. aestivum*	*A. thaliana*	Drought, salinity, and cold	Mao et al. (2014)
	MLNAC5	*Miscanthus lutarioriparius*	*A. thaliana*	Drought and cold	Yang et al. (2015)
	TaNAC29	*T. aestivum*	*A. thaliana*	Drought and salinity	Huang et al. (2015)
	CarNAC4	*Cicer arietinum*	*A. thaliana*	Drought and salinity	Yu et al. (2016)
	ThNAC13	*T. hispida*	*A. thaliana*	Salinity and osmotic stress	Wang et al. (2017)
	EsNAC1	*Eutrema salsugineum*	*A. thaliana*	Salinity and oxidative stress	Liu et al. (2018a)
	SlNAC8	*Suaeda liaotungensis*	*A. thaliana*	Drought and salinity	Wu et al. (2018)
	PgNAC21	*Pennisetum glaucum*	*A. thaliana*	Salinity	Shinde et al. (2019)
bZIP	*GmbZIP44, GmbZIP62, GmbZIP78, GmbZIP132*	*G. max*	*A. thaliana*	Salinity and freezing	Liao et al. (2008)

(continued)

Table 2 (continued)

Transcription factor type/subfamily	Gene	Donor plant	Accepter plant	Stress type	References
	ABP9	*Z. mays*	*A. thaliana*	Drought, cold and salinity	Zhang et al. (2011)
	GmbZIP1	*G. max*	*A. thaliana*	Drought, cold and salinity	Gao et al. (2011)
	ZmbZIP72	*Z. mays*	*A. thaliana*	Drought and salinity	Ying et al. (2012)
	ThbZIP1	*T. hispida*	*A. thaliana*	Drought and salinity	Ji et al. (2013)
	TabZIP60	*T. aestivum*	*A. thaliana*	Drought, salinity, and low temperature	Zhang et al. (2015)
	BnaABF2	*B. napus*	*A. thaliana*	Drought and salinity	Zhao et al. (2016a)
	VqbZIP39	*Vitis quinquangularis*	*A. thaliana*	Drought and salinity	Tu et al. (2016)
CDR	*CcCDR*	*Cajanus cajan*	*A. thaliana*	Drought, cold and salinity	Tamirisa et al. (2014)
GRAS	*VaPAT1*	*V. amurensis*	*A. thaliana*	Drought, cold and high salinity	Yuan et al. (2016)
LEA/DHNs	*HbDHN1, HbDHN2*	*H. brasiliensis*	*A. thaliana*	Drought and osmotic stress	Cao et al. (2017)
MT2A	*PdMT2A*	*Phoenix dactylifera*	*A. thaliana*	Drought, salinity and oxidative stress	Patankar et al. (2019)
HD-Zip	*JcHDZ07*	*Jatropha curcas*	*A. thaliana*	Salinity	Tang et al. (2019)

(continued)

Table 2 (continued)

Transcription factor type/subfamily	Gene	Donor plant	Accepter plant	Stress type	References
SnRK2	*MpSnRK2.10*	*Malus prunifolia*	*A. thaliana*	Drought	Shao et al. (2019)
RBG	*OsRBGD3*	*O. sativa*	*A. thaliana*	Cold	Lenka et al. (2019)
KLCR	*GhKLCR1*	*G. hirsutum*	*A. thaliana*	Drought	Li et al. (2019b)

Table 3 *Arabidopsis* drought stress-inducible TFs and protein functions. Modified from Shinozaki and Yamaguchi-Shinozaki (2007)

Drought stress	
Functional proteins	Regulatory proteins
• LEA proteins • Chaperones • Detoxification enzymes • Proteases • Water channels • Transporters • Enzymes for osmolyte biosynthesis	• ABA biosynthesis • Phosphatases • Phospholipid metabolism • Protein kinases • Transcription factors: HB, bZIP, MYC, NAC, MYB, DREB2, AREB

These include numerous phosphatases, TFs, protein kinases, and enzymes help in phospholipid metabolisms. Proteins and TFs might direct the stress-inducible genes expression, both supportively and autonomously and may establish gene networks in *Arabidopsis*, as shown in Table 3 and Fig. 2 (Shinozaki and Yamaguchi-Shinozaki 2007).

In *Hevea brasiliensis*, *HbDHN1* and *HbDHN2* (dehydrin genes) formerly identified, induced in epicormic shoots at wounding position and recognized as an important sign of laticifer differentiation. In *Arabidopsis*, *HbDHNs* overexpression presented higher tolerance to drought, osmotic stresses. Therefore, together, *HbDHNs* affect physiological processes and upgrading *Arabidopsis* tolerance toward numerous environmental stresses (Cao et al. 2017).

Sumoylation (post-translational modifications) influence the plant's cellular process via SUMO conjugation (small ubiquitin-like modifier) to locate substrate proteins. SUMO conjugation regulates the major cellular junctions' response to different environmental abiotic stresses. SIZ1, a SUMO E3 ligase, is a smoothing key stage in the sumoylation pathway. It has studied the overexpression of Os*SIZ1* rice gene *A. thaliana* leads to enhanced tolerance for numerous environmental stresses. For instance, plants overexpressing Os*SIZ1* displayed increased drought tolerance and produced a high yield of seeds in various stressed situations (Mishra et al. 2018). In recent research, it was revealed that γ-aminobutyric acid (GABA) accumulation under drought sign of stress-specific reaction also also buildup their prompts in response to stomatol opening regulation thus avoids water loss in *Arabidopsis* (Mekonnen et al. 2016). In *Arabidopsis*, *TSPO* (tryptophan-rich sensory protein) promoter boosts the expression of acetic acid biosynthesis genes, confirming extended existence in drought stress (Rasheed et al. 2018).

Another factor, osmotic stress caused by salinity and drought in plants, affecting plant growth and development. In *A. thaliana*, phytase overexpression improves the growth of plants to osmotic strain (Belgaroui et al. 2018). For tolerance of osmotic stress in *Arabidopsis*, ABA plays a vital role in regulating leaf starch degradation. Starch swiftly assemble with the synergistic action of a-AMYLASE3 (AMY3) as well as b-AMYLASE1 (BAM1) which stimulates stomatal opening in guard cells.

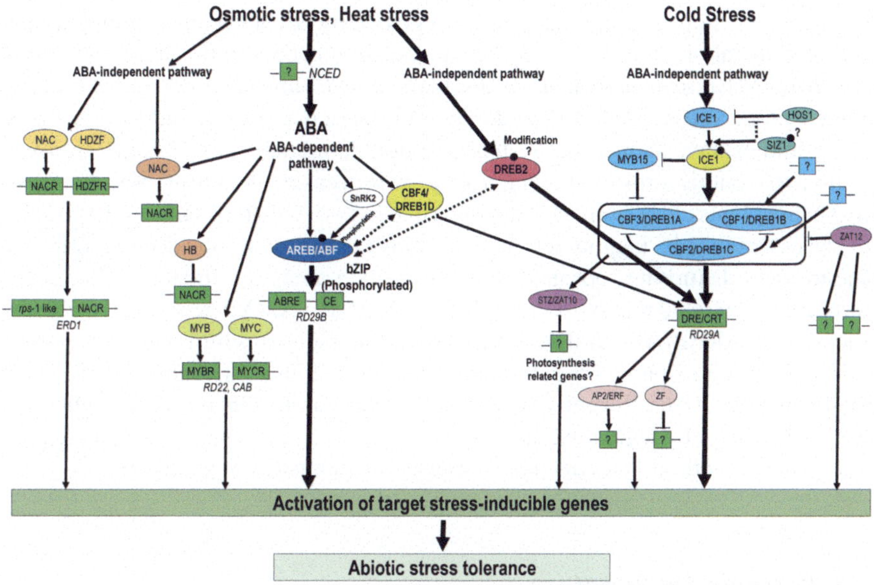

Fig. 2 A schematic representation of the cross-talk transcriptional network among *cis*-acting elements and transcription factors of ABA-independent and ABA-dependent pathways during the perception of abiotic stresses. The small filled black dotes indicating the post-translational modifications of transcriptions factors in response to stress signals for their activation via phosphorylation. Bold black arrows correspond to the major signaling pathways, which leads the regulation of many downstream genes. Black dotted arrow showing the protein–protein interactions. Two different DRE/CRT-binding proteins, CBF1/DREB1B, CBF2/DREB1C, and CBF3/DREB1A, distinguish two different signal transduction pathways in response to cold and drought stresses, respectively. A transcriptional activator, ICE1 (Inducer of CBF Expression 1), functions upstream of the DRE/DREB regulon. DREB1/CBF genes are induced by cold stress, whereas DREB2 is induced by dehydration and salt stress. Both work in ABA-independent pathways. ICE1 is an MYC-type transcription factor and binds to *cis*-elements in the promoter of CBF3/DREB1A to induce its expression. CBF3/DREB1A is an AP2-type transcription factor to regulate the expression of cold-regulated genes and cold tolerance. CBF4/DREB1D is a DRE *cis*-element binding factor that is ABA-dependent. HOS1 function as a negative regulator of ICE1. The black dot on HOS1 and SIZ1 indicate the sumoylation modification by SIZ1 of the ICE1 transcription factor. The dotted black line from SIZ1 to HOS1 represents competition for binding places on the ICE1 transcription factor, whereas SIZ1 blocks the access of HOS1 to the ubiquitination sites on the ICE1. SnRK2 is involved in ABA signalling and phosphorylation. The small blue squares with question marks indicate the putative MYC ICE1-like transcription factors, which may play a role in the activation of CBF1/DREB1B, CBF2/DREB1C, and CBF3/DREB1A. The small green squares represent the *cis*-elements present in stress-responsive genes. The green squares with question marks correspond to the putative *cis*-elements in the promoter of stress-responsive genes

During the day, starch accumulates and is degraded by BAM3 at night to support heterotrophic metabolism in leaves. Starch degradation occurs in the presence of light through BAM1 (stress-activated) to discharge sugar plus sugar-derived osmolytes during osmotic stress. AMY3 also takes part in starch degradation induced by stress. Recent studies have shown that *A. thaliana* double mutant's amy3 bam1 are oversensitive to osmotic stress displaying root growth weakened. During osmotic stress, amy3 bam1 plants close stomata at the same extents as a wild type but could not mobilize leaves starch. 14C tagging revealed that amy3 bam1 plants had decreased carbon trade to roots disturbing root growth and osmolyte amassing in stress. By genetic approaches, we found that ABA controls the BAM1 and AMY3 activity in *Arabidopsis* leaves via AREB/ABF-SnRK2 kinase signaling pathways during osmotic stress, certifying carbon resource for continuous plant growth in a fluctuating environment (Thalmann et al. 2016). Recently, heterologous expression of increased drought tolerance in strawberry by regulating the ABA content, stomatal closure and by increasing the activites of antioxidant enzymes in transgenic plants (Li et al. 2020).

3.2 Extreme Temperature

According to current climate predictions, in the next 50–100 years, temperatures might increase from 3 to 5 °C, and crop plants may encounter increased weather disasters that have severe consequences (Dolzblasz and Dołzbłasz 2018; Teixeira et al. 2013). The temperature could have overwhelming effects on the metabolism of plants, uncoupling physiological processes, as well as distracting cellular homeostasis (Wahid 2007; Raza et al. 2020). Stress-induced cellular changes to result in an increased buildup of toxic cellular complexes similar to ROS that leads to extreme aldehyde accumulation (Singh et al. 2013; Stiti et al. 2011). In reply to heat, cold, desiccation, and dehydration, aldehydes produced and intermediates in various fundamental metabolic pathways. Aldehydes may destroy proteins and nucleic acid functions by reacting with them which leads to cell death. *ALDH* enzymes can be used to remove aldehydes (Zhao et al. 2017). In *Arabidopsis*, *ALDH* genes stress-related members have been explored (Stiti et al. 2011). A family of enzymes, aldehyde dehydrogenase (ALDH), takes part in the metabolism of plants and is involved in aldehyde homeostasis via eliminating noxious aldehydes. Under the heat stress conditions, ALDH3I1 and ALDH7B4 are strongly induced in *Arabidopsis*. Increased ALDH7B4 expression also detected in heat mutual with salinity, wounding, and dehydration. Therefore, *ALDH* genes play a vital part in defending *Arabidopsis* plants from increased temperature harm as well as shared stressed conditions (Zhao et al. 2017).

A study has discovered, mild rise in environmental temperature encourages the *Arabidopsis* seedlings development through producing auxin plant hormone. A high temperature upgrades the fast accretion of TIR1 auxin co-receptor dependent of HSP90 (chaperone). SGT1 co-chaperone and HSP90 network with TIR, TIR1, i.e. HSP90 client. HSP90 activity affects the TIR1 degradation as well as stimulatingly faults in auxin-mediated plant growth progressions at low and high temperatures. So, SGT1 and HSP90 integrate auxin signaling and temperature to modify plant growth in changing temperature (Wang et al. 2016b).

In *Arabidopsis*, putative intracellular auxin carrier PIN-LIKES (PILS) is located at the endoplasmic reticulum (ER) and transports intracellular auxin into the ER lumen. PILS proteins like PILS5, PILS3, and PILS2 lowers the auxin signaling via sequestering the auxin to ER. Thus, PILS proteins join in environmental cues to induce auxin signaling minima. A temperature-sensitive PILS6 regulates the auxin nuclear availability, contributing to the rise of root growth and nuclear auxin signaling (Feraru et al. 2019). *Arabidopsis* has 21 HSF (von Koskull-Döring et al. 2007), *Arabidopsis* HSF pathway activation is complex and numerous downstream TFs tangled (Schramm et al. 2008). In response to elevated temperature, transcriptome cause dynamic and rapid H2A.Z nucleosome activation at the target sites and dependent on the HSFA1 clade of *Arabidopsis* HSFs. Due to transcriptional cascade different downstream stress-responsive TFs activates, triggering transcriptome alterations to elevated temperature (Cortijo et al. 2017). Genome-wide association mapping has discovered that developmental stage-specific 4-QTLs (quantitative trait loci) are intensely related through heat response in *Arabidopsis* (Bac-Molenaar et al. 2015).

Arabidopsis thaliana has shown increased freezing tolerance named as cold acclimation to low temperature. CBF (C-repeat binding factor) pathways has a key role in freezing tolerance in *Arabidopsis* (Thomashow 2010). In a recent study, it has found that when shifts *Arabidopsis* plants from a warm temperature to cold then CBF1, CBF2, and CBF3 induced within minutes also called DREB1b, DREB1c, and DREB1a, respectively. These encode AP2/ERF TFs which identify C-repeat (CRT)/dehydration-responsive element (DRE) DNA regulatory element available in the CBF-targeted gene promoters. Around a hundred cold-regulated genes (COR) expressed from any CBF1, CBF2, or CBF3 constitutive overexpression results in an increased freezing tolerance referred to as CBF regulon (Park et al. 2015). CBF regulon raises tolerance to freeze while the mechanisms not understood entirely but have genes that encode cryoprotective agents (Hughes et al. 2013), enzymes take part in cryoprotectants generation (low molecular weight) (Kaplan et al. 2007). Vogel et al. (2005) found that ZAT12, a zinc finger TF encoded by *ZAT12*, has genes in shared with CBF2 regulon and ZAT12 constitutive overexpression resulted in 9 or 15 COR genes induction and repression, respectively. Seven COR genes were also a member of CBF2 regulon, from which four induced and three repressed by CBF2 overexpression. So, in *Arabidopsis*, the low-temperature regulatory network is greatly interrelated, and CBF regulon expression comprises wide co-regulation by further TFs (Park et al. 2015; Fig. 2).

3.3 Salinity

Salinization considered a major concern for worldwide agriculture. Salinity stress affects greater than 800 Mha overall land (Munns and Tester 2008). Roots hardly absorb water in the presence of high salt concentration and ions Na^+ and Cl^- stop the growth of the plant through reducing photosynthesis and ruining metabolic processes (Deinlein et al. 2014). In plants, salt response interpretation then the progress of salinity stress tolerance are critical requirements to feed the growing population. Antiporter system SOS pathway covers SOS1, SOS2, SOS3 remove sodium from the cytosol. *A. thaliana* SOS1 (*AtSOS1*) encodes Na^+/H^+ antiporter plasma membrane which plays an important role in Na^+ efflux. An *AtSOS2/AtSOS3* kinase complex involved in the phosphorylation and activation of *AtSOS1*. *AtSOS2* which encodes Ser/Thr protein kinase act together with SOS3 encoding EF-hand Ca^{2+}-binding protein. For salt tolerance, *AtSOS3* behaves as a calcium sensor and activates *AtSOS2*. Thus, the development of SOS1-dependent sodium removal from cytosol stands crucial for upgrading salt stress tolerance in *Arabidopsis* (Sako et al. 2015). Genome-wide modifications in gene expression are necessary for adaptation to environmental pressure (Kim et al. 2015). Histone H4 acetylation, as well as histone H3Ser10 phosphorylation, revealed to increase in high salinity in *Arabidopsis* T87cells and Tobacco BY2 (Sokol et al. 2007). In *Arabidopsis*, histone deacetylase (HDAC) inhibitor cyclo (-L-2-amino-8-hydroxamido-suberoyl-aminoisobutylyl-L-phenylalanyl-D-prolyl-), known Ky-2 increased the resistance to salinity. Centered on the HDAC inhibitory activity and trichostatin A (TSA) structure and chlamydocin, Ky-2 proposed as an inhibitor of HDAC. Ky-2 produced alterations in transcription of different salt-responsive genes and metabolites accumulation like polyamines and proline. Under salinity stress, *Arabidopsis* improved acetylation with Ky-2 and produced *AtSOS1* upregulation leading to reduce Na^+ accumulation (Sako et al. 2015).

According to a recent study, different chemical combinations used to increase plants stress resistance (Savvides et al. 2016). Further research revealed salinity stress resistance enhanced by applying exogenous chemical compounds such as methyl jasmonate, salicylic acid, and strigolactone (Khan et al. 2015; Van Ha et al. 2014). Epigenetic inhibitors like suberoylanilide hydroxamic acid and Ky-2 (Patanun et al. 2017; Sako et al. 2015), and further chemical compounds such as polyamines and nitroprusside (Savvides et al. 2016), furthermore increased salt resistance. In *A. thaliana* and rice, ethanol application improves salinity stress tolerance. Under the salt stress situation, ethanol up-regulated the ROS expression signaling-related genes linked by high salinity resistance. Among them, few encode ROS scavengers plus TFs (*AtZAT10* and *AtZAT12*). When exposed to increased salinity, *A. thaliana* cytosolic ascorbate peroxidases (APX) action improved by ethanol. Henceforth, ethanol application boosted high salinity stress resistance through increasing ROS detoxification, also regulating ROS-related genes (Nguyen et al. 2017). *VrDREB2A* (DREB-binding TF) overexpression activated the various downstream genes expression then increased resistance to high salt with no retardation of plant growth (Chen et al. 2016). A class II WRKY-TF, *TaWRKY93* brought through ABA and NaCl and enhanced salinity by increasing osmotic stress tolerance. *Arabidopsis* has improved salt resistance by ABA-dependent plus independent pathway (Qin et al. 2015). *Arabidopsis lipoxygenase3* (LOX3) is an enzyme that takes part in the production of JA, functions in the response of salt stress. Transcriptional analysis revealed that LOS3 intensely made during salt dealing. LOX3 mutant showed salt hypersensitivity once matched with wild type (Ding et al. 2016).

Osmotic stress occurs in the plant due to drought, and salinity affects the crop yield throughout the world. Plants reply to osmotic stress through alteration of the different cellular processes, including stress-responsive genes activation, membrane transport regulation across vacuolar membranes (tonoplast) and plasma membrane to retain water as well as ionic homeostasis. In response to certain environmental stimulus, Ca^{2+} signature generated and received by different Ca^{2+} sensors; calmodulin (CaM), calcineurin B-like proteins (CBLs), and Ca^{2+}-dependent protein kinases (CDPKs) and Ca^{2+}-binding proteins to start numerous cellular responses (Pandey et al. 2015). CBL and CIPKs (CBL-interacting protein kinases) form a complex signaling network working in various plant stress responses (Luan 2009). In *Arabidopsis*, calcium sensors CBL2 and CBL3 discovered act together with CIPK21 and mark this kinase to the tonoplast. Under salt stress conditions, CIPK21 preferential localization to the vacuolar membrane detected when co-expressed with both CBL2 and CBL3. In *Arabidopsis*, CIPK21 mediates responses to salt pressure through regulating ion as well as water homeostasis across tonoplast (Pandey et al. 2015).

In plants, ERFs (ethylene-responsive factors) cover significant transcription factors family and show major parts in stress responses and developmental practices. *AcERF2* is an ethylene-responsive factor and its overexpression by CaMV35S promoter results in decreased ROS and malondialdehyde (MDA) accumulation, and improved activities of antioxidant enzymes plus swift stomatal closure in *Arabidopsis* osmotic treatment. *Arabidopsis* plants with *ACERF2* overexpression found to be ABA hypersensitive during germination, primary root elongation, seedling formation, and showed a significant osmotic stress resistance (Sun et al. 2018b). Aquaporin (AQP) proteins show an important part in plants towards abiotic stress. While in plants, defined roles of the aquaporin genes during abiotic stress not understood fully. An AQP gene: *ThPIP2;5* of subgroup PIP2 has cloned and characterized by the *Tamarix hispida*. *ThPIP2;5* overexpressing *Arabidopsis* plants exhibited a better seed germination rate then enhanced root growth plus fresh weight gain in osmotic and salt tolerance, showing abiotic stress tolerance improved by *ThPIP2;5*. *ThPIP2;5* overexpression in transgenic *Arabidopsis* plus *Tamarix* plants showed enhanced ROS-scavenging ability, antioxidant enzyme activities and proline contents, lowered MDA and H_2O_2 contents, and decreased electrolyte leakage proportions as equated to control plants in reply to salt plus osmotic stress. These data recommended that *ThPIP2;5* shows a significant part in abiotic stress resistance in transgenic plants via lowering ROS accumulation and growing the antioxidant activities (Wang et al. 2018a). In *A. thaliana*, grapevine *VvABF2* gene overexpression could improve osmotic stress resistance but completely needed for ABA signaling (Liu et al. 2019).

3.4 Waterlogging

In plants, abiotic stresses are sometimes short-lived, and rescue phase next stress removal is critical. Main abiotic stress, waterlogging affect agriculture and plant biodiversity adversely (Yeung et al. 2018; Raza et al. 2020). Unexpected waterlogging upsets the growth of plants then development by reducing oxygen resource (Hinz et al. 2010; Voesenek and Bailey-Serres 2015). Plants developed different morphological also metabolic versions to live in brief phases of thorough or half-submergence and disappear hypoxia (Sairam et al. 2009; Thirunavukkarasu et al. 2013; Yin et al. 2009). In response to waterlogging, oxygen scarcity behaves as a primary signal (Milroy and Bange 2013). Under low-oxygen stress conditions, plant ethanolic fermentation activated and classically linked with waterlogging stress. The first pyruvate converted into acetaldehyde in the presence of pyruvate decarboxylase and then acetaldehyde changes to ethanol by alcohol dehydrogenase. Evidence revealed NADPH oxidase is a principal home of H_2O_2. Ten NADPH (*AtrbohA-J*) oxidase genes have found in *Arabidopsis*. In response to hypoxic stress, AtrbohD and AtrbohF performed functions by regulating various hypoxia-inducible genes expression also Ca^{2+} signaling. Waterlogging stress-induced AtrbohD expression in *Arabidopsis* and AtrbohD mutation directed to clear *Arabidopsis* sensitivity toward waterlogging. Furthermore, waterlogging increased the alcohol dehydrogenase activity (ADH), H_2O_2 accumulation, and ADH1 expression considerably decreased in two mutant lines of AtrbohD. In conclusion, AtrbohD is essential for *Arabidopsis* waterlogging tolerance (Sun et al. 2018a). In transgenic *Arabidopsis*, kiwifruit gene *AdPDC1* overexpression showed enhanced resistance to waterlogging. Three different genes *RESPIRATORY BURST OXIDASE HOMOLOG D*, *SENESCENCE-ASSOCIATED GENE 113*, and *ORESARA1* have functions in a regulatory network comprising ROS burst, ABA, and ethylene. During submergence recovery, this network regulates ROS homeostasis, stomatal aperture, plus chlorophyll degradation. In *A. thaliana*, flooding resistance improved by these signaling networks (Yeung et al. 2018).

According to recent work, *HaHB11* proposed as a biotechnological tool to develop *Arabidopsis* biomass, harvest, and overflowing tolerance. *HaHB11* belongs to sunflower homeodomain-leucine zipper I subfamily of TF and *HaHB11* sunflower gene expression made in flooding. Transgenic *Arabidopsis HaHB11* displayed wider stems, larger rosettes than wild type, and significantly enhanced biomass than control. Transgenic *Arabidopsis HaHB11* had enhanced waterlogging stress tolerance (Cabello et al. 2016). Another sunflower TF, *HaWRKY76* confer tolerance to both submergence and dehydration with a yield penalty. When *Arabidopsis* plants transformed with *HaWRKY76* cDNA with the regulator of the constitutive promoter, transgenic plants presented improved biomass, sucrose content, and seed production than control plants. Furthermore, they showed an enhanced tolerance to waterlogging stress. Thus, *HaWRKY76* behaves as a prospective biotechnological tool that increases crop harvest as well as flood, drought tolerance (Raineri et al. 2015). By combing fine genetic mapping, single nucleotide polymorphism (SNP) haplotyping and whole-genome sequencing identified that *ARN6.1* was responsible for waterlogging stress tolerance by increasing the formation of the adventitious root in cucumber line Zaoer-N. Through multiple pieces of evidence, it has been found that *CsARN6.1* is the utmost possible candidate for ARN6.1 encoding AAA ATPase. In *Arabidopsis*, the ectopic expression of *CsARN6.1* showed lateral root development and better rooting ability associated with waterlogging stress tolerance (Xu et al. 2018).

In eukaryotes, autophagy acts as a highly conserved process involving decomposition or degradation of intracellular toxic constituents induced by endoplasmic stress and accumulation of ROS. Autophagy's role in the stimulation of programmed cell death (PCD) is still under infancy. During waterlogging, ROS contribute to the PCD regulation through autophagy. Waterlogging stimulates the transcription of hypoxia-responsive genes, and RBOH mediated ROS generation (respiratory burst oxidase homolog). Furthermore, autophagy-related gene (ATG), as well as the number of autophagosomes, increased under waterlogging. First, autophagy appeared at the root stele then autophagosomes occurred at other locations in roots. Therefore, during waterlogging autophagy occurred in the roots of *Arabidopsis* has an attenuating effect on PDC in the roots (Guan 2019). To deal with a diversity of stressed situations, different bacterial genes have recognized and transformed to crop plants. From *Pseudomonas veronii-KJ* acdS gene isolated and cloned into *E. coli* for the recombinant 1-aminocyclopropane-1-carboxylate (ACC) deaminase expression. To alleviate the adverse effects of waterlogging and salt, *Arabidopsis* plants transformed with bacterial acdS gene to produce ACC deaminase (Jung et al. 2018). A form of hemoglobin present in Vitreoscilla (Gram-negative bacterium) called Vitreoscilla hemoglobin (VHb) overexpressed in *Arabidopsis* plants and improved resistance to waterlogging stress (Du et al. 2016).

3.5 Heavy Metals Stress

Different abiotic stresses have a vast impression on plant growth as well as crop production. Among these different environmental stresses, heavy metal has found to be the drastic environmental hazard that adversely disturbs plant growth because of its extended soil persistence, also threatens human health by entering the food chain (Hasanuzzaman et al. 2020). Plants have developed a different mechanism to survive against heavy metal stress like trafficking, uptake regulation, and heavy metals chelation to lowers the toxicity (Kim and Kang 2018). Low-molecular weight Cys-rich proteins; metallothionein (MT) are well known because of their part in the heavy metals sequestration like mercury (Mg) and cadmium (Cd). MTs also has a chief role in the regulation of homeostasis of essential metals like copper (Cu) and zinc (Zn). For sequestration of heavy metals, MTs cysteine residues directly or indirectly involved in removing ROS, which may accumulate at the toxic levels during biotic and abiotic stresses (Patankar et al. 2019). In *Arabidopsis* as well rice, expression of 7 and 10 MT genes have been studied in response combined salt and heavy metal stress. *OsMT1a*, *OsMT1b*, *OsMT1c*, *OsMT1g*, and *OsMT2a* enhanced in several heavy metals. Six *Arabidopsis* MTs increased Zn as well as Cu tolerance in Zn- and Cu-sensitive yeast mutants, and transgenic plants which overexpress some genes including *AtMT2a*, *AtMT3*, *BjMT2*, or *PsMT* increased Cd, Cu, and Zn resistance (Kim and Kang 2018). *A. thaliana* transgenic plants were established to enhance resistance and accretion of metalloids and heavy metals through *AsPCS1* and *GSH1* overexpression from garlic and baker's yeast. Binding peptides, i.e. GSH and phytochelatins (PCs) take part in ion chelation of heavy metal. Studies have shown that single-gene transgenic lines exhibited greater resistance and stored further arsenic (As) plus Cd when equated with wild type. While dual-gene transformants have shown considerably greater resistance and stored more As and Cd than single-gene transgenic plants. *AsPCS1* and *GSH1* immediate overexpression ensued high overall PCs assembly in transgenic *Arabidopsis* (Guo et al. 2008). Subgroup genes of basic helix-loop-helix (bHLH) IVc: *bHLH34/104/115* are the main regulators of iron (Fe) homeostasis in *Arabidopsis*. In *Arabidopsis*, *bHLH104* found to be tangled in positive regulation of Cd tolerance. *bHLH104* identified as a positive regulator for Cd tolerance recommends that this may be used in new approaches to engineer crops for Cd stress adaptation (Yao et al. 2018).

In plants, stress-associated protein (SAP), family members identified as imparting resistance to numerous environmental stresses. In *A. thaliana*, 14 SAP genes having ANI, A20, and Cys2–His2 zinc finger domains. SAP family member, *AtSAP13* has two AN1 zinc finger domain and Cys2–His2 domain. *AtSAP13* transcripts exhibited up-regulation to ABA, Cd, plus salt-stressed (Dixit et al. 2018). Wheat heavy metal ATPase2 (*TaHMA2*) is involved in the transport of Cd and Zn across the membrane. *TaHMA2* involved in translocation of Cd as well as Zn and tolerance in *Arabidopsis* (Qiao et al. 2018). Glutathione-*S*-transferases (GSTs) enzymes are well known because of their role in xenobiotic compounds detoxifications. Based on their sequence identity, soluble GSTs have classified into numerous classes, i.e., tau, phi, zeta, theta, lambda, dehydroascorbate reductase (DHAR), elongation factor 1 gamma (EF1G), and hydroquinone dehalogenase (TCHQD). Studies have shown that rice tau class *OsGSTU30* has a major part for Cr (VI) detoxification. *A. thaliana* transgenic plants have developed through the transformation of *OsGSTU30* gene isolated from rice variety and transgenic lines overexpressing *OsGSTU30* controls regulatory network resulting in drought and heavy metal stress tolerance (Srivastava et al. 2019).

ATP-binding cassette (ABC) family stands among main protein families found in the living organisms. These transporter proteins contain numerous sub-states comprising ions, lipids, carbohydrates, antibiotics, xenobiotics, drugs, and heavy metals. *A. thaliana* has around 130 ABC proteins, but substrates specificities, as well as precise functions of the most transporters, are still in doubt. Certain *Arabidopsis* ABC transporters recently discovered are involved in the detoxification processes plus plant growth and development. *AtATM3*, an ABC transporter of *Arabidopsis*, is the mitochondrial protein that takes part in Fe–S clusters biogenesis and plant homeostasis. Recently, it was found that *AtATM3* gene expression was up-regulated in the *Arabidopsis* roots when treated with Cd and lead (Pb), indicating that this gene involved in heavy metals tolerance (Kim et al. 2006).

4 Regulation of Abiotic Stress-Responsive Transcription Factors in *A. Thaliana*

The gene pool is regulated and controlled by TFs that are chief regulatory factors. A single TF controls various target genes expression via attachment of TF to the specific *cis*-acting component inside the promoter site of the corresponding gene of interest. The transcription regulation system of such category is termed as Regulon. Some of these important TFs respond to drought, low temperature, and high salinity stress, including ERF, MYC, MYB, bZIP, WRKY, and NAC TFs (Fig. 2). In *Arabidopsis,* various such main regulons were identified that had a vigorous response toward abiotic stress. DREB1 or CBF along with DREB2 regulons plays a function and an important part in ABA-independent gene expression. On the other hand, the ABRE (ABA response element) binding protein or the ABF (ABRE-binding factor) regulon have functionality over the gene expression of the ABA-dependent gene expression (Fig. 2). Besides the accumulation of these foremost pathways, various additional regulons are also playing a part in gene expression of the abiotic stress response. Those regulons are; MYC/MYB and NAC regulons (Chinnusamy et al. 2004; Nakashima et al. 2009b, 2014; Nuruzzaman et al. 2013). A schematic representation of the cross-talk transcriptional network among *cis*-acting elements and TFs of ABA-independent and -dependent pathways during the perception of abiotic stresses is shown in Fig. 2.

The DRE (dehydrator response element) comprises of fundamental sequence, i.e. A/GCCGAC which was recognized to be a *cis*-acting element, an ABA-independent in *Arabidopsis*; that is essential for the controlling drought stress reactions, great saline conditions, and cold-related stressors (Yamaguchi-Shinozaki and Shinozaki 1994). Related *cis* substitute elements are also part of cold-inducible genes which are known as CRT (C-repeat), and short-temperature response element; and they both contain DRE as their essential design (Baker et al. 1994; Yamaguchi-Shinozaki and Shinozaki 1994). The identification of CRTs and DRE was made in *Arabidopsis* and grasses having various promotor sites of stress-inducing genes. The cDNA of *Arabidopsis* which encodes the CBF or DREB1 and DREB2 were sequestered with the help of yeast one hybridizing screening (Stockinger et al. 1997; Liu et al. 1998). The mentioned proteins comprise of motifs of AP2 (APETALA2)/ERF (ethylene response element-binding factor). They are precisely attached to the CRT/DRE sequence; thus they activate the gene transcription determined by CRT/DRE sequence present in *A. thaliana*. The AP2/ERF motif is explicit to plants beside it is utilized as a DNA-binding domain. There are six DREB1/CBF genes along with eight DREB2 genes in the genome of *A. thaliana* (Sakuma et al. 2002). Moreover, the other genes present are DREB1A, DREB1B, and DREB1C. These genes are promptly and ephemerally persuaded through cold, whereas high saline stress and dehydration dents have much influence over them (Liu et al. 1998; Fig. 2).

Contrastingly, on the other hand, the genes DREB2A and DREB2B are prompted over dehydration and other saline conditions rather than cold temperature conditions (Nakashima et al. 2000; Fig. 2). The transcription level for other genes, i.e. three DREB1 and six DREB2 were found truncated for such stress circumstances (Sakuma et al. 2002). The main TFs of *Arabidopsis* required for inducing cold gene are DREB1A—DREB1B—DREB1C, whereas the main TFs for inducing high drought and saline expressing genes are DREB2A—DREB2B. Some researchers have signified a cross-link under osmosis stress conditions amid the pathways of DREB1/CBF with that of DREB2 as the other genes DREB1F/DDF1, DREB1D/CBF4 and DREB1E/DDF2 have inadequate prompt effect under osmotic stress (Novillo et al. 2004; Yamaguchi-Shinozaki and Shinozaki 2005; Fujita et al. 2011; Medina et al. 2011; Akhtar et al. 2012). The overexpression of DREB1A/CBF3 or the DREB1B/CBF1 genes under the influence of CaMV (cauliflower mosaic virus) 35S promoter in transgenic *Arabidopsis* exhibited that they have are high tolerant under high-saline conditions and freezing temperatures (Nakashima et al. 2000; Kitashiba et al. 2004; Jaglo-Ottosen et al. 1998; Liu et al. 1998;). This phenomenon suggested that the target sites for DREBs/CBFs are numerous gene sites. *Arabidopsis* protoplasts induced DREB2A's domain analysis exposed that it is center region has a negative regulatory domain and by deleting that region, DREB2A becomes DREB2A-CA, i.e. constitutively active. Several phenotype-related changes were made in the transgenic *Arabidopsis* due to DREB2A-CA overexpression, which includes, drought endurance, dwarfism, and many stresses inducing downstream genes up-regulation (Qin et al. 2008). In addition to these, other changes that were induced due to overexpressing of DREB2A constitutively active includes; expression of HS (Heat shock)-related genes along with enriched heat endurance and tolerance in transgenic plants as well as transgenic *Arabidopsis* (Sakuma et al. 2006b; Lim et al. 2007; Yoshida et al. 2011).

The cDNA of *Arabidopsis* was isolated with the help of yeast one-hybrid system which encodes bZIP TFs and called as AREB or ABFs (Jakoby et al. 2002; Yamaguchi-Shinozaki and Shinozaki 2006). There are 75 discrete bZIP TFs in the genome of *Arabidopsis*, apart from that thirteen members having conserved domains are categorized as sub-family of AREB/ABFs (Jakoby et al. 2002; Nakashima and Yamaguchi-Shinozaki 2009). The *Arabidopsis* genes that express themselves in vegetative issues instead of seeds during saline and dehydrated conditions are ABF3, AREB1/ABF2, and AREB2/ABF4 (Fujita et al. 2005). It was reported that the overexpression of AREB1 active form in transgenic *Arabidopsis* exhibited increased drought endurance and hyper-sensitivity of ABA. Moreover, up-regulation of 8 genes also occurred that signify two groups. They were ABA dehydration stress induces regulation genes and the LEA class of genes along with histones H1 (linker) and AAA ATPase (Fujita et al. 2005; Roychoudhury and Paul 2012; Barbosa et al. 2013).

Various protein kinases in *Arabidopsis* were identified, i.e. type 2 SNF1 (SnRK2 type), OSTI/SRK2E as ABA activates protein kinase (Nakashima et al. 2009a). They remained too revealed toward facilitating the regulation of stomata opening and to purpose up-stream of the ABA response gene expression. Hyper-osmosis activated the nine SnRK2 genes, and ABA in *Arabidopsis* triggers five *SnRK2* genes established that ABA activated SnRK2 protein kinase; this also includes *SnRK2-2/SRK2D*, *SnRK2-3/SRK21*, and *SnRK2-6/SRK2E*, phosphoryl the AREB1 polypeptide (Furihata et al. 2006; Yoshida et al. 2006). In recent times, it was established that two protein kinases, *SnRK2-2/SRK2D* plus *SnRK2-3/SRK2I*, regulate the reactions of ABA during germination of a seed, growth of the seedling, and dormancy in *A. thaliana*. The writers recommended that the contrivance of *SnRK2-2* plus *SnRK2-3* exploit during ABA indicating includes the initiation of ABRE-derived gene expression from side to side phosphor-relations of AREB/ABFs (Fujii et al. 2007; Feng et al. 2014;).

Some of the added pathways in *Arabidopsis* for gene expression are there whose regulation via NAC (*N*-acetylcysteine) TFs in the dehydration stress occurs (Fig. 2). There are two diverse *Arabidopsis cis*-acting essentials which are crucial for the introduction of genetic factor manifestation by desiccation in ERD1 (Early Response to Dehydration1) this translates a Clp pro-tease regulation subunit, i.e. ClpD (Simpson et al. 2003). There are three coding DNA that encodes the NAC-type protein, i.e. ANAC19 (AT1G52890)–ANAC055 (ATNAC3–AT3G15500) besides ANAC072 (RD26–AT4G27410). They were sequestered utilizing yeast 1 hybrid selection then were established to attach with one of the *cis* components of ERD1 promoter (Tran et al. 2004). NAC type protein is specific TFs for each plant, and they exhibit different relative developments in plants for both responses, i.e. biotic and abiotic. The first NAC protein in *Arabidopsis* encoded by cDNA was stated to be RD26 (responsive to dehydration 26) gene (Fujita et al. 2004). Taking, for instance, drought, ABA, high saline conditions and cold are the inducing agents for the expression of *OsNAC6* (Nakashima et al. 2007). There was a great resemblance among the sequences of *OsNAC6* with that of stress response NAC proteins (ANAC019, ANAC055, and ANAC072 (RD26) of *Arabidopsis*. It was established that the SNAC gene group that are NAC-type stress-responsive TFs plays an important role in enduring the stressors related to the environment, i.e. drought (Nakashima et al. 2012).

TFs related to stress existed among the communal stress reaction genes, together with five WRKY associates, three NAM, four MYB, three HSF, and two AP2 with the SCL13 TF. WRKY11 inter-relates with that of calmodulin and turned out to be an adverse manager of basal resistance in *Arabidopsis* (Journot-Catalino et al. 2006). The function of light signaling was to be performed by SCL13 (Torres-Galea et al. 2006). The WRKYs had an effect against the necro-trophic pathogens and not against the bio-trophic ones. The damage caused by necrotrophic pathogens is of physical level influenced by abiotic factors. There is very less evidence existing for other TFs even though the iso-forms of NAM, AP2, MYB, and HSF that are not part of N12 were previously found to have a link with pathways related to stress. The mutual stress reaction genes characterize constituents of cross-link amongst abiotic plus biotic stress reaction contrivances by recognizing genes, such as TFs WRKY, NHL3, in addition to PUB17. Certainly, the *Arabidopsis* transformed BOS1 unveiled conceded confrontation against Botrytis cinerea pathogen as well as abridged forbearance to scarcity, elevated saline conditions plus stress caused by oxidation (Luo et al. 2010; Mengiste et al. 2003).

Post-transcriptional instruction is additionally equal to gene expression intonation that is exemplified by four assemblies of progressions: mRNA (pre-messenger), capping, splicing, and poly-adenylating, i.e. (processing), nuclei-cytoplasmic transferring of mRNA, turn-over, and constancy of mRNA as well as translation of mRNA

(Floris et al. 2009). Substitute intertwining is extensively identified to normalize plant gene expression imperiled to different squat and elevated temperatures (Zhao et al. 2015). For instance, in *A. thaliana*, a pre-mRNA splicing factor encoded gene, i.e. STA1 (STABILIZED1) is of paramount value in the low temperature (cold) stressful scenarios (Lee et al. 2006).

Ever since the initial twentieth century, numerous news has associated that the small RNAs are liked with the abiotic stress response. This shows that post-transcriptional instruction of gene expression has a significant value among these occurrences (Barciszewska-Pacak et al. 2015; Pegler et al. 2019). Small RNAs comprising of 20–25 nucleotides are administered as of non-coding dsRNA predecessors by RNA-seq of DCL (DICER-LIKE) family. This also facilitates sequences of gene muzzling contrivances. Among the mechanisms is the one who slices messenger RNAs or else averts the translation via facilitation of micro-RNAs having 21 nucleotides. The role of the micro-RNAs in regard to stress responses was guessed with the discovery of regulation of stress by micro-RNA levels along with stress linked genes identification. Functional-level investigations have established that numerous micro RNAs of plants are found to have a very essential part in resistance of plants against abiotic stressors (Sunkar and Zhu 2004; Sunkar et al. 2006; Lin et al. 2008b; Yan et al. 2016; Pegler et al. 2019;).

At the post-translation stage of phosphorylation, the sumoylation and ubiquitination of proteins are methods that are essential for the variation of plant reaction toward the abiotic stressors. Phosphorylation, as well as de-phosphorylation, has paramount roles against stress due to abiotic conditions. Numerous signal transduction cataracts molded by MAPKs and SnRKs are triggered when the water deficiency and osmosis related stress via specific phosphorylation of remains (Hrabak et al. 2003; Zhu 2016). Amongst them, SnRK2 proteins are known to have a role in ABA-dependent reactions against water discrepancy such as the closing of stomata (Yoshida et al. 2006).

The drought tolerance can be enhanced (due to increased closure of stomata because of ABA) by the XERICO gene up-regulation which encodes the H_2 type of zinc finger as well as an E3 ubiquitin ligase (Ko et al. 2006). XERICO controls the ABA levels by augmenting the transcriptional state of the *AtNCED3* (ABA biosynthetic gene). The conclusions specify that the protein dilapidation facilitated by the ubiquitin/proteasome pathway has an important part in ABA homeostatic and reaction (Raab et al. 2009; Kang et al. 2011). A simulation was also described to contribute to reactions to phosphate malnourishment toward the tolerance from high and low temperatures. A rise in t intensities of SUMO-protein conjugations could also be noticed in H_2O destitute plants (Kim et al. 2017a; Mishra et al. 2018; Tomanov et al. 2018).

Likewise, it is predictable that numerous regions, i.e. post-transcription parameter connecting alterations of proteins, deprivation of proteins, and breakdown of RNA would appear (Prall et al. 2019). The example of the application of microRNA against stress resistance importance could be taken into account. The detection of micro RNAs that plays a part in regulated stress responses besides discerning the prospective usage of such micro RNAs to restrain or even rise the stress resistance in plants is said to be a wide field of research (Barciszewska-Pacak et al. 2015). Sunkar et al. (2006) have spawned transgenic *A. thaliana* plants as an example; having over-expressed miR398 resistant type of a plastic Cu/Zn-SOD; CSD2 (Copper zinc Super Oxide Dismutase), in addition, they have established that transgenic plants gather additional CSD2 messenger RNA as compared to plants. Moreover, it also overexpresses a consistent CSD2 and is subsequently much more forbearing against elevated light, additional oxidative-type stress, and different heavy metals. The findings obtained from the experiments proposed that to understand stress tolerance manipulation in the plants, we must have expanded knowledge about post-translation-level gene regulation. This would also suggest an upgraded approach to wangle crop plants with boosted stress forbearance. The intensive activities of the transcription, post-transcription, and post-translation procedures have ensured those temporal and spatial applicable arrangements of downstream gene expression. Eventually, determination associated with the proteome and transcriptome of stress exposed plants to shift on adaptive reaction has also been ensured. The detailed comprehension of interaction between these three regulation mechanisms is critical for the comprehension of the molecular procedures prevailing plant variation to the atmosphere along with the development of plant against stress forbearance.

5 Conclusion and Future Perspectives

Due to the number of collective databases for *A. thaliana* obtained after dissected plant reactions toward numerous stress conditions that are abiotic nature, the combined genome research study has become possible. Regulation of many plant genes that are responsive to stressors due to abiotic factors, i.e. high saline conditions, high temperature, dehydrated conditions, and cold temperature, etc., is accomplished. Those genes are regarded to be essential for the tolerance of stress due to the regulation of these stress-related genes. Monocot rice plant and Dicot *Arabidopsis* plant are said to have mutual mechanisms for gene expression regulation. This fact was brought to light after several biological, genetic, and molecular studies. In the gene expression regulation due to abiotic stressors, the important role is played by TFs. The importance of TFs in stress-tolerant genetic engineering are due to the fact that, when a single TF is overexpressed; this can direct up-regulation of extensive range related to stress-responsive genes monetized by TFs. There are some major TFs (DREB/CBF, AP2/ERF, bZIP, WRKY, NAC, etc.) that have been extensively studied in a model plant for multiple abiotic stresses. However, there is a need to work on other TFs to explore their mechanisms in abiotic stress tolerance.

Various scientists have used innovative genetic engineering approaches over the last decades in some major crops and have exhibited auspicious results for comprehending molecular contrivances of tolerance related to abiotic stress. The verdicts obtained from such researches have highlighted the fact that more studies should be carried out focusing on the physiology, metabolic, and molecular characteristics of stress forbearance to assist the crop development having an innate ability to endure abiotic related stress. This, in return, would not only support the stabilization of crop production but also subsidize in nutrition and food security issues faced by under develop countries and tropically semi-arid regions. The impeding studies should focus on the dichotomization of detailed stress-related components and mark out different elements involved in the stress signaling transduction pathway. The study should be continued, and the advantages should be obtained from the present, and forthcoming techniques and methodologies in elementary and pragmatic research must propose valued gears. These tools would help out in the evaluation of transcriptome analysis, reflecting a complex biological system alongside numerous concurrent environment-related conditions. However, Fig. 3 documented the importance of basic research in a model plant and the results are translating into other crop plants which really helps out other major crops to grow in the field to feed millions, worldwide, using modern breeding and biotechnological platforms, under stressful environment.

Arabidopsis thaliana research has been furnishing a useful understanding of all aspects of modern molecular biology. Remarkable advancement has been attained in basic research efforts, like cloning of disease tolerance genes, cloning of abiotic stress tolerance genes, hormone biogenesis, and flowering. Likewise, the functions of thousands of genes have been evaluated in *Arabidopsis*, and these findings started to be applied to other crops. *A. thaliana* is supposed to continue to be the most fabulous studied model plant, and its resources will continue to enhance both in numbers and refinement.

As a final point, the cooperative and accommodating exertions of physiologists, plant molecular biologists, and breeders are mandatory for the engendering of stress forbearing reap plants via genetic engineering. It is anticipated that impending these communal determinations and outcomes of combined lessons will completely subsidize to ecological food assembly in the realm and will aid to preclude international-scale conservational mutilation that affects the starting of abiotic stress.

Acknowledgements We are grateful to all the researchers whose contributions have been cited in this chapter, which helped us to prepare this chapter.

References

Abe H, Urao T, Ito T, Seki M, Shinozaki K, Yamaguchi-Shinozaki K (2003) *Arabidopsis AtMYC2* (bHLH) and *AtMYB2* (MYB) function as transcriptional activators in abscisic acid signaling. Plant Cell 15:63–78

Abogadallah GM, Nada RM, Malinowski R, Quick P (2011) Overexpression of HARDY, an AP2/ERF gene from *Arabidopsis*, improves drought and salt tolerance by reducing transpiration and sodium uptake in transgenic *Trifolium alexandrinum* L. Planta 233:1265–1276

Agarwal M, Hao Y, Kapoor A, Dong C-H, Fujii H, Zheng X, Zhu J-K (2006) A R2R3 type MYB transcription factor is involved in the cold regulation of CBF genes and in acquired freezing tolerance. J Biol Chem 281:37636–37645

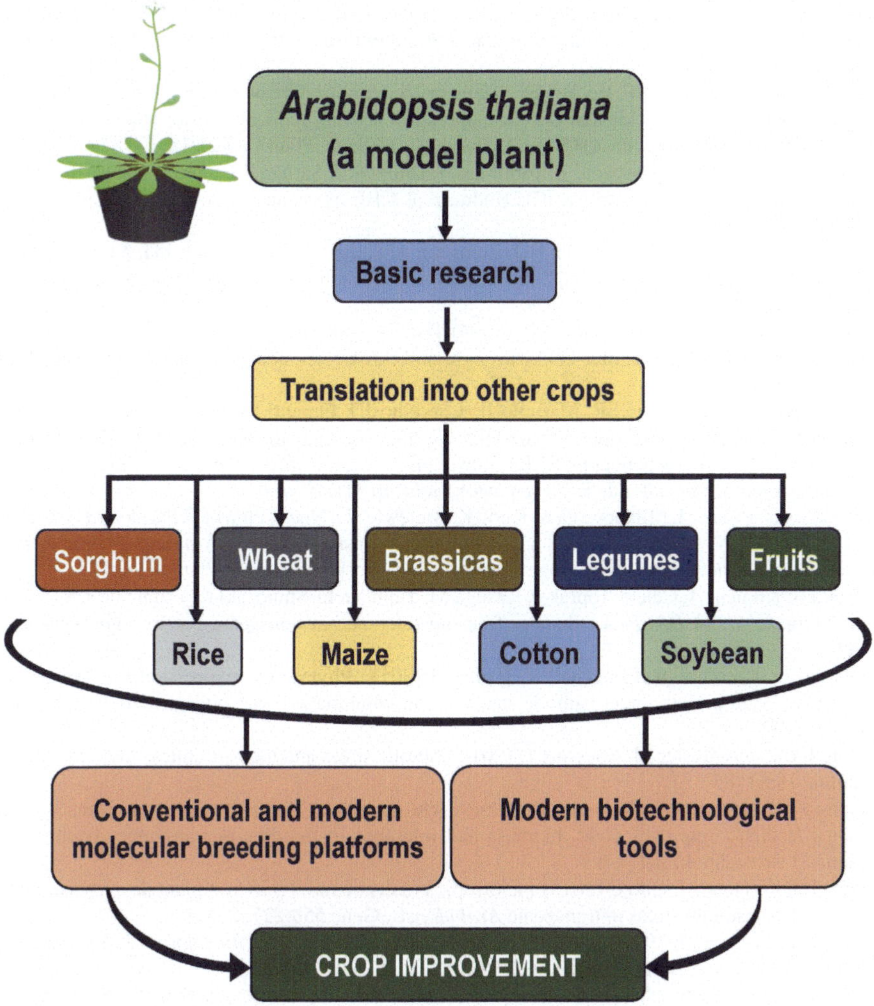

Fig. 3 A simplest model showing the importance of basic research in a model plant

Akhtar M, Jaiswal A, Taj G, Jaiswal J, Qureshi M, Singh N (2012) DREB1/CBF transcription factors: their structure, function and role in abiotic stress tolerance in plants. J Genet 91:385–395

Apel K, Hirt H (2004) Reactive oxygen species: metabolism, oxidative stress, and signal transduction. Annu Rev Plant Biol 55:373–399

Assunção AG, Herrero E, Lin YF, Huettel B, Talukdar S, Smaczniak C, Immink RG, Van Eldik M, Fiers M, Schat H, Aarts MG (2010) *Arabidopsis thaliana* transcription factors bZIP19 and bZIP23 regulate the adaptation to zinc deficiency. Proc Natl Acad Sci USA 107:10296–10301

Atkinson NJ, Urwin PE (2012) The interaction of plant biotic and abiotic stresses: from genes to the field. J Exp Bot 63:3523–3543

Babitha K, Ramu S, Pruthvi V, Mahesh P, Nataraja KN, Udayakumar M (2013) Co-expression of *AtbHLH17* and *AtWRKY28* confers resistance to abiotic stress in *Arabidopsis*. Transgenic Res 22:327–341

Bac-Molenaar JA, Fradin EF, Becker FF, Rienstra JA, van der Schoot J, Vreugdenhil D, Keurentjes JJ (2015) Genome-wide association mapping of fertility reduction upon heat stress reveals developmental stage-specific QTLs in *Arabidopsis thaliana*. Plant Cell 27:1857–1874

Badawi M, Reddy YV, Agharbaoui Z, Tominaga Y, Danyluk J, Sarhan F, Houde M (2008) Structure and functional analysis of wheat ICE (inducer of CBF expression) genes. Plant Cell Physiol 49:1237–1249

Bahieldin A, Atef A, Edris S, Gadalla NO, Ramadan AM, Hassan SM, Al Attas SG, Al-Kordy MA, Al-Hajar AS, Sabir JS, Nasr ME (2018) Multifunctional activities of *ERF109* as affected by salt stress in *Arabidopsis*. Sci Rep 8:6403

Baker SS, Wilhelm KS, Thomashow MF (1994) The 5′-region of *Arabidopsis thaliana* cor15a has *cis*-acting elements that confer cold-, drought-and ABA-regulated gene expression. Plant Mol Biol 24:701–713

Barbosa EG, Leite JP, Marin SR, Marinho JP, Carvalho JD, Fuganti-Pagliarini R, Farias JR, Neumaier N, Marcelino-Guimarães FC, de Oliveira MC, Yamaguchi-Shinozaki K (2013) Overexpression of the ABA-dependent AREB1 transcription factor from *Arabidopsis thaliana* improves soybean tolerance to water deficit. Plant Mol Biol Rep 31:719–730

Barciszewska-Pacak M, Milanowska K, Knop K, Bielewicz D, Nuc P, Plewka P, Pacak AM, Vazquez F, Karlowski W, Jarmolowski A, Szweykowska-Kulinska Z (2015) *Arabidopsis* microRNA expression regulation in a wide range of abiotic stress responses. Front Plant Sci 6:410

Behnam B, Kikuchi A, Celebi-Toprak F, Kasuga M, Yamaguchi-Shinozaki K, Watanabe KN (2007) Arabidopsis *rd29A:DREB1A* enhances freezing tolerance in transgenic potato. Plant Cell Rep 26:1275–1282

Belgaroui N, Lacombe B, Rouached H, Hanin M (2018) Phytase overexpression in *Arabidopsis* improves plant growth under osmotic stress and in combination with phosphate deficiency. Sci Rep 8:1137

Bowne J, Bacic A, Tester M, Roessner U (2018) Abiotic stress and metabolomics. Annu Plant Rev Online 15:61–85

Cabello JV, Giacomelli JI, Piattoni CV, Iglesias AA, Chan RL (2016) The sunflower transcription factor *HaHB11* improves yield, biomass and tolerance to flooding in transgenic *Arabidopsis* plants. J Biotechnol 222:73–83

Cai H, Tian S, Dong H, Guo C (2015) Pleiotropic effects of *TaMYB3R1* on plant development and response to osmotic stress in transgenic *Arabidopsis*. Gene 558:227–234

Cao B, Shu L, Li A (2019) Functional characterization of *LkERF-B2* for improved salt tolerance ability in *Arabidopsis thaliana*. Biotech 9: 263

Cao Y, Xiang X, Geng M, You Q, Huang X (2017) Effect of *HbDHN1* and *HbDHN2* genes on abiotic stress responses in *Arabidopsis*. Front Plant Sci 8:470

Chen H, Lai Z, Shi J, Xiao Y, Chen Z, Xu X (2010) Roles of *Arabidopsis* WRKY18, WRKY40 and WRKY60 transcription factors in plant responses to abscisic acid and abiotic stress. BMC Plant Biol 10:281

Chen H, Liu L, Wang L, Wang S, Cheng X (2016) *VrDREB2A*, a DREB-binding transcription factor from *Vigna radiata*, increased drought and high-salt tolerance in transgenic *Arabidopsis thaliana*. J Plant Res 129:263–273

Chen L-T, Wu K (2010) Role of histone deacetylases HDA6 and HDA19 in ABA and abiotic stress response. Plant Signal Behav 5:1318–1320

Chen M, Xu Z, Xia L, Li L, Cheng X, Dong J, Wang Q, Ma Y (2008) Cold-induced modulation and functional analyses of the DRE-binding transcription factor gene, *GmDREB3*, in soybean (*Glycine max* L.). J Exp Bot 60:121–135

Chen T, Yang Q, Zhang X, Ding W, Gruber M (2012) An alfalfa (*Medicago sativa* L.) ethylene response factor gene, *MsERF11*, enhances salt tolerance in transgenic *Arabidopsis*. Plant Cell Rep 31:1737–1746

Cheng L, Li X, Huang X, Ma T, Liang Y, Ma X, Peng X, Jia J, Chen S, Chen Y, Deng B (2013a) Overexpression of sheepgrass R1-MYB transcription factor *LcMYB1* confers salt tolerance in transgenic *Arabidopsis*. Plant Physiol Biochem 70:252–260

Cheng M-C, Liao P-M, Kuo W-W, Lin T-P (2013b) The Arabidopsis ETHYLENE RESPONSE FACTOR1 regulates abiotic stress-responsive gene expression by binding to different *cis*-acting elements in response to different stress signals. Plant Physiol 162:1566–1582

Cheng M, Hsieh E, Chen J, Chen Y, Lin T (2012) Arabidopsis RGLG2, functioning as a RING E3 ligase, interacts with *AtERF53* and negatively regulates the plant drought stress response. Plant Physiol 158:363–375

Cheong YH, Chang H-S, Gupta R, Wang X, Zhu T, Luan S (2002) Transcriptional profiling reveals novel interactions between wounding, pathogen, abiotic stress, and hormonal responses in *Arabidopsis*. Plant Physiol 129:661–677

Chinnusamy V, Ohta M, Kanrar S, Lee B-h, Hong X, Agarwal M, Zhu J-K (2003) ICE1: a regulator of cold-induced transcriptome and freezing tolerance in *Arabidopsis*. Genes Dev 17:1043–1054

Chinnusamy V, Schumaker K, Zhu JK (2004) Molecular genetic perspectives on cross-talk and specificity in abiotic stress signalling in plants. J Exp Bot 55:225–236

Choudhury FK, Rivero RM, Blumwald E, Mittler R (2017) Reactive oxygen species, abiotic stress and stress combination. Plant J 90:856–867

Clauw P, Coppens F, De Beuf K, Dhondt S, Van Daele T, Maleux K, Storme V, Clement L, Gonzalez N, Inzé D (2015) Leaf responses to mild drought stress in natural variants of *Arabidopsis*. Plant Physiol 167:800–816

Coolen S, Proietti S, Hickman R, Davila Olivas NH, Huang PP, Van Verk MC, Van Pelt JA, Wittenberg AH, De Vos M, Prins M, Van Loon JJ (2016) Transcriptome dynamics of *Arabidopsis* during sequential biotic and abiotic stresses. Plant J 86:249–267

Cortijo S, Charoensawan V, Brestovitsky A, Buning R, Ravarani C, Rhodes D, van Noort J, Jaeger KE, Wigge PA (2017) Transcriptional regulation of the ambient temperature response by H2A.Z nucleosomes and HSF1 transcription factors in *Arabidopsis*. Mol Plant 10:1258–1273

Craigon DJ, James N, Okyere J, Higgins J, Jotham J, May S (2004) NASCArrays: a repository for microarray data generated by NASC's transcriptomics service. Nucleic Acids Res 32:D575–D577

Davila Olivas NH, Kruijer W, Gort G, Wijnen CL, van Loon JJ, Dicke M (2017) Genome-wide association analysis reveals distinct genetic architectures for single and combined stress responses in Arabidopsis thaliana. New Phytol 213:838–851

Deinlein U, Stephan AB, Horie T, Luo W, Xu G, Schroeder JI (2014) Plant salt-tolerance mechanisms. Trends Plant Sci 19:371–379

Ding H, Lai J, Wu Q, Zhang S, Chen L, Dai YS, Wang C, Du J, Xiao S, Yang C (2016) Jasmonate complements the function of *Arabidopsis* lipoxygenase3 in salinity stress response. Plant Sci 244:1–7

Ding Z, Li S, An X, Liu X, Qin H, Wang D (2009) Transgenic expression of MYB15 confers enhanced sensitivity to abscisic acid and improved drought tolerance in *Arabidopsis thaliana*. J Genet Genomics 36:17–29

Dixit A, Tomar P, Vaine E, Abdullah H, Hazen S, Dhankher OP (2018) A stress-associated protein, AtSAP13, from *Arabidopsis thaliana* provides tolerance to multiple abiotic stresses. Plant, Cell Environ 41:1171–1185

Dolzblasz A, Dołzbłasz S (2018) Arabidopsis high temperature stress research. Acta Soc Bot Pol 87(3):3594

Du H, Shen X, Huang Y, Huang M, Zhang Z (2016) Overexpression of Vitreoscilla hemoglobin increases waterlogging tolerance in *Arabidopsis* and maize. BMC Plant Biol 16:35

Fang RJ, Li AQ, Tian RN, Zhang WJ, Zou AL, Wu FY, Liao YH, Wang XM, Pang YJ, Yang RW, Lu GH (2019) Heterologous overexpression of *Lithospermum erythrorhizon LeERF-1* gene increases drought and pathogen resistance in Arabidopsis. Acta Physiol Plant 41:24

Feng CZ, Chen Y, Wang C, Kong YH, Wu WH, Chen YF (2014) Arabidopsis RAV1 transcription factor, phosphorylated by SnRK2 kinases, regulates the expression of ABI3, ABI4, and ABI5 during seed germination and early seedling development. Plant J 80:654–668

Feraru E, Feraru MI, Barbez E, Waidmann S, Sun L, Gaidora A, Kleine-Vehn J (2019) PILS6 is a temperature-sensitive regulator of nuclear auxin input and organ growth in *Arabidopsis thaliana*. Proc Natl Acad Sci USA 116:3893–3898

Fleta-Soriano E, Munné-Bosch S (2016) Stress memory and the inevitable effects of drought: a physiological perspective. Front Plant Sci 7:143

Floris M, Mahgoub H, Lanet E, Robaglia C, Menand B (2009) Post-transcriptional regulation of gene expression in plants during abiotic stress. Int J Mol Sci 10:3168–3185

Fujii H, Verslues PE, Zhu J-K (2007) Identification of two protein kinases required for abscisic acid regulation of seed germination, root growth, and gene expression in *Arabidopsis*. Plant Cell 19:485–494

Fujita M, Fujita Y, Maruyama K, Seki M, Hiratsu K, Ohme-Takagi M, Tran LS, Yamaguchi-Shinozaki K, Shinozaki K (2004) A dehydration-induced NAC protein, RD26, is involved in a novel ABA-dependent stress-signaling pathway. Plant J 39:863–876

Fujita Y, Fujita M, Satoh R, Maruyama K, Parvez MM, Seki M, Hiratsu K, Ohme-Takagi M, Shinozaki K, Yamaguchi-Shinozaki K (2005) AREB1 is a transcription activator of novel ABRE-dependent ABA signaling that enhances drought stress tolerance in *Arabidopsis*. Plant Cell 17:3470–3488

Fujita Y, Fujita M, Shinozaki K, Yamaguchi-Shinozaki K (2011) ABA-mediated transcriptional regulation in response to osmotic stress in plants. J Plant Res 124:509–525F

Furihata T, Maruyama K, Fujita Y, Umezawa T, Yoshida R, Shinozaki K, Yamaguchi-Shinozaki K (2006) Abscisic acid-dependent multisite phosphorylation regulates the activity of a transcription activator AREB1. Proc Natl Acad Sci USA 103:1988–1993

Fursova OV, Pogorelko GV, Tarasov VA (2009) Identification of ICE2, a gene involved in cold acclimation which determines freezing tolerance in *Arabidopsis thaliana*. Gene 429:98–103

Gao F, Zhou J, Deng RY, Zhao HX, Li CL, Chen H, Suzuki T, Park SU, Wu Q (2017) Overexpression of a tartary buckwheat R2R3-MYB transcription factor gene, *FtMYB9*, enhances tolerance to drought and salt stresses in transgenic *Arabidopsis*. J Plant Physiol 214:81–90

Gao SQ, Chen M, Xu ZS, Zhao CP, Li L, Xu HJ, Tang YM, Zhao X, Ma YZ (2011) The soybean *GmbZIP1* transcription factor enhances multiple abiotic stress tolerances in transgenic plants. Plant Mol Biol 75:537–553

Gilmour SJ, Sebolt AM, Salazar MP, Everard JD, Thomashow MF (2000) Overexpression of the *Arabidopsis* CBF3 transcriptional activator mimics multiple biochemical changes associated with cold acclimation. Plant Physiol 124:1854–1865

Gough NR (2002) Science's signal transduction knowledge environment: the connections maps database. Ann NY Acad Sci 971:585–587

Grover A, Kapoor A, Lakshmi OS, Agarwal S, Sahi C, Katiyar-Agarwal S, Agarwal M, Dubey H (2001) Understanding molecular alphabets of the plant abiotic stress responses. Curr Sci 206–216

Guan B (2019) Effect of waterlogging-induced autophagy on programmed cell death in Arabidopsis roots. Front Plant Sci 10:468

Guo J, Dai X, Xu W, Ma M (2008) Overexpressing GSH1 and *AsPCS1* simultaneously increases the tolerance and accumulation of cadmium and arsenic in *Arabidopsis thaliana*. Chemosphere 72:1020–1026

Haake V, Cook D, Riechmann J, Pineda O, Thomashow MF, Zhang JZ (2002) Transcription factor CBF4 is a regulator of drought adaptation in Arabidopsis. Plant Physiol 130:639–648

Hao YJ, Wei W, Song QX, Chen HW, Zhang YQ, Wang F, Zou HF, Lei G, Tian AG, Zhang WK, Ma B (2011) Soybean NAC transcription factors promote abiotic stress tolerance and lateral root formation in transgenic plants. Plant J 68:302–313

Harb A, Krishnan A, Ambavaram MM, Pereira A (2010) Molecular and physiological analysis of drought stress in *Arabidopsis* reveals early responses leading to acclimation in plant growth. Plant Physiol 154:1254–1271

Hasanuzzaman M, Fujita M (2012) Heavy metals in the environment: Current status, toxic effects on plants and possible phytoremediation. In: Anjum NA, Pereira MA, Ahmad I, Duarte AC, Umar

S, Khan NA (eds) Phytotechnologies: remediation of environmental contaminants. CRC Press, Boca Raton, pp 7–73

Hasanuzzaman M, Hossain MA, Silva JAT, Fujita M (2012) Plant responses and tolerance to abiotic oxidative stress: antioxidant defense is a key factor. In: Bandi V, Shanker AK, Shanker C, Mandapaka M (eds) Crop stress and its management: perspectives and strategies. Springer, New York, pp 261–316

Hasanuzzaman M, Nahar K, Fujita M, Ahmad P, Chandna R, Prasad MNV, Ozturk M (2013a) Enhancing plant productivity under salt stress: relevance of poly-omics. In: Ahmad P, Azooz MM, Prasad MNV (eds) Salt stress in plants-omics, signaling and responses. Springer, Berlin, pp 113–156

Hasanuzzaman M, Nahar K, Fujita M (2013b) Plant response to salt stress and role of exogenous protectants to mitigate salt-induced damages. In: Ahmad P, Azooz MM, Prasad MNV (eds) Ecophysiology and responses of plants under salt stress. Springer NY, USA, pp 25–87

Hasanuzzaman M, Borhannuddin Bhuyan MHM, Raza A, Hawrylak-Nowak B, Matraszek-Gawron R, Al Mahmud J, Nahar K, Fujita M, (2020) Selenium in Plants: Boon or Bane?. Environ Exp Bot 104170 https://doi.org/10.1016/j.envexpbot.2020.104170

Hinz M, Wilson IW, Yang J, Buerstenbinder K, Llewellyn D, Dennis ES, Sauter M, Dolferus R (2010) Arabidopsis RAP2.2: an ethylene response transcription factor that is important for hypoxia survival. Plant Physiol 153:757–772

Hirayama T, Shinozaki K (2010) Research on plant abiotic stress responses in the post-genome era: past, present and future. Plant J 61:1041–1052

Hossain MA, Bhattacharjee S, Armin SM, Qian P, Xin W, Li HY, Burritt DJ, Fujita M, Tran LS (2015) Hydrogen peroxide priming modulates abiotic oxidative stress tolerance: insights from ROS detoxification and scavenging. Front Plant Sci 6:420

Hrabak EM, Chan CW, Gribskov M, Harper JF, Choi JH, Halford N, Kudla J, Luan S, Nimmo HG, Sussman MR, Thomas M (2003) The *Arabidopsis* CDPK-SnRK superfamily of protein kinases. Plant Physiol 132:666–680

Hsieh E-J, Cheng M-C, Lin T-P (2013) Functional characterization of an abiotic stress-inducible transcription factor *AtERF53* in *Arabidopsis thaliana*. Plant Mol Biol 82:223–237

Hu X, Xu X, Li C (2018) Ectopic expression of the *LoERF017* transcription factor from *Larix olgensis* Henry enhances salt and osmotic-stress tolerance in *Arabidopsis thaliana*. Plant Biotechnol Rep 12:93–104

Huang Q, Wang Y, Li B, Chang J, Chen M, Li K, Yang G, He G (2015) *TaNAC29*, a NAC transcription factor from wheat, enhances salt and drought tolerance in transgenic Arabidopsis. BMC Plant Biol 15:268

Hughes SL, Schart V, Malcolmson J, Hogarth KA, Martynowicz DM, Tralman-Baker E, Patel SN, Graether SP (2013) The importance of size and disorder in the cryoprotective effects of dehydrins. Plant Physiol 163:1376–1386

Hundertmark M, Hincha DK (2008) LEA (late embryogenesis abundant) proteins and their encoding genes in *Arabidopsis thaliana*. BMC Genom 9:118

Initiative AG (2000) Analysis of the genome sequence of the flowering plant *Arabidopsis thaliana*. Nature 408:796

Inupakutika MA, Sengupta S, Devireddy AR, Azad RK, Mittler R (2016) The evolution of reactive oxygen species metabolism. J Exp Bot 14: erw382

Jaglo-Ottosen KR, Gilmour SJ, Zarka DG, Schabenberger O, Thomashow MF (1998) Arabidopsis CBF1 overexpression induces COR genes and enhances freezing tolerance. Science 280:104–106

Jakoby M, Weisshaar B, Dröge-Laser W, Vicente-Carbajosa J, Tiedemann J, Kroj T, Parcy F (2002) bZIP transcription factors in *Arabidopsis*. Trends Plant Sci 7:106–111

Jensen MK, Kjaersgaard T, Nielsen MM, Galberg P, Petersen K, O'shea C C, Skriver K (2010) The *Arabidopsis thaliana* NAC transcription factor family: structure–function relationships and determinants of ANAC019 stress signalling. Biochem J 426:183–196

Ji X, Liu G, Liu Y, Zheng L, Nie X, Wang Y (2013) The bZIP protein from *Tamarix hispida*, *ThbZIP1*, is ACGT elements binding factor that enhances abiotic stress signaling in transgenic *Arabidopsis*. BMC Plant Biol 13:151

Jiang Y, Deyholos MK (2009) Functional characterization of Arabidopsis NaCl-inducible WRKY25 and WRKY33 transcription factors in abiotic stresses. Plant Mol Biol 69:91–105

Jin X, Xue Y, Wang R, Xu R, Bian L, Zhu B, Han H, Peng R, Yao Q (2013) Transcription factor *OsAP21* gene increases salt/drought tolerance in transgenic *Arabidopsis thaliana*. Mol Biol Rep 40:1743–1752

Journot-Catalino N, Somssich IE, Roby D, Kroj T (2006) The transcription factors WRKY11 and WRKY17 act as negative regulators of basal resistance in *Arabidopsis thaliana*. Plant Cell 18:3289–3302

Jung C, Seo JS, Han SW, Koo YJ, Kim CH, Song SI, Nahm BH, Do Choi Y, Cheong JJ (2008) Overexpression of *AtMYB44* enhances stomatal closure to confer abiotic stress tolerance in transgenic Arabidopsis. Plant Physiol 146:623–635

Jung H, Ali S, Kim JY, Kim W-C (2018) Transgenic *Arabidopsis* expressing acdS gene of *Pseudomonas veronii*-KJ alleviate the adverse effects of salt and water-logging stress. Plant Breed Biotech 6:221–232

Jung J, Won SY, Suh SC, Kim H, Wing R, Jeong Y, Hwang I, Kim M (2007) The barley ERF-type transcription factor *HvRAF* confers enhanced pathogen resistance and salt tolerance in Arabidopsis. Planta 225:575–588

Kang J-y, Choi H-i, Im M-y, Kim SY (2002) Arabidopsis basic leucine zipper proteins that mediate stress-responsive abscisic acid signaling. Plant Cell 14:343–357

Kang M, Fokar M, Abdelmageed H, Allen RD (2011) Arabidopsis SAP5 functions as a positive regulator of stress responses and exhibits E3 ubiquitin ligase activity. Plant Mol Biol 75:451–466

Kaplan F, Kopka J, Sung DY, Zhao W, Popp M, Porat R, Guy CL (2007) Transcript and metabolite profiling during cold acclimation of Arabidopsis reveals an intricate relationship of cold-regulated gene expression with modifications in metabolite content. Plant J 50:967–981

Karaba A, Dixit S, Greco R, Aharoni A, Trijatmiko KR, Marsch-Martinez N, Krishnan A, Nataraja KN, Udayakumar M, Pereira A (2007) Improvement of water use efficiency in rice by expression of *HARDY*, an *Arabidopsis* drought and salt tolerance gene. Proc Natl Acad Sci U S A 104:15270–15275

Khan MIR, Fatma M, Per TS, Anjum NA, Khan NA (2015) Salicylic acid-induced abiotic stress tolerance and underlying mechanisms in plants. Front Plant Sci 6:462

Kim S, Kang JY, Cho DI, Park JH, Kim SY (2004) ABF2, an ABRE-binding bZIP factor, is an essential component of glucose signaling and its overexpression affects multiple stress tolerance. Plant J 40:75–87

Kim D-Y, Bovet L, Kushnir S, Noh EW, Martinoia E, Lee Y (2006) *AtATM3* is involved in heavy metal resistance in Arabidopsis. Plant Physiol 140:922–932

Kim J-M, Sasaki T, Ueda M, Sako K, Seki M (2015) Chromatin changes in response to drought, salinity, heat, and cold stresses in plants. Front Plant Sci 6:114

Kim JY, Song JT, Seo HS (2017a) Post-translational modifications of Arabidopsis E3 SUMO ligase AtSIZ1 are controlled by environmental conditions. FEBS Open Bio 7:1622–1634

Kim MJ, Ruzicka D, Shin R, Schachtman DP (2012) The *Arabidopsis* AP2/ERF transcription factor RAP2.11 modulates plant response to low-potassium conditions. Mol Plant 5:1042–1057

Kim SH, Kim HS, Bahk S, An J, Yoo Y, Kim J-Y, Chung WS (2017b) Phosphorylation of the transcriptional repressor MYB15 by mitogen-activated protein kinase 6 is required for freezing tolerance in *Arabidopsis*. Nucleic Acids Res 45:6613–6627

Kim SY (2006) The role of ABF family bZIP class transcription factors in stress response. Physiol Plant 126:519–527

Kim Y-O, Kang H (2018) Comparative expression analysis of genes encoding metallothioneins in response to heavy metals and abiotic stresses in rice (*Oryza sativa*) and *Arabidopsis thaliana*. Biosci Biotechnol Biochem 82:1656–1665

Kitashiba H, Ishizaka T, Isuzugawa K, Nishimura K, Suzuki T (2004) Expression of a sweet cherry DREB1/CBF ortholog in *Arabidopsis* confers salt and freezing tolerance. J Plant Physiol 161:1171–1176

Ko JH, Yang SH, Han KH (2006) Upregulation of an Arabidopsis RING-H2 gene, XERICO, confers drought tolerance through increased abscisic acid biosynthesis. Plant J 47:343–355

Kranner I, Minibayeva FV, Beckett RP, Seal CE (2010) What is stress? Concepts, definitions and applications in seed science. New Phytol 188:655–673

Krishnaswamy S, Verma S, Rahman MH, Kav NN (2011) Functional characterization of four APETALA2-family genes (RAP2. 6, RAP2. 6L, DREB19 and DREB26) in Arabidopsis. Plant Mol Biol 75:107–127

Lämke J, Bäurle I (2017) Epigenetic and chromatin-based mechanisms in environmental stress adaptation and stress memory in plants. Genom Biol 18:124

Larcher W (2003) Physiological plant ecology: ecophysiology and stress physiology of functional groups, 4th edn. Springer, New York, p 513

Le BH, Wagmaister JA, Kawashima T, Bui AQ, Harada JJ, Goldberg RB (2007) Using genomics to study legume seed development. Plant Physiol 144:562–574

Lee B-h, Kapoor A, Zhu J, Zhu J-K (2006) STABILIZED1, a stress-upregulated nuclear protein, is required for pre-mRNA splicing, mRNA turnover, and stress tolerance in Arabidopsis. Plant Cell 18:1736–1749

Lee S-Y, Hwang EY, Seok H-Y, Tarte VN, Jeong MS, Jang SB, Moon Y-H (2015) Arabidopsis *AtERF71/HRE2* functions as transcriptional activator via *cis*-acting GCC box or DRE/CRT element and is involved in root development through regulation of root cell expansion. Plant Cell Rep 34:223–231

Lenka SK, Singh AK, Muthusamy SK, Smita S, Chinnusamy V, Bansal KC (2019) Heterologous expression of rice RNA-binding glycine-rich (RBG) gene *OsRBGD3* in transgenic *Arabidopsis thaliana* confers cold stress tolerance. Funct Plant Biol 46:482–491

Liao Y, Zou HF, Wei W, Hao YJ, Tian AG, Huang J et al (2008) Soybean GmbZIP44, GmbZIP62 and GmbZIP78 genes function as negative regulator of ABA signaling and confer salt and freezing tolerance in transgenic *Arabidopsis*. Planta 228:225–240

Li B, Fan R, Guo S, Wang P, Zhu X, Fan Y, Chen Y, He K, Kumar A, Shi J, Wang Y (2019a) The Arabidopsis MYB transcription factor, *MYB111* modulates salt responses by regulating flavonoid biosynthesis. Environ Exper Bot: 103807

Li H, Gao Y, Xu H, Dai Y, Deng D, Chen J (2013) *ZmWRKY33*, a WRKY maize transcription factor conferring enhanced salt stress tolerances in Arabidopsis. Plant Growth Regul 70:207–216

Li J, Wang Y, Yu B, Song Q, Liu Y, Chen TH, Li G, Yang X (2018) Ectopic expression of *StCBF1* and *ScCBF1* have different functions in response to freezing and drought stresses in Arabidopsis. Plant Sci 270:221–233

Li J, Yu D, Qanmber G, Lu L, Wang L, Zheng L, Liu Z, Wu H, Liu X, Chen Q, Li F (2019b) *GhKLCR1*, a kinesin light chain-related gene, induces drought-stress sensitivity in Arabidopsis. Sci China Life Sci 62:63–75

Li L, Sheen J (2016) Dynamic and diverse sugar signaling. Curr Opin Plant Biol 33:116–125

Li S, Fu Q, Chen L, Huang W, Yu D (2011) *Arabidopsis thaliana* WRKY25, WRKY26, and WRKY33 coordinate induction of plant thermotolerance. Planta 233:1237–1252

Li S, Fu Q, Huang W, Yu D (2009) Functional analysis of an *Arabidopsis* transcription factor WRKY25 in heat stress. Plant Cell Rep 28:683–693

Li S, Zhou X, Chen L, Huang W, Yu D (2010) Functional characterization of *Arabidopsis thaliana* WRKY39 in heat stress. Mol Cells 29:475–483

Li X, Liang Y, Gao B, Mijiti M, Bozorov TA, Yang H, Zhang D, Wood AJ (2019c) *ScDREB10*, an A-5c type of DREB gene of the desert moss *Syntrichia caninervis*, confers osmotic and salt tolerances to Arabidopsis. Genes 10:146

Li Y, Zhang H, Zhang Q, Liu Q, Zhai H, Zhao N, He S (2019d) An AP2/ERF gene, *IbRAP2-12*, from sweetpotato is involved in salt and drought tolerance in transgenic Arabidopsis. Plant Sci 281:19–30

Li M, Yang Y, Raza A, Yin S, Wang H, Zhang Y, Dong J, Wang G, Zhong C, Zhang H, Liu J, (2020) Heterologous Expression of *Arabidopsis* rty Enhances Drought Tolerance in Strawberry (*Fragaria*× *ananassa* Duch.). https://doi.org/10.21203/rs.3.rs-34565/v1

Lichtenthaler HK (1988) In vivo chlorophyll fluorescence as a tool for stress detection in plants. In: Applications of chlorophyll fluorescence in photosynthesis research, stress physiology, hydrobiology and remote sensing. Springer, pp 129–142

Lichtenthaler HK (1996) Vegetation stress: an introduction to the stress concept in plants. J Plant Physiol 148:4–14

Lim CJ, Hwang JE, Chen H, Hong JK, Yang KA, Choi MS, Lee KO, Chung WS, Lee SY, Lim CO (2007) Over-expression of the Arabidopsis DRE/CRT-binding transcription factor DREB2C enhances thermotolerance. Biochem Biophys Res Commun 362:431–436

Lin R-C, Park H-J, Wang H-Y (2008a) Role of *Arabidopsis* RAP2.4 in regulating light-and ethylene-mediated developmental processes and drought stress tolerance. Mol Plant 1:42–57

Lin S-I, Chiang S-F, Lin W-Y, Chen J-W, Tseng C-Y, Wu P-C, Chiou T-J (2008b) Regulatory network of microRNA399 and PHO_2 by systemic signaling. Plant Physiol 147:732–746

Lippold F, Sanchez DH, Musialak M, Schlereth A, Scheible W-R, Hincha DK, Udvardi MK (2009) *AtMyb41* regulates transcriptional and metabolic responses to osmotic stress in Arabidopsis. Plant Physiol 149:1761–1772

Lisso J, Altmann T, Müssig C (2006) The AtNFXL1 gene encodes a NF-X1 type zinc finger protein required for growth under salt stress. FEBS Lett 580:4851–4856

Liu C, Sun Q, Zhao L, Li Z, Peng Z, Zhang J (2018a) Heterologous expression of the transcription factor *EsNAC1* in *Arabidopsis* enhances abiotic stress resistance and retards growth by regulating the expression of different target genes. Front Plant Sci 9:1495

Liu H, Yang W, Liu D, Han Y, Zhang A, Li S (2011) Ectopic expression of a grapevine transcription factor *VvWRKY11* contributes to osmotic stress tolerance in Arabidopsis. Mol Biol Rep 38:417–427

Liu J, Chu J, Ma C, Jiang Y, Ma Y, Xiong J, Cheng Z-M (2019) Overexpression of an ABA-dependent grapevine bZIP transcription factor, *VvABF2*, enhances osmotic stress in Arabidopsis. Plant Cell Rep 1–10

Liu J, Srivastava R, Howell SH (2008) Stress-induced expression of an activated form of AtbZIP17 provides protection from salt stress in Arabidopsis. Plant, Cell Environ 31:1735–1743

Liu Q, Kasuga M, Sakuma Y, Abe H, Miura S, Yamaguchi-Shinozaki K, Shinozaki K (1998) Two transcription factors, DREB1 and DREB2, with an EREBP/AP2 DNA binding domain separate two cellular signal transduction pathways in drought-and low-temperature-responsive gene expression, respectively, in Arabidopsis. Plant Cell 10:1391–1406

Liu S, Lv Z, Liu Y, Li L, Zhang L (2018b) Network analysis of ABA-dependent and ABA-independent drought responsive genes in *Arabidopsis thaliana*. Genet Mol Biol 41:624–637

Lu M, Ying S, Zhang D-F, Shi Y-S, Song Y-C, Wang T-Y, Li Y (2012) A maize stress-responsive NAC transcription factor, *ZmSNAC1*, confers enhanced tolerance to dehydration in transgenic Arabidopsis. Plant Cell Rep 31:1701–1711

Luan S (2009) The CBL–CIPK network in plant calcium signaling. Trends Plant Sci 14:37–42

Luo H, Laluk K, Lai Z, Veronese P, Song F, Mengiste T (2010) The Arabidopsis Botrytis Susceptible1 Interactor defines a subclass of RING E3 ligases that regulate pathogen and stress responses. Plant Physio 154:1766–1782

Luo JS, Yang Y, Gu T, Wu Z, Zhang Z (2019) The Arabidopsis defensin gene *AtPDF2.5* mediates cadmium tolerance and accumulation. Plant, Cell Environ 42:2681–2695

Luo X, Bai X, Sun X, Zhu D, Liu B, Ji W, Cai H, Cao L, Wu J, Hu M, Liu X (2013) Expression of wild soybean WRKY20 in *Arabidopsis* enhances drought tolerance and regulates ABA signalling. J Exp Bot 64:2155–2169

Lv Q, Lan Y, Shi Y, Wang H, Pan X, Li P, Shi T (2016a) AtPID: a genome-scale resource for genotype–phenotype associations in Arabidopsis. Nucleic Acids Res 45:D1060–D1063

Lv Y, Fu S, Chen S, Zhang W, Qi C (2016b) Ethylene response factor *BnERF2*-like (ERF2.4) from *Brassica napus* L. enhances submergence tolerance and alleviates oxidative damage caused by submergence in *Arabidopsis thaliana*. Crop J 4:199–211

Mao X, Chen S, Li A, Zhai C, Jing R (2014) Novel NAC transcription factor *TaNAC67* confers enhanced multi-abiotic stress tolerances in Arabidopsis. PLoS ONE 9:e84359

Mao X, Zhang H, Qian X, Li A, Zhao G, Jing R (2012) *TaNAC2*, a NAC-type wheat transcription factor conferring enhanced multiple abiotic stress tolerances in Arabidopsis. J Exp Bot 63:2933–2946

Matiolli CC et al (2011) The Arabidopsis bZIP gene *AtbZIP63* is a sensitive integrator of transient abscisic acid and glucose signals. Plant Physiol 157:692–705

Matsukura S, Mizoi J, Yoshida T, Todaka D, Ito Y, Maruyama K, Shinozaki K, Yamaguchi-Shinozaki K (2010) Comprehensive analysis of rice DREB2-type genes that encode transcription factors involved in the expression of abiotic stress-responsive genes. Mol Genet Genomics 283:185–196

Medina J, Catalá R, Salinas J (2011) The CBFs: three Arabidopsis transcription factors to cold acclimate. Plant Sci 180:3–11

Mekonnen DW, Flügge U-I, Ludewig F (2016) Gamma-aminobutyric acid depletion affects stomata closure and drought tolerance of *Arabidopsis thaliana*. Plant Sci 245:25–34

Mengiste T, Chen X, Salmeron J, Dietrich R (2003) The BOTRYTIS SUSCEPTIBLE1 gene encodes an R2R3MYB transcription factor protein that is required for biotic and abiotic stress responses in Arabidopsis. Plant Cell 15:2551–2565

Milroy S, Bange M (2013) Reduction in radiation use efficiency of cotton (*Gossypium hirsutum* L.) under repeated transient waterlogging in the field. Field Crops Res 140:51–58

Mishra N, Srivastava AP, Esmaeili N, Hu W, Shen G (2018) Overexpression of the rice gene *OsSIZ1* in Arabidopsis improves drought-, heat-, and salt-tolerance simultaneously. PLoS ONE 13:e0201716

Miura K, Jin JB, Lee J, Yoo CY, Stirm V, Miura T, Ashworth EN, Bressan RA, Yun DJ, Hasegawa PM (2007) SIZ1-mediated sumoylation of ICE1 controls CBF3/DREB1A expression and freezing tolerance in *Arabidopsis*. Plant Cell 19:1403–1414

Munns R, Tester M (2008) Mechanisms of salinity tolerance. Annu Rev Plant Biol 59:651–681

Nahar K, Hasanuzzaman M, Alam MM, Rahman A, Suzuki T, Fujita M (2016) Polyamine and nitric oxide crosstalk: antagonistic effects on cadmium toxicity in mung bean plants through upregulating the metal detoxification, antioxidant defense, and methylglyoxal detoxification systems. Ecotoxicol Environ Saf 126:245–255

Nahar K, Hasanuzzaman M, Suzuki T, Fujita M (2017) Polyamine-induced aluminium tolerance in mung bean: a study on antioxidant defense and methylglyoxal detoxification systems. Ecotoxicology 26:58–73

Nakashima K, Fujita Y, Kanamori N, Katagiri T, Umezawa T, Kidokoro S, Maruyama K, Yoshida T, Ishiyama K, Kobayashi M, Shinozaki K (2009a) Three Arabidopsis SnRK2 protein kinases, SRK2D/SnRK2.2, SRK2E/SnRK2.6/OST1 and SRK2I/SnRK2.3, involved in ABA signaling are essential for the control of seed development and dormancy. Plant Cell Physiol 50:1345–1363

Nakashima K, Ito Y, Yamaguchi-Shinozaki K (2009b) Transcriptional regulatory networks in response to abiotic stresses in Arabidopsis and grasses. Plant Physiol 149:88–95

Nakashima K, Shinwari ZK, Sakuma Y, Seki M, Miura S, Shinozaki K, Yamaguchi-Shinozaki K (2000) Organization and expression of two Arabidopsis DREB2 genes encoding DRE-binding proteins involved in dehydration-and high-salinity-responsive gene expression. Plant Mol Biol 42:657–665

Nakashima K, Takasaki H, Mizoi J, Shinozaki K, Yamaguchi-Shinozaki K (2012) NAC transcription factors in plant abiotic stress responses. BBA-Gene Regul Mech 1819:97–103

Nakashima K, Tran LS, Van Nguyen D, Fujita M, Maruyama K, Todaka D, Ito Y, Hayashi N, Shinozaki K, Yamaguchi-Shinozaki K (2007) Functional analysis of a NAC-type transcription factor OsNAC6 involved in abiotic and biotic stress-responsive gene expression in rice. Plant J 51:617–630

Nakashima K, Yamaguchi-Shinozaki K (2009) Promoters and transcription factors in abiotic stress-responsive gene expression. In: Abiotic stress adaptation in plants. Springer, pp 199–216

Nakashima K, Yamaguchi-Shinozaki K (2013) ABA signaling in stress-response and seed development. Plant Cell Rep 32:959–970

Nakashima K, Yamaguchi-Shinozaki K, Shinozaki K (2014) The transcriptional regulatory network in the drought response and its crosstalk in abiotic stress responses including drought, cold, and heat. Front Plant Sci 5:170

Nelson DE, Repetti PP, Adams TR, Creelman RA, Wu J, Warner DC, Anstrom DC, Bensen RJ, Castiglioni PP, Donnarummo MG, Hinchey BS (2007) Plant nuclear factor Y (NF-Y) B subunits confer drought tolerance and lead to improved corn yields on water-limited acres. Proc Natl Acad Sci USA 104:16450–16455

Nguyen HM, Sako K, Matsui A, Suzuki Y, Mostofa MG, Ha CV, Tanaka M, Tran LS, Habu Y, Seki M (2017) Ethanol enhances high-salinity stress tolerance by detoxifying reactive oxygen species in *Arabidopsis thaliana* and rice. Front Plant Sci 8:1001

Nishizawa A, Yabuta Y, Shigeoka S (2008) Galactinol and raffinose constitute a novel function to protect plants from oxidative damage. Plant Physiol 147:1251–1263

Novillo F, Alonso JM, Ecker JR, Salinas J (2004) CBF2/DREB1C is a negative regulator of CBF1/DREB1B and CBF3/DREB1A expression and plays a central role in stress tolerance in Arabidopsis. Proc Natl Acad Sci USA 101:3985–3990

Nuruzzaman M, Sharoni AM, Kikuchi S (2013) Roles of NAC transcription factors in the regulation of biotic and abiotic stress responses in plants. Front Microbiol 4:248

Pachauri R.K, Allen MR, Barros VR, Broome J, Cramer W, Christ R, van Ypersele J-P (2014) Climate change 2014 synthesis report contribution of working groups I, II, and III to the fifth assessment report of the Intergovernmental Panel on Climate Change. IPCC

Pandey GK, Kanwar P, Singh A, Steinhorst L, Pandey A, Yadav AK, Tokas I, Sanyal SK, Kim BG, Lee SC, Cheong YH (2015) Calcineurin B-like protein-interacting protein kinase CIPK21 regulates osmotic and salt stress responses in *Arabidopsis*. Plant Physiol 169:780–792

Park HY, Seok HY, Woo DH, Lee SY, Tarte VN, Lee EH, Lee CH, Moon YH (2011) *AtERF71/HRE2* transcription factor mediates osmotic stress response as well as hypoxia response in Arabidopsis. Biochem Biophys Res Commun 414:135–141

Park S, Lee CM, Doherty CJ, Gilmour SJ, Kim Y, Thomashow MF (2015) Regulation of the Arabidopsis CBF regulon by a complex low-temperature regulatory network. Plant J 82:193–207

Patankar HV, Al-Harrasi I, Al Kharusi L, Jana GA, Al-Yahyai R, Sunkar R, Yaish MW (2019) Overexpression of a metallothionein 2a gene from date palm confers abiotic stress tolerance to yeast and *Arabidopsis thaliana*. Int J Mol Sci 20:2871

Patanun O, Ueda M, Itouga M, Kato Y, Utsumi Y, Matsui A, Tanaka M, Utsumi C, Sakakibara H, Yoshida M, Narangajavana J (2017) The histone deacetylase inhibitor suberoylanilide hydroxamic acid alleviates salinity stress in cassava. Front Plant Sci 7:2039

Pegler JL, Oultram JM, Grof CP, Eamens AL (2019) Profiling the abiotic stress responsive microRNA landscape of *Arabidopsis thaliana*. Plants 8:58

Peng X, Zhang L, Zhang L, Liu Z, Cheng L, Yang Y, Shen S, Chen S, Liu G (2013) The transcriptional factor *LcDREB2* cooperates with *LcSAMDC2* to contribute to salt tolerance in *Leymus chinensis*. Plant Cell Tiss Org 113:245–256

Pereira A (2016) Plant abiotic stress challenges from the changing environment. Front Plant Sci 7:1123

Phukan UJ, Jeena GS, Tripathi V, Shukla RK (2018) Ma RAP 2–4, a waterlogging-responsive ERF from Mentha, regulates bidirectional sugar transporter AtSWEET10 to modulate stress response in Arabidopsis. Plant Biotechnol J 16:221–233

Prall W, Sharma B, Gregory BD (2019) Transcription is just the beginning of gene expression regulation: the functional significance of RNA-binding proteins to post-transcriptional processes in plants. Plant Cell Physiol 67: 1–14

Pruthvi V, Narasimhan R, Nataraja KN (2014) Simultaneous expression of abiotic stress responsive transcription factors, *AtDREB2A*, *AtHB7* and *AtABF3* improves salinity and drought tolerance in peanut (*Arachis hypogaea* L.). PLoS One 9: e111152

Qiao K, Gong L, Tian Y, Wang H, Chai T (2018) The metal-binding domain of wheat heavy metal ATPase2 (*TaHMA2*) is involved in zinc/cadmium tolerance and translocation in *Arabidopsis*. Plant Cell Rep 37:1343–1352

Qin F, Sakuma Y, Tran LS, Maruyama K, Kidokoro S, Fujita Y, Fujita M, Umezawa T, Sawano Y, Miyazono KI, Tanokura M (2008) Arabidopsis DREB2A-interacting proteins function as RING E3 ligases and negatively regulate plant drought stress–responsive gene expression. Plant Cell 20:1693–1707

Qin Y, Wang M, Tian Y, He W, Han L, Xia G (2012) Over-expression of *TaMYB33* encoding a novel wheat MYB transcription factor increases salt and drought tolerance in *Arabidopsis*. Mol Biol Rep 39:7183–7192

Qin Y, Tian Y, Han L, Yang X (2013) Constitutive expression of a salinity-induced wheat WRKY transcription factor enhances salinity and ionic stress tolerance in transgenic *Arabidopsis thaliana*. Biochem Biophys Res Commun 441:476–481

Qin Y, Tian Y, Liu X (2015) A wheat salinity-induced WRKY transcription factor *TaWRKY93* confers multiple abiotic stress tolerance in *Arabidopsis thaliana*. Biochem Biophys Res Commun 464:428–433

Qiu Q-S, Guo Y, Dietrich MA, Schumaker KS, Zhu J-K (2002) Regulation of SOS1, a plasma membrane Na^+/H^+ exchanger in *Arabidopsis thaliana*, by SOS2 and SOS3. Proc Natl Acad Sci USA 99:8436–8441

Qiu Y, Yu D (2009) Over-expression of the stress-induced *OsWRKY45* enhances disease resistance and drought tolerance in *Arabidopsis*. Environ Exp Bot 65:35–47

Qu T, Liu R, Wang W, An L, Chen T, Liu G, Zhao Z (2011) Brassinosteroids regulate pectin methylesterase activity and *AtPME41* expression in *Arabidopsis* under chilling stress. Cryobiology 63:111–117

Raab S, Drechsel G, Zarepour M, Hartung W, Koshiba T, Bittner F, Hoth S (2009) Identification of a novel E3 ubiquitin ligase that is required for suppression of premature senescence in *Arabidopsis*. Plant J 59:39–51

Raineri J, Ribichich KF, Chan RL (2015) The sunflower transcription factor *HaWRKY76* confers drought and flood tolerance to *Arabidopsis thaliana* plants without yield penalty. Plant Cell Rep 34:2065–2080

Rasheed S, Bashir K, Kim JM, Ando M, Tanaka M, Seki M (2018) The modulation of acetic acid pathway genes in *Arabidopsis* improves survival under drought stress. Sci Rep 8:7831

Rasmussen S, Barah P, Suarez-Rodriguez MC, Bressendorff S, Friis P, Costantino P, Bones AM, Nielsen HB, Mundy J (2013) Transcriptome responses to combinations of stresses in *Arabidopsis*. Plant Physiol 161:1783–1794

Raza A, Razzaq A, Mehmood SS, Zou X, Zhang X, Lv Y, Xu J (2019a) Impact of climate change on crops adaptation and strategies to tackle its outcome: a review. Plants 8:34

Raza A, Mehmood SS, Tabassum J, Batool R (2019b) Targeting plant hormones to develop abiotic stress resistance in wheat. In: Hasanuzzaman M, Nahar K, Hossain M. (eds)Wheat production in changing environments. Springer, Singapore, pp 557–577

Raza A, Mehmood SS, Shah T, Zou X, Yan L, Zhang X, Khan RS (2019c) Applications of molecular markers to develop resistance against abiotic stresses in wheat. In: Hasanuzzaman M, Nahar K, Hossain M. (eds) Wheat production in changing environments. Springer, Singapore, pp 393–420

Raza A, Ashraf F, Zou X, Zhang X, Tosif H (2020) Plant Adaptation and tolerance to environmental stresses: mechanisms and perspectives. In Hasanuzzaman M. (eds) Plant ecophysiology and adaptation under climate change: mechanisms and perspectives I. Springer, Singapore, pp. 117–145

Razzaq A, Sadia B, Raza A, Khalid Hameed M, Saleem F (2019) Metabolomics: a way forward for crop improvement. Metabolomics 9:303

Rhee SY, Beavis W, Berardini TZ, Chen G, Dixon D, Doyle A, Garcia-Hernandez M, Huala E, Lander G, Montoya M, Miller N (2003) The arabidopsis information resource (TAIR): a model organism database providing a centralized, curated gateway to *Arabidopsis* biology, research materials and community. Nucleic Acid Res 31:224–228

Rizhsky L, Liang H, Shuman J, Shulaev V, Davletova S, Mittler R (2004) When defense pathways collide. The response of *Arabidopsis* to a combination of drought and heat stress. Plant Physiol 134:1683–1696

Rolland F, Baena-Gonzalez E, Sheen J (2006) Sugar sensing and signaling in plants: conserved and novel mechanisms. Annu Rev Plant Biol 57:675–709

Roychoudhury A, Paul A (2012) Abscisic acid-inducible genes during salinity and drought stress. Adv in Med Biol 51:1–78

Sairam RK, Dharmar K, Chinnusamy V, Meena RC (2009) Waterlogging-induced increase in sugar mobilization, fermentation, and related gene expression in the roots of mung bean (*Vigna radiata*). J Plant Physiol 166:602–616

Sako K, Kim JM, Matsui A, Nakamura K, Tanaka M, Kobayashi M, Saito K, Nishino N, Kusano M, Taji T, Yoshida M (2015) Ky-2, a histone deacetylase inhibitor, enhances high-salinity stress tolerance in *Arabidopsis thaliana*. Plant Cell Physiol 57:776–783

Sakuma Y, Liu Q, Dubouzet JG, Abe H, Shinozaki K, Yamaguchi-Shinozaki K (2002) DNA-binding specificity of the ERF/AP2 domain of *Arabidopsis* DREBs, transcription factors involved in dehydration-and cold-inducible gene expression. Biochem Biophys Res Commun 290:998–1009

Sakuma Y, Maruyama K, Osakabe Y, Qin F, Seki M, Shinozaki K, Yamaguchi-Shinozaki K (2006a) Functional analysis of an *Arabidopsis* transcription factor, DREB2A, involved in drought-responsive gene expression. Plant Cell 18:1292–1309

Sakuma Y, Maruyama K, Qin F, Osakabe Y, Shinozaki K, Yamaguchi-Shinozaki K (2006b) Dual function of an *Arabidopsis* transcription factor DREB2A in water-stress-responsive and heat-stress-responsive gene expression. Proc Natl Acad Sci USA 103:18822–18827

Salim N, Raza A (2020) Nutrient use efficiency (NUE) for sustainable wheat production: a review. J Plant Nutr 43:297–315

Savvides A, Ali S, Tester M, Fotopoulos V (2016) Chemical priming of plants against multiple abiotic stresses: mission possible? Trends Plant Sci 21:329–340

Schramm F, Larkindale J, Kiehlmann E, Ganguli A, Englich G, Vierling E, Von Koskull-Döring P (2008) A cascade of transcription factor DREB2A and heat stress transcription factor *HsfA3* regulates the heat stress response of *Arabidopsis*. Plant J 53:264–274

Seki M, Narusaka M, Ishida J, Nanjo T, Fujita M, Oono Y, Kamiya A, Nakajima M, Enju A, Sakurai T, Satou M (2002) Monitoring the expression profiles of 7000 Arabidopsis genes under drought, cold and high-salinity stresses using a full-length cDNA microarray. Plant J 31:279–292

Seo JS, Sohn HB, Noh K, Jung C, An JH, Donovan CM, Somers DA, Kim DI, Jeong SC, Kim CG, Kim HM (2012) Expression of the Arabidopsis *AtMYB44* gene confers drought/salt-stress tolerance in transgenic soybean. Mol Breed 29:601–608

Seo PJ, Xiang F, Qiao M, Park JY, Lee YN, Kim SG, Lee YH, Park WJ, Park CM (2009) The MYB96 transcription factor mediates abscisic acid signaling during drought stress response in Arabidopsis. Plant Physiol 151:275–289

Sewelam N, Kazan K, Thomas-Hall SR, Kidd BN, Manners JM, Schenk PM (2013) Ethylene response factor 6 is a regulator of reactive oxygen species signaling in *Arabidopsis*. PLoS ONE 8:e70289

Shahjee HM, Banerjee K, Ahmad F (2002) Comparative analysis of naturally occurring L-amino acid osmolytes and their D-isomers on protection of *Escherichia coli* against environmental stresses. J Biosci 27:515–520

Shao Y, Zhang X, van Nocker S, Gong X, Ma F (2019) Overexpression of a protein kinase gene *MpSnRK2.10* from *Malus prunifolia* confers tolerance to drought stress in transgenic *Arabidopsis thaliana* and apple. Gene 692:26–34

Shi H, Lee B-h, Wu S-J, Zhu J-K (2003) Overexpression of a plasma membrane Na^+/H^+ antiporter gene improves salt tolerance in *Arabidopsis thaliana*. Nat Biotechnol 21:81

Shinde H, Dudhate A, Tsugama D, Gupta SK, Liu S, Takano T (2019) Pearl millet stress-responsive NAC transcription factor *PgNAC21* enhances salinity stress tolerance in *Arabidopsis*. Plant Physiol Biochem 135:546–553

Shinozaki K, Yamaguchi-Shinozaki K (2007) Gene networks involved in drought stress response and tolerance. J Exp Bot 58:221–227

Siddiqua M, Nassuth A (2011) Vitis CBF1 and Vitis CBF4 differ in their effect on *Arabidopsis* abiotic stress tolerance, development and gene expression. Plant, Cell Environ 34:1345–1359

Silveira AB, Gauer L, Tomaz JP, Cardoso PR, Carmello-Guerreiro S, Vincentz M (2007) The Arabidopsis *AtbZIP9* protein fused to the VP16 transcriptional activation domain alters leaf and vascular development. Plant Sci 172:1148–1156

Simpson SD, Nakashima K, Narusaka Y, Seki M, Shinozaki K, Yamaguchi-Shinozaki K (2003) Two different novel cis-acting elements of erd1, a clpA homologous Arabidopsis gene function in induction by dehydration stress and dark-induced senescence. Plant J 33:259–270

Singh S, Brocker C, Koppaka V, Chen Y, Jackson BC, Matsumoto A, Thompson DC, Vasiliou V (2013) Aldehyde dehydrogenases in cellular responses to oxidative/electrophilicstress. Free Radic Biol Med 56:89–101

Skinner JS, von Zitzewitz J, Szűcs P, Marquez-Cedillo L, Filichkin T, Amundsen K, Stockinger EJ, Thomashow MF, Chen TH, Hayes PM (2005) Structural, functional, and phylogenetic characterization of a large CBF gene family in barley. Plant Mol Biol 59:533–551

Sokol A, Kwiatkowska A, Jerzmanowski A, Prymakowska-Bosak M (2007) Up-regulation of stress-inducible genes in tobacco and *Arabidopsis* cells in response to abiotic stresses and ABA treatment correlates with dynamic changes in histone H3 and H4 modifications. Planta 227:245–254

Song Y, Jing S, Yu D (2009) Overexpression of the stress-induced *OsWRKY08* improves osmotic stress tolerance in *Arabidopsis*. Chinese Sci Bull 54:4671–4678

Song Y, Lv J, Qiu N, Bai Y, Yang N, Dong W (2019) The constitutive expression of alfalfa *MsMYB2L* enhances salinity and drought tolerance of *Arabidopsis thaliana*. Plant Physiol Biochem 141:300–305

Srivastava D, Verma G, Chauhan AS, Pande V, Chakrabarty D (2019) Rice (*Oryza sativa* L.) tau class glutathione S-transferase (*OsGSTU30*) overexpression in *Arabidopsis thaliana* modulates a regulatory network leading to heavy metal and drought stress tolerance. Metallomics 11:375–389

Stiti N, Missihoun TD, Kotchoni S, Kirch H-H, Bartels D (2011) Aldehyde dehydrogenases in *Arabidopsis thaliana*: biochemical requirements, metabolic pathways, and functional analysis. Front Plant Sci 2:65

Stockinger EJ, Gilmour SJ, Thomashow MF (1997) *Arabidopsis thaliana* CBF1 encodes an AP2 domain-containing transcriptional activator that binds to the C-repeat/DRE, a cis-acting DNA regulatory element that stimulates transcription in response to low temperature and water deficit. Proc Natl Acad Sci USA 94:1035–1040

Su LT, Li JW, Liu DQ, Zhai Y, Zhang HJ, Li XW, Zhang QL, Wang Y, Wang QY (2014) A novel MYB transcription factor, *GmMYBJ1*, from soybean confers drought and cold tolerance in *Arabidopsis thaliana*. Gene 538:46–55

Sun L, Ma L, He S, Hao F (2018a) *AtrbohD* functions downstream of ROP2 and positively regulates waterlogging response in *Arabidopsis*. Plant Signal Behav 13:e1513300

Sun X, Li Y, Cai H, Bai X, Ji W, Ding X, Zhu Y (2012) The *Arabidopsis AtbZIP1* transcription factor is a positive regulator of plant tolerance to salt, osmotic and drought stresses. J Plant Res 125:429–438

Sun X, Yu G, Li J, Liu J, Wang X, Zhu G, Zhang X, Pan H (2018b) *AcERF2*, an ethylene-responsive factor of Atriplex canescens, positively modulates osmotic and disease resistance in *Arabidopsis thaliana*. Plant Sci 274:32–43

Sun X, Zhu Z, Zhang L, Fang L, Zhang J, Wang Q, Li S, Liang Z, Xin H (2019) Overexpression of ethylene response factors *VaERF080* and *VaERF087* from *Vitis amurensis* enhances cold tolerance in *Arabidopsis*. Sci Hortic 243:320–326

Sun Z-M, Zhou M-L, Xiao X-G, Tang Y-X, Wu Y-M (2014) Genome-wide analysis of AP2/ERF family genes from *Lotus corniculatus* shows *LcERF054* enhances salt tolerance. Funct Integr Genomic 14:453–466

Sunkar R, Kapoor A, Zhu J-K (2006) Posttranscriptional induction of two Cu/Zn superoxide dismutase genes in *Arabidopsis* is mediated by downregulation of miR398 and important for oxidative stress tolerance. Plant Cell 18:2051–2065

Sunkar R, Zhu J-K (2004) Novel and stress-regulated microRNAs and other small RNAs from *Arabidopsis*. Plant Cell 16:2001–2019

Tada Y (2019) The HKT transporter gene from *Arabidopsis*, *AtHKT1;1*, is dominantly expressed in shoot vascular tissue and root tips and is mild salt stress-responsive. Plants 8:204

Takahashi F, Shinozaki K (2019) Long-distance signaling in plant stress response. Curr Opin Plant Biol 47:106–111

Tamirisa S, Vudem DR, Khareedu VR (2014) Overexpression of pigeonpea stress-induced cold and drought regulatory gene (CcCDR) confers drought, salt, and cold tolerance in *Arabidopsis*. J Exp Bot 65:4769–4781

Tang Y, Bao X, Wang S, Liu Y, Tan J, Yang M, Zhang M, Dai R, Yu X (2019) A physic nut stress-responsive HD-Zip transcription factor, *JcHDZ07*, confers enhanced sensitivity to salinity stress in transgenic *Arabidopsis*. Front Plant Sci 10:942

Teixeira EI, Fischer G, Van Velthuizen H, Walter C, Ewert F (2013) Global hot-spots of heat stress on agricultural crops due to climate change. Agric For Meteorol 170:206–215

Thalmann M, Pazmino D, Seung D, Horrer D, Nigro A, Meier T, Kölling K, Pfeifhofer HW, Zeeman SC, Santelia D (2016) Regulation of leaf starch degradation by abscisic acid is important for osmotic stress tolerance in plants. Plant Cell 28:1860–1878

Thirunavukkarasu N, Hossain F, Mohan S, Shiriga K, Mittal S, Sharma R, Singh RK, Gupta HS (2013) Genome-wide expression of transcriptomes and their co-expression pattern in subtropical maize (*Zea mays* L.) under waterlogging stress. PLoS One 8: e70433

Thomashow MF (2010) Molecular basis of plant cold acclimation: insights gained from studying the CBF cold response pathway. Plant Physiol 154:571–577

To TK, Nakaminami K, Kim JM, Morosawa T, Ishida J, Tanaka M, Yokoyama S, Shinozaki K, Seki M (2011) *Arabidopsis* HDA6 is required for freezing tolerance. Biochem Biophys Res Commun 406:414–419

Tomanov K, Nukarinen E, Vicente J, Mendiondo GM, Winter N, Nehlin L, Weckwerth W, Holdsworth MJ, Teige M, Bachmair A (2018) Sumoylation and phosphorylation: hidden and overt links. J Exp Bot 69:4583–4590

Tong Z, Hong B, Yang Y, Li Q, Ma N, Ma C, Gao J (2009) Overexpression of two chrysanthemum *DgDREB1* group genes causing delayed flowering or dwarfism in *Arabidopsis*. Plant Mol Biol 71:115–129

Torres-Galea P, Huang L-F, Chua N-H, Bolle C (2006) The GRAS protein SCL13 is a positive regulator of phytochrome-dependent red light signaling, but can also modulate phytochrome A responses. Mol Genet Genomics 276:13–30

Tran LS, Nakashima K, Sakuma Y, Simpson SD, Fujita Y, Maruyama K, Fujita M, Seki M, Shinozaki K, Yamaguchi-Shinozaki K (2004) Isolation and functional analysis of *Arabidopsis* stress-inducible NAC transcription factors that bind to a drought-responsive cis-element in the early responsive to dehydration stress 1 promoter. Plant Cell 16:2481–2498

Tran L (2004) Functional analysis of *Arabidopsis* NAC transcription factors controlling expression of erd1 gene under drought stress. Plant Cell 16:2481–2498

Tu M, Wang X, Huang L, Guo R, Zhang H, Cai J, Wang X (2016) Expression of a grape bZIP transcription factor, *VqbZIP39*, in transgenic *Arabidopsis thaliana* confers tolerance of multiple abiotic stresses. Plant Cell Tiss Org 125:537–551

Urao T, Yakubov B, Satoh R, Yamaguchi-Shinozaki K, Seki M, Hirayama T, Shinozaki K (1999) A transmembrane hybrid-type histidine kinase in *Arabidopsis* functions as an osmosensor. Plant Cell 11:1743–1754

Usadel B, Nagel A, Thimm O, Redestig H, Blaesing OE, Palacios-Rojas N, Selbig J, Hannemann J, Piques MC, Steinhauser D, Scheible WR (2005) Extension of the visualization tool MapMan to allow statistical analysis of arrays, display of coresponding genes, and comparison with known responses. Plant Physiol 138:1195–1204

Van Ha C, Leyva-González MA, Osakabe Y, Tran UT, Nishiyama R, Watanabe Y, Tanaka M, Seki M, Yamaguchi S, Van Dong N, Yamaguchi-Shinozaki K (2014) Positive regulatory role of strigolactone in plant responses to drought and salt stress. Proc Natl Acad Sci USA 111:851–856

Vera-Estrella R, Barkla BJ, Bohnert HJ, Pantoja O (2004) Novel regulation of aquaporins during osmotic stress. Plant Physiol 135:2318–2329

Verslues PE, Agarwal M, Katiyar-Agarwal S, Zhu J, Zhu JK (2006) Methods and concepts in quantifying resistance to drought, salt and freezing, abiotic stresses that affect plant water status. Plant J 45:523–539

Vinocur B, Altman A (2005) Recent advances in engineering plant tolerance to abiotic stress: achievements and limitations. Curr Opin Biotech 16:123–132

Voesenek LA, Bailey-Serres J (2015) Flood adaptive traits and processes: an overview. New Phytol 206:57–73

Vogel JT, Zarka DG, Van Buskirk HA, Fowler SG, Thomashow MF (2005) Roles of the CBF2 and ZAT12 transcription factors in configuring the low temperature transcriptome of *Arabidopsis*. Plant J 41:195–211

von Koskull-Döring P, Scharf K-D, Nover L (2007) The diversity of plant heat stress transcription factors. Trends Plant Sci 12:452–457

Wahid A (2007) Physiological implications of metabolite biosynthesis for net assimilation and heat-stress tolerance of sugarcane (*Saccharum officinarum*) sprouts. J Plant Res 120:219–228

Wang F, Kong W, Wong G, Fu L, Peng R, Li Z, Yao Q (2016a) *AtMYB12* regulates flavonoids accumulation and abiotic stress tolerance in transgenic *Arabidopsis thaliana*. Mol Genet Genomics 291:1545–1559

Wang L, Li Z, Lu M, Wang Y (2017) *ThNAC13*, a NAC transcription factor from *Tamarix hispida*, confers salt and osmotic stress tolerance to transgenic *tamarix* and *Arabidopsis*. Front Plant Sci 8:635

Wang L, Zhang C, Wang Y, Wang Y, Yang C, Lu M, Wang C (2018a) *Tamarix hispida* aquaporin *ThPIP2;5* confers salt and osmotic stress tolerance to transgenic Tamarix and *Arabidopsis*. Environ Exp Bot 152:158–166

Wang R, Zhang Y, Kieffer M, Yu H, Kepinski S, Estelle M (2016b) HSP90 regulates temperature-dependent seedling growth in *Arabidopsis* by stabilizing the auxin co-receptor F-box protein TIR1. Nat Commun 7:10269

Wang S, Zhou B, Yao W, Jiang T (2018b) *PsnERF75* transcription factor from *Populus simonii* × *P. nigra* confers salt tolerance in transgenic *Arabidopsis*. J Plant Biol 61:61–71

Wang W, Vinocur B, Shoseyov O, Altman A (2004) Role of plant heat-shock proteins and molecular chaperones in the abiotic stress response. Trends Plant Sci 9:244–252

Wang Z, Zhao X, Wang B, Liu E, Chen N, Zhang W, Liu H (2016c) Overexpression of an *Arabidopsis* heterogeneous nuclear ribonucleoprotein gene, *AtRNP1*, affects plant growth and reduces plant tolerance to drought and salt stresses. Biochem Biophys Res Commun 472:353–359

Welling A, Palva ET (2008) Involvement of CBF transcription factors in winter hardiness in birch. Plant Physiol 147:1199–1211

Wu D, Sun Y, Wang H, Shi H, Su M, Shan H, Li T, Li Q (2018) The *SlNAC8* gene of the halophyte *Suaeda liaotungensis* enhances drought and salt stress tolerance in transgenic *Arabidopsis thaliana*. Gene 662:10–20

Xiong AS, Jiang HH, Zhuang J, Peng RH, Jin XF, Zhu B, Wang F, Zhang J, Yao QH (2013) Expression and function of a modified AP2/ERF transcription factor from *Brassica napus* enhances cold tolerance in transgenic *Arabidopsis*. Mol Biotechnol 53:198–206

Xu X, Ji J, Xu Q, Qi X, Weng Y, Chen X (2018) The major-effect quantitative trait locus CsARN6.1 encodes an AAA ATP ase domain-containing protein that is associated with waterlogging stress tolerance by promoting adventitious root formation. Plant J 93:917–930

Xu ZS, Ni ZY, Li ZY, Li LC, Chen M, Gao DY, Yu XD, Liu P, Ma YZ (2009) Isolation and functional characterization of *HvDREB1*—a gene encoding a dehydration-responsive element binding protein in *Hordeum vulgare*. J Plant Res 122:121–130

Xu ZY, Kim SY, Kim DH, Dong T, Park Y, Jin JB, Joo SH, Kim SK, Hong JC, Hwang D, Hwang I (2013) The *Arabidopsis* NAC transcription factor *ANAC096* cooperates with bZIP-type transcription factors in dehydration and osmotic stress responses. Plant Cell 25:4708–4724

Yamaguchi-Shinozaki K, Shinozaki K (1994) A novel cis-acting element in an *Arabidopsis* gene is involved in responsiveness to drought, low-temperature, or high-salt stress. Plant Cell 6:251–264

Yamaguchi-Shinozaki K, Shinozaki K (2005) Organization of cis-acting regulatory elements in osmotic-and cold-stress-responsive promoters. Trends Plant Sci 10:88–94

Yamaguchi-Shinozaki K, Shinozaki K (2006) Transcriptional regulatory networks in cellular responses and tolerance to dehydration and cold stresses. Annu Rev Plant Biol 57:781–803

Yamori W (2016) Photosynthetic response to fluctuating environments and photoprotective strategies under abiotic stress. J Plant Res 129:379–395

Yan J, Zhao C, Zhou J, Yang Y, Wang P, Zhu X, Tang G, Bressan RA, Zhu JK (2016) The miR165/166 mediated regulatory module plays critical roles in ABA homeostasis and response in *Arabidopsis thaliana*. PLoS Genet 12:e1006416

Yang O, Popova OV, Süthoff U, Lüking I, Dietz K-J, Golldack D (2009a) The *Arabidopsis* basic leucine zipper transcription factor *AtbZIP24* regulates complex transcriptional networks involved in abiotic stress resistance. Gene 436:45–55

Yang S, Wang S, Liu X, Yu Y, Yue L, Wang X, Hao D (2009b) Four divergent Arabidopsis ethylene-responsive element-binding factor domains bind to a target DNA motif with a universal CG step core recognition and different flanking bases preference. FEBS J 276:7177–7186

Yang W, Liu XD, Chi XJ, Wu CA, Li YZ, Song LL, Liu XM, Wang YF, Wang FW, Zhang C, Liu Y (2011) Dwarf apple *MbDREB1* enhances plant tolerance to low temperature, drought, and salt stress via both ABA-dependent and ABA-independent pathways. Planta 233:219–229

Yang X, Wang X, Ji L, Yi Z, Fu C, Ran J, Hu R, Zhou G (2015) Overexpression of a *Miscanthus lutarioriparius* NAC gene *MlNAC5* confers enhanced drought and cold tolerance in *Arabidopsis*. Plant Cell Rep 34:943–958

Yao X, Cai Y, Yu D, Liang G (2018) *bHLH104* confers tolerance to cadmium stress in *Arabidopsis thaliana*. J Integr Plant Biol 60:691–702

Yeung E, van Veen H, Vashisht D, Paiva AL, Hummel M, Rankenberg T, Steffens B, Steffen-Heins A, Sauter M, de Vries M, Schuurink RC (2018) A stress recovery signaling network for enhanced flooding tolerance in *Arabidopsis thaliana*. Proc Natl Acad Sci USA 115:E6085–E6094

Yin D, Chen S, Chen F, Guan Z, Fang W (2009) Morphological and physiological responses of two chrysanthemum cultivars differing in their tolerance to waterlogging. Environ Exp Bot 67:87–93

Yin Y, Jiang X, Ren M, Xue M, Nan D, Wang Z, Xing Y, Wang M (2018) *AmDREB2C*, from *Ammopiptanthus mongolicus*, enhances abiotic stress tolerance and regulates fatty acid composition in transgenic *Arabidopsis*. Plant Physiol Bioch 130:517–528

Ying S, Zhang D-F, Fu J, Shi Y-S, Song Y-C, Wang T-Y, Li Y (2012) Cloning and characterization of a maize bZIP transcription factor, *ZmbZIP72*, confers drought and salt tolerance in transgenic *Arabidopsis*. Planta 235:253–266

Yokotani N, Ichikawa T, Kondou Y, Matsui M, Hirochika H, Iwabuchi M, Oda K (2009) Tolerance to various environmental stresses conferred by the salt-responsive rice gene *ONAC063* in transgenic *Arabidopsis*. Planta 229:1065–1075

Yoshida R, Umezawa T, Mizoguchi T, Takahashi S, Takahashi F, Shinozaki K (2006) The regulatory domain of SRK2E/OST1/SnRK2.6 interacts with ABI1 and integrates abscisic acid (ABA) and osmotic stress signals controlling stomatal closure in *Arabidopsis*. J Biol Chem 281:5310–5318

Yoshida T, Ohama N, Nakajima J, Kidokoro S, Mizoi J, Nakashima K, Maruyama K, Kim JM, Seki M, Todaka D, Osakabe Y (2011) Arabidopsis *HsfA1* transcription factors function as the main positive regulators in heat shock-responsive gene expression. Mol Genet Genomics 286:321–332

Yoshida T, Fujita Y, Maruyama K, Mogami J, Todaka D, Shinozaki K et al (2015) Four *Arabidopsis* AREB/ABF transcription factors function predominantly in gene expression downstream of

SnRK2 kinases in abscisic acid signalling in response to osmotic stress. Plant, Cell Environ 38:35–49

Yu X, Liu Y, Wang S, Tao Y, Wang Z, Shu Y, Peng H, Mijiti A, Wang Z, Zhang H, Ma H (2016) *CarNAC4*, a NAC-type chickpea transcription factor conferring enhanced drought and salt stress tolerances in *Arabidopsis*. Plant Cell Rep 35:613–627

Yu Y, Bi C, Wang Q, Ni Z (2019) Overexpression of *TaSIM* provides increased drought stress tolerance in transgenic *Arabidopsis*. Biochem Biophy Res Co 512:66–71

Yuan Y, Fang L, Karungo SK, Zhang L, Gao Y, Li S, Xin H (2016) Overexpression of *VaPAT1*, a GRAS transcription factor from *Vitis amurensis*, confers abiotic stress tolerance in *Arabidopsis*. Plant Cell Rep 35:655–666

Zhang J-Y, Huang S-N, Wang G, Xuan J-P, Guo Z-R (2016) Overexpression of *Actinidia deliciosa* pyruvate decarboxylase 1 gene enhances waterlogging stress in transgenic *Arabidopsis thaliana*. Plant Physiol Biochem 106:244–252

Zhang JZ, Creelman RA, Zhu J-K (2004a) From laboratory to field. Using information from *Arabidopsis* to engineer salt, cold, and drought tolerance in crops. Plant Physiol 135:615–621

Zhang L, Liu G, Zhao G, Xia C, Jia J, Liu X, Kong X (2014) Characterization of a wheat R2R3-MYB transcription factor gene, *TaMYB19*, involved in enhanced abiotic stresses in *Arabidopsis*. Plant Cell Physiol 55:1802–1812

Zhang L, Zhang L, Xia C, Zhao G, Liu J, Jia J, Kong X (2015) A novel wheat bZIP transcription factor, *TabZIP60*, confers multiple abiotic stress tolerances in transgenic *Arabidopsis*. Physiol Plant 153:538–554

Zhang X, Fowler SG, Cheng H, Lou Y, Rhee SY, Stockinger EJ, Thomashow MF (2004b) Freezing-sensitive tomato has a functional CBF cold response pathway, but a CBF regulon that differs from that of freezing-tolerant Arabidopsis. Plant J 39:905–919

Zhang X, Wang L, Meng H, Wen H, Fan Y, Zhao J (2011) Maize ABP9 enhances tolerance to multiple stresses in transgenic *Arabidopsis* by modulating ABA signaling and cellular levels of reactive oxygen species. Plant Mol Biol 75:365–378

Zhang X, Wollenweber B, Jiang D, Liu F, Zhao J (2008) Water deficits and heat shock effects on photosynthesis of a transgenic *Arabidopsis thaliana* constitutively expressing ABP9, a bZIP transcription factor. J Exp Bot 59:839–848

Zhao B-Y, Hu Y-F, Li J-J, Yao X (2016a) *BnaABF2*, a bZIP transcription factor from rapeseed (*Brassica napus* L.), enhances drought and salt tolerance in transgenic *Arabidopsis*. Bot Stud 57:12

Zhao C, Lang Z, Zhu J-K (2015) Cold responsive gene transcription becomes more complex. Trends Plant Sci 20:466–468

Zhao C, Zhang Z, Xie S, Si T, Li Y, Zhu J-K (2016b) Mutational evidence for the critical role of CBF transcription factors in cold acclimation in *Arabidopsis*. Plant Physiol 171:2744–2759

Zhao H, Bughrara SS (2008) Isolation and characterization of cold-regulated transcriptional activator *LpCBF3* gene from perennial ryegrass (*Lolium perenne L.*). Mol Genet Genomics 279:585–594

Zhao J, Missihoun TD, Bartels D (2017) The role of *Arabidopsis* aldehyde dehydrogenase genes in response to high temperature and stress combinations. J Exp Bot 68:4295–4308

Zhao Y, Yang Z, Ding Y, Liu L, Han X, Zhan J, Wei X, Diao Y, Qin W, Wang P, Liu P (2019) Overexpression of an R2R3 MYB Gene, *GhMYB73*, increases tolerance to salt stress in transgenic *Arabidopsis*. Plant Sci 286:28–36

Zhong H, Guo Q-Q, Chen L, Ren F, Wang Q-Q, Zheng Y, Li X-B (2012) Two *Brassica napus* genes encoding NAC transcription factors are involved in response to high-salinity stress. Plant Cell Rep 31:1991–2003

Zhong L, Chen D, Min D, Li W, Xu Z, Zhou Y, Li L, Chen M, Ma Y (2015) *AtTGA4*, a bZIP transcription factor, confers drought resistance by enhancing nitrate transport and assimilation in *Arabidopsis thaliana*. Biochem Biophys Res Commun 457:433–439

Zhou L, Wang N-N, Gong S-Y, Lu R, Li Y, Li X-B (2015) Overexpression of a cotton (*Gossypium hirsutum*) WRKY gene, *GhWRKY34*, in *Arabidopsis* enhances salt-tolerance of the transgenic plants. Plant Physiol Bioch 96:311–320

Zhou QY, Tian AG, Zou HF, Xie ZM, Lei G, Huang J, Wang CM, Wang HW, Zhang JS, Chen SY (2008) Soybean WRKY-type transcription factor genes, GmWRKY13, GmWRKY21, and GmWRKY54, confer differential tolerance to abiotic stresses in transgenic Arabidopsis plants. Plant Biotechnol J 6:486–503

Zhu D, Wu Z, Cao G, Li J, Wei J, Tsuge T, Gu H, Aoyama T, Qu LJ (2014) TRANSLUCENT GREEN, an ERF family transcription factor, controls water balance in *Arabidopsis* by activating the expression of aquaporin genes. Mol Plant 7:601–615

Zhu J-K (2016) Abiotic stress signaling and responses in plants. Cell 167:313–324

Zhu Q, Zhang J, Gao X, Tong J, Xiao L, Li W, Zhang H (2010) The *Arabidopsis* AP2/ERF transcription factor RAP2.6 participates in ABA, salt and osmotic stress responses. Gene 457:1–12

Zimmermann P, Hirsch-Hoffmann M, Hennig L, Gruissem W (2004) GENEVESTIGATOR. *Arabidopsis* microarray database and analysis toolbox. Plant Physiol 136:2621–2632

Newly Revealed Promising Gene Pools of Neglected *Brassica* Species to Improve Stress-Tolerant Crops

Mohammad Mafakheri and Mojtaba Kordrostami

Abstract To efficiently address the threatened food security worldwide and the gap between supply and demand for agricultural products from a growing population on the earth, novel gene pools are necessities to enable plant breeders to develop crops capable of providing high quality and nutritionally rich food products under the changing climate era with increasing the incidence of environmental abiotic and biotic stresses. *Brassica*, as an essential source of crops worldwide with extensive gene pool which the close relatives of domesticated species, crop wild relatives, known for highly applicable gene pool have been exploited by plant breeders. While the resistance of current cultivars to the damaging insect pests and pathogens is deteriorating due to constant evolution and favorable environmental requirements for their reproduction of pests, besides increasing in abiotic stresses, hence obtaining the field knowledge on potential resistance wild species, landraces, accessions, and CWR is highly required, since it is a prior step to utilize the current advancements in biotechnological tools that allowing a cost-efficient and reliable screening and sequencing of a large number of plant materials for the gene(s) in target in a time-saving manner. The development of agronomically important traits is mainly time-consuming and laborious because they are multigene-regulated. The gene pool of *Brassica* crops is narrow, thus exploiting and probing for potential gens among wild genetic resources is inevitable, which in *Brassica*, *Brassica villosa*, *Brassica incana*, and *Brassica montana* as wild relative of important vegetable crops manifested as a rich source for agronomic traits in particular insect-pest resistance, among amphidiploid species *Brassica carinata* has been observed to be a promising source of salt and drought tolerance gens and *Brassica tournefortii* has shown compatible potentials for drought tolerance. Further genome evaluation leads to the identification of urgently needed quantitative trait loci (QTL) and alleles in wild species and CWR, and the large-scale collaborations on the application of genomics to reveal the untapped wild species can

M. Mafakheri
Department of Horticultural Sciences, Faculty of Agricultural Sciences, University of Guilan, P.O. Box 41635-1314, Rasht, Iran

M. Kordrostami (✉)
Nuclear Agriculture Research School, Nuclear Science and Technology Research Institute (NSTRI), Karaj, Iran
e-mail: kordrostami009@gmail.com

M. Hasanuzzaman (ed.), *The Plant Family Brassicaceae*,
https://doi.org/10.1007/978-981-15-6345-4_4

greatly contribute in the acceleration of sustainable utilization of genetic resources and reach a balance in agricultural production.

Keywords Genome evaluation · QTL · Gene pool · *Brassica*

1 Introduction

The Earth is in a situation that supplying adequate food products is a matter of impossibility without the utilization of productive crop cultivars as well as insecticides and fungicides (Abouhussein 2012; Béné et al. 2015; Vosman et al. 2016). The exacerbation in challenge is foreseen by the exponential increase in human population and radical weather fluctuation which the former has led to intensive agriculture method that requires plenty of chemical fertilizers and insecticides, herbicides, and fungicides to operate, while the latter has imposed and increased the incidence of biotic and abiotic stresses (Hasanuzzaman and Fujita 2012; Taiz 2013; Josephson et al. 2014; Obaisi 2017; Hasanuzzaman et al. 2020). The projects on the earth population are around 10 billion by 2050, picture this rise in population growth is accompanied by losing arable lands to salinization on a large scale which the solely achievable, reliable, and cost-efficient strategy to address these challenges in agriculture sector is the development of tolerant crop cultivars (DeLucia et al. 2012; Cubasch et al. 2013; Gray and Brady 2016; Hansen et al. 2016; Fanzo et al. 2018). In this case, *Brassica* a known genus of mustard family includes a number of nutritionally economically important crops with closeness in taxonomy and significant diversity in morphology. Having over 100 wild species and weedy relatives provides unique genetic resources for the development of resistance against environmental stresses in *Brassica* crops (Zhang et al. 2014; Kumar et al. 2015; Šamec and Salopek-Sondi 2019). The genus commonly classified into oilseed, fodder crop, vegetable and condiment crops (Cartea et al. 2011) which canola (*Brassica napus* L.) is the third main crop for industrial source of edible oil (Wang et al. 2011; Hegelund et al. 2018), swede (*Brassica napobrassica* L.), turnip (*Brassica rapa* subsp. *Rapa* L.), kale (*Brassica oleracea* var. *sabellica* L.), and forage rapeseed (*Brassica napus* L.) are important forage crops in particular in the U.S and Europe (Westwood and Mulcock 2012; Vosman et al. 2016), vegetable crops in this genus are three important species *B. oleracea* (e.g., kale, cauliflower, broccoli, cabbage, Brussels sprouts, Savoy cabbage), *B. juncea* (e.g., several mustard varieties), and *B. rapa* (e.g., turnip, Chinese cabbage, pak choi) (Cartea et al. 2011; Warwick 2011; Bhandari and Kwak 2015), and finally a widely cultivated condiment crop from *Brassica* is black mustard (*B. napus*) (Borpatragohain et al. 2019). A capable species from this genus as a potential oil crop, *Brassica carinata* (Ethiopian rape), during the last two decades has attracted the attention of scholars (Warwick et al. 2006; Kaur and Sharma 2016; Hossain et al. 2018). Given the enormous diversity in brassica genus in addition to related wild relatives and lesser-known species, all make brassica an exceptional untapped genetic resource for important urgently needed commercial and agronomical characteristics, additionally, a number of biodiversity databases related to *Brassica* are accessible to assist genetic scholars, managers of genetic resources, and plant breeders to appropriately opting taxa for evolutionary relationship studies and screening for agronomically important traits (Warwick 2011; Chen et al. 2018; El-Esawi 2018; Rao et al. 2019). In this chapter, we attempt to summarize the limited published data on monitoring and utilizing Brassica germplasm, landraces, accessions, and CWR to develop biotic or abiotic stress resistance varieties.

2 Developing Insect-Pest and Pathogen Resistance Crops Varieties

Climate change has the capability to generate an ecological upheaval that could have increased the outbreak of insect pests, as a result, the application of insecticides may no longer be a viable option owing to acceleration in a number of generation per year and consequently ineffective of commercial pesticides in a rather short period. Additionally, the population of natural enemies of pests may negatively be affected by changes in climate and further facilitate the conditions in the favor of pests (DeLucia et al. 2012; Vosman et al. 2016; Rehman and Kumar 2018; Singh et al. 2018; Xu et al. 2019). Two of the large and commercially known species *B. oleracea*, and *B. rapa* have several of important vegetable crops and close to half million hectares cultivated area with vegetable crops where they have a significant annual crop loss to pests such as cabbage thrips, root flies, aphids, caterpillars, and whiteflies, especially cabbage aphid (*Brevicoryne brassicae*) and whitefly (*Aleyrodes proletella*) that have a devastating impact on brussels sprouts, kale, and savoy cabbage. These pests are sap-sucking from the phloem on the underside of leaves, hence in addition to disruption of growth, secretion of honeydew which is a media favorable of molds, application of effective control is also difficult, therefore, resistance cultivars can be highly advantageous (Gulidov and Poehling 2013; Springate 2016, 2017). Defensive mechanisms in plants are physical or chemical which can play their role upon the occurrence of attack or some are permeant as physical barriers. Some specific evolved characteristics in leaves such as a wax layer or leaf toughness can prohibit the feeding or settlement of leaf-eating insect pests. Similarly, deterrence essential oils that repel attackers or proteins that their digestion negatively affects the herbivores are other parts of this arms race that plants have against insect pests (Broekgaarden et al. 2011; Mithöfer and Boland 2012; War et al. 2012; Mitchell et al. 2016). From the defensive secondary chemicals in species of *Brassica*, Glucosinolates, are extensively investigated compounds, which their enzymatic hydrolysis at the time cell disruption leads to the generation of toxic constituents, for instance, isothiocyanates (Cole 1997; Halkier and Gershenzon 2006; Poelman et al. 2009). Nonetheless, specialist herbivore insects mostly have the capability to enzymatically detoxify glucosinolates or their byproducts (Ratzka et al. 2002; Bruce 2014). Hence, resistance in *Brassica* against specialized herbivores is possibly more dependent on other compounds or proteins rather than glucosinolates, which to date, some responsible genes for encoding them have been detected (Smith and Clement 2012; De Vos and VanDoorn 2013). The initiation step in developing insect-resistant varieties is the need to determine resistance sources that more often than not are wild species of the same genus, crop wild relatives (CWR), or landraces (LR). After the identification of the source, the putative genes required to be detected to perform an efficient transformation to the crop of interest (Broekgaarden et al. 2011).

In this case, a group of scholars under a program named "The PGR Secure project" (Vosman et al. 2016) conducted a comprehensive series of studies to developed resistance varieties in *Brassica oleracea* and *Chinese cabbage* to cabbage whitefly and aphid, their endeavors summarized here. Pelgrom et al. (2012) screened a sum of 432 accessions from wild species and landraces of *B. oleracea* and some CWR to identify potential cabbage whitefly and aphid resistance accession(s). Using simple field tests in which the infested plants with those pests utilized to exclude the vulnerable accessions in a low-cost and time-saving manner and facilitated the monitoring of a number of accessions (Pelgrom et al. 2015). The resistance group opted at the first stage again tested under the greenhouse atmosphere. Numerous accessions among landraces of *B. oleracea* var. capitate (heading cabbage) found, which formerly a cultivar of this cabbage, (cv.) Rivera, insect-resistance was reported (Broekgaarden et al. 2009, 2011), the outcomes to this point showed that (cv.) Rivera is not exceptional and the resistance can be identified in landraces of heading cabbage. Similarly, (Pelgrom et al. 2015) in less-known species of *Brassica* including *B. villosa, B. incana,* and *B. montana* witnessed the resistance. While the resistance accessions against whiteflies detected in the species of *Brassica*, the resistance to aphid found to be in CWR, which all indicates the essentiality of considering unknown species as well as CWR in order to take a larger scope and increase the chance of identifying the resistance source. From the accessions of *B. villosa* and *B. incana*, some had specific morphological differences such as the higher intensity of trichomes that possibly coffers the resistance (Maxted et al. 2012). To locate the resistance factor in the identified accessions, the electrical penetration graph or PEG which is a simple system composed of a partial circuit that completed solely when an insect begins to enter its stylet into the veins to find proper host for feeding, in this project they found the phloem where the resistance factor is located since on resistance accessions of *B. villosa, B. incana,* and *B. montana* cabbage aphids and whiteflies had a long period of sustained phloem sap ingestion in contrary to sensitive accessions of *B. oleracea.* Therefore, the anatomical differences between resistance and vulnerable accessions had no association with resistance (Broekgaarden et al. 2011). Such valuable data could greatly contribute in understand the location of resistance which further assists in developing a multi mechanism resistance with higher effectiveness as well as stability. To pinpoint the underlying molecular relationship involve in resistance to the pests of interest molecular markers are a significant help that could accelerate the whole process to another level, to this end, single nucleotide polymorphism (SNP) utilized to identify the informative SNP markers in wild *Brassica* species, CRW, and *B. oleracea* with the aim of RNAseq analyzing, as result, the distribution pattern of the SNPs on the various pseudochromosomes and scaffolds of the *B. oleracea* genome obtained. The acquired array can be utilized extensively in studies such as mapping of QTL in *B. oleracea* and CWR, association mapping in *B. oleracea*, and finally association analysis among species, subspecies, varieties, and landraces.

In an effective resistance of host plant to sap-sucking pests numerous mechanisms can be involved (Broekgaarden et al. 2011). Using QTL mapping the physical position on chromosome associated with resistance to whitefly what identified. From the screening of the accessions, the parents, resistance, and sensitivity were selected. As a result of crossing cv. Rivera (resistant to whitefly) with cv. Christmas Drumhead (vulnerable) an F2 population was created. From this population a number of individuals identified with resistance-related anatomical characteristics (time of head formation, leaf wax layer, and leaf toughness), and low rates of whitefly adult survival and oviposition. With SNP markers the genotyping of the population was carried out. The whitefly resistance-associated putative QTLs for adult survival and oviposition rate were identified. A major QTL located for the "wax layer," explaining over 50% of its variation. No co-localization between the morphological features and adult survival or oviposition rate was observed. Irrespective of significant resistance of cv. Rivera against the cabbage whitefly, absence of effective QTLs for survival and oviposition rate was witnessed, which supports the possibility of multigene control of tolerance in this variety or present of a basically different defensive mechanism. In another cross between a completely resistant accession of *B. incana* to whitefly and a sensitive cultivar of *B. oleracea* which the obtained F1 was backcrossed with the parent, *B. incana*, that the mapping of this population for whitefly resistance focused on a single locus which 57% of the variation of whitefly adult survival and 82% of oviposition rate explained. The existence of trichomes on the exact same locus was mapped, that with a significant negative correlation for the existence of trichomes, all lead to this conclusion the trichomes presence is probably accountable for the recorded resistance, it needs to be taken into consideration that maybe a gene responsible for conferring the resistance is tightly linked to the gene regulating the presence/absence of the trichomes. To clarify this ambiguity comprehensive studies are required.

Complementary studies on resistance of cv. Rivera to the most damaging insect-pest of *Brassica* worldwide, cabbage whitefly, followed by Broekgaarden et al. (2018) through carrying out a comprehensive investigation exploiting several omics techniques: transcriptome, metabolome profiling, and genetic mapping to unmask the underlying molecular mechanism involved in this resistance to whitefly. To understand the effect of age in cv. Rivera, transcriptome profiles of susceptible (young) and resistant (older) employing RNA sequencing were assessed, it was observed that in older plants a number of processes associated with defense mechanisms were overexpressed. While the concentration of phytohormones jasmonic acid (JA) decremented with no age dependency as the plants infested by whitefly, abscisic acid (ABA) on the contrary increased dependent on the plant age in reaction to whitefly infestation, that is, the concentration of ABA in old plants enhanced whereas lessened in young plants. The biosynthesis of Auxin was considerably lower at older plants and higher in young ones regardless of the insect infestation (Khan et al. 2014), while glucosinolate concentration increased in older plants. QTL mapping indicated the presence of QTL intervals with large effects on chromosome 2 and 9 for oviposition rate and whitefly adult survival, respectively. The overexpressed genes in older plants found on the identified QTL intervals observed to be orthologous of ABA signaling-related genes in Arabidopsis, emphasizing the regulatory role of ABA in resistance against whiteflies.

The essentiality of wild species to procure resistance gens is crystal clear and requires prior knowledge on resistance source and underlying mechanisms, in which Subramanian et al. (2005) to develop resistance in oilseed crops of *Brassica* against blackleg caused by a deadly fungal pathogen, *Leptosphaeria maculans*, carried out an interspecific cross of rapeseed (*B. napus*) as a sensitive parent and Ethiopian rapeseed (*B. carinata*) the resistant parent. Proteome analysis revealed the protein bases of resistance, in resistance progenies infected with *L. maculans* enzymatic antioxidant (superoxide dismutase), key enzymes such as nitrate reductase, and carbonic anhydrase, as well as photosynthetic enzymes, enhanced significantly. Explorations in *Brassica* germplasm regarding identifying resistance genes for insect-pest or pathogens have been ample which reflects the major demand from breeders of *Brassica* and economical significance of crop losses in *Brassica* crops worldwide.

3 Probing for Salt and Drought Tolerance Resources

One of the highly complicated multigene-regulated agronomic characteristics in various plant species is tolerance to salinity stress (Zhang et al. 2014; Lang et al. 2017; Lopez et al. 2018). The interspecific, as well as intraspecific variation for salt stress within Brassica, is expansive (Kumar et al. 2009; Zhang et al. 2014). The tolerance mechanisms vary greatly as the intensity of resistance during ontological phases. However, these mechanisms in a specific plant species are operational only in an especial window of time. Additionally, the influence mechanism of one process is likely to rule out the impact of ongoing processes at a specific time (Yeo 1998; Carvajal et al. 1999; Hussain et al. 2018). Resistance to salinity has a significant dependency on developmental stages which means that resistance at one developmental phase may not necessarily reflect tolerance in the other stages, as case in point in several crops such as barley, corn, rice, tomato, and wheat, enhancement in salt resistance observed as plants age (Kazachkova et al. 2013; Julkowska 2019). To make the matter more complex, amphidiploid species have a higher capability in survival against stresses than diploid parents (Zhu et al. 2016). Several studies have undertaken the salt stress tolerance in amphidiploid species of *Brassica* in comparison to diploid species, Ashraf et al. (2001) reported that the amphidiploid species including *B. carinata*, *B. juncea*, and *B. napus* had a superior ability to tolerance salt when compared with diploid species, *B. campestries*, *B. nigra*, and *B. oleraceae*. Further, Ashraf et al. (1989) evaluated the response of *B. carinata, B. juncea, B. napus, and B. campestries* when they received four types of various salts CaCl2, MgCl2, NaCl, and Na2SO4, they found NaCl more toxic, comparably. Also, the germination rate of all four species reduced significantly and indiscriminately except for *B. napus*, which had higher seedling growth and germination rate under salt stress.

On the other hand, there are contradictory results on salinity resistance of *B. carinata*, He and Cramer (1992) compared the resistance of *B. napus*, *B. campestris*, *B. nigra*, *B. juncea*, *B. oleracea*, and *B. carinata* to seawater and interestingly, *B. carinata* found to have susceptibility to salt while *B. napus* was highly tolerant and the rest were moderately resistance. Correspondingly, Kumar et al. (2009) as a result of their assessments on various species of *Brassica* for salt tolerance source, found *B. carinata* is sensitive or at last moderately salt tolerance.

Additionally, the higher ability of *B. carinata* and *B. napus* to withstand salinity in comparison to *B. campestries* suggested by Ashraf and McNeilly (1990). Following studies on salt tolerance of *B. carinata* and *B. napus* Ashraf (2001) subjected these two species to salt stress in comparative study with sensitive diploid species which the resistant amphidiploid species produced a considerably higher biomass in addition to several another salt tolerance-associated physiological responses including higher photosynthesis rate, reduced stomatal conductance, and greater water use efficiency. The capability of plants to cope with salt stress and the ability to discriminate themselves from sensitive genotypes all indicate the existence of physiological and genetical bases for resistance to salinity.

Drought resistance is one of the quantitative complex traits, thus unmasking the complete genetic control over responses of the plant to drought stress is considerably difficult (Jiang et al. 2017; Oladosu et al. 2019). More, the absence of proper selection measures also impeded the development of varieties resistant to drought stress. There a few published research on screening wild species of *Brassica* to discover drought resistance source, Rashidi et al. (2017) reported a relatively large-scale study on 36 genotypes of oilseed rape derived from a group of seven Brassica such as *B. napus*, *B. rapa*, *B. juncea*, *B. carinata*, *B. oleracea*, *B. nigra*, and *Brassica fruticulosa* subjected them to water deficiency. The seed yield in all water-stressed genotypes declined. The highest tolerance and productivity found to be in *B. napus*, as it has been frequently reported (Kagale et al. 2007; Ashraf and Ali 2008; Dalal et al. 2009; Zhang et al. 2014; Wang et al. 2017; Channaoui et al. 2019). The moderate level drought tolerance observed in *B. fruticulosa*. Their applicable measures also reported in this study to select genotypes with high yield under water stress including mean productivity (MP), geometric mean productivity (GMP), and stress tolerance index (STI). Another Brassica species are known for its salt tolerance, *B. carinata*, also in this study observed to be drought-tolerant as well. Considering yield as a measure, highly negatively affected species were *B. rapa* and *B. oleracea*. Other wild species of *Brassica* observed to have the capability to resistance drought, *B. tournefortii* reported showing superior ecological performance under drought conditions in open field in comparison with several brassica species (Guo 2015). Analogous results from *B. tournefortii* also recorded by Prakash and Bhat (2007). Further, *B. tournefortii* and B. carinata are potential thermotolerances (Salisbury and Barbetti 2011; Singh and Chandra 2005) that further molecular studies may unveil the gene(s) in control of abiotic stress tolerance in these species. The huge diversity in *Brassica* germplasm beside the presence of contrasting wild species in *Brassica* is significantly valuable owing to numerous benefits that such an opportunity can have for breeders.

The entire advantages of wild species described in details above without the knowledge of cytogenetic variation of cultivated and wild species of *Brassica*, the affinity of their chromosomes and interspecific crossability would be useless (Choudhary et al. 2002; Prakash et al. 2009), also identified and helps the breeder to identify the desired alleles and to control the desirable traits and ultimately to transfer them to the cultivars (Bing et al. 1995; Ana et al. 2009; Bocianowski et al. 2018). In order to study the cytogenetic diversity of rapeseed cultivars and two wild species including wild cabbage and Mediterranean mustard Dehdari (2012) conducted a chromosomal study that showed high intraspecies similarity so that different rapeseed cultivars, both spring, and autumn were not significantly different in relation to the total length of chromosomes number 1–19. For most of the other measured characteristics, the difference was not significant in most cases. Mustard masses showed no significant difference in measured chromosomal characteristics except for the short arm length ratio of chromosome 3. This result clearly shows that all the Mediterranean mustard studied despite being from different regions with water conditions and completely different climates were collected but did not have much chromosomal variation. Among the many traits measured, only the ratio of the total length of chromosome 7 and the short arm ratio of chromosome 6 were significantly different between the two wild cabbage masses. Results of analysis of variance and cluster analysis of the studied species based on the characteristics of intrachromosomal asymmetry index, interchromosomal asymmetry index, percentage of total shape, maximum and minimum relative length difference, and the number of mitotic chromosomes indicated cross-species similarities and the possibility of successful hybridization between the selected genotypes.

4 Conclusion and Future Prospects

The germplasm of wild *Brassica* is capable to provide essential agronomic and economic traits and new crops. Significant endeavors have been made to explore the *Brassica* germplasm to find desirable genetic materials that could be imported into breeding projects. Considering the rise in demand for agricultural products as a consequence of the growth in population on the earth in addition to the importance of *Brassica* crops as a vegetable, oil, animal feed, and condiment information on untapped germplasms of *Brassica* is pivotal to fuel the breeding programs. Several wild species of *Brassica* have found to have significant capabilities from resistance against insect pests and pathogens to tolerance of abiotic stresses. Of course, many of studies still are in the first step and generating information from field screening of potential wild species, CWR, landraces, and accessions of crops and a few took advantage of biotechnological toolbox, but the background data are cornerstone of future projects relying on using gene sequencing techniques, in a nutshell, it is a combination of studies at macro or micro scale that makes progress possible in developing new cultivars that can cope with biotic or abiotic stresses under changing climate.

Over the recent decades, a great deal of information has been published and shared on databases, however, the absence of systematic holistic approaches to screen extensively diverse *Brassica* germplasm for agronomically important QTLs is vivid. The presented data so far is partial and fragmentary, although they are highly invaluable, to a have a bigger picture, conducting comprehensive studies is highly needed. There is ongoing degradation of genetic resources owing to intensive agriculture and climate change, hence attempts to explore germplasms should be accelerated. Identification of QTLs and their precise location on the chromosome can significantly facilitate the transformation of candidate genes into elite lines through gen stacking.

References

Abouhussein S (2012) Climate change and its impact on the productivity and quality of vegetable crops. J Appl Sci Res 8:4359–4383

Ana M-J, Ankica K-S, Dejana S-P, Radovan M, Nikola H (2009) Phenotypic and molecular evaluation of genetic diversity of rapeseed (*Brassica napus* L.) genotypes. Afric J Biotechnol 8(19):4835–4844

Ashraf M (2001) Relationships between growth and gas exchange characteristics in some salt-tolerant amphidiploid Brassica species in relation to their diploid parents. Environ Exp Bot 45:155–163

Ashraf M, Ali Q (2008) Relative membrane permeability and activities of some antioxidant enzymes as the key determinants of salt tolerance in canola (*Brassica napus* L.). Environ Exp Bot 63:266–273

Ashraf M, Bokhari M, Mehmood S (1989) Effects of four different salts on germination and seedling growth of four *Brassica* species. Biologia 35:173–187

Ashraf M, McNeilly T (1990) Responses of four Brassica species to sodium chloride. Environ Exp Bot 30:475–487

Ashraf M, Nazir N, McNeilly T (2001) Comparative salt tolerance of amphidiploid and diploid *Brassica* species. Plant Sci 160:683–689

Béné C, Barange M, Subasinghe R, Pinstrup-Andersen P, Merino G, Hemre G-I, Williams M (2015) Feeding 9 billion by 2050–putting fish back on the menu. Food Sec 7:261–274

Bhandari S, Kwak J-H (2015) Chemical composition and antioxidant activity in different tissues of *Brassica* vegetables. Molecules 20:1228–1243

Bing D, Downey R, Rakow G (1995) An evaluation of the potential of intergeneric gene transfer between *Brassica napus* and *Sinapis arvensis*. Plant Breed 114:481–484

Bocianowski J, Niemann J, Nowosad K (2018) Genotype-by-environment interaction for seed quality traits in interspecific cross-derived *Brassica* lines using additive main effects and multiplicative interaction model. Euphytica 215:7. https://doi.org/10.1007/s10681-018-2328-7

Borpatragohain P, Rose TJ, Liu L, Raymond CA, Barkla BJ, King GJ (2019) Seed glucosinolate yield is maximized by higher rates of sulfur nutrition than required for seed yield in condiment mustard (*Brassica juncea* L.). PloS One 14:e0213429

Broekgaarden C et al (2018) Combining QTL mapping with transcriptome and metabolome profiling reveals a possible role for ABA signaling in resistance against the cabbage whitefly in cabbage Plos One 13:e0206103. https://doi.org/10.1371/journal.pone.0206103

Broekgaarden C, Poelman EH, Voorrips RE, Dicke M, Vosman B (2009) Intraspecific variation in herbivore community composition and transcriptional profiles in field-grown *Brassica oleracea* cultivars. J Exp Bot 61:807–819. https://doi.org/10.1093/jxb/erp347

Broekgaarden C, Snoeren TA, Dicke M, Vosman B (2011) Exploiting natural variation to identify insect-resistance genes. Plant Biotechnol J 9:819–825. https://doi.org/10.1111/j.1467-7652.2011.00635.x

Bruce T (2014) Glucosinolates in oilseed rape: secondary metabolites that influence interactions with herbivores and their natural enemies. Ann Appl Biol 164:348–353

Cartea M, Lema M, Francisco M, Velasco P (2011) Basic information on vegetables Brassica crops. In: Sadowski J, Kole C (eds) Genetics, genomics and breeding of vegetable brassicas. CRC Press, pp 1–34

Carvajal M, Martínez V, Alcaraz CF (1999) Physiological function of water channels as affected by salinity in roots of paprika pepper. Physiol Plant 105:95–101. https://doi.org/10.1034/j.1399-3054.1999.105115.x

Channaoui S, El Idrissi IS, Mazouz H, Nabloussi A (2019) Reaction of some rapeseed (*Brassica napus* L.) genotypes to different drought stress levels during germination and seedling growth stages. OCL 26:23

Chen H, Al-Shehbaz IA, Yue J, Sun H (2018) New insights into the taxonomy of tribe Euclidieae (*Brassicaceae*), evidence from nrITS sequence data. PhytoKeys 100:125–139

Choudhary B, Joshi P, Rao SR (2002) Cytogenetics of *Brassica juncea* × *Brassica rapa* hybrids and patterns of variation in the hybrid derivatives. Plant Breed 121:292–296

Cole R (1997) The relative importance of glucosinolates and amino acids to the development of two aphid pests *Brevicoryne brassicae* and *Myzus persicae* on wild and cultivated *Brassica* species. Entomol Exp Appl 85:121–133

Cubasch U, Wuebbles D, Chen D, Facchini MC, Frame D, Mahowald N, Winther J-G (2013) Introduction. In: Plattner K, Tignor M, Allen SK, Boschung J, Nauels A, Xia Y, Bex V, Midgley PM (eds) Climate change 2013: the physical science basis. Contribution of working group I to the fifth assessment report of the intergovernmental panel on climate change. Cambridge University Press, Cambridge, UK, and New York. http://www.climatechange2013org/images/report/WG1AR5_Chapter01_FINALpdf

Dalal M, Tayal D, Chinnusamy V, Bansal KC (2009) Abiotic stress and ABA-inducible Group 4 LEA from Brassica napus plays a key role in salt and drought tolerance. J Biotech 139:137–145

De Vos M, VanDoorn A (2013) Resistance to sap-sucking insects in modern-day agriculture. Front Plant Sci 4:222

Dehdari A (2012) Cytogenetical evaluation of canola cultivars and two wild species of *Brassica.* J Molec Cell Res 26:446–461

DeLucia EH, Nabity PD, Zavala JA, Berenbaum MR (2012) Climate change: resetting plant-insect interactions. Plant Physiol 160:1677–1685

El-Esawi MA (2018) *Brassica* germplasm: characterization, breeding and utilization. IntechOpen, USA

Fanzo J, Davis C, McLaren R, Choufani J (2018) The effect of climate change across food systems: implications for nutrition outcomes. Glob Food Secur 18:12–19

Gray SB, Brady SM (2016) Plant developmental responses to climate change. Dev Biol 419:64–77

Gulidov S, Poehling H-M (2013) Control of aphids and whiteflies on Brussels sprouts by means of UV-absorbing plastic films. J Plant Dis Protect 120:122–130

Guo Y (2015) Screening for drought tolerance in *Brassica rapa* L.: from genetic variation to gene expression. The University of Western Australia

Halkier BA, Gershenzon J (2006) Biology and biochemistry of glucosinolates. Annu Rev Plant Biol 57:303–333

Hansen J et al (2016) Ice melt, sea level rise and superstorms: evidence from paleoclimate data, climate modeling, and modern observations that 2 C global warming could be dangerous. Atmos Chem Phys 16:3761–3812

Hasanuzzaman M, Fujita M (2012) Heavy metals in the environment: current status, toxic effects on plants and possible phytoremediation. In: Anjum NA, Pereira MA, Ahmad I, Duarte AC, Umar S, Khan NA (eds) Phytotechnologies: remediation of environmental contaminants. CRC Press, Boca Raton, pp 7–73

Hasanuzzaman M, Mohsin SM, Bhuyan MHMB, Bhuiyan TF, Anee TI, Masud AAC, Nahar K (2020) Phytotoxicity, environmental and health hazards of herbicides: challenges and ways forward. In: Prasad MNV (ed) Agrochemicals detection, treatment and remediation: pesticides and chemical fertilizers. Butterworth-Heinemann, London, pp 55–99

He T, Cramer GR (1992) Growth and mineral nutrition of six rapid-cycling *Brassica* species in response to seawater salinity. Plant Soil 139:285–294. https://doi.org/10.1007/BF00009320

Hegelund JN, Liang C, Lauridsen UB, Kemp O, Lutken H, Muller R (2018) Increasing genetic variability in oilseed rape (*Brassica napus*)—genotypes and phenotypes of oilseed rape transformed by wild type *Agrobacterium rhizogenes*. Plant Sci 271:20–26

Hossain Z et al (2018) Agronomic responses of Brassica carinata to herbicide, seeding rate, and nitrogen on the Northern Great Plains. Crop Sci 58:2633–2643

Hussain HA, Hussain S, Khaliq A, Ashraf U, Anjum SA, Men S, Wang L (2018) Chilling and drought stresses in crop plants: implications, cross talk, and potential management opportunities. Front Plant Sci 9. https://doi.org/10.3389/fpls.2018.00393

Jiang Y et al (2017) Quantitative trait loci associated with drought tolerance in *Brachypodium distachyon*. Front Plant Sci 8. https://doi.org/10.3389/fpls.2017.00811

Josephson AL, Ricker-Gilbert J, Florax RJGM (2014) How does population density influence agricultural intensification and productivity? Evidence from Ethiopia. Food Policy 48:142–152

Julkowska MM (2019) Changing gears in salt stress tolerance: *AtCAMTA6*'s role in transcriptional regulation of ion transport. Plant Physiol 180:684–685

Kagale S, Divi UK, Krochko JE, Keller WA, Krishna P (2007) Brassinosteroid confers tolerance in *Arabidopsis thaliana* and *Brassica napus* to a range of abiotic stresses. Planta 225:353–364. https://doi.org/10.1007/s00425-006-0361-6

Kaur M, Sharma VSP (2016) Quality of component crops as influenced by intercropping of canola oilseed rape (*Brassica napus*) and Ethiopian mustard (*Brassica carinata*) with Indian rape (*Brassica rapa* var. Toria). J Oilseed Brassica 1:38–44

Kazachkova Y, Batushansky A, Cisneros A, Tel-Zur N, Fait A, Barak S (2013) Growth platform-dependent and -independent phenotypic and metabolic responses of *Arabidopsis* and its halophytic relative, Eutrema salsugineum, to salt stress. Plant Physiol 162:1583–1598

Khan M, Rozhon W, Poppenberger B (2014) The role of hormones in the aging of plants—a mini-review. Gerontology 60:49–55

Kumar G, Purty RS, Sharma MP, Singla-Pareek SL, Pareek A (2009) Physiological responses among *Brassica* species under salinity stress show strong correlation with transcript abundance for SOS pathway-related genes. J Plant Physiol 166:507–520. https://doi.org/10.1016/j.jplph.2008.08.001

Kumar M, Choi J, Kumari N, Pareek A, Kim S (2015) Molecular breeding in Brassica for salt tolerance: importance of microsatellite (SSR) markers for molecular breeding in *Brassica*. Front Plant Sci 6. https://doi.org/10.3389/fpls.2015.00688

Lang L et al (2017) Quantitative trait locus mapping of salt tolerance and identification of salt-tolerant genes in *Brassica napus* L. Front Plant Sci 8:1000. https://doi.org/10.3389/fpls.2017.01000

Lopez C, Orazaly M, Mozzoni L, Korth KL, Chen P (2018) Quantitative trait loci for salt tolerance in soybean. J Crop Improv 32:766–780. https://doi.org/10.1080/15427528.2018.1516261

Maxted N, Kell S, Ford-Lloyd B, Dulloo E, Toledo Á (2012) Toward the systematic conservation of global crop wild relative diversity. Crop Sci 52:774–785

Mitchell C, Brennan RM, Graham J, Karley AJ (2016) Plant defense against herbivorous pests: exploiting resistance and tolerance traits for sustainable crop protection. Front Plant Sci 7. https://doi.org/10.3389/fpls.2016.01132

Mithöfer A, Boland W (2012) Plant defense against herbivores: chemical aspects. Ann Rev Plant Biol 63:431–450

Obaisi A (2017) Overpopulation: a threat to sustainable agriculture and food security in developing countries? A review. https://doi.org/10.13140/RG.2.2.20613.04325

Oladosu Y et al (2019) Drought resistance in rice from conventional to molecular breeding: a review. Int J Mol Sci 20:3519. https://doi.org/10.3390/ijms20143519

Pelgrom K et al (2012) Looking for resistance to phloem feeders in *Brassica olerace*. Crop Wild Relat 8:12–14

Pelgrom KTB, Broekgaarden C, Voorrips RE, Bas N, Visser RGF, Vosman B (2015) Host plant resistance towards the cabbage whitefly in *Brassica oleracea* and its wild relatives. Euphytica 202:297–306. https://doi.org/10.1007/s10681-014-1306-y

Poelman EH, Dam NM, Loon JJA, Vet LE, Dicke M (2009) Chemical diversity in *Brassica oleracea* affects biodiversity of insect herbivores. Ecology 90:1863–1877

Prakash S, Bhat S (2007) Contribution of wild crucifers in *Brassica* improvement: past accomplishment and future perspectives. In: Proceedings of GCIRC 12th international rapeseed congress, 2007, pp 213–215

Prakash S, Bhat S, Quiros C, Kirti P, Chopra V (2009) Brassica and its close allies: cytogenetics and evolution. Plant Breed Rev 31:21–187

Rao M, Bhat S, Gupta N, Chamola R, Singh N, Bhatacharya R (2019) NRCPB rapa 8 (IC374272)(IC0623820; INGR17050), a Yellow Sarson (*Brassica rapa*) germplasm with potential as a parent for resynthesis of *B. juncea*. Indian J Plant Genet Res 32:257

Rashidi F, Majidi M, Pirboveiry M (2017) Response of different species of *Brassica* to water deficit. Inter J Plant Prod 11:1–16

Ratzka A, Vogel H, Kliebenstein DJ, Mitchell-Olds T, Kroymann J (2002) Disarming the mustard oil bomb. Proc Natl Acad Sci USA 99:11223–11228

Rehman SA, Kumar R (2018) Scenario of insect pests under changing climatic situations. Inter J Chem Stud 6:77–81

Salisbury PA, Barbetti MJ (2011) Breeding oilseed Brassica for climate change. Crop adaptation to climate change 448–463

Šamec D, Salopek-Sondi B (2019) Cruciferous (*Brassicaceae*) vegetables. In: Nonvitamin and nonmineral nutritional supplements. Elsevier, pp 195–202

Singh B, Singh B, Singh P (2018) Breeding cauliflower: a review. Inter J Veget Sci 24:58–84

Singh D, Chandra N (2005) Oilseed *Brassica* improvement. In: Haryana state: genetic resources. 14th Australian research assembly on Brassicas, Port Lincoln, proceedings. SARDI, primary industries and resources South Australia, pp 47–51

Smith CM, Clement SL (2012) Molecular bases of plant resistance to arthropods. Ann Rev Entom 57:309–328

Springate S (2016) The cabbage whitefly, Aleyrodes proletella: causes of outbreaks and potential solutions. Doctoral dissertation, University of Greenwich

Springate S (2017) The cabbage whitefly *Aleyrodes proletella* and its natural enemies on wild cabbage *Brassica oleracea* on the Kent coast. Trans Kent Field Club 20:42

Subramanian B, Bansal VK, Kav NN (2005) Proteome-level investigation of *Brassica carinata*-derived resistance to *Leptosphaeria maculans*. J Agric Food Chem 53:313–324. https://doi.org/10.1021/jf048922z

Taiz L (2013) Agriculture, plant physiology, and human population growth: past, present, and future. Theor Exp Plant Physiol 25:167–181

Vosman B et al (2016) Using phenomics and genomics to unlock landrace and wild relative diversity for crop improvement. In: Enhancing crop genepool use: capturing wild relative and landrace diversity for crop improvement. CABI, pp 1–9

Wang N et al (2011) Flowering time variation in oilseed rape (*Brassica napus* L.) is associated with allelic variation in the FRIGIDA homologue BnaA.FRI.a. J Exp Bot 62:5641–5658. https://doi.org/10.1093/jxb/err249

Wang P, Yang C, Chen H, Song C, Zhang X, Wang D (2017) Transcriptomic basis for drought-resistance in *Brassica napus* L. Sci Rep 7:40532. https://doi.org/10.1038/srep40532

War AR, Paulraj MG, Ahmad T, Buhroo AA, Hussain B, Ignacimuthu S, Sharma HC (2012) Mechanisms of plant defense against insect herbivores. Plant Signal Behav 7:1306–1320. https://doi.org/10.4161/psb.21663

Warwick SI (2011) *Brassicaceae* in agriculture. In: Genetics and genomics of the Brassicaceae. Springer, pp 33–65

Warwick SI, Gugel RK, McDonald T, Falk KC (2006) Genetic variation of Ethiopian mustard (*Brassica carinata* A. Braun) germplasm in Western Canada. Genet Res Crop Evol 53:297–312. https://doi.org/10.1007/s10722-004-6108-y

Westwood C, Mulcock H (2012) Nutritional evaluation of five species of forage brassica. In: Proceedings of the New Zealand Grassland Association, 2012. New Zealand Grassland Association, pp 31–38

Xu H, Xie H, Wu S, Wang Z, He K (2019) Effects of elevated CO_2 and increased N fertilization on plant secondary metabolites and chewing insect fitness. Front Plant Sci 10:739

Yeo A (1998) Molecular biology of salt tolerance in the context of whole-plant physiology. J Exp Bot 49:915–929. https://doi.org/10.1093/jxb/49.323.915

Zhang X, Lu G, Long W, Zou X, Li F, Nishio T (2014) Recent progress in drought and salt tolerance studies in *Brassica* crops. Breed Sci 64:60–73. https://doi.org/10.1270/jsbbs.64.60

Zhu M et al (2016) Molecular and systems approaches towards drought-tolerant canola crops. New Phytol 210:1169–1189. https://doi.org/10.1111/nph.13866

Enhancement of Abiotic Stress Tolerance in *Camelina sativa*: Conventional Breeding and Biotechnology

Larysa V. Nishchenko and Mirza Hasanuzzaman

Abstract *Camelina* (*Camelina sativa* (L.) Crantz] (*Brassicaceae*) is a plant currently being researched as an alternative oilseed crop. The interest in *Camelina* is growing rapidly because the species has several agronomic advantages: high productivity, low-input agriculture, cold weather tolerance, short growing season (85–100 days), grows well in semi-arid regions, and low-fertility or saline soils. In addition, *Camelina* has been reported to be naturally resistant to diseases such as *Alternaria brassicae* and insect pests. Also, *Camelina* seed has higher oil content (>40%) with beneficial fatty acids composition that makes it suitable for industrial and nutritional applications. Therefore, it is a potential oil crop for sustainable agriculture. Different abiotic stresses cause disturbances in physiological and biochemical processes in plants. Due to the influence of these factors, fluctuations in gross harvest are observed. Therefore, an important task during the cultivation of plants is to obtain high productivity of crops, regardless of climatic conditions. The most effective farming can be achieved by integrating conventional breeding, physiology, and molecular biology. There are some achievements in this. So, this chapter summarizes advances in improving the resistance of *Camelina* to adverse environmental factors, viz. drought, salinity, extreme temperatures, and toxic metals.

Keywords Brassicaceae, oilseed crops · Stress tolerance · Plant breeding · Future crops

L. V. Nishchenko (✉)
Institute of Food Biotechnology and Genomics, National Academy of Sciences of Ukraine, Kiev, Ukraine
e-mail: 10lesya0916@gmail.com

M. Hasanuzzaman
Department of Agronomy, Faculty of Agriculture, Sher-e-Bangla Agricultural University, Dhaka 1207, Bangladesh

M. Hasanuzzaman (ed.), *The Plant Family Brassicaceae*,
https://doi.org/10.1007/978-981-15-6345-4_5

1 Introduction

Camelina is a crop that is being actively investigated as a production base for biofuel. This is mainly because its seeds are high in oil (30–40% of oil by dry weight) combined with several outstanding agronomic features. Unsaturated fatty acids include approximately 90% of the total seed oils among which the essential polyunsaturated fatty acids constitute over 50% of total fatty acids (35–40% a-linolenic acid (18:3n-3), 20–25% linoleic acid (18:2n-6), 15% eicosanoic acid (20:1n-9), and 3% erucic acid (22:1n-9)) (Lu and Kang 2008). It was demonstrated that *Camelina* oil is more stable to oxidation compared to oils from other crops (Abramovic and Abram 2005). Because of those properties *Camelina* oil is quit suitable for many applications including production of biofuel, manufacture of soaps, cosmetics, oil for lamp, varnishes, safe paints, and using for nutrition purposes (Zubr 1997; Moser and Vaughn 2010). Regarding the food application eating of *Camelina* oil can be healthy for human as a valuable source of omega-3 fatty acids (Hrastar et al. 2009; Lu and Kang 2008), moreover, investigations have displayed that consuming of Camellia oil reduced cholesterol in blood serum (Karvonen et al. 2002). The *Camelina* cold-pressed meal is rich in omega-3 fatty acids and amino acids combined with a relatively low glucosinolate content (Berhow et al. 2013) making it suitable for animal feed (Russo and Reggiani 2015). Several investigations display that feeding *Camelina* oil or meal is beneficial for some animals including fish (Berti et al. 2016; Betancor et al. 2017). When it comes to agronomic qualities, *Camelina* has some benefits. It is a short life cycle (85–100 days), high yield potential and requires fewer inputs than other oilseed crops (Steppuhn et al. 2010), adaptability to marginal soils, relatively frost and drought resistant (Iskandarov et al. 2014). Besides this, *Camelina* was shown as less susceptible to pathogens such as *Alternaria* (Séguin-Swartz et al. 2009), *Sclerotinia* (Eynck et al. 2012), and to some insect pests (Soroka et al. 2015). All together these characteristics show *Camelina* as a suitable candidate for sustainable cropping systems.

2 Achievements in Improving the *Camelina* Resistance to Various Abiotic Stresses

As a sessile organism, plants are often exposed to various abiotic stresses which leads to physiological and biochemical changes in plants (Hasanuzzaman et al. 2012, 2013a, b; Zingaretti et al. 2013). Abiotic stresses, such as drought, salinity, toxic metal/metalloid, extreme temperatures, toxic metals, etc., reduce crop yields significantly (Golldack et al. 2011; Hasanuzzaman et al. 2012; Hasanuzzaman and Fujita 2012). Up to 70% decline in productivity of major crops worldwide has been attributed to abiotic stresses (Mittler 2006). Despite its overall resistance, *Camelina* is sensitive to bacterial and fungal pathogens, which cause wilting (*Fusarium*), root rot (*Pythium*), and downy mildew (*Peronospora*) (Zakharchenko et al. 2013) which also leads to crops productivity losses worldwide. Because of those, there is a necessity to investigate and understand the physiological and molecular responses of plants toward the stresses of different origins. Moreover, identification of the stress regulatory genes and employing in plants toward stress tolerance should be explored. To enhance crop productivity and crop yield under stresses, *Camelina* needs to increase stress tolerance through genetic engineering and traditional breeding, which is the best viable option. There are several achievements in the improvement of stress tolerance of *Camelina.* Some of them have already been described in detail in the Sainger's article (Sainger et al. 2017). In this chapter, we have summarized the achievements in increasing the resistance of *Camelina* to stresses of various origins.

2.1 Drought

In general, *Camelina* has a relatively low water use requirement and has a natural resistance to drought. Perhaps this is due to the ability of *Camelina* to extract soil moisture deep in the soil profile, which was determined to be about 1.4 m. (Hunsaker et al. 2011) But despite this, it was shown that drought significantly affects the growth, physiological and biochemical processes in *Camelina.* During water deficiency, plants undergo various changes such as decreased photosynthesis, stomatal conductance, and transpiration rate leading to lower crop yields (Waraich et al. 2017). The first step in improving plant stability is to study the physiological and biochemical changes in plants during stress. Changes in gene activity during stress are also should be investigated. All this knowledge is effectively used to improve the drought resistance of plants. For example, one study showed that increasing the accumulation of leaf cuticular wax in *Camelina* can reduce non-stomatal water loss and, as a result, increase its resistance to drought. To do this, the ways of wax biosynthesis in *Camelina* should be studied, as well as the genes that are involved in this process should be determined. The wax components have been detected that can be applied as suitable biomarkers for screening and breeding drought-resistant plants of *Camelina.* Also, resistant plants can be created through modifying cuticle composition and properties using a transgenic approach (Tomasi et al. 2018). In another study, the significant single nucleotide polymorphisms (SNP) have been detected which were located in genes connected to cuticular wax accumulation, which can also be applied for molecular breeding and selection of drought-resistant *Camelina* plants (Luo et al. 2019).

Also, another approach to improving plant stability is to study the transcription factors of genes that are involved in response to stress. Such factors have been discovered in *Camelina* (Kanth et al. 2015). It was shown that some modulating factors, for example, miRNAs interact with target genes and regulate their expression during the drought in *Camelina* plants. The six miRNAs which were strongly upregulated during exposure to drought in comparison with normal conditions were discovered (Subburaj et al. 2016). Expression of CsalncRNAs (Long non-coding RNAs) (a total of 5390) that were detected in the *Camelina* genome was also studied during water deficiency. It was shown that CsalncRNAs expression was significantly increased, which indicates their involvement in response to drought (Subburaj et al. 2018).

Also, there are already some advances in the development of drought-resistant *Camelina* plants using a transgenic approach. Lee et al. (2014) found that overexpression of the *Arabidopsis MYB96* gene upregulated the *Camelina* wax biosynthetic genes (*CsKCS2, CsKCS6, CsKCR1-1, CsKCR1-2, CsECR,* and *CsMAH1*) which resulted in 50% increase in wax biosynthesis. This has led to an increase in drought resistance of transgenic *Camelina* plants compared to wild-type plants.

Arabidopsis G-protein subunit 3 (*AGG3*) overexpression using two types of promoters (CaMV35S and seed-specific glycinin promoter) was studied in *Camelina* during exposure to some stresses (Roy Choudhury et al. 2014). For instance, *Camelina* plants overexpressing this gene exhibited elevated photosynthetic parameters, and these seeds were less sensitive to drought at the early stages.

2.2 Salinity

It has been reported that *Camelina* can grow well on highly saline soils (Budin et al. 1995). Later, it was shown that *Camelina* is sensitive to salinity stress (Khalid et al. 2015). It was demonstrated that the salinization negatively affects the growth of *Camelina,* so the height and weights of the roots and shoots were reduced, as well as the chlorophyll content, and as a result, photosynthesis rate was decreased. It was also shown that the regulation of certain genes was changed under saline conditions (Heydarian et al. 2018a). Another study showed that during salt stress, the height of *Camelina* plants decreased by 57%, and the yield was also significantly reduced (Ullah et al. 2018). In addition, salinization negatively affects the germination of seeds in Camelina, as well as their oil content (Steppuhn et al. 2009). In order to obtain stable crop yield of *Camelina* it is important to investigate the mechanisms of resistance of *Camelina* to salt stress and to increase the stability of this crop.

There are already several achievements in this. The effect of salt stress on *Camelina* was studied, as well as changes in the expression of various metabolites-related genes during salt stress (Heydarian et al. 2018a). These metabolites reduce the loss of water in plants during salt stress. Also, the expression of genes involved in nitrate uptake and nitrogen assimilation was reduced. Using these techniques, the stability of *Camelina* plants to salt stress can be increased. In the following study, it was found that the expression of *acdS* also affected the expression of some genes involved in responses to stresses in *Camelina* (Heydarian et al. 2018b). These results can be used for the cultivation of *Camelina* on marginal lands.

2.3 Extreme Temperatures

Cold stress negatively affects plant growth and crop productivity (Chinnusamy et al. 2010). *Camelina* is believed to be insensitive to cold stress. But it was shown that some varieties better withstand low temperatures, while others are sensitive to them. In order to identify varieties with a high or low level of tolerance to freezing stress 136 doubled haploid (DH) lines of *Camelina* have been investigated under freezing stress conditions (−14 °C for 6 h) (Soorni et al. 2017). Three lines have been identified as freezing tolerant. In addition, a protocol was developed for large-scale screening of *Camelina* germplasm under freezing stress conditions.

By introducing a gene isoprene synthase (*PcISPS*) from *Populus* × *canescens* into the *Camelina* genome, plants with increased the isoprene emission have been made. This can increase the stability of *Camelina*, as the plants that emit isoprene are tolerant to high temperature and high light intensities (Rossi et al. 2014).

2.4 Toxic Metals

Heavy-metal pollution soil negatively affects plant growth. Excessive levels of heavy metal can lead to leaf chlorosis and necrosis, photosynthesis disorder, and a decrease in the rate of seed germination. In this regard, there is a need, on the one hand, to increase the resistance of plants to heavy metals, and on the other hand, to look for plants that are able to accumulate heavy metals and thus clean the soil (phytoremediation).

In order to investigate genes involved in response to heavy-metal expression patterns of eight *HMA* genes in *Camelina* have been analyzed (Park et al. 2014), *CsHMA3* genes were expressed in all organs, especially after Pb treatment. Transgenic *Camelina* plants expressing a 35S promoter:: *CsHMA3* have been generated using a gateway cloning system. Transgenic plants that grew on media containing heavy metals (Cd, Pb, Zn, Co) were investigated. It was observed that transgenic plants showed better resistance Cd, Pb, and Zn and had improved physiological and yield attributed. These plants also accumulated a high amount of metals. In addition, it was found that *Camelina* is more resistant to heavy metals and is able to accumulate more metals in shoots compared with rapeseed. These results allow us to suggest that *Camelina* can be used for phytoremediation. Park et al. (2015) reported that *CsHMA3* overexpressed plants contained higher unsaturated fatty acids than wild type. Moreover, ROS generation was also lower under metal stress (Park et al. 2015). All these data indicate that *CsHMA3* transgenic *Camelina* can be used for cultivation on marginal land for biofuel production.

3 Conclusions

The interest in *C. sativa* as an alternative oilseed crop has grown rapidly in recent years because it is a fast-growing crop with a short life cycle and high yield potential. In addition, *Camelina* is relatively resistant to the effects of several stressful environmental factors. But despite its general stability, *Camelina* is sensitive to the effects of certain stresses, as a result of which the crop yield is significantly reduced. During the cultivation of oilseeds, it is important not only the ability of plants to survive under stress but also to give a high yield. In this review, some knowledge on the study of the resistance of *Camelina* and data on some genes that are involved in the response of *Camelina* to stress were summarized. Existing achievements in improving the resistance of this species to stresses, such as drought, salinity, extreme temperatures, heavy metals influence, and effects of pathogenic fungi have also been described. A generalization of the available information will help to better understand the mechanisms of sustainability of camellias, as well as having knowledge can be used to improve the efficiency of growing this crop under stressful environmental conditions.

Acknowledgements Larysa V. Nishchenko is grateful for financial support to the National Academy of Science of Ukraine, Grant UkrISTEI № 0115U005022.

References

Abramovic H, Abram V (2005) Physico-chemical properties, composition and oxidative stability of *Camelina sativa* oil. Food Technol Biotechnol 43:63–70

Berhow MA, Polat U, Glinski JA, Glensk M, Vaughn SF, Isbell T, Gardner C (2013) Optimized analysis and quantification of glucosinolates from *Camelina sativa* seeds by reverse-phase liquid chromatography. Ind Crop Prod 43:119–125

Berti M, Gesch R, Eynck C, Anderson J, Cermak S (2016) Camelina uses, genetics, genomics, production, and management. Ind Crops Prod 94:690–710

Betancor MB, Li K, Sprague M, Bardal T, Sayanova O, Usher S, Tocher DR (2017) An oil containing EPA and DHA from transgenic *Camelina sativa* to replace marine fish oil in feeds for Atlantic salmon (Salmo salar L.): effects on intestinal transcriptome, histology, tissue fatty acid profiles and plasma biochemistry. PloS One 12:e0175415

Budin JT, Breene WM, Putnam DH (1995) Some compositional properties of Camelina (*Camelina sativa L. Crantz*) seeds and oils. J Am Oil Chemists Soc 72:309–315

Chinnusamy V, Zhu JK, Sunkar R (2010) Gene regulation during cold stress acclimation in plants. In: Plant stress tolerance. Humana Press, pp 39–55

Eynck C, Séguin-Swartz G, Clarke WE, Parkin IA (2012) Monolignol biosynthesis is associated with resistance to *Sclerotinia sclerotiorum* in *Camelina sativa*. Mol Plant Pathol 13:887–899

Golldack D, Lüking I, Yang O (2011) Plant tolerance to drought and salinity: stress regulating transcription factors and their functional significance in the cellular transcriptional network. Plant Cell Rep 30:1383–1391

Hasanuzzaman M, Fujita M (2012) Heavy metals in the environment: current status, toxic effects on plants and possible phytoremediation. In: Anjum NA, Pereira MA, Ahmad I, Duarte AC, Umar

S, Khan NA (eds) Phytotechnologies: remediation of environmental contaminants. CRC Press, Boca Raton, pp 7–73
Hasanuzzaman M, Hossain MA, Teixeira da Silva JA, Fujita M (2012) Plant responses and tolerance to abiotic oxidative stress: antioxidant defense is a key factor. In: Bandi V, Shanker AK, Shanker C, Mandapaka M (eds) Crop stress and its management: perspectives and strategies. Springer, New York, pp 261–316
Hasanuzzaman M, Nahar K, Fujita M, Ahmad P, Chandna R, Prasad MNV, Ozturk M (2013a) Enhancing plant productivity under salt stress—relevance of poly-omics. In: Ahmad P, Azooz MM, Prasad MNV (eds) Salt stress in plants: omics, signaling and responses. Springer, Berlin, pp 113–156
Hasanuzzaman M, Nahar K, Fujita M (2013b) Plant response to salt stress and role of exogenous protectants to mitigate salt-induced damages. In: Ahmad P, Azooz MM, Prasad MNV (eds) Ecophysiology and responses of plants under salt stress. Springer, New York, pp 25–87
Heydarian Z, Gruber M, Glick BR, Hegedus DD (2018a) Gene expression patterns in roots of *Camelina sativa* with enhanced salinity tolerance arising from inoculation of soil with plant growth promoting bacteria producing 1-aminocyclopropane-1-carboxylate deaminase or expression the corresponding acdS gene. Front Microbiol 9
Heydarian Z, Yu M, Gruber M, Coutu C, Robinson SJ, Hegedus DD (2018b) Changes in gene expression in *Camelina sativa* roots and vegetative tissues in response to salinity stress. Sci Rep 8:9804
Hrastar R, Petrisic MG, Ogrinc N, Kosir IJ (2009) Fatty acid and stable carbon isotope characterization of *Camelina sativa* oil: implications for authentication. J Agric Food Chem 57:579–585
Hunsaker DJ, French AN, Clarke TR, El-Shikha DM (2011) Water use, crop coefficients, and irrigation management criteria for Camelina production in arid regions. Irrig Sci 29:27–43
Iskandarov U, Kim HJ, Cahoon EB (2014) Camelina: an emerging oilseed platform for advanced biofuels and bio-based materials. Plants and bioenergy. Springer, New York, pp 131–140
Kanth BK, Kumari S, Choi SH, Ha HJ, Lee GJ (2015) Generation and analysis of expressed sequence tags (ESTs) of *Camelina sativa* to mine drought stress-responsive genes. Biochem Biophys Res Commun 467:83–93
Karvonen HM, Aro A, Tapola NS, Salminen I, Uusitupa MI, Sarkkinen ES (2002) Effect of [alpha]-linolenic acid [ndash] rich *Camelina sativa* oil on serum fatty acid composition and serum lipids in hypercholesterolemic subjects. Metab Clin Exp 51:1253–1260
Khalid H, Kumari M, Grover A, Nasim M (2015) Salinity stress tolerance of Camelina investigated in vitro. Sci Agric Bohem 46:137–144
Lee SB, Kim H, Kim RJ, Suh MC (2014) Overexpression of Arabidopsis *MYB96* confers drought resistance in *Camelina sativa* via cuticular wax accumulation. Plant Cell Rep 33:1535–1546
Lu C, Kang J (2008) Generation of transgenic plants of a potential oilseed crop *Camelina sativa* by Agrobacterium-mediated transformation. Plant Cell Rep 27:273–278
Luo Z, Tomasi P, Fahlgren N, Abdel-Haleem H (2019) Genome-wide association study (GWAS) of leaf cuticular wax components in *Camelina sativa* identifies genetic loci related to intracellular wax transport. BMC Plant Biol 19:187
Mittler R (2006) Abiotic stress, the field environment and stress combination. Trends Plant Sci 11:15–19
Moser BR, Vaughn SF (2010) Evaluation of alkyl esters from *Camelina sativa* oil as biodiesel and as blend components in ultra low-sulfur diesel fuel. Biores Technol 101:646–653
Park W, Feng Y, Ahn SJ (2014) Alteration of leaf shape, improved metal tolerance, and productivity of seed by overexpression of *CsHMA3* in *Camelina sativa*. Biotechnol Biofuel 7:96
Park W, Feng Y, Kim H, Suh MC, Ahn SJ (2015) Changes in fatty acid content and composition between wild type and *CsHMA3* overexpressing *Camelina sativa* under heavy-metal stress. Plant Cell Rep 34:1489–1498
Rossi L, Borghi M, Xie D (2014) Characterization of transgenic *Camelina plants* emitting isoprene. In: Proceedings of the 53rd annual meeting of the phytochemical society of North America

Roy Choudhury S, Riesselman AJ, Pandey S (2014) Constitutive or seed-specific overexpression of Arabidopsis G-protein γ subunit 3 (AGG 3) results in increased seed and oil production and improved stress tolerance in Camelina sativa. Plant Biotechnol J 12:49–59

Russo R, Reggiani R (2015) Seed protein in *Camelina sativa (L.) Crantz* var. Calena. Int J Plant Soil Sci 8:1–6

Sainger M, Jaiwal A, Sainger PA, Chaudhary D, Jaiwal R, Jaiwal PK (2017) Advances in genetic improvement of *Camelina sativa* for biofuel and industrial bio-products. Renew Sustain Energy Rev 68:623–637

Séguin-Swartz G, Eynck C, Gugel RK, Strelkov SE, Olivier CY, Li JL, Falk KC (2009) Diseases of *Camelina sativa* (false flax). Can J Plant Pathol 31:375–386

Soorni J, Kazemitabar SK, Kahrizi D, Dehestani A, Bagheri N (2017) Screening of Camelina (*Camelina sativa L.*) doubled haploid lines for freezing tolerance in the seedling stage. Genetika 49:173–181

Soroka J, Olivier C, Grenkow L, Séguin-Swartz G (2015) Interactions between *Camelina sativa (Brassicaceae)* and insect pests of canola. Can Entomol 147:193–214

Steppuhn H, Falk KC, Zhou R (2010) Emergence, height, grain yield and oil content of Camelina and canola grown in saline media. Can J Soil Sci 90:151–164

Steppuhn H, Wall KG, Falk KC, Zhou R, Brandt SK (2009) How well does Camelina tolerate root zone salinity? In: Soils and crops workshop

Subburaj S, Jeon Y, Tu L, Jin YT, Kumari S, Lee GJ (2018) Genome-wide identification, functional prediction and expression profiling of long non-coding RNAs in *Camelina sativa*. Plant Growth Regul 86:49–63

Subburaj S, Kim AY, Lee S, Kim KN, Suh MC, Kim GJ, Lee GJ (2016) Identification of novel stress-induced microRNAs and their targets in *Camelina sativa* using computational approach. Plant Biotechnol Rep 10:155–169

Tomasi P, Dyer JM, Jenks MA, Abdel-Haleem H (2018) Characterization of leaf cuticular wax classes and constituents in a spring *Camelina sativa* diversity panel. Ind Crop Prod 112:247–251

Ullah MA, Mahmood IA, Zaman B (2018) *In vitro* saline sodic status of *Camelina sativa* cv. Blaine creek. Horticult Int J 2:179–183

Waraich EA, Ahmed Z, Ahmad R, Saifullah Shahbaz M, Ehsanullah (2017) Modulation in growth, development, and yield of *Camelina sativa* by nitrogen application under water stress conditions. J Plant Nut 40:726–735

Zakharchenko NS, Kalyaeva MA, Buryanov YI (2013) Expression of cecropin *P1* gene increases resistance of *Camelina sativa* (L.) plants to microbial phytopathogenes. Russ J Genet 49:523–529

Zingaretti SM, Inácio MC, de Matos Pereira L, Paz TA, de Castro França S (2013) Water stress and agriculture. In: Responses of organisms to water stress. IntechOpen

Zubr J (1997) Oil-seed crop: *Camelina sativa*. Ind Crop Product 6:113–119

Brassicaceae Plants Response and Tolerance to Salinity

Subhankar Mondal and Koushik Chakraborty

Abstract Increasing soil salinity is one of the most important global issues for recent agriculture, which can potentially cut down the total crop productivity. Brassicas or different members of *Brassicaceae* family are very important components of human food. Interestingly, the members of *Brassicaceae* family have diverse genetic composition and thus their response and adaptation to salinity differs greatly. Hence, it is necessary to know the morphological, physiological, biochemical, hormonal and molecular responses of *Brassica* for sustainable production in saline environment. Due to its high economic and nutritional importance, characterization and identification of salt-tolerant genotypes in each of the cultivated species of *Brassicaceae* is an essential approach for maintaining food production and security. Several techniques are available to identify the tolerant donors of *Brassica.* However, identification through key morpho-physiological traits along with established molecular markers is still a universal technique. Due to the diverse nature of different members of *Brassicaceae* family, its salt tolerance response is not universal. Rather, from highly glycophytic to extreme halophytic responses were reported within *Brassicaceae* family. Understandably, the physiological and molecular responses also vary greatly between different species present in the family. In general, it was reported by several researchers that most of the amphidiploid species in *Brassicaceae* family were comparatively more salt-tolerant than the diploid members. Looking at the agronomic importance of different Brassicas, over the years several conventional and advanced breeding approaches were performed to improve salt tolerance in individual *Brassica* species, despite their differential critical levels of salt tolerance. Genetic interventions either by transgenics or through genome editing approaches were also exercised to incorporate/modify key regulatory genes viz. *SOS*, *HKT*, *NHX* and *LEA* gene families into different members of *Brassicaceae* in order to increase their salt tolerance potential. In this compilation, we focussed on gathering related information, which would help in better understanding of differential responses of different members of *Brassicaceae* family towards salt stress.

S. Mondal · K. Chakraborty (✉)
Division of Crop Physiology & Biochemistry, ICAR National Rice Research Institute, Cuttack 753006, India
e-mail: koushikiari@gmail.com; Koushik.Chakraborty@icar.gov.in

M. Hasanuzzaman (ed.), *The Plant Family Brassicaceae*,
https://doi.org/10.1007/978-981-15-6345-4_6

Keywords Brassicas · Canola · Ionic imbalance · Mustard · SOS pathway · Salt stress

1 Introduction

Members of *Brassicaceae* are one of the important contributors to human daily needs and used as vegetables, oil, condiments, etc. Among different oilseed species, members of *Brassica* occupy third place in vegetable oil production (Ashraf and McNeilly 2004; Purty et al. 2008). In general, *Brassica campestris* and *Brassica napus* are the most common oil-producing crops in *Brassica* genus and cultivated for commercial oil production (Ashraf and McNeilly 2004). The family Brassicaceae belongs to the Division magnoliophyta, Class magnoliopsida, Subclass dilleniidae and Order capparales (Cronquist 1968) with the type Genus *Brassica*, also known as mustard family. Earlier the genus was referred as cruciferae due to its typical cruciform flowers, where sepals and petals are crossly placed and six stamens are arranged in two whorls of which, 2 are short and 4 are long and the fruits are mainly siliques. The origin and evolution of Brassicaceae plant family showed that it is divided into two major groups—(i) Aethionemaeae and (ii) the core Brassicaceae. Molecular studies based on the chloroplast *ndhF*, ITS DNAand *nad4 intron 1* showed core Brassicaceae group contains 3 major lineages—lineage-I, lineage-II, and lineage-III, which is represented by 13, 4 and 7 tribes, respectively. The members of the lineage I and III are characterized by branched trichomes, while members of lineage-II showed simple trichome (Franzke et al. 2010). The genus *Brassica* belongs to lineage-II and are characterized by simple trichome. Fossil records showed that the family Brassicaceae has originated nearly 19 million years ago. The oldest pollen of Brassicaceae member was found from the middle Miocene epoch of Cenozoic era (approx. 16 million years ago). Studies with Brassicaceae also showed that the genus *Brassica* was evolved from the ancient member *Sinapidendron* through *Diplotaxis-Erucastrum* complex at Miocene. The family Brassicaceae maintains a wide diversity, contains 435 genera and nearly 3675 species (Purty et al. 2008).

Among different types of genera and species, some members of Brassicaceae contribute importantly in oilseed production, while some are used as model organism (*Arabidopsis thaliana*) for different scientific purposes. In general, the members of Brassicaceae showed a moderately tolerant character to salinity. But, some members like *Thellungiella halophile* show extreme tolerance to salt due to high environmental pressure and evolution. On the contrary, some members belong to genus *Brassica* are economically and nutritionally important and cultivated as food crops viz. *B. campestris*, *Brassica rapa*, *Brassica juncea*, *Brassica nigra* and *Brassica oleracea* . Studies with different species of *Brassica* showed that the growth and development of different *Brassica* members are affected by different kinds of abiotic stresses like drought, salinity, high temperature, etc. Among different kinds of abiotic stresses, soil salinization is an important constraint for agriculture. The level of salt is continuously increasing in the agricultural field due to poor irrigation practices. Recent reports showed that nearly 0.8 billion hector lands were affected by the increasing salinity (Qadir et al. 2014; Shahid et al. 2018).

Increasing soil salinity is associated with rapid water loss from plants. To protect themselves from osmotic stress, plants synthesize a variety of osmoticums like proline, glycine betaine, sugar compounds like trehalose, etc. High soil salinity is a major problem for growth and normal life cycle of *Brassica*, affecting total biomass and yield. Members of Brassicaceae family showed various levels of tolerance against salt stress based on their genetic organization and evolution. In general, normal glycophytic plant growth and development severely affected if the salt concentration exceeds 25 mM (Munns et al. 2019). However, the concept was completely different and more complicated for halophytes, where salt is essential for maintaining the growth, water potential and ionic homeostasis. During stress, morphological traits like root and shoot growth; physiological parameters like leaf water potential, chlorophyll content, photosynthetic rate and membrane integrity; biochemical factors like osmolyte accumulation, antioxidant enzyme activities and molecular responses like expressional changes in salt responsive genes are important attributes for identifying the tolerance level present between inter- and intra-specific members of Brassicaceae family. In this compilation, we shall try to summarize the morphological, physiological and biochemical responses showed by the different members of Brassicaceae family under saline environment.

2 Different Species of *Brassica* Genus and Their Response to Salinity

Brassica is one of the most important genera of the mustard family (Brassicaceae), where members like *B. rapa*, *B. juncea*, *B. oleracea*, *B. napus*, *B. carinata*, *B. nigra*and *B. rupestris* are agriculturally important and used as vegetables, oilseeds, condiments and biodiesel production. Interestingly, members of *Brassica* showed an evolutionary relationship among them, which is known as U's-triangle and identified by a Japanese botanist Woo Jang-Choon. As per the Us'-triangle ancestral three members of *Brassica* genus are diploid and generally showed 3 different kinds of genome organization (*B. rapa*-AA, *B. nigra*-BB and *B. oleracea*-CC) and from there 3 other species of *Brassica* genus are derived by chromosome combinations and evolution (*B. juncea*-AABB, *B. carinata*-BBCC and *B. napus*-AACC). As most of the members of the *Brassica* genus are agriculturally important, it's essential to understand the differential response of different species of *Brassica* to various abiotic stresses. Among different types of abiotic stresses salinity is one of the most important constraints for agriculture and previous studies showed that different members of *Brassica* respond differently in saline environment (Kumar et al. 2009; Chakraborty et al. 2012b, 2016b). Inter-specific evaluation of *Brassica* genotypes suggested that the amphidiploid members of the family viz. *B. juncea* (AABB), *B. carinata* (BBCC) and *B. napus* (AACC) showed an improved level of tolerance than their ancestral diploid species like *B. rapa* (AA), *B. nigra* (BB) and *B. oleracea* (CC) (Malik 1990; Kumar 1995; Purty et al. 2008). Further cytogenetical evidences showed that the superior salt tolerant behaviour of amphidiploid species comes from the process of natural hybridization and crossing over belonging to two diploid ancestral genomes AA (*B. rapa*) and CC (*B. oleracea*) (Ashraf et al. 2001). Ashraf and McNeilly (2004) reported that significant inter and intra-specific variation was present in different members of *Brassica* and in general, tolerant members maintained a high cellular potassium (K^+) and calcium (Ca^{2+}) level to withstand stress.

3 Different Halophytic Members in *Brassicaceae* Family

Among the different members of Brassicaceae, *Thellungiella salsuginea* is a known halophyte and showed high degree of tolerance to high salinity and drought stress. This genotype is extensively used as an important source for the salinity tolerant genes. Some important genes like *ThNHX1* (Vacuolar Na^+/H^+ antiporter), *ThSOS1* (a plasma membrene H^+ mediated antiporter of Na^+), *TsLEA1* late embryogenesis abundant (*LEA*), *TsTIP1* tonoplast aquaporin genes were identified from *Thellungiella halophila* and over-expressed in *Arabidopsis thaliana.* The transgenic lines of *Arabidopsis* over-expressing the genes from *T. halophila* showed an improved level of salt tolerance (Wu et al. 2009; Oh et al. 2009a, b). Similarly, some other important genes like *TsVP* (H^+-PPase) were also transformed in *Gossypium* sp. and *Nicotiana tabacum* (Zhang et al. 2012; Wang et al. 2014). The transgenic lines over-expressing the *TsVP* construct from *Thellungiella* showed an improved level of salt tolerance (Gao et al. 2006; Lv et al. 2008). Recently, *Thellungiella parvula* was identified as a model halophyte that can be used for basic studies concerning salinity and also an important source for novel gene discovery (Orsini et al. 2010).

Besides the genus *Thellungiella, Cakile maritima* is also identified as another potential halophyte that can tolerate high level of salt (Boyd and Barbour 1986; Megdiche et al. 2009). Megdiche et al. (2009) identified and characterized the *Cystatin* gene from *C. maritima* which encodes a 221 amino acid sequence and produces a protein with molecular mass of 25KDa. Furthermore, studies indicated that the expression of *Cystatin* gene was high under severe salinity and water deficit conditions. Other reported halophytic members of the Brassicaceae family are *Eruca vesicaria*–cultivated as a crop under saline conditions–(Ashraf and Noor 1993; Ashraf 1994); *Lesquerella fendleri* is considered as a moderately tolerant for salinity and can survive under the electrical conductance of 6.9 dS m^{-1} (Dierig et al. 2001, 2004); *Lobularia maritima* is a halophyte, which is able to maintain low Na^+/K^+ ratio under high salt conditions, by the improved activity of *NHX1*, *PIP2;1*, and *V-ATPase* (Popova et al. 2008) and *Raphanus raphanistrum* ssp. *maritimus*—wild radish, which can tolerate a high level of salinity.

4 Basic Responses of *Brassicaceae* Plant Family to Salinity Stress

Salinity affects plants in two different ways.— Initially it is osmotic stress, which reduced the water conductance status in the plant cells and changes the overall plant water and nutritional status (Munns and Tester 2008). The second response is associated with the accumulation of high concentrations of sodium to upper plant parts, which ultimately reaches the toxic level and strongly puts a negative effect on the growth and development of plants by changing the normal status of potassium, calcium and essential micronutrients. Reports showed that salinity affects the normal function of organelles like chloroplast and mitochondria. Due to its negative effects on photosynthesis and respiration process, it produces reactive species of dioxygen (ROS) like H_2O_2 and other free radicals of oxygen, which causes significant oxidative damage in plants resulting in cell death and necrosis (Munns and Tester 2008). As salinity is a polygenic trait, which can interact with several genes that are involved in various biological pathways, so plant response to stress is also distinctly different from one species to other. The response towards salinity in different species can be characterized by attributes like plant biomass, growth, along with the response of physiological, biochemical and molecular responses that are altered due to salt stress (Fig. 1).

4.1 Plant Biomass and Growth

Total plant biomass and growth are an important character for identifying a plant's sensitivity to salt stress. Not only for *Brassicaceae* family of plants, but in general significant reduction of growth and biomass was highest for most sensitive genotype, while maintaining a considerable of plant biomass is a sign of tolerance. Previous studies with three different members of Brassicaceae-*B. napus*, *B. juncea* and *B. oleracea* showed that reduction of fresh weight and dry weight occurred under salinity and magnitude of reduction was correlated with the amount of salt accumulation and overall salinity tolerance (Chakraborty et al. 2016a). Recent studies with three different members of *Brassicaceae* with different doses of sodium showed that significant inhibition of growth associated with sensitive *B. oleracea* var *caitata* and *B. rapa*, while the reduction was not significant in tolerant *B. oleracea* var *acephala* plants (Pavlovic et al. 2019).

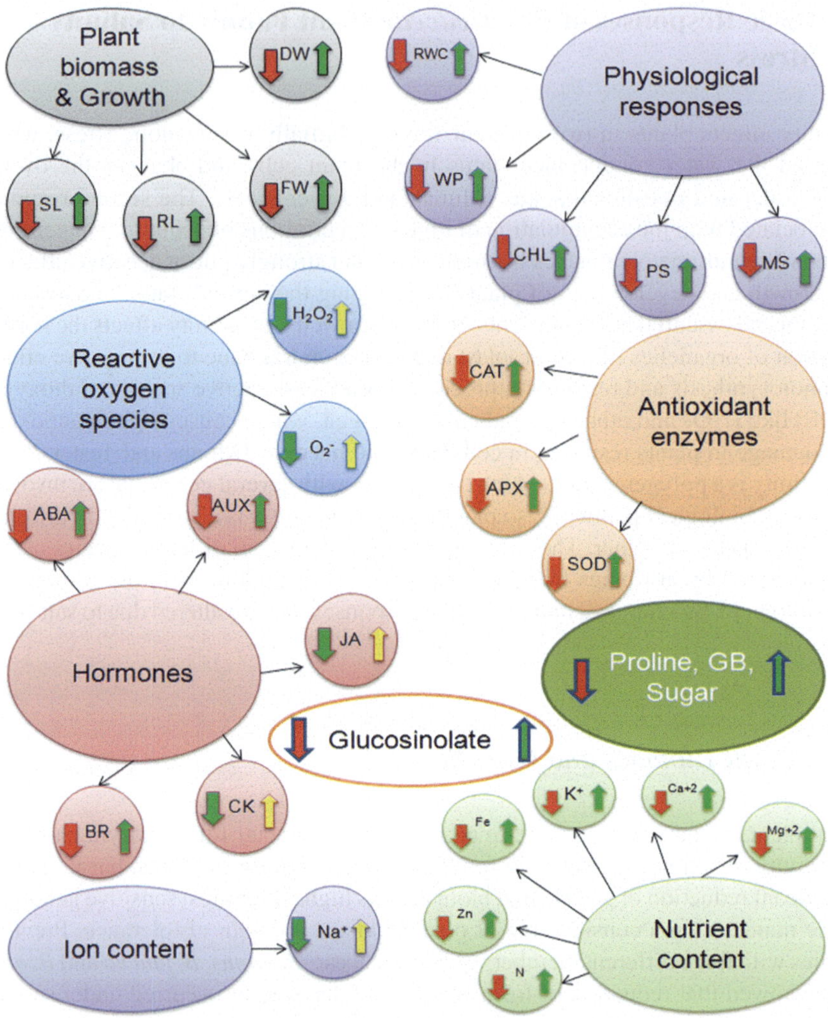

Fig. 1 Morphological, physiological and biochemical responses showed by the tolerant and sensitive genotypes of *Brassica*, under salt stress. Morphological traits like DW, FW, SL and RL; physiological parameters like RWC, WP, CHL, PS and MS; antioxidant enzymes like CAT, APX and SOD; osmolytes like proline, GB and sugar; hormones like ABA, AUX and BR; inorganic nutrients like N, K, Ca, Zn, Mg and Fe; glucosinolate level induced/maintained for tolerant genotypes are represented by an upward green arrow, while reduction and leakage for the sensitive genotypes are represented by a red arrow. Reactive oxygen species like H_2O_2 and O_2^-, hormones like JA, CK and Na^+ level maintenance/reduction/minute elevation are characteristic for tolerant genotypes and represented by a downward green arrow, while induction of these are characteristics of sensitive genotypes and represented by a yellow arrow. Here, DW: Dry weight; FW: Fresh weight; SL: Shoot length and RL: Root length; RWC: Relative water content; WP: Water content; CHL: Chlorophyll content; PS: Photosynthesis; MS: Membrane stability; H_2O_2: Hydrogen peroxide; O_2: Free radicals of oxygen; CAT: Catalase; APX: Ascorbate peroxidase; SOD: Superoxide dismutase; ABA: Abscisic acid; AUX: Auxin; JA: Jasmonic acid; CK: Cytokinin; BR: Brassinosteroid; GB: Glycine betaine; Na^+: Sodium; K^+: Potassium; Ca^{2+}: Calcium; Mg^{2+}: Magnesium; N: Nitrogen; Fe: Iron and Zn: Zinc

4.2 *Leaf Chlorophyll Content, Photosynthetic Efficiency and Transpiration Rate*

Along with morphological traits, physiological markers like leaf area, leaf chlorophyll content, relative water content, permeability index, water potential, photosynthetic efficiency, transpiration rate and stomatal conductance are also important factors to identify a plant's sensitivity towards salinity stress. Previous studies showed that leaf area and leaf number was significantly decreased under salt stress (Jamil et al. 2007). Chakraborty et al. (2016c) reported that salinity significantly reduced the total chlorophyll content in different species of *Brassica.* The degradation of chlorophyll pigments and other important attributes like chlorophyll fluorescence, photosynthetic efficiency, photochemical efficiency and non-photochemical efficiency was significant for the sensitive members, while reduction was non-significant for tolerant members. Comparative study between glycophytic and halophytic members of *Brassicaceae* family showed that chlorophyll degradation was highest in *Arabidopsis,* while non-significant degradation of chlorophyll was observed in the halophyte *Thellungiella.* Salt stress induced stomatal closure observed in *Arabidopsis*; while in halophytic *Thellungiella* no stomatal closure was observed (Stepien and Johnson 2009). Similarly, the chlorophyll fluorescence parameters like PSII efficiency, photochemical efficiency and non-photochemical efficiency were unchanged for *Thellungiella,* while in *Arabidopsis* significant amount of reduction was observed (Stepien and Johnson 2009). Chakraborty et al. (2016d) with three different members of *Brassica* showed that leaf chlorophyll content, photosynthetic efficiency, transpiration rate and stomatal conductance were decreased under stress. As per the reports, the degradation of chlorophyll was least in tolerant *B. napus* followed by moderately tolerant *B. juncea* and most in susceptible *B. oleracea.* Similarly, transpiration, photosynthesis and stomatal conductance were reduced under salt stress; the level of reduction is most important to identify the intra-specific differences in tolerance level present in different members of *Brassica* sp. Kumar et al. (2015) cloned a *lectin* gene from a member of leguminosae chickpea and successfully transformed the *lectin* gene in *B. juncea* plants, transgenic plants over-expressing the *lectin* gene showed an improved resistance against leaf necrosis disease caused by *Alternaria brasicae.* Further investigations with the transgenic plants showed that the gene also providing salt tolerance by maintaining a high amount of chlorophyll under stress.

4.3 *Stomatal Regulation Under Salinity*

Photosynthesis and transpiration are regulated by the stomata, which help plants to exchange CO_2 and water vapour between leaf and atmosphere. Under saline conditions, in general the rate of stomatal transpiration decreases. Under the influence of osmotic stress, excessive water loss is very important; hence closure of stomata happens to be the most immediate response by plants. As the stomata are also an important site for CO_2 influx, the net photosynthesis generally declines under saline condition. Ulfat et al. (2007) reported that net CO_2 assimilation reduced due to imposition of salt stress in *B. napus*. Comparative analysis in between a sensitive and a tolerant genotype showed that the number of guard cells and stomatal aperture size differed in tolerant and sensitive genotype. Further, studies with halophytes showed that guard cells of halophytic members can able to retain function even when they were grown in high concentrations of salt (Robinson et al. 1997). Noreen et al. (2010) reported that only the tolerant accessions (Peela shajalum and Neela shajalum) of *B. rapa* were able to maintain the gas exchange parameters like photosynthetic efficiency and stomatal conductance under high concentration of salt stress.

4.4 *Biochemical Responses*

4.4.1 Osmotic Stress-Related Marker Compounds in *Brassicaceae*

The reduction of water content, chlorophyll pigments, fresh weight and dry weights are the most common responses of salinity stress. To retain a normal water level in the cells and maintaining the cellular homeostasis, plants synthesize different types of osmolytes like proline, sugars and glycine betaine. Accumulation of these compounds helps to maintain the cellular water balance and thus, the functions of several proteins are intact in the cell. Along with, plasma membrane integrity and electrolyte leakage was restored by the enhanced production of osmolytes. Among different kinds of osmoprotectants, amount of proline inthe cell is an important factor for identifying the magnitude of tolerance in *B. juncea* genotypes. Madan et al. (1995) showed that proline accumulation and activity of the metabolic enzymes that are involved in proline biosynthesis pathway like *P5CR* (Pyrroline 5-carboxylate reductase) and ornithine aminotransferase were induced under stress, whereas, the activity of a proline breakdown enzyme proline oxidase was reduced under salt stress. Later, Chakraborty et al. (2012a) showed that a higher amount of proline was maintained in the tolerant genotypes of *Brassica* by the induced activity of *P5CS* gene. Further, comparative analysis between a glycophytic *A. thaliana* and a halophytic member *T. halophila* showed that the level of an important amino acids like proline was significantly high in *Thellungiella* as compared to *Arabidopsis*, which possibly helps *Thellungiella* to withstand salt stress (Ghars et al. 2007). Similarto this study, later a study with another important halophyte *C. maritima* also showed the same result, where induced accumulation of proline in the cell played an important role in maintaining the membrane stability and cellular homeostasis under high salinity (Megdiche et al. 2007). Murakeözy et al. (2003) showed that the induced cytoplasmic level of proline can potentially reduce the osmotic potential of the cell, which is around 50% for *Lepidium crassifolium*. Along with proline, evidences also revealed that the level of glycine betaine (a quaternary ammonium compounds synthesized in plants from choline) and trehalose in the cell is also important for determining the tolerance level . Chakraborty et al. (2012a) showed that tolerant genotypes of *B. juncea*—CS52 and CS54 were able to synthesize high amount of proline, glycine betaine and trehalose under stress and induced activity of these compounds are important for their higher tolerance under saline environment. Along with, another two key enzymes of osmolyte biosynthetic pathway (*BADH*—Betaine aldehyde dehydrogenase and *T6PS*—Trehalose 6-phosphate synthase)that are involved in glycine betaine and trehalose biosynthesis pathway, over-expressed in the tolerant genotypes of *B. juncea* (CS52 and CS54), indicates induced activity of these genes are also important for providing tolerance to *Brassica* against salt and oxidative stress (Chakraborty and Sairam 2017).

It is well known that among different kinds of osmoprotectants glycine betaine is most efficient in terms of energy management; it helps plants to survive under osmotic stress as well as oxidative stress, synthesized by the activity of the two enzymes *choline monooxygenase* (*CodA*) and *betaine aldehyde dehydrogenase* (*betA*). Contribution of glycine betaine for maintaining the cellular homeostasis was functionally proved by several scientist. The biotechnological approach by introducing 1.9 kb *choline oxidase* (*COX*) gene into *B. napus* showed an improved level of salt tolerance. For that 1.9 kb gene (*COX*) was at first amplified from *Arthrobacter pascens* and ligated into a binary vector *pHS993* and *pHS724.* The construct was cloned under the regulation of a constitutive promoter *CaMV35S* and finally transformed into the plants by *Agrobacterium*-mediated transformation technique. The transgenic plants were able to synthesize an enhanced amount of glycine betaine and had better germination capability under salt stress as compared to control plants of *Brassica* (Huang et al. 2000). Wang et al. (2010) showed that the *CodA* gene from *Arthrobacter tumifaciens* also improved the salinity tolerance ability in *B. campestris* by producing enhanced level of glycine betaine. Similarly, Prasad et al. (2000) transformed the *B. juncea* plants with a bacterial *CodA* gene and successfully improved the salt tolerance capacity of transgenic *B. juncea* plants. Bhattacharya et al. (2004) isolated the *betA* gene from bacteria and successfully transformed the gene into the *B. oleracea* by *Agrobacterium*-mediated transformation technique; the transgenic plants overexpressing the *betA* gene constitutively showed an improved glycine betaine production and had a better tolerance to salt stress.

4.4.2 Oxidative Stress Scavenging Compounds in *Brassicaceae*

Myriad of evidences showed that salt stress is associated with osmotic and oxidative stress. Due to salt stress, accumulation of H_2O_2 increases, which in turn interacts with unsaturated fatty acids of the plasma membrane and changes its integrity. Production of H_2O_2 aggravates the peroxidation of membrane lipids (another indicator of oxidative stress), estimated by the production of a biochemical marker compound Malondialdehyde (a product of lipid peroxidation). To protect the cells from the oxidative damage plants generally synthesizes several enzymes like catalase, ascorbate peroxidase, superoxide dismutase, etc. Induced activities of these enzymes were considered to be a signature attribute for salt-tolerant genotype. Previous study (Noreen et al.

2010) with five different cultivars of *B. rapa* showed that the activity of these antioxidant enzymes was high under stress for tolerant genotypes like Neela Shajilum and Peela Shajilum. Similarly, Ashraf and Ali (2007) reported, the activity of antioxidant enzymes like catalase, peroxidase and superoxide dismutase was highest in tolerant genotypes and might be considered as the principal factors for salt tolerance in *B. napus*. Mittal et al. (2012) stated a strong relationship is present in between photosystem II and antioxidant enzymes, and showed that better photosynthetic performance of *B. juncea* cv. Bio 902 under high salinity was directly related to the activities of the oxidative stress scavenging enzymes like catalase, ascorbate peroxidase and superoxide dismutase present in the cell. Along with the level of antioxidant enzymes, the ratio of different antioxidants like superoxide dismutase and ascorbate peroxidase was an important factor for survivability under salt stress. Based on the report by Joshi et al. (2011) the highest activity of antioxidant enzymes was found in the tolerant *B. juncea* CS52 genotype. Tolerant genotypes of *B. juncea* CS52 were able to maintain a low ratio of superoxide dismutase to ascorbate peroxide compared to the susceptible genotypes like RLM-198. Molecular basis and biotechnological approach showed that over-expression of an important antioxidant enzyme ascorbate peroxidase (*APX1*) can potentially improve the salinity tolerance. For that, *APX1* gene was cloned from *Hordeum vulgare* and transformed into *A. thaliana* by *Agrobacterium*-mediated gene transformation technique. Transgenic *Arabidopsis* plants over-expressing the *HvAPX1* construct showed an improved level of tolerance to salt stress and significantly lowered the amount of H_2O_2 under saline conditions (Xu et al. 2008). To improve the antioxidant capacity in *Brassica,* Metwali et al. (2012) incorporated two important antioxidant genes—*SOD* (superoxide dismutase) and *APX* (ascorbate peroxidase) jointly into *B. oleracea* plants by *Agrobacterium*-mediated transformation. Transgenic plants over-expressing the genes in a constitutive manner showed an improved level of tolerance under saline conditions. Apart from the genes related to antioxidant enzyme, the LEA group of proteins protect plants from multiple abiotic stresses. Park et al. (2005) characterized an *LEA* gene from *B. napus* (*BnLEA4-1*), cloned into a binary vector *pGI121,* under the regulation of *CaMV35S* promoter. The whole construct was transferred into *B. campestris* by means of *Agrobacterium*-mediated transformation technique. Transgenic plants over-expressing the *BnLEA4-1* gene showed an improved level of tolerance against salt by enhanced production of antioxidant enzymes. Similarly, Saha et al. (2016) cloned an LEA domain containing gene from *A. thaliana LEA4-1* (*AtLEA4-1*) and transformed the gene into *B. juncea* plants constitutively, transgenic *Brassica* plants over-expressing the transgene showed an improved salinity tolerance by maintaining chlorophyll content, as well as increased its ability to synthesize higher amount of antioxidant enzymes as compared to control plants. Most recently, Chen et al. (2019) cloned a *Syntaxin* gene (*GmSYP24*) from *Glycine max* and successfully transformed the gene into *A. thaliana.* Transgenic *Arabidopsis* lines over-expressing the *GmSYP24* showed enhanced tolerance to salinity. The Syntaxin protein contains an LEA2 domain that interacts with ABA to control stomatal responsive genes, also important for maintaining the high expression of *POD* and *SOD* genes under stress compared to control plants. Sun et al. (2014) transformed a constituted gene named *YHem*1 into the *B. napus* for increasing the production of 5-ALA (5-Aminolevulinic acid). Transgenic *B. napus* plants over-expressing the *YHem1* gene accrued an improved level of salt tolerance where antioxidant enzyme activity and chlorophyll content were higher in transgenic lines and transgenic plants can sustain salt concentration at the level of 200 mM.

4.4.3 Role of Glucosinolates in Salt Tolerance in *Brassicaceae* Family

Glucosinolates are a type of secondary metabolite present specifically in the members of Brassicaceae. Glucosinolates are generally used as a defensive compound against pathogens. Apart from its biotic stress tolerance, reports showed that, glucosinolates also help in abiotic stress tolerance in *Brassica* and other members of Brassicaceae. Previous reports of Lopez-Berenguer et al. (2009) showed that *B. oleracea* var *Italica* significantly increased the glucosinolates under salt stress, therefore, showed a strong relationship between glucosinolates content and salt tolerance. Martinez-Ballesta et al. (2015) in his study with *A. thaliana*, concluded glucosinolates are important for maintaining overall plant water status under salt stress. Molecular analysis showed that two genes are involved in glucosinolate biosynthesis pathway viz., *MYB28* and *MYB29*have played role under salt stress. Comparative analysis with a wild and double mutant line (*myb28myb29*) showed that the osmotic damage was severe in double mutant lines of *Arabidopsis*. Wild type plants with functional *MYB28* and *MYB29* gene showed increased glucosinolates production/accumulation under salt stress to maintain water transport status and limits destructive effects of osmotic stress. While, no increment was observed in double mutant lines of *Arabidopsis*, and osmotic damage was most severe for *myb28myb29* lines. Later, Pang et al. (2012) reported different types of glucosinolates accumulated in different parts of *T. halophila* as a response to salt stress.

4.4.4 Role of Hormones in Salt Tolerance in *Brassicaceae* Family

Growth, development and every mechanism in a plant are highly regulated by the endogenous phytohormones like (ABA—Abscisic acid, IAA—Indole 3-acetic acid, GA—Gibberellic acid, JA—Jasmonic acid, SA—Salisyclic acid, BR—Brassinosteroids and Triazoles). Phytohormones act like mediators for environmental responses (Pavlovic et al. 2019). It is well known that the stomatal movement is highly dependent on ABA contentand an elevated level of ABA under stress can provide salt tolerance in *Brassica* (He and Cramer 1996). It was also reported to promote stomatal closure and minimizing the amount of water loss during stress (Ryu and Cho 2015). Recently, Pavlovic et al. (2019) showed that important hormones like ABA were positively correlated with the level of tolerance in *B. rapa* ssp. *pekinensis*, *B. oleracea* var. *capitata* and *B. oleracea* var. *acephala*, while JA was found to be negatively correlated. Apart from ABA, auxin level in the cell is another important factor for growth.

The amount of auxin in the cells is also important for salinity response and tolerance. Previous reports on *Arabidopsis thaliana* showed that, YUCCA pathway gene *YUC6* when over-expressed under the regulation of an oxidative stress-responsive promoter SWAP2, the plant showed better performance against oxidative stress. Reports also showed that auxin potentially controls the expression of some stress-responsive genes like *DREB2A*, *DREB2B*, *RD22*, *RD29A*, which are important for abiotic stress tolerance (Bieleach et al. 2017). In *Arabidopsis*, reduction of cytokinin phytohormone was related with better tolerance. Study with CK-deficient *IPT* gene knockout lines (where *ipt 1,3,5,7* genes are mutated) showed better performance during salt stress. Mutated lines of *IPT* gene showed an improved tolerance by induced activity of several genes and proteins belonging to *DREB*, *ZFHD*, *SOS*, *LEA* and *NAC* family genes, while, downregulating several genes that are involved in photosynthesis pathway (Nishiyama et al. 2012). Previous reports also stated that a molecular crosstalk present between ABA and CK phytohormones and CKs were found to be the negative regulator for salt stress (Nishiyama et al. 2011) and initiate a response that cause delay for the stomatal closure and movement. Studies related to jasmonates revealed that tolerant cultivars maintained a low level of jasmonates under stress, which effectively helped to reduce the level of sodium inside cells. In general, studies with BR showed that the increased BR content in *Arabidopsis* can potentially increase the overall salt tolerance (Krishna 2003). Most recently, Pavlovic et al. (2019) showed that among the members of Brassica family salt tolerance was positively related with the level of BR precursor compound typhasterol. Molecular mechanism showed that BR activates BIN2 dependent inhibition of Mitogen-activated protein kinase (MAPK) protein and reduce stomatal development. Low stomatal density in high saline conditions effectively enhances the stress tolerance in *Arabidopsis* (Kim et al. 2012) by minimizing the water loss during stress.

4.5 Alteration in Ionic Balance in Brassicaceae *Under Salinity*

Normal cellular function is highly regulated by the avilability of ions and nutrients . Salinity potentially changed the ionic and nutritional homeostasis of the cell as it strongly competes with potassium and calcium ion uptake. A multitude of evidences showed that sodium caused the K^+ efflux or leakage. Therefore, actual level of sodium, potassium and calcium present in the cell under stress is another important factor to identify the plant's tolerance.. From the previous studies, it is clear that low amount of sodium in the cell and maintenance of high K^+/Na^+ ratio is considered a signature attribute for identifying the salt tolerance genotypes (Shabala and Pottosin 2014). Studies on rapeseed showed that maintainance of a high K^+ in the cell had played a legitimate role in plant survival (Rameeh et al. 2012). Das et al. (1994) reported a strong correlation between Na^+ influx and K^+ efflux is present in *B. campestris*. Along with the amount of sodium and potassium, the ratio of K^+/Na^+

was also important under stress. The amphidiploid species of *Brassica* were able to maintain a high K^+/Na^+ ratio under stress compared to the diploid species of *Brassica* (Ashraf and McNeilly 2004). Chakraborty et al. (2012b) reported maintenance of low Na^+/K^+ ratio was the main discriminating factor behind the tolerant nature of *B. juncea* cultivars viz. CS52 and CS54. Ashraf et al. (2001) reported that amphidiploid species are able to maintain a high K^+/Na^+ ratio in their root and shoot, which directly associated with the better performance of amphidiploids under salinity than diploid species of *Brassica*. Umar et al. (2011) showed that external application of potassium also improves the salt tolerance capability , where application of excess potassium in the soil effectively improved the salinity tolerance efficiency by improving photosynthetic traits, antioxidant enzymes in *B. campestris*. Similarly, Chakraborty et al. (2016c) showed that, an external application of potassium also ameliorated tolerance level in sensitive and tolerant genotypes in other crops. Molecular basis behind salt accumulation was identified in *Brassica*. Transporters like NSCC, AKT and HAK preferred to transport sodium under stress to increase the cytoplasmic concentration of salt. To maintain the proper concentration of sodium in the cell, plants upregulate the expression of several transporters belonging to the family SOS (Salt overly sensitive), HKT (High-affinity K^+ transporter), PM-ATPase (Plasma membrane ATPase) and PM-PPase (Plasma membrane pyrophosphatase). Kumar et al. (2009) showed that, a low Na^+/K^+ ratio was maintained in amphidiploid member of *B. juncea* by induced activity of SOS family members.

Besides, exclusion of sodium by induced activities of SOS and HKT transporters, some studies with *NHX1* gene family members showed that storing sodium into the vacuole could also be an alternative approach for maintaining cellular ionic homeostasis under saline environment. Transgenic *B. napus* plants showed an improved tolerance when *NHX1* gene was over-expressed under the regulation of a constitutive promoter. Previous studies also showed that *NHX1* gene from glycophyte and halophyte can potentially increase the level of tolerance when over-expressed constitutively. Though in both cases tolerance level was elevated, the signature difference was identified based on the intensity of tolerance showed by the plants containing glycophytic *NHX* and halophytic *NHX* genes (Kronzucker and Britto 2011; Mishra and Tanna 2017). Previous studies on salinity also showed that high concentration of sodium in the cell can influence calcium efflux process. As per the reports of Collins et al. (2007), higher concentrations of calcium in *B. campestris* can also provide enhanced tolerance to salt stress. Previous studies showed that sodium can compete with calcium by removing calcium from plasma membrane and cause membrane damage.

Apart from the maintenance of enzyme activity and plasma membrane integrity, calcium acts as an important signalling molecule under stress, which orchestrates the signal transduction network into the whole plant under stress. Reports showed that elevation of cytosolic calcium level was observed after immediate exposure of salt stress. He and Cramer (1992) working with six different species of *Brassica* reported that among different types of micro and macronutrient calcium played a central regulatory role for awakening the salinity response in the different members belonging to *Brassica* species. According to Knight (2000), in *Arabidopsis* several different types of proteins were induced and involved in calcium signalling under salt stress. The calcium response further activates the proteins, which actively regulates other important processes like protein regulation, protein–protein interaction, etc. The elevated level of calcium was also important to initiate the Salt overly sensitive (SOS) pathway genes in plants. Previous reports showed that, under salt stress, expressions of the two calcium transporters are significantly changed that belongs to the *CAX* and *ACA* gene families. Bose et al. (2011) showed that the *CAX* and *ACA* gene family members work exclusively for maintaining the cytoplasmic level of calcium, and stores the excess calcium into organelles like vacuole and endoplasmic reticulum. Boursiac et al. (2010) showed that in *A. thaliana* disruption of an important member of *ACA* gene can potentially activate the salicylic acid response to initiate the programmed cell death process.

4.6 Changes in Nutritional Contents in Brassicaceae *Under Salinity*

Normally, plant growth and development is highly dependent on the bio-availability of the macro and micronutrients present in the soil. Wide range of evidences showed that salinity effectively changed the nutritional status (macronutrient like nitrogen, phosphate and micronutrients like zinc, iron and manganese) inside plant cells other than K^+ and Ca^{2+}. Study with *Brassica*, in general, showed salinity potentially decreased essential macronutrients except phosphate (Lopez-Berenguer et al. 2009). Chakraborty et al. (2016e), showed that in *Brassica* spp. total nitrogen content decreased in the leaves under salinity stress, alongside it also effectively lowered down the contents of important micronutrients viz. iron, zinc and manganese in root, stem and leaves. However, they also showed that tolerant genotypes were able to retain higher contents of nitrogen and other micronutrients under stress. Previous studies also showed that the activities of nitrate reductase (an important enzyme for N-uptake and metabolism) were reduced under stress in *B. napus* at the concentration of 150 mM and 200 mM NaCl (Bybordi and Ebrahimian 2011). Therefore, based on the activity of the enzyme it was clear that nitrate influx is significantly decreased at high concentrations of salt. Along with nitrogen, *B. napus* showed a significant increase in phosphate concentration. The activities of ACP (acid phosphatase) and ALP (alkaline phosphatase) enzymes were up-regulated under salt stress (Bybordi and Ebrahimian 2011). Retention of nutrients under stress can be beneficial for survival and helpful for maintaining the cellular homeostasis. Previous reports showed that the external application of nitrogen or other micronutrients improves plants' performance under salinity stress. Siddiqui et al. (2011) reported that associated application of nitrogen and sulphur in the medium can improve the response under stress, by effectively increasing the morphological, physiological characters and enzyme activities of *B. juncea*. Similarly, Iqbal et al. (2015) showed that the external application of nitrogen can potentially improve the tolerance level in *B.*

juncea by regulating a hormone ethylene and proline biosynthesis. Siddiqui et al. (2008) showed that the application of nitrogen along with gibberellin enhanced the salinity tolerance in *B. juncea* plants by regulating antioxidant enzymes, proline and glycine betaine synthesis. Besides, some important micronutrients like iron, zinc and manganese content were reduced under stress in case of sensitive genotypes of *Brassica*. Tuncturk et al. (2011) showed that micronutrients like iron, manganese, copper were highly accumulated in roots of tolerant *B. napus*, also, iron, copper and manganese were significantly increased in all plant parts of *B. napus* after imposition of salt stress. However, there was no significant amount of increase in the Zn content. Goel and Singh (2015), in his study with *B. juncea* showed that nitrogen uptake and its assimilation pathway genes like NRT (Nitrate transporter), AMT (Ammonium transporter), NR (Nitrate reductase), NiR (Nitrite reductase), GS (Glutamine synthetase), GOGAT (Glutamate synthase), GDH (Glutamate dehydrogenase) and ASN (Asparagines synthetase) down-regulated under salt stress. As per the reports, *AMT1.2*, *AMT2*, *GS1.1*, *GDH1* and *ASN2* genes were downregulated and *NRT1.1*, *NRT1.5* and GDH2 were upregulated immediately after exposure to an hour of stress, while expression of *NRT1.1*, *NRT2.1*, *NiR1*, *AMT2*, *GDH1* and *ASN2* genes was repressed after 24 h of salt treatment.

4.7 Molecular Responses

4.7.1 Genes for Salinity Tolerance

Plants generally use 2 different kinds of mechanisms to cope with the destructive effects of salt stress. I. Sodium exclusion strategy—which directly excludes sodium in the soil. II. Sodium compartmentation strategy—where cells generally store sodium into the vacuoles to maintain a non-toxic level of sodium in the cytoplasm. Among different kinds of mechanisms, the Na^+ exclusion occurs mainly by the activities of the SOS gene family members like SOS1, SOS2 and SOS3, HKT1 and 2, and plasma membrane located H^+ pumps.

The SOS (Salt Overly Sensitive) Gene Family Members

Shi et al. (2000) first characterized the *AtSOS1* gene as a potential Na^+/H^+ transporter, which located in the plasma membrane of *A. thaliana* and helps to exclude sodium out of the cell under salt stress, thus maintaining low concentrations of sodium inside the cell. The activity of *SOS1* gene is highly regulated by the action of SOS2–SOS3 complex. SOS2 is a protein kinase belonging to the SnRK3 family (sucrose non-fermenting-1-related protein kinase-3), while SOS3 encodes a calcium-binding protein that perceives the Ca^{2+}-signal. The calcium concentration increases in the cell due to hike in cellular sodium content and activates SOS3-like calcium-binding protein 8 (also known as Calcineurin B-like CBL10), which further binds with SOS2 to activate SOS2–SOS3 complex and finally activates the SOS1 through phosphorylation process. Contribution of SOS1 for maintaining the cellular homeostasis and ion-exclusion was functionally proved by using *sos1* mutant lines of *A. thaliana* (Shi et al. 2002). Reports showed that SOS1 is involved in excluding sodium from the cytosol to surrounding medium of epidermal cells and from vascular tissues to the surrounding parenchyma cells when over-expressed (Na^+/H^+ antiporter gene-*AtSOS1*) in *Arabidopsis* in a constitutive manner. For that, the *SOS1* gene was cloned under the regulation of a constitutive promoter *CaMV35S*. Transgenic plants over-expressing the *AtSOS1* gene had improved salt tolerance and were able to maintain low Na^+ concentration under high salt concentrations. Therefore, by limiting the accumulation of sodium in cells, it is also possible to improve the magnitude of salt tolerance (Fig. 2).

But, with electrophysiological evidences Shabala et al. (2005) reported that the activity of the *SOS1* gene is not solely dependent on the SOS2–SOS3 complex, rather it could be a possible target of phospholipase D (PLD) as well. Exposure of salt stress induces the PLDα1 in *Arabidopsis* plant, which in turn activates a secondary messenger molecule phosphatidic acid (PA). PA further activates the Mitogen-Activated Protein Kinase 6 (MAPK6), which directly targets and activates the SOS1 (Yu et al. 2010) to exclude Na^+ outside cytosol and maintain cellular homeostasis. Previous studies depicted that the gene *SOS2* played a central role under salt stress, where it interacts with various enzymes and calcium channel proteins (*CAX1*) and vacuolar H^+ pumps (*V-ATPase*) (Cheng et al. 2004). In addition, SOS2 also interacts with the genes that are involved in the ABA biosynthesis pathway like *ABI2*. Qiu et al. (2004) showed in the mutants of *SOS2*, the activity of a tonoplast localized Na^+/H^+ channel is highly repressed under salt stress. Therefore, itprovided evidence that the gene *SOS2* definitely had some interaction with the *AtNHX* gene. Most recently, two new SOS members were identified in *Arabidopsis* named SOS4 and SOS5. SOS4 involved in the synthesis of vitamin and SOS5 plays an important role in salt stress for maintaining the cell expansion and retaining normal growth of root (Shi et al. 2003).

The expression pattern of different SOS family genes was studied in *B. juncea* cv CS52 and *B. napus* by Kumar et al. (2009). As per the reports, amphidiploid *B. juncea* and diploid *B. napus* maintained a distinct transcript abundance pattern of SOS to withstand stress. Similarly, Chakraborty et al. (2012b) showed, in *B. juncea* and *B. campestris* induction of three different kinds of SOS family members (*SOS1*, *SOS2* and *SOS3*) was different and expression of SOS was high in the salt-tolerant genotype *B. juncea*.

Figure 2 shows that at high external sodium concentrations, sodium moves in the cell by NSCC (Non specific cation channels) channels and increase the inner concentration of sodium. Elevated level of sodium is associated with the immediate release of calcium by the activity of vacuolar CAX and ACA gene families. Further, calcium activates the inactive SOS3 protein and the activated SOS3 binds with SOS2 to activate SOS1 protein-(H^+ mediated Na^+ antiporter), which excludes sodium out of the cell. Recent reports showed that elevated level of sodium also activates PLDα1 protein, which activates MAPK6 protein with the help of phosphatidic acid to regulate the expression of SOS1. Plasma membrane localized AHA members help to create

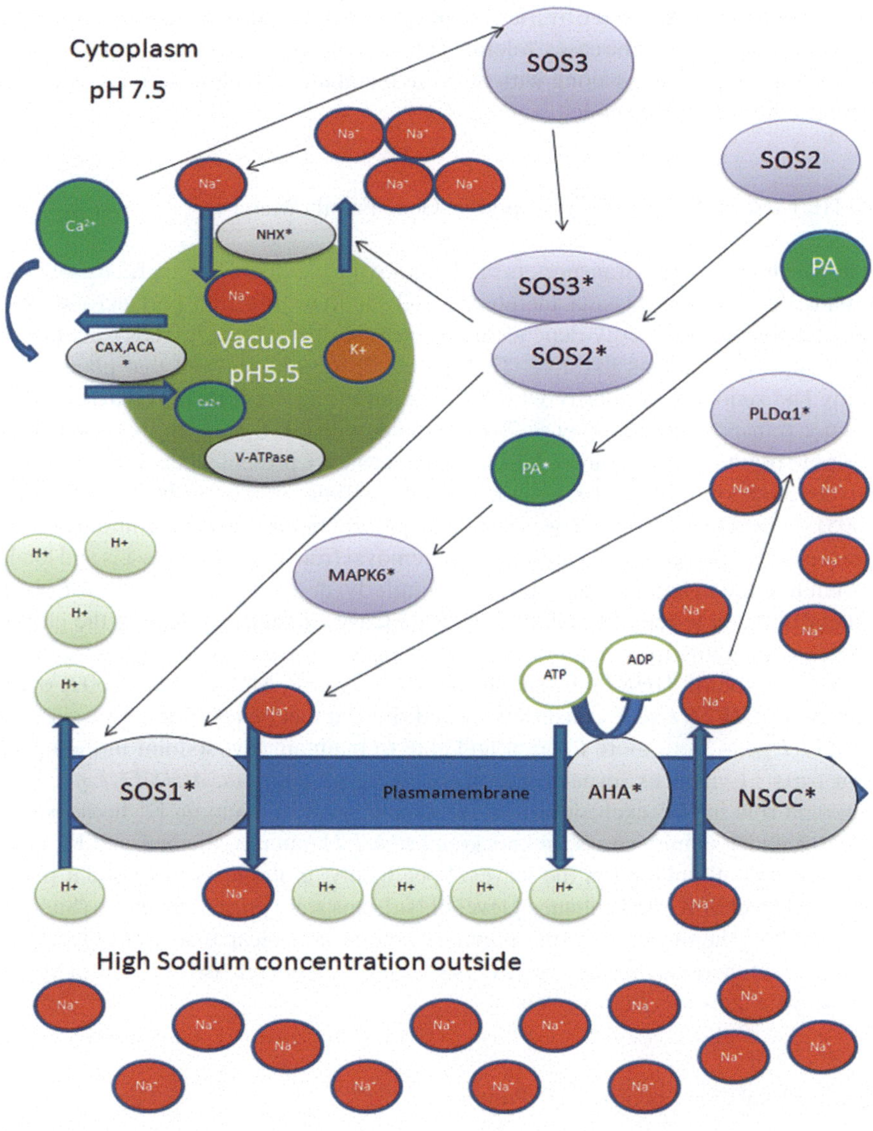

Fig. 2 The *SOS* pathway and *NHX* gene functions in *Arabidopsis* for maintaining the cellular homeostasis under salt stress. Here '*' denotes activated form of the protein and PA: Phosphatidic acid; MAPK6: Mitogen-Activated Protein Kinase 6

necessary proton motive force by transferring protons (H^+) out of the cell in an energy-dependent manner. In some cases, tolerant genotypes also showed the induced activity of vacuolar NHX proteins along with SOS1 to maintain the cellular homeostasis and storing sodium inside vacuole.

The HKT (High-Affinity K^+ Transporter) Gene Family Member

High-affinity potassium transporters of plants are the members of integral membrane proteins, localized in plasma membrane and facilitate ion transport across plasmamembrane in an energy-dependent manner. Multiple types of HKT members are present in monocots like rice (*Oryza sativa*). Reports showed that plant HKT family had sequence similarity with bacterial and fungal TrKH/TrKG/ktrB classes of cation transporters (Waters et al. 2013). Studies with HKT gene family showed that it helps to provide salt tolerance. Myriad evidences were present where HKT member actively transports Na^+ and K^+ to maintain the cellular homeostasis. Sunarpi et al. (2005) showed that *AtHKT1.1* was primarily present in the plasma membrane of the xylem parenchyma cells and facilitates Na^+ removal from the xylem vessels and loads sodium into the xylem parenchyma cells. Similarly, in *Arabidopsis AtHKT1.1* plays a major function to remove sodium ions from leaves, thereby protecting the photosynthetic apparatus from salinity induced damage (Horie et al. 2009). To validate the contribution of *AtHKT1.1* gene in salinity tolerance, Moller et al. (2009) overexpressed *AtHKT1.1* gene constitutively and showed that plants over-expressing the *AtHKT1.1* gene were more tolerant and able to maintain low sodium in the upper plant parts . Evidences revealed that in *Arabidopsis thaliana*, *AtHKT1.1* plays an important role in Na^+ exclusion from leaves and helps to maintain K^+ homeostasis in leaves under salinity stress. In knockout *(athkt1.1)* mutants, the Na^+ and K^+ level were significantly altered in the leaves. Specifically, in the leaves of mutant plant, a reduced level of K^+ and enhanced level of Na^+ ions were present in leaves (Sunarpi et al. 2005). Therefore, from the previous reports it is clear that *AtHKT1;1* gene played dual functions in cells - remove Na^+ along with the release of K^+ from xylem parenchyma cells to xylem vessels.

Plasma Membrane Localized H^+ Pumps

Proton gradient is important for transporting the ions across the plasma membrane of plant cells. Evidences showed that in the plasma membrane of plants, two different types of proton pumps were present, which were designated as P-type H^+-ATPase and P-type H^+-PPase. Both these pumps help to transport protons out of the cells by phosphorylated aspartyl formation to utilize energy and phosphate group from ATP (Fuglsang et al. 2011). Reports showed that in *A. thaliana* genome, 11 different types of *AHA* genes are present. Among different types of AHA isoforms, AHA1 and AHA2 are most abundant in the case of plants and expressed predominantly in leaves and roots (Fuglsang et al. 2011). Knockout lines of both *aha1-aha2* mutants are lethal for survival as reported by Haruta et al. (2010). The activities of these genes were directly related to the activity of *AtSOS1* gene. Members of Plasma membrane located H^+ transporters create a proton motive force, which is actually needed during the exclusion of sodium by the activity of *AtSOS1*. As reported earlier, the combination of both genes can successfully enhance the salt tolerance (Zhou et al. 2015). Most recently, Fan et al. (2019) cloned two genes from a halophyte *Sesuvium portulacastrum* and jointly over-expressed in *A. thaliana* plants. For transgenic plant development, *SpSOS1* and *SpAHA1* genes were cloned in a binary vector *pCAMBIA1304* under the regulation of a constitutive promoter and transformed through *Agrobacterium*-mediated technique. Transgenic lines over-expressing both the genes imparted an improved tolerance to salt. Previous reports also showed that in plants two different types of H^+-PPase pumps were present: type I and type II. Based on the cytosolic concentration of K^+ and $Ca^{2+,}$ these H^+ pumps respond differentially. In general, type I pumps were sensitive to cytosolic K^+ concentration, whereas type II pumps were extremely Ca^{2+} sensitive (Fuglsang et al. 2011). The second type of mechanism mainly focused on compartment and storing the sodium into vacuoles to maintain the cellular homeostasis.

The NHX Type Cation/H^+ Antiporters

In *A. thaliana* 8 types of NHX family genes are reported, of which two are endosomal (*NHX5*, *NHX6*) (Bassil et al. 2011), four are vacuolar *(NHX1-NHX4)* (Apse and Blumwald 2007), and NHX7 and NHX8 are located at plasma membrane (Shi et al. 2000). Among different types of NHX members, NHX1 and NHX2 are most abundant and play an important role in maintaining the cellular homeostasis and intracellular pH in *A. thaliana* (Yokoi et al. 2002). NHX1 is a cation/H^+ transporter that helps to compartmentalize sodium into the vacuole to maintain a balance in the cytosol. The *NHX1* and *NHX2* genes also help to exchange K^+ with vacuole and play a major role in intra-vacuolar exchange (Bassil et al. 2011). Earlier studies with NHX genes showed that expression of the *AtNHX1* gene was more specifically induced in the leaves compared to roots under salt stress (Quintero et al. 2000). Characterization and validation through transgenics revealed that the majority of NHX members are the valuable contributors for salinity tolerance. To improve salt tolerance in *Arabidopsis* Apse et al. (1999) transformed the vacuolar gene *AtNHX*1 in *Arabidopsis* in a constitutive manner. Transgenic plants over-expressing the gene showed an improved level of tolerance under salt stress and were able to retain growth and development up to 200 mM of NaCl concentration. Similarly, Zhang et al. (2001) used *Arabidopsis AtNHX1* gene and transformed the gene into the *B. napus* plant under the regulation of a constitutive promoter. Reports showed that transgenic *B. napus* plants overexpressing the *AtNHX1* gene could tolerate up to 100 mM concentration of salt. Evidences also present where NHX from distant members could potentially increase the salinity tolerance in *Brassica* spp. Rajagopal et al. (2007) cloned *NHX1* gene from *Pennisetum glaucum* and transformed in *B. juncea* by *Agrobacterium*-mediated method. Transgenic plants over-expressing the *PgNHX1* gene showed better performance when exposed to salt stress and can withstand up to 300 mM NaCl concentration.

Vacuolar H^+ Pumps

The activity of the vacuolar H^+ pumps is highly associated with the activities of the genes present in the vacuolar-membrane like *NHX1* (Popova and Golldack 2007; Krebs et al. 2010). In *A. thaliana* two types of proton pumps were located in vacuole viz. vacuolar H^+-ATPase (*V-ATPase*) and vacuolar H^+ pyrophosphatase (*V-PPase*). Wide ranges of evidences were found where over-expression of a vacuolar H^+-PPase provides improved salinity tolerance. Gaxiola et al. (1999) cloned and over-expressed the *Arabidopsis AVP1* gene constitutively. Reports showed plants over-expressing the *AVP1* construct have improved tolerance to salt. Similarly, yeast cells over-expressing the *AtAVP1* gene showed an improved level of salt tolerance. Kumar et al. (2013) cloned vacuolar–PPase gene (*AVP1*) from *Arabidopsis* and successfully transformed it into sugarcane. Transgenic sugarcane plants over-expressing the *Arabidopsis AVP1* gene showed an improved tolerance to drought and salinity.

5 Conclusion and Future Perspectives

Taken together, the studies carried out by various researchers confirmed that there are huge variability of salinity specific responses in different members of *Brassicaceae* family. By and large majority of the members of *Brassicaceae* family are salt sensitive having a threshold salt tolerance level of 50 mM (of NaCl), but at the same time there is existence of some halophytic members like *Thellungiella* in this family which can tolerate extremely high concentration of salt. Studies showed diploid members of this family are comparatively more sensitive to salinity as against the amphidiploid species. Till date, various morphological, physiological, biochemical, and molecular studies were performed with glycophytic and halophytic members of Brassicaceae, which revealed retention of growth and biomass (shoot length, root length, dry weight, and fresh weight) under stress, were most legitimate and differentiating factor for salt tolerance in Brassicaceae family. Other physiological responses like retention of water potential, relative water content, total chlorophyll, and membrane integrity were found to be quite useful to identify the salt-tolerant genotypes. Importantly, studies showed that massive induction of glucosinolates was associated with salt tolerance in Brassicaceae. Generally, tolerant members of the family could maintain greater cellular homeostasis under salt stress by producing higher amounts of proline, glycine betaine, and other organic compounds. It was observed that different phytohormones like ABA, IAA, and BR were elevated under stress while CK and JA were decreased under stress to prevent the water loss and retain the growth in Brassica under stress. Alongside, high selectivity of ions (K^+ over Na^+) during upward transportation and low Na^+/K^+ ratio supported by the activity of *SOS*, *HKT*, and *PM-ATPase* genes were found to be crucial for differential salt tolerance in different members of *Brassicaceae* family. Although, a good amount of work was carried out to know the salinity specific response of different members of *Brassicaceae*, most were oriented in glycophytic species and there is scope to have a clearer understanding of salt tolerance strategies in halophytic members of this family. More precisely, the harmony of different salt tolerance mechanisms (like ion exclusion, sequestration, and secretion) was not very clear in halophytic species. Also, studies on the role of different phytohormones in imparting salt tolerance need greater attention for effective stress adaptation strategy in these crops.

References

Apse MP, Aharon GS, Snedden WA, Blumwald E (1999) Salt tolerance conferred by overexpression of a vacuolar Na^+/H^+ antiport in *Arabidopsis*. Science 285(5431):1256–1258

Apse MP, Blumwald E (2007) Na^+ transport in plants. FEBS Lett 581:2247–2254

Ashraf M, Ali Q (2007) Relative membrane permeability and activities of some antioxidant enzymes as the key determinants of salt tolerance in canola (*Brassica napus* L.). Environ Exp Bot 63: 266–273

Ashraf M, McNeilly T (2004) Salinity tolerance in Brassica oilseeds. Crit Rev Plant Sci 23:157–174

Ashraf M, McNeilly T, Nazir M (2001) Comparative salt tolerance of amphidiploid and diploid *Brassica* species. Plant Sci 160:683–689

Ashraf M, Noor R (1993) Growth and pattern of ion uptake in *Eruca sativa* Mill. under salt stress. Angew Bot 67:17–21

Ashraf M (1994) Organic substances responsible for salt tolerance in *Eruca sativa*. Biol Plant (Prague) 36:255–259

Bassil E, Tajima H, Liang Y, Ohto M, Ushijima K, Nakano R et al (2011) The Arabidopsis Na^+/H^+ antiporters NHX1 and NHX2 control vacuolar pH and K^+ homeostasis to regulate growth, flower development, and reproduction. Plant Cell. https://doi.org/10.1105/tpc.111.089581

Bhattacharya RC, Maheswari M, Dineshkumar V, Kriti PB, Bhat SR, Chopra VL (2004) Transformation of *Brassica oleracea* var. capitata with bacterial betA gene enhances tolerance to salt stress. Sci Hortic 100:215–227

Bieleach A, Hrtyan M, Tognetti VB (2017) Plant under stress: involvement of auxin and cytokinin. Int J Mol Sci 18(7):1427

Bose J, Pottosin I, Shabala SS, Palmgren MG, Shabala S (2011) Calcium efflux systems in stress signalling and adaptation in plants. Front Plant Sci 2:85

Boursiac Y, Lee SM, Romanowsky S, Blank R, Sladek C, Chung WS, Harper JF (2010) Disruption of the vacuolar calcium-ATPases in Arabidopsis results in the activation of a salicylic acid-dependent programmed cell death pathway. Plant Physiol 154(3):1158–1171

Boyd RS, Barbour MG (1986) Relative salt tolerance of *Cakile edentula* (Brassicaceae) from Lacustrine and marine beaches. Amer J Bot 73(2):236–241

Bybordi A, Ebrahimian E (2011) Effect of salinity stress on activity of enzymes involved in nitrogen and phosphorous metabolism case study: Canola (*Brassica napus* L.). Asian J Agric Res 5(3): 208–214

Chakraborty K, Bhaduri D, Meena HN, Kalariya K (2016a) External potassium (K^+) application improves salinity tolerance by promoting Na^+-exclusion, K^+-accumulation and osmotic adjustment in contrasting peanut cultivars. Plant Physiol Biochem 103:143–153

Chakraborty K, Bishi SK, Goswami N, Singh AL, Zala PV (2016b) Differential fine-regulation of enzyme driven ROS detoxification network imparts salt tolerance in contrasting peanut genotypes. Environ Exp Bot 128:79–90

Chakraborty K, Bose J, Shabala L, Eyles A, Shabala S (2016c) Evaluating relative contribution of osmo- and tissue-tolerance mechanisms towards salinity stress tolerance in three Brassica species. Physiol Plant 158:135–151

Chakraborty K, Bose J, Shabala L, Shabala S (2016d) Difference in root K^+ retention ability and reduced sensitivity of K^+-permeable channels to reactive oxygen species confer differential salt tolerance in three *Brassica* species. J Exp Bot 67:4611–4625

Chakraborty K, Sairam RK (2017) Induced-expression of osmolyte biosynthesis pathway genes improved salt and oxidative stress tolerance in *Brassica* species. Indian J Exp Biol 55:711–721

Chakraborty K, Sairam RK, Bhaduri D (2016e) Effects of different levels of soil salinity on yield attributes accumulation of nitrogen, and micronutrients in *Brassica* Spp. J Plant Nutr 39:1026–1037

Chakraborty K, Sairam RK, Bhattacharya RC (2012a) Salinity induced expression of pyrrolline-5-carboxylate synthetase determine salinity tolerance in *Brassica* spp. Acta Physiol Plant 34:1935–1941

Chakraborty K, Sairam RK, Bhattacharya RC (2012b) Differential expression of salt overly sensitive pathway genes determines salinity stress tolerance in *Brassica* genotypes. Plant Physiol Biochem 51:90–101

Chen L, Fang Y, Zhang C, Hao Q, Cao D, Yuan S et al (2019) GmSYP24, a putative syntaxin gene, confers osmotic/drought, salt stress tolerances and ABA signal pathway. Sci Reports 9

Cheng NH, Pittman JK, Zhu JK, Hirschi KD (2004) The protein kinase SOS2 activates the *Arabidopsis* H^+/Ca^{2+} antiporter CAX1 to integrate calcium transport and salt tolerance. J Biol Chem 279:2922–2926

Collins RP, Harris PJC, Bateman MJ, Henderson J (2007) Effect of calcium and potassium nutrition on yield, ion content and salt tolerance of *Brassica campestris* (rapa). J Plant Nutrition 31(8):1461–1481

Cronquist A (1968) The evolution and classification of flowering plants. London: Nelson

Das S, Chandra M, Ghosh B (1994) Response of *Brassica campestris* cultivars to salinity stress. Plant Physiol Biochem 21(2):76–79

Dierig DA, Shannon MC, Grieve CM (2001) Registration of WCL-SL1 salt tolerant *Lesquerella fendleri* germplasm. Crop Sci 41:604–605

Dierig DA, Tomasi PM, Salywon AM, Ray DT (2004) Improvement in hydroxy fatty acid seed oil content and other traits from interspecific hybrids of three *Lesquerella* species: *Lesquerella fendleri*, *L. pallida*, and *L. lindheimeri*. Euphytica 139:199–206

Fan Y, Yin X, Xie Q, Xia Y, Wang Z, Song J et al (2019) Co-expression of *SpSOS1* and *SpAHA1* in transgenic *Arabidopsis* plants improves salinity tolerance. BMC Plant Biol. https://doi.org/10.1186/s12870-019-1680-7

Franzke A, Lysak MA, Al-Shehbaz IA, Koch MA, Mummenhoff K (2010) Cabbage family affairs: the evolutionary history of Brassicaceae. Trend Plant Sci 16:1360–1385

Fuglsang AT, Paez-Valencia J, Gaxiola RA (2011) Plant proton pumps: regulatory circuits involving H^+-ATPase and H^+-PPase. Transporters Pumps Plant Signal. https://doi.org/10.1007/978-3-642-14369-4_2

Gao F, Gao Q, Duan X, Yue G, Yang A, Zhang J (2006) Cloning of an H + -PPase gene from *Thellungiella halophila* and its heterologous expression to improve tobacco salt tolerance. J Exp Bot 57(12):3259–3270

Gaxiola RA, Rao R, Sherman A, Grisafi P, Alper SL, Fink GR (1999) The *Arabidopsis thaliana* proton transporters, AtNhx1 and Avp1, can function in cation detoxification in yeast. Proc Natl Acad Sci U S A 96(4):1480–1485

Ghars MA, Parre E, Debez A, Bordenave M, Richard L, Leport L et al (2007) Comparative salt tolerance analysis between *Arabidopsis thaliana* and *Thellungiella halophila* with special emphasis to K^+/Na^+ selectivity and proline accumulation. J Plant Physiol 158:588–599

Goel P, Singh AK (2015) Abiotic stresses downregulate key genes involved in nitrogen uptake and assimilation in *Brassica juncea* L. PLoS ONE. https://doi.org/10.1371/journal.pone.0143645

Haruta M, Burch HL, Nelson RB, Barrett-Wilt G, Kline KG, Mohsin SB, Young JC, Otegui MS, Sussman MR (2010) Molecular characterization of mutant Arabidopsis plants with reduced plasma membrane proton pump activity. J Biol Chem 285(23):17918–17929
He T, Cramer GR (1992) Growth and mineral nutrition of six rapid-cycling *Brassica* species in response to seawater salinity. Plant Soil 139:285–294
He T, Cramer GR (1996) Abscisic acid concentrations are correlated with leaf area reductions in two salt-stressed rapid-cycling *Brassica* species. Plant Soil 179(1):25–33
Horie T, Hauser F, Schroeder JI (2009) HKT transporter-mediated salinity resistance mechanisms in *Arabidopsis* and monocot crop plants. Trends Plant Sci 14:660–668
Huang J, Hirji R, Adam L et al (2000) Genetic engineering of glycinebetaine production toward enhancing stress tolerance in plants: metabolic limitations. Plant Physiol 122(3):747–756. https://doi.org/10.1104/pp.122.3.747
Iqbal N, Umar S, Khan NA (2015) Nitrogen availability regulates proline and ethylene production and alleviates salinity stress in mustard (*Brassica juncea*). J Plant Physiol 178:84–91
Jamil M, Rehman S, Lee KJ, Kim JM, Rha ES (2007) Salinity reduced growth PS2 phytochemistry and chlorophyll content in radish. Sci Agric. https://doi.org/10.1590/S0103-90162007000200002
Joshi PK, Saxena SC, Arora S (2011) Characterization of *Brassica juncea* antioxidant potential under salinity stress. Acta Physiol Plant 33(3):811–822
Kim TW, Michniewicz M, Bergmann DC, Wang ZY (2012) Brassinosteroid regulates stomatal development by GSK3-mediated inhibition of a MAPK pathway. Nature 482:419–422
Knight H (2000) Calcium signaling during abiotic stress in plants. Int Rev Cytol 195:269–324
Krebs M, Beyhl D, Görlich E, Al-Rasheid KA, Marten I, Stierhof YD, Hedrich R, Schumacher K (2010) Arabidopsis V-ATPase activity at the tonoplast is required for efficient nutrient storage but not for sodium accumulation. Proc Natl Acad Sci U S A 107(7):3251–3256
Krishna P (2003) Brassinosteroid-mediated stress responses. J Plant Growth Regul 22:289–297
Kronzucker HJ, Britto DT (2011) Sodium transport in plants: a critical review. New Phytol 189:54–81
Kumar D (1995) Salt tolerance in oilseed brassicas present status and future prospects. Plant Breed Abst 65:1438–1447
Kumar D, Shekhar S, Bisht S, Kumar V, Varma A, Kumar M (2015) Ectopic overexpression of lectin in transgenic *Brassica juncea* plants exhibit resistance to fungal phytopathogen and showed allivation to salt and drought stress. J Bioeng Biomed Sci 5:1
Kumar G, Purty RS, Sharma MP, Singla-Pareek SL, Pareek A (2009) Physiological responses among *Brassica* species under salinity stress show strong correlation with transcript abundance for SOS pathway-related genes. J Plant Physiol 166(5):507–520
Kumar T, Uzma, Khan MR, Abbas Z, Ali GM (2013) Genetic improvement of sugarcane for drought and salinity stress tolerance using *Arabidopsis* vacuolar pyrophosphatase (AVP1). Gene Mol Biotechnol 56(3): 199–209
Lopez-Berenguer C, Martinez-Ballesta MC, Moreno DA, Carvajal M, Garcia-Viguera C (2009) Growing hardier crops for better health: salinity tolerance and the nutritional value of Broccoli. J Agric Food Chem 57(2):572–578
Lv S, Zhang K, Gao Q, Lian L, Song Y, Zhang J (2008) Overexpression of an H^+-PPase gene from *Thellungiella halophila* in cotton enhances salt tolerance and improves growth and photosynthetic performance. Plant Cell Physiol 49(8):1150–1164
Madan S, Nainawatee HS, Jain RK, Chowdhury JB (1995) Proline and proline metabolising enzymes in in-vitro selected NaCl-tolerant *Brassica juncea* L. under salt stress. Ann Bot (Lond) 76:51–57
Malik RS (1990) Prospects for *Brassica carinata* as an oilseed crop in India. Exp Agric 26:125–129
Martinez-Ballesta M, Moreno-Fernandez DA, Castejon D, Ochando C, Morandi PA, Carvajal M (2015) The impact of the absence of aliphatic glucosinolates on water transport under salt stress in *Arabidopsis thaliana*. Front Plant Sci 6:524
Megdiche W, Amor NB, Debez A, Hessini K, Ksouri R, Zuily-Fodil Y, Abdelly C (2007) Salt tolerance of the annual halophyte *Cakile maritima* as affected by the provenance and the developmental stage. Acta Physiol Plant 29:375–384

Megdiche W, Passaquet C, Zorrig W, Fodil YZ, Abdelly C (2009) Molecular cloning and characterization of novel cystatin gene in leaves *Cakile maritima* halophyte. J Plant Physiol 166:739–749

Metwali EMR, Fuller MP, Jellings AJ (2012) Agrobacterium mediated transformation of anti-stress genes into cauliflower (*Brassica oleracea* var. botrytis L.). 2. Transformation and confirmation of stress tolerance. Aust J Basic Appl Sci 6(5): 31–39

Mishra A, Tanna B (2017) Halophytes: potential resources for salt stress tolerance genes and promoters. Front Plant Sci 8:829

Mittal S, Kumari N, Sharma V (2012) Differential response of salt stress on *Brassica juncea*: photosynthetic performance, pigment, proline, D1 and antioxidant enzymes. Plant Physiol Biochem 54:17–26

Moller IS, Gilliham M, Jha D, Mayo GM, Roy SJ, Coates JC, Haseloff J, Tester M (2009) Shoot Na^+ exclusion and increased salinity tolerance engineered by cell type-specific alteration of Na^+ transport in *Arabidopsis*. Plant Cell 21:2163–2178

Munns R, Passioura JB, Colmer TD, Byrt SC (2019) Osmotic adjustment and energy limitations to plant growth in saline soil. New Phytol

Munns R, Tester M (2008) Mechanisms of salinity tolerance. Annu Rev Plant Biol 59:651–681

Murakeözy EP, Nagy Z, Duhazé C, Bouchereau A, Tuba Z (2003) Seasonal changes in the levels of compatible osmolytes in three halophytic species of inland saline vegetation in Hungary. J Plant Physiol 160(4): 395–401

Nishiyama R, Le DT, Watanabe Y, Matsui A, Tanaka M, Seki M, Yamaguchi-Shinozaki K, Shinozaki K, Tran LS (2012) Transcriptome analyses of a salt-tolerant cytokinin-deficient mutant reveal differential regulation of salt stress response by cytokinin deficiency. PLoS ONE 7:e32124

Nishiyama R, Watanabe Y, Fujita Y et al (2011) Analysis of cytokinin mutants and regulation of cytokinin metabolic genes reveals important regulatory roles of cytokinins in drought, salt and abscisic acid responses, and abscisic acid biosynthesis. Plant Cell 23(6):2169–2183. https://doi.org/10.1105/tpc.111.087395

Noreen Z, Ashraf M, Akram NA (2010) Salt-induced regulation of some key antioxidant enzymes and physio-biochemical phenomena in five diverse cultivars of turnip (*Brassica rapa* L.). J Agron Crop Sci 196(4): 273–285

Oh D, Leidi E, Zhang Q, Hwang SM, Li Y, Quintero FJ et al (2009a) Loss of halophytism by interference with *SOS1* expression. Plant Physiol 151:210–222

Oh D, Zahir A, Yun D, Bressan RA, Bohnert HJ (2009b) SOS1 and Halophytism. Plant Signal Behav 4(11):1081–1083

Orsini F, D'Urzo MP, Inan G, Serra S, Oh D, Mickelbart MV et al (2010) A comparative study of salt tolerance parameters in 11 wild relatives of *Arabidopsis thaliana*. J Exp Bot 61(13):3787–3798

Pang Q, Guo J, Chen S, Chen Y, Zhang L, Fei M, Jin S, Li M, Wang Y, Yan X (2012) Effect of salt treatment on the glucosinolate-myrosinase system in *Thellungiella salsuginea*. Plant Soil 355:363–374

Park BJ, Akira LZK, Kameya T (2005) Genetic improvement of Chinese cabbage for salt and drought tolerance by constitutive expression of a *B. napus* LEA gene. Plant Sci 169:553–558

Pavlovic I, Mlinaric S, Tarkowska D, Oklestkova J, Novak O, Lepedus H, Bok VV, Brkanac SR, Strnad M, Salopek-Sondi B (2019) Early *Brassica* crops responses to salinity stress: a comparative analysis between Chinese cabbage, white cabbage and kale. Front Plant Sci. https://doi.org/10.3389/fpls.2019.00450

Popova OD, Golldack D (2007) In the halotolerant *Lobularia maritima* (Brassicaceae) salt adaptation correlates with activation of the vacuolar H^+-ATPase and the vacuolar Na^+/H^+ antiporter. J Plant Physiol 164(10):1278–1288

Popova OV, Yang O, Dietx KJ, Golldack D (2008) Differential transcript regulation in *Arabidopis thaliana* and the halotolerant *Lobularia maritima* indicates genes with potential function in plant salt adaptation. Gene 423(2):142–148

Prasad KVSK, Sharmila P, Kumar PA, Saradhi PP (2000) Transformation of *Brassica juncea* (L.) Czern with bacterial coda gene enhances its tolerance to salt stress. Mol Breeding 6:489–499

Purty RS, Kumar G, Singla-Pareek SL, Pareek A (2008) Towards salinity tolerance in *Brassica*: an overview. Physiol Mol Biol Plants 14(1–2):39–49

Qadir M, Quillérou E, Nangia V, Murtaza G, Singh M, Thomas RJ et al (2014) Economics of salt-induced land degradation and restoration. Nat Resour Forum 38:282–295

Qiu QS, Guo Y, Quintero FJ, Pardo JM, Schumaker KS, Zhu JK (2004) Regulation of vacuolar Na^+/H^+ exchange in *Arabidopsis thaliana* by the salt-overly-sensitive (SOS) pathway. J Biol Chem 279:207–215

Quintero FJ, Blatt MR, Pardo JM (2000) Functional conservation between yeast and plant endosomal Na^+/H^+ antiporters. FEBS Lett 471(2–3): 224–228

Rajagopal D, Agarwal P, Tyagi W, Singla-Pareek SL, Reddy MK, Sopory SK (2007) *Pennisetum glaucum* Na^+/H^+ antiporter confers high level of salinity tolerance in transgenic *Brassica juncea*. Mol. Breed 19:137–151

Rameeh V, Cherati A, Abbaszadeh F (2012) Relationship between seed yield and shoot ions at vegetative and reproductive stages of rapeseed genotypes under saline environment. Int J Plant Res 2(3):61–64

Robinson MF, Very A, Sanders D, Mansfield TA (1997) How can stomata contribute to salt tolerance? Ann Bot 80(4):387–393

Ryu H, Cho YG (2015) Plant hormones in salt stress tolerance. J Plant Biol 58:147–155

Saha B, Mishra S, Awasthi JP, Sahoo L, Panda SK (2016) Enhanced drought and salinity tolerance in transgenic mustard [*Brassica juncea* (L.) Czern & Coss.] overexpressing *Arabidopsis* group 4 late embryogenesis abundant gene (AtLEA4-1). Environ Exp Bot 128:99–111

Shabala L, Cuin TA, Newman IA, Shabala S (2005) Salinity-induced ion flux patterns from the excised roots of *Arabidopsis sos* mutants. Planta 222:1041–1050

Shabala S, Pottosin I (2014) Regulation of potassium transport in plant under hostile conditions: implications for abiotic and biotic stress tolerance. Physiol Plant 151(3):257–279

Shahid SA, Zaman M, Heng L (2018) Soil salinity: historical perspectives and a world overview of the problem. In: Guideline for salinity assessment, mitigation and adaptation using nuclear and related techniques. Springer, Cham, pp 43–53

Shi H, Kim Y, Guo Y, Stevenson B, Zhu J (2003) The Arabidopsis *SOS5* locus encodes a putative cell surface adhesion protein and is required for normal cell expansion. Plant Cell 15(1):19–32

Shi HZ, Ishitani M, Kim CS, Zhu JK (2000) The Arabidopsis thaliana salt tolerance gene SOS1 encodes a putative Na^+/H^+ antiporter. Proc Nat Acad Sci USA 97:6896–6901

Shi HZ, Quintero FJ, Pardo JM, Zhu JK (2002) The putative plasma membrane Na^+/H^+ antiporter SOS_1 controls long-distance Na^+ transport in plants. Plant Cell 14(2):465–477

Siddiqui MH, Khan MN, Mohammad F, Khan MMA (2008) Role of nitrogen and gibberellin (GA3) in the regulation of enzyme activities and in osmoprotectant accumulation in *Brassica juncea* L. under salt stress. J Agron Crop Sci 194:214–224

Siddiqui MH, Mohammad F, Masrooor M, Khan A et al (2011) Cumulative effect of nitrogen and sulphur on *Brassica juncea* L. genotypes under NaCl stress. Protoplasma 249:139–153

Stepien P, Johnson GN (2009) Contrasting responses of photosynthesis to salt stress in the glycophyte Arabidopsis and the halophyte *Thellungiella*: role of the plastid terminal oxidase as an alternative electron sink. Plant Physiol 149(2):1154–1165

Sun XE, Feng XX, Li C, Zhang ZP, Wang LJ (2014) Study on salt tolerance with YHem1 transgenic canola (*Brassica napus*). Physiol Plant 154(2):223–242

Sunarpi, Horie T, Motoda J, Kubo M, Yang H, Yoda K et al (2005) Enhanced salt tolerance mediated by AtHKT1 transporter-induced Na^+ unloading from xylem parenchyma cells

Tuncturk M, Tuncturk R, Yildirin B, Ciftci V (2011) Changes of micronutrients, dry weight and plant development in canola (*Brassica napus* L.) cultivars under stress. Afr J Biotechnol 10(19): 3726–3730

Ulfat M, Athar H, Ashraf M, Akram NA, Jamil A (2007) Appraisal of physiological and biochemical selection criteria for evaluation of salt tolerance in canola (*Brassica napus* L.). Pak J Bot 39(5): 1593–1603

Umar S, Diva I, Anjum NA, Iqbal M, Ahmad I, Pereira E (2011) Potassium-induced alleviation of salinity stress in *Brassica campestris* L. Cent Eur J Biol 6(6):1054–1063
Wang LL, Chen AP, Zhong NQ, Liu N, Wu XM, Wang F, Yang CL, Romero MF, Xia GX (2014) The *Thellungiella salsuginea* tonoplast aquaporin TsTIP1;2 functions in protection against multiple abiotic stresses. Plant Cell Physiol 55(1):148–161
Wang Q, Xu W, Xue Q, Su W (2010) Transgenic *Brassica chinensis* plants expressing a bacterial coda gene exhibit enhanced tolerance to extreme temperature and high salinity. J Zhejiang Univ Sci B 11(11):851–861
Waters S, Gilliham M, Hrmova M (2013) Plant high-affinity potassium (HKT) transporters involved in salinity tolerance: structural insights to probe differences in ion selectivity. Int J Mol Sci 14(4):7660–7680
Wu C, Gao X, Kong X, Zhao Y, Zhang H (2009) Molecular cloning and functional analysis of a Na^+/H^+ antiporter gene *ThNHX1* from a halophytic plant *Thellungiella halophila*. Plant Mol Biol Rep 27:1. https://doi.org/10.1007/s11105-008-0048-1
Xu W, Shi W, Veda A, Takabe T (2008) Mechanisms of salt tolerance in transgenic *Arabidopsis thaliana* carrying a peroxisomal ascorbate peroxidase gene from barley. Pedosphere 18(4):486–495
Yokoi S et al (2002) Differential expression and function of *Arabidopsis thaliana NHX* Na^+/H^+ antiporters in the salt stress response. Plant J 30:529–539
Yu L, Nie J, Cao C, Jin Y, Yan M, Wang F, Liu J et al (2010) Phosphatidic acid mediates salt stress response by regulation of MPK6 in *Arabidopsis thaliana*. New Phytol 188:762–773
Zhang H, Hodson JN, William JP, Blumwald E (2001) Engineering salt-tolerant *Brassica* plants: characterisation of yield and seed oil quality in transgenic plants with increased vacuolar sodium accumulation. Proc Natl Acad Sci U S A 98:12832–12836
Zhang Y, Li Y, Lai J, Zhang H, Liu Y, Liang L, Xie Q (2012) Ectopic expression of a LEA protein gene TsLEA1 from *Thellungiella salsuginea* confers salt-tolerance in yeast and Arabidopsis. Mol Biol Rep 39(4):4627–4633
Zhou Y, Yin X, Duan R, Hao G, Guo J, Jiang X (2015) SpAHA1 and SpSOS1 coordinate in transgenic yeast to improve salt tolerance. PLoS ONE 10(9):e0137447. https://doi.org/10.1371/journal.pone.0137447

Brassicaceae Plants Response and Tolerance to Drought Stress: Physiological and Molecular Interventions

Ali Raza, Sidra Charagh, Ali Razzaq, Rida Javed, Rao Sohail Ahmad Khan, and Mirza Hasanuzzaman

Abstract The plant family Brassicaceae (Cruciferae) is among the largest angiospermic family belonging to the order Brassicales. It consists of 12–15 tribes with 338–360 genera and about 3709 species distributed all around the globe in all continents. Plants usually grown in fields; their development and growth are mostly affected by environmental stresses, including both biotic and abiotic stresses. From abiotic stresses, mostly plants face drought and salinity stress more than others. Out of these two, agriculture is mostly affected by drought stress. Almost 40% of world's agricultural land area is affected by drought stress. Drought is one of the major environmental stress factor that alters morphological, biochemical also physiological characteristics of all Brassicaceae plants. Drought also lowers the productivity of various central *Brassica* spp. via altering normal plant growth as well as their development. In this chapter, we have discussed the physiological and biochemical responses and tolerance of Brassicaceae plants to drought stress. Further, we also describe the most challenging strategies to overcome drought stress in Brassicaceae plants.

Keywords Abiotic stress · Brassicaceae · Biochemical mechanisms · Drought · Oilseed crops · Physiological responses · Stress tolerance

A. Raza (✉)
Oil Crops Research Institute, Chinese Academy of Agricultural Sciences (CAAS), Wuhan 430062, China
e-mail: alirazamughal143@gmail.com

S. Charagh · A. Razzaq · R. Javed · R. S. A. Khan
Centre of Agricultural Biochemistry and Biotechnology (CABB), University of Agriculture, Faisalabad 38040, Pakistan

M. Hasanuzzaman
Department of Agronomy, Faculty of Agriculture, Sher-e-Bangla Agricultural University, Dhaka 1207, Bangladesh

M. Hasanuzzaman (ed.), *The Plant Family Brassicaceae*,
https://doi.org/10.1007/978-981-15-6345-4_7

1 Introduction

The plant family Brassicaceae is also known as dicots family (Al-Shehbaz et al. 2006). Worldwide, this family is also called Cruciferae and comprises of mainly 338 genera and 3709 species (GIDIK et al. 2019). It includes approximately 100 species of vegetables which includes *Brassica juncea* L. (mustard), *Brassica rapa* L. (turnip rape), *Brassica napus* L. (rapeseed), and *Brassica oleracea* L. (cabbage) (Jan et al. 2018). These crops are all important not only economically but also gained importance industrially because these crops are used for condiment, oil, and fodder. One of the earliest cultivated members of Brassicaceae is Mustard. The seed of mustards was date back for cultivation overused 5000 years ago (Rahman et al. 2018). Oilseed rape, also known as Canola, being an oil crop, is the most important member of this family (Alagoz and Toorchi 2018; Raza et al. 2019a). *Camelina sativa*, *Brassica carinata*, *Eruca vesicaria*, and *Crambe abyssinica* is considered as biodiesel and protein crop whereas, *B. oleracea* is called as vegetable crop among the members of Brassicaceae family (Warwick et al. 2007).

The knowledge about the breeding of crops such as *C. sativa*, *B. oleracea*, and *B. napus*, is provided by wild species of Brassicaceae which were present in natural habitats. The study of a gene of these members of family Brassicaceae, helped in the conservation and production of new crops (Gidik et al. 2016). Brassicaceae vegetables, being a source of antioxidants, contains a high level of glucosinolate and phenolic content (Jahangir et al. 2009). These compounds prevent cardiovascular diseases and cancer (Moreno et al. 2006). Besides these preventive roles, anti-nutritional effects of glucosinolates, polyphenols, tannins, and S-methyl cysteine sulphoxide was also reported from vegetables of Brassicaceae family (Griffiths et al. 1998).

Plants usually grown in fields; their development and growth are mostly affected by environmental stresses, which includes salinity, drought, low, and high temperature. From abiotic stresses, mostly plants face drought and salinity stress more than others. Out of these two, agriculture is mostly affected by drought stress. Almost 40% of world's agricultural land area is affected by drought problem (Trenberth et al. 2014; Raza et al. 2019b, 2019c, 2019d; Razzaq et al. 2019a). Water shortages affected crop production greatly (Sprouts 2018). The mechanism of opening and closing of stomata and leave growth affects badly due to unavailability or less availability of water (Saeed et al. 2016). Drought affects the leaf growth of leaves of plants, due to this, the rate of photosynthesis decreased (Zhang et al. 2004). The effect of drought on the development of leaves is more evident on cereals than oil crops (Manivannan et al. 2008). Canola plant becomes unable to expand due to water shortage, results in shoot length reduction (Ashraf et al. 2013). The reduction of growth of shoot was also reported in soybean by (Specht et al. 2001).

Drought stress resulted in the reduction of enzyme activities which controls metabolism of plants (Xu et al. 2008), resulted in the biomass reduction of crops (Farooq et al. 2009). Due to huge loss of crops by drought stress, various methods were adopted previously to make crops tolerant to drought and increase its productivity. For this purpose, the in vitro technique proved to be effective in dealing with drought stress for plants of family Brassicaceae (Jan et al. 2018). Sucrose, PEG, and mannitol are known as osmotic stress agents, were used by researchers during in vitro selection. These compounds can make the osmatic potential of nutrient solution low without producing any toxicity to plants (Hassan et al. 2004). *Brassica juncea* becomes drought tolerant by using in vitro selection by (Gangopadhyay et al. 1997). In this chapter, we have discussed the physiological and biochemical responses and tolerance of Brassicaceae plants to drought stress.

2 Physiological and Biochemical Mechanisms to Study Drought Stress Response

Drought tolerance occurs due to different cellular, physiological, and molecular mechanisms. These processes either induce or repress the gene functions which then cause osmolyte accumulation, improvement of the antioxidant system, reduction in transpiration and inhibition of growth of roots and shoots (Panjabi-Sabharwal et al. 2010; Hasanuzzaman et al. 2012, 2014; Raza et al. 2020). Hence, Fig. 1 showing the physiological, biochemical, and molecular responses under drought stress. Abscisic acid (ABA), a phytohormone, present abundantly where water supply is too short

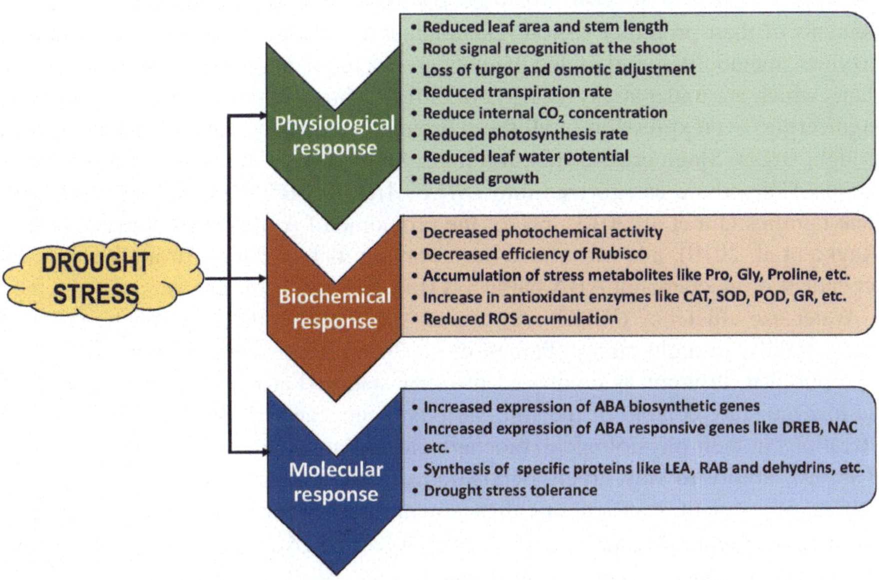

Fig. 1 Summary of physiological, biochemical, and molecular responses under drought stress

and resulted in stomatal closure and induction of gene expression which are stress-related (Yang et al. 2011). ABA-independent system is also involved in the expression of genes involved in drought stress (Aguado et al. 2014). In response to abiotic stress, metabolic proteins are produced. These metabolic proteins involve regulatory proteins (kinases) and protein involved in osmoprotectant's synthesis—transcription factors (TFs) or kinases response towards abiotic stress (Chaves and Oliveira 2004). The promoter region of a target gene contains *cis*-acting elements on which TFs attached with DNA domain hat. They regulate the gene expression by activating or repressing the RNA polymerase's activity (Riechmann et al. 2000). There is a group of a gene, called regulon are controlled by various types of TFs. Four types of regulons were identified in plants towards abiotic stress, which are ABA-dependent and ABA independent. *CBF/DREB* regulon, *NAC*, and *ZF-HD* regulon are ABA independent, whereas *AREB/ABF* regulon and *MYC* regulon are ABA dependent (Fig. 2; Saibo et al. 2008).

There is a signaling pathway of abiotic stress, which includes perception of signals, transduction, and responsiveness. Physiological and metabolic reactions become activated during this pathway (Liu et al. 2014b). Stress stimulus perceived by plant cells through sensors is located at the cell membrane. Inositol phosphate, calcium ions, ROS, sugars, and cyclic nucleotides are second messengers, and with the help of these messengers, intracellular ones become activated by receiver signals (Bhargava and Sawant 2013). Protein kinases regulate phosphorylation, whereas phosphatases regulate dephosphorylation during the pathway of signal transduction (Huang et al. 2012). The suppression or activation of TFs by phosphatases or kinases, played its role in the regulation of downstream gene's expression, by its interaction with *cis*-elements of their promoter region (Danquah et al. 2014). TFs, functional proteins, enzymes, metabolites, and molecular chaperones regulate the signaling pathways in plants which are under stress (Song et al. 2013). Through the approaches of genetic engineering, overexpression of all these proteins helps the plants to survive against drought stress (Singh et al. 2015). In Plants, up to 10% of the genes from genomes encode TFs and are categorized into *DREB*, *AREB*, *NAC*, *MYB*, *WRKY*, and *bZIP* gene families (Jin et al. 2013). 6% of the proteome of *Arabidopsis* consists of TFs (Rayko et al. 2010), and most have been playing its role during drought stress, by dependent or independent ABA pathways (Fig. 2; Joshi et al. 2016).

Water use efficiency (WUE) and growth stage govern the response by crops to stress mainly, drought stress (Pareek et al. 2009). Reproductive stage of crop is susceptible to drought as compared to other stages (Farooq et al. 2017). Due to the low moisture content of the soil, various plants can cope with this problem by alterations in their physiological, biochemical, and molecular mechanisms. Rolling of leaves, ability to stay green, deposition of epicuticular wax, stomatal closing, osmotic adjustment, root length enhancement, and resistance of photoinhibition is the clear symptoms of crops towards drought tolerance (Khazaei et al. 2013). Various biochemical changes occur due to drought stress, such as accumulation of ROS, osmolytes, antioxidant enzymes, and species at high level (Kholová et al. 2011). The maintenance of membrane stability and activities of enzymes under osmotic stress is controlled by these osmolytes (Singh et al. 2015).

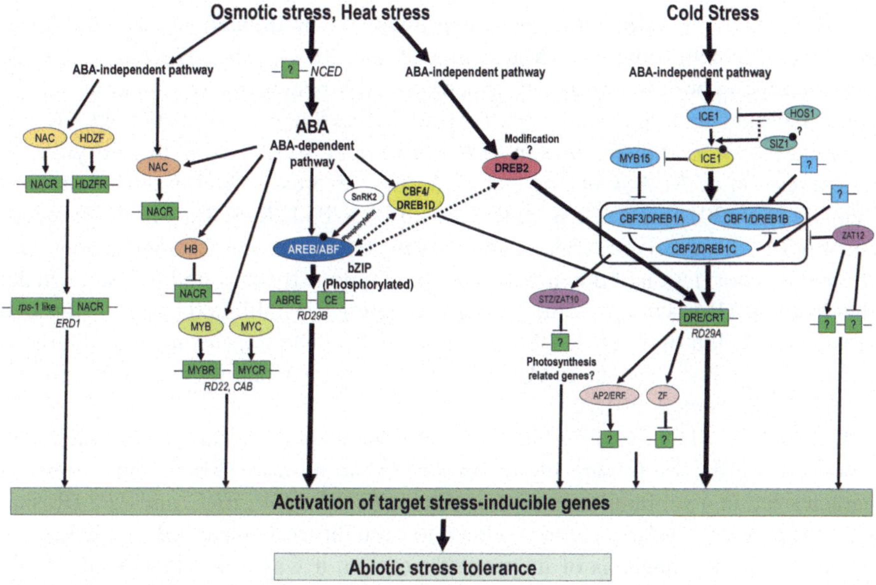

Fig. 2 A schematic representation of the crosstalk transcriptional network among *cis*-acting elements and transcription factors of ABA-independent and ABA-dependent pathways during the perception of abiotic stresses. The small filled black dotes indicating the post-translational modifications of transcriptions factors in response to stress signals for their activation via phosphorylation. Bold black arrows correspond to the major signaling pathways, which leads to the regulation of many downstream genes. Black dotted arrow showing the protein-protein interactions. Two different *DRE/CRT*-binding proteins, *CBF1/DREB1B*, *CBF2/DREB1C*, and *CBF3/DREB1A*, distinguish two different signal transduction pathways in response to cold and drought stresses, respectively. A transcriptional activator, ICE1 (Inducer of *CBF* Expression 1), functions upstream of the *DRE/DREB* regulon. *DREB1/CBF* genes are induced by cold stress, whereas *DREB2* is induced by dehydration and salt stress. Both work in ABA-independent pathways. ICE1 is an MYC-type transcription factor and binds to *cis*-elements in the promoter of *CBF3/DREB1A* to induce its expression. *CBF3/DREB1A* is an *AP2*-type transcription factor to regulate the expression of cold-regulated genes and cold tolerance. *CBF4/DREB1D* is a *DRE* cis-element binding factor that is ABA-dependent. HOS1 function as a negative regulator of ICE1. The black dot on HOS1 and SIZ1 indicate the sumoylation modification by SIZ1 of the ICE1 transcription factor. The dotted black line from SIZ1 to HOS1 represents competition for binding places on the ICE1 transcription factor. Whereas, SIZ1 blocks the access of HOS1 to the ubiquitination sites on the ICE1. SnRK2 is involved in ABA signaling and phosphorylation. The small blue squares with question marks indicate the putative *MYC* ICE1-like transcription factors which may play a role in the activation of *CBF1/DREB1B*, *CBF2/DREB1C*, and *CBF3/DREB1A*. The small green squares represent the *cis*-elements present in stress-responsive genes. The green squares with question marks correspond to the putative *cis*-elements in the promoter of stress-responsive genes

During ABA signaling, the gene expression is modulated by *AREB* (ABA-responsive element binding) (Maruyama et al. 2011). The *ABRE cis*-element controlled the expression of genes, which are ABA-responsive (Fujita et al. 2013). Vascular tissues and guard cells synthesized ABA (Bauer et al. 2013). *NCED* is a major enzyme, coded by 5 genes in *Arabidopsis*, and is involved in the biosynthesis and expression of ABA. Out of 5 genes, *NCED3* expressed highly during drought (Behnam et al. 2013). *PP2Cs* is attached with PYR/PYL/RCARs, which are receptors of ABA and perceived ABA, resulting in the release of *SnRK2s* (Miyakawa et al. 2013). Phosphorylation of downstream proteins is done by activated *SnRK2s*. On the other hand, if ABA is not present, the ABA signaling was inhibited by *PP2C* through dephosphorylation of *SnRK2s* (Yoshida et al. 2010). The improvement in tolerance mechanism against drought in *Arabidopsis* is reported due to overexpression of *AREB1* (Yoshida et al. 2015).

AP2/ERFBP TFs have a role in plant hormone responses, cell proliferation, and response towards abiotic and biotic stresses (Sharoni et al. 2010). This family is characterized in 4 subfamilies, which includes, *DREB*, *RAV*, *ERF*, and *AP2* (Rashid et al. 2012). In *Arabidopsis*, genes related to drought stress have ABA-independent expression. By the analysis of its promoter region, it was found that A/GCCGAC sequence was found with *cis*-element (Lucas et al. 2011). *DREB2A* and *DREB2B* are 2 genes which played their role against drought, in few members of the Brassicaceae family (Nakashima et al. 2014). By the overexpression of *OsDREB2B* gene in transgenic *Arabidopsis*, the expression of *DREB2A* gene also increased, resulting in the improvement of tolerance against drought (Matsukura et al. 2010). The regulation of *DREB2A* gene under normal conditions occurred through two mechanisms, i.e., the inhibition of expression of *DREB2A* by *GRF7* and *DRIP1* and *DRIP2* assisted the targeted proteolysis which is mediated by 26S proteasome (Singh and Laxmi 2015).

The *NAC* family is plant-specific and largest gene family (Shao et al. 2015). The activation of "EARLY RESPONSIVE TO DEHYDRATION STRESS 1 gene" is mitigated by TFs of *NAC* and *ZF-HD* and is revealed by analysis of the promoter of *ERD1* (Tran et al. 2007). In *Arabidopsis*, tolerance against drought achieved by overexpression of *NAC* genes. Activation of the *ERD1* gene is not necessary for this (Tran et al. 2007). The overexpression of *ANACO72*, *ANACO55*, and *ANACO19* genes also helped in drought tolerance up to a great extent (Liu et al. 2014a). It was observed that due to overexpression of "*TaNAC2* and *TaNAC9*" genes not only enhanced drought tolerance but also increased the level of transcripts of a gene which are stress-responsive and helps in the improvement of physiological parameters (Huang et al. 2015). *bZIP* TFs family regulates the development and growth of plants along with response against drought stress. It regulates the expressions of a gene in an ABA-dependent manner (Llorca et al. 2014). Zhang et al. 2015b reported that due to overexpression of *TabZIP60*, improved the tolerance against drought in *Arabidopsis*.

DRE/CRT motif is present in the promoter region of a gene, responsible for the drought. It represents itself as a binding site for *DREB/CBF* TF (ABA-independent). This site is served as a coupling element in gene expression of ABA-dependent pathway for *ABRE* (Singh and Laxmi 2015). The crosstalk between both pathways signaling pathways (ABA-dependent and Independent) exists due to the interaction of *DREB2C*, *DREB1A/CBF3*, and *DREB2A* proteins with *AREB/ABF* (Lee et al. 2010). At the transcriptional level, the *DREB2A* gene activated due to *SnRK2s* and *AREB/ABF* TFs (Kim et al. 2011). *NACs* and *AREB/ABFs* both also have similar interaction. *ANAC096* interacted with *ABF4/AREB2* and *ABF2/AREB1* during osmotic and dehydration stress (Xu et al. 2013). *Arabidopsis ERF1* regulate the gene expression of a transcript by coupling with 2 different *cis*-elements, due to this correlation among *AP2/ERFs* and *DREB/CBFs* was confirmed at the transcript level (Cheng et al. 2013). TFs played an important role in the activation of several genes which may encode for protein which were involved in detoxification, ROS scavenging, osmolyte protection, and ubiquitination (Puranik et al. 2012). *AREB2/ABF4*, *AREB1*, and *ABF3* (Yoshida et al. 2015), *FeDREB1* (Fang et al. 2015), *TaNAC2* (Mao et al. 2012), and *OsAREB1* (Jin et al. 2010) were the gene studied for drought tolerance and reported enhanced tolerance against drought by overexpressing these genes in member of family Brassicaceae.

3 Past and Current Status of Drought Stress in Brassicaceae Plants

Drought is one of the major environmental stress factor that alters morphological, biochemical also physiological characteristics of all Brassicaceae plants. General layout of drought stress signaling mechanisms is described in Fig. 3. Drought also lowers the productivity of various central *Brassica* spp. via altering normal plant growth as well as their development. This chapter will discuss the drought impact on diverse growth phases further physio-morphological processes of *Brassica* spp. like *B. juncea*, *B. carinata*, *Brassica nigra*, *Brassica campestris*, *B. napus*, and *B. oleracea* (Jan et al. 2017). Modern genomic, as well as classical genetic methodologies linked with defined phenotyping, have increased the genes unraveling further metabolic pathways conferring crops drought tolerance. In *Brassica*, several quantitative trait locis (QTLs) provided by QTL-mapping of drought. Several drought tolerant genes have been isolated, and few of them confirmed having potential for plant tolerance through genetic improvement (Zhang et al. 2014). Nevertheless, there is limited work has been done to develop drought stress tolerance Brassicaceae plants expect some major *Brassica* species and some other plants. There is a need to explore the drought stress mechanisms for a maximum number of plants from Brassicaceae family. However, a list of drought stress-responsive TFs has been given in Table 1.

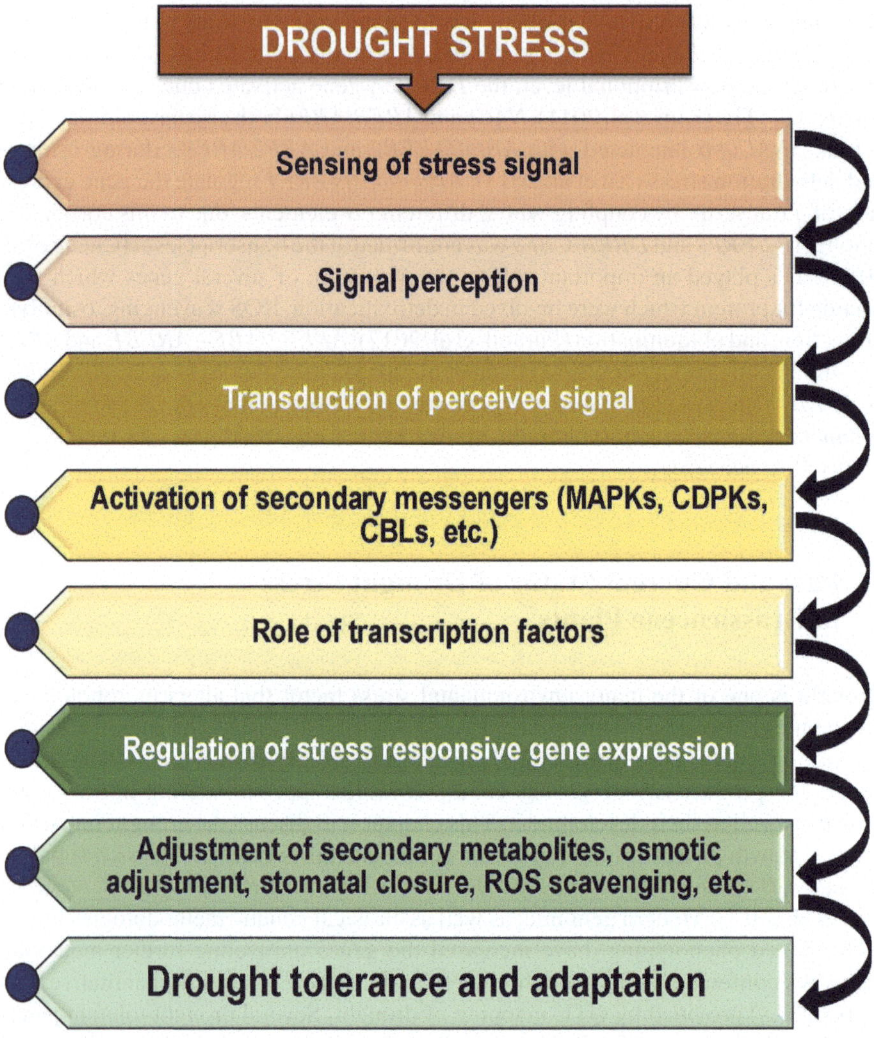

Fig. 3 General layout of drought stress signaling mechanisms

3.1 Brassica napus

Oilseed rape (*B. napus*) growing all over the globe, has diverse variety of consumptions comprising animal feed, plant oil as well as fuel substitute (Abdallah et al. 2010). Oilseed rape appeared as a superb rotation crop that suppresses pathogens; soil-born type, also limits the mineral nutrients leaching into aquifer (Lee et al. 2016). Oilseed rape yield is growing globally, in 2007 FAO reported 47 million tons of oilseed rape production which increased up to 58.4 million tons in 2010–2011 (Saeidnia and Gohari 2012). Canola very sensitive against drought stress and

Table 1 List of drought stress-responsive TFs/genes from Brassicaceae plants

Plant	Gene	References
B. napus	*AtDWF4*	Sahni et al. (2016)
	BnaABI1	Babula-Skowrońska et al. (2015)
	BnWRKY57	Shirazi et al. (2019)
	LEA3	Liang et al. (2019)
	VOC	Liang et al. (2019)
	C2H2 zinc finger proteins	Wang et al. (2019)
B. oleracea	*AOP2*	Podda et al. (2019)
	AtEDT1/HDG11	Zhu et al. (2016b)
	BrDSR28	Park et al. (2017b)
B. campestris	GR2 gene	Zhang et al. (2018)
	Rboh C; Rboh F genes	Zhang et al. (2019)
	LEA gene	Park et al. (2005)
B. juncea	*AtLEA4-1*	Saha et al. (2016)
	Glyoxalase I (Gly I)	Rajwanshi et al. (2016)
	Check pea lectin gene	Kumar et al. (2015)
	γ-TMT	Yusuf et al. (2010)
A. thaliana	*AtMYB44*	Jung et al. (2008)
	DREB1A	Kasuga et al. (2004)
	LeGSTU2	Xu et al. (2015)
	AtMYB12	Wang et al. (2016)
	AtGolS2	Selvaraj et al. (2017)
	VrDREB2A	Chen et al. (2016)
	MlNAC9	Zhao et al. (2016)
	FtMYB9	Gao et al. (2017)
	BrLAS	Li et al. (2018b)
	BrRH22	Nawaz et al. (2018)
Raphanus sativus	*RsWRKY*	Karanja et al. (2017a)
	RsNAC145	Karanja et al. (2017b)
	RsAP2/ERF	Karanja et al. (2019)
E. vesicaria sativa	*EvDREB2A*	Huang et al. (2016)

drought is considered a main yield-limiting cause for canola. Drought can inhibit the establishment of plant seedling that results in worse plant densities and lowered productivity (Zhang et al. 2015a). This problem can be solved by developing canola cultivars that maintain oil production by tolerating water stress. In canola, genomic-assisted breeding comprises description of functional genes for water stress-tolerant as well as markers connected through these genes (Xue et al. 2013). Exogenous

Melatonin (MT) application has been found to protect *B. napus* seedlings by alleviating drought stress and improves seedlings growth. MT holds osmotic activity adjusting, anti-oxidative and growth-inducing characteristics that make them valuable for drought-acclimatization (Li et al. 2018a). Currently has been found that plant growth regulators (PGRs) used broadly to maintain plant survival against environmental stresses. Nitric oxide (NO) 0.02 mM and 5-aminolevulinic acid 0.895 mM applied on two cultivars of canola; Cyclone and Dunkeld, under water-deficit environment. Drought stress considerably lowered the plant growth, soluble proteins, chlorophyll pigments, and relative water contents, while increased malondialdehyde (MDA), proline, glycine betaine, and catalyse (CAT) as well as peroxidase enzymes activities in both varieties. PGRs application resulted in improved growth, GB, CAT activity, chl-a plus soluble proteins, whereas lowered the relative membrane permeability (RMP), MDA then peroxidase (POD) activity. These findings suggested that exogenous NO and ALA application enhanced drought tolerance of canola through up-regulation of the oxidative defense system, osmoprotectant accumulation, and lowering lipid peroxidation (Akram et al. 2018).

Silicon (Si) also has major role in metabolic and physiological processes of plants, exogenous Si application (1 mM SiO_2) meaningfully improved resistance of *B. napus* in drought condition (10 and 20% PEG). Si guards photosynthetic pigments also lessened the oxidative stress in *B. napus* by increasing glutathione (GSH) and ascorbate (AsA) pool; activities of anti-oxidant enzymes like glutathione-*S*-transferase (GST), CAT, glutathione reductase (GR), ascorbate peroxidase (APX), and proline content. Therefore, exogenous application of silicon enhanced antioxidant defense system and glyoxalase enzymes activities under drought condition (Hasanuzzaman et al. 2018). Transgenic *B. napus* have developed through overexpression of *Arabidopsis* Brassinosteroid (BR) gene *AtDWF4* and scored stress-response phenotypes as well as growth. Transgenic *B. napus* plants displayed improved seed productivity and increased oil content/plant then significantly increased drought tolerance, heat stress plus fungal pathogens resistance, i.e., *Sclerotinia sclerotiorum* and *Leptosphaeria maculans*. So, BR-related genes enhance tolerance against biotic as well as abiotic stresses and enhanced crop productivity (Sahni et al. 2016).

Drought boosted the exogenous levels of ABA and jasmonic acid (JA). Pretreatment with salicylic acid (SA) alleviates stress induce by drought through increased genes expression for proline synthesis (*P5CS1*, *P5CS2*, and *P5CR*), NADPH oxidase, also increased redox regulating genes expression (*TRXh5* and *GRXC9*). Salicylic acid-mediated stress responses overlapped by increased *NPR1* and *PR-1* expression, which antagonistically depress the JA as well as ABA related genes (*NCED3*, *MYC2*, and *PDF1.2*). These showed SA modulated NPR1 dependent signaling pathway plus proline synthesis are an integrative process of redox control in *B. napus* in drought (Lee et al. 2019). Transgenic *B. napus* plants have developed by overexpression of *Arabidopsis thaliana ABI1* orthologue and to learn resultant fluctuations under drought stress and growing effect of *BnaABI1* genes in drought situations. Against drought response, *ABI1* behaves as hub of protein regulator and *BnaABI1* take part in dehydration-related stress responses in *B. napus* (Babula-Skowrońska et al. 2015).

The latest finding suggested C2H2 zinc finger proteins behaves as main transcriptional regulators to wide range of stressed situations like high temperature, drought, salinity, and oxidative stress in plant responses (Wang et al. 2019). Transgenic *B. napus* plants overexpressed with drought-responsive genes like vicinal oxygen chelate (*VOC*) and late embryogenesis rich in group3; *LEA3* enhanced tolerance to drought stress and increased rapeseed oil contents by lowering ROS and improving photosynthetic efficiency (Liang et al. 2019). TFs *WRKY* has major role in regulating numerous biological processes of plants like abiotic stress responses. In *B. napus* (rapeseed), *WRKY* genes have appeared as prospective target for enhancing stress tolerance through transgenic breeding. Researchers have cloned and characterized *WRKY57* (*BnWRKY57*) *from B. napus* and studied *BnWRKY57* expression pattern under drought as well as salt stresses two *B. napus* cultivars. *BnWRKY57* gene encodes a hydrophilic polypeptide of 293 amino acids. *BnWRKY57* gene promoter enclosed *cis*-regulatory elements which tangled in biotic plus abiotic stress replies proposing this gene has role in modulating various signaling pathways. Therefore in *B. napus, BnWRKY57* gene might give adaptive responses against salt and drought stresses (Shirazi et al. 2019).

3.2 *Brassica oleracea*

Brassica oleracea is the most important agronomically and diploid crop species covering several crop; cauliflower, Brussels sprout, kale, cabbage, kohlrabi, and cabbage. *Brassica* crops grown because of their inflorescence, stems, leaves, and axillary buds, also present incredible morphological diversity (Golicz et al. 2016). Besides their morphological discrepancy, these cultivars along with their correlated wild species have a collection of secondary metabolites which have been involved in supporting human health like different compounds functions as antioxidant and anti-carcinogens (Parkin et al. 2014). The *Brassica* crops are rich in different secondary metabolites like glucosinolates along with byproducts named isothiocyanates, nitriles, thiocyanates, epithionitriles which have been acknowledged to play a part in plant defense to the pathogen violence. It has been reported that isothiocyanates have a role in cancer prevention as well as immunomodulatory and neuroprotective effects (Traka and Mithen 2009; Giacoppo et al. 2013). In reply to environmental strains like salt and drought, secondary metabolites synthesis helpful for signaling, lowering, and coping with the stressed environment in chloroplast generated by photosynthetic imbalance. During plant growing stage, water deficit may be used as a sustainable practice for production of curly kale having increased concentration of bioactive metabolites; important for promoting health. Drought stress enhanced the trans-2-hexenal, tocopherol, and phytol content while the content of chlorophyll. Furthermore, drought also increased anti-oxidant activity and expression with *AOP2* gene associated with aliphatic alkenyl glucosinolates synthesis and *TGG1*, *TGGE*, *PEN2* genes encodes myrosinases; enzyme for glucosinolate breakdown (Podda et al. 2019).

Recent studies have stated that NO protects broccoli plants from abiotic stresses by alleviating their adverse effects. Broccoli seeds soaked in a solution of NO 0.02 mM and two water-stress levels applied (control, 100% field capacity (FC0, 60% F) to *B. oleracea* (4-week old). After three-week irrigation of drought stress, NO (002 mM) applied to broccoli plants, and leaf samples collected after 12-days foliar application to calculate anti-oxidant as well as photosynthetic activities and other biochemical parameters. Water-stress resulted in reduced shoot length, chlorophyll content, and betaine, while it increased hydrogen peroxide, ascorbic acid (AsA), SOD, and CAT enzymes activities. NO which applied exogenously resulted in increased fresh also dry shoot biomass, GB, chlorophyll content, shoot length, total phenolics and soluble proteins, and POD and SOD enzymes activities in water deficiency in broccoli plants. In this way, NO is more effective for boosting drought tolerance and help broccoli plants by up-regulating oxidative defense under water-deficit situation (Munawar et al. 2019). Cauliflower (*B. oleracea*) widely consumed as vegetable in various countries but their productivity is severely affected by drought. Pretreatment of seeds with AsA has an active role in alleviating adverse drought stress of vegetable crops by increasing CAT and SOD activities, GB, and total phenolic content (Latif et al. 2016).

In red cabbage plants, zinc (Zn) deficiency effects have been studied under drought condition. Plants growth impaired because of Zn deficiency in drought than with watered environment. Zn content decline drastically due to drought and resulted in photosynthetic apparatus damage in leaves deficit with Zn than leaves having sufficient Zn. Anti-oxidant enzymes activities except SOD increased in zinc-deficient conditions, whereas drought increased all enzymes activities containment with an accumulation of H_2O_2 and MDA (Hajiboland and Amirazad 2010). In *B. oleracea*, Zn biofortification increased the essential amino acid accumulation, flavonols, gluciberin as well as aliphatic plus indolic glucosinolates (Barrameda-Medina et al. 2017). Improving osmotic stress as well as drought tolerance without decreasing the crop yield has been a major challenge in crop improvement. A protein *AtEDT1/HDG11* (*Arabidopsis ENHANCED DROUGHT TOLERANCE1/HOMEODOMAIN GLABROUS11*) belongs to class IV HD-Zip family has a role in drought resistance in pepper, *Arabidopsis*, plus rice. Transgenic Chinese kale with *AtEDT1/HDG11* gene enhanced resistance to drought plus osmotic stress, also exhibit long hypocotyls, dense root system, tall stems than untransformed kale plants. Transgenic kale plant leaves have proline content, and ROS was increased significantly after osmotic, drought stress than wild type. Overexpression of *AtEDT1/HDG11* also resulted in ABA hypersensitivity and induced stomatal closure. Therefore, *AtEDT1/HDG11* has role in Chinese kale of boosting abiotic stress tolerance by ABA- and auxin-mediated signaling (Zhu et al. 2016b). In Chinese cabbage, *BrDSR28* (*B. rapa* drought stress resistance) gene expression increased above four-fold than wild type (Park et al. 2017b).

3.3 *Brassica juncea*

Brassica juncea is among the most important vegetable oil sources worldwide, land higher than 6 million hectares cultivated with *B. juncea* and more vulnerable to abiotic stresses that affect their overall productivity (Singh et al. 2009). Rapid industrialization and exponentially increased population caused a decline of arable land. It is essential to work on the improvement of the mustard crop to fulfill the increased oilseed demand (Dutta et al. 2008). Late Embryogenesis Abundant Protein (LEA) expression has been studied previously involved in alleviating osmotic stress through stabilization of water status, cytosolic structures protection as well as cell by improving membrane plus protein stability. Researchers transformed and over-expressed mustard with *AtLEA4-1* protein from *Arabidopsis* 4LEA protein group, resulting transgenic lines showed improved salinity and drought tolerance. This study suggests that LEA proteins functions as molecular chaperone and antioxidant conferring tolerance to abiotic stresses (Saha et al. 2016). Glyoxalase tangled in converting cytotoxic methylglyoxal into D-lactate by S-D-lactoylglutathione. It has *Glyoxalase I* (*Gly I*) plus *Glyoxalase II* (*Gly II*) enzymes as well as GSH that behaves as cofactor through substrate anchoring into the two enzymes active site. Plants overexpressed either alone/in combination with *Gly I* or *Gly II* identified to give tolerance against various environmental stresses. *Brassica juncea* plants (R and S lines) transformed with *Glyoxalase I* (*Gly I*) gene and expressed with rd29A stress-inducible promoter or CaMV 35S constitutive promoter. Both *B. juncea* lines R and S presented increased resistance to drought as well as salt stress than control plants. Therefore, *Gly I* gene overexpression in the presence of rd29A stress-inducible promoter appeared best choice to increase drought, heavy metals as well as salt resistance in transgenic *B. juncea* (Rajwanshi et al. 2016).

Plants secondary metabolites lectins having ubiquitous nature belongs to heterogeneous proteins which reversibly and specifically bind to carbohydrates and works in defense system to different biotic plus abiotic stresses. Transgenic *B. juncea* plants overexpressed with chickpea lectins confers resistance against fungal pathogens as well as salt and drought stress. In addition, lectin gene expression imparts high proline content, water holding capacity, plus chlorophyll retention along low membrane damage and electrolyte leakage reduction than control plants (Kumar et al. 2015). Abiotic stresses; drought, high temperature, and salinity adversely affect nitrogen (N) uptake as well as assimilation in plants. However, *B. juncea*, knowledge about N-pathway genes regulation in reply to environmental stresses at the transcriptional level is very limited. From *B. juncea*, genes that encode ammonium transporters (AMT), nitrate and nitrite reductase (NR, NiR), nitrate transporters (NRT), glutamine synthetase (GS), glutamate synthase (GOGAT), asparagines synthetase, and glutamate dehydrogenase (GDH) clones. After 1 h then 24 h of 42 °C heat, 150 mM salt; NaCl, and osmotic stresses (250 mM Mannitol) expression of these genes studied. Many of these genes encode enzymes and nitrate transporters that involved in nitrogen assimilation and remobilization, which down-regulated in abiotic stresses.

Many transgenic plants overexpressing N-pathway genes have presented improved resistance to environmental stresses (Goel and Singh 2015).

Vitamin E includes tocopherols, which are lipid-soluble antioxidant produced by few cyanobacteria and plants. It has been reported that γ-tocopherol methyltransferase (γ-*TMT*) gene overexpression in transgenic *B. juncea* from *Arabidopsis* showed 6-fold increased α-tocopherol level; most active tocopherol form. In reply to numerous abiotic stresses; drought, salinity, osmotic stress, and heavy metals, tocopherol levels increased in *B. juncea.* Shoot growth and seed germination measurements have revealed that overexpression of the γ-*TMT* gene in transgenic *B. juncea* enhanced resistance to several environmental strains (Yusuf et al. 2010). Transgenic *B. juncea* enriched with α-tocopherol confers resistance to drought, salt, osmotic, as well as heavy metal pressures (Kumar et al. 2013).

3.4 *Brassica campestris*

Brassica campestris (*B. rapa*) contains turnip, turnip rape, fast plants, field mustard, and turnip mustard; broadly cultivated as root and leaf vegetable, as well as an oilseed. *Brassica* vegetables provide soluble fibers, vitamin C, and numerous compounds having anticancer properties, which modulates innate immune system response by anticancer, antibacterial, and antiviral activity. Researchers have found that ethanol extracted from *B. campestris* leaves are highly effective while, methanol, ethyl acetate, and petroleum ether extracts from *B. campestris* root, leaves and stem showed antibacterial properties to different bacterial strains like *Escherichia coli*, *Staphylococcus aureus*, *Bacillus cereus*, *Staphylococcus epidermidis*, and *Pseudomonas aeruginosa* (Agrawal et al. 2013). Glutathione reductase gene (GR2) obtained from *B. campestris* through cDNA ends (RACE) isolation. GR2 has 2073 bp of full-length cDNA as well as an open reading frame of 1692 bp, a protein having 563 amino acids encoded by GR2 of 60.7 kDa molecular mass having 7.9 isoelectric points. PCR results have indicated that GR2 expressed in *B. campestris* leaves, roots and stems, among these GR2 expression was highest in leaves. In response to drought stress, GR1, GR2 transcripts level as well as GR activity increased. These findings revealed that GR has a particular part in coping with various stresses in *B. campestris.* Before giving drought stress, ABA applied to plants indicated that GR1, GR2 expression levels increased significantly than single stress indicated that abscisic acid up-regulated the transcription and activity of GR1 and GR2 gene (Zhang et al. 2018).

Two *B. campestris* seedlings; Tianyou 2 and Longyou 6 treated with NADPH oxidase inhibitors (DPI, IMD), H_2O_2 scavenger (DMTU), and MAPKK inhibitors (U0126), to investigate salt as well as drought stress on activities of antioxidant enzymes and RbohC plus RbohF genes expression. The findings showed antioxidant enzymes; SOD, CAT, APX, GR plus H_2O_2 accumulation and expression of RbohC plus RbohF genes also increased but the accumulation of $O_2^{\cdot-}$ lowered (Zhang et al.

2019). Chinese cabbage (*B. campestris*) transformed with late embryogenesis abundant gene (LEA) isolated from *B. napus* by LBA4404 *Agro* strain having pIG121-LEA binary vector that contained LEA protein gene associated with CaMV promoter as well as terminator sequences and applied with *NPTII* gene (neomycin phosphotransferase II). In response to drought stress, transgenic cabbage plants exhibited improved growth ability. The improved tolerance was indicated by delaying damage symptoms development resulted by stress (Park et al. 2005). Pakchoi (*B. campestris*) among the utmost widespread vegetables, cultivated in China over northern to southern range. Increased nitrogen as fertilizer is used to ensure the yield and cost-effectiveness form vegetables cultivation. To find the drought stress effects on nitrogen, growth status of pakchoi at varying levels of nitrogen, nitrogen accumulation changes, and enzymes that involved in the accumulation of nitrogen were studied. Polyethylene glycol (PEG) was added to induce drought in hydroponic culture. A modified solution of nutrient was applied to pakchoi seedling with different concentrations of nitrogen ($N1 = 2$, $N_2 = 9$, and $N_3 = 18$ mM $NaNO_3$) as well as osmotic potential ($W1 = 0$, $W2 = 60$, $W3 = 120$ g L^{-1} PEG 6000). Seven days drought stress resulted in lowered transpiration rate, shot biomass also nitrogen concentration. Increased nitrogen availability lightens the drought stress through enhancing amino acids contents in roots, improving nitrate reductase, and glutamine synthetase activities that lessen moisture limitations. These findings indicated that pakchoi (*B. campestris*) seedling treated with higher nitrogen levels resulted in improved growth performance in drought stress (Xiong et al. 2018).

3.5 *Brassica carinata*

Brassica carinata (Ethiopian mustard) 25 different strain response were assessed to drought stress under greenhouse conditions in a pot experiment. Plants from all lines (4-week old) were exposed to drought stress (0 or 2 cycles) twice wilting then re-watering. Drought stress resulted in shoot fresh plus a dry mass reduction in all lines. However, 5-strains C90-1191, C90-1203, C90-1210, 77-1271, and P48-3 found superior compared with leftover lines in shoot fresh as well as dry mass production in drought situations. Two lines, C90-1205, and 4007-A classified as moderate tolerant due to their performance in two growth variables. While all leftover lines put in the sensitive category of drought, the capability of water retention (WRC) and osmotic adjustment did not appear as a successful criterion for strain discrimination because drought-resistant, moderately resistant and sensitive lines contained similar osmotic adjustment and WRC values. *Brassica carinata* germplasm drought tolerance improvement is considerably high value by selection and breeding processes (Ashraf and Sharif 1998). Researchers have analyzed the *B. carinata* response to drought stress and the following rehydration by measuring plant growth, chlorophyll concentration and fluorescence, water status, net transpiration, and photosynthetic rate as well as stomatal conductance. Plants were kept for 15 days in drought stress then subjected for 5 days to routine irrigation with 100% FC (field capacity), and next analyzed the possible loss extent caused by stress as well as gain by rehydration. Drought stress resulted in reduced root and shoot growth rate, reduced number and dimensions of leaves as well as biomass accumulation in various plants parts. Increased drought stress also lowered chlorophyll content, relative water content, chlorophyll fluorescence, net transpiration, and photosynthetic rate as well as stomatal conductance. Partial or complete recovery observed after rehydration for all studied parameters. As a result, *B. carinata* uses morpho-physiological drought avoidance strategy (Husen et al. 2014).

Four species of *B.* plants; *B. campestris* (Sarson), *B. juncea* (Brown mustard), *B. napus* (Oilseed rape), and *B. carinata* (Ethiopian mustard) exposed to 24 days drought stress and next assessed their relative drought resistance. These species showed close relationship for water content plus biomass production toward repeated cycles of drought stress. *Brassica carinata* as compared to other *Brassica* species produced considerably lower dry and fresh biomass as well as reduced water content, leaf wax surface, while more significant leaf water potential upon wilting also osmotic potential when rehydration applied. *Brassica napus* showed greater dry and fresh biomass also high chlorophyll, protein, and water content in roots and shot; however, lesser osmotic leaf water potentials. These results showed *B. napus* appeared as most drought-resistant, *B. campestris* and *B. juncea* intermediate while, *B. carinata* sensitive for drought stress. The results suggest a possible element of *Brassica* species drought resistance is an osmotic adjustment (Ashraf and Mehmood 1990).

3.6 *Arabidopsis thaliana*

TFs take part in gene expression regulation, changes plus drive various abiotic stress responses and developmental processes. Around 1600, TFs have been recognized in *Arabidopsis* genome representing almost 6% of genes and categorized into numerous relatives centered on DNA binding domain structures. In *Arabidopsis*, *AtMYB44* belongs to TFs family; R2R3 MYB sub-group22 and upon treatment with ABA resulted in accumulation of *AtMYB44* in 30 min. Various environmental stresses also activate the gene like salinity, low temperature, and drought. *Arabidopsis* plants with overexpression of *AtMYB44* sensitive to ABA than wild type and showed reduced water loss and enhanced drought stress resistance (Jung et al. 2008). Transgenic *Arabidopsis* plants overexpressing *DREB1A* (*CBF3*) in the presence of CaMV35S promoter displayed strong stress gene expression and enhanced drought tolerance (Kasuga et al. 2004). A *DREB*-binding transcription factor *VrDREB2A* from *Vigna radiata* has shown enhanced resistance against salt plus drought stress in *Arabidopsis* (Chen et al. 2016). GST characterized as detoxification enzymes take part in stress tolerance. Although in tomato, data is limited regarding GST gene function involved in abiotic strain response. Currently, researchers have cloned *LeGSTU2* gene and transferred into *Arabidopsis* by *Agrobacterium tumefaciens* mediated transformation. These transgenic *Arabidopsis* revealed improved resistance to drought pressure (Xu et al. 2015).

GRAS proteins lie in plants specific transcription factors family also have functions against various abiotic stress signals and physiological processes. A gene of GRAS family, i.e., *PeSCL7* from poplar used to improve resistance against drought plus salt in *Arabidopsis* (Ma et al. 2010). *BrLAS* a gene from *B. rapa* that belongs to GRAS TFs family was expressed primarily in the axillary stems and roots then localized in nucleus of protoplast cells (*B. rapa*). *Arabidopsis* plants overexpressed with *BrLAS* gene showed delayed flowering time, senescence, and blotting also reduced fertility. *Arabidopsis* transgenic plants also exhibited improved tolerance for drought and lowered ROS accretion with improved activities of antioxidant enzymes in drought conditions than wild type plants (Li et al. 2018b). In Chinese cabbage (*B. rapa*), *BrRH22* is a DEAD-box RNA helicase (RHs) (chloroplast targeted) that has functions in environmental stress responses. In drought and salt stress *BrRH22* expression increased while UV stress decreased *BrRH22* expression. *BrRH22* has RNA chaperone activity and localized in the chloroplast. *Arabidopsis* plants expressing *BrRH22* increased germination also plantlet growth in high drought stress by reducing AB13, AB14, and AB15 levels (Nawaz et al. 2018). TFs; *NAC* has significant roles in environmental stress responses in plants. At present, limited data is available regarding *NAC* gene function in *Miscanthus lutarioriparius* in stress response. *MlNAC9* gene overexpressed in *Arabidopsis* plants conferring ABA hypersensitivity during stages of seed germination and root elongation also showed enhanced resistance for drought stress. Under drought conditions, various anti-oxidant enzyme activities increased significantly such as CAT, SOD, and POD while lowered MDA content in transgenic lines (Zhao et al. 2016). Transgenic *Arabidopsis* overexpressed with galactinol synthase (*AtGolS2*) resulted in increased raffinose and galactinol content in leaves also showed improved drought stress tolerance (Selvaraj et al. 2017). Recently, *ArtRTY* gene from *Arabidopsis* enhances drought tolerance in strawberry (Li et al. 2020).

3.7 *Raphanus sativus* (Common Radish)

One of the broader TFs is *WRKY* that related to various biological activities necessary for the survival of plant-like mechanisms for control response for environmental strains such as drought, salinity, heat also heavy metals. *Raphanus sativus* is root vegetable crop, and that is why the investigation of the *WRKY* TF expression and characterization is imperative in radish. Currently, from radish genome database 126 *WRKY* putative genes retrieved. *RsWRKY* proteins have zinc finger motif and domains (highly conserved) confirmed by their protein sequences and annotation scrutiny. qRT-PCR analysis revealed the important difference of 35 *RsWRKY* genes in various abiotic stress conditions, connecting that these genes could be stress-responsive. Out of 126 *WRKY* TFs recognized from radish 35 showed expression patterns induced by environmental stresses. These findings provide baseline for improving the radish yield as well as quality by increasing resistance against abiotic stress (Karanja et al. 2017a). Another larger family of TF is *NAC* proteins play role in biological processes like expansion and cell differentiation and hormone signaling to several biotic also abiotic stresses. Around 172 *RsNACs* identified from genome of *R. sativus* including 17 members for membrane-bound. RT-qPCR analysis identified 21 *NAC* genes responsive to drought from radish genome. *RsNAC023* and *RsNAC080* genes expressed differentially and positively to all strains except abscisic acid whereas *RsNAC145* further dynamically responded to drought stress than other genes. These findings provide candidate genes and valuable knowledge for refining abiotic stress resistance in common radish (Karanja et al. 2017b). Around 247 *AP2/ERF* genes have been identified in radish genome and superfamily *AP2/ERF* categorized into five groups. In *RsAP2/ERF* genes expansion, tandem duplication has vital role. Out of 247 *RsAP2/ERF* genes, 21 expressed in drought except for abscisic acid and heavy metals representing their significant characters in environmental pressure response and stress tolerance (Karanja et al. 2019).

A natural disaccharides Trehalose has been used to improve a variety of crop plants for stress tolerance. Radish plant pretreatment with 25 mM trehalose induced drought tolerance under water-stressed conditions and increased the chlorophyll content, photosynthesis rate, WUE, proline contents, and SOD enzyme activities in radish cultivars (Akram et al. 2016). Researchers have studied the 28-homobrassinolide and 24-epibrassinolide effects on growth of radish seedlings placed in water stress induced by 15% w/v PEG (polyethylene glycol). Supplementation with brassinosteroids lowered the water stress inhibitory effects on germination and growth of seed. Brassinosteroids stimulates the growth of radish seedlings in water-deficit environment by elevating the nucleic acid plus soluble protein levels and lowered RNase activities. PEG alleviate the osmotic stress by increasing actions of CAT, SOD and APX enzymes (Mahesh et al. 2013). Noman et al., 2018 studied zinc-chelated lysine (Zn-Lys) effects on radish plants germination and yield in water-deficit conditions. Radish seeds first preconditioned with Zn-Lys 0, 1.5, 3, 4.5, or 6 mg kg^{-1} grown, these priming treatment considerably improved seed germination in water stress and plants developed from this primed seed showed enhancements in plant biomass, nutritional quality, photosynthetic leaf pigments, and final roots production. Moreover, peroxidase and SOD enzymes activities also increased but lowered MDA content. 3–4.5 mg kg^{-1} of Zn-Lys regimes appeared sufficient for drought tolerance induction and nutritional quality. These suggest the Zn-Lyn kept a possible balance of translocation and nutrients uptake through inhibiting drought-induced membrane lipid peroxidation (Noman et al. 2018).

3.8 *Eruca vesicaria sativa*

Eruca vesicaria sativa is one of the drought-resistant species of Cruciferae and *DREB2A* (dehydration responsive element binding protein 2A) take part in abiotic stress replies specifically drought. The qRT-PCR analysis revealed the increased *EvDREB2A* expression in leaves than hypocotyls and roots and decreased expression in flower bud. *EvDREB2A* can be used to makes crops tolerant to drought stress (Huang et al. 2016).

3.9 *Sinapis alba*

Sinapis alba has various necessary agronomically important characteristics like drought tolerant. Researchers have performed the *S. alba* transcriptional profiling genome-wide in the drought stressed environment following re-watering, to find out the candidate genes which take part in drought resistance with the help of Illumina deep sequencing technology. Then comparative analysis has discovered various alterations at gene expression level referable to drought strain, caused 248 genes up-regulation and 309 genes down-regulation. Gene ontology analysis discovered these differentially expressed genes generally tangled in cell division as well as metabolic plus catabolic processes (Dong et al. 2012).

3.10 *Anastatica hierochuntica*

Anastatica hierochuntica is a well-known species of Brassicaceae family due to their hygrochasy seed dispersal mechanism (Eshel et al. 2017). Recently found that *A. hierophantica* have strangely short nonphotochemical quenching (NPQ) plus high PSII electron transport rate (ETR). Currently, it was examined that these unusual behaviors are due to x net photosynthetic rate (P_N) inhibition and lack of CO_2. Photoprotective and photosynthetic reply of *A. hierochuntica, as well as sunflower (Helianthus annuus)*, compared with drought or low CO_2 in a photosynthetic inhibition state. *A. hierochuntica* showed approximately half of NPQ values, and twice values of ETR of *H. annuus* in nonsteady state of low CO_2. Whereas, long term P_N inhibition in drought produced a similar rise in NPQ and fell in ETR in both *H. annuus* and *A. hierochuntica.* These findings suggest that in *A. hierochuntica* low NPQ as well as high ETR not related to drought response conditions (Eppel and Rachmilevitch 2016).

3.11 *Raphanus raphanistrum*

Raphanus raphanistrum (wild radish) seeds morphometric characteristics and dormancy expression caused by water stress have been examined. In completely randomized design, experimentation was carried with five water levels (control, S4, S5, S6, and S8) and four replicates. Restrained irrigation caused water pressure then transferring pots in rain shelter at various stages of growth from flowering onset (S4) to flowering completion (S8). Pods (wild radish) were collected from all treatments and tested the seed dormancy for all treatment with pod and without pod. Water stress affected the pod morphometric features like seed weight and pod wall weight. The pod that sampled earlier had thicker walls as well as higher pod segment weight. Pods that collected later have lower dormancy level. These suggest that water stress influenced the seed dormancy expression during reproductive development (Eslami et al. 2012).

3.12 *Diplotaxis tenuifolia*

Diplotaxis tenuifolia, wild rocket (Brassicaceae) demand as well production has improved in Mediterranean countries and has been suggested as a useful insectary plant for agro-ecosystem management. These wild rocket has the most prolonged flowering period and attracts pollinators (bees, hoverflies). Scientists have studied the effects of drought, moderate deficit irrigation (MDI), and severe deficit irrigation (SDI) on *D. tenuifolia* appeal toward pollinators. The results revealed that *D. tenuifolia* tolerate four-day drought in late spring-summer conditions without affecting flowers as well as pollinators attractiveness. However, 8 days or more extended drought period reduced the flower number as well as pollinators attractiveness. Additionally, results revealed that water use efficiency could be improved through regulated deficit irrigation (RDI) without upsetting floral development and *D. tenuifolia* attractiveness to hoverflies and bees (Barbir et al. 2014).

4 Strategies to Overcome Drought Stress

In modern agriculture, breeding of drought tolerance cultivars has to gain significant importance. Recent advances in genomic approaches paved the way for deciphering drought stress mechanism in Brassicaceae plants. Various modern breeding strategies have been successfully applied to tackle drought syndrome in Brassicaceae plants. In this section, we briefly described modern genomic-assisted breeding strategies to develop drought tolerance Brassicaceae cultivars.

5 Genome-Wide Association Studies

Genome-wide association studies (GWAS) has been extensively applied for elucidating the drought tolerance response of various Brassicaceae plants (EL-SODA et al. 2015; Zhu et al. 2016a). GWAS can be used to reveal the function of novels genes linked with drought tolerance and identify the drought-associated loci in genome-wide studies (Exposito-Alonso et al. 2018). In general, GWAS merges single-nucleotide polymorphism (SNP) and phenotypic data from a large number of accessions to discover many important loci having correlations in allelic frequency to the environment (Lasky et al. 2012). Hence, GWAS is a modern breeding strategy to cope with drought stress syndrome in many *Brassica* species. *Arabidopsis thaliana* is an important member of the Brassicaceae family and an excellent model organism for plant research. GWAS has been extensively executed in such *A. thaliana*, which shows natural genetic variation for investigation of drought tolerance mechanism (Verslues et al. 2014). For example, GWAS was carried out to study the natural variation in 324 *Arabidopsis* accessions subjected to drought stress for biomass accumulation. Six quantitative trait loci (QTLs) was found to control the drought stress mechanism (Bac-Molenaar et al. 2016). In another experiment, El-Soda and colleagues analyzed numerous root and shoot growth-related traits in *Arabidopsis* under drought stress. GWAS mapping and recombinant inbred line (RIL) techniques were used to identify drought-responsive QTLs in diverse genotypes of *Arabidopsis*. The results showed that the discovery of 58 SNPs linked loci showing QTLs by environment interaction (Q × E), while 16 SNPs were not displaying Q × E (EL-SODA et al. 2015). Proline accumulation is vital to combat stress tolerance, and *Arabidopsis* shows natural variation in proline contents under drought stress conditions. GWAS was conducted for natural accessions and publicly access SNPs data to investigate the variations in proline accumulation under drought stress. Metrics tools were used to identify the candidate genomic regions that comprised of novel genes controlling the proline content accumulation. Reverse genetics studies of numerous genomic regions found new proline associated genes such as nucleoside triphosphate hydrolase, MADS-box protein, protein phosphatase 2A, thioredoxins, ribosomal protein RPL24A, mitochondrial protease LON1 and many universal stress-related proteins such as A domain protein (Verslues et al. 2014).

SNPs aided with whole-genome sequencing have been applied to study the genetic architecture of *Brassica* species, like the construction of linkage maps, assessment of genetic variation, taxonomy, phylogenetic investigation, estimation of population structure and GWAS (Bancroft et al. 2011; Huang et al. 2013). SNPs markers have numerous, advantages due to their abundance in genomes and are co-dominant which provide high-throughput automation (Tsuchihashi and Dracopoli 2002). There are many reported studies on GWAS to decipher the genomic regions of several *Brassica* cultivars to attain drought tolerance. *Brassica napus* is a vital oilseed crop; its yield and plant densities are adversely affected by drought stress. For this, 140 canola accessions were subjected to drought stress by using PEG for phenotyping analysis. Genetic variability and phenotypic variations showed that the ratio of root to shoot length was considered as a preferable sign for drought resistance response. Genotyping was done via 25,495 *Brassica* SNPs, and GWAS detected 16 QTLs strongly associated with drought stress.

Moreover, combining the QTL loci identified from GWAS with several expressed genes by RNA sequencing revealed 79 candidate genes, of which 8 genes were regulated the drought stress via gene ontology of *A. thaliana*. These findings give detailed information about the regulatory mechanism for drought tolerance in *B. napus* (Zhang et al. 2015a). In 2016 Fletcher and co-workers conducted GWAS to identify the polymorphism linked with drought tolerance QTLs in canola. Re-sequencing was performed for the mapping population to investigate genetic variance across QTL regions and within the genome. Illumina HiSeq platform was used to sequence parental genotypes and construct a reference assembly to explain the genetic diversity. High-density SNPs were used to characterize genome-wide patterns. Molecular marker analysis was executed for numerous of the identified polymorphic QTLs, which ultimately provide a better understanding to dissect the genetic mechanism underlying drought tolerance in *B. napus* (Fletcher et al. 2016).

To conclude, the only limited number of research work has been published for important Brassicaceae plants in order to develop drought tolerant cultivars. Progress in modern tools for genomic-assisted breeding for other *Brassica* species will help the scientists to use high-throughput SNPs and sequencing platforms for cost-effective and speedy breeding in the future. Besides, genomic selection (GS) and genotyping by sequencing (GBS) will provide a more efficient way to study the genetic variation under drought stress and can improve the quality and yield of Brassicaceae plants.

6 CRISPR/Cas9 Based Genome Editing

Continuous advancement in modern breeding techniques permits scientist to develop resilient climate crops with increased yield and better acclamation in unfavorable climatic conditions (Raza et al. 2019b). Clustered regularly interspaced short palindromic repeats (CRISPR/Cas9) is an advanced genome editing technique, which has been widely applied for crop improvement in modern breeding programs (Okuzaki et al. 2018; Raza et al. 2019b). CRISPR/Cas9 based targeted gene editing is opening the new horizons in plant biology and provide a precise, cost-effective, and efficient system for genome manipulations of many organisms from several kingdoms. This remarkable genome editing toolkit was isolated from prokaryotic organisms like archaea and bacteria. CRISPR/Cas9 system is consist of a single guide RNA (sgRNA) for targeting the desired DNA sequence and Cas9 protein which act like scissors. A specific PAM sequence in the seed sequence is detected by Cas9 protein and cuts the sequence at 3–4 base pairs upstream. The double standard breaks (DSBs) are repaired via homology-directed repair (HDR) or non-homologous end-joining (NHEJ) mechanisms (Hsu et al. 2013; Razzaq et al. 2019b). The CRISPR/Cas9 toolkit has demonstrated its simplicity, versatility, and efficiency in several crop species by successfully targeted gene editing (Murovec et al. 2018; Razzaq et al. 2019b). In combination with other speed breeding platforms, it will be revolutionized the crop breeding in the near future.

CRISPR/Cas9 system has emerged as a powerful tool for improving numerous crop to develop climate-resilient species. Although *Brassica* species has excellent commercial value, modern genomic-assisted breeding techniques are still lacking. There are only limited events of genome editing have been reported in Brassicaceae plants (Braatz et al. 2017; Kirchner et al. 2017; Lawrenson et al. 2015; Murovec et al. 2018). In these studies, CRISPR/Cas9 mediated genome editing was executed in *B. napus* and *B. oleracea* to develop knockout mutants (Braatz et al. 2017; Lawrenson et al. 2015), but comprehensive understanding of genetic characteristics of these knockout and mutation patterns needs to be studied more carefully. In addition, there is no method available of CRISPR/Cas9 based targeted genome editing for drought stress tolerance in important oilseed crops and other plants of Brassicaceae family. Continuous efforts are still making to develop precise and accurate methods for manipulating the genome of *Brassica* crops. Polyploidy is the main hurdle to manipulate the genome of most *Brassica* species. Because of gene redundancy modifying any specific trait by mutagenesis is very ineffective (Braatz et al. 2017).

On the other hand, there are some studies published for achieving drought stress tolerance in *Arabidopsis*. In 2017 Park and colleagues conducted an experiment by using the CRISPR/Cas9 system to improve the drought stress tolerance in *Arabidopsis*. To enhance the endogenous transcriptional levels of *Arabidopsis vacuolar* H^+*-pyrophosphatase* (*AVP1*) *and anthocyanin pigment 1* (*PAP1*), CRISPR/Cas9 activation toolbox was used. The expression level of *AVP1* was elevated 2–5 folds in engineered plants and showed greater leaf surface area, number of leaves and increased drought tolerance as compared to wild type (Park et al. 2017a). A plant produced some unique redox enzymes called glutaredoxins that are responsible for plant growth and root development about nutrient sensing. CRISPR/Cas9 technique was used to achieve targeted mutagenesis in *Arabidopsis* by disrupting *AtGRXS3/4/5/7/8*, *AtGRXS6*, and *AtGRXS11* genes. Agrobacterium-mediated transformation was done for the delivery of CRISPR machinery, and mutants were screened. Sanger sequencing method was applied to confirm the small insertion-deletion mutations in genes. PCR screening was performed to detect the targeted mutagenesis in engineered plants. The results showed increase root growth, greater nitrogen use efficiency, and drought resistance in *Arabidopsis* (Fernandez et al. 2018).

The plant responds to drought stress by triggering several regulating mechanisms leading to crucial changes in the developmental and physiological processes. Deciphering the role of genes regulating the stress response can be used to develop drought tolerant plants. A unique responsive element of ABA pathway like *AREB1/ABF2* binding factor is a major factor for regulating drought tolerance. CRISPR/Cas9 based genome editing was executed to study the CRISPR activation (CRISPRa) system that mutates *AREB1* to elucidate its function under drought stress conditions in *Arabidopsis*. CRISPRa dCas9^{HAT} toolkit was used to trigger the expression of *AREB1*, and mutated plants indicated dwarf trait. Moreover, qRT-PCR was conducted to calculate the expression level of both responsive factors and found that both *RD29A* and *AREB1* showed an elevated level of a gene in controlled plants. The plants produced by this approach revealed more chlorophyll concentration and rapid stomatal aperture in drought stress conditions, which allowed increased tolerance against drought stress. Hence, it was concluded that CRISPRa dCas9^{HAT} is an excellent tool to develop drought tolerant plants (Paixão et al. 2019).

In the future, CRISPR/Cas9 tools can be used to improve the economic importance of oilseed and other Brassicaceae plants. There is the prime need to design novel cargo vector system for targeted delivery of CRISPR/Cas9 toolbox. More efforts are required to develop accurate and efficient ways to manipulate the genomes of Brassicaceae plants and produce drought stress-tolerant plants.

7 Conclusion and Outlooks

In plants, stress could be explained as which can originate change from the optimized developmental conditions and absolute growth. One of the crucial environmental stress factors is drought, which alters the growth of plants as well as their development—stress like drought results in different biochemical changes, morphological plus physiological that aids to maintain plants growth. In climatic areas, soil water scarcity and drought might be chronic through low water availability; unpredictable and random as a result of weather conditions fluctuations during the plant growth period. With the increase in water scarcity and climate changes, the drought effects also increased. The genes that have direct or indirect involvement in exerting abiotic stress in a plant are regulated by the inducible transcription factors regulating the function and regulatory proteins.

Nevertheless, there is limited work has been done to develop drought stress tolerance Brassicaceae plants expect some major *Brassica* species and some other plants. There is a need to explore the drought stress mechanisms for a maximum number of plants from the Brassicaceae family.

In modern agriculture, breeding of drought tolerance cultivars has to gain significant importance. Recent advances in genomic approaches paved the way for deciphering drought stress mechanism in Brassicaceae plants. Various modern breeding strategies have been successfully applied to tackle drought syndrome in Brassicaceae plants. In the future, GS and GBS will provide a more efficient way to study the genetic variation under drought stress and can improve the quality and yield of Brassicaceae plants. CRISPR/Cas9 tools can be used to improve the economic importance of oilseed and other Brassicaceae plants. Together with genome editing tools, OMICS approaches can lead researchers to identify candidate genes and regulators for stress response parameters that may improve the yield and drought stress tolerance via transgenic and modern plant breeding strategies, such as speed breeding. More efforts are required to develop accurate and efficient ways to manipulate the genomes of Brassicaceae plants and produce drought stress-tolerant plants.

Acknowledgements The authors are grateful to all the researchers whose excellent work has been cited and helped us to prepare this chapter.

References

Abdallah M, Dubousset L, Meuriot F, Etienne P, Avice J, Ourry A (2010) Effect of mineral sulphur availability on nitrogen and sulphur uptake and remobilization during the vegetative growth of *Brassica napus* L. J Exp Bot 61:2635–2646

Agrawal MK, Rathore D, Goyal S, Varma A, Varma A (2013) Antibacterial efficacy of *Brassica campestris* root, stem and leaves extracts. Int J Adv Res 1:131–135

Aguado A, Capote N, Romero F, Dodd IC, Colmenero-Flores JM (2014) Physiological and gene expression responses of sunflower (*Helianthus annuus* L.) plants differ according to irrigation placement. Plant Sci 227:37–44

Akram NA, Iqbal M, Muhammad A, Ashraf M, Al-Qurainy F, Shafiq S (2018) Aminolevulinic acid and nitric oxide regulate oxidative defense and secondary metabolisms in canola (*Brassica napus* L.) under drought stress. Protoplasma 255:163–174

Akram NA, Waseem M, Ameen R, Ashraf M (2016) Trehalose pretreatment induces drought tolerance in radish (*Raphanus sativus* L.) plants: some key physio-biochemical traits. Acta Physiol Plant 38:3

Al-Shehbaz I, Beilstein MA, Kellogg E (2006) Systematics and phylogeny of the *Brassicaceae* (Cruciferae): an overview. Plant Syst Evol 259:89–120

Alagoz SM, Toorchi M (2018) An investigation of some key morpho-physiological attributes and leaf proteome profile in canola (*Brassica napus* L.) under salinity stress. Pak J Bot 50:847–852

Ashraf M, Mehmood S (1990) Response of four *Brassica* species to drought stress. Environ Exp Bot 30:93–100

Ashraf M, Shahbaz M, Ali Q (2013) Drought-induced modulation in growth and mineral nutrients in canola (*Brassica napus* L.). Pak J Bot 45:93–98

Ashraf M, Sharif R (1998) Assessment of inter-cultivar/line variation of drought resistance in a potential oil-seed crop, Ethiopian mustard (*Brassica carinata* A. Br.). Arch Agron Soil Sci 43:251–265

Babula-Skowrońska D, Ludwików A, Cieśla A, Olejnik A, Cegielska-Taras T, Bartkowiak-Broda I, Sadowski J (2015) Involvement of genes encoding ABI1 protein phosphatases in the response of *Brassica napus* L. to drought stress. Plant Mol Biol 88:445–457

Bac-Molenaar JA, Granier C, Keurentjes JJ, Vreugdenhil D (2016) Genome-wide association mapping of time-dependent growth responses to moderate drought stress in Arabidopsis. Plant, Cell Environ 39:88–102

Bancroft I, Morgan C, Fraser F, Higgins J, Wells R, Clissold L, Baker D, Long Y, Meng J, Wang X, Liu S (2011) Dissecting the genome of the polyploid crop oilseed rape by transcriptome sequencing. Nat Biotechnol 29:762

Barbir J, Dorado J, Fernández-Quintanilla C, Blanusa T, Maksimovic C, Badenes-Pérez FR (2014) Impact of drought and regulated deficit irrigation on attractiveness of *Diplotaxis tenuifolia* (*Brassicaceae*) to bees and hoverflies. Impact of drought and regulated deficit irrigation on attractiveness of *Diplotaxis tenuifolia* (*Brassicaceae*) to bees and hoverflies 100:13–16

Barrameda-Medina Y, Blasco B, Lentini M, Esposito S, Baenas N, Moreno DA, Ruiz JM (2017) Zinc biofortification improves phytochemicals and amino-acidic profile in *Brassica oleracea* cv. Bronco. Plant Sci 258:45–51

Bauer H, Ache P, Lautner S, Fromm J, Hartung W, Al-Rasheid KA, Sonnewald S, Sonnewald U, Kneitz S, Lachmann N, Mendel RR (2013) The stomatal response to reduced relative humidity requires guard cell-autonomous ABA synthesis. Curr Biol 23:53–57

Behnam BA, Iuchi SA, Fujita MI, Fujita YA, Takasaki HI, Osakabe YU, Yamaguchi-Shinozaki KA, Kobayashi MA, Shinozaki KA (2013) Characterization of the promoter region of an *Arabidopsis* gene for 9-cis-epoxycarotenoid dioxygenase involved in dehydration-inducible transcription. DNA Res 20:315–324

Bhargava S, Sawant K (2013) Drought stress adaptation: metabolic adjustment and regulation of gene expression. Plant Breed 132:21–32

Braatz J, Harloff H-J, Mascher M, Stein N, Himmelbach A, Jung C (2017) CRISPR-Cas9 targeted mutagenesis leads to simultaneous modification of different homoeologous gene copies in polyploid oilseed rape (*Brassica napus*). Plant Physiol 174:935–942

Chaves M, Oliveira M (2004) Mechanisms underlying plant resilience to water deficits: prospects for water-saving agriculture. J Exp Bot 55:2365–2384

Chen H, Liu L, Wang L, Wang S, Cheng X (2016) *VrDREB2A*, a *DREB*-binding transcription factor from *Vigna radiata*, increased drought and high-salt tolerance in transgenic *Arabidopsis thaliana*. J Plant Res 129:263–273

Cheng L, Li X, Huang X, Ma T, Liang Y, Ma X, Peng X, Jia J, Chen S, Chen Y, Deng B (2013) Overexpression of sheepgrass R1-MYB transcription factor *LcMYB1* confers salt tolerance in transgenic *Arabidopsis*. Plant Physiol Biochem 70:252–260

Danquah A, de Zelicourt A, Colcombet J, Hirt H (2014) The role of ABA and MAPK signaling pathways in plant abiotic stress responses. Biotechnol Adv 32:40–52

Dong CH, Li C, Yan XH, Huang SM, Huang JY, Wang LJ, Guo RX, Lu GY, Zhang XK, Fang XP, Wei WH (2012) Gene expression profiling of Sinapis alba leaves under drought stress and rewatering growth conditions with Illumina deep sequencing. Mol Biol Rep 39:5851–5857

Dutta I, Saha P, Das S (2008) Efficient Agrobacterium-mediated genetic transformation of oilseed mustard [*Brassica juncea* (L.) Czern.] using leaf piece explants. Vitro Cell Develop Biol Plant 44:401–411

El-Soda M, Kruijer W, Malosetti M, Koornneef M, Aarts MG (2015) Quantitative trait loci and candidate genes underlying genotype by environment interaction in the response of *Arabidopsis thaliana* to drought. Plant, Cell Environ 38:585–599

Eppel A, Rachmilevitch S (2016) Photosynthesis and photoprotection under drought in the annual desert plant Anastatica hierochuntica. Photosynthetica 54:143–147

Eshel G, Shaked R, Kazachkova Y, Khan A, Eppel A, Cisneros A, Acuna T, Gutterman Y, Tel-Zur N, Rachmilevitch S, Fait A (2017) Anastatica hierochuntica, an *Arabidopsis* desert relative, is tolerant to multiple abiotic stresses and exhibits species-specific and common stress tolerance strategies with its halophytic relative, Eutrema (Thellungiella) salsugineum. Front Plant Sci 7:1992

Eslami S, Gill G, McDonald G (2012) Effect of water stress during seed development on morphometric characteristics and dormancy of wild radish (*Raphanus raphanistrum* L.) seeds. Int J Plant Prod 4:159–168

Exposito-Alonso M, Vasseur F, Ding W, Wang G, Burbano HA, Weigel D (2018) Genomic basis and evolutionary potential for extreme drought adaptation in *Arabidopsis thaliana*. Nat Ecol Evol 2:352

Fang Z, Zhang X, Gao J, Wang P, Xu X, Liu Z, Shen S, Feng B (2015) A buckwheat (*Fagopyrum esculentum*) *DRE*-binding transcription factor gene, *FeDREB1*, enhances freezing and drought tolerance of transgenic *Arabidopsis*. Plant Mol Biol Rep 33:1510–1525

Farooq M, Gogoi N, Barthakur S, Baroowa B, Bharadwaj N, Alghamdi SS, Siddique K (2017) Drought stress in grain legumes during reproduction and grain filling. J Agron Crop Sci 203:81–102

Farooq M, Wahid A, Kobayashi N, Fujita D, Basra S (2009) Plant drought stress: effects, mechanisms and management. In: Sustainable agriculture. Springer, pp 153–188

Fernandez F, Sanchez K, Escobar M (2018) Identification of CRISPR-Cas9 Mutants in *Arabidopsis* Glutaredoxin Genes *AtGRXS11*, *AtGRXS6*, and *AtGRXS3/4/5/7/8* G ene Cluster. FASEB J 32:lb20-lb20

Fletcher RS, Herrmann D, Mullen JL, Li Q, Schrider DR, Price N, Lin J, Grogan K, Kern A, McKay JK (2016) Identification of polymorphisms associated with drought adaptation QTL in *Brassica napus* by resequencing. Genes Genome Genet 6:793–803

Fujita Y, Yoshida T, Yamaguchi-Shinozaki K (2013) Pivotal role of the *AREB/ABF*-SnRK2 pathway in ABRE-mediated transcription in response to osmotic stress in plants. Physiol Plant 147:15–27

Gangopadhyay G, Basu S, Gupta S (1997) In vitro selection and physiological characterization of NaCl-and mannitol-adapted callus lines in *Brassica juncea*. Plant Cell Tiss Org Cult 50:161–169

Gao F, Zhou J, Deng RY, Zhao HX, Li CL, Chen H, Suzuki T, Park SU, Wu Q (2017) Overexpression of a tartary buckwheat R2R3-MYB transcription factor gene, *FtMYB9*, enhances tolerance to drought and salt stresses in transgenic *Arabidopsis*. J Plant Physiol 214:81–90

Giacoppo S, Galuppo M, Iori R, De Nicola GR, Cassata G, Bramanti P, Mazzon E (2013) Protective role of (RS)-glucoraphanin bioactivated with myrosinase in an experimental model of multiple sclerosis. CNS Neurosci Ther 19:577–584

Gidik B, Gül V, Sefali A (2019) A study of wild plant species of *Brassicaceae* family in bayburt region of Turkey. Pak J Bot 51:681–687

Gidik B, Onemli F, Cabi E (2016) Determination of wild plant species of *Brassicaceae* family in Turkish Thrace. Biol Divers Conserv 9:100–105

Goel P, Singh AK (2015) Abiotic stresses downregulate key genes involved in nitrogen uptake and assimilation in *Brassica juncea* L. PLoS ONE 10:e0143645

Golicz AA, Bayer PE, Barker GC, Edger PP, Kim H, Martinez PA, Chan CK, Severn-Ellis A, McCombie WR, Parkin IA, Paterson AH (2016) The pangenome of an agronomically important crop plant *Brassica oleracea*. Nat Commun 7:13390

Griffiths D, Birch A, Hillman J (1998) Antinutritional compounds in the Brasi analysis, biosynthesis, chemistry and dietary effects. J Horti Sci Biotechnol 73:1–18

Hajiboland R, Amirazad H (2010) Drought tolerance in Zn-deficient red cabbage (*Brassica oleracea* L. var. capitata f. rubra) plants. Horti Sci 37:88–98

Hasanuzzaman M, Hossain MA, Teixeira da Silva JA, Fujita M (2012) Plant responses and tolerance to abiotic oxidative stress: antioxidant defense is a key factor. In: Bandi V, Shanker AK, Shanker C, Mandapaka M (eds) Crop stress and its management: perspectives and strategies. Springer, New York, pp 261–316

Hasanuzzaman M, Nahar K, Gill SS, Fujita M (2014) Drought stress responses in plants, oxidative stress and antioxidant defense. In: Tuteja N, Gill SS (eds) Climate change and plant abiotic stress tolerance, vol 1—Biological techniques, 1st edn, Wiley, Weinheim, pp 209–250

Hasanuzzaman M, Nahar K, Anee T, Khan M, Fujita M (2018) Silicon-mediated regulation of antioxidant defense and glyoxalase systems confers drought stress tolerance in *Brassica napus* L. South Afr J Bot 115:50–57

Hassan NM, Serag MS, El-Feky FM (2004) Changes in nitrogen content and protein profiles following in vitro selection of NaCl resistant mung bean and tomato. Acta Physiol Plant 26:165

Hsu PD, Scott DA, Weinstein JA, Ran FA, Konermann S, Agarwala V, Li Y, Fine EJ, Wu X, Shalem O, Cradick TJ (2013) DNA targeting specificity of RNA-guided Cas9 nucleases. Nat Biotechnol 31:827

Huang BL, Zhang XK, Li YY, Li DY, Ma MY, Cai DT, Wu WH, Huang BQ (2016) Cloning and characterization of the dehydration-responsive element-binding protein 2A gene in *Eruca vesicaria* subsp sativa. Genet Mol Res 15:23–29

Huang GT, Ma SL, Bai LP, Zhang L, Ma H, Jia P, Liu J, Zhong M, Guo ZF (2012) Signal transduction during cold, salt, and drought stresses in plants. Mol Biol Rep 39:969–987

Huang Q, Wang Y, Li B, Chang J, Chen M, Li K, Yang G, He G (2015) *TaNAC29*, a *NAC* transcription factor from wheat, enhances salt and drought tolerance in transgenic *Arabidopsis*. BMC Plant Biol 15:268

Huang S, Deng L, Guan M, Li J, Lu K, Wang H, Fu D, Mason AS, Liu S, Hua W (2013) Identification of genome-wide single nucleotide polymorphisms in allopolyploid crop *Brassica napus*. BMC Genom 14:717

Husen A, Iqbal M, Aref IM (2014) Growth, water status, and leaf characteristics of *Brassica carinata* under drought and rehydration conditions. Brazilian J Bot 37:217–227

Jahangir M, Abdel-Farid IB, Kim HK, Choi YH, Verpoorte R (2009) Healthy and unhealthy plants: the effect of stress on the metabolism of *Brassicaceae*. Environ Exp Bot 67:23–33

Jan N, Qazi HA, Ramzan S, John R (2018) Developing stress-tolerant plants through in vitro tissue culture: family *Brassicaceae*. In: biotechnologies of crop improvement, vol 1. Springer, pp 327–372

Jan SA, Bibi N, Shinwari ZK, Rabbani MA, Ullah S, Qadir A, Khan N (2017) Impact of salt, drought, heat and frost stresses on morpho-biochemical and physiological properties of *Brassica* species: an updated review. J Rural Dev Agri 2:1–10

Jin J, Zhang H, Kong L, Gao G, Luo J (2013) PlantTFDB 3.0: a portal for the functional and evolutionary study of plant transcription factors. Nucleic Acids Res 42:D1182–D1187

Jin X-F, Xiong A-S, Peng R-H, Liu J-G, Gao F, Chen J-M, Yao Q-H (2010) *OsAREB1*, an ABRE-binding protein responding to ABA and glucose, has multiple functions in *Arabidopsis*. BMB Rep 43:34–39

Joshi R, Wani SH, Singh B, Bohra A, Dar ZA, Lone AA, Pareek A, Singla-Pareek SL (2016) Transcription factors and plants response to drought stress: current understanding and future directions. Front Plant Sci 7:1029

Jung C, Seo JS, Han SW, Koo YJ, Kim CH, Song SI, Nahm BH, Do Choi Y, Cheong JJ (2008) Over-expression of *AtMYB44* enhances stomatal closure to confer abiotic stress tolerance in transgenic *Arabidopsis*. Plant Physiol 146:623–635

Karanja BK, Fan L, Xu L, Wang Y, Zhu X, Tang M, Wang R, Zhang F, Muleke EM, Liu L (2017a) Genome-wide characterization of the *WRKY* gene family in radish (*Raphanus sativus* L.) reveals its critical functions under different abiotic stresses. Plant Cell Rep 36:1757–1773

Karanja BK, Xu L, Wang Y, Muleke EM, Jabir BM, Xie Y, Zhu X, Cheng W, Liu L (2017b) Genome-wide characterization and expression profiling of *NAC* transcription factor genes under abiotic stresses in radish (*Raphanus sativus* L.). PeerJ 5:e4172

Karanja BK, Xu L, Wang Y, Tang M, Muleke EMm, Dong J, Liu L (2019) Genome-wide characterization of the *AP2/ERF* gene family in radish (*Raphanus sativus* L.): unveiling evolution and patterns in response to abiotic stresses. Gene 144048

Kasuga M, Miura S, Shinozaki K, Yamaguchi-Shinozaki K (2004) A combination of the *Arabidopsis DREB1A* gene and stress-inducible rd29A promoter improved drought-and low-temperature stress tolerance in tobacco by gene transfer. Plant Cell Physiol 45:346–350

Khazaei H, Street K, Bari A, Mackay M, Stoddard FL (2013) The FIGS (Focused Identification of Germplasm Strategy) approach identifies traits related to drought adaptation in Vicia faba genetic resources. PLoS ONE 8:e63107

Kholová J, Hash CT, Kočová M, Vadez V (2011) Does a terminal drought tolerance QTL contribute to differences in ROS scavenging enzymes and photosynthetic pigments in pearl millet exposed to drought? Environ Exp Bot 71:99–106

Kim JS, Mizoi J, Yoshida T, Fujita Y, Nakajima J, Ohori T, Todaka D, Nakashima K, Hirayama T, Shinozaki K, Yamaguchi-Shinozaki K (2011) An ABRE promoter sequence is involved in osmotic stress-responsive expression of the *DREB2A* gene, which encodes a transcription factor regulating drought-inducible genes in *Arabidopsis*. Plant Cell Physiol 52:2136–2146

Kirchner TW, Niehaus M, Debener T, Schenk MK, Herde M (2017) Efficient generation of mutations mediated by CRISPR/Cas9 in the hairy root transformation system of *Brassica carinata*. PLoS ONE 12:e0185429

Kumar D, Shekhar S, Bisht S, Kumar V, Varma A, Kumar M (2015) Ectopic overexpression of lectin in transgenic *Brassica juncea* plants exhibit resistance to fungal phytopathogen and showed alleviation to salt and drought stress. J Bioeng Biomed Sci 5:147

Kumar D, Yusuf MA, Singh P, Sardar M, Sarin NB (2013) Modulation of antioxidant machinery in α-tocopherol-enriched transgenic *Brassica juncea* plants tolerant to abiotic stress conditions. Protoplasma 250:1079–1089

Lasky JR, Des Marais DL, McKAY JK, Richards JH, Juenger TE, Keitt TH (2012) Characterizing genomic variation of *Arabidopsis thaliana*: the roles of geography and climate. Mol Ecol 21:5512–5529

Latif M, Akram N, Ashraf M (2016) Regulation of some biochemical attributes in drought-stressed cauliflower (*Brassica oleracea* L.) by seed pre-treatment with ascorbic acid. J Horti Sci Biotechnol 91:129–137

Lawrenson T, Shorinola O, Stacey N, Li C, Østergaard L, Patron N, Uauy C, Harwood W (2015) Induction of targeted, heritable mutations in barley and *Brassica oleracea* using RNA-guided Cas9 nuclease. Genome Biol 16:258

Lee B-R, Islam MT, Park S-H, Jung H-i, Bae D-W, Kim T-H (2019) Characterization of salicylic acid-mediated modulation of the drought stress responses: reactive oxygen species, proline, and redox state in *Brassica napus*. Environ Exp Bot 157:1–10

Lee B-R, Zaman R, Avice J-C, Ourry A, Kim T-H (2016) Sulfur use efficiency is a significant determinant of drought stress tolerance in relation to photosynthetic activity in *Brassica napus* cultivars. Front Plant Sci 7:459

Lee S-j, Kang J-y, Park H-J, Kim MD, Bae MS, Choi H-i, Kim SY (2010) *DREB2C* interacts with *ABF2*, a *bZIP* protein regulating abscisic acid-responsive gene expression, and its overexpression affects abscisic acid sensitivity. Plant Physiol 153:716–727

Li M, Yang Y, Raza A, Yin S, Wang H, Zhang Y, Dong J, Wang G, Zhong C, Zhang H, Liu J (2020) Heterologous Expression of *Arabidopsis* rty Enhances Drought Tolerance in Strawberry (*Fragaria*× *ananassa* Duch.). https://doi.org/10.21203/rs.3.rs-34565/v1

Li J, Zeng L, Cheng Y, Lu G, Fu G, Ma H, Liu Q, Zhang X, Zou X, Li C (2018a) Exogenous melatonin alleviates damage from drought stress in *Brassica napus* L. (rapeseed) seedlings. Acta Physiol Plant 40:43

Li P, Zhang B, Su T, Li P, Xin X, Wang W, Zhao X, Yu Y, Zhang D, Yu S, Zhang F (2018b) BrLAS, a GRAS transcription factor from *Brassica rapa*, is involved in drought stress tolerance in transgenic *Arabidopsis*. Front Plant Sci 9:1792

Liang Y, Kang K, Gan L, Ning S, Xiong J, Song S, Xi L, Lai S, Yin Y, Gu J, Xiang J (2019) Drought-responsive genes, late embryogenesis abundant group3 (LEA 3) and vicinal oxygen chelate, function in lipid accumulation in *Brassica napus* and *Arabidopsis* mainly via enhancing photosynthetic efficiency and reducing ROS. Plant Biotechnol J 17:2123–2142

Liu G, Li X, Jin S, Liu X, Zhu L, Nie Y, Zhang X (2014a) Overexpression of rice *NAC* gene *SNAC1* improves drought and salt tolerance by enhancing root development and reducing transpiration rate in transgenic cotton. PLoS ONE 9:e86895

Liu J-H, Peng T, Dai W (2014b) Critical cis-acting elements and interacting transcription factors: key players associated with abiotic stress responses in plants. Plant Mol Biol Rep 32:303–317

Llorca CM, Potschin M, Zentgraf U (2014) *bZIPs* and *WRKYs*: two large transcription factor families executing two different functional strategies. Front Plant Sci 5:169

Lucas S, Durmaz E, Akpınar BA, Budak H (2011) The drought response displayed by a *DRE*-binding protein from Triticum dicoccoides. Plant Physiol Biochem 49:346–351

Ma H-S, Liang D, Shuai P, Xia X-L, Yin W-L (2010) The salt-and drought-inducible poplar GRAS protein SCL7 confers salt and drought tolerance in *Arabidopsis thaliana*. J Exp Bot 61:4011–4019

Mahesh K, Balaraju P, Ramakrishna B, Rao SSR (2013) Effect of brassinosteroids on germination and seedling growth of radish (*Raphanus sativus* L.) under PEG-6000 induced water stress. Am J Plant Sci 4:2305

Manivannan P, Jaleel CA, Somasundaram R, Panneerselvam R (2008) Osmoregulation and antioxidant metabolism in drought-stressed *Helianthus annuus* under triadimefon drenching. C R Biol 331:418–425

Mao X, Zhang H, Qian X, Li A, Zhao G, Jing R (2012) *TaNAC2*, a *NAC*-type wheat transcription factor conferring enhanced multiple abiotic stress tolerances in *Arabidopsis*. J Exp Bot 63:2933–2946

Maruyama KY, Todaka DA, Mizoi JU, Yoshida TA, Kidokoro SA, Matsukura SA, Takasaki HI, Sakurai TE, Yamamoto YY, Yoshiwara KY, Kojima MI (2011) Identification of cis-acting promoter elements in cold-and dehydration-induced transcriptional pathways in Arabidopsis, rice, and soybean. DNA Res 19:37–49

Matsukura S, Mizoi J, Yoshida T, Todaka D, Ito Y, Maruyama K, Shinozaki K, Yamaguchi-Shinozaki K (2010) Comprehensive analysis of rice *DREB2*-type genes that encode transcription factors involved in the expression of abiotic stress-responsive genes. Mol Genet Genomics 283:185–196

Miyakawa T, Fujita Y, Yamaguchi-Shinozaki K, Tanokura M (2013) Structure and function of abscisic acid receptors. Trends Plant Sci 18:259–266

Moreno DA, Carvajal M, López-Berenguer C, García-Viguera C (2006) Chemical and biological characterisation of nutraceutical compounds of broccoli. J Pharm Biomed Anal 41:1508–1522

Munawar A, Akram NA, Ahmad A, Ashraf M (2019) Nitric oxide regulates oxidative defense system, key metabolites and growth of broccoli (*Brassica oleracea* L.) plants under water limited conditions. Sci Hortic 254:7–13

Murovec J, Guček K, Bohanec B, Avbelj M, Jerala R (2018) DNA-free genome editing of *Brassica oleracea* and *B. rapa* protoplasts using CRISPR-Cas9 ribonucleoprotein complexes. Front Plant Sci 9:1594

Nakashima K, Yamaguchi-Shinozaki K, Shinozaki K (2014) The transcriptional regulatory network in the drought response and its crosstalk in abiotic stress responses including drought, cold, and heat. Front Plant Sci 5:170

Nawaz G, Lee K, Park SJ, Kim Y-O, Kang H (2018) A chloroplast-targeted cabbage DEAD-box RNA helicase *BrRH22* confers abiotic stress tolerance to transgenic *Arabidopsis* plants by affecting translation of chloroplast transcripts. Plant Physiol Biochem 127:336–342

Noman A, Ali Q, Maqsood J, Iqbal N, Javed MT, Rasool N, Naseem J (2018) Deciphering physio-biochemical, yield, and nutritional quality attributes of water-stressed radish (*Raphanus sativus* L.) plants grown from Zn-Lys primed seeds. Chemosphere 195:175–189

Okuzaki A, Ogawa T, Koizuka C, Kaneko K, Inaba M, Imamura J, Koizuka N (2018) CRISPR/Cas9-mediated genome editing of the fatty acid desaturase 2 gene in *Brassica napus*. Plant Physiol Biochem 131:63–69

Paixão JF, Gillet FX, Ribeiro TP, Bournaud C, Lourenço-Tessutti IT, Noriega DD, de Melo BP, de Almeida-Engler J, Grossi-de-Sa MF (2019) Improved drought stress tolerance in *Arabidopsis* by CRISPR/dCas9 fusion with a Histone AcetylTransferase. Sci Rep 9:8080

Panjabi-Sabharwal V, Karan R, Khan T, Pareek A (2010) Abiotic stress adaptation in plants. In: Pareek A, Sopory SK, Bohnert HJ, Govindjee (eds) Physiological, molecular and genomic foundation. Springer, Dordrecht, the Netherlands, pp 177–198

Pareek A, Sopory SK, Bohnert H (2009) Abiotic stress adaptation in plants. Springer, Dordrecht, Switzerland

Park B-J, Liu Z, Kanno A, Kameya T (2005) Genetic improvement of Chinese cabbage for salt and drought tolerance by constitutive expression of a *B. napus* LEA gene. Plant Sci 169:553–558

Park J-J, Dempewolf E, Zhang W, Wang Z-Y (2017a) RNA-guided transcriptional activation via CRISPR/dCas9 mimics overexpression phenotypes in Arabidopsis. PLoS ONE 12:e0179410

Park J-S, Yu J-G, Park Y-D (2017b) Characterization of a drought tolerance-related gene of Chinese cabbage in a transgenic tobacco plant. Hortic Environ Biotechnol 58:48–55

Parkin IA, Koh C, Tang H, Robinson SJ, Kagale S, Clarke WE, Town CD, Nixon J, Krishnakumar V, Bidwell SL, Denoeud F (2014) Transcriptome and methylome profiling reveals relics of genome dominance in the mesopolyploid *Brassica oleracea*. Genome Biol 15:R77

Podda A, Pollastri S, Bartolini P, Pisuttu C, Pellegrini E, Nali C, Cencetti G, Michelozzi M, Frassinetti S, Giorgetti L, Fineschi S (2019) Drought stress modulates secondary metabolites in *Brassica oleracea* L. convar. acephala (DC) Alef, var. sabellica L. J Sci Food Agric 99:5533–5540

Puranik S, Sahu PP, Srivastava PS, Prasad M (2012) *NAC* proteins: regulation and role in stress tolerance. Trends Plant Sci 17:369–381

Rahman M, Khatun A, Liu L, Barkla B (2018) *Brassicaceae* mustards: traditional and agronomic uses in Australia and New Zealand. Molecules 23:231

Rajwanshi R, Kumar D, Yusuf MA, DebRoy S, Sarin NB (2016) Stress-inducible overexpression of Glyoxalase I is preferable to its constitutive overexpression for abiotic stress tolerance in transgenic *Brassica juncea*. Mol Breed 36:76

Rashid M, Guangyuan H, Guangxiao Y, Hussain J, Xu Y (2012) *AP2/ERF* transcription factor in rice: genome-wide canvas and syntenic relationships between monocots and eudicots. Evol Bioinform 8:EBO. S9369

Rayko E, Maumus F, Maheswari U, Jabbari K, Bowler C (2010) Transcription factor families inferred from genome sequences of photosynthetic stramenopiles. New Phytol 188:52–66

Raza A, Ashraf F, Zou X, Zhang X, Tosif H (2020) Plant Adaptation and Tolerance to Environmental Stresses: Mechanisms and Perspectives. In Hasanuzzaman M. (eds) Plant Ecophysiology and Adaptation under Climate Change: Mechanisms and Perspectives I. Springer, Singapore, pp 117–145

Raza A, Mehmood SS, Ashraf F, Khan RSA (2019a) Genetic diversity analysis of *Brassica* species using PCR-based SSR markers. Gesunde Pflanz 71:1–7

Raza A, Mehmood SS, Shah T, Zou X, Yan L, Zhang X, Khan RS (2019d) Applications of molecular markers to develop resistance against abiotic stresses in wheat. In: Hasanuzzaman M, Nahar K, Hossain M (eds) Wheat production in changing environments. Springer, Singapore, pp 393–420

Raza A, Mehmood SS, Tabassum J, Batool R (2019c) Targeting plant hormones to develop abiotic stress resistance in wheat. In: Hasanuzzaman M, Nahar K, Hossain M (eds)Wheat production in changing environments. Springer, Singapore, pp 557–577

Raza A, Razzaq A, Mehmood SS, Zou X, Zhang X, Lv Y, Xu J (2019b) Impact of climate change on crops adaptation and strategies to tackle its outcome: a review. Plants 8:34

Razzaq A, Sadia B, Raza A, Khalid Hameed M, Saleem F (2019a) Metabolomics: a way forward for crop improvement. Metabolomics 9:303

Razzaq A, Saleem F, Kanwal M, Mustafa G, Yousaf S, Imran Arshad HM, Hameed MK, Khan MS, Joyia FA (2019b) Modern trends in plant genome editing: an inclusive review of the CRISPR/Cas9 toolbox. Int J Mol Sci 20:4045

Riechmann JL, Heard J, Martin G, Reuber L, Jiang CZ, Keddie J, Adam L, Pineda O, Ratcliffe OJ, Samaha RR, Creelman R (2000) *Arabidopsis* transcription factors: genome-wide comparative analysis among eukaryotes. Science 290:2105–2110

Saeed M, Noshin I, Abida A, Naveed R, Roomina M, Fatima B, Nazima B (2016) Effect of drought stress on *Brassica* crops and its mitigation by inoculation of PGPR. Int J Biosci 9:282–291

Saeidnia S, Gohari AR (2012) Importance of *Brassica napus* as a medicinal food plant. J Med Plant Res 6:2700–2703

Saha B, Mishra S, Awasthi JP, Sahoo L, Panda SK (2016) Enhanced drought and salinity tolerance in transgenic mustard [*Brassica juncea* (L.) Czern & Coss.] overexpressing *Arabidopsis* group 4 late embryogenesis abundant gene (*AtLEA4-1*). Environ Exp Bot 128:99–111

Sahni S, Prasad BD, Liu Q, Grbic V, Sharpe A, Singh SP, Krishna P (2016) Overexpression of the brassinosteroid biosynthetic gene DWF4 in *Brassica napus* simultaneously increases seed yield and stress tolerance. Sci Rep 6:28298

Saibo NJ, Lourenço T, Oliveira MM (2008) Transcription factors and regulation of photosynthetic and related metabolism under environmental stresses. Ann Bot 103:609–623

Selvaraj MG, Ishizaki T, Valencia M, Ogawa S, Dedicova B, Ogata T, Yoshiwara K, Maruyama K, Kusano M, Saito K, Takahashi F (2017) Overexpression of an *Arabidopsis thaliana* galactinol synthase gene improves drought tolerance in transgenic rice and increased grain yield in the field. Plant Biotechnol J 15:1465–1477

Shao H, Wang H, Tang X (2015) *NAC* transcription factors in plant multiple abiotic stress responses: progress and prospects. Front Plant Sci 6:902

Sharoni AM, Nuruzzaman M, Satoh K, Shimizu T, Kondoh H, Sasaya T, Choi IR, Omura T, Kikuchi S (2010) Gene structures, classification and expression models of the *AP2/EREBP* transcription factor family in rice. Plant Cell Physiol 52:344–360

Shirazi FA, Razi H, Niazi A, Alemzadeh A (2019) Molecular cloning and expression analysis of a stress-responsive *WRKY* transcription factor gene, *BnWRKY57*, from *Brassica napus*. Plant Omics 12:37

Singh B, Bohra A, Mishra S, Joshi R, Pandey S (2015) Embracing new-generation 'omics' tools to improve drought tolerance in cereal and food-legume crops. Biol Plant 59:413–428

Singh D, Laxmi A (2015) Transcriptional regulation of drought response: a tortuous network of transcriptional factors. Front Plant Sci 6:895

Singh VV, Verma V, Pareek AK, Mathur M, Yadav R, Goyal P, Thakur AK, Singh YP, Koundal KR, Bansal KC, Mishra AK (2009) Optimization and development of regeneration and transformation protocol in Indian mustard using lectin gene from chickpea [*Cicer arietinum* (L.)]. J Plant Breed Crop Sci 1:306–310

Song X, Li Y, Hou X (2013) Genome-wide analysis of the *AP2/ERF* transcription factor superfamily in Chinese cabbage (*Brassica rapa* ssp. pekinensis). BMC Genomics 14:573

Specht JE, Chase K, Macrander M, Graef GL, Chung J, Markwell JP, Germann M, Orf JH, Lark KG (2001) Soybean response to water. Crop Sci 41:493–509

SPROUTS B (2018) Effect of mannitol stress on morphological, biochemical and polyphenol parameters in broccoli sprouts (*Brassica oleracea* var. italica). Appl Ecol Environ Res 16:2043–2058

Traka M, Mithen R (2009) Glucosinolates, isothiocyanates and human health. Phytochem Rev 8:269–282

Tran LS, Nakashima K, Sakuma Y, Osakabe Y, Qin F, Simpson SD, Maruyama K, Fujita Y, Shinozaki K, Yamaguchi-Shinozaki K (2007) Co-expression of the stress-inducible zinc finger homeodomain ZFHD1 and *NAC* transcription factors enhances expression of the ERD1 gene in Arabidopsis. Plant J 49:46–63

Trenberth KE, Dai A, Van Der Schrier G, Jones PD, Barichivich J, Briffa KR, Sheffield J (2014) Global warming and changes in drought. Nat Clim Chang 4:17

Tsuchihashi Z, Dracopoli N (2002) Progress in high throughput SNP genotyping methods. Pharmacogenomics J 2:103

Verslues PE, Lasky JR, Juenger TE, Liu T-W, Kumar MN (2014) Genome-wide association mapping combined with reverse genetics identifies new effectors of low water potential-induced proline accumulation in Arabidopsis. Plant Physiol 164:144–159

Wang F, Kong W, Wong G, Fu L, Peng R, Li Z, Yao Q (2016) *AtMYB12* regulates flavonoids accumulation and abiotic stress tolerance in transgenic *Arabidopsis thaliana*. Mol Genet Genomics 291:1545–1559

Wang K, Ding Y, Cai C, Chen Z, Zhu C (2019) The role of C2H2 zinc finger proteins in plant responses to abiotic stresses. Physiol Plant 165:690–700

Warwick S, Gugel R, Gomez-Campo C, James T (2007) Genetic variation in *Eruca vesicaria* (L.) Cav. Plant Genet Resour 5:142–153

Xiong X, Chang L, Khalid M, Zhang J, Huang D (2018) Alleviation of Drought Stress by Nitrogen Application in *Brassica campestris* ssp. Chinensis L. Agronomy 8:66

Xu D-Q, Huang J, Guo S-Q, Yang X, Bao Y-M, Tang H-J, Zhang H-S (2008) Overexpression of a TFIIIA-type zinc finger protein gene ZFP252 enhances drought and salt tolerance in rice (*Oryza sativa* L.). FEBS Lett 582:1037–1043

Xu J, Xing X-J, Tian Y-S, Peng R-H, Xue Y, Zhao W, Yao Q-H (2015) Transgenic *Arabidopsis* plants expressing tomato glutathione S-transferase showed enhanced resistance to salt and drought stress. PLoS ONE 10:e0136960

Xu ZY, Kim SY, Kim DH, Dong T, Park Y, Jin JB, Joo SH, Kim SK, Hong JC, Hwang D, Hwang I (2013) The *Arabidopsis NAC* transcription factor *ANAC096* cooperates with *bZIP*-type transcription factors in dehydration and osmotic stress responses. Plant Cell 25:4708–4724

Xue Y, Warburton ML, Sawkins M, Zhang X, Setter T, Xu Y, Grudloyma P, Gethi J, Ribaut JM, Li W, Zhang X (2013) Genome-wide association analysis for nine agronomic traits in maize under well-watered and water-stressed conditions. Theor Appl Genet 126:2587–2596

Yang W, Liu XD, Chi XJ, Wu CA, Li YZ, Song LL, Liu XM, Wang YF, Wang FW, Zhang C, Liu Y (2011) Dwarf apple *MbDREB1* enhances plant tolerance to low temperature, drought, and salt stress via both ABA-dependent and ABA-independent pathways. Planta 233:219–229

Yoshida T, Fujita Y, Maruyama K, Mogami J, Todaka D, Shinozaki K, Yamaguchi-Shinozaki K (2015) Four *Arabidopsis AREB/ABF* transcription factors function predominantly in gene expression downstream of SnRK2 kinases in abscisic acid signalling in response to osmotic stress. Plant, Cell Environ 38:35–49

Yoshida T, Fujita Y, Sayama H, Kidokoro S, Maruyama K, Mizoi J, Shinozaki K, Yamaguchi-Shinozaki K (2010) *AREB1*, *AREB2*, and *ABF3* are master transcription factors that cooperatively regulate ABRE-dependent ABA signaling involved in drought stress tolerance and require ABA for full activation. Plant J 61:672–685

Yusuf MA, Kumar D, Rajwanshi R, Strasser RJ, Tsimilli-Michael M, Sarin NB (2010) Overexpression of γ-tocopherol methyl transferase gene in transgenic *Brassica juncea* plants alleviates abiotic stress: physiological and chlorophyll a fluorescence measurements. Biochim Biophys Acta (BBA) 1797:1428–1438

Zhang J, Mason AS, Wu J, Liu S, Zhang X, Luo T, Redden R, Batley J, Hu L, Yan G (2015a) Identification of putative candidate genes for water stress tolerance in canola (*Brassica napus*). Front Plant Sci 6:1058

Zhang L, Zhang L, Xia C, Zhao G, Liu J, Jia J, Kong X (2015b) A novel wheat *bZIP* transcription factor, *TabZIP60*, confers multiple abiotic stress tolerances in transgenic *Arabidopsis*. Physiol Plant 153:538–554

Zhang M, Duan L, Zhai Z, Li J, Tian X, Wang B, He Z, Li Z (2004) Effects of plant growth regulators on water deficit-induced yield loss in soybean. In: Proceedings of the 4th international crop science congress, Brisbane, Australia, pp 252–256

Zhang T, Li Q, Diao Z, Li P, Wang J, Zheng S (2019) Effects of salt and drought stresses on antioxidant system and RbohC and RbohF genes expression in *Brassica campestris*. J Appl Ecol 30:969–978

Zhang T, Nie T, Sun W, Shi Z, Wang J (2018) Effects of diverse stresses on gene expression and enzyme activity of glutathione reductase in *Brassica campestris*. J Appl Ecol 29:213–222

Zhang X, Lu G, Long W, Zou X, Li F, Nishio T (2014) Recent progress in drought and salt tolerance studies in *Brassica* crops. Breed Sci 64:60–73

Zhao X, Yang X, Pei S, He G, Wang X, Tang Q, Jia C, Lu Y, Hu R, Zhou G (2016) The Miscanthus *NAC* transcription factor *MlNAC9* enhances abiotic stress tolerance in transgenic *Arabidopsis*. Gene 586:158–169

Zhu M, Monroe JG, Suhail Y, Villiers F, Mullen J, Pater D, Hauser F, Jeon BW, Bader JS, Kwak JM, Schroeder JI (2016a) Molecular and systems approaches towards drought-tolerant canola crops. New Phytol 210:1169–1189

Zhu Z, Sun B, Xu X, Chen H, Zou L, Chen G, Cao B, Chen C, Lei J (2016b) Overexpression of *AtEDT1/HDG11* in Chinese kale (*Brassica oleracea* var. alboglabra) enhances drought and osmotic stress tolerance. Front Plant Sci 7:1285

Rapeseed: Biology and Physiological Responses to Drought Stress

Mojtaba Kordrostami and Mohammad Mafakheri

Abstract Among crops, oilseeds are the second largest food source in the world after cereals. Rapeseed is the third most important oil plant in the world and its extracted oil is used for human and industrial purposes due to its fatty acid composition. Its oil is one of the highest quality edible oils due to its good composition of unsaturated fatty acids and low percentage of saturated fatty acids. Rapeseed accounts for 14.7% of total vegetable oil production in the world, and countries such as China, Canada, India, France, England, Germany, Poland, Sweden, Czech Republic, Pakistan, and Bangladesh are the major producers of rapeseed. Most of the rapeseed production in the world is done in dryland conditions, so the plants response to water stress is crucial. In recent years, canola cultivation has declined due to the lack of rainfall and drought. Thus, one of the important factors that threaten the successful development and cultivation of rapeseed in the world is drought stress. This chapter deals with the study of rapeseed and the effect of drought on its morphological and physiological characteristics.

Keywords Rapeseed · Proline · Oil · Yield

1 Introduction

Oilseeds are the second largest food reserves in the world after cereals (Bernard 2011). These crops, in addition to having rich reserves of fatty acids, also contain protein (McKevith 2005). Rapeseed is one of the world's major oilseeds in recent decades (Beta and Isaak 2016a, b). *Brassica napus* L. (rapeseed) is one of the most important crops grown in the world and has the highest annual cultivation area among the most important vegetable oils globally and is ranked second after soybean

M. Kordrostami (✉)
Nuclear Agriculture Research School, Nuclear Science and Technology Research Institute (NSTRI), Karaj, Iran
e-mail: kordrostami009@gmail.com

M. Mafakheri
Department of Horticultural Sciences, Faculty of Agricultural Sciences, University of Guilan, 41635-1314 Rasht, Iran

M. Hasanuzzaman (ed.), *The Plant Family Brassicaceae*,
https://doi.org/10.1007/978-981-15-6345-4_8

(Hamzei 2011). Rapeseed has high yield potential and high oil content (40–45%) among oilseeds (Wittkop et al. 2009). The level of canola cultivation in the world has increased from 8.2 million hectares in 1970 to more than 34.2 million hectares by 2017 (FAO-Stat 2017). The average yield of this plant is 1–2 tons per hectare, particularly in temperate and cold regions, due to having a longer growth period, it has a higher yield (Wrigley et al. 2004). The yield of rapeseed depends on cultivar capacity, weather conditions, soil type, and crop management (Farre et al. 2001). Genetic and agronomic factors are also determinant of growth and development, and consequently of grain yield (Beta and Isaak 2016a). Adhering to the planting date in terms of adapting the appropriate ambient temperature for germination, providing favorable environmental conditions to increase flowering period and preventing seed filling with early spring heat will increase yield of canola (Jaime et al. 2018).

Rapeseed is an ancient plant and available information indicates that it was cultivated in India 2000 years BC (Gupta 2007). At that time, rapeseed oil was used as a feed oil, edible oil, and herbaceous plant for animal nutrition (Eskin and McDonald 1991). Nowadays, with the use of new varieties, the quality and quantity of oil of this plant has been increased for edible use and is considered as one of the most valuable oilseed crops, especially in temperate regions of the world, due to its diverse and varied uses (Li and Olsen 2016). Canola is the third largest supplier of vegetable oil after soybean and palm oil. It accounts for about 7.14% of the total vegetable oil production in the world (McVetty and Duncan 2015). Erucic acid oil in cultivars is used in soap making, plastics industries, as well as lubricants in industrial machines, and jet engines (due to high heat tolerance) (Nath et al. 2016). The most important producing countries are China, Canada, India, and European countries (France, England, Germany), which together account for 89.4% of total production (Edwards et al. 2011).

2 Cytogenetics of Rapeseed

The botanical relationships between oilseed rape varieties were identified in the 1930s. Basic cytological studies indicate three diploid species, *B. oleracea* n = 9, C; *B. compestris* n = 18, A; and *B. nigra* n = 8, B as well as three amphidiploid species, *B. napus* n = 19, AC; *B. juncea* n = 18, AB; and *B. corinata* n = 17, BC derived from a cross between the abovementioned diploid species (Morinaga 1934). Understanding *Brassica* species relationships will enable plant breeders to produce artificial amphidiploid using diploid species as well as the transfer of genes controlling the desired properties through interspecific crossing (Xu 2010). Nagaharu (1935) demonstrated the relationships between the diploids and the amphidiploid by synthesizing amphidiploid which is schematically illustrated in the U triangle.

3 Botanical Properties of Rapeseed

Rapeseed (*B. napus* subsp. *napus*), is a member of the family Brassicaceae and is an amphidiploid species derived from a cross between *B. oleracea* and *B. campestris* in nature. It is an annual herbaceous plant divided into two types: spring and autumn. The morphological characteristics of this plant are as follows: The rapeseed has a vertical, often long, spindle-shaped main root, the upper part of which is 1–3 cm in diameter and penetrates to a depth of 80 cm. It also has numerous lateral roots that are usually horizontal and penetrate less deeply into the soil. Depth of penetration and expansion of the root system plays an important role in plant drought tolerance and optimum use of moisture stored in the soil. It also protects the plant at high altitude with dense cultivation against strong winds. In heavy clay soils, the root penetration depth is quite limited.

The rapeseed produces the main stem from which many branches branch out. At the end of winter, the main stem begins to elongate and after the flowering of the main stem, the branches begin to elongate. Its degree of branching depends on variety, environment, plant nutrition, farming techniques, and so on. For example, plant density has a significant effect on the amount of branching and height at which the main stem begins to branch. But mainly the lateral branches form in the middle and upper part of the main stem and 8–10 lateral branches branch out from the main stem. The main stem has a nearly circular cross-section. It is vertical and its color is light green which turns yellow over time. Stem height varies from 50–200 cm in different varieties but is usually 80–150 cm in height. Rapeseed is often well regenerated and if it is provided with sufficient nutrients, it can offset the effects of the low number of plants by creating multiple branches.

The leaves are greenish-blue in color and the veins appear in the text. The rosette leaves and the lower leaves of the stem have little trichomes, but the upper and middle leaves lack trichomes, petioles, and lobes, and the edges may be serrated or flat. The leaves are heart-shaped and cover one-third to one-half of the stems at the junction with the stem. Rapeseed leaves are placed alternately on the stems. The number of leaves of the main stem varies from 5–12 in the spring type and up to 40 in the autumn type. The amount of leaf production is related to the flowering period. Leaf fall is common in tropical regions due to some factors such as pests. If leaf fall occurs at the beginning of flowering, it has a negative effect on the final yield but has no significant effect after flowering (Leterme 1988)

4 Qualitative Characteristics of Grain, Meal, and Oil

Canola seed contains 40–45% of oil. Also, 12–16% of rapeseed grains make up the crust. Rapeseed meal accounts for approximately 50–58% of the total dry matter of grains, with 43–38% of protein (Durkee and Thivierge 1975). Fiber constitutes 13% of rapeseed meal. Canola protein is a good combination of essential and nonessential amino acids; So that the amount of lysine, methionine, cysteine, threonine, and tryptophan amino acids are comparable to cereal grains (Diosday et al. 1984). Canola oil with low saturated fatty acids (less than 4% palmitic acid) and relatively high oleic acid (60%) and linoleic acid (10%), is among the most common oils after olive oil, and with the exception of soybean oil, it is the only edible oil that contains significant amounts of alpha-linolenic acid. It also has a good balance in the ratio of linoleic acid to linolenic acid (2:1) (Shahidi 2005). It has been shown that there is a relationship between oil content and protein content of grain and fiber with canola seed color. The seeds with dark brown color had 39.9% oil, 39.3% protein, and 13.7% crude fiber. Seeds with yellowish-brown color have 42.2% oil, 41% protein, and 11.5% crude fiber. Finally, the seeds with yellow color had 45.5% oil, 42.1% protein, and 8.9% crude fiber.

5 Environmental Stressors

The distribution, growth, and production of plants are always restricted by the occurrence of natural disasters in the form of biotic and abiotic stresses (Mahajan and Tuteja 2005). Environmental stresses are one of the most important factors in reducing crop production in the world (Dat et al. 2000). Drought, high and low temperatures, soil salinity, pests, and diseases and weeds adversely affect germination, plant growth, and ultimately crop production (Van den Berg and Zeng 2006).

6 Drought Stress

In agriculture, stress is referred to as deviation from normal growth conditions and involves changes in all physiological processes at different levels whose effects can be reversible and, if continued, may become permanent and irreversible. Drought is one of the most important abiotic stresses for crops (Hasanuzzaman and Fujita 2011; Hasanuzzaman et al. 2012; Bruce et al. 2002). Drought stress or in other words the limitation of available water is the main limiting factor for agricultural production (Geerts and Raes 2009). More than 45% of agricultural land is permanently exposed to drought and 38% of the world's population resides in these areas (Ashraf and Foolad 2007). So in the future, most efforts will be toward producing more products under dehydration conditions (Masoud 2007). In other words, more yield must be produced per drop of water (Zhao et al. 2007). Results of various studies show that plant physiological responses to drought and heat vary depending on species, region, time of occurrence, and severity of these stresses (Wright et al. 1995). Under drought stress conditions, as the soil dries, its osmotic potential becomes more negative. Plants will be able to continue to absorb water from the soil as long as their water potential is lower than the soil water potential. Therefore, plants reduce the harmful effects of dehydration by increasing metabolism and osmotic regulation

(Kordrostami et al. 2019). Drought stress can prevent or even stop one or more physiological activities such as transpiration, photosynthesis, elongation of tissues and organs, or enzymatic activities of the cell (Hasanuzzaman and Fujita 2011; Alam et al. 2014a, b; Hasanuzzaman et al. 2014; Nahar et al. 2017; Hasanuzzaman et al. 2018; Kordrostami and Rabiei 2019a). In addition to the severity of the stress and its duration, the growth stage in which the plant is stressed is also important (in terms of its impact on plant growth and yield) (Karafyllidis et al. 1996). Drought resistance is achieved by multiple and often independent morphological and physiological properties, whose interaction has not yet been adequately studied (Hua et al. 1997). In harsh environmental conditions, the amount and rate of photosynthetic storage or assimilate transfer depend on the reaction rate of the plant, and the efficiency of the plant's enzymatic-hormonal and vascular system and the interaction of these factors through the rate and duration of grain filling play a key role in grain yield stability (Papakosta and Gagianas 1991). Under drought stress conditions, partial closure of the stomata reduces transpiration more than photosynthesis and thus increases water use efficiency, but severe stress causes complete closure of the stomata and water use efficiency due to lower photosynthesis, decreases yield (Kordrostami and Rabiei 2019a). Much phenotypic and genetic variation have been reported for the grain yield of *Brassica* species under drought stress conditions (Enjalbert et al. 2013).

Under stress conditions, dry matter depletion can be reduced cellular turgidity due to decreased leaf area of the plant as well as reduced photosynthetic rate due to biochemical constraints caused by water scarcity such as reduction of photosynthetic pigments, especially chlorophylls (Lawlor and Cornic 2002).

Drought affects the plant's performance through the following three mechanisms: (A) Decrease in photosynthetic active radiation (PAR) uptake by canopy resulting from leaf area depletion results in widespread wilting under extreme stress conditions and eventually premature aging of plant leaves (Earl and Davis 2003). (B) Reduction of radiation use efficiency; this reduction can be measured by measuring the dry matter accumulated in the unit of light absorbed over a specific time period (Stone et al. 2001). (C) Immediate decrease in carbon gas exchange absorbed per unit light (Kramer 1983).

Since plants cannot escape from different environmental stresses, they need mechanisms to identify and respond to stresses (Ghasemi et al. 2018; Kordrostami and Rabiei 2019b). One of these mechanisms is an osmotic adjustment. Osmotic adjustment is a type of adaptation to water scarcity stress that, through the accumulation of soluble substances within cells, can lead to the maintenance of cell turgor and its dependent processes at low water potentials (Kordrostami and Rabiei 2019a). This regulation is achieved by producing most of the different types of organic matter such as proline, protein, betaine, and soluble sugars in roots and shoots (Kordrostami et al. 2019). Osmotic adjustment by storing soluble solutes such as sugars, organic acids, and ions (especially K^+) by cells is a process that can lower the cell's water potential without lowering the pressure potential (Abdel Latef et al. 2019).

7 Drought Resistance Mechanisms

Drought resistance is defined as the ability of a plant to survive long-term drought and is usually applied in three ways (Levitt and Levitt 1987): (A) drought escape; (B) drought avoidance; (C) drought tolerance. Drought resistance is defined as the ability of a cultivar to produce more economically (or better life-sustaining) than other cultivars in situations where these cultivars are exposed to soil or atmospheric drought (Kordrostami and Rabiei 2019a).

Drought escape is an important strategy for plant phenology modification in areas where short growing season and late drought stress are dominant. Rapid shading development is essential for the proper use of environmental factors such as light, water, and temperature, and for areas where drought stress decreases during the growing season, stable varieties can be produced by selecting early varieties. Thus, the short-term growth period in areas with the possibility of early drought is considered an advantage, which reduces water consumption due to reduced leaf area (Arrandeau 1989). The drought escape mechanism is adopted by plants that complete their life cycles during the wet season before drought begins.

Plants can prevent drought by utilizing deep and well-developed roots, proper shade structure and surface, leaf angle and movement, thick cuticle, leaf surface adjustment and stomatal closure in hot and dry hours, and osmotic pressure. The ability of a plant to withstand rainless periods by keeping its water levels high is called drought avoidance (Levitt and Levitt 1987). The ability of a plant to survive periods of no rain and tolerate water deficiency in tissues is called drought tolerance, which is typically associated with osmotic adjustment. Soluble substances such as sugars, amino acids, and ammonium compounds protect the membrane and enzymes from damage depending on the cell's ability to maintain the cell membrane under normal conditions and to prevent alteration of the nature of the protein. Soluble compounds such as sugars, amino acids, and ammonium compounds protect the membrane and enzymes from damage depending on the cell's ability to maintain the cell membrane under normal conditions and to prevent alteration of the protein structure. The plant may survive in spite of its internal moisture, which is termed the drought-tolerant mechanism, and the plant with this mechanism is able to regenerate and grow by providing moisture. Drought tolerance is one of the most important plants' mechanisms to cope with drought stress (Kordrostami and Rabiei 2019a). Water shortage tolerance means the least reduction in all plant life processes compared to when the plant has enough water (Levitt and Levitt 1987).

Correlations between traits and morphology indicate that most plants prefer the mechanism of drought escape through higher growth rate, early flowering, and early maturity. However, in the case of severe drought, this interpretation is reversed because, although the plants may be alive, they cannot enter the reproductive stage and produce a yield in terms of drought (Chugh et al. 2011). In such situations, drought avoidance is important, so that the plant can withstand periods of no rain or drought by maintaining a turgor state by altering the root system or regulating leaf area. However, a combination of avoidance and tolerance mechanisms is needed

(Khan et al. 2018). Stress tolerant genotypes are selected in two ways: direct (the measurement of yield) and indirect (the measurement of morphological and physiological traits related to stress tolerance). One of the important issues in evaluating cultivars for drought tolerance is the quantitative measurement of drought tolerance indices. Fernandez (1993) divided the plant response into four groups when studied the performance of mung bean (*Vigna radiata* L.) genotypes in two environments (drought stress and non-stress conditions), high yield genotypes in both environments (Group A), high yield genotypes under normal conditions (Group B), high yield genotypes under stress conditions (Group C), and genotypes with low performance in both environments (Group D).

8 Effects of Drought Stress on Morphological Characteristics of Rapeseed

It is believed that plant height per se does not have a specific effect on aquatic relationships in the plant and determining the appropriate height for the plant under drought stress conditions is of interest with regard to other agronomic traits (Blum and Sullivan 1986). The measured canopy temperature of the plant during the first few days of drought can be used as a selection index for drought tolerance in oily species of *Brassica* (Kumar and Singh 1998). The canopy temperature corresponds to the drought stress, so that, following a decrease in available soil water, the water potential of the plant is reduced and its transpiration is reduced, and, based on the energy balance at the leaf surface, the decrease in transpiration results in an increase in canopy temperature (Cárcova et al. 1998).

9 Effect of Drought Stress on Seed Oil and Protein Content

Oil is the most valuable component of rapeseed and although its amount and composition are largely genetically determined (Fieldsend et al. 1991), it is significantly affected by environmental conditions such as heat and drought (Gunasekera et al. 2006; McCartney et al. 2004). It was observed that oil content and grain weight increased with increasing irrigation rate, so environmental conditions, especially soil moisture, could accurately determine the relationship between grain oil content and grain weight for use in breeding programs (Krogman and Hobbs 1975). In a study by Jensen et al. (1996), it was found that mild drought stress had little effect on oil content, but in severe stress, oil content decreased. Greenhouse analysis showed that drought stress before and after flowering increased glucosinolate content in rapeseed (Masoud 2007). Fanaei et al. (2014) and Daneshmand et al. (2008) reported that there was a significant correlation between oil yield and grain yield and oil content (%), and the cultivars with high grain yield had approximately the same high oil yield per unit area. Downey (1983) reported that drought and high temperatures reduced the unsaturated fatty acids in rapeseed oil. In another report by Rao and Mendham (1991), the effect of supplemental irrigation (an indicator of improved drought stress) on oil content is shown in *B. rapa* and *B. napus* (increased from 46.3 to 48.9% and from 47.4 to 51%, respectively). Increasing water reduces the average amount of oil (Potfer et al. 1988). In arid regions, a positive correlation was observed between rainfall after pollination and oil content (Pritchard et al. 2000).

Water deficiency causes the expression of specific genes and the expression of these genes is associated with adaptation responses of plants under stress (Shinwari 1999). Water stress due to changes in gene expression in plants, in spite of the decrease in total protein synthesis, produces specific proteins that protect cells and the plant from stress. These proteins include chaperones, LEAs, osmotins and antifreeze proteins, RNA-binding proteins, key enzymes for the biosynthesis of osmolytes, water channel proteins, proline and sugar carriers, ROS enzymes, and various proteases (Kordrostami et al. 2019). LEA proteins, mRNA-binding proteins, and chaperones are involved in protecting macromolecules such as enzymes, lipids, and mRNA from dehydration (Yamaguchi-Shinozaki and Shinozaki 2001). Production of other proteins has also been reported against water deficit stress such as responsive to ABA, dehydrins RABs, and vegetative storage proteins in various plants (Artlip and Funkhouser 1995). Shabani et al. (2010), with a two-year trial on the effect of water deficit stress on rapeseed, reported that drought stress was inversely correlated with seed yield. The effect of drought on canola protein percentage also varied, but some reports have reported an increase in seed protein content under drought stress conditions. However, a negative correlation between oil content and grain protein was reported with emphasis on the oil content and protein content of rapeseed (Si et al. 2003).

10 Effect of Drought Stress on Physiological Traits

One of the important factors on canola yield under drought stress conditions is the effect of early aging and chlorophyll content on seed storage material. The elimination of chlorophyll reduces the production of sap. The findings of Kaiser (1987) also show a sharp decrease in RWC (leaf relative water content) under severe stress which causes cell death. Mild stress increases and severe stress decreases chlorophyll content per leaf area.

Sugars can maintain high turgor pressure as an immune system in times of stress (Kishor et al. 2005). In the study of the effect of stress on diploid and amphiploid rapeseed cultivars, it was concluded that leaf water-soluble sugars were higher in the tolerant cultivars in water deficit stress. But in extreme stress due to the decrease in extracellular water content, the soluble sugars content in susceptible cultivars is higher (Ahanger et al. 2017). Soluble sugars act as osmotic regulators, stabilizing cell membranes, and maintaining cellular turgor. In fact, in plants that accumulate soluble sugars in response to drought stress, osmotic adjustment can be done better.

The amount of soluble sugar and starch increases when stress occurs, and as the stress continues, the amount of sugar decreases due to consumption and starch becomes simpler components, and its amount decreases.

Proline is one of the osmotic regulators found in many higher plants and usually accumulates in large quantities in response to environmental stresses. Rapid breakdown of proline after stress conditions have ended may itself provide the factors required for mitochondrial oxidative phosphorylation and ATP production to repair stress-induced injuries. The special role of proline in drought-tolerant plants has been demonstrated, and its accumulation in the tolerant plants is more than sensitive ones (Bates et al. 1973). Under moderate or severe stress conditions, proline concentrations increase with respect to other amino acids. Proline acts as a nitrogen storage tank or soluble substance that reduces the osmotic potential of the cytoplasm and helps the plant to tolerate stress. The results of Mirzaee et al. (2013) showed that drought stress increased the proline concentration of leaves and roots of SLM-046 and Hyola 308 cultivars, but under stress conditions, proline content in SLM-046 was higher than Hayola 308.

11 Effect of Drought Stress on Yield and Yield Components

Rapeseed yield depends on density, number of siliques per plant, number of grains per siliques, and grain weight (Angadi et al. 2003). Flowering stage and silique formation are the most sensitive stages to drought stress (Sadras and Milroy 1996), which can be due to the susceptibility of pollen grains development, pollination, and fertilization to stress (Hall 1992). Blum and Sullivan (1986) found that drought stress at the reproductive stage reduced canola reproductive organs including the number of siliques per plant, number of seeds per silique, and grain weight. Masoud (2007) observed that water deficit during flowering to maturity stage reduced grain yield, biological yield, and number of silique per plant, but the number of grains per silique was not affected. A study by Jensen et al. (1996) showed that late-season drought stress in *B. napus* and *B. rapa* caused more than half of the siliques to fall, while the remaining siliques had more seeds with more weight. Sierts et al. (1987) reported that the growing silique wall competes strongly with the developing seeds for the uptake of photosynthetic materials, which increases with increasing environmental stresses, leading to a decrease in canola seed yield through the siliques fall. The existence of a linear correlation between the number of true siliques with cumulative dry matter to the end of the flowering stage in *B. napus* indicates that this ratio could be an indicator of the effect of stress on the yield of rapeseed. Sims et al. (1993) reported that the yield of rapeseed in Montana increased with an increasing amount of available water. Decreasing water content in rapeseed at the flowering stage decreases the number of siliques per plant, but supplemental irrigation of rapeseed increases silique number by prolonging flowering periods, which is due to higher leaf area at the flowering period (Mendham and Salisbury 1995). In a study, drought stress was investigated on yield and yield components in three rapeseed and mustard cultivars. The results showed that the number of silique in rapeseed is one of the components sensitive to drought stress (Wright et al. 1995). The number of silique and the number of seeds per silique decreased with increasing moisture deficiency. It seems that continued stress during flowering and silique development stages leads to a lack of fertilization and silique formation, and to some extent, an increased percentage of aborted siliques and finally siliques fall. Drought stress at flowering stage also significantly reduces the number of siliques per plant, which is due to the deficiency in the production and supply of photosynthetic materials at the time of stress and insufficient photosynthetic materials to be appropriately allocated to produced and growing siliques, resulting in their loss and reduction of this trait in rapeseed. In general, the persistence of drought stress during the grain filling period by accelerating aging reduces grain yield.

According to Rao and Mendham (1991), grain weight, which is mainly determined at the stage of grain filling, is most influenced by genetic structure, but environmental factors also influence it. However, the contribution of assimilate transfer to grain filling under drought conditions increases. Effects of drought stress on grain filling rate and duration and weight loss have been reported and some have found it to be effective on grain growth rate (Haq et al. 2014).

Harvest index is a measure of the seed weight to the total weight of plant ratio and usually high-yielded rapeseed have higher harvest index (Sana et al. 2003). This index showed a significant positive correlation with drought tolerance of chickpea genotypes under drought conditions (Sánchez et al. 1998). The results of Jensen et al. (1996) on canola showed that water deficit at vegetative growth, flowering, and grain filling stages significantly reduced harvest index. In a study by Shabani et al. (2010), control treatment (full irrigation) and stress treatment at the flowering stage had the highest and rainfed treatment had the lowest harvest index.

12 Conclusion

Rapeseed with extensive genetic resources and high diversity for traits such as cold tolerance, salt tolerance, high yield, and high oil and protein content can be effective in freeing countries from oil imports. One of the goals of plant breeding is to increase yield under various environmental stresses including water scarcity (Clarke et al. 1992). The most severe effects of drought stress are observed when water scarcity occurs during flowering and accelerates grain ripening (Gan et al. 2004; Masoud 2007). Grain yield, as the most important index for selection of drought-tolerant cultivars, is influenced by many genetic and environmental factors, making it difficult to obtain suitable genotypes. Drought-tolerant varieties are relatively resistant to drought and have no significant yield loss (Clarke et al. 1992). One of the ways to cope with drought stress is to breed tolerant and premature plants and it is important to understand how each plant or genotype copes with stress (Khan et al. 2018).

References

Abdel Latef A, Kordrostami M, Zakir A, Zaki H, Saleh OM (2019) Eustress with H_2O_2 facilitates plant growth by improving tolerance to salt stress in two wheat cultivars. Plants 8:303

Ahanger M, Akram N, Ashraf M, Alyemeni M, Wijaya L, Ahmad P (2017) Plant responses to environmental stresses—from gene to biotechnology. AoB Plants 9:plx025

Alam MM, Hasanuzzaman M, Nahar K, Fujita M (2014a) Exogenous jasmonic acid modulates the physiology, antioxidant defense and glyoxalase system in imparting drought stress tolerance in different Brassica species. Plant Biotechnol Rep 8(3):279–293. https://doi.org/10.1007/s11816-014-0321-8

Alam MM, Hasanuzzaman M, Nahar K, Fujita M (2014b) Alleviation of osmotic stress in *Brassica* napus, B. campestris, and *B. juncea* by ascorbic acid application. Biol Plant 58(4):697–708. https://doi.org/10.1007/s10535-014-0447-0

Angadi S, Cutforth H, McConkey B, Gan Y (2003) Yield adjustment by canola grown at different plant populations under semiarid conditions. Crop Sci 43:1358–1366

Arrandeau MA (1989) Breeding strategies for drought resistance. In: Baker EWG (ed) Drought resistance in cereals. CAB International, UK, pp 107–116

Artlip T, Funkhouser E (1995) Protein synthetic responses to environmental stresses. Marcel Dekker, Handbook of plant and crop physiology New York, pp 627–644

Ashraf M, Foolad M (2007) Roles of glycine betaine and proline in improving plant abiotic stress resistance. Environ Exp Bot 59:206–216

Bates L, Waldren R, Teare I (1973) Rapid determination of free proline for water-stress studies. Plant Soil 39:205–207

Bernard JK (2011) Oilseed and oilseed meal. Encyclopedia of dairy sciences 2:349–355

Beta T, Isaak C (2016a) Grain production and consumption: overview. In: Wrigley C (ed) Encyclopedia of food grains. Academic Press, UK, Oxford

Beta T, Isaak C (2016b) Grains around the world. In: Wrigley C (ed) Encyclopedia of food grains. Oxford, Academic Press, UK, pp 349–428

Blum A, Sullivan C (1986) The comparative drought resistance of landraces of sorghum and millet from dry and humid regions. Ann Bot 57:835–846

Bruce WB, Edmeades GO, Barker TC (2002) Molecular and physiological approaches to maize improvement for drought tolerance. J Exp Bot 53:13–25

Cárcova J, Maddonni G, Ghersa C (1998) Crop water stress index of three maize hybrids grown in soils with different quality. Field Crops Res 55:165–174

Chugh V, Kaur N, Gupta AK (2011) Evaluation of oxidative stress tolerance in maize (*Zea mays* L.) seedlings in response to drought. Indian J Biochem Biophys 48:47–53

Clarke JM, DePauw RM, Townley-Smith TF (1992) Evaluation of methods for quantification of drought tolerance in wheat. Crop Sci 32:723–728

Daneshmand A, Shiranirad A, Nourmohammadi G, Zareei G, Daneshian J (2008) Effect of irrigation regiems and nitrogen levels on seed yield and seed quality of two rapeseed (*Brassica napus* L.) cultivars. Iranian J Crop Sci 10(3):244–261

Dat J, Vandenabeele S, Vranová E, Van Montagu M, Inzé D, Van Breusegem F (2000) Dual action of the active oxygen species during plant stress responses. Cell Mol Life Sci 57:779–795

Diosday L, Tzeng YM, Rubin L (1984) Preparation of rapeseed protein concentrates and isolates using ultrafiltration. J Food Sci 49:768–770

Downey R (1983) Origin and description of the *Brassica* oilseed In: Kramer, JKG, FD Sauer and WJ Pigden, (eds). High and low erucic acid rapeseed oils production, usage, chemistry and toxicological evaluation. Academic press, Toronto, Canada

Durkee A, Thivierge P (1975) Bound phenolic acids in *Brassica* and *Sinapis* oilseeds. J Food Sci 40:820–822

Earl HJ, Davis RF (2003) Effect of drought stress on leaf and whole canopy radiation use efficiency and yield of maize. Agron J 95:688–696

Edwards D, Batley J, Parkin I, Kole C (2011) Genetics, genomics and breeding of oilseed *Brassicas*. CRC Press

Enjalbert J-N, Zheng S, Johnson JJ, Mullen JL, Byrne PF, McKay JK (2013) *Brassicaceae* germplasm diversity for agronomic and seed quality traits under drought stress. Ind Crops Prod 47:176–185

Eskin N, McDonald B (1991) Canola oil. Nutr Bull 16:138–146

Fanaei H, Rad M, Ghassemi M (2014) Evaluation of seed yield, yield components and tolerance to drought stress of spring canola genotypes. Seed Plant Improv J 30:269–287

FAO-Stat (2017) Food and agriculture organization of the united nations. FAOSTAT database

Farre I, Robertson M, Walton G, Asseng S (2001) Yield analysis of canola in a variable environment. In: 12th Biennial Australian research assembly on *Brassicas* Victoria. Geelong, Australia. 2–5 Oct. Australian assembly on *Brassicas*. Horsham, Australia, pp 130–134

Fernandez GC (1993) Effective selection criteria for assessing stress tolerance. In: Kuo, CG (ed) Proceedings of the international symposium on adaptation of vegetables and other food crops in temperature and water stres., AVRDC Publication, Tainan, pp 257–270

Fieldsend J, Murray F, Bilsborrow P, Milford G, Evans E (1991) Glucosinolate accumulation during seed development in winter sown oilseed rape (*B. napus*). In: Proceedings of 8th international rapeseed congress, pp 686–694

Gan Y, Angadi S, Potts D, McDonald C (2004) Canola and mustard response to short periods of high temperature and drought stresses at different growth stages. Can J Plant Sci 84:697–704

Geerts S, Raes D (2009) Deficit irrigation as an on-farm strategy to maximize crop water productivity in dry areas. Agric Water Manag 96:1275–1284

Ghasemi S, Kumleh HH, Kordrostami M (2018) Changes in the expression of some genes involved in the biosynthesis of secondary metabolites in *Cuminum cyminum* L. under UV stress. Protoplasma 256(1):279–290

Gunasekera C, Martin L, Siddique K, Walton G (2006) Genotype by environment interactions of Indian mustard (*Brassica juncea* L.) and canola (*Brassica napus* L.) in Mediterranean-type environments: II. Oil and protein concentrations in seed. Eur J Agron 25:13–21

Gupta SK (2007) Advances in botanical research: rapeseed breeding, vol 45. Academic press

Hall AE (1992) Breeding for heat tolerance. Plant Breed Rev 10:129–168

Hamzei J (2011) Seed, oil, and protein yields of canola under combinations of irrigation and nitrogen application. Agron J 103:1152–1158

Haq TU, Ali A, Nadeem SM, Maqbool MM, Ibrahim M (2014) Performance of canola cultivars under drought stress induced by withholding irrigation at different growth stages. Soil Environ 33:43–50

Hasanuzzaman M, Nahar K, Gill SS, Fujita M (2014) Drought stress responses in plants, oxidative stress and antioxidant defense. In: Tuteja N, Gill SS (eds) Climate change and plant abiotic stress tolerance. Biological techniques, vol 1, 1st edn. Wiley, Weinheim, pp 209–250

Hasanuzzaman M, Hossain MA, Teixeira da Silva JA, Fujita M (2012) Plant responses and tolerance to abiotic oxidative stress: antioxidant defense is a key factor. In: Bandi V, Shanker AK, Shanker C, Mandapaka M (eds) Crop stress and its management: perspectives and strategies. Springer, New York, pp 261–316

Hasanuzzaman M, Fujita M (2011) Selenium pretreatment upregulates the antioxidant defense and methylglyoxal detoxification system and confers enhanced tolerance to drought stress in rapeseed seedlings. Biol Trace Elem Res 143(3):1758–1776

Hasanuzzaman M, Nahar K, Rahman A, Inafuku M, Oku H, Fujita M (2018) Exogenous nitric oxide donor and arginine provide protection against short-term drought stress in wheat seedlings. Physiol Mol Biol Plants 24(6):993–1004. https://doi.org/10.1007/s12298-018-0531-6

Hua X-J, Van de Cotte B, Van Montagu M, Verbruggen N (1997) Developmental regulation of pyrroline-5-carboxylate reductase gene expression in *Arabidopsis*. Plant Physiol 114:1215–1224

Jaime R, Alcántara JM, Manzaneda AJ, Rey PJ (2018) Climate change decreases suitable areas for rapeseed cultivation in Europe but provides new opportunities for white mustard as an alternative oilseed for biofuel production. PLoS ONE 13:e0207124

Jensen C, Mogensen V, Mortensen G, Fieldsend J, Milford G, Andersen MN, Thage J (1996) Seed glucosinolate, oil and protein contents of field-grown rape (*Brassica napus* L.) affected by soil drying and evaporative demand. Field Crops Res 47:93–105

Kaiser WM (1987) Effects of water deficit on photosynthetic capacity. Physiol Plant 71:142–149

Karafyllidis D, Stavropoulos N, Georgakis D (1996) The effect of water stress on the yielding capacity of potato crops and subsequent performance of seed tubers. Potato Res 39:153–163

Khan A, Pan X, Najeeb U, Tan DKY, Fahad S, Zahoor R, Luo H (2018) Coping with drought: stress and adaptive mechanisms, and management through cultural and molecular alternatives in cotton as vital constituents for plant stress resilience and fitness. Biol Res 51:47

Kishor PK et al (2005) Regulation of proline biosynthesis, degradation, uptake and transport in higher plants: its implications in plant growth and abiotic stress tolerance. Curr Sci 88:424–438

Kordrostami M, Rabiei B (2019a) Breeding for improved crop resistance to osmotic stress. In: Handbook of plant and crop stress, 4th edn. CRC Press, pp 593–602

Kordrostami M, Rabiei B (2019b) Salinity stress tolerance in plants: physiological, molecular, and biotechnological approaches. In: Hasanuzzaman M, Hakeem KR, Nahar K, Alharby HF (eds) Plant abiotic stress tolerance: agronomic, molecular and biotechnological approaches. https://doi.org/10.1007/978-3-030-06118-0_4

Kordrostami M, Rabiei B, Ebadi AA (2019) Oxidative stress in plants: production, metabolism, and biological roles of reactive oxygen species. In: Handbook of plant and crop stress, 4th edn. CRC Press, pp 85–92

Kramer PJ (1983) Problems in water relations of plants and cells. In: International review of cytology, Elsevier, vol 85. pp 253–286

Krogman K, Hobbs E (1975) Yield and morphological response of rape (*Brassica campestris* L. cv. Span) to irrigation and fertilizer treatments. Can J Plant Sci 55:903–909

Kumar A, Singh D (1998) Use of physiological indices as a screening technique for drought tolerance in oilseed *Brassica* Species. Ann Bot 81:413–420

Lawlor DW, Cornic G (2002) Photosynthetic carbon assimilation and associated metabolism in relation to water deficits in higher plants. Plant, Cell Environ 25:275–294

Leterme P (1988) Modelisation du fonctionnement du peuplement de colza d'hiver en fin de cycle: elaboration des composantes finales du rendement. Colza: Physiologie et Elaboaration du Rendement CETIOM:124–129

Levitt J, Levitt J (1987) Responses of plants to environmental stresses, vol 2. Academic Press, New York

Li LF, Olsen K (2016) To have and to hold: selection for seed and fruit retention during crop domestication. In: Current topics in developmental biology, vol 119. Elsevier, pp 63–109

Mahajan S, Tuteja N (2005) Cold, salinity and drought stresses: an overview. Arch Biochem Biophys 444:139–158

Masoud S (2007) The effects of water deficit during growth stages of canola (*Brassica napus* L.). Am-Eurasian J Sustain Agric 2:417–422

McCartney C, Scarth R, McVetty P, Daun J (2004) Genotypic and environmental effects on saturated fatty acid concentration of canola grown in Manitoba. Can J Plant Sci 84:749–756

McKevith B (2005) Nutritional aspects of oilseeds. Nutr Bull 30:13–26

McVetty P, Duncan R (2015) Canola, Rapeseed, and Mustard: for biofuels and bioproducts. In: Cruz VMV, Dierig DA (eds) Industrial crops. Handbook of plant breeding, vol 9. Springer, New York, NY

Mendham N, Salisbury P (1995) Physiology: crop development, growth and yield. In: Kimber D, McGregor D (eds) Brassiva oilseeds: production and utilization, CAB international. Wallingford, Oxon

Mirzaee M, Moieni A, Ghanati F (2013) Effect of drought stress on proline and soluble sugar content in canola (*Brassica napus* L.) seedlings. Iran J Biol 26:90–98

Morinaga T (1934) Interspecific hybridization in *Brassica*. Cytologia 6:62–67

Nagaharu U (1935) Genome analysis in *Brassica* with special reference to the experimental formation of *B. napus* and peculiar mode of fertilization. Jpn J Bot 7:389–452

Nahar K, Hasanuzzaman M, Alam MM, Rahman A, Mahmud JA, Suzuki T, Fujita M (2017) Insights into spermine-induced combined high temperature and drought tolerance in mung bean: osmoregulation and role of antioxidant defense and methylglyoxal detoxification system. Protoplasma. 254(1):445–460

Nath UK, Kim H-T, Khatun K, Park J-I, Kang K-K, Nou I-S (2016) Modification of fatty acid profiles of rapeseed (*Brassica napus* L.) oil for using as food, industrial feed-stock and biodiesel. Plant Breed Biotech 4:123–134

Papakosta DK, Gagianas A (1991) Nitrogen and dry matter accumulation, remobilization, and losses for mediterranean wheat during grain filling. Agron J 83:864–870

Potfer J, Merrien A, Vartanian N (1988) Etude in situ du système racinaire du colza de printemps en condition de sécheresse. Phisiologie et elaboration du rendement du colza dhiver Ed Centre Technique Interprofessionel des Oléagineaux Métropolitains (CETIOM), París:47–53

Pritchard F, Eagles H, Norton R, Salisbury P, Nicolas M (2000) Environmental effects on seed composition of Victorian canola. Aus J Exp Agric 40:679–685

Rao M, Mendham N (1991) Soil–plant–water relations of oilseed rape (*Brassica napus* and *B. campestris*). J Agric Sci 117:197–205

Sadras V, Milroy S (1996) Soil-water thresholds for the responses of leaf expansion and gas exchange: a review. Field Crops Res 47:253–266

Sana M, Ali A, Malik MA, Saleem MF, Rafiq M (2003) Comparative yield potential and oil contents of different canola. Pak J Agron 2:1–7

Sánchez FJ, Ma Manzanares, de Andres EF, Tenorio JL, Ayerbe L (1998) Turgor maintenance, osmotic adjustment and soluble sugar and proline accumulation in 49 pea cultivars in response to water stress. Field Crops Res 59:225–235

Shabani A, Haghighi A, Sepaskhah A, Emam Y, Honar T (2010) Effect of water stress on grain yield, yield components and quality of winter rapeseed (*Brasica napus* L.) cv. Licord. Iran J Crop Sci 12

Shahidi F (2005) Bailey's industrial oil and fat products, vol 1. Wiley, New Jersey

Shinwari Z (1999) Function and regulation of genes that are induced by dehydration stress. Biosci Agric 5:39–47

Si P, Mailer RJ, Galwey N, Turner DW (2003) Influence of genotype and environment on oil and protein concentrations of canola (*Brassica napus* L.) grown across southern Australia Australian. J Agric Res 54:397–407

Sierts HP, Geisler G, Leon J, Diepenbrock W (1987) Stability of yield components from winter oilseed rape (*Brassica napus* L.). J Agron Crop Sci 158:107–113

Sims J et al (1993) Canola variety yield trials. Montana State Univ Ag Expt Stat, *Bozeman*. Montana Agresearch 10:15–20

Stone P, Wilson D, Reid J, Gillespie R (2001) Water deficit effects on sweet corn. I. Water use, radiation use efficiency, growth, and yield. Aust J Agric Res 52:103–113

Van den Berg L, Zeng Y (2006) Response of South African indigenous grass species to drought stress induced by polyethylene glycol (PEG) 6000. South Afric J Bot 72:284–286

Wittkop B, Snowdon R, Friedt W (2009) Status and perspectives of breeding for enhanced yield and quality of oilseed crops for Europe. Euphytica 170:131

Wright P, Morgan J, Jessop R, Cass A (1995) Comparative adaptation of canola (*Brassica napus*) and Indian mustard (*B. juncea*) to soil water deficits: yield and yield components. Field Crops Res 42:1–13

Wrigley CW, Corke H, Walker CE (2004) Encyclopedia of grain science, vol 1. Elsevier Academic Press, Oxford

Xu Y (2010) Molecular plant breeding. CABI Publishing

Yamaguchi-Shinozaki K, Shinozaki K (2001) Improving plant drought, salt and freezing tolerance by gene transfer of a single stress-inducible transcription factor. In: Novartis foundation symposium. Wiley, pp 176–185

Zhao T-J, Liu Y, Yan Y-B, Feng F, Liu W-Q, Zhou H-M (2007) Identification of the amino acids crucial for the activities of drought responsive element binding factors (DREBs) of *Brassica napus*. FEBS Lett 581:3044–3050

Responses and Tolerance of Brassicas to High Temperature

Pushp Sharma

Abstract Crop plants depict high productivity when grown in ambient temperatures, optimum for various growth and physio-metabolic processes. Temperatures higher than optimum (5–7 °C) decreases rate and duration of metabolic process and therefore yield. *Brassica* species are known to be heat-susceptible at seedling emergence if planted at greater than 30 °C mean day/night temperature. Light textures arid soils show very high seedling mortality, leading to lower plant strands. Hence, tolerance to heat at seedling emergence stage in *Brassica* will facilitate an extended planting period increasing the cropped area and utilization of conserved soil moisture received due to late rains. Therefore, under resource constraint farming systems, it is desirable to breed for heat tolerance toward critical sensitive growth stages. High temperature-induced yield losses could be due to accelerated senescence, reduced photosynthetic rates but increased respiration and inhibition of metabolic processes of seed development like starch synthesis. Crop species including *Brassica* adapt to heat stress membrane stability and accumulation. Due to intensive cropping systems planting of Rapeseed mustard gets delayed and the crop experiences high temperature ($\geq$30 °C) during seed filling stage, hampering translocation and resulting in higher shriveled seeds leading to yield penalties. Improved genotypes have been developed in *Brassica juncea* that can tolerate high temperature during germination and later growth stages and should be preferred over other high yielding varieties. Two foliar applications of salicylic acid improved seed filling in *B. juncea*. Compatible solutes are small organic molecules that help in osmotic adjustment under stressful environment added by upregulation of antioxidative enzymes and non-enzymatic biomolecules, the later stabilizes the proteins and lipids of cell membranes thus maintaining osmoregulation by scavenging ROS. Foliar application of the compatible solutes/osmoprotectants can mitigate or reduce the negative effect of various stresses. Genes and the QTLs can be identified associated with complex traits of heat stress by the recent available tools like Genome-wide association studies (GWAS), genotyping by sequencing (GBS) and the next-generation sequencing (NGS).

P. Sharma (✉)
Department of Plant Breeding and Genetics, Punjab Agricultural University, Ludhiana 141004, India
e-mail: pushp20@pau.edu

M. Hasanuzzaman (ed.), *The Plant Family Brassicaceae*,
https://doi.org/10.1007/978-981-15-6345-4_9

Keywords Heat tolerance · Osmoregulation · Climate change · Oxidative stress · Antioxidant defense · Photosynthetic efficiency

Abbreviations

Pn	Net photosynthesis
gs	Stomatal conductance
Tr	Transpiration
Ci	Intracellular CO_2
PSII	Quantum efficiency of photo system II photochemistry
LAI	Leaf area index
PAR	Photosynthetic active radiations
CT	Canopy temperature
CTD	Canopy temperature depression
FC	Field capacity
RWC	Relative water content
WSD	Water saturation deficit
WRC	Water retention capacity
Fo	Minimal fluorescence
Fm	Maximal fluorescence
ARDM	Amount of remobilized dry matter
REE	Remobilization efficiency
RP	Remobilization percentage
SOD	Superoxide dismutase
POX	Peroxidase
CAT	Catalase
GR	Glutathione reductase
GM	Geometric mean
HTI	Heat tolerance index
HSI	Heat susceptibility index
PAU	Punjab Agricultural University
RRS	Regional research station
SDS-PAGE	Sodium dodecyl sulfate-polyacrylamide gel electrophoresis
IAA	Indole acetic acid
GA	Gibberalic acid
ABA	Abscisic acid
SA	Salicylic acid
MDA	Malondialdehyde
24EBL	24-epibrassinolide
HI	Harvest index
BY	Biological yield
HSPs	Heat shock proteins
QTL	Quantitative trait loci

SNPS	Single nucleotide polymorphisms
NGS	Next-generation sequencing
GBS	Genotype by sequencing
GWAS	Genome wide association studies

1 Introduction

In India, among the oilseeds, Rapeseed mustard holds significant importance. Breeding for high yields under stress condition is the major thrust area for the breeders. Heat stress is the temperature higher than the optimal which leads to yield reductions worldwide (Wahid 2007). Therefore, it is considered a major constraint to sustain crop production. Many areas of the world witness heat stress due to increased temperature causing problems in agricultural production (Hasanuzzaman et al. 2012a, 2013b). Heat stress would affect survival, growth attributes and physiological processes of plant species (Hasanuzzaman et al. 2013b, c, 2019; Nahar et al. 2015a, b). The effects are mainly dependent upon temperature, plant species, and their metabolic processes. Cell and tissue death usually occurs when plant tissue surface temperature reaches 48–50 °C. At high temperature, seedlings are more prone to injuries due to direct contact with the soil surface. Soil temperature is generally 3–4 °C higher than the ambient temperature (Sharma unpublished). Reproductive processes are especially sensitive to high temperature, which causes flower abscission, pollen mortality, and poor fruit set (Kaushal et al. 2016). However, such lethal temperature would occur for brief periods, when surviving the heat stress become paramount. Temperature and the duration of exposure define the detrimental effects on crop species (Bita and Gerats 2013). Heat is known to hasten the development of plant specifically, organ size, biomass and yield by shortening seed filling duration leading to a decline in seed weight. Accelerated development may also lead to premature senescence. The impact of high temperature is mostly seen on germination, different phases of growth and development critical being reproductive stage culminating to decline in yield (Fig. 1). Further, physiology and metabolism as affected by heat stress will elucidate the potential mechanisms and various tools for modulating stress. Genetic variations available in the germplasm, land races and wild species for the component traits will pave way for crop improvement. Brief or frequent episodes of heat stress are accompanied by varied changes in morphology, physiology, anatomy and also in biochemical metabolism which alters growth and development ultimately reducing yield. Two types of injuries as reported by Kaushal et al. (2016) are inflicted by high temperature (i) increased membrane fluidity due to protein aggregation and denaturation generally named as direct injury) (ii) enzymes of organelles like chloroplast and mitochondria gets inactivated limiting protein synthesis but accelerating degradation of proteins responsible for disrupting membrane stability referred as indirect injury (Howarth 2005).

High temperature stress	
▪ Seed germination ➢ Seedling mortality ➢ Dry matter declines ▪ Vegetative phase reduced ➢ Phenophaes shortened ▪ Water loss (RWC) ➢ Pn reduced ➢ Dry matter partitioning affected ➢ Remobilization efficiency reduced ▪ Oxidative stress ▪ Seed yield reduced ▪ Oil content /oil yield reduced ▪ Quality affected	▪ Growth parameters and yield components reduced ➢ Seed filling reduced ➢ Translocation affected ➢ More shrivelled seeds ➢ Lesser mature seeds ➢ Forced maturity ▪ Yield reduction

Fig. 1 Effect of high temperature stress on growth and development of Rapeseed mustard

Indian mustard (*B. juncea*), is mostly cultivated as Rabi/winter season crop and according to GOI (2017) its contribution is about 28.6% toward edible oils. *B. juncea* is a C_3 plant and is prone to climatic vagaries leading to a drastic decline in productivity. Rainfed farming in India accounts for 30.7% area and Indian mustard requires 80–240 mm water, mostly available from conserved moisture thereby making this crop well suited in the rainfed cropping system. Crop encounters high temperatures at two stages, initially at the germination stage and second at seed filling commonly referred as terminal stage due to delayed planting under intensive cropping system. In addition to heat stress other abiotic stresses like low light, salinity, dry spells, frost, etc., which further lowers the yields. Rapeseed-mustard is photosynthetically efficient at 15–20 °C. Early temperature rise at the seedling stage and during the reproductive phase limits the rate of photosynthesis. High temperature not only affects germination but increased seedling mortality thereby reducing the crop stand as reported by Azharudheen et al. (2013). Harvesting of rice and cotton in the multiple cropping systems is mainly responsible for delayed planting of mustard in India. So the reproductive phase is exposed to heat stress (Chauhan et al. 2009). Post-anthesis or the seed filling stage mostly encounters terminal heat stress which hampers rate of photosynthesis, translocation of assimilates to the developing seeds/sinks further limiting the synthesis but also triggers the disease incidence along with insect–pest infestation (Tuteja et al. 1996) thus decreasing seed size and seed yield. As reported earlier, late sowing shortens the vegetative stage, delays onset of flowering with a decreased period for seed development (Srivastava and Balkrishna 2003; Dhaliwal et al. 2007). Impact of high temperature on growth parameters i.e., plant height, length of the main raceme, branch number, yield attributes like siliquae on main raceme and seed weight along with seed yield has been reported by our laboratory (Sharma 2014a; Sharma and Sardana 2016). Earlier findings of Young et al. (2004)

and a recent study by Priya (2020) have reported the negative impact of terminal heat stress on seed filling in *B. juncea.* Due to climatic change and increasing temperature at the terminal stage cultivation of Indian mustard is becoming a serious threat to sustaining productivity. There is a dire need to develop heat-tolerant varieties of *B. juncea* as reported by Woods et al. (1991) and later by Chauhan et al. (2009) for cultivation in warm and semi-arid regions in India but also worldwide. Research on abiotic stresses in All India coordinated Research Project on Rapeseed mustard (DRMR 2013, 2015, 2019) has identified some heat-tolerant genotypes like PM25 and RH749 which can tolerate heat shocks (45 °C $\pm$ 1 °C for 4 h) daily for 4 days in controlled conditions and also when field-grown in end September and early October when soil and ambient temperatures are high, NPJ211 for drought and terminal heat stress whereas RH 749, NPJ207 and NPJ213 for terminal heat stress ($\geq$30 °C) based on the physiological trails in different agroclimatic zones in India.

2 Phenological Traits

Thermal timing is regarded as the influence of temperature in plant development. Ontogeny of leaf starting from leaf appearance, rate of elongation and duration of elongation phase increased with high temperature. When the reproductive development ceases, an increase in leaf number occurs without any enhancement in photosynthetic rates. Plant growth predominately is due to cell division followed by cell growth. Both the processes are stimulated by high temperature but the role of temperature in the expansion of leaf area and related dynamics needs further investigation. Initiation of flowering, fruiting, durations of flowering and fruiting are altered by high temperature. Within the optimal range however, time from emergence to initiation of reproductive phase is shorter.

During the development, many processes are prone to high temperature till the embryo formation which initiates with micro and mega sporogenesis, anthesis, post-anthesis includes pollination, fertilization. Fertilization may fail or embryos may abort if any of the processes fail resulting in lesser seeds or grains and ultimately reduced yields. Therefore, during crop development, the reproductive period is more critical if the stresses occur during this phase. Floral abortion enhances if stress occurs before anthesis and even at the time of anthesis with a declining number of seeds in most of the field crops. If temperature stress occurs at the flowering stage, pollen fertility declines poor dehiscence and viability with fewer pollen grains germinating, ovule formation, functioning retarded, hampering seed growth and seed set. Pollen viability and seed setting are vulnerable to even short episodes of heat stress.

Three phasic life cycle of mustard plant includes (i) seed sowing to emergence (ii) seedling emergence to full bloom/flowering (iii) flowering till maturity (Boomiraja et al. 2010). Phenophases are reduced accordingly time to reach maturity is shortened or high temperature hastens maturity. Rise in temperature by 1–2 °C than the optimum reduced the grain filling period thus impacting yield attributes in cereals (Nahar et al. 2010). A decrease in flower buds and abortion has been reported by

Hemantaranjan et al. (2018) even with a small episode of high temperature. Late planted Indian mustard (*B. juncea*) has early onset of flowering due to terminal heat stress according to Chauhan et al. (2009) and later by Rameeh (2012). Sharma and Sardana (2013) reported that the late sown genotypes of *B. juncea* had shorter flowering duration and completed the life cycle in 140 days as compared to normal sown where flowering duration was of 148 days. The time period sowing to heading was minimum in heat-stressed wheat crop (30 °C for 15 days) over non-stressed plants (Balla et al. 2013). Germination, seedling mortality and less number of plants leads to reduced plant population if the high temperature is prevalent during sowing (Azharudheen et al. 2013; Chauhan et al. 2009). *Brassica rapa* sown at 3 sowing dates revealed that 1st October sowing took lesser days for emergence (6.4 days) than 15 October (7.4 days) trailed by 30 October (12.4 days) sowing owing to favorable soil and air temperature on first planted date which declined with late planting. Delayed planting reduced crop duration by shortening vegetative and reproductive phase according to Solanki and Mundra (2015) and endorsed by Alam et al. (2015) where vegetative growth phase of the early sown *B. juncea* witnessed cool temperature and flowered late. In canola cultivars of *Brassica napus* when planted at the first optimum date emerged/germinated in a short time whereas flowering duration was advanced. The trend was reversed in late sowing where days to seedling emergence increased but decreased flowering duration according to Khayat et al. (2015). With the anticipated extreme temperatures owing to increased temperature due to climate change productivity will be enormously impacted. In all the species, pollination is greatly liable to temperature extremities (Hatfield and Prueger 2015). Dehiscence of anthers and endosperm re-absorption highly reduced pollen load with heat stress (Peterson and Murphy 2015). Vegetative and reproductive phases were influenced by different sowing dates in *B. juncea* where vegetative phase was extended by 4–5 days while the reproductive phase was reduced by 8–10 days in late planting over timely sown crop as per the finding of Solanki and Mundra (2015). Delay in sowing at 15 days interval from 15 to 30 October and then 15 November significantly delayed the emergence and required significantly more number of days for flowering but attained early physiological maturity (Kaur H et al. 2018; Kaur L et al. 2018). Late planting affected phenology by delaying flowering initiation and completion leading to a late transition from vegetative to flowering stage owing to low temperature during this phase. Flowering stage in the timely planted crop had an average temperature of 12.9 and 15.7 °C in the late-planted resulting in shortening of flowering duration by 9.5 days at PAU, Punjab Agricultural University, Ludhiana. At RRS, Regional research station, Abohar flowering duration was delayed by 6.9 days where timely sown crop faced 11.9 °C and late sown experienced 14.0 °C average temperature. However, siliquing duration did not vary significantly at the two locations according to the recent study of Priya (2020) in a set of forty-nine advanced breeding material. Late planting in *Brassica campestris* decreased the duration of phenological phases (Akhter et al. 2015). Sowing of Indian mustard (*B. juncea*) on 15 November attained early maturity (131.8 days) comparable to crop sown on 26 October (140.6 days) and 5 November (133.8 days). It may be due to high temperature which reduced the efficiency of photosynthetic translocation and capacity of nutrients uptake and ultimately leads to a reduction in various growth phases and maturity (Khan et al. 2018). Rate and duration of grain/seed filling relate to grain/seed weight or seed size however heat stress accelerated seed/grain filling but limited the duration of grain/seed filling (Girousse et al. 2018). Anthesis and maturity time recorded in days were reduced by drought to 10% and 14% but with high temperature to 16% and 20%, respectively, as reported by Qaseem et al. (2019).

3 Physiological Traits

3.1 Photosynthetic Pigments

Early-onset of leaf senescence due to high temperature controlled the alteration in chlorophyll a/b ratio and chlorophyll decline. Temperature stress produced ROS which degraded the main photosynthetic pigments. In the chloroplast, photosystem II is present on the thylakoid membranes which are temperature sensitive. Thylakoid membranes are disrupted under high temperatures with a decline in activity of PS II (Kim and Portis 2005). Oxygen evolving complex gets interrupted with temperature which disturbs the flow of electrons to the acceptors of the photosystem II reaction center. PSI system is usually more conserved under high-temperature stress than PSII. Either decreased inflow of CO_2 into the mesophyll tissue of leaves is controlled by the closure of stomata or the weakening of metabolic activities with heat stress influences the photosynthetic capacity. Further, with stress caused by impaired photosynthetic electron transport, Rubisco protein added by the reduced enzymatic activity of Rubisco acts as carboxylase/oxygenase. A decrease in photosynthesis was mainly due to a decline in stomatal conductance in the early period of stress. Cs and Pn are mostly down-regulated by stress owing to reduced Rubisco activase enzyme. However, high temperature increased catalytic activity of Rubisco and lowered the affinity between Rubisco enzyme and CO_2 with enhanced oxygenase activity thus net photosynthetic (Pn) rate declined. Respiration rates are generally known to increase with the increasing temperatures but the mechanism controlling and regulating respiration under temperature stress is still not as much known. Therefore understanding these responses is important because photosynthesis is temporal i.e., during day time and confined to only green tissues mainly leaves/pods/siliquae whereas respiration is day and night process of all the organs. Changes in the processes associated with photosynthesis (Fig. 2) correlated well with thermostability and seed yield (Mathur et al. 2014). Investigations have cited the adverse effect of photosynthesis leading to growth restraint in *B. juncea* (Hayat et al. 2009). Photosynthetic restraint may be due to loss of chlorophyll, electron transport and thermostability of PSII (Sinsawat et al. 2004). Growth and development of parental and gametophytic tissues are affected as the photosynthetic rates declined thus restricting the availability of assimilates and depleting the energy reserves causing starvation of tissues (Young et al. 2004). Physiological traits were significantly affected by sowing time

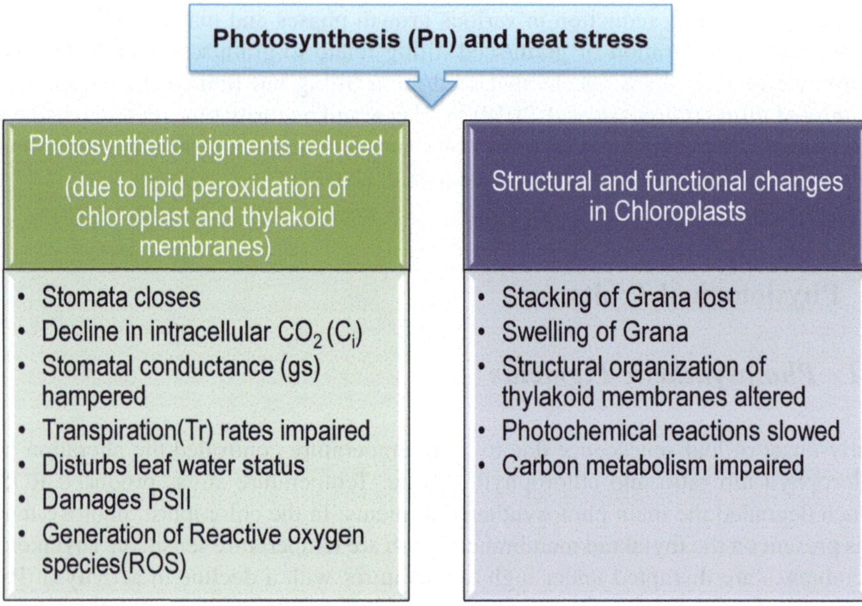

Fig. 2 Heat stress in relation to photosynthetic efficiency

under irrigated condition in *B. juncea* according to Lakra et al. (2018). Detailed study on the chlorophyll, ascorbate and other phytochemicals have been carried out by Krumbein et al. (2005) in *Brassica* species. The enzymes of chlorophyll biosynthesis are inhibited by high temperatures. Heat stress alters the carbon metabolism due to changes in the enzymatic activities which affects the photosynthetic capacity in C_3 and C_4 plants. Thylakoid lamellae of chloroplast where the photochemical reaction occurs and stroma for carbon metabolism are inflicted by heat injuries as per the findings of Wang et al (2010). In *B. juncea*, a decline in chlorophyll content was observed with delayed planting in comparison to normal planting. Carotenoid content also showed decline trends with increased temperature (Kumar et al. 2013). The reduction in photosynthesis resulting from heat stress was mainly due to reduced leaf area, functioning of photosynthetic machinery, and the onset of early senescence culminating in production losses. Photosynthetic pigments especially chlorophyll a (28%), chlorophyll b (24%) and total chlorophyll (25%) suffered decline to variable extent at seedling stage with heat shock of 40 °C, suggesting heat stress caused the structural damages in chloroplast (Gosavi et al. 2014). Reducing trend is supported by the findings of Bali et al (2016) in *B. juncea* upon copper stress and with increasing Cu concentration chlorophyll and carotenoids were reduced. Chlorophyll content was higher (Kavita and Pandey 2017) in tolerant genotypes of *B. juncea*. Jangid and Srivastava (2018) reported that chlorophyll content decreased in plant with successive delayed sowings as compared to optimum sowing. With the growth and development of crop plants toward maturity and also with late planting, canopy temperature increased. Non-destructive assessment of chlorophyll content has been widely reported by SPAD meter (Soil Plant Analysis Development) which determines the chlorophyll content by measuring the absorbance of the leaf and light transmission in different ranges. Leaf SPAD values indicate the ratio of thickness to greenness determined by transmitted light in infrared and red ranges, respectively. Cen et al. (2006) reported decrease in chlorophyll pigment as oilseed rape matured but different results were for SPAD readings because of the nutrients in the leaves which are transported to fruit/siliquae and the nutrient content gets depleted gradually in the leaves. Hayat et al. (2009) reported decreased SPAD by 17.5% in *Brassica* seedling exposed to heat stress over control. Chouksey et al. (2014) reported the reduction in SPAD value at later growth stages (90 and 120 DAS) in comparison to early stages (30 and 60 DAS) in Indian mustard. SPAD values and chlorophyll pigments have decreased at different locations in the late-planted genotypes (DRMR 2015, 2019) of Indian mustard. Each successive fortnight delay in sowing from 15 to 30 October and further to November 15 significantly reduced leaf area index, SPAD value when recorded at two stages i.e., 40 and 120 DAS and photosynthetic active radiations intercepted at 40 and 80 DAS in *B. napus* (Kaur H et al. 2018; Kaur L et al. 2018). Photosynthetic pigments, SPAD value, increased from vegetative to flowering and declined at siliquing stage in the advanced set of breeding lines indicated the best expression of genotypes at flowering stage (Priya 2020). The decline in total chlorophyll content was 14.7%, 13.1%, and 24% at vegetative, flowering and siliquing stage respectively. However, carotenoids were reduced to 15.5, 13.7, and 29.2% in the successive stages. Leaf greenness was also measured by SPAD meter with similar trends reported at different growth stages in *B. juncea* genotypes. In the temperature gradient tunnels, impairment in senescence-related metabolism contrasting results was recorded. Pigments and SPAD values increased in the genotypes planted in the tunnels as compared to control/open field planted genotypes at flowering and siliquing stages.

3.2 Leaf Area Index and Photosynthetic Active Radiation

LAI denotes the leaf area to ground area and PAR as the radiations between 400 and 700 nm wavelength of the natural incoming sunlight that are absorbed by green plants and used for photosynthesis to produce plant biomass. Crop canopy intercepts PAR according to the total LA. LA denotes size for light-harvesting and the capacity of the crop for conversion into assimilates or assimilatory capacity. The crop will deflect more LAI and PAR, if growth is rapid. The lower value for LAI during the vegetative phase, higher during flowering and moderate during siliquing was reported by Panda et al. (2004) in Indian mustard. Early sown crops efficiently converted PAR into dry matter (Singh and Singh 2006). Terminal heat stress according to Shengri et al. (2012) decreased LAI from 3.7 in optimum condition to 2.5 in stressed *B. napus*. LAI of *B. juncea* increased by 49.1% at 75 DAS over 45 DAS with 50% shading (Sharma et al. 2012a). Genotypes from different agroclimatic zones were tested at different locations for terminal heat stress revealed a decline in LAI and PAR interception in the late-planted genotypes (DRMR 2019). The increasing trend of PAR and LAI in *B. juncea* at 30 DAS and 90 DAS and decreased thereafter according to the findings of Chouksey et al. (2014). PAR interception and LAI improved gradually from vegetative to flowering and declined at siliquing stage LAI reduced by 34.5, 39.6, and 48.2% and PAR interception by 35.8, 34.2, and 30.6% with respect to growth stages with late planting (Priya 2020) and earlier by Kaur and Sharma (2015b) due to water restrictions regimes in two species of *B. juncea a*nd *B. napus.*

3.3 Canopy Temperature

Lower canopy temperature (CT) denotes the cooler environment of the crop due to transpiration which was strongly associated with yield in drought and heat stress because of a common genetic basis (Pinto et al. 2010). Canopy temperature was associated with deeper roots as indicated by the rooting behavior in moisture deficit and temperature stress related studies (Sharma et al. 2015). Genetic variation in stomatal conductance under heat is also linked with the cooler canopy thereby canopy temperature can be selected as physiological marker trait for improving tolerance to temperature (Rhythm 2019). Canopy air temperature depression, CATD or CTD usually computed by subtracting canopy temperature (CT) from the ambient temperature (Rosyara et al. 2008) which results in negative value as the air temperature is higher than the plant canopy. By ignoring the negative value higher value was observed under well-irrigated and non-stressed crops. Additionally, CTD is greatly affected by vapor pressure deficit (Vpd) with lesser impact of solar radiation, ambient temperature and speed with which wind blows (Sharma and Sardana 2017). Environmental factors including the moisture content of the soil, wind along with evaporation, cloudy weather, ambient temperature, RH, low light intensity along with the biological factors like transpiration and metabolism of crop plants greatly affects temperature depression. The higher canopy temperature was reported in water stress conditions as compared to non-stressed control condition in *Brassicas*. Probably when the stomata close transpiration declines and respiration increases. Canopy temperature was higher in normal planted cultivars but CTD was higher under late planting (Anonymous 2019; DRMR 2019). However, contradictory results were reported by Priya (2020) where canopy temperature (CT) was 4.1 °C higher under late planting whereas canopy temperature depression (CTD) was higher in timely planted genotypes. Late-planted genotypes had higher canopy temperature depression indicating cooler canopies. CTD is highly correlated with yield and is identified as escape mechanism. Evaluation of mustard genotypes revealed both escape and true tolerance. Certain genotypes maintained elevated leaf temperature with not much decline in yield that can be exploited during breeding for high temperature stress (Anonymous 2017).

3.4 Plant Water Status

Change in the air temperature is responsible for variations in the water status of the plant. Martinez-Ballesta et al. (2015) suggested that under elevated temperature the hydraulic conductivity increased which resulted in increased aquaporin activity; hence membrane became more permeable as water flows through pores. The reason lies with the heat stress wherein the gradient, driving force for water flow into the flowers or the grains gets disrupted which dehydrates the flower and grains leading to increased membrane permeability. Water deficit occurs in the plants during the daytime due to increased transpiration which decreases the water potential and disrupts the physiological processes (Tsukaguchi et al. 2003; Blum 2005; Kauser et al. 2006). Water potential (Ψ) and the relative water content (RWC) in leaves of Indian mustard genotypes declined with increased leaf temperature eventually decreasing the photosynthetic efficiency has been documented by Kumar et al. (2018a, b). Earlier study of Ram et al. (2014) reported that the germplasm accession which had higher leaf RWC were good indicator of heat and drought resistance. Aneja et al. (2015) studied the rapid decline in RWC when Indian mustard was subjected to drought conditions where osmotic potential became more negative in stressed genotypes. RWC significantly decreased (Kaur 2017) with an increase in heat stress and water saturation deficit (WSD) increased with an increase in heat stress in mustard seedling. The balance between the leaf water supply and transpiration rates is manifested by excised leaf water loss (ELWL). Genotypes with lesser ESWL under stress could maintain leaf water balance which seems to be attributed to stress tolerance indicator (Kaur and Sharma 2015c). Genotypes maintained higher relative water content (RWC) and possessed greater water retention capacity (WRC) under irrigated condition whereas higher water saturation deficit (WSD) was in water-stressed mustard as reported by Tasmina et al. (2016). Late planting reduced RWC to variable extent with an increased water saturation deficit in *B. juncea* (Anonymous 2015, 2019). Late sown *B. juncea* (November 21) had the lowest leaf water potential, osmotic potential, and relative water content whereas optimum sown crop (October 16) had high osmotic and water potentials along with relative water content. Further, in the late-planted mustard crop on 21 November witnessed low temperature which slowed down the physiological processes, decreased water absorption besides poor root growth. The temperature was favorable on 16 October sowing for root growth because of higher RWC along with water and osmotic potentials (Kumar et al. 2013). RWC declined under late planting by 23.0% whereas water saturation deficit and relative saturation deficit increased due to delayed sowing by 16.7% and 15.4%, respectively, indicating higher transpiration rates/water loss (Sharma and Sardana 2017; Priya 2020).

3.5 Membrane Stability and Injury

The structure and function of the membranes are disrupted by high temperatures leading to increased fluidity. The functioning of the membrane in the plant cell is important for photosynthesis and respiration. Commonly used assays for testing membrane damage are the cell membrane stability assay and the chlorophyll fluorescence. Leakage of ions from the cell leads to loss of membrane integrity as membrane-bound enzyme activity gets inhibited under high temperature. Thylakoid membrane damage (Fo/Fm) can be assayed by measuring the ground state of chlorophyll a fluorescence (Fo) and maximum (Fm) using chloro-fluorometer. Thylakoid membranes and plasma membranes are prone to damage by temperature and *Brassica* genotypes are known to differ in their response (Sharma 2015). These differential responses are sensed by changes in chlorophyll fluorescence traits (Sharma and Sardana 2018). Carbon isotopic discrimination studies relate to the ratio of stable carbon isotopes ($^{13}C/^{12}C$) in the plant tissue to be analyzed especially 3rd and 4th leaf on the main raceme of *B. juncea*. Stable carbon isotope discrimination differences were positively correlated with high-temperature tolerance in many crop plants. Further QTLs associated with isotopic discrimination studies can reveal the correlation and contribution of Cs and Pn under temperature stress (Sharma et al. 2012a, b). Persistent cellular functions of the membranes under stress are essential for respiration and photosynthesis. For improving the selection efficiency physiological and biochemical marker screens/traits should complement the breeding methodology. Empirical selection in breeding programs is liable for the loss of heat-tolerant genes (Hemantaranjan et al. 2018). Membrane stability under temperature stress is the reliable selection criteria for heat tolerance and the genetic variations existing in the genotypes can be tapped and used in the breeding programs to develop thermotolerant varieties to be planted in high-temperature prone environments. In the plant cell, plasma membranes are sensitive, liable to heat, and prone to injuries. The phase transition from solid-gel structure to flexible liquid-crystalline structure occurs when the membranes sustain injury due to high temperature. Membrane fluidity increases due to the denaturation of proteins and increase in unsaturated fatty acids. Loss of membrane integrity due to injury inflicted by heat causes electrolyte leakage consisting of both organic and inorganic ions from the tissue (Salvucci and Craft-Brander 2004). Age of the tissue/plant, sample, stage of development, season, and plant species affect the electrolyte leakage. Indian mustard when sown under diverse sowing conditions showed an adverse effect on membrane stability index (MSI) only in the late planting of *Brassica* crop (Kumar et al. 2013). The membrane stability in *B. juncea* ranged from 5.2 to 45.4% and a declining trend was observed under heat stress (Ram et al. 2014). Some researchers have supported similar result findings in Indian mustard (Kavita and Pandey 2017; Wilson et al. 2014; Aneja et al. 2015). Late sowing disrupts membrane stability due to terminal heat stress as indicated by electrolyte leakage studies (Anonymous 2015, 2017, 2018). Recent findings of Priya (2020) revealed reduced stability of membrane in *B. juncea* by 36.3% and increased membrane injury by 18.1% in late sowing. Membrane stability index, acquired thermotolerance, and

RWC had a significant association with seed yield in heat-stressed Indian mustard genotypes along with the stress indices by Singh et al. (2016).

4 Assimilate Partitioning

High-temperature stress decreased the crop growing season by increasing the crop development rate and a direct association exists between temperature and developmental rate former varying from base to optimum temperature. Temperature rise above the optimum considerably affects the rate of development with the decline of photoassimilate production which is more likely to be limiting under high temperature if the stress occurs during the seed filling stage. The reduced photosynthate production will result in a reduced transport of carbon to the developing seeds/sinks when the demand for assimilates is greatest. The aerial dry mass and seed yield are strongly associated which influences harvest index. Yield and yield components remain, however, the most effective tools for stress evaluation. Heat stress not only slows down the assimilate process but during seed filling the uptake of mineral elements as well. Transport of assimilates to the developing kernel and those stored in the vegetative parts are remobilized. So both the direct and remobilized assimilates add to the seed yield. High temperature during the reproductive phase or grain/seed filling significantly impacts the rate and duration of grain/seed filling. Therefore source-sink relationship is critical at these two stages for selecting germplasm which could maintain high yields under temperature stress (Abdelrahman et al. 2019). There are two carbon sources to be used for seed filling firstly assimilates from current photosynthesis and secondly, the stored reserves that are translocated to the seeds from leaves and stem (Yuan et al. 2017). The main source of carbon for seed/grain filling lies in the carbohydrate reserves in the stem with the inhibition of current photosynthesis under stress. Under mild stress, even the current photo assimilates may be inadequate for normal seed/grain filling (Blum 1998). This mechanism depends on the accumulation of reserves before flowering and the remobilization of the reserves during grain filling. Different stresses reduced the assimilates for seed/grain filling which induces the enhanced mobilization of stem reserves to be utilized by the developing grains/seeds. In the stressed crop plants the major source of transported assimilates about 75–100% to the grain yield are the stored carbohydrates. Carbohydrates metabolism especially sucrose synthesis and its utilization has a strong relation with mobilization of stored reserves from both leaves and stem (Prasad et al. 2008). Leaf senescence increased with heat stress while duration of grain filling decreased which further obstructed translocation of photoassimilates to grain, affecting synthesis and accumulation in the grain lowering the kernel weight and yield (Kaur and Behl 2010). The positive association of biomass with grain yield revealed that dry matter is the vital component for the formation of grain and its filling. This could be one way to achieve higher yield under stress. For improved dry matter translocation to the grains, enhanced biomass at anthesis or efficient dry matter translocation is required. Dry matter decreased with delayed planting of *B.*

napus (Kaur H et al. 2018; Kaur L et al. 2018) and in *B. juncea* (DRMR 2018, 2019). High reserve mobilization in the genotypes exhibited yield stability under varied stressed environments. Current photosynthesis contributes to the grain filling and with the limiting assimilates under stress the increased remobilization of stem reserves strengthens the post grain filling. Average stem reserve mobilization under stress condition ranged from 13.3 to 24.9% in a timely sown and 21.3 to 34.8% in late sown (Ram et al. 2018). In another study under normal condition pre-anthesis stem reserves contributed about 11.9–15.1% and 18.7–21.3% to the grain weight in the heat-tolerant and sensitive cultivars, respectively (Sikder and Paul 2010). The effect of continuous heat stress at 30/25 °C on source-sink relation in wheat was studied by Hutsch et al. (2019) during the vegetative phase. Harvesting at four phenological stages starting from three-node stages, initiation of flowering, grain filling, and maturity. Heat stress reduced the developmental phases and aerial dry matter or biomass varying between 19 and 41%. Total shoot dry matter declined by 77% at grain filling and 58% at maturity. Results also revealed that assimilates were not restricted signifying source was not a limiting factor. Sink capacity decreased with heat stress as lesser and smaller kernels were produced. Active transport of hexoses across the plasma membrane to the developing kernel was the main sink limiting step during grain/seed filling (Baghel and Srivastava 2010). Over the years dry matter accumulation declined in stem by 26.7 and 46.4% whereas in leaves by 38.0 and 41.2% at the flowering stage. Stem dry weight was reduced by 47.6 and 23.3% in late planting during the two crop seasons at maturity. The demand for utilizing stem reserve is represented by the amount of remobilized dry matter (ARDM) which was higher (44.2%) in late-planted genotypes in comparison to the genotypes planted at the optimum time. The remobilization efficiency (REE) is the indicator of stem mobilization reserve which increased by 62.5% in the late-planted genotypes; however, stem reserve utilization to the seed or remobilizing percentage (REP) again increased by 83.1% in late planting. The elite genotypes showed higher remobilization and higher stem reserve utilization. Grain filling is the critical stage with regard to terminal heat stress therefore the capability of these genotypes to use stem reserve exists which seems to be an important aspect under abiotic stress particularly under heat and drought stress (Priya 2020).

5 Biochemical Traits

5.1 Total Carbohydrates

Sugar is regarded as a signaling molecule in different abiotic stresses and the increase in total soluble sugars or carbohydrates imparted protection under temperature stress, therefore considered as an adaptive mechanism. Mohamed and Abdel-Hamid (2013) reported the enhanced soluble sugars imparted tolerance to genotypes which were rated as tolerant to heat stress over other genotypes. Starch and soluble sugar estimated in the grains after anthesis and at maturity indicated maximum starch content at 35 days after sowing in field condition in both tolerant and sensitive cultivars whereas under elevated temperature conditions maximum at 30 days after anthesis in tolerant and at 25 days after anthesis in sensitive cultivars. A similar trend was observed in soluble sugar content at the anthesis stage. Starch synthesis ended at maturity due to an ample amount of soluble sugar under elevated temperature in the grains. Soluble sugar could not be converted into starch due to some biochemical constraints. Grain development was affected by heat stress leading to immature grains probably due to enzymatic dysfunction associated with starch synthesis from sugars (Khatun and Ahmed 2015). Upon heat shock, total soluble sugars increased while starch content decreased due to breakdown of starch and conversion to soluble carbohydrates. Identification and categorization of genotypes as tolerant was attributed to higher total soluble sugars and starch relative to the sensitive ones in *B. campestris* (Yuan et al. 2017). Carbohydrates in leaves were reduced by 18.4% and in stem by 27.1% at the flowering stage, however, 44.1% reduction was observed in stem at maturity stage. Mobilization of carbohydrates under terminal heat stress ($\geq$30 °C) in the late planting endorsed the decline of carbohydrates being utilized for seed filling (Priya 2020). Increase in total soluble sugars (28.7%), reducing (33.8%) and non-reducing sugars (20.2%) were recorded under rainfed over irrigated regime (Rhythm 2019) and earlier by Kaur and Sharma (2015a) and in *B. napus* (Kaur 2012) whereas under heat stress only reducing sugars enhanced considerably in mustard cultivars than total soluble and non-reducing sugars (Sharma unpublished data).

Various stresses including biotic as well have a common path of metabolism referred to as oxidative stress which when regulated confers tolerance to the abiotic and biotic stresses in crop plants (Hasanuzzaman et al. 2012b). Toxic reactive oxygen species generated include hydrogen peroxide (H_2O_2), singlet oxygen ($^1O_2$), hydroxyl radical (OH^-), and superoxide radical (O_2^-) causes metabolic oxidative stress as reported by Vasseur et al. (2011). Leakage of electrons invariably from electron transport occurring during the metabolic processes of mitochondria, chloroplasts and at sensor plasma membranes contributes to ROS generation (Blokhina and Fagerstedt 2010). The defense system in the plants comprising of enzymatic and non-enzymatic component referred to as antioxidant system checks the ROS being produced under extreme temperatures and is also helpful in maintaining the redox homeostasis (Hasanuzzaman et al. 2013a). Several studies indicated upregulation of antioxidative defense system including both enzymatic and non-enzymatic antioxidants but the degree and type of expression of the antioxidative defense system depends upon the plant type, intensity and duration of the stresses.

5.2 *Antoxidative Enzymes*

When ROS generation overcomes the scavenging abilities of the defensive antioxidant system consequently leading to extensive cell damage culminating even to cell death under extreme conditions. Under such circumstances, manipulation of defense-related genes along with the exogenous application of protectants in the form of foliar sprays can overcome and regulate the antioxidative defense mechanism induced by heat stress oxidative damage. Numerous research studies positively correlated heat tolerance with the increased antioxidants in the plants (Gill and Tuteja 2010; Hasanuzzaman et al. 2019). Protective role of antioxidant enzymes in heat-stressed wheat was evaluated at three stages viz. vegetative, anthesis, and 15 days after anthesis. Significant increased activities of SOD, PX, and CAT while GR and POX decreased at all the stages in late and very late sowings over normal sowing (Almeselmani et al. 2006).

At all the studied growth stages activities of APX, SOD, CAT, GR, and POX increased in heat-tolerant wheat cultivars than the susceptible ones. Drought stress enhanced the SOD (6.60 units/mg protein) and POX activity (3.90 units/mg protein) in 30% FC (Field capacity) as compared to 60% FC in *B. napus*. Plant's ability to overcome oxidative stress is partially determined by the SOD activity followed by the upregulation of the other down-regulated antioxidative enzymes (Abedi and Pakniyat 2010). CAT and SOD are the most active enzymes activated in response to stressed environment (Wojciechowska et al. 2013). The activity of antioxidant enzymes particularly SOD, CAT and POX increased under heat stress in *B. juncea* but the increase was significantly higher in the tolerant genotypes (Rani B et al. 2013, Rani T et al. 2013). In response to oxidative stress, SOD is considered as the first line of defense according to Zhang et al. (2015) Superoxide dismutase catalyzed the partitioning of O_2^- into ordinary molecular O_2 or into damaging H_2O_2 which is detoxified by peroxidase and catalase. Catalase neutralizes the excessive ROS generated by stress whereas peroxidase fine-tunes the modulation of ROS signaling. Rate of antioxidative enzymes in *B. juncea* in response to salt and temperature stress of 44 °C were studied by Kaur H et al. (2018), Kaur L et al. 2018) and the results revealed enhanced activity of SOD by 61.6%, CAT by 64.7%, and APX by 61.6% over control. Under shading stress, SOD activity increase by 26.4%, CAT by 12.9%, and POX by 25.6% as compared to sunlight control in *B. juncea* (Kaur H et al. 2018; Kaur L et al. 2018). A similar trend was seen in *Brassica oleracea* by Soengas et al. (2018). An increase in antioxidative enzymes in different *Brassica* species and also under different abiotic stresses has been reported from our laboratory viz. *Brassica carinata* (Kaur 2016), cold tolerance (Kaur H et al. 2018; Kaur L et al. 2018), salt stress in *B. juncea* (Kannu 2019) and drought (Rhythm 2019). Recently, high enzyme activity under late planting in research trails over the years has been reported by Priya (2020) where CAT increased by 1.2 folds, SOD by 1.1, and POX by 1.7 at Punjab agricultural university, Ludhiana while at second location i.e., regional research station Abohar the respective increase was 1.2, 1.7, and 4.6 folds due to terminal heat stress. Antioxidative enzymes increased to variable extent in *B. juncea* cultivars planted simultaneously in temperature gradient tunnels.

5.3 Non-enzymatic Antioxidative Biomolecules

The antioxidant metabolites mainly consist of three groups including phenols, carotenoids, and ascorbic acids along with tocopherols. Under elevated stress conditions, these metabolites scavenge the ROS. Chakraborty and Pradhan (2010) studied biochemical response to elevated high-temperature stress and reported that ascorbic acid content increased at temperature 30 and 35 °C. Tocopherols are valued antioxidants because of their mode of action as they react with polyunsaturated acyl group and protect polyunsaturated fatty acids from degradation by lipid peroxidation by not only scavenging lipid peroxy radicals but also quenching ROS. Tocopherols in this process contribute their phenolic hydrogen to lipid-free radical, thus neutralizing the radical and terminating the autocatalytic lipid peroxidation thereby protecting cell membranes. Tocopherol radicals thus produced are less reactive; more stable can be reconverted to tocopherol on reacting with other antioxidants like ascorbates or glutathione (Fritshe et al. 2018). Seeds of heat-tolerant genotype in different crops are enriched with an array of antioxidants like tocopherol, flavonoids, phenylpropanoids, and ascorbates precursors (Chebrolu et al. 2016). Research conducted in our laboratory again revealed enhanced tocopherols and ascorbates in *B. juncea* under low light stress (Kaur H et al. 2018; Kaur L et al. 2018), drought (Rhythm 2019), saline condition (Kannu 2019) and recently by Priya (2020) where tocopherols increased by 9.2 mg and ascorbic acid by 2.01 mg on the basis of dry matter (g) where late sowing faced terminal high-temperature stress.

6 Yield and Yield Attributes

Expression of genotype is governed by the environment and the interaction of genotype × environment with seed yield as quantitative trait (Gunasekera et al. 2006). Association studies along heat susceptibility and tolerance indices for identifying high-temperature tolerance have been established in Indian mustard (*B. juncea*) cultivars by Sharma and Sardana (2015, 2019). Three *Brassica* species were germinated in the growth chambers to visualize the impact of short temperature stress on reproductive phasic development and yield by Angadi et al. (2000). Day/night temperature for the experiment was 35/15 °C which proved detrimental to the reproductive phase for all the three species at different developmental stages studied. Genotypic performance under stress and low stress were evaluated for stress indices which revealed that consistent heat tolerance was relatively in the genotype with high geometric mean (GM) added by stress tolerance index (STI) and lower stress susceptibility index (SSI). The significant impact of sowing dates has been reported by research studies on yielding abilities of *B. juncea* (Shivani and Kumar 2002; Rafiei et al. 2011; Patel et al. 2013; Singh and Singh 2014; Devi 2017) and in canola cultivars of *B. napus* (Khayat 2015; Balodis and Gaile 2016). To evaluate the impact of planting times on yield and component traits, three canola cultivars (Bulbul-98, Zafar-2000, and Rainbow) were planted on 15, 30 October and 15 November considered as early, late, and very late sowings, respectively, by Abdul et al. (2013). 30 October sown canola cultivars had maximum seed size, biomass, HI and seed yield but minimum when the same canola cultivars were sown on 15 November. The negative correlation between yields of canola cultivars and delayed plantings was linked with genetic variations and heat stress at seed filling. Zafar-2000 was higher yielder (2142.5 kg ha^{-1}) and low yield was in Rainbow (2014.4 kg ha^{-1}) due to genetic variation added by varietals character affected by heat stress. Earlier, a similar trend in seed yield, yield attributes and oil content was reported by Sardana and Sharma (2012) in *B. juncea* cultivars with sowing dates. Kumar and Kudesia (2014) studied the yield contributing character of oleiferous *Brassicas* (*B. juncea* (Brown and yellow), *B. carinata* and *B. campestris*) revealing that *B. juncea* with harvest index of 41.3% which was due to lesser plant dry weight, higher seed weight (0.501 g), seed yield/plant (9.24 g), more siliquae $plant^{-1}$ (180) along with seed number $siliqua^{-1}$ (15). To visualize the impact of terminal high temperature on seed filling caused by delayed sowing, Indian mustard was planted on 26 October and one month thereafter on 26 November. The results were supportive of the earlier findings resulting in decreased yield and yield attributes especially branch number, seeds $siliqua^{-1}$ and seed size in November sowing over October sowing. A maximum average temperature >30 °C during seed filling was noticed in the late sown November crop. Stress intensity varied from 0.11 to 0.82 while tolerance ranged from 1.05 to 1.68. Photosynthesis was greatly impacted by heat stress at the terminal stage resulting in the drop of Pn from 9.8 to 48.3% over normally cultivated genotypes (Singh et al. 2014). Canola cultivars (Hyola 401, PP401, RGS003, and Option 500) when planted at two dates 6 and 21 in the month of November and December inferred maximum total siliquae, 1000 seed weight, biomass, yield and HI in the 6 November sown crop. Grain yield was highest in first planting (2611.6 kg ha^{-1}) followed by a subsequent decline of 13.7, 31.4 and 41.9%, respectively (Khayat et al. 2015). Mustard variety, BARI Sarisha 14 was planted in October twice on 25 and 30 dates and four times in the month of November on 5,10, 14, and 15 by Alam et al. (2014, 2015) to demonstrate the significant impact of planting time on total siliquae $plant^{-1}$, seed weight and seed yield. These three traits decreased considerably with a delay in planting after 30 October. The highest seed yield of 1.5 ton ha^{-1} was recorded in 30 October planting and 1.0 ton ha^{-1} from 15 November sowing. Gawariya et al. (2015) planted 24 mustard varieties in October at 3 dates 1, 16, 31 and 15 November and observed better growth due to primary and secondary branch number, higher dry matter accumulation, component yield trait especially total siliquae $plant^{-1}$ and seed yield of 2013 kg ha^{-1} in 1st October sowing relative to 31st October and 15th November late plantings. Further, late planted mustard cultivars on 15 November had lesser primary and secondary branches, siliqua length and seeds $siliqua^{-1}$. Studies of Kumari et al. (2012) and later by Kumar P et al. (2017), Kumar Y et al. (2017) endorsed that 5th November sowings reduced test weight, seed yield $plant^{-1}$, biomass $plant^{-1}$ and harvest index as compared to 25th October sown crop. The declining trend with moisture stress in Indian mustard genotypes was reported by Sharma et al. (2011) in yield and component traits. Oil content was affected by the duration of the seed filling period, which was maximized when crop was exposed to a lower temperature. However, a linear negative relationship existed between air temperature and oil content during the seed filling period. High temperature hastened the plant development rate with reduced duration of seed filling and decreased oil content (Faraji 2012; Sharma and Sardana 2013, 2014, 2018) and later by Kumar et al. (2018a, b) in Indian mustard when sown at 23 September, 16 October, and 21 November. Khan et al. (2018) indicated a reduction in the number of seeds, thousand-grain weights and fewer tillers per plant was due to reduced photosynthesis, accelerated flag leaf senescence ultimately grain yield owing to high temperature. Rate of seed filling and seed filling duration controlled the seed size. Reduced number of endosperm cells and decreased starch synthesis which may be due to limited assimilate supply to developing seeds or impeded biosynthetic process during the development of seed are the consequences of heat stress encountered by seed filling stage due to late planting (Barnabas et al. 2008). Further, a positive association between the grain weight and rate of grain filling due to temperature variation was endorsed by Wu et al. (2018), which consequently was related to the number of grains. Planting dates and nitrogen doses had significant impact on growth and productivity (Kaur 2012; Kaur H et al. 2018; Kaur L et al. 2018) has been demonstrated in canola oilseed rape (*B. napus*). Crop sown on 15 October produced 4.1% more seed yield (2476 kg ha^{-1}) and 27.3% more stover yield (9458 kg ha^{-1}) than 30 October sown crop, which in turn produced 36.1% and 36.4% more seed and stover yields than 15 November sown crop. Effect of dose (100 and 125 kg) and time of application (two or three splits) of nitrogen (N) on initiation or completion of emergence was inconspicuous but application of 125 kg ha^{-1} of N delayed the maturity, increased plant height (mean over time of application) at all growth stages over 100 kg ha^{-1} of N, whereas its effect on other growth parameters was variable (Kaur H et al. 2018; Kaur L et al. 2018). Significant interaction for fresh weight and seed yield among the genotypes of *Brassica* species during heat, drought and combination of both the stresses has been reported by Chen et al. (2019) where fresh weight declined by 48.2% with drought, 11.4% with heat and 69.4% when both heat and drought stresses occurred simultaneously. Seed yield also decreased by 40.8%, 57.3%, and 57.3% in drought, heat and combination of drought and heat stress, respectively. In oilseeds, seed development stage exposed to different temperature array in safflower and castor bean showed no change in the fatty acid composition but declined the amount of unsaturated fatty acid (Sardana et al. 2008, 2014) and later similar results were endorsed by Patel et al. (2017) in *B. juncea* and also by Rashid et al. (2017) in *B. napus*. Late planting after 27 October decreased total siliquae $plant^{-1}$, seeds $siliqua^{-1}$, seed size, biomass, seed yield, oil content, oil yield along with oleic and linoleic acid but increased linolenic acid by 23.3% and erucic acid by 48.7% (Nazeri et al. 2018). Decrease of 59.1% in seed yield and 55.9% in oil yield earlier too has been reported by late planting (Hassan et al. 2005). Asghari et al. (2018) studied the performance of 11 canola cultivars at different sowing dates where delayed sowing checked the plant height, 1000 seed weight, oil content, and yield reduction of 33.6% over the first sowing date. Research material of Koscielny et al. (2018) comprised of ten inbred lines of *B napus*, a commercial cultivar of *B. napus* and *B. juncea* which were open-pollinated and sown in the greenhouse

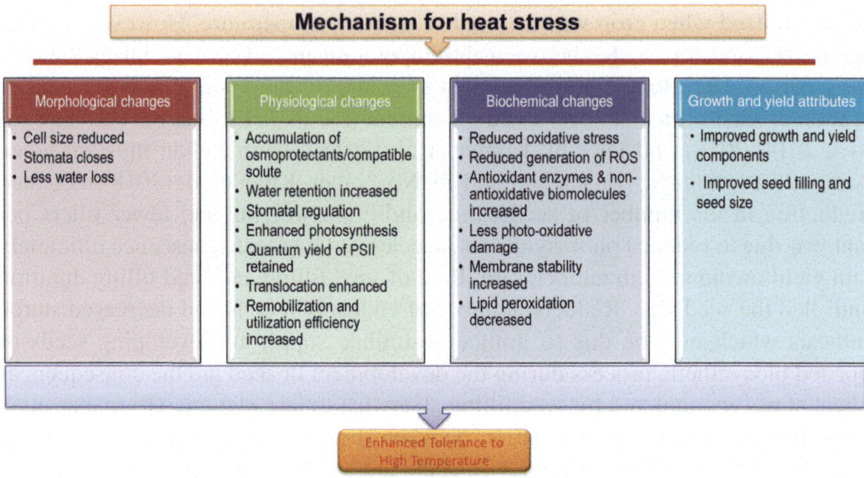

Fig. 3 Mechanism conferring heat tolerance in Rapeseed–mustard

then were subjected to day/night temperature regimes of 22 °C/10 °C in control and 31 °C/14 °C for high temperature. Days were more for the appearance of first and last flower in *B. juncea* genotypes in control. Flowering duration, number of total flowers, total pods, pod: flower, total seeds, seeds pod^{-1}, biomass, yield and HI were lesser in heat-stressed genotypes in comparison to control (Dipti et al. 2016). Germplasm of *B. juncea* and *B. napus* has been evaluated for heat stress at seedling and terminal stages and characterized for morpho-physiological and heat resistance parameters (Sharma et al. 2011a, b, 2013, 2014). Genetic variation for heat tolerance in *B. juncea* was confirmed for identified traits by Sharma et al. (2015) and mechanism depicted in Fig. 3.

A diversified set of 500 Indian mustard (*B. juncea*) germplasm including accession from different countries, landraces, released varieties, Introgressed lines developed at Punjab Agricultural University, Ludhiana were screened at two locations PAU and Abohar for seed yield and yield attributes for two years with two sowing dates. Yield depression and heat susceptibility index were the selection criteria for identifying differential responses in a panel of 96 genotypes (Sandhu 2019) which was further used for GWAS by Sandhu et al. (2019). A significant reduction in the seed yield under terminal heat stress along with plant height, number of pods on the main shoot and seed weight endorsed the detrimental effect of high temperature on growth and development. Genetic variations characterized the germplasm, identified the donors for heat tolerance and the role of wild species in conferring abiotic stress tolerance from the developed introgression lines in *B. juncea*. The elevated temperature at the terminal stage had a great impact on growth attributes at two locations as recently reported by Priya (2020) in a set of 48 advanced breeding lines. Reduction in plant height by 21.0% and 9.2%, length of the main raceme by 32.5% and 16.0%, primary branches by 23.0% and 20.8%, secondary branches by 23.0% and 38.2%, biomass to the tune of 35.0% and 25.2% and seed yield 40.9% and 32.2% over the years at PAU and RRS, Abohar, respectively. Late planting reduced the oil content resulting in the reduction of unsaturated fatty acid as revealed by fatty acid profiling (Sharma and Sardana 2014). Selection criteria of seed yield reduction (<30%), lower HSI ($\leq$1) and higher HTI ($\geq$0.6) rated genotypes as tolerant to terminal heat stress. Seed yield had a positive association with phenological traits like flowering duration, siliquing duration, reproductive phase and maturity when planted at two locations during two crop seasons. Positive and significance correlation of seed yield with BY (r = 0.745**), HI (r = 0.569**), HSI (r = 0.455**), HTI (r = 0.894**) at PAU while at second location the trend was significant for BY (r = 0.883**), HI (r = 0.433**), HSI (r = 0.744**), HTI (r = 0.939**) under optimum planting condition whereas BY (r = 0.593**) and HTI (r = 893**) showed positive association in delayed planting at PAU and similar trend existed at RRS, Abohar for BY (r = 0.782**) and HTI (r = 0.892**).

7 Protein Profiling

Morphological characters recorded at the time of maturity over the years to explore genetic diversity are regularly prone to environmental fluctuation therefore cannot be considered as marker traits. To record genetic diversity and develop phylogenetic relation in the available genetic resources, both molecular and biochemical markers are widely used. Polypeptides or protein profiling serves as a useful genetic marker system with expressed gene diversity and has been applied for gene mapping, gene regulation, development genetics, evolution, screening for mutants and cultivars identification (Jangid et al. 2017; Kumar P et al. 2017, Kumar Y et al. 2017). SDS PAGE is one of the simple and reliable techniques for genetic diversity analysis and cultivar identification. Planting location has shown to contribute to genetic diversity in morphological characters (Pal et al. 2019) which in turn can affect the genes responsible for a particular trait. Many of diversity studies were done for different species of *Brassicas* from different geographical regions (Sadia et al. 2009). With more stress on quality traits and oil content the germplasm in *Brassica* eroded (Hasan et al. 2006). Subsequently, for breeding in *Brassicas* genetic difference is narrow for many valuable traits. Effective germplasm conservation in this important oleiferous crop along with breeding strategies will be more precise to assess genetic diversity and phylogenetic relationships (Moghaieb et al. 2014). Indian mustard seedlings (*B. juncea* L.) were subjected to heat stress of 45 °C and on revival variability in banding pattern was recorded. Further, thermotolerant and thermosensitive genotypes two each were screened for heat tolerance on the basis of 50% seedling mortality and banding pattern of protein as reported by Rani et al. (2015). A major band of 53.12 kDa and minor band of 100, 89.12, 74.13, 46.76, and 38.9 kDa in thermotolerant genotypes and protein bands of molecular weight 25.79 and 30.7 kDa in thermosensitive genotypes appeared under high-temperature stress which disappeared when stress was relieved. Mawlong et al. (2017) studied peptide polymorphism in seed protein under N_0 and

N_{80} doses. Results revealed that in N_0 condition, a total of 166 bands (86.5%) were scored ranging from 10 to 71 kDa in all genotypes, of which 118 showed polymorphic loci with 71.08% polymorphism. In N_{80} condition, a range of 170 bands (88.5%) was found among 24 genotypes, of which 74 bands showed 38.5% polymorphism. The protein band ranged from 18.1 and 31.2 kDa belonged to cruciferin in *B. juncea*. EI-Esawi (2015) studied total seed protein profiling in *B. napus* and reported a total of 18 protein band ranging from 16 to 46 kDa. SDS PAGE protein profiling (Priya 2020) showed differential responses of genotypes for banding patterns. The intense banding pattern indicated protein structure integrity was maintained in genotypes planted at the optimum time, however light banding pattern was due to denaturation of protein in late-planted genotypes. Most of the peptides bands ranged from 10 to 75 kDa. Physiological and biochemical marker traits identified at the reproductive stage suggested variations for yield and the component traits which may be used for cultivar screening. Improved varieties with heat and drought tolerance are possible with improved marker traits in accordance with the principal component analysis in late sown wheat genotypes (Saleem et al. 2017; Qaseem et al. 2019). Identification of stable genotype in the varied environments and their interaction indicated an additional effect on yield of *B. napus* as demonstrated by multiplicative interaction analysis by Ramesh (2014) and later by Bibi et al. 2018). Cluster analysis with the biplot analysis was used to classify the sunflower hybrids (Tyagi and Dhillon 2019). In Indian mustard genotypes, the variance for clustered mean was calculated for various quantitative characters of mustard and results revealed wider genetic diversity in the genotypes due to greater distance between two clusters (Kumar and Pandey 2013). Several heat-inducible genes encoding for HSPs are increased under heat stress for protection and survival by preventing protein denaturation, through protein folding acting as chaperones for preserving stability and functioning of proteins according to Saini (2017).

8 Hormonal Regulation

Foliar sprays of growth-promoting substances, signaling molecules, osmoprotectants and other biomolecules favored growth under high temperatures because of their antoxidative and growth-promoting effects. Proline, glycine betaine and trehalose are the accumulated osmolytes in response to various abiotic stresses including tolerance to heat. On the contrary, heat-sensitive cultivars could not accumulate osmoprotectants conferring heat tolerance therefore, foliar sprays of osmolytes can improve tolerance to high temperature. Reduced H_2O_2 production, accumulation of soluble sugars, improved K^+ and Ca^{2+} protected the developing tissues from heat stress with the application of proline. The literature cites the application of phytohormones quite effective in mitigating heat stress. When the seeds of *B. juncea* were soaked in different plant hormones like IAA and GA @100 μM, 50 and 100 μM kinetin and 0.5 and 1 μM ABA mitigated the effect of high temperature (47 ± 0.5 °C) as demonstrated by Chhabra et al. (2009). Growth promoters and retardants were active in alleviating heat stress and these effects were a noteworthy observation. The role of growth-regulating hormone in alleviating temperature stress at a concentration that otherwise could prove toxic for the growth of seedling. Plant hormone, salicylic acid protected effectively Indian mustard genotypes in high temperature conditions (Sharma 2014b). Foliar spray of salicylic acid @ 100 ppm twice in an assembly of 25 Indian mustard genotypes at initiation of flowering and ten days thereafter improved growth predominantly by enhancing siliquae on the main raceme and length of the siliqua whereas seed filling improved due to 3.2% increase in total seeds siliqua^{-1} and 2.5% of fully developed seeds as compared to unsprayed control. Improved seed yield in RGN-145 and RH-0116 was due to decreased shriveled seeds and enhanced fully developed seeds resulting in effective partitioning of assimilate as indicated by improved HI (Sharma et al. 2013). The oxidative damages by ABA treated mustard seedlings under heat stress was lesser than control and was linked to decreased malondialdehyde and H_2O_2 content. Fluridone, an inhibitor of ABA biosynthesis, reverted the inhibition made by ABA which proposed the role of this inhibitor in mitigating damages owing to heat (unpublished data). Pre-treatment of maize seedlings with different concentrations of plant growth regulators imparted tolerance to heat (Sharma and Basra 2004a) and further concerning heat shock treatment in maize seedlings with salicylic acid and its interaction with growth regulators has been reported by Sharma and Basra (2004b). Application of SA @ 10 and 20 μM at high temperature (40–50 °C) enhanced CAT and POX antioxidative abilities of *Brassica* species (Kaur et al. 2009). 24–epibrassinolide were applied in different concentrations of mustard (*B. juncea*) seedlings by Kumar et al. (2012) to visualize the impact on growth and antioxidant enzymes. 24EBL treated seedlings displayed better growth, enhanced protein content and upregulation of antioxidative enzymes particularly SOD, CAT, POX thus imparting tolerance to high temperature-induced oxidative stress. A spray of salicylic acid at different seeding dates proved to be beneficial for phenology, growth, yield and other indices in *Brassica* species as reported by Muhal and Solanki (2015).

9 Conclusion and Future Thrust Areas

Climatic conditions particularly temperature and photoperiod impact growth and reproduction directly as physiological limitations and including competition for resources referred to as indirect ecological influences are collectively responsible for the distribution of species. Over the past decades, changes in the climatic conditions including temperature was instrumental not only in the distribution and abundance of a wide range of crop species including phenological and physiological traits. Shift in geographical location to a greater extent and therefore survivals of species are expected as a result of climate change. Species with higher migration rates and fitness have greater chances of survival and dominance, one promising strategy is to improve existing genotypes and develop new genotypes that can tolerate abiotic stresses particularly high-temperature stress. Targeted and controlled breeding programs generally result in narrow genetic diversity specifically tolerance to high temperature. In the process of domestication or evolution crops go through a genetic bottleneck that leads to more uniformity ensuring more vulnerability to abiotic and biotic stresses. Genotype development needs sources of new genes and tolerance traits. Exploration of wild species, relatives of cultivated species, available landraces, germplasm, developed introgression lines can provide that opportunity and play a critical role in developing climate-smart genotypes. Wild species have retained noble traits for abiotic stresses which can be used as donors in the cultivated *Brassicas* breeding programs. Therefore, this source of diversity needs to be tapped, preserved, expanded and finally utilized in developing heat-tolerant varieties.

Phenotyping techniques categorize tolerance, escape, or avoidance as heat stress responses essential for developing heat-tolerant varieties in the crops to be cultivated in the future hot climates. Signaling due to high-temperature episodes and subsequent expression of specific genes involved in temperature tolerance calls for a better understanding of developing heat-tolerant species. However, seed yield is a quantitative trait and further, the yield attributing component traits obscures the direct phenotypic selection. Several studies have identified QTLs for seed yield and its component traits in *B. juncea* (Kaur et al. 2014; Akhatar and Banga 2015; Dhaka et al. 2017). However, no study has been reported for SNPs identification with SY and its component traits in response to high temperature. Therefore, it is desirable to screen germplasm variability for these traits to be used in breeding practices. With the advances in genetic tools like QTL mapping and transgenic methods, further advancement of genetic material calls for research in understanding the underlying mechanism of environmental interactions. Genetic base for heat tolerance can be significantly expanded via phenotyping approaches for biochemical means like metabolomics, the promising strategy leading to the identification of biochemical markers to supplement breeding efforts. Aerial sensor technology along with advanced ground techniques may aid field phenotyping with the widening of improved gene pool required in the breeding strategies for abiotic stresses. Multiple stresses interact in the field trials mainly high temperature or drought stress, which are complex to understand individually. The research strategy for producing cultivars which can withstand hot and dry spell should consider need-based characters or genes stacking for specific environment involving advanced techniques. Integration of advanced high-throughput approaches like microarray, genomics, and proteomics during different growth stages and also under stress will pave way in future for developing thermotolerant varieties and transgenic *Brassica* plants. Methods to reduce genome complexity, genotyping by sequencing (GBS), next-generation sequencing (NGS) are the recent genotyping tools. Across any genome of interest among any individual number in a population, thousands of markers can be revealed in a few weeks by GBS as demonstrated by Davey et al. (2011). Identification of QTLs/genes underlying complex traits is possible by Genome-wide association studies (Zhu et al. 2008; Shilpa et al. 2019).

References

Abdelrahman M, Burritt DJ, Gupta A, Tsujimoto H, Tran LSP (2019) Heat stress effects on source-sink relationship and metabolome dynamics in wheat. J Expt Bot 4:1–13

Abdul S, Cheema MA, Wahid MA, Saleem MF, Ghaffari MA, Hussain S, Arshad MS (2013) Effect of sowing time on seed yield and oil contents of canola varieties. J Glob Innov Agric Soc Sci 1(1):1–4

Abedi T, Pakniyat H (2010) Antioxidant enzyme changes in response to drought stress in ten cultivars of oilseed rape (*Brassica napus* L.). Czech. J Genet Plant Breed 46(1):24–34

Akhatar J, Banga SS (2015) Genome-wide association mapping for grain yield components and root traits in *Brassica juncea* (L.) Czern & Coss. Mol Breed 35(1):48

Akhter MT, Mannan MA, Kundu PB, Paul NK (2015) Effects of different sowing dates on the phenology and accumulated heat units in three rapeseed (*Brassica campestris* L.) varieties. Bangladesh J Bot 44(1):94–101

Alam M, Begum F, Roy P (2014) Yield and yield attributes of rapeseed-mustard (*Brassica*) genotypes grown under late sown condition. Bangladesh J Agril Res 39:311–336

Alam MJ, Ahmed KS, Mollah MRA, Tareq MZ, Alam J (2015) Effect of planting dates on the yield of mustard seed. Int J Appl Sci Biotechnol 3(4):651–654

Almeselmani M, Deshmukh SR, Kushwaha SR, Singh TP (2006) Protective role of antioxidant enzymes under high temperature stress. Plant Sci 171:382–388

Aneja B, Yadav NR, Kumar N, Yadav RC (2015) Hsp transcript induction is correlated with physiological changes under drought stress in Indian mustard. Physiol Mol Biol Plants 21(3):305–316

Angadi SV, Cutforth HW, Miller PR, McConkey BG, Entz MH, Brandt SA, Volkmar KM (2000) Response of three *Brassica* species to high temperature stress during reproductive growth. Can J Plant Sci 80:693–701

Anonymous (2015) Annual progress report rapeseed mustard, oilseed section. Department of Plant Breeding and Genetics, Punjab Agricultural University, Ludhiana, pp 111–126

Anonymous (2017) Annual progress report rapeseed mustard, oilseed section. Department of Plant Breeding and Genetics, Punjab Agricultural University, Ludhiana, pp 1–26

Anonymous (2018) Area and average yield under rape seed and mustard in Punjab (1960–61 to 2017–18). ENVIS Centre, Punjab, Ministry of Environment, Forests & Climate Change, Govt. of India

Anonymous (2019) Annual progress report rapeseed mustard, oilseed section. Department of Plant Breeding and Genetics, Punjab Agricultural University, Ludhiana, pp 120–131

Asghari BAH, Heravan EM, Alizadeh B, Abad HHS, Madani H (2018) Oil content, seed yield and morphological changes of canola cultivars in response to different sowing dates. Crop Res 53(2):38–44

Azharudheen TPM, Yadava DK, Singh N, Vasudev S, Prabhu KV (2013) Screening of Indian mustard (*Brassica juncea* L. Czern & Coss.) germplasm for seedling thermo-tolerance using a new screening protocol. Afr J Agric Res 8:4755–4760

Baghel RS, Srivastava SBL (2010) Assessment of mustard genotypes for thermo-tolerance at seed development stage. Ind J Plant Physiol 15:36–43

Bali S, Poonam Kohli SK, Kaur H, Bhardwaj R (2016) Improvement in photosynthetic efficiency of *Brassica juncea* under copper stress by plant steroid hormone. J Chem Pharma Res 8(5):464–470

Balla K, Karsai I, Bencze SZ, Kiss T, Veisz O (2013) Effect of heat stress on the physiological processes of wheat. Acta Agron Hung 61:1–12

Balodis O, Gaile Z (2016) Sowing date and rate effect on winter oilseed rape (*Brassica napus* L.) yield component's formation. Proc Latvian Acad Sci 70:384–392

Barnabas B, Jager K, Feher A (2008) Effect of drought and heat stress on reproductive processes in cereals. Plant Cell Environ 31:11–38

Bibi T, Mustafa HSB, Mahmoud T, Hameed A, Ali Q (2018) Multivariate analysis for adaptability and yield stability of rapeseed (*Brassica napus* L.) strains in different agro-climatic zones. Genetika 50(2):369–378

Bita CE, Gerats T (2013) Plant tolerance to high temperature in a changing environment: scientific fundamentals and production of heat tolerance crops. Front Plant 4:273–296

Blokhina O, Fagerstedt KV (2010) Reactive oxygen species and nitric oxide in plant mitochondria: origin and redundant regulatory systems. Physiol Plant 138(4):447–462

Blum (1998) Improving wheat grain filling under stress by stem reserve mobilization. Euphytica 100:77–83

Blum (2005) Drought resistance, water use efficiency and yield potential-are they compatible, dissonant, or mutually exclusive. Aust J Agric Res 56:1159–1168

Boomiraja K, Chakrabartia B, Aggarwala PK, Choudharya R, Chanderb S (2010) Assessing the vulnerability of Indian mustard to climate change. Agril Ecosyst Environ 138:265–273

Cen H, Shao Y, Song H, He Y (2006) Nondestructive estimation of rape nitrogen status using SPAD chlorophyll Meter ICSP 2006

Chakraborty U, Pradhan D (2010) Biochemical responses of lentil (*Lentil culinaris* Medik.) to elevated temperature stress. Res J Pharma Bio Chem Sci 1(3):575–585

Chauhan JS, Meena ML, Saini MK, Meena DR, Singh M, Meena SS, Singh KH (2009) Heat stress effects on morpho-physiological characters of Indian mustard (*Brassica juncea* L.). In: 16th Australian Research Assembly on Brassicas, Ballarat Victoria, pp 91–97

Chebrolu KK, Fritschi FB, Ye SQ, Krishnan HB, Smith JR, Gillman JD (2016) Impact of heat stress during seed development on soybean seed metabolome. Metabolomics 12:1–14

Chen S, Gua Y, Sirault X, Stefanova K, Saradadevi R, Turner NC, Nelson MN, Furbank RT, Siddique KHM, Cowling WA (2019) Non-destructive phenomics tools for the prediction of heat and drought tolerance at anthesis in *Brassica* species. Plant Phenomics 2:1–16

Chhabra ML, Dhawan A, Sangwan N, Dawan K, Singh D (2009) Phytohormones induced amelioration of high temperature stress in *Brassica juncea* (L.). In: Proceedings of 16th Australian Research Assembly on Brassicas, Ballarart, Australia, pp 10–14

Chouksey H, Sardana V, Tiwari AK, Singh A (2014) Lai, Par and Spad value of cultivars of Indian mustard (*Brassica juncea)* as influenced phosphorus levels. Int J Curr Micro Appl Sci 3(9):322–329

Davey JW, Hohenlohe PA, Etter PD, Boone JQ, Catchen JM, Blaxter ML (2011) Genome-wide genetic marker discovery and genotyping using next-generation sequencing. Natl Rev Genet 12(7):499

Devi S (2017) Effect of sowing date on flowering and seed set of mustard (*Brassica juncea* L.). J Ento Zoo Stud 5(5):1534–1537

Dhaka N, Rout K, Yadava SK, Sodhi YS, Gupta V, Pental D, Pradhan AK (2017) Genetic dissection of seed weight by QTL analysis and detection of allelic variation in Indian and east European gene pool lines of *Brassica juncea.* Theor Appl Genet 130(2):293–307

Dhaliwal K, Hundal SS, Chahal SK (2007) Agro climatic indices of Indian mustard (*Brassica juncea*) under Punjab conditions. Ind J Agric Sci 77:88–91

Dipti, Chawla V, Priyanka (2016) Correlation and path coefficient analysis for yields contributing parameters in Indian mustard [*Brassica juncea* (L.) Czern and Coss.]. Am J Pharm Tech Res 6(2):665–667

DRMR (2013) Directorate of Rapeseed-Mustard Research, Bharatpur, Rajasthan. www.drmr.res.in/about_rm.html

DRMR (2015) Vision2050. Directorate of Rapeseed-Mustard Research, Bharatpur, Rajasthan. http://www.icar.org.in/vision%202050%20DRMR%20Rajasthan.pdf

DRMR (2018) Annual progress report of AICRP on Rapeseed mustard, Bharatpur, pp 1–19

DRMR (2019) Annual progress report of AICRP on Rapeseed mustard, Bharatpur, pp 1–30

El-Esawi MA (2015) Genetic diversity and phylogenetic relationship of *Brassica napus* L. as revealed by protein profiling and SSR markers. Egypt J Expt Bot 11(2):245–256

Faraji A (2012) Oil concentration in Canola (*Brassica napus* L.) as a function of environment conditions during seed filling period. Int J Plant Prod 6(3):267–277

Fritshe S, Wang X, Jung C (2018) Recent advances in our understanding of tocopherol biosynthesis in plants: an overview of key genes, functions, and breeding of Vitamin K improved crop. Antioxidants 6:1–18

Gawariya SC, Chopra NK, Chopra N, Harika AS (2015) Effect of date of sowing and crop geometry on growth and yield parameters of forage mustard (var. *Chinese Cabbage).* Afr J Agric Res 10(33):3292–3295

Gill SS, Tuteja N (2010) Reactive oxygen species and antioxidant machinery in abiotic stress tolerance in crop plants. Plant Physiol Biochem 48:909–930

Girousse C, Roche J, Guerin C, Gouis Le, Balzegue S, Mouzeyar S, Bouzidi MF (2018) Co-expression network and phenotypic analysis identify metabolic pathways associated with the effect of warming on grain yield component in wheat. PLOS One 13:160–175

GOI (Government of India) (2017) In: Third advance estimates of production of food grains for 2016–17. Agricultural Statistics Division, Department of Agriculture Cooperation & Farmers welfare, GOI, New Delhi. http://agricoop.nic.in/statistics/advance-estimate-apy-english

Gosavi GU, Jadhav AS, Kale AA, Gadakh SR, Pawar BD, Chimote VP (2014) Effect of heat stress on proline, chlorophyll content, heat shock protein and antioxidant enzymes activity in sorghum (*Sorghum bicolor*) at seedling stage. Ind J Biotechnol 14:356–363

Gunasekera CP, Martin LD, Siddique KHM, Walton GH (2006) Genotypes by environment interaction of Indian mustard (*Brassica juncea* L.) and Rapeseed (*B napus L.*) in Mediterranean type environment. Crop growth and seed yield. Eur J Agron 25:1–12

Hasan M, Seyis F, Badani AG, Pons-Kuhnemann J, Friedt W, Luhu W, Snowdon RJ (2006) Analysis of genetic diversity in the *Brassica napus* L. gene pool using SSR markers. Genet Res Crop Evo 53(4):793–802

Hasanuzzaman M, Hossain MA, Teixeira da Silva JA, Fujita M (2012a) Plant responses and tolerance to abiotic oxidative stress: antioxidant defense is a key factor. In: Bandi V, Shanker AK, Shanker C, Mandapaka M (eds) Crop stress and its management: perspectives and strategies. Springer, New York, pp 261–316

Hasanuzzaman M, Hossaina MA, Fujita M (2012b) Exogenous selenium pretreatment protects rapeseed seedling from cadmium-induced oxidative stress by up-regulating the antioxidant defense and methyl detoxification system. Biol Trace Element Res 4:24–38

Hasanuzzaman M, Nahar K, Alam MM, Roychowdhury R, Fujita M (2013a) Physiological biochemical and molecular mechanisms of heat stress tolerance in plants. Int J Mol Sci 14:9643–9684

Hasanuzzaman M, Nahar K, Alam MM, Roychowdhury R, Fujita M (2013b) Physiological, biochemical, and molecular mechanisms of heat stress tolerance in plants. Int J Mol Sci 14(5):9643–9684

Hasanuzzaman M, Nahar K, Fujita M (2013c) Extreme temperatures, oxidative stress and antioxidant defense in plants. In: Vahdati K, Leslie C (eds) Abiotic stress—plant responses and applications in agriculture. InTech, Rijeka, pp 169–205

Hasanuzzaman M, Nahar K, Khan MIR, Mahmud JA, Alam MM, Fujita M (2019) Regulation of reactive oxygen species metabolism and glyoxalase systems by exogenous osmolytes confers thermotolerance in *Brassica napus*. Gesunde Pflanzen 72:3–16

Hassan FU, Ali H, Cheema MA (2005) Effects of environmental variation on oil content and fatty acid composition of canola cultivars. J Res Sci 16:65–72

Hatfield JL, Prueger JH (2015) Temperature extremes: effect on plant growth and development. Weather Climate Extremes 10:4–10

Hayat S, Masood A, Yusuf M, Fariduddin Q, Ahmad A (2009) Growth of Indian mustard (*Brassica juncea* L.) in response to salicylic acid under high-temperature stress. Braz J Plant Physiol 21:187–195

Hemantaranjan A, Malik CP, Bhanu AN (2018) Physiology of heat stress and tolerance mechanism—an overview. J Plant Sci Res 33(1):55–68

Howarth CJ (2005) Genetic improvements of tolerance to high temperature. In: Ashraf M, Harris PJC (eds) Abiotic stresses: plant resistance through breeding and molecular approaches. Howarth Press, New York, NY, pp 277–300
Hutsch BW, Jahn D, Schubert S (2019) Grain yield of wheat (*Triticum aestivum* L.) under long-term heat stress is sink-limited with stronger inhibition of kernel setting than grain filling. J Agro Crop Sci 205:22–32
Jangid KK, Srivastava JP (2018) Canopy temperature depression and stay green: Major component for identifying terminal heat stress resistant genotypes of wheat. Int J Pure Appl Biosci 6(3):374–381
Jangid K, Krishan KR, Parasher N, Bhatt B, Gaur A, Kumar V (2017) Assessment of genetic diversity for storage seed protein in a set of released varieties and their diallelic set of crosses in Indian mustard [*Brassica juncea* (L.) Czern & Coss] using SDS PAGE. Int J Curr Micro Appl Sci 6(8):2241–2245
Kannu P (2019) Performance of Indian mustard (*Brassica juncea* L. Czern and Coss) genotypes under saline conditions. MSc thesis, Punjab Agricultural University, Ludhiana, India
Kaur S (2012) Identification and characterization of morpho-physiological traits associated with response to water stress in *Brassicas*. MSc thesis, Punjab Agricultural University, Ludhiana, India
Kaur H (2016) Physiological implications of determinate plant growth habit in Ethiopian mustard (*Brassica carinata* A. Braun). MSc thesis, Punjab Agricultural University, Ludhiana, India
Kaur K (2017) Photosynthetic traits and antioxidative enzymes under low light stress in Indian mustard (*Brassica juncea* L.) genotypes. MSc thesis, Punjab Agricultural University, Ludhiana, India
Kaur V, Behl RK (2010) Grain yield in wheat as affected by short periods of high temperature, drought and their interaction during pre-and post an thesis stage. Cer Res Comm 38:514–520
Kaur S, Sharma P (2015a) Biochemical characterization of Indian mustard (*B. juncea* L.) genotypes in response to moisture stress and irrigation modules. J Oilseed Brassica 6(2):265–272
Kaur S, Sharma P (2015b) Correlation between leaf traits and moisture availability in Indian mustard (*Brassica juncea* L.). Appl Bio Res 17(1):174–184
Kaur S, Sharma P (2015c) Physiological responses of Indian mustard (*Brassica juncea* L.) to different moisture regimes. The Bioscan 10(3):1259–1267
Kaur P, Ghai N, Sangha MK (2009) Induction of thermo tolerance through heat acclimation and salicylic acid in Brassica species. Afr J Biotechnol 8:619–625
Kaur H, Gupta S, Kumar N, Akhatar J, Banga SS (2014) Progression of molecular and phenotypic diversification in resynthesized *Brassica juncea* (L) gene pool with determinate inflorescence. Euphytica 199:325–338
Kaur H, Sirhindi G, Bhardwaj R, Alyemeni MN, Siddique KHM, Ahmad P (2018) 28-homobrassinolide regulates antioxidant enzymes activities and gene expression in response to salt- and temperature-induced oxidative stress in *Brassica juncea*. Sci Rep 8:1–13
Kaur L, Sardana V, Sharma P (2018) Effect of sowing dates and nitrogen application on growth and productivity of canola oilseed rape (*Brassica napus*). J oilseed Brassica 9(2):114–121
Kauser R, Athar HR, Asraf M (2006). Chlorophyll fluorescence: a potential indicator for rapid assessment of water stress tolerance in canola (*Brassica napus* L.) Pak J Bot 38:1501–1509
Kaushal N, Bhnadari K, Siddique HM, Nayyar H (2016) Food crops face rising temperatures: an overview of responses, adaptive mechanisms, and approaches to improves heat tolerance. Cogent Food Agric 2:1–42
Kavita PA (2017) Physiological attributes for screening of Indian mustard (*Brassica juncea* L. Czern and Coss) genotypes during terminal heat stress. Int J Curr Micro Appl Sci 6(9):2908–2913
Khan A, Khaliq I, Ahmad M, Ahmed HGMD, Khan AG, Farooq MS (2018) Comparative performance of spring wheat (*Triticum Aestivum* L.) through heat stress indices. Pak J Bot 50(2):481–488
Khatun S, Ahmed JU (2015) Response of elevated temperature on carbohydrates accumulation and grain yield in different wheat cultivars. Bangladesh J Agril Res 40(2):205–215

Khayat M (2015) Measuring yield and its components of canola genotypes in different planting date. Ind J Fund Appl Life Sci 5:48–55

Khayat M, Rahnama A, Lorzadeh S, Lack S (2015) Physiological indices, phenological characteristics and trait evaluation of canola genotypes response to different planting dates. Proc Natl Acad Sci India 2016 Sect B: Bio Sci 88(1)

Kim K, Portis AR (2005) Temperature dependence of photosynthesis in *Arabidopsis* plants with modifications in Rubisco activase and membrane fluidity. Plant Cell Physiol 46:522–530

Koscielny CB, Hazebroek J, Duncan RW (2018) Phenotypic and metabolic variation among spring *Brassica napus* genotypes during heat stress. Crop Pasture Sci 69:284–295

Krumbein AB, Schonhof I, Schreiner M (2005) Composition and contents of phytochemicals (glucosinolates, carotenoids and chlorophylls) and ascorbic acid in selected Brassica species (*B juncea, B. rapa* subsp. nipposinica var. chinoleifera, *B. rapa* subsp. chinensis and *B. rapa* subsp. Rapa). J Appl Bot Food Qual 79(3):168–174

Kumar S, Kudesia R (2014) Study of yield contributing characters of oleiferous *Brassicas* in Bundelkhand region. J Global Biosci 3(5):931–934

Kumar B, Pandey A (2013) Diversity analysis in Indian mustard (*Brassica juncea* L. Czern and Coss). Madras Agric J 100:62–66

Kumar S, Sirhindi G, Bharadwaj R, Kumar M, Arora P (2012) Role of 24-epibrassinolide in amelioration of high temperature stress through antioxidant defense system in *Brassica juncea* L. Plant Stress 6:55–58

Kumar S, Sairam RK, Prabhu KV (2013) Physiological traits for heat temperature stress tolerance in *Brassica juncea*. Ind J Plant Physiol 18(1):89–93

Kumar P, Dolker R, Manjunatha G, Pallavi HM (2017) Molecular finger printing and assessment of genetic variations among advanced breeding lines of *Moringa oleifera* L. by using seed protein. RAPD and cytochrome P 450 based markers. South Afr J Bot 111:60–67

Kumar Y, Singh R, Singh D, Kumar A, Dhaka AK (2017) Influence of weather parameters on yield and yield attributes of Mustard (*Brassica juncea*) at Hisar condition. Environ Ecosyt 35(2C):1274–1280

Kumar A, Kumar M, Gill P, Dharamvir, Kumar N (2018a) Physiological and Biochemical responses of Indian mustard (*Brassica juncea* L.) genotypes of different sowing dates. Int J Curr Micro Appl Sci 7(12):2794–2801

Kumar A, Lal M, Mohan N, Kumar N (2018b) Effect of different sowing dates on yield and yield attributes of Indian mustard (*Brassica juncea* L.) genotypes. Int J Pure Appl Biosci 6(2):848–883

Kumari A, Singh RP, Yashpal (2012) Productivity, nutrient uptake and economics of mustard hybrid (*Brassica juncea)* under different planting time and row spacing. Ind J Agron 57(1):61–67

Lakra RK, Alam P, Nayar A (2018) Effect of sowing time and crop geometry on productivity of mustard (*Brassica juncea)* under irrigated conditions of Jharkhand. Int J Curr Micro Appl Sci (7):777–781

Martinez-Ballesta MC, Perez LL, Muries B, Azcarate MO, Carvajal M (2015) Climate change and plant water balance: the role of Aquaporins—a review In: Lichtfouse E (ed) Climate change, intercropping, pest control and beneficial microorganism. Sustain Agric Rev 71–90

Mathur C, Agrawal D, Jajoo A (2014) Photosynthesis: response of high temperature stress. J Photochem Photo Biol 137:116–126

Mawlong I, Reema R, Kumar MSS, Kandpal BK, Premi OP (2017) Peptides polymorphism under recommended dose of nitrogen fertilization in *Brassica juncea*. J Oilseeds Res 34(4):217–225

Moghaieb RE, Mohammed EH, Youssief SS (2014) Genetic diversity among some canola cultivars as revealed by RAPD, SSR and AFLP analysis. Biotechnology 4(4):403–410

Mohamed HI, Abdel-Hamid AME (2013) Molecular and biochemical studies for heat tolerance on four cotton genotypes. Romania Biotechnol Lett 18(6):8823–8831

Muhal, Solanki (2015) Effect of seeding dates and salicylic acid foliar spray on growth yield, phenology and agro meteorological indices of *Brassica* species. J Oilseed Brassica 6(1):183–190

Nahar K, Ahamed KU, Fujita M (2010) Phenological variation and its relation with yield in several wheat (*Triticum aestivum* L.) cultivars under normal and late sown mediated heat stress condition. Natl Sci Biol 2:51–56

Nahar K, Hasanuzzaman M, Alam MM, Fujita M (2015a) Exogenous glutathione confers high temperature stress tolerance in mung bean (*Vigna radiata* L.) by modulating antioxidant defense and methylglyoxal detoxification system. Environ Exp Bot 112:44–54

Nahar K, Hasanuzzaman M, Fujita M (2015b) Heat stress responses and thermotolerance in soybean. In: Miransari M (ed) Abiotic and biotic stresses in soybean production: soybean production, vol 1, 1st ed. Academic Press/Elsevier Inc, New York, USA, pp 261–284

Nazeri P, Rad AHS, Valad Abadi SA, Mirakhori M, Masoule EH (2018) Effect of sowing dates and late season water deficit stress on quantitative and qualitative traits of canola cultivar. Outlook Agric 47(4):291–297

Pal S, Dubey N, Avinashe H, Khan S, Reddy PJ (2019) Estimation of genetic variability, correlation and path analysis for yield and yield contributing characters in Indian mustard (*Brassica juncea* L.). J Pharma Phytochem 4:102–105

Panda BB, Shivay YS, Bandyopadhyaya SK (2004) Growth and development of Indian mustard *(Brassica juncea)* under different levels of irrigation and dates of sowing. Ind J Plant Physiol 9(4):419–425

Patel N, Tyagi PK, Shukla KC (2013) Influence of sowing dates and varieties on total biomass and its partitioning into different plant parts of Indian mustard (*Brassica juncea* L.). Technofame 2:12–16

Patel A, Singh AK, Singh AV, Sharma A, Raghuvanshi N, Kuma A (2017) Effect of different sowing dates on growth, yield and quality of various Indian mustard (*Brassica juncea* L.) varieties. Int J Curr Micro Appl Sci 4:71–77

Peterson A, Murphy KM (2015) Quinoa cultivation for temperate North America: Consideration and areas for investigation. In: Murphy K, Matanguihan J (eds) Quinoa: improvement and sustainable production. Wiley, Hoboken, NJ, pp 173–192

Pinto RS, Reynolds MP, Mathews KL, McIntyre CL, Olivares-Villegas JJ, Chapman SC (2010) Heat and drought adaptive QTL in a wheat population designed to minimize cofounding agronomic effects. Theo Appl Genet 121:1001–1021

Prasad PVV, Pisopati SR, Ristic Z, Bukovnik U, Fritz AK (2008) Impact of night time temperature on physiology and growth of spring wheat. Crop Sci 48:2372–2380

Priya (2020) Characterization of morpho-physiological and biochemical traits for heat tolerance in Indian mustard (*Brassica juncea* L. Czern & Coss). Dissertation submitted to Punjab Agricultural University, Ludhiana

Qaseem MF, Qureshi R, Shaheen H (2019) Effects of pre-anthesis drought, heat and their combination on the growth, yield and physiology of diverse wheat (*Triticum aestivum L.*) genotypes varying in sensitivity to heat and drought stress. Sci Reports 9(6955):1–12

Rafiei S, Delkhosh B, Shirani Rad AH, Zandi P (2011) Effect of sowing dates and irrigation regimes on agronomic traits of Indian mustard in semi-arid area of Takestan. J Am Sci 7:721–728

Ram B, Meena HS, Singh VV, Singh BK, Nanjundan J, Kumar A, Singh SP, Bhogal NS, Singh D (2014) High temperature stress tolerance in Indian mustard (*Brassica juncea*) germplasm as evaluated by membrane stability index and excised-leaf water loss techniques. J Oilseed Brassica 5(2):149–157

Ram K, Kumar R, Sunita MR (2018) Stem reserve mobilization in relation to yield under different drought and high temperature stress conditions in wheat (*Triticum aestivum* L) genotypes. Int J Curr Micro Sci 7(4):3695–3704

Rameeh V (2012) Correlation analysis in different planting dates of rapeseed varieties. J Agric Sci 7(2):76–84

Ramesh V (2014) Multivariate regression analysis of yield associated traits in Rapeseed (*Brassica napus* L.) genotypes. Adv Agric 3:4–15

Rani B, Dhawan K, Jain V, Chhabra ML, Singh D (2013) High temperature induced changes in antioxidative enzymes in *Brassica juncea* (L) Czern & Coss. http://www.australianoilseeds.com/data/assets/pdffile/0003/6861/46
Rani T, Yadav RC, Yadav NR, Rani A, Singh D (2013) A genetic transformation in oilseed Brassica—a review. Ind J Agril Sci 83:367–373
Rani B, Kumari N, Jain V, Dhawan K, Avtar R (2015) Heat stress induced changes in protein profile of Indian mustard (*Brassica juncea* L.). J Oilseed Brassica 6(2):302–305
Rashid M, Hampton JG, Rolston MP, Khan KM, Seville DJ (2017) Heat stress during seed development affects forage brassica (*Brassica napus* L.) seed quality. J Agron Crop Sci 204:147–154
Rhythm (2019) Differential response of Indian mustard (*Brassica juncea* L. Czern & Coss) genotypes under rainfed and irrigated conditions. MSc thesis submitted to Punjab Agricultural University, Ludhiana
Rosyara UR, Vromman D, Duveiller E (2008) Canopy temperature depression as an indicator of correlative measure of spot blotch resistance and heat stress tolerance in spring wheat. J Pl Path 9(1):103–107
Sadia M, Malik SA, Rabbani MA, Pearce SR (2009) Electrophoretic characterization and the relationship between some *Brassica* species. Electron J Bio 5:1–4
Saini N (2017) Study on the effect of heat shock on stress responsive gene expression in Indian mustard (*Brassica juncea.* L). MSc thesis, Indira Gandhi Krishi Vishwavidyalaya, Raipur
Saleem N, Jan SA, Atif MJ, Khurshid H, Khan SA, Addullah M, Jahanzaib M, Ahmed H, Ullah SF, Iqbal A, Naqi S, Ilyas M, Ali N, Rabbani MA (2017) Multivariate based variability within diverse Indian mustard (*Brassica juncea* L.) genotypes. Open J Genet 7:69–83
Salvucci ME, Craft-Brander SJ (2004) Inhibition of photosynthesis by heat stress. The activation state of Rubisco as a limiting factor in photosynthesis. Physiol Plant 120:179–186
Sandhu SK (2019) Selection indices for improving selection efficiency in Indian mustard. J Crop Improv 33(1):25–41
Sandhu SK, Pal L, Kaur J, Bhatia D (2019) Genome wide association studies for yield and its component traits under terminal heat stress in Indian mustard (*Brassica juncea* L.). Euphytica 215(188):1–15
Sardana V, Sharma P (2012) Seed yield and oil content of Indian mustard (*Brassica juncea* l.) as influenced by sowing time. Poster paper presented in the 15th Punjab Science Congress held at GNDU, Amritsar, February 7–9, 2012. Abstracts No. 122, pp 16–17
Sardana V, Sangha MK, Atwal AK, Sheoran P (2008) Influence of sowing dates on thermal requirement, productivity and oil quality of Indian mustard, *Brassica juncea* (L.) Czern and Coss cultivars. J Oilseeds Res 25(1):82–84
Sardana V, Sharma P, Banga S, Kaur G, Brar KS (2014) Effect of sowing dates on productivity of low erucic and high erucic Indian mustard (*Brassica juncea*) cultivars. Theme III: Crop management challenges and possibilities. Abstract: 2nd National Brassica Conference on Brassicas for Addressing Edible Oil and Nutritional Security organized by Society for Rapeseed Mustard Research, Bharatpur at Punjab Agricultural University Ludhiana during February 14–16, 2014, p 104
Sharma P (2014a) Heat tolerance in Indian mustard (*Brassica juncea*) at seedling and terminal stage. Ann Pl Soil Res 16(3):224–228
Sharma P (2014b) Salicylic acid: a novel plant growth regulator-role in physiological processes and abiotic stresses under changing environments. In: Tuteja N, Gill SS (eds) Climate change and plant abiotic stress tolerance. Wiley-VCH Verlag Gmbh & Co KGaA Boschstr 12 69469 Weinheim, Germany, vol 2, pp 939–990
Sharma P (2015) Chlorophyll fluorescence parameters, SPAD chlorophyll and yield in Brassica cultivars. J Oilseed Brassica 6(2):249–256
Sharma P, Basra AS (2004a) Establishment of thermo tolerance in maize seedlings by plant growth regulators. Ann Pl Soil Res 6(2):184–188

Sharma P, Basra AS (2004b) Regulation of heat shock tolerance in maize seedlings by salicylic acid and its interaction with growth regulators. Crop Improv 31(2):119–127

Sharma P, Sardana V (2013) Screening of Indian mustard (*Brassica juncea*) for thermo tolerance at seedling and terminal stages. J Oilseed Brassica 4(2):61–67

Sharma P, Sardana V (2014) High temperature effect at terminal stage in mustard genotypes. In: Proceedings of national symposium on crop improvement for inclusive sustainable development, November 7–9, 2014 held at Punjab Agricultural University, Ludhiana, pp 428–430

Sharma P, Sardana V (2015) Correlation and heat susceptible index analysis for high temperature tolerance in Indian mustard (*Brassica juncea* L.) Phytomorphology 65(1–2):75–83

Sharma P, Sardana V (2016) Effect of low light on photosynthetic characteristics and yield in *Brassica juncea*. In: National Conference on Basic and Applied Researches in Plants and Microbes, 03–05,November 2016 held at Punjabi University, Patiala D0–13 pp 131

Sharma P, Sardana V (2017) Morphological and physiological responses of Indian mustard (*Brassica juncea* L.) genotypes to low light stress. In 3rdNational Brassica Conference (NBC2017) Enhancing Oilseed Brassica Production through climate smart technologies February 16–18, 2017, ICAR-IARI, New Delhi, India pp 221–222

Sharma P, Sardana V (2018) Physiological indices for heat tolerance in Indian mustard (*Brassica juncea*). In: 21st Punjab science congress on scientific advances for inclusive development and environmental protection, 7–9 February 2018 held at Punjab Agricultural University, Ludhiana AP-39, p 39

Sharma P, Sardana V (2019) Morpho-physiological and biochemical characterization of Indian mustard (*Brassica juncea* L.) under abiotic stresses. In: the 4th National Brassica Conference on Innovative approaches in oilseeds Brassica towards self-sufficiency held at Chandra Shekhar Azad University of Agriculture and Technology, Kanpur during February 1–3, 2019, p 43

Sharma P, Sardana V, Banga S, Salisbury P, Banga SS (2011a) Morpho-physiological responses of oilseed rape (*Brassica napus L.*) genotypes to drought stress. In: Proceedings of 13th International Rapeseed Congress held at Prague, Czech Republic, 5–9 June 2011, pp 238–240

Sharma P, Sardana V, Banga SS (2011b) Differential tolerance of Indian mustard (*Brassica juncea*) genotypes to high temperature at seedling and terminal stage. In: International conference preparing agriculture for climate change, February 6–8, 2011 held at PAU Ludhiana Abst no II B-30

Sharma P, Sardana V, Banga SS (2012a) Thermo tolerance responses of selected Brassica genotypes. In: Proceedings of 1st National Brassica conference on production barriers and technological options in oilseed Brassica held at CCS Haryana Agricultural University Hisar from 2–3 March 2012, pp 32–33

Sharma P, Sardana V, Banga SS (2012b) Shading effect on photosynthesis, dry matter accumulation and yield of Indian mustard (*Brassica juncea).* Crop Improv 130:27–29

Sharma P, Sardana V, Banga SS (2013) Effect of Salicylic acid on growth and seed filling in Indian mustard (*Brassica juncea*) under high temperature stress. Vegetos 26(1):243–248

Sharma P, Sardana V, Banga SS (2014) Identification of morpho-physiological and heat resistance characteristics in Indian mustard under terminal heat stress. Theme IV: Crop physiology and nutritional quality. Abstract: 2nd National Brassica Conference on Brassicas for Addressing Edible Oil and Nutritional Security organized by Society for Rapeseed Mustard Research, Bharatpur at Punjab Agricultural University Ludhiana during February 14–16, 2014, pp 120–121

Sharma P, Kaur S, Sardana V, Salisbury P, Banga S, Banga SS (2015a) Characterization of morpho-physiological responses associated with water stress tolerance in *Brassica napus*. In: 14th International Rapeseed Congress held at Saskatoon, Canada, 5–9 July 2015 under the Theme D: Abiotic stress on Monday 6, July 2015

Sharma P, Sardana V, Banga SS (2015b) Genetic variation for heat tolerance in *Brassica juncea*. In: Proceedings of 14th International Rapeseed Congress held at Saskatoon, Canada, 5–9 July 2015, Poster no. 508

Shengri AH, Gharineh MH, Bakhshandeh A, Fathi G, Karmollachaab (2012) Effect of terminal heat stress and nitrogen on morpho-physiological characteristics, ANUE and yield of canola (*Brassica napus* L.). Int J Agril Crop Sci 4(6):284–288

Shilpa G, Akhatar J, Kaur P, Sharma A, Sharma P, Mittal M, Bharti B, Banga SS (2019) Genetic analyses of nitrogen assimilation enzymes in *Brassica juncea* (L.) Czern & Coss. Mol Biol Rep. https://doi.org/10.1007/s11033-019-04878-5 (online)

Shivani, Kumar S (2002) Response of Indian mustard (*Brassica juncea*) to sowing date and row spacing in mid hills of Sikkim under rainfed conditions. Ind J Agron 47:405–410

Sikder S, Paul NK (2010) Effect of post anthesis heat stress on stem reserves mobilization, canopy temperature depression and floret sterility of wheat cultivars. Bangladesh J Bot 39(1):51–55

Singh RB, Singh RN (2006) Spray schedule for the management of Alternaria blight and white rust of Indian mustard (*B. juncea*) under different dates of sowing. Ind J Agril Sci 76:575–579

Singh PK, Singh AK (2014) Effect of different dates of sowing and irrigation scheduling on growth and yield of mustard (*Brassica juncea* L.). J Prog Agric 5:18–21

Singh AK, Singh RR, Singh AK, Singh PK (2014) Influence of dates of sowing and irrigation scheduling on growth and yield of mustard (*Brassica juncea* L.). Int J Farm Sci 4:80–85

Singh V, Pant U, Bhajan R (2016) Physiological traits versus seed yield derived parameters based heat stress study in Indian mustard. J Breed Genet 48(2):127–135

Sinsawat V, Leipner J, Stamp P, Frachcboud Y (2004) Effect of heat stress on the photosynthetic apparatus in maize (*Zea mays* L.) grown at control or high temperature. Env Expt Bot 52:123–129

Soengas P, Rodriguez VM, Velasco P, Cartea ME (2018) Effect of temperature stress on antioxidant defenses in *Brassica oleracea*. Epubl 3(5):5237–5243. https://doi.org/10.1021/acsomega.8b00242

Solanki NS, Mundra SL (2015) Phenology and productivity of mustard (*Brassica juncea* L.) under varying sowing environment and irrigation levels. Ann Agril Res 36(3):312–317

Srivastava JP, Balkrishna (2003) Environmental parameters influencing phenological development of mustard in relation to yield. Ind J Plant Physiol 8:349–353

Tasmina T, Khan AR, Karim A, Akter N, Islam R (2016) Physiological changes of wheat varieties under water deficit conditions. Bangladesh Agron J 19(2):105–114

Tsukaguchi T, Kawamitsu Y, Takeda H, Suzuki K, Egawa Y (2003) Water status of flower buds and leaves as affected by high temperature in heat tolerant and heat-sensitive cultivars of Snap bean (*Phaseolus vulgaris* L.). Plant Prod Sci 6:4–27

Tuteja SS, Lakpale R, Tripathi RS (1996) Effect of date of sowing, nitrogen level and nitrogen splitting on mustard (*B. juncea*). Adv Plant Sci 9:167–168

Tyagi V, Dhillon SK (2019) Water use efficient sunflower hybrids having diverse cytoplasmic background. Helia 42(71):187–201

Vasseur F, Pantin F, Vile D (2011) Changes in light intensity reveal a major role for carbon balance in *Arabidopsis* responses to high temperature. Plant Cell Environ 34:1563–1576

Wahid A (2007) Physiological implications of metabolites biosynthesis in net assimilation and heat stress tolerance of sugarcane (*Saccharum officinarum*) sprouts. J Pl Res 120:219–228

Wang LJ, Fan L, Loescher W, Duan W, Liu GJ, Cheng JS, Li SH (2010) Salicylic acid alleviates decreases in photosynthesis under heat stress and accelerates recovery in grapevine leaves. BMC Plant Bio 10:28–34

Wilson RA, Sanga MK, Banga SS, Atwal AK, Gupta S (2014) Heat stress tolerance in relation to oxidative stress and antioxidant in *Brassica juncea*. J Environ Biol 35:383–387

Wojciechowska R, Hanus-Fajerska EJ, Kolton A, Kaminska I, Grabowska A, Kunicki E (2013) The effect of seedling chilling on glutathione content, catalase and peroxidase activity in *Brassica oleracea* L. var. italic. Acta Soc Bot Pol 82:243–248

Woods DL, Capcara JJ, Downey RK (1991) The potential of mustard (*Brassica juncea* L.) Coss) as an edible oil crop on the Canadian Prairies. Can J Pl Sci 71(1):195–198

Wu X, Tang Y, Li C, Chun W (2018) Characterization of the rate and duration of grain filling in wheat in south western China. Pl Prod Sci 21(4):358–369

Young LW, Ron WW, Peta C (2004) High temperature stress of *Brassica napus* during flowering reduces micro and mega gametophyte fertility, induces fruit abortion and disrupts seed production. J Expt Bot 55:485–495
Yuan L, Tang L, Zhu S, Hou J, Chen G, Liu F (2017) Influence of heat stress on leaf morphology and nitrogen-carbohydrates metabolism in two wucai (*Brassica campestris* L.) genotypes. Acta Soc Bot Pol 86(2):1–16
Zhang J, Jiang F, Yang P, Li J, Yan G, Hu L (2015) Response of canola (*Brassica napus* L.) cultivar under contrasting temperature regimes during early seedling growth stage as revealed by multiple physiological criteria. Acta Physiol Plant 37:7–13
Zhu C, Gore M, Buckler ES, Yu J (2008) Status and prospects of association mapping in plants. Plant Genome 1(1):5–20

Brassicaceae Plants Response and Tolerance to Waterlogging and Flood

Mrinalini Prasad and Rajiv Ranjan

Abstract Waterlogging is the most severe worldwide problems in agriculture. The presence of excess amount than its optimum requirement is called as waterlogging. In the waterlogged/flooded condition, water gets filled in the pores of the soil, so the oxygen concentration decreases in the soil. Oxygen deficiency decreases growth and survival of plants growing in it also induces stomatal closening. The flooding frequently induces stomatal closing mostly in C3 plants. Most of the crop plants are more susceptible to flooding and every year significant yield reduction accounts due to flooding. The present book chapter highlights the recent development and approaches to improve crop resistance to flooding of Brassicaceae plants. Most of the studies were carried out on maize, barley, and soybean. The study and research on the crop like brassica plants has just been started. Some of the promising traits that are important to enhance crop flooding tolerance are anatomical adaptations, i.e., formation of aerenchyma, the barrier formation against radial oxygen loss, and the adventitious roots development. Metabolic adaptations might be helpful to improve waterlogging tolerance as well, but further studies are required in this direction. The various approaches like marker assisted selection, quantitative trait loci analyses, or genome-wide association studies in combination with specific tolerance traits need to be addressed. The use of flooding-tolerant varieties/cultivars of the crop of interest might be helpful in finding useful tolerance traits to be used in plant breeding.

Keywords Flood · Waterlogging · Adaptation · Aerenchyma · ROS

1 Introduction

Across the world due to increasing food demand, scientists are trying to produce more food crops in waste land in general areas where water, soil condition, and temperature are not suitable for crop production. Waterlogging is one of the most regular phenomenon can be also known as flood, submergence, soil saturation, hypoxia,

M. Prasad (✉) · R. Ranjan
Faculty of Science, Department of Botany, Dayalbagh Educational Institute (Deemed University), Dayalbagh, Agra-5, India
e-mail: mrinu.prasad.raj@gmail.com

M. Hasanuzzaman (ed.), *The Plant Family Brassicaceae*,
https://doi.org/10.1007/978-981-15-6345-4_10

anoxia. Generally, two types of flood are reported in the field first waterlogging in which root and some parts of shoot goes under water and second is complete submergence where total plant goes under water (Hasanuzzaman et al. 2017; Anee et al. 2019). Waterlogging is a perennial problem in low laying rainfed areas. The main reason for damage in waterlogging condition is lack of oxygen supply to the plants that hampers nutrients and water uptake as a results plants show wilting.

Flooding is creating misbalance in ecosystems and therefore all living creatures are developing or adapting a tolerance regarding high water level and enhance the capacity for gas exchange into the environment is important to avoid hypoxia inside the plant. Oxygen concentration is low in soil (hypoxia) and no presence of oxygen (anoxia) has an effect on nutrient uptake, growth regulation, carbohydrate reaction, respiration, photosynthesis (Ferreira et al. 2008; Table 1). Shortage of oxygen in roots develops a change in anaerobic and aerobic respiration which reduced the utilization of organic compounds and accumulation of ethanol (Halcomb 2003). In oxygen deficit condition plants shift its metabolism anaerobic from aerobic mode. Plant can tolerate waterlogging conditions by various mechanisms like increased availability of dissolved sugar, aerenchyma formation, higher level of ethylene production, greater activity of glycolytic pathway and fermentation enzymes and involvement of antioxidant defense mechanism to adapt oxidative stress caused by waterlogging.

Brassica napus L. and relatives rape (*Brassica napus* L.) belong to Brassicaceae family, is a main oilseed crop that can also be used as both food and feed. It is an allotetraploid species and has complex genetic structure that derived from the two diploid species *B. rapa* and *B. oleracea.* It has broad range of cultivars ranging from winter to spring types. *Brassica napus* L is more susceptible to waterlogging, in contrast to other Brassica species (Ashraf and Mehmood 1990). This is at least in part due to its inability to form aerenchyma (Voesenek et al. 1999), similar to its parent *B. rapa* (Daugherty et al. 1994). There are various effects caused by flood as described in detail.

Table 1 Plant development for the period of submergence and waterlogging condition

S. No.		Short term		Long term	
		Escape	Quiescence	Escape	Quiescence
1.	Photosynthesis	−−	−−	+	−−
2.	O_2 levels	−−	−	+	−
3.	Metabolic activity	++	+	−	+
4.	Reserve mobilization	++	−+	−	−+
5.	Glycolytic flux	++	−	−+	−
6.	Shoot elongation	++	−	−	−
7.	Carbon starvation	++	−+	−−	−+
8.	Adventitious root	++	−	++	−

2 General Effect of Flood

Due to flood, plants leave accumulate starch in leaves, phloem transport decline, reduced carbohydrates in root (Liao and Lin 2001). As we all know that the oxygen is the most important part of life and Reactive Oxygen Species (ROS) is used in various cellular activities in normal condition (Ashraf and Akram 2009). When the plants having abiotic stress like flood then the ROS level is enhanced and it damaged many cellular/ metabolic activity like photosynthesis II, calvine cycle. ROS develops unstable free radicals which react with other cellular molecules and produce more unstable radicals (Foyer and Halliwell 1976). ROS is generally produced by the reaction between molecules of aerobic system with ionizing radiations and specific organelles where ROS produces in chloroplasts, mitochondria, and Peroxisomes (Sairam and Srivastva 2002). In the flooded condition, ethylene accumulated inside the plant shows the harsh condition of plant (Voesenek and Sasidharan 2013) and in other hand oxygen availability. Both the gases generate a balance in inner region between production, consumption, and diffusion activity. Meristems and roots easily experience low oxygen condition due to waterlogging and activity of photosynthetic tissues are dependent on oxygen availability. Ethylene used an activator for the ABA, GA, and auxin to stimulate adventitious rooting in different plants (Vidoz et al. 2010; Dawood et al. 2016; Steffens et al. 2006). However, root also need ethylene to persuade for ROS construction in the epithelial cell and leading to their cell death to allow root penetration (Steffens and Sauter 2009) and successive raise in ROS leading to the cell death (Rajhi et al. 2011; Yamauchi et al. 2011; Steffens et al. 2011).

2.1 Effect of Soil on Waterlogging

When the water flows on the soil surface faster than absorption then it's called waterlogging of submergence and water level increased by the climate change. Due to this excess water flow surface run-off increased, made worse flooding problem, soil aeration reduced, denitrification and leaching are increased (Belford et al. 1992; Rienk et al. 2002). The soil damage was occurred not only by the flow of water but also by the rate of water flow on the land and it is especially affected by soil structure. The properties of soil play an important role in the limiting of water flow, i.e., pore size, pore volume, interconnection of soil, proportion of each size class and stability. According to Greenland (1977) classification, more than 50 μm pore size is help in the drain water under gravity, aerobic respiration, and root penetration. And pore size with 50–0.5 μm can grasp water against the flow but frail for root to explore it and less than 0.2 μm pore size have strongly hold water so that neither root nor gravity can extract it (Biswas and Kalra 2018). Classification of soil pores according to Greenland:

A. Residual (pore <0.5 μm diameter, Water not extractable, Always filled with water and 23%),
B. Storage (50–0.5 μm diameter, Extractable water, Water or gas filled and 38%), and
C. Transmission (Pores >50 μm diameter, Drain under gravity, gas filled at field capacity and 3%).

2.2 *Physiological Effect*

Due to flood, plant shows the stomata stress and limited water stress and it creates a deficiency of internal water (Parent et al. 2008a, b). Low oxygen level causes decrease in hydraulic conductivity and photosynthetic rate (Else et al. 2001; Ashraf et al. 2011). And this condition is responsible for stomata closing (Ashraf and Arfan 2005), reduced chlorophyll contents, demolition of chlorophyll (Jackson and Lambers 1992; Huang et al. 1994; Ashraf et al. 2011), decreased leaf area (Malik et al. 2001) but increase the concentration on reactive oxygen species (ROS). ROS influences the harm of metabolic reactions, cellular molecules, and different metabolites, i.e., lipids, proteins, pigments, DNA, etc. (Ashraf 2009). The ROS having free radicals (unstable condition) in the form of unpaired electrons and they produced more free radicals after reaction with other cellular molecules (Foyer and Halliwell 1976; Hideg 1997). Long-drawn-out contact of plants to waterlogging could produce root injuries which limit the photosynthesis and hamper the biochemical reactions like Ribulose Bisphosphate Carboxylase (RuBPC) activity, glycollate oxidase, and phosphoglycollate (Yordanova and Popova 2001), destruction of chloroplast membrane which hinders the photosynthesis efficiency and electron transport (Titarenko 2000), few examples are *Lolium perenne* (McFarlane et al. 2003), *Lycopersicon esculentum* (Bradford and Yang 1980; Jackson 1990) *Pisum sativum* (Zhang and Davies 1987), and *Triticum aestivum* (Trought and Drew 1980). Flooding can reduce the oxygen influx to the roots (Vartapetian and Jackson 1997) and hinder in the diffusion of gases breakdown such as ethylene (Arshad and Frankenberger 1990) or CO_2 produced by microbes of soil and root. High accumulation of ethylene reduced the extension of root (Boru et al. 2003) and carbon dioxide converted to bicarbonate ions which create a condition of high lime that affects to the plant like chlorosis, iron unavailability, etc. Due to shortage of oxygen, some biochemical changes were occurred in soil (Ponnamperuma 1972) because microbes consume inorganic ions in a place of free electron which is an acceptor of oxygen to generate energy. At the initial stage of anaerobic reduction, nitrates convert to nitrite, nitrous oxide and nitrogen gas and it results in reduced quantity of nitrate around root. And other reduces oxides are formed which affects the enzymatic activity of the cells they are Mn^{4+} and Fe^{3+} oxidized to highly soluble Mn^{2+} and Fe^{2+} (Laanbroek 1990). If the duration of food was increased then the anaerobic bacteria alter SO_4^{2-} to H_2S which is a poison for respiratory enzyme and it reduced the iron quantity in soil. The effect of flood on soil is the most destructive and harmful. Most bacterial activity increases in the soil, e.g., methogenic bacterial converts carbon dioxide to methane, increase soil borne fungal diseases by *Gliocladium roseum* in case of germination seeds (Yanar et al. 1997), alfalfa infection in vegetables by phytophthora and trees by pythium (Walker 1991).

3 Adaptation

Due to flood, plant develops aerenchyma cells for the longitudinal connection of the plant and air to create a gas space, which supply an easy passage of aerial gas exchange over long distance inside the plant (Takahashi et al. 2014). These spaces create a connection between flooded root tip and aerial surface (Steffens and Rasmussen 2016) and for adventitious root hypocotyls or stems are build a link into the anaerobic substance. However, aerenchymas are protected from oxygen leakage into the closer anaerobic soil, which significantly recovers flooding tolerance (Abiko et al. 2012). This system is efficient when the plant in waterlogged condition where plant shoot gets in touch of air then flows down to the root. But in complete submergence condition shoot does not make a contact with air or oxygen and the after effect root also compromise. In that case, few wetland plant species develop an aerial shoot contact, vertical elongation of leaves, petiole and internodes develop snorkel. This approach was observed in few rice varieties and other plants (Jackson 2007). In other hand, plant produces an inactive condition in which it reduces the energy and carbon utilization to make reserves last a long time and maintain cellular integrity (Voesenek and Bailey-Serres 2015; Akman et al. 2012). In case of submergence condition, both short and long term plants are affected but in long term escaping plants will have recovered aerial or O_2 level contact (Table 2). Photosynthesis is also decreased by flooding, however during escape approach few photosynthesis can regain the establishment of the oxygen level contact. Primarily, escaping submerged plants will have high metabolic action by which reserve mobilization and glycolytic flux increases. The escaping plants are strongly suffering and quiescence plants are having low energy for activity. An aqueous condition hinders the ethylene to quit the plant tissues through gas diffusion (Fig. 1). The ethylene entrapment has become an important point from which help in modification of root architecture and construction of adventitious roots. It also becomes a pool of hormones and epidermal cell was dead

Table 2 The different scenarios encountered by plants in front to increasing levels of water excess, ranging from waterlogging to complete submergence (Striker 2012)

Waterlogging	Flooding	
	Partial submergence	Complete submergence
When the root system is under anaerobic conditions	Complete roots are immersed in water while just a portion of the shoot is dipped by water	Complete plants are under the water level. Water depth and turbidity are important factors

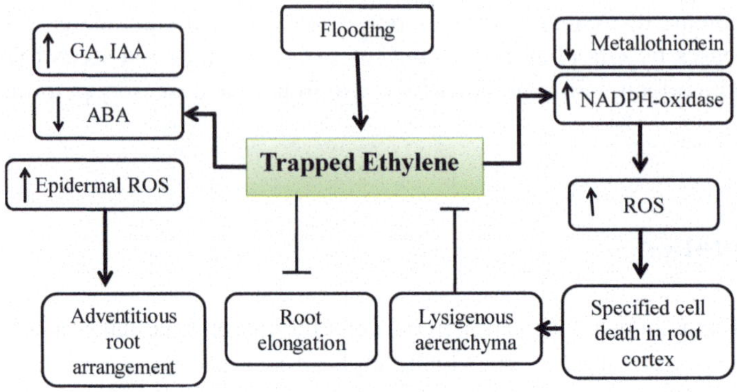

Fig. 1 Entrapment of ethylene in waterlogged condition

due to ROS induced. However, ethylene helps in the reduction of metallothionein and raise NADPH oxidase.

3.1 Physiological Adaptation

At the first point, plant responses to waterlogging are the diminution in stomata conductance, partial water uptake (Parent et al. 2008a, b). Due to damage of root, the plant faces low oxygen condition which leads to the decline in photosynthetic rate, reduction in chlorophyll contents, and many more problems related to leaf. Long exposure of plant to waterlogging condition developed root injuries and it changes some biochemical reactions in photosynthesis such as restrict the activity of Ribulose Bisphosphate Carboxylase (RuBPC), glycollate oxidase, phosphoglycollate, destruction of chloroplast membrane, hinder electron transport, and reduction in quantum energy (Fv/Fm) of PSII (Photosystem II) (Pociecha et al. 2008). Lysigenous aerenchymas are formed by apoptosis of cortex cells which engaged in the fall of antioxidant activity due to ethylene. Some inert oxygen is not affected by ethylene but few genes are involved in this activity (Shiono et al. 2014). Ethylene also helps the plant to escape on the water surface which is the downstream signaling process found in rice using VII ERFs group (van Veen et al. 2013, 2014; Hattori et al. 2009). As the after effect of flood, ethylene is increased which restricts the root elongation but by the help of aerenchyma cells unnecessary ethylene is removed. Few plants are unsuccessful in the development of aerenchyma are facing lessening in root growth. Ethylene signaling plays an important role in the development of root in adverse condition. Excess flood tolerant species are lost their ability to react with ethylene (Pierik et al. 2006; Voesenek et al. 2015; Loreti et al. 2016).

3.2 Anatomical Adaptation

Aerenchyma tissue found in the root cortex system which corrects the aerial spaces from aerial shoot to the submerged root for the transportation of oxygen (Colmer and Greenway 2005). By the arrangement of these tissue with cortex cells are separated in four types of root structure and there shape—*Apium* (irregular), and *Rumex* (honeycomb), graminaceous (bicycle wheel), cyperaceous (spider web) (Justin and Armstrong 1987). Lysigeny is the pattern of aerenchyma tissue in which disintrigate and death of cells present in the cortex region and in schizogeny expansion of intracellular spaces into radial sectors (Seago et al. 2005).

3.3 Morphological Adaptation

Waterlogging causes many morphological and anatomical changes in the submerged plants (Table 3). It develops hypertrophied lenticels (Yamamoto et al. 1995; Parelle et al. 2006), radical cell division, hypertropical growth, formation of adventitious roots (Malik et al. 2001), and the most important adaptation in the root cortex is the development of lacunae gas spaces (aerenchyma) (Evans 2004). In the submerged condition the plant develops a replacement root systems and their mechanisms are 1. Formation of the outgrowth of previously existing root primordial in the shoot base (Jackson et al. 1981), 2. Initiation of a new root system which creates root primordial and their outgrowth (Jackson and Armstrong 1999; Shimamura et al. 2007), and 3. Insertion of roots at the soil surface connecting of the root system as seen in woody species (Pereira and Kozlowski 1977) and herbaceous species (Gibberd et al. 2001). First two mechanisms were focused on ethylene which is developed the activity in presence of auxin. The lenticels are helping in the downward diffusion of CO_2 and in the production of ethanol, CO_2, and CH_4. It is acclimatized by two different ways first is development of aerenchyma which creates highly separation in specific pattern of cells and produced abiotic stress and regulate external stimulus. It also occurs cell death as a result of separation is called shizogeny. And in case of second adaptation, enlargement of aerenchyma called as Isogeny which is partially stop working of cortex. This adaptation of cells becomes cell death and again formation of external stimulus just like abiotic stress (Pellinen et al. 1999). The first adaptation is to avoid the oxygen deficiency inside the plant during submerged condition then it improves by morphologically modification of gas exchange by the surrounding. Under submerged condition some plants develop leaves in upward direction and few enhance shoot growth for the contact with air. This is called "escape strategy" and it develops growth regulation in petioles.

Table 3 Comparison of leaf traits influencing gas exchange and photosynthesis by terrestrial wetland plants when under water and submerged aquatic plants (Pedersen et al. 2013)

Leaf traits for	Terrestrial wetland plants	Submerged aquatic plants
Morphology		
Leaf size	Medium-large	Small-medium
Dissected/lobed	Rare	Common
Strap-shaped	Rare	Common
Leaf thickness	Moderate-thick	Thin
Hairs/trichomes	Rare	Absent
Surface hydrophobicity resulting in leaf gas films	Common	Absent
Anatomy		
Stomata	Always present	Absent/non-functional
Cuticle	Always present	Absent/highly educed
Chloroplasts in epidermal cells	Only in guard cells	Common
Aerenchyma	Variable	Variable
Supporting fibres	Always present	Rare
Porosity of lamina	High in thick, low in thin lamina	High in thick, low in thin lamina
Photosynthetic pathway		
C3	Common	Common
C4	Rare	Rare (but uncertain)
CAM	Absent	Rare
Use of HCO_3	Absent	Common

4 Mechanism of Root Tips Adaptation

Under waterlogging condition, the survival of plants depends upon the sufficient capacity and limited speed of internal transport of oxygen. A porous tissue is present on the internal system of plants which creates gas-filled spaces and help in the movement of gases called aerenchyma. In some species aerenchyma can help to control aeration of roots up to 300 mm and respiration. The aerenchyma can be increased due to waterlogging and submergence condition. The absence of oxygen in the soil has an effect on root tips and the damage caused to the root is due to insufficient supply of ATP and byproducts of anaerobic metabolism. In anoxia condition plant tissues decreased the rate of production by 65–97% (Gibbs and Greenway 2003). Roots generate ATP by using glycolysis pathway which converts pyruvic acid into ethanolic and lactic acid fermentation (Table 4). At the starting hours, metabolic processes are censored due to less availability of ATP. Cell death can be neglected if small amount of present energy can be utilized for the formation of important anaerobic protein (Xia et al. 1995). The decrease of membrane integrity is due to imbalance of ATP by

Table 4 Possible causes of plant injury by hypoxia and anoxia, and mechanisms of adaptation

Causes of injury	Mechanisms of adaptation	References
Self-poisoning by fermentation end-products (ethanol)	Avoidance of fermentation end product (ethanol) accumulation	McManmon and Crawford (1971), Crawford (1978)
High sensitivity of plant root cells to oxygen deficiency	Avoidance of root anaerobiosis by oxygen translocation from shoots. Aerenchyma formation. Fast under-water elongation. Adventitious rooting	Vartepetian et al. (1970), Vartapetian (1973), Webb and Armstrong (1983), Ap Rees and Wilson (1984), Armstrong et al. (1991), McPherson (1939), Musgrave et al. (1972), Jackson and Drew (1984)
Cytoplasmic acidosis	Avoidance of cytoplasmic acidosis	Davies et al. (1974), Davies (1980), Roberts et al. (1982, 1984, 1985)
Energy starvation	Avoidance of energy starvation by efficient substrate provision and ATP production	Vartapetian et al. (1976, 1978a), Saglio et al. (1980, 1988), Andrews and Pomeroy (1981, 1983), Mocquot et al. (1981), Webb and Armstrong (1983), Johnson et al. (1989, 1994), Waters et al. (1991), Perata et al. (1992a, b, 1993), Sachs (1993), Subbaiah et al. (1994), Vartapetian and Poljakova (1994), Xia et al. (1995)
Post-anoxic damage from free radicals	Detoxification of free radicals	Monk et al. (1987), Crawford et al. (1994), Pfister-Sieber and Brandle (1994)
Impedance to the supply of water, inorganic minerals, and hormones from damaged roots to shoots	Modifying root–shoot relationships to decrease shoot damage caused by injured roots	Kramer and Jackson (1954), Jackson and Drew (1984), Drew (1991), Jackson (1993), Vartapetian and Jackson (1997)

the activity of lipid hydrolysis which is affected by lipolytic acyl hydrolase (Rawyler et al. 1999). In an anaerobic condition, scarcity of sugar is due to starch breakdown and sugar utilized by the roots reflects the less survival. By this process, rice seedlings are germinated without oxygen as a result of α-amylase coding gene which is activated in anaerobic condition (Loreti et al. 2003). Roots seedling are longer time stay alive because of external glucose supply (Webb and Armstrong 1983). The cause of death of plants is caused by less glycolysis rate, less supply of ATP, absence of molecular oxygen, and many more (Vartapetian et al. 1978a, b). It is also an important factor for cell death. Protons as byproduct of anaerobic metabolism are creating acidic surrounding in cytoplasm and vacuole (Waisel and Eshel 2002). Another toxic compound is acetaldehyde, which generally converts to ethanol in presence of enzyme. Excess production of acetaldehyde is generating cell damage (Boamfa et al. 2003). Rapid death of root tips is an acclimatization response (Subbaiah and Sachs 2003). An absence of oxygen and ATP, plant cells cannot protect them adjacent to accomplishment of active oxygen species after the flood (VanToai and Bolles 1991).

5 Antioxidant Defense Mechanism of Plants

Plants have a quality to reduce the effects of ROS (reactive oxygen species) by generation of different antioxidants like catalase (CAT), peroxidase (POD), ascorbate peroxidase (APX), glutathione reductase (GR), superoxide dismutase (SOD), glutathione, tocopherols, ascorbic acid, and carotenoids (Gupta et al. 2005; Hasanuzzaman et al. 2019). For example, in case of mungbean, waterlogging condition decreased SOD, GR, APX and CAT (Ahmed et al. 2002) and in corn plant (Yan et al. 1996).

5.1 Nutrient Uptake Effect

Stresses can affect the plant growth and nutrient uptake like salinity, drought, and flood. In case of waterlogging the tolerant plant reduces nutrient uptake, plant growth rate, and root–shoot ratio. Under submerged condition, the quantity of soluble carbohydrates like fructose, glucose, sucrose are increased in shoots shows that photosynthesis does not limit nutrient uptake. In general, both the N and Mn create low redox potential in soil that encourages denitrification of NO_3 and produces Mn^{2+}. It also reduces the selectivity of K^+/Na^+ uptake in root and hinders the transport of K^+ in shoot. It causes lacking in N, K, Mg, Ca, and P.

5.2 Biochemical Approach

Plant growth regulators (PGRs) like gibberellic (GA) and cytokinins (CK) are declined in the roots and improve the intensity of ethylene and ABA. Under waterlogging condition quantity of auxins decreased and precursor (Aminocyclopropane-1-Carboxylic Acid) of ethylene biosynthesis are enhanced. Some after effects of flood on PGRs and their consequences on different plants like decrease auxins shows swelling of stem and cortex of cell, reduced GA illustrate cell elongation and stem enlargement, ABA amplification reduced the rate of gas exchange during photosynthesis, transpiration, and respiration and at the end increased ethylene shows epinasty of leaves. Binding of oxygen to the ubiquitous in plant is another way but the binding is too compact for recognition of oxygen (Dordas et al. 2003). Genes coding for anaerobic protein (Hypoxically Induced Proteins or HIPs) are enhanced in post and transcriptional level when oxygen is in deficient condition and it is needed for adaptations (Subbaiah and Sachs 2003; Dolferus et al. 2003; Baxter-Burrell et al. 2003).

6 Mechanism of Arial Shoots Against Waterlogging

Soil waterlogging and submerged roots are directly affecting the shoot system; nitrate uptake is reduced due to deficiency of oxygen. Juvenile leaves are taking the nutrient from the older leaves for their activity (Drew and Sisworo 1977).

Some important systems of drainage water

There are few suggestions of different systems which can be helpful for the drainage of water and reduced the effect of waterlogging condition.

- Re-designing or re-structuring the area of the field
- Developing the surface area of drainage
- Increasing the system of subsurface drainage
- Rising a new and improved structure of underground drainage system
- Gives a sufficient drainage system around the root system.
- Supply some growth retardant (500 ppm Cycocel) by spray system for reducing apical growth
- For leaf area some spray are used 1% KCl + 2% DAP (MOP)
- Nipping terminal buds for capturing apical dominance
- NAA is an important one which helps in the control of premature fall of buds, flower, and early maturation of fruits and pods.
- To improve photosynthesis rate 0.5 ppm of Brassinolide is supplied
- 100 ppm Salicylic acid was foliar spray which used for the increasing stem reserve under high moisture stress
- Under critical stress condition 0.3% Boric acid + 0.5% $ZnSO_4$ + 0.5 and % $FeSO_4$ + 1.0% was foliar spray.

7 Conventional Breeding for Tolerance

Many research institutes have tried to develop submergence tolerant varieties with desired agronomic characters. In China, as a part of experiment, 18 cultivars were compared with different seed coat color and seed viability was tested after 24 h of submergence and interestingly, yellow-colored cultivars were found to be mainly susceptible to submergence at the seed stage, and dark-colored cultivars were mostly tolerant, including Zhongshuang 9 and 10 (Zhang et al. 2008a, b). These cultivars also were used for other screens of waterlogging tolerance. Further screens were performed with more Chinese lines at post developmental stages (Zou et al. 2013a, b; Zou et al. 2014; Xu et al. 2015), confirmed previous tolerant lines (Zhongshuang 9 and 10), and identified further lines to be tolerant (Huayouza 9, Ningyou 12, Xiangyou 13,) and susceptible (Yuhuang 1, Zhongyouza 3, Zhongshuang 8). When genetic analysis was carried out it was suggested that two genes were involved in waterlogging tolerance of Zhongshuang 9, but no QTL has been associated with this tolerance (Cong et al. 2009). In separate study, a cross between six cultivars differing in waterlogging tolerance was conducted and it was reported that additive and non-additive effects pointing to many genes or alleles involved in which Zhongshuang 9 being the most elite cultivar (Cheng et al. 2010). Further, an independent group was conducted a study used a double haploid population between two lines different from the ones discussed above with high and low waterlogging and drought tolerance (Li et al. 2014). Mapping of quantitative trait loci associated (QTL) was studied with waterlogging tolerance and drought resistance during the seedling stage in oilseed rape (*Brassica napus*). The group were identified at least 11 different QTL for waterlogging tolerance in this species implying a complex regulation. In another effort to associate waterlogging tolerance with physiological traits suggested the significance of the antioxidant system. Further analyses reported a difference in nitrate metabolism between two Chinese cultivars under waterlogging condition (Yu et al. 2014). Moreover, a correlation of low ethanolic fermentation and higher waterlogging tolerance was reported by Xu et al. (2016).

Till date there is little information available on the molecular response of rape seed to waterlogging or submergence, making it hard to find a basis to select candidate genes in the QTL regions. In first attempt the aim was to solve this problem conducted on two Chinese cultivars with contrasting waterlogging tolerance, Zhongshuang 9 and GH01 (Zou et al. 2013a, b; Zou et al. 2015). The plant roots to 12 h of waterlogging were taken and its transcriptional response studied using RNAseq, and many genes were reported and some responded to the stress treatment, mainly similar in both genotypes. Though, there were minor differences between the two cultivars, might be related to the plant hormone abscisic acid, but it has yet to be explored further. But, full QTL analysis of these two lines was not yet conducted. Another transcriptomic analysis for shoot responses after 36 and 72 h of waterlogging was evaluated, revealing down regulation of photosynthesis (Lee et al. 2014), as was also shown for soybean (Chen et al. 2016). The genus Brassica includes the diploid species *B. nigra*, *B. rapa*, and *B. oleracea*, and the allotretraploid species *B. napus*, *B. juncea*, *and B. carinata* (Zhang et al. 2016). Very few literature is known on waterlogging tolerance of other Brassica species, in separate study it was described that *B. juncea and B. carinata* are more waterlogging tolerant than rape (Ashraf and Mehmood 1990). A direct comparative relationship between rape and diploid Brassica species under waterlogging would be most useful. A comprehensive study was conducted with two populations of *B. rapa* with different waterlogging tolerance suggested the significance of carbohydrate supply to roots as a potential factor for waterlogging tolerance (Daugherty and Musgrave 1994). One study was done on *B. oleracea* at the seed stage and reported that seed germination of Brassica species is mainly dependent on oxygen. A QTL study was performed between two *B. oleracea* cultivars, a sensitive Chinese cultivar (A12DHd) and a more tolerant calabrese cultivar (GDDH33), suggested that three QTLs are associated with seed viability under low oxygen concentrations (Finch-Savage et al. 2005), however associated genes were not yet identified.

The relative more closer to the rape and the well-studied model Brassicaceae *Arabidopsis thaliana* (L.) Heynh. might be another way to exploit existing knowledge in improving rape waterlogging tolerance. First, ecotype identified on *Arabidopsis* has established differences in submergence tolerance (Vashisht et al. 2011; Van Veen et al. 2016; Akman et al. 2017; Yeung et al. 2018). Although, *Arabidopsis* is rather flooding-sensitive, the analysis of flooding-tolerant Brassicaceae from the *Rorippa* genus (Akman et al. 2012; Sasidharan et al. 2013) might be more appropriate.

8 Molecular Strategies for Improving Waterlogging Tolerance in Plants

Scientists across the globe are actively involved in making the plants tolerant to flooding stress by application of nutrient and plant hormones externally. In recent past, Ashraf et al. (2011) reported that exogenous application of potassium in soil and as foliar spray improved the adverse effects of waterlogging on cotton plants. Earlier Ashraf and Rehman (1999) reported that application of nitrate in soil proved useful in mitigating the damaging effects of waterlogging on various physiological attributes of maize. Similarly, in 2009, Yiu et al. observed that exogenous application of spermidine and spermine facilitates several biochemical and physiological and biochemicals adaptations in anion when exposed to flooding stress. In addition to that application of uniconazole externally was also helpful in reducing the damaging effects of waterlogging in oil seed rape plants (Webb and Fletcher 1996; Zhou et al. 1997). Therefore, the application of these organic and inorganic compounds imparts an excellent platform for inducing tolerance to flooding stress. The waterlogging causes substantial changes in gene expression, which coordinate morphological and metabolic adaptations to the stress (Bailey-Serres and Voesenek 2010; Mustroph et al. 2014). Group VII ethylene response factors (ERF-VIIs) is a class of ERF transcription factors (TFs) that regulate the expression of a wide range of genes involved in adaptive responses to flooding and low oxygen. In *Arabidopsis*, five ERF-VII genes, viz., HRE1, HRE2, RAP2.2, RAP2.3, and RAP2.12, were recognized as important regulators for flooding and low oxygen tolerance. These ERF-VIIs regulate a similar set of hypoxia-responsive genes, but constitutively expressed RAP2.2, RAP2.3, and RAP2.12 are more powerful activators than stress-inducible HRE1 and HRE2 as cited in literature (Bui et al. 2015; Gasch et al. 2016).

To solve the problem it is very important to understand the physiology of flooding tolerance/sensitivity and identify genes in particular condition. In this regard, researchers across the globe conducted study on reverse genetics to find out the genes expressed in low oxygen treated plant roots. The sense/antisense over expression of the said genes in transgenic plants was used to identify any contribution they make to flooding tolerance.

After many investigations, around 20 newly synthesized polypeptides (Sachs et al. 1980) are reported upon switching from aerobic to anaerobic condition in roots. These atrial natriuretic peptides (ANPs) were identified via cDNA cloning (Table 5). These ANPs include enzyme (PDC, LDH, ADH, Ala AT) were involved in fermentation pathways. Among them, enzymes of glycolytic pathway and sugar degradation pathways are important ANPs.

The key genes derived cDNA library constructed from mRNA of an aerobically induced of Arabidopsis were classified into 4 groups, viz, diverse metabolic genes described earlier as ANPs, signal transduction components., proteins related to defense mechanism, and finally proteins involved in DNA structure and post transcriptional regulation. At molecular level, various approaches have been proposed to improve the plant growth in anaerobic conditions. First approach is based on fermentation pathways. Either by enhancing the expression of genes encoding the fermentation pathways enzymes or by making the expression of these pathways constitutive, it might be helpful for the plants to survive with initial low oxygen condition. Another approach is to alter the expression or function of AtMyB2 transcription factor. Both approaches might have a beneficial effect in switching on longer term adaptation to low oxygen stress.

Now a days Next Generation Technology has become a powerful tool, various microRNAs were reported, uncovering a diversity of biological functions in response to various biotic and abiotic stresses in plants. miRNAs have an active role in response to flooding in plants as revealed by genome-wide studies in response to hypoxia (Moldovan et al. 2009; Licausi et al. 2011), waterlogging (Liu et al. 2012; Zhai et al. 2013), and submergence (Zhang et al. 2008a, b; Jeong et al. 2013; Jin et al. 2017; Li et al. 2017; Franke et al. 2018). Despite the fact that reports of miRNA's responsiveness to flooding stress are still inadequate in comparison to information on other abiotic stress, existing literatures have revealed that miRNAs regulate four main lines of response to flooding stress including morphological adaptation, management of energy supply, control of flowering, and oxidative stress response.

9 Conclusion

Waterlogging is one of the major limitation for sustainable and integrated agriculture. Its effects were observed both on the entire plant as well as, cellular levels. There is the urgent requirement to screen available germplasm for waterlogging tolerance and use the genes responsible for inducing tolerance in brassica crops so as to make them resistant as well. Waterlogging causes deficiency of several essential nutrients in the field. Thus, the application of these nutrient or other plant hormones could be used externally to alleviate the adverse effects of waterlogging/flood. QTL discovery for flood tolerant cultivars in brassica plants are under process. The comprehensive knowledge on the overall crop response to the stress at

Table 5 List of all known proteins that are actively synthesized (protein level) and genes which are induced (RNA level) by low oxygen conditions

Atrial natriuretic peptide	Detection level	Function	References
Sucrose synthase	RNA, protein	Sucrose breakdown	Springer et al. (1986), Martin et al. (1993)
α-Amylase	RNA, protein	Sucrose breakdown	Perata et al. (1993)
Glucose-6-phosphate isomerase	RNA, protein	Glycolysis	Kelley and Freeling (1984a), Sachs et al. (1996)
Pyrophosphate-dependent phosphofructokinase	Protein	Glycolysis	Botha and Botha (1991), Mertens (1991)
Hexokinase	Protein	Glycolysis	Bouny and Saglio (1996), Fox et al. (1998)
Fructose-1,6-bisphosphate aldolase	RNA, protein	Glycolysis	Kelly and Freeling (1984b), Dennis et al. (1988)
Glyceraldehyde-3-phosphate dehydrogenase	RNA, protein	Glycolysis	Sachs et al. (1996)
Enolase	RNA, protein	Glycolysis	Sachs et al. (1996)
Alcohol dehydrogenase	RNA, protein	Alcohol fermentation	Freeling and Bennett (1985)
Pyruvate decarboxylase	RNA, protein	Alcohol fermentation	Kelley (1989)
Lactate dehydrogenase	RNA, protein	Lactic acid fermentation	Hoffman et al. (1986), Germain et al. (1997)
Alanine aminotransferase	RNA, protein	Alanine fermentation	Good and Crosby (1989)
Glutamine synthase	Protein	Nitrogen metabolism	Mattana et al. (1994a)
Nitrate reductase	Protein	Nitrogen metabolism	Mattana et al. (1994b)
Nitrate reductase	Protein	Nitrogen metabolism	Mattana et al. (1994b)
Formate dehydrogenase	RNA, protein	Cl metabolism	Hourton- Cabassa et al. (1998), Suzuki et al. (1998)
Xyloglucan Endotransglycosylase	RNA	Cell wall loosening	Sachs et al. (1996)
1-aminocyclopropane-1-carboxylic acid synthase	RNA	Ethylene biosynthesis	Olson et al. (1995)

(continued)

Table 5 (continued)

Atrial natriuretic peptide	Detection level	Function	References
Haemoglobin, AtHgbl	RNA	Oxygen storage	Trevaskis et al. (1997)
Vacuolar H^+-translocating pyrophosphatase	RNA	Cytosolic acidosis	Carystinos et al. (1995)
Cytosolic pyruvate orthophosphate dikinase	Protein, RNA	Anoxic CO_2 fixation	Moons et al. (1998)
AtMYB2	RNA	Transcription factor	Hoeren et al. (1998)
Myb7 (rice, X89605)	RNA	Transcription factor	Menguzzato et al. (1995)
G-Box binding factor Gbf1	RNA	Transcription factor	De Vetten and Ferl (1995)
Calcium-dependent protein kinase (CDPK)	Protein	Protein kinase	Morello et al. (1994)

the transcriptomic, proteomic, and metabolomic level would definitely be useful to understand the molecular mechanisms that might provide the basis for underlying tolerance traits.

Acknowledgements Authors are thankful to Director, Dayalbagh Educational Institute (Deemed University), and Head, Department of Botany for providing infrastructure.

References

Abiko T, Kotula L, Shiono K, Colmer TD, Nakazono M (2012) Enhanced formation of aerenchyma and induction of a barrier to radial oxygen loss in adventitious roots of Zea nicaraguensis contribute to its waterlogging tolerance as compared with maize (*Zea mays* ssp. mays). Plant Cell Environ 35(9):1618–1630

Ahmed S, Nawata E, Hosokawa M, Domae Y, Sakuratani T (2002) Alterations in photosynthesis and some antioxidant enzymatic activities of mungbean subjected to waterlogging. Plant Sci 163(1):117–123

Akman M, Bhikharie AV, McLean EH, Boonman A, Visser EJ, Schranz ME, van Tienderen PH (2012) Wait or escape? Contrasting submergence tolerance strategies of *Rorippa amphibia, Rorippa sylvestris* and their hybrid. Ann Bot 109(7):1263–1276

Akman M, Kleine R, van Tienderen PH, Schranz EM (2017) Identification of the submergence tolerance QTL Come Quick Drowning1 (CQD1) in *Arabidopsis thaliana*. J Hered 108(3):308–317

Andrews CJ, Pomeroy MK (1981) The effect of flooding pretreatment on cold hardiness and survival of winter cereals in ice encasement. Canad J Plant Sci 61(3):507–513

Andrews CJ, Pomeroy MK (1983) The influence of flooding pretreatment on metabolic changes in winter cereal seedlings during ice encasement. Canad J Bot 61(1):142–147

Anee TI, Nahar K, Rahman A, Mahmud JA, Bhuiyan TF, Alam MU, Fujita M, Hasanuzzaman M (2019) Oxidative damage and antioxidant defense in sesamum indicum after different waterlogging durations. Plants 8:196

Ap Rees T, Wilson PM (1984) Effects of reduced supply of oxygen on the metabolism of roots of *Glyceria maxima* and *Pisum sativum*. Zeit Pflan Physiol 114(5):493–503

Armstrong WI, Armstrong JE, Beckett PM, Justin SH (1991) Convective gas-flows in wetland plant aeration. SPB Acad Pub Netherlands 283:283–302

Arshad M, Frankenberger WT (1990) Production and stability of ethylene in soil. Bio Ferti Soil 10(1):29–34

Ashraf M (2009) Biotechnological approach of improving plant salt tolerance using antioxidants as markers. Biotech advan 27(1):84–93

Ashraf M, Akram NA (2009) Improving salinity tolerance of plants through conventional breeding and genetic engineering: an analytical comparison. Biotech advan 27(6):744–752

Ashraf M, Arfan M (2005) Gas exchange characteristics and water relations in two cultivars of *Hibiscus esculentus* under waterlogging. Biol Plant 49(3):459–462

Ashraf M, Mehmood S (1990) Effects of waterlogging on growth and some physiological parameters of four Brassica species. Plant Soil 121(2):203–209

Ashraf M, Rehman H (1999) Mineral nutrient status of corn in relation to nitrate and long-term waterlogging. J Plant Nutri 22(8):1253–1268

Ashraf MA, Ahmad MS, Ashraf M, Al-Qurainy F, Ashraf MY (2011) Alleviation of waterlogging stress in upland cotton (*Gossypium hirsutum* L.) by exogenous application of potassium in soil and as a foliar spray. Crop Pastu Sci 62(1):25–38

Bailey-Serres J, Voesenek LA (2010) Life in the balance: a signaling network controlling survival of flooding. Cur Opin Plant Biol 13(5):489–494

Baxter-Burrell AI, Chang R, Springer P, Bailey-Serres JU (2003) Gene and enhancer trap transposable elements reveal oxygen deprivation-regulated genes and their complex patterns of expression in Arabidopsis. Annals Bot 91(2):129–141

Belford RK, Dracup M, Tennant D (1992) Limitations to growth and yield of cereal and lupin crops on duplex soils. Aust J Exp Agri 32(7):929–945

Biswas JC, Kalra N (2018) Effect of waterlogging and submergence on crop physiology and growth of different crops and its remedies: Bangladesh perspectives. Saudi J Eng Technol 3(6):315–329

Boamfa EI, Ram PC, Jackson MB, Reuss J, Harren FJ (2003) Dynamic aspects of alcoholic fermentation of rice seedlings in response to anaerobiosis and to complete submergence: relationship to submergence tolerance. Annals Bot 91(2):279–290

Boru G, van Ginkel M, Trethowan RM, Boersma L, Kronstad WE (2003) Oxygen use from solution by wheat genotypes differing in tolerance to waterlogging. Euphyt 132(2):151–158

Botha AM, Botha FC (1991) Effect of anoxia on the expression and molecular form of the pyrophosphate dependent phosphofructokinase. Plant Cell Physiol 32(8):1299–1302

Bouny JM, Saglio PH (1996) Glycolytic flux and hexokinase activities in anoxic maize root tips acclimated by hypoxic pretreatment. Plant Physiol 111(1):187–194

Bradford KJ, Yang SF (1980) Xylem transport of 1-aminocyclopropane-1-carboxylic acid, an ethylene precursor, in waterlogged tomato plants. Plant Physiol 65(2):322–326

Bui LT, Giuntoli B, Kosmacz M, Parlanti S, Licausi F (2015) Constitutively expressed ERF-VII transcription factors redundantly activate the core anaerobic response in *Arabidopsis thaliana*. Plant Sci 236:37–43

Carystinos GD, MacDonald HR, Monroy AF, Dhindsa RS, Poole RJ (1995) Vacuolar H^+-translocating pyrophosphatase is induced by anoxia or chilling in seedlings of rice. Plant Physiol 108(2):641–649

Chen W, Yao Q, Patil GB, Agarwal G, Deshmukh RK, Lin L, Wang B, Wang Y, Prince SJ, Song L, Xu D (2016) Identification and comparative analysis of differential gene expression in soybean leaf tissue under drought and flooding stress revealed by RNA-seq. Front Plant Sci 7:1044

Cheng Y, Min GU, Ye CO, Zou CS, Zhang XK, Wang HZ (2010) Combining ability and genetic effects of germination traits of *Brassica napus* L. under waterlogging stress condition. Agric Sci China 9(7):951–957

Colmer TD, Greenway H (2005) Oxygen transport, respiration, and anaerobic carbohydrate catabolism in roots in flooded soils. Plant Respir 137–158. Springer

Cong Y, Cheng Y, Zou C, Zhang X, Wang H (2009) Genetic analysis of waterlogging tolerance for germinated seeds of rapeseed (*Brassica napus* L.) with mixed model of major gene plus polygene. Acta Agron Sin 35(8):1462–1467

Crawford RMM (1978) Metabolic adaptation to anoxia. In: Hook DD, Crawford RMM (eds) Plant life in anaerobic environments. Ann Arbor Science, Ann Arbor, pp 119–136

Daugherty CJ, Musgrave ME (1994) Characterization of populations of rapid-cycling Brassica rapa L. selected for differential waterlogging tolerance. J Exp Bot 45(3):385–392

Daugherty CJ, Matthews SW, Musgrave ME (1994) Structural changes in rapid-cycling *Brassica rapa* selected for differential waterlogging tolerance. Can J Bot 72(9):1322–1328

Davies DD (1980) Anaerobic metabolism and the production of organic acids. Metabol Respir 581–611

Davies DD, Grego S, Kenworthy P (1974) The control of the production of lactate and ethanol by higher plants. Planta 118(4):297–310

Dawood T, Yang X, Visser EJ, Te Beek TA, Kensche PR, Cristescu SM, Lee S, Floková K, Nguyen D, Mariani C, Rieu I (2016) A co-opted hormonal cascade activates dormant adventitious root primordia upon flooding in *Solanum dulcamara*. Plant Physiol 170(4):2351–2364

De Vetten NC, Ferl RJ (1995) Characterization of a maize G-box binding factor that is induced by hypoxia. Plant J 7(4):589–601

Dennis ES, Gerlach WL, Walker JC, Lavin M, Peacock WJ (1988) Anaerobically regulated aldolase gene of maize: a chimaeric origin? J Mol Biol 202(4):759–767

Dolferus R, Klok EJ, Delessert C, Wilson S, Ismond KP, Good AG, Peacock WJ, Dennis ES (2003) Enhancing the anaerobic response. Annals Bot 91(2):111–117

Dordas C, Rivoal J, HILL RD (2003) Plant haemoglobins, nitric oxide and hypoxic stress. Annals Bot 91(2):173–178

Drew MC (1991) Oxygen deficiency in the root environment and plant mineral nutrition. Plant Life Oxy Depriv 303–316

Drew MC, Sisworo EJ (1977) Early effects of flooding on nitrogen deficiency and leaf chlorosis in barley. New Phytol 79(3):567–571

Else MA, Coupland D, Dutton L, Jackson MB (2001) Decreased root hydraulic conductivity reduces leaf water potential, initiates stomatal closure and slows leaf expansion in flooded plants of castor oil (*Ricinus communis*) despite diminished delivery of ABA from the roots to shoots in xylem sap. Physiol Plant 111(1):46–54

Evans DE (2004) Aerenchyma formation. New Phytol 161(1):35–49

Ferreira JL, Coelho CH, Magalhães PC, Santána GC, Borém A (2008) Evaluation of mineral content in maize under flooding. Embrapa Milho e Sorgo-Artigo em periódico indexado (ALICE). Crop Breed Appl Biotech 8:134–140

Finch-Savage WE, Côme D, Lynn JR, Corbineau F (2005) Sensitivity of *Brassica oleracea* seed germination to hypoxia: a QTL analysis. Plant Sci 169(4):753–759

Fox TC, Green BJ, Kennedy RA, Rumpho ME (1998) Changes in hexokinase activity in *Echinochloa phyllopogon* and *Echinochloa crus-pavonis* in response to abiotic stress. Plant Physiol 118(4):1403–1409

Foyer CH, Halliwell B (1976) The presence of glutathione and glutathione reductase in chloroplasts: a proposed role in ascorbic acid metabolism. Planta 133(1):21–25

Franke KR, Schmidt SA, Park S, Jeong DH, Accerbi M, Green PJ (2018) Analysis of *Brachypodium* miRNA targets: evidence for diverse control during stress and conservation in bioenergy crops. BMC Genom 19(1):547

Freeling M, Bennett DC (1985) Maize adhl. Ann Rev Genet 19(1):297–323

Gasch P, Fundinger M, Müller JT, Lee T, Bailey-Serres J, Mustroph A (2016) Redundant ERF-VII transcription factors bind to an evolutionarily conserved cis-motif to regulate hypoxia-responsive gene expression in *Arabidopsis*. Plant Cell 28(1):160–180

Germain V, Raymond P, Ricard B (1997) Differential expression of two tomato lactate dehydrogenase genes in response to oxygen deficit. Plant Mol Biol 35(6):711–721

Gibberd MR, Gray JD, Cocks PS, Colmer TD (2001) Waterlogging tolerance among a diverse range of *Trifolium* accessions is related to root porosity, lateral root formation and 'aerotropic rooting'. Annals Bot 88(4):579–589

Gibbs J, Greenway H (2003) Mechanisms of anoxia tolerance in plants. I. Growth, survival and anaerobic catabolism. Funct Plant Bio 30(1):1–47

Good AG, Crosby WL (1989) Anaerobic induction of alanine aminotransferase in barley root tissue. Plant Physiol 90(4):1305–1309

Greenland DJ (1977) Soil damage by intensive arable cultivation: temporary or permanent?. Philosophical Transactions of the Royal Society of London. Bio Sci 281(980):193–208

Gupta KJ, Stoimenova M, Kaiser WM (2005) In higher plants, only root mitochondria, but not leaf mitochondria reduce nitrite to NO, in vitro and in situ. J Exp Bot 56(420):2601–2609

Halcomb M (2003) Too much water kills more plants than too little water. The University of Tennessee Extension

Hasanuzzaman M, Mahmud JA, Nahar K, Inafuku M, Oku H, Fujita M (2017) Plant responses, adaptation and ROS metabolism in plants exposed to waterlogging stress. In: Khan MIR, Khan NA, Ismail AM (eds) reactive oxygen species and antioxidant systems: role and regulation under abiotic stress. Springer Nature, Singapore, pp 257–281

Hasanuzzaman M, Bhuyan MHMB, Anee TI, Parvin K, Nahar K, Mahmud JA, Fujita M (2019) Regulation of ascorbate-glutathione pathway in mitigating oxidative damage in plants under abiotic stress. Antioxidants 8(9):384. https://doi.org/10.3390/antiox8090384

Hattori Y, Nagai K, Furukawa S, Song XJ, Kawano R, Sakakibara H, Wu J, Matsumoto T, Yoshimura A, Kitano H, Matsuoka M (2009) The ethylene response factors SNORKEL1 and SNORKEL2 allow rice to adapt to deep water. Nature 460(7258):1026–1030

Hideg E (1997) Free radical production in photosynthesis under stress conditions. Hand Photosynth 911–930

Hoeren FU, Dolferus R, Wu Y, Peacock WJ, Dennis ES (1998) Evidence for a role for AtMYB2 in the induction of the Arabidopsis alcohol dehydrogenase gene (ADH1) by low oxygen. Genet 149(2):479–490

Hoffman NE, Bent AF, Hanson AD (1986) Induction of lactate dehydrogenase isozymes by oxygen deficit in barley root tissue. Plant Physiol 82(3):658–663

Hourton-Cabassa C, Ambard-Bretteville F, Moreau F, de Virville JD, Rémy R, des Francs-Small CC (1998) Stress induction of mitochondrial formate dehydrogenase in potato leaves. Plant Physiol 116(2):627–635

Huang B, Johnson JW, NeSmith DS, Bridges DC (1994) Root and shoot growth of wheat genotypes in response to hypoxia and subsequent resumption of aeration. Crop Sci 34(6):1538–1544

Jackson MB (1990) Hormones and developmental change in plants subjected to submergence or soil waterlogging. Aquat Bot 38(1):49–72

Jackson MB (1993) Are plant hormones involved in root to shoot communication? Adv Bot Res 19:103–118

Jackson MB (2007) Ethylene-promoted elongation: an adaptation to submergence stress. Ann Bot 101(2):229–248

Jackson MB, Armstrong W (1999) Formation of aerenchyma and the processes of plant ventilation in relation to soil flooding and submergence. Plant Biol 1(03):274–287

Jackson MB, Drew MC (1984) Effect of flooding on herbaceous plants. Flood Plant Growth 47–128

Jackson DD, Lambers H (1992) Plant life under oxygen deprivation, ecology, physiology and biochemistry. Edited by MB

Jackson MB, Drew MC, Giffard SC (1981) Effects of applying etnylene to the root system of *Zea mays* on growth and nutrient concentration in relation to flooding tolerance. Physiol Plant 52(1):23–28

Jeong DH, Schmidt SA, Rymarquis LA, Park S, Ganssmann M, German MA, Accerbi M, Zhai J, Fahlgren N, Fox SE, Garvin DF (2013) Parallel analysis of RNA ends enhances global investigation of microRNAs and target RNAs of *Brachypodium distachyon*. Genom biol 14(12):R145

Jin Q, Xu Y, Mattson N, Li X, Wang B, Zhang X, Jiang H, Liu X, Wang Y, Yao D (2017) Identification of submergence-responsive microRNAs and their targets reveals complex miRNA-mediated regulatory networks in lotus (*Nelumbo nucifera* Gaertn). Front Plant Sci 18(8):6

Johnson J, Cobb BG, Drew MC (1989) Hypoxic induction of anoxia tolerance in root tips of *Zea mays*. Plant Physiol 91(3):837–841

Johnson JR, Cobb BG, Drew MC (1994) Hypoxic induction of anoxia tolerance in roots of Adh1 null Zea mays L. Plant Physiol 105(1):61–67

Justin SH, Armstrong W (1987) The anatomical characteristics of roots and plant response to soil flooding. Phytol 465–495

Kelley PM (1989) Maize pyruvate decarboxylase mRNA is induced anaerobically. Plant Mol Biol 13(2):213–222

Kelley PM, Freeling M (1984a) Anaerobic expression of maize fructose-1,6-diphosphate aldolase. J Bio Chem 259(22):14180–14183

Kelley PM, Freeling M (1984b) Anaerobic expression of maize glucose phosphate isomerase I. J Bio Chem 259(1):673–677

Kramer PJ, Jackson WT (1954) Causes of injury to flooded tobacco plants. Plant Physiol 29(3):241–245

Laanbroek HJ (1990) Bacterial cycling of minerals that affect plant growth in waterlogged soils: a review. Aquat Bot 38(1):109–125

Lee YH, Kim KS, Jang YS, Hwang JH, Lee DH, Choi IH (2014) Global gene expression responses to waterlogging in leaves of rape seedlings. Plant Cell Rep 33(2):289–299

Li Z, Mei S, Mei Z, Liu X, Fu T, Zhou G, Tu J (2014) Mapping of QTL associated with waterlogging tolerance and drought resistance during the seedling stage in oilseed rape (*Brassica napus*). Euphyt 197(3):341–353

Li G, Deng Y, Geng Y, Zhou C, Wang Y, Zhang W, Song Z, Gao L, Yang J (2017) Differentially expressed microRNAs and target genes associated with plastic internode elongation in Alternanthera philoxeroides in contrasting hydrological habitats. Fronti Plant Sci 8:2078

Liao CT, Lin CH (2001) Physiological adaptation of crop plants to flooding stress. Proceedings of the National Science Council, Republic of China. Part B Life Sci 25(3):148–157

Licausi F, Weits DA, Pant BD, Scheible WR, Geigenberger P, van Dongen JT (2011) Hypoxia responsive gene expression is mediated by various subsets of transcription factors and miRNAs that are determined by the actual oxygen availability. Phytol 190(2):442–456

Liu Z, Kumari S, Zhang L, Zheng Y, Ware D (2012) Characterization of miRNAs in response to short-term waterlogging in three inbred lines of *Zea mays*. PLoS ONE 7(6):e39786

Loreti E, Yamaguchi J, Alpi A, Perata P (2003) Sugar modulation of α-amylase genes under anoxia. Ann Bot 91(2):143–148

Loreti E, van Veen H, Perata P (2016) Plant responses to flooding stress. Curr Opin Plant Biol 33:64–71

Malik AI, Colmer TD, Lambers H, Schortemeyer M (2001) Changes in physiological and morphological traits of roots and shoots of wheat in response to different depths of waterlogging. Funct Plant Bio 28(11):1121–1131

Martin T, Frommer WB, Salanoubat M, Willmitzer L (1993) Expression of an Arabidopsis sucrose synthase gene indicates a role in metabolization of sucrose both during phloem loading and in sink organs. Plant J 4(2):367–377

Mattana M, Bertani A, Reggiani R (1994a) Expression of glutamine synthetase during the anaerobic germination of *Oryza sativa* L. Planta 195(1):147–149

Mattana M, Coraggio I, Bertani A, Reggiani R (1994b) Expression of the enzymes of nitrate reduction during the anaerobic germination of rice. Plant Physiol 106(4):1605–1608

McFarlane NM, Ciavarella TA, Smith KF (2003) The effects of waterlogging on growth, photosynthesis and biomass allocation in perennial ryegrass (*Lolium perenne* L.) genotypes with contrasting root development. J Agric Sci 141(2):241–248

McManmon M, Crawford RM (1971) A metabolic theory of flooding tolerance: the significance of enzyme distribution and behaviour. Phytol 70(2):299–306

McPherson DC (1939) Cortical air spaces in the roots of Zea mays L. Phytol 38(3):190–202
Menguzzato E, Valle G, Coraggio I (1995) A myb encoding cDNA (X89605) expressed in anaerobic rice presents features suggesting post transcriptional regulation. Plant Physiol 95–104
Mertens E (1991) Pyrophosphate-dependent phosphofructokinase, an anaerobic glycolytic enzyme? FEBS Lett 285(1):1–5
Mocquot B, Prat C, Mouches C, Pradet A (1981) Effect of anoxia on energy charge and protein synthesis in rice embryo. Plant Physiol 68(3):636–640
Moldovan D, Spriggs A, Yang J, Pogson BJ, Dennis ES, Wilson IW (2009) Hypoxia-responsive microRNAs and trans-acting small interfering RNAs in *Arabidopsis*. J Exp Bot 61(1):165–177
Monk LS, Fagerstedt KV, Crawford RM (1987) Superoxide dismutase as an anaerobic polypeptide: a key factor in recovery from oxygen deprivation in *Iris pseudacorus*? Plant Physiol 85(4):1016–1020
Moons A, Valcke R, Van M (1998) Low-oxygen stress and water deficit induce cytosolic pyruvate orthophosphate dikinase (PPDK) expression in roots of rice, a C3 plant. Plant J 15(1):89–98
Morello L, Gianì S, Breviario D (1994) The influence of anaerobiosis on membrane-associated rice (*O. sativa* L.) protein kinase activities. J Plant Physiol 144(4–5):500–504
Musgrave A, Jackson MB, Ling E (1972) Callitriche stem elongation is controlled by ethylene and gibberellin. Nat New Bio 238(81):93
Mustroph A, Hess N, Sasidharan R (2014) Hypoxic energy metabolism and PPi as an alternative energy currency. Low Oxyg Stress Plant 165–184
Olson DC, Oetiker JH, Yang SF (1995) Analysis of LE-ACS3, a 1-aminocyclopropane-1-carboxylic acid synthase gene expressed during flooding in the roots of tomato plants. J Bio Chem 270(23):14056–14061
Parelle J, Brendel O, Bodénès C, Berveiller D, Dizengremel P, Jolivet Y, Dreyer E (2006) Differences in morphological and physiological responses to water-logging between two sympatric oak species (*Quercus petraea* [Matt.] Liebl., *Quercus robur* L.). Ann Fores Sci 63(8):849–859
Parent C, Berger A, Folzer H, Dat J, Crevècoeur M, Badot PM, Capelli N (2008a) A novel nonsymbiotic hemoglobin from oak: cellular and tissue specificity of gene expression. Phytol 177(1):142–154
Parent C, Capelli N, Dat J (2008b) Reactive oxygen species, stress and cell death in plants. Compt Rend Biol 331(4):255–261
Pedersen O, Colmer TD, Sand-Jensen KA (2013) Underwater photosynthesis of submerged plants–recent advances and methods. Front Plant Sci 4:140
Pellinen R, Palva T, KangasjaÈrvi J (1999) Subcellular localization of ozone-induced hydrogen peroxide production in birch (Betula pendula) leaf cells. Plant J 20(3):349–356
Perata P, Pozueta-Romero J, Akazawa T, Yamaguchi J (1992a) Effect of anoxia on starch breakdown in rice and wheat seeds. Plant 188(4):611–618
Perata P, Vernieri P, Armellini D, Bugnoli M, Tognoni F, Alpi A (1992b) Immunological detection of acetaldehyde-protein adducts in ethanol-treated carrot cells. Plant Physiol 98(3):913–918
Perata P, Geshi N, Yamaguchi J, Akazawa T (1993) Effect of anoxia on the induction of α-amylase in cereal seeds. Plant 191(3):402–408
Pereira JS, Kozlowski TT (1977) Variations among woody angiosperms in response to flooding. Physiol Plant 41(3):184–192
Pfister-Sieber M, Brändle R (1994) Aspects of plant behaviour under anoxia and post-anoxia. Proc R Soc Edinb Sect B Bio Sci 102:313–324
Pierik R, Tholen D, Poorter H, Visser EJ, Voesenek LA (2006) The Janus face of ethylene: growth inhibition and stimulation. Trends Plant Sci 11(4):176–183
Pociecha E, Kościelniak J, Filek W (2008) Effects of root flooding and stage of development on the growth and photosynthesis of field bean (*Vicia faba* L. minor). Acta Physiol Plant. 30(4):529
Ponnamperuma FN (1972) The chemistry of submerged soils. Adv Agron 24:29–96)
Rajhi I, Yamauchi T, Takahashi H, Nishiuchi S, Shiono K, Watanabe R, Mliki A, Nagamura Y, Tsutsumi N, Nishizawa NK, Nakazono M (2011) Identification of genes expressed in maize root

cortical cells during lysigenous aerenchyma formation using laser microdissection and microarray analyses. Phyto 190(2):351–368

Rawyler A, Pavelic D, Gianinazzi C, Oberson J, Braendle R (1999) Membrane lipid integrity relies on a threshold of ATP production rate in potato cell cultures submitted to anoxia. Plant Phys 120(1):293–300

Rienk R, Ploeg D, Machulla GA, Hermsmeyer DI, Ilsemann J, Gieska MA (2002) Changes in land use and the growing number of flash floods in Germany. Agric Eff Gr Surf Waters Rese Edge Sci Soc 273:317

Roberts JK, Wemmer D, Ray PM, Jardetzky O (1982) Regulation of cytoplasmic and vacuolar pH in maize root tips under different experimental conditions. Plant Physiol 69(6):1344–1347

Roberts JK, Callis J, Jardetzky O, Walbot V, Freeling M (1984) Cytoplasmic acidosis as a determinant of flooding intolerance in plants. Proc Natl Acad Sci 81(19):6029

Roberts JK, Andrade FH, Anderson IC (1985) Further evidence that cytoplasmic acidosis is a determinant of flooding intolerance in plants. Plant Physiol 77(2):492–494

Sachs MM (1993) Molecular genetic basis of metabolic adaptation to anoxia in maize and its possible utility for improving tolerance of crops to soil waterlogging. Interact Stress Plants Chang Clim 375–393

Sachs MM, Freeling M, Okimoto R (1980) The anaerobic proteins of maize. Cell 20(3):761–767

Sachs MM, Subbaiah CC, Saab IN (1996) Anaerobic gene expression and flooding tolerance in maize. J Exp Bot 47(1):1–5

Saglio PH, Raymond P, Pradet A (1980) Metabolic activity and energy charge of excised maize root tips under anoxia: control by soluble sugars. Plant Physiol 66(6):1053–1057

Saglio PH, Drew MC, Pradet A (1988) Metabolic acclimation to anoxia induced by low (2–4 kPa partial pressure) oxygen pretreatment (hypoxia) in root tips of *Zea mays*. Plant Physiol 86(1):61–66

Sairam RK, Srivastava GC (2002) Changes in antioxidant activity in sub-cellular fractions of tolerant and susceptible wheat genotypes in response to long term salt stress. Plant Sci 162(6):897–904

Sasidharan R, Mustroph A, Boonman A, Akman M, Ammerlaan AM, Breit T, Schranz ME, Voesenek LA, van Tienderen PH (2013) Root transcript profiling of two Rorippa species reveals gene clusters associated with extreme submergence tolerance. Plant Physiol 163(3):1277–1292

Seago JRJL, Marsh LC, Stevens KJ, Soukup A, Votrubova O, Enstone DE (2005) A re-examination of the root cortex in wetland flowering plants with respect to aerenchyma. Ann Bot 96(4):565–579

Shimamura S, Yoshida S, Mochizuki T (2007) Cortical aerenchyma formation in hypocotyl and adventitious roots of *Luffa cylindrica* subjected to soil flooding. Ann Bot 100(7):1431–1439

Shiono K, Yamauchi T, Yamazaki S, Mohanty B, Malik AI, Nagamura Y, Nishizawa NK, Tsutsumi N, Colmer TD, Nakazono M (2014) Microarray analysis of laser-microdissected tissues indicates the biosynthesis of suberin in the outer part of roots during formation of a barrier to radial oxygen loss in rice (*Oryza sativa*). J Exp Bot 65(17):4795–4806

Springer B, Werr W, Starlinger P, Bennett DC, Zokolica M, Freeling M (1986) The Shrunken gene on chromosome 9 of *Zea mays* L is expressed in various plant tissues and encodes an anaerobic protein. Mol Gener Gene 205(3):461–468

Steffens B, Rasmussen A (2016) The physiology of adventitious roots. Plant Physiol 170(2):603–617

Steffens B, Sauter M (2009) Epidermal cell death in rice is confined to cells with a distinct molecular identity and is mediated by ethylene and H_2O_2 through an autoamplified signal pathway. Plant Cell 21(1):184–196

Steffens B, Wang J, Sauter M (2006) Interactions between ethylene, gibberellin and abscisic acid regulate emergence and growth rate of adventitious roots in deepwater rice. Plant 223(3):604–612

Steffens B, Geske T, Sauter M (2011) Aerenchyma formation in the rice stem and its promotion by H_2O_2. Phytol 190(2):369–378

Striker GG (2012) Flooding stress on plants: anatomical, morphological and physiological responses. Botany 1:3–28

Subbaiah CC, Sachs MM (2003) Molecular and cellular adaptations of maize to flooding stress. Ann Bot 91(2):119–127

Subbaiah CC, Zhang J, Sachs MM (1994) Involvement of intracellular calcium in anaerobic gene expression and survival of maize seedlings. Plant Physiol 105(1):369–376

Suzuki K, Itai R, Suzuki K, Nakanishi H, Nishizawa NK, Yoshimura E, Mori S (1998) Formate dehydrogenase, an enzyme of anaerobic metabolism, is induced by iron deficiency in barley roots. Plant Physiol 116(2):725–732

Takahashi H, Yamauchi T, Colmer TD, Nakazono M (2014) Aerenchyma formation in plants. Low Oxy Stress Plant 247–265

Titarenko T (2000) Test parameters of revealing the degree of fruit plants tolerance to the root hypoxia caused by flooding of soil. Plant Physiol Biochem 115

Trevaskis B, Watts RA, Andersson CR, Llewellyn DJ, Hargrove MS, Olson JS, Dennis ES, Peacock WJ (1997) Two hemoglobin genes in *Arabidopsis thaliana*: the evolutionary origins of leghemoglobins. Proc Natl Acad Sci 94(22):12230–12234

Trought MC, Drew MC (1980) The development of waterlogging damage in wheat seedlings (*Triticum aestivum* L.). Plant Soil 54(1):77–94

van Veen H, Mustroph A, Barding GA, Vergeer-van Eijk M, Welschen-Evertman RA, Pedersen O, Visser EJ, Larive CK, Pierik R, Bailey-Serres J, Voesenek LA (2013) Two Rumex species from contrasting hydrological niches regulate flooding tolerance through distinct mechanisms. Plant Cell 25(11):4691–4707

van Veen H, Akman M, Jamar DC, Vreugdenhil D, Kooiker M, van Tienderen P, Voesenek LA, Schranz ME, Sasidharan R (2014) Group VII E thylene R esponse F actor diversification and regulation in four species from flood-prone environments. Plant Cell Environ 37(10):2421–2432

Van Veen H, Vashisht D, Akman M, Girke T, Mustroph A, Reinen E, Hartman S, Kooiker M, van Tienderen P, Schranz ME, Bailey-Serres J (2016) Transcriptomes of eight *Arabidopsis thaliana* accessions reveal core conserved, genotype-and organ-specific responses to flooding stress. Plant Physiol 172(2):668–689

VanToai TT, Bolles CS (1991) Postanoxic injury in soybean (*Glycine max*) seedlings. Plant Physiol 97(2):588–592

Vartapetian BB (1973) Aeration of roots in relation to molecular oxygen transport in plants. Plant Respon Clim Fact Proceed Upps Symp 227–231

Vartapetian BB, Jackson MB (1997) Plant adaptations to anaerobic stress. Anna Bot 79(1):3–20

Vartapetian BB, Poljakova LI (1994) Blocking of anaerobic protein synthesis destabilizes dramatically plant mitochondrial membrane ultrastructure. Biochem Mol Bio Int 33(2):405–409

Vartapetian BB, Andreeva IN, Maslova IP, Davtian NG (1970) The oxygen and ultrastructure of root cells. Agrochim 15:1–19

Vartapetian BB, Andreeva IN, Kozlova GI (1976) The resistance to anoxia and the mitochondrial fine structure of rice seedlings. Protoplasm 88(2–4):215–224

Vartapetian BB, Andreeva IN, Nuritdinov N (1978a) Plant cells under oxygen stress. In: Hook DD, Crawford RMM (eds) Plant life in anaerobic environments. Ann Arbor Sci, Ann Arbor, pp 13–88

Vartapetian BB, Mazliak P, Lance C (1978b) Lipid biosynthesis in rice coleoptiles grown in the presence or in the absence of oxygen. Plant Sci Let 13(4):321–328

Vashisht D, Hesselink A, Pierik R, Ammerlaan JM, Bailey-Serres J, Visser EJ, Pedersen O, Van Zanten M, Vreugdenhil D, Jamar DC, Voesenek LA (2011) Natural variation of submergence tolerance among *Arabidopsis thaliana* accessions. New Phytol 190(2):299–310

Vidoz ML, Loreti E, Mensuali A, Alpi A, Perata P (2010) Hormonal interplay during adventitious root formation in flooded tomato plants. Plant J 63(4):551–562

Voesenek LA, Bailey-Serres J (2015) Flood adaptive traits and processes: an overview. Phytol 206(1):57–73

Voesenek LA, Sasidharan R (2013) Ethylene–and oxygen signalling–drive plant survival during flooding. Plant Bio 15(3):426–435

Voesenek LA, Armstrong W, Bögemann GM, Colmer TD, McDonald MP (1999) A lack of aerenchyma and high rates of radial oxygen loss from the root base contribute to the waterlogging intolerance of *Brassica napus*. Funct Plant Bio 26(1):87–93

Voesenek LA, Pierik R, Sasidharan R (2015) Plant life without ethylene. Trends Plant Sci 20(12):783–786

Waisel Y, Eshel A (2002) Functional diversity of various constituents of a single root system. Plant Root 271–296

Walker GE (1991) Chemical, physical and biological control of carrot seedling diseases. Plant Soil 136(1):31–39

Waters I, Morrell S, Greenway H, Colmer TD (1991) Effects of anoxia on wheat seedlings: II. Influence of O_2 supply prior to anoxia on tolerance to anoxia, alcoholic fermentation, and sugar levels. J Exp Bot 42(11):1437–1447

Webb T, Armstrong W (1983) The effects of anoxia and carbohydrates on the growth and viability of rice, pea and pumpkin roots. J Exp Bot 34(5):579–603

Webb JA, Fletcher RA (1996) Paclobutrazol protects wheat seedlings from injury due to waterlogging. Plant Grow Regul 18(3):201–206

Xia JH, Saglio P, Roberts JK (1995) Nucleotide levels do not critically determine survival of maize root tips acclimated to a low-oxygen environment. Plant Physiol 108(2):589–595

Xu M, Ma H, Zeng L, Cheng Y, Lu G, Xu J, Zhang X, Zou X (2015) The effect of waterlogging on yield and seed quality at the early flowering stage in *Brassica napus* L. Field Crop Res 180:238–245

Xu B, Cheng Y, Zou X, Zhang X (2016) Ethanol content in plants of *Brassica napus* L. correlated with waterlogging tolerance index and regulated by lactate dehydrogenase and citrate synthase. Acta Physiol Plant 38(3):81

Yamamoto F, Sakata T, Terazawa K (1995) Physiological, morphological and anatomical responses of Fraxinus mandshurica seedlings to flooding. Tree Physiol 15(11):713–719

Yamauchi T, Rajhi I, Nakazono M (2011) Lysigenous aerenchyma formation in maize root is confined to cortical cells by regulation of genes related to generation and scavenging of reactive oxygen species. Plant Signal Beh 6(5):759–761

Yan B, Dai Q, Liu X, Huang S, Wang Z (1996) Flooding-induced membrane damage, lipid oxidation and activated oxygen generation in corn leaves. Plant Soil 179(2):261–268

Yanar Y, Lipps PE, Deep IW (1997) Effect of soil saturation duration and soil water content on root rot of maize caused by *Pythium arrhenomanes*. Plant Dis 81(5):475–480

Yeung E, van Veen H, Vashisht D, Paiva AL, Hummel M, Rankenberg T, Steffens B, Steffen-Heins A, Sauter M, de Vries M, Schuurink RC (2018) A stress recovery signaling network for enhanced flooding tolerance in Arabidopsis thaliana. Proc Nat Acad Sci 115(26):E6085–E6094

Yiu JC, Liu CW, Fang DY, Lai YS (2009) Waterlogging tolerance of Welsh onion (*Allium fistulosum* L.) enhanced by exogenous spermidine and spermine. Plant Physiol Biochem 47(8):710–716

Yordanova RY, Popova LP (2001) Photosynthetic response of barley plants to soil flooding. Photosynth 39(4):515–520

Yu CB, Xie YY, Hou JJ, Fu YQ, Hong S, Xing LI (2014) Response of nitrate metabolism in seedlings of oilseed rape (*Brassica napus* L.) to low oxygen stress. J Integr Agric 13(11):2416–2423

Zhai L, Liu Z, Zou X, Jiang Y, Qiu F, Zheng Y, Zhang Z (2013) Genome-wide identification and analysis of microRNA responding to long-term waterlogging in crown roots of maize seedlings. Physiol Plant 147(2):181–193

Zhang J, Davies WJ (1987) ABA in roots and leaves of flooded pea plants. J Exp Bot 38(4):649–659

Zhang XK, Chen J, Chen L, Wang HZ, Li JN (2008a) Imbibition behavior and flooding tolerance of rapeseed seed (*Brassica napus* L.) with different testa color. Gene Resour Crop Evol 55(8):1175–1184

Zhang Z, Wei L, Zou X, Tao Y, Liu Z, Zheng Y (2008b) Submergence-responsive microRNAs are potentially involved in the regulation of morphological and metabolic adaptations in maize root cells. Ann Bot 102(4):509–519

Zhang X, Liu T, Li X, Duan M, Wang J, Qiu Y, Wang H, Song J, Shen D (2016) Interspecific hybridization, polyploidization, and backcross of *Brassica oleracea* var. alboglabra with B. rapa var. purpurea morphologically recapitulate the evolution of Brassica vegetables. Sci Rep 6:18618

Zhou W, Zhao D, Lin X (1997) Effects of waterlogging on nitrogen accumulation and alleviation of waterlogging damage by application of nitrogen fertilizer and mixtalol in winter rape (*Brassica napus* L.). J Plant Grow Regul 16(1):47–53

Zou X, Tan X, Hu C, Zeng L, Lu G, Fu G, Cheng Y, Zhang X (2013a) The transcriptome of *Brassica napus* L. roots under waterlogging at the seedling stage. Int J Mol Sci 14(2):2637–2651

Zou XL, Cong Y, Cheng Y, Lu GY, Zhang XK (2013b) Screening and identification of waterlogging tolerant rapeseed (*Brassica napus* L) during Germination stage. In 2013 third international conference on intelligent system design and engineering applications, pp 1248–1253

Zou X, Hu C, Zeng L, Cheng Y, Xu M, Zhang X (2014) A comparison of screening methods to identify waterlogging tolerance in the field in *Brassica napus* L. during plant ontogeny. PloS one 9(3):e89731

Zou XL, Liu ZE, Lu GY, Cheng Y, Xu JS, Zhang XK (2015) Comparison of transcriptomes undergoing waterlogging at the seedling stage between tolerant and sensitive varieties of *Brassica napus* L. J Integr Agric 14(9):1723–1734

Brassicaceae Plants Response and Tolerance to Nutrient Deficiencies

K. S. Karthika, Prabha Susan Philip, and S. Neenu

Abstract The plant family Brassicaceae commonly known as the cabbage family or mustard family has several agriculturally important crops. The major vegetable crops include cauliflower, cabbage, broccoli, kale, Brussels sprouts, turnip, radish, etc., and major oilseed crops are mustard and canola. Every essential nutrient has an inevitable role to play in a plant's growth and development; however some may be critical towards some crops such as in providing resistance to pests and diseases, or in terms of improving quality of the crop. Among macronutrients, S is one of them, very much essential to the plants of Brassicaceae family as it has got prominent effect on oilseed crops. Striking a perfect balance between N and S is important in growth and development of Brassicaceae plants. An understanding on the role of nutrients in the growth and development of plants belonging to Brassicaceae family is essential to identify their requirements for nutrients. This would provide an insight into identification of the symptoms exhibited as a result of nutrient deficiencies. The chapter deals with the nutrient deficiencies in Brassicaceae plants, and their responses and tolerance to stresses such as nutrient deficiencies. Hence, a better knowledge on these would aid in enhancing productivity of Brassicaceae plants by properly managing nutrient deficiencies and related abiotic stresses.

Keywords Brassicaceae · Critical concentration · Plant nutrition · Deficiencies · Symptoms

K. S. Karthika (✉)
ICAR-National Bureau of Soil Survey and Land Use Planning, Regional Centre, Hebbal, Bangalore 560 024, India
e-mail: kskavukattu@gmail.com

P. S. Philip
ICAR-National Bureau of Soil Survey and Land Use Planning, Regional Centre, Delhi 110 012, India

S. Neenu
ICAR-Central Plantation Crops Research Institute, Kasaragod 671 124, Kerala, India

M. Hasanuzzaman (ed.), *The Plant Family Brassicaceae*,
https://doi.org/10.1007/978-981-15-6345-4_11

1 Introduction

Brassicaceae, (formerly Cruciferae) the mustard family of flowering plants (order Brassicales) is composed of 338 genera and some 3,700 species. Most of the plants in this family possess huge economic importance. The family includes many vegetable crops and the genus *Brassica* is the major one consisting of several leafy vegetable crops such as cabbage, Chinese cabbage, broccoli; root vegetable crop such as turnip, major oilseed crops such as rapeseed and mustard (Collett et al. 2014; Spragg 2016). Oleiferous brassicas include rapeseed and mustard which are the major oilseed crops belonging to this family. Soils that are well-drained having a pH of 5.5–8.5 are found to be the most ideal for *Brassica* species as poorly drained soils hinder the growth and establishment. Brassicaceae plants are mainly herbs which can be annuals, biennials, or perennials. The leaves are mostly simple and alternately arranged, and many are peppery-flavoured. The flowers have four petals and four sepals each which appear in the form of a cross known as cruciform; referring the plants as "crucifers" or "cruciferous". Almost all Brassicaceae members are C3 plants except Moricandia which has a hybrid system between C3 and C4. *Brassica* crops require about 25% more nitrogen, phosphorus, potassium, and about five times more sulphur compared to most other grain crops. Cabbage, cauliflower, radish, watercress, horseradish, oil crops, etc., are the cultivated plant species coming under this family. Some of the important members of the family are mentioned in Table 1. In this chapter, we discuss the role of essential nutrients in the growth and development of Brassicaceae plants, their critical concentration (Table 2), physiology and mechanisms and how the deficiencies of essential nutrients are exhibited in the form of different symptoms.

2 Nutrition

Essential nutrients are those nutrients that fulfil the criteria developed by Arnon and Stout (1939). Accordingly, there are 17 nutrients that are considered essential for higher plants and these include carbon (C), hydrogen (H), oxygen (O); primary nutrients—nitrogen (N), phosphorus(P), potassium (K); secondary nutrients—calcium (Ca), magnesium (Mg), sulphur (S); and micronutrients—iron (Fe), manganese (Mn), zinc (Zn), copper (Cu), boron (B), molybdenum (Mo), nickel (Ni), and chlorine (Cl). The C, H, and O are derived from the environment; however other elements need to be provided from the soil or through the fertilizers to plants.

It is very important that we understand the critical concentration of each essential nutrient in the soil as well as plant as any deviation be it deficiency or excess will affect the crop growth and development. Deficiency of a particular nutrient reduces the crop production and excess concentration results in toxic symptoms. The plant family Brassicaceae is one of the well-known plant families with *Brassica, Arabidopsis,* and *Alyssum* as representative plants. Cabbage, cauliflower, radish,

Table 1 Important members of family Brassicaceae

Common name	Scientific name	Uses
Cabbage	*Brassica oleracea var. capitata*	As vegetable
Broccoli	*B. oleracea var. italica*	
Cauliflower	*B. oleracea var. botrytis*	
Kale/leaf cabbage	*B. oleracea var. sabellica*	
Brussels sprout	*B. oleracea var. gemmifera*	
Collard greens	*B. oleracea var. acephala*	
Savoy cabbage	*B. oleracea var. sabauda*	
Kohlrabi/German turnip	*B. oleracea var. gonglodes*	
Gan Ian	*B. oleracea var. alboglabra*	
Turnip	*B. rapa subsp. rapa*	
Napa cabbage	*B. rapa subsp. pekinensis*	
Bok choy	*B. rapa subsp. chinensis*	
Rappini	*B. ruvo*	
Arugula/Rocket salad	*Eruca vesicaria*	
Garden cress	*Lepidium sativum*	
Watercress	*Nasturtium officinale*	
Radish	*Raphanus sativus*	
Horseradish	*Armoracia rusticana*	As spice/oilseed crops
Wasabi/Japanese horseradish	*Eutrema japonicum*	
White mustard	*Sinapis alba*	
Indian/brown mustard	*B. juncea*	
Black mustard	*B. nigra*	
Garlic mustard	*Allairia petiolata*	
White ball mustard	*Calepina irregularis*	
Ball mustard	*Neslia paniculata*	
Treacle mustard	*Erysimum repandum*	
Hedge mustard	*Sisymbrium officinale*	
Asian hedge mustard	*Sisymbrium orientale*	
Smooth mustard	*Sisymbrium crysimoides*	
Rapeseed/oilseed rape	*B. napus*	For vegetable oil
Woad	*Isatis tinctoria*	Dye
Aethionema	*Aethionema grandiflorium*	As Ornamental plant
Alyssum	*Lobularia maritima*	
Honesty	*Lunaria annua*	
Arabidopsis	*Arabidopsis thaliana*	As model plant for molecular studies

Table 2 Critical nutrient concentration, nutrient content, and nutrient uptake in Brassicaceae crops

Critical nutrient concentration (CNC) for 90% yield										
	N	P	K	Mg	S	Mn	Zn	Cu	B	Mo
	%					$\mu g\ g^{-1}$				
Oilseed rape	3.50	0.30	2.50	0.20	0.50	30.0	20.0	5.0	25.0	0.30
Nutrient content(kg t^{-1})										
	N	P_2O_5	K_2O	Ca	Mg	S	Cu	Mn	Zn	
Cabbage	2.90	0.80	2.90	0.40	0.20	1.0	0.001	0.002	0.002	
Total nutrient uptake (kg t^{-1})										
		N	P_2O_5	K_2O	Fe	Mn	Zn	Cu		
Cauliflower(curd)		4.0	2.0	4.0						
Cabbage (head)		3.5	1.3	4.2						
Mustard (Tandon 2004)					1.12	0.1	0.1	0.017		
Relative tolerance to exchangeable sodium percentage in soils (Range of ESP[a]) (Tyagi 2000; Gupta and Abrol 1990)										
Rapeseed, mustard		30–50								

[a]Relative crop yields are only 50% of the maximum in the alkalinity range indicated

watercress, horseradish, oil crops, etc., are the cultivated plant species coming under this family. Almost all Brassicaceae members are C3 plants except Moricandia which has a hybrid system between C3 and C4. *Brassica* crops require about 25% more nitrogen, phosphorus, potassium, and about five times more sulphur compared to most other grain crops. However, studies also confirm that the requirement of cabbage for S was the minimum among the macronutrients and it followed the order N > ca > K > Mg > P > S. Higher levels of primary nutrients, S, Mn, and B are required by oilseed rape (*B. napus* L.) for optimum yield (Grant and Bailey 1993).

3 Nitrogen

Nitrogen is a major component in many of the plant cell components and is directly involved in processes like plant photosynthesis, protein synthesis signifying its importance as a primary nutrient in plant growth. It is a key constituent of amino acids, nucleic acids, nucleotides, proteins, chlorophyll, ribosomes, chromosomes, genes, and a component of all enzymes. Since it is constituent of chlorophyll, nitrogen is responsible for the dark green colour to plants and it helps in the vegetative growth of crop-plants (Karthika et al. 2018). Nitrogen plays a key role in the production of starch and amino acids in leaves. Since it helps in the production of protein and amino acids it improves the quality of leafy vegetables, fodders, and food-grains. Generally plants contain about 1–5% nitrogen by dry weight. Brassicaceae crops to attain optimum growth with better quality inflorescences require comparatively higher amounts of N (Coulombe et al. 1998).

The critical N concentration was defined as that critical plant N concentration (PNCc) in the dry matter mass required to attain the maximum instantaneous growth rate (Lemaire and Gastal 1997). It refers to the optimal growth, though it also signifies the susceptibility to physiological disorder or diseases. Critical concentration of N in a generic C3 species was recorded to be 4.8% N, DW-0.34 (Lemaire and Gastal 1997), for generic *Brassica* species was 5.2% N, DW-0.178 (Greenwood and Stone 1998), and for white cabbage, it was recorded to contain 5.1% N, DW-0.33 (Ekbladh and Writter 2010). In cabbage (cv. Rare Ball Hybrid) 0.51% was found to be the critical concentration of N.

3.1 Nitrogen Deficiency, Response, and Stress Tolerance

Brassica plants require more nitrogen and hence available nitrogen is a major limiting factor for attaining higher yields in these crops. Nitrogen in sufficient quantities imparts a dark green colour in *Brassica* plants. Mineral N is crucial for production and yield of oleiferous brassicas, which mainly denote rapeseed and mustard (Abdallah et al. 2010; Anjum et al. 2012). Lack of adequate nitrogen leads to reduced production of chlorophyll, the green colour imparting pigment resulting in yellowing of older leaves. In general, nitrogen deficiency results in smaller plant size, early flowering and fruiting. A deficiency in N, decreases vegetative growth during the initial stages of crop development and reduces the leaf area index, thereby lessening the duration of photosynthetic activity resulting in decreased productivity. Nitrogen deficiency in broccoli leads to paling of foliage and heads. It results in the formation of very small loose heads and bolting is also seen. Stunted plant growth and delayed crop development are resulted as a result of N deficiency in soil.

In cabbage, N deficiency results in chlorotic foliage. Even though the shape of the leaves remains the same the growth is severely affected and delayed development of the plant occurs. Under conditions of severe deficiency the head will not be formed in cabbage. The shoot and root of cabbage recorded lower contents of N, Ca, and Mg under conditions of N deficiency (Sung et al. 2018).

In mustard, the deficiency is exhibited as reduced vigour, pale green, and smaller leaves. The flowering is reduced and the overall growth of the plant gets severely restricted. In cauliflower yellowing of the older leaves and red discolouration beginning at the margin of the leaves are the main symptoms of nitrogen deficiency. In cauliflower, it results in the development of small curd or button referred as buttoning. In rapeseed, leaf blades become narrower during N starvation. The stems become thinner giving better resistance to lodging. The younger leaves turn pale green and the older ones turn yellowish green in colour.

However, excess of nitrogen causes more vegetative growth and lodging before seed-production. These plants are more prone to insect pests and diseases. It increases the length of reproductive period, delay the ripening and quality.

3.2 Tolerance Mechanisms to N Stress

Broccoli, when grown under N stress conditions recorded a rise in total glucosinolates, if S was not limiting (Schonhof et al. 2007). This is due to the interrelationship between N and S assimilatory pathways. Deficiency of N suppresses S pathway and vice versa (Koprivova et al. 2000). Limited S allows the accumulation of O-acetyl serine and limited N inhibits its accumulation indicating the antagonistic effect (Leustek et al. 2000). Nitrogen fertilization affects the contents of aliphatic and indolic glucosinolates in *Eruca sativa* L. (Omirou et al. 2009). Aliphatic glucosinolates responded negatively and indolic glucosinolates responded positively to N fertilization (Ballesta et al. 2013). In cabbage, it was found that N stress resulted in increased levels of inositol in shoots and glucose, fructose in shoots indicating a general increase in sugar accumulation, and decreased levels of sugar in roots especially fructose, inositol levels in roots, absolute succinate and malate, and most of the amino acids (Sung et al. 2018).

4 Phosphorus

Phosphorus is an essential nutrient involved in the processes of photosynthesis, energy storage and transfer in the form of AMP, ADT, ATP, NADP or NADPH, respiration, cell division and enlargement, etc. It is very much important for root formation and development which prevents P sufficient plants from lodging and improves the water use efficiency by the production of more root structures. It also helps in flowering, fruiting, seed formation, in hastening maturity, and survival of plants in harsh winter. The phosphorus requirement for the optimum growth of plant is 0.3–0.5% (dry weight basis) in the vegetative stage and when the concentration reaches beyond 1% toxicity symptoms may arise. In *Brassica* crops, P encourages early flowering and ripening as well as development of roots. It is mostly important in early stages of growth. *Brassica* crops require higher P supply to increase the content of high energy containing oil.

4.1 Phosphorus Deficiency, Response, and Stress Tolerance

In rapeseed, a deficiency in P hinders the shoot and root growth, whereas an adequate supply of P increases the concentration of oil and protein (Bailey and Grant 1990). In short season areas, higher P leads to earlier maturity of rape plants. Supply of P is restricted by the cold and drought conditions. In *Brassica* plants, growth retardation forms the main symptom of P deficiency with no discolouration of leaves. Phosphorus deficiency in cabbage was exhibited by changes in leaf shape particularly by narrowing the leaf (Sung et al. 2018). In broccoli, the deficiency leads to red discolouration on the edges of the older leaves, with a yellow-green transition zone and the leaf size remain unaffected. In case of rapeseed older leaves show reddish/purplish discolouration. In cauliflower older leaves are pale bluish green with purple discolouration. Underside of the leaves turn red colour. In cabbage, P stress conditions recorded an increase in glucose levels, inositol levels in shoots, and an eight-fold increase in citrate concentrations. However, it also recorded a decrease in fructose levels, soluble sugar levels in roots, inositol levels in roots, absolute succinate and malate levels, and amino acids (Sung et al. 2018).

4.2 Tolerance Mechanisms to P Stress

Root hairs in oilseed rape possess a higher P influx rate, contributing towards increased efficiency in P uptake (Fohse et al. 1991). Rape excretes organic acids in the root zone enabling scavenging of P by converting insoluble forms into available forms. The P status in soil, availability of moisture and yield determines the response to P. When the nutrient availability is limited in a soil, the capacity of soils to overcome unfavourable environmental conditions lead to optimum plant growth and development (Lynch 1995). General recommendation of N and P_2O_5 are in a ratio of 2:1.

5 Potassium

Potassium plays a significant role in photosynthesis and food production. It acts as a catalyst in at least 60 different enzymes in the plants. It regulates stomatal movement of the plant and helps to regulate water within plant cells and loss of water by transpiration and reduces the negative effects of salts in the plants. It is helpful in the transportation of photosynthates. It enhances resistance power against pests and diseases and lodging of the plant. Potassium helps to improve the quality of seeds and fruits by converting starch into sugars. It accelerates ripening of the crop due to its involvement in carbohydrate metabolism and breakdown and translocation of starch. It also helps in improving the storage life of vegetables and fruits. It gives strength to roots, stem, and branches of the plant thereby reduce the lodging of the plant.

5.1 Potassium Deficiency, Response, and Stress Tolerance

The form in which potassium is absorbed by the plant is potassium ions (K^+). Potassium deficiency symptoms initially appear in older leaves of the plant as it is mobile inside the plant system. The most important symptom of potassium deficiency is scorching or firing along the margin which may fall out in later stages of deficiency. In cauliflower, due to potassium deficiency leaf margins become yellow to light brown and later dark brown in colour and turn necrotic. If deficiency persists, the margin withers and curls up. The leaves produced are less waxy in nature. At severe deficiency, a gradual progress of necrosis is seen from the margins to the middle of the leaves. In broccoli, older leaves become yellow to brown with well defined border zones. Since broccoli has a high requirement of K the deficiency affects the plant severely. In cabbage, deformation of the head occurs due to K deficiency. Initially the older leaves show light brown and withered necrotic border zone and later necrosis progresses from the edge to the middle. Leaf margins will be cupped upward resulting in deformed heads. In rapeseed, before the actual symptoms occur, K deficiency results in growth and yield reductions. Often younger leaves remain small and growth is stopped in later stages. A bluish green colouration of the younger leaves also occurs.

5.2 Tolerance Mechanisms to K Stress

Potassium deficiency increased oxylipins and glucosinolate levels in *Arabidospis thaliana*. The glucosinolates were more in roots than shoots. Glucosinolates play a significant role in enhancing defense mechanisms in the plant during K deficiency in soil (Troufflard et al. 2010). In *Arabidopsis thaliana,* for improving resistance to potassium deficiency in soils and for improved root hair and primary root growth, ethylene was found to be responsible (Jung et al. 2009). During potassium deficiency, ethylene acts upstream of reactive oxygen species. A marker that is recognized for potassium deficiency is the expression of *High-Affinity K + Transporter5*. This is also related to ethylene signaling indicating its dependency. However, there are some responses out of K deprivation which are ethylene independent which is explained by *ethylene insensitive2-1* (*ein2-1*) mutant ensuring the probability that ethylene alone can never be a mediator to K stress response. As explained, the stimulation of production of reactive oxygen species by ethylene signaling under K stress plays an important role in bringing the changes in root morphology and whole plant tolerance. Potassium stress conditions recorded a marked increase in the levels of quinate, a precursor of the shikimate pathway (Sung et al. 2018).

6 Secondary Nutrients

The secondary nutrients involved in plant nutrition are calcium, magnesium, and sulphur. These are required in lesser amounts than that of the primary nutrients, however essential for a plant to complete its life cycle. However, Brassicaceae has a higher requirement for S especially, oleiferous brassicas.

6.1 Calcium

Calcium concentration in the shoot ranges from 0.1 to 5% of dry weight for plant growth. Calcium is an immobile element in plants. Calcium pectate is a constituent in cell wall indicating the role of Ca in maintenance of structure of cell wall and membrane. Calcium makes cell wall stronger, enhances the nutrient uptake, imparts resistance to the plant tissue, and maintains normal development of the root system (Hirschi 2004). Calcium, as a second messenger, is responsible for several functions like root growth, pollen tube growth and fertilization, responses to biotic and abiotic stress, etc. There are two types of plant species based on their adaptation, i.e. calcifuges and calcicoles. Calcifuges are those plants occur in soils with low pH and Ca and calcicoles are those occur in calcareous soils. Brassicaceae members are mainly coming under the group calcicoles, which have a high calcium concentration. They are also termed as calciotrophs, where the Ca accumulation is stimulated by increasing external calcium concentration. Cabbage has a higher requirement for Ca. A Ca concentration of 1.1% on dry weight basis is critical in reducing the head yield of cabbage by 50%. However, a Ca content >4% is excessive for the growth of cabbage.

6.1.1 Calcium Deficiency, Response, and Stress Tolerance

Deficiency of Ca may arise when the levels of Ca is low. This could be due to an imbalance in soil nutrients as a result of weather conditions of high rainfall, low soil moisture, or increased levels of N or K which reduces Ca availability.

In Brassicaceae crops, calcium deficiency is mainly exhibited as tipburn of cabbage, cauliflower, Brussels sprouts, and Chinese cabbage (Napa cabbage). In cabbage and Chinese cabbage, tipburn refers to the breakdown of plant tissue inside the head. Calcium deficiency may occur as narrowing of leaves in cabbage (Sung et al. 2018). In cabbage, however, the most prominent symptom is "tip burn" which means necrotic lesions on the tip and margins of the leaf. In later stages the necrosis progresses inward and affect the entire foliage. Calcium deficiency occurring in the early stages of growth, results in stunted plant growth with no formation of the head. In Brussels sprouts, tipburn happens in individual sprouts and in cauliflower on the inner wrapper leaves. In cauliflower it results in necrotic leaf margins and "claw-like" deformation of the youngest leaf and ultimately the curd dies off. Studies also confirm that tipburn could be considered as a physiological disorder too, which is aggravated by a deficient supply of Ca resulting in collapse and death of tissues. Tipburn in cauliflower may be followed by a bacterial rot on the heads, though a few cultivars are comparatively free of problems. Calcium deficiency in broccoli starts on young leaves as necrotic lesions on the margins. Growth of the entire plant is affected due to calcium deficiency.

Liming is a recommended practice to overcome the problems of tipburn, as it helps to raise the soil pH. Maintaining adequate soil moisture and limiting the application of ammoniacal forms of nitrogen could be helpful in preventing tipburn. Choosing a resistant cultivar is another option to overcome the tipburn.

Emblem, Vantage Point, Bobcat, Padok, Cecile, Royal Vantage, Blue Vantage, Quick Start, Thunderhead, Viceroy, Superstar, Platinum Dynasty, Solid Blue 780 are a few Green cabbage varieties with higher resistance to tipburn. Red cabbage is less susceptible to tipburn.

6.1.2 Tolerance Mechanisms to Ca Stress

Calcium inhibits cell wall degrading enzymes produced by fungi and bacteria and hence low level of Ca in the plant tissue leads to higher degradation and infection. Calcium ions by certain mechanisms alleviate ionic stress in plants and are able to prevent subsequent damage. In *Arabidopsis*, an increased stress resistance to conditions of Ca stress was found as a result of overexpression of proteins such as CAX1(Ca^{2+}/H^+ antiporters) and CRT (Ca-binding protein) (Pittman and Hirsch 2001). In Chinese cabbage, cytoplasmic Ca^{2+} fluctuation-induced downstream signaling, salicylic acid signaling events were found to be the plant defense responses under Ca deprived conditions (Su et al. 2015). Quinate, (a precursor of the shikimate pathway) levels increased under conditions of Ca stress (Sung et al. 2018). Calcium plays an important role in salt tolerance. Salt tolerance of *B. campestris* (rapa) decreased with increased concentration of Ca^{2+}, K^+, and Ca^{2+}+K^+ in seeds of *B. campestris* (rapa) (Collins et al. 2008). Calcium protects the plants from Al toxicity due to the protective behaviour of Ca ions (Kochian 1995). Calcium transport in the plant takes places mostly through the xylem, along with water. Hence calcium uptake and transpiration rate of the plant share a direct relationship. It helps to improve the stomatal function and stimulate the heat shock proteins. It is utilized for cell division and formation, nitrogen metabolism, nutrient signaling, plant immunity signaling, etc. Calcium toxicity in the rhizosphere prevents germination of seeds and reduces plant growth rate.

6.2 Magnesium

Magnesium in soils ranges from 0.05 to 0.5% and in plants from 0.1 to 0.4% on dry weight basis. The availability of Mg to plants is limited as it is a chemical constituent of minerals. Magnesium plays an indispensable role in chlorophyll production. Magnesium is involved in improving ion utilization in plants, activation of many enzymes, in many metabolic processes and reactions like photophosphorylation, protein synthesis, photosynthetic carbon-di-oxide fixation, partitioning and utilization of photoassimilates, generation of reactive oxygen species, phloem loading, photo-oxidation of leaf tissue, etc. The element influences earliness and uniform maturity of fruits. It improves the utilization and movement of phosphorus in the plant system. It has a role in translocation of starch and regulation of other nutrient uptake. Magnesium is very important in oilseed crops. In cabbage, a critical Mg content of 0.15% on dry weight basis results in a cabbage head yield reduction by 50% due to its deficiency. However, a concentration >1.5% is excessive for the growth of cabbage. For the normal growth of cabbage, an adequate Ca: Mg ratio is essential.

6.2.1 Magnesium Deficiency, Response, and Stress Tolerance

Magnesium being mobile in the plant system, its deficiency symptoms begin to appear in older leaves. Magnesium deficiency leads to degradation of the chlorophyll pigment and leaves become yellowish in colour with green veins, leads to interveinal chlorosis. The leaves may turn pink to light red and may curl upward along the margins. Several processes in plant physiology are dependent on magnesium, hence its deficiency will interrupt all those processes.

In broccoli, the older leaves show chlorosis between the veins progress from the margin to the middle of the leaflet. The veins remain green and white and severe deficiency leads to necrotic spots between the veins. In cabbage, purplish coloured leaves develop leading to irregular chlorosis due to Mg deficiency. Small veins turn chlorotic but the bigger veins remain green and white respectively. The interveinal zones curl up. The inner leaves show necrotic areas in their edges. The leaf size reduces resulting in narrowing of leaves (Sung et al. 2018). In moderate deficiency, growth reduction is very slight but severe deficiency leads to retarded growth. In cauliflower, Mg deficiency leads to chlorosis and severe deficiency leads to necrotic spots between veins and older leaves become stiff and prematurely fall down. In cabbage, it was reported that Mg stress resulted in increased sucrose levels and decreased glucose and fructose levels in shoots. However, the roots recorded a decrease in soluble sugars and inositol levels. The levels of quinate and amino acids increased in shoots and roots of cabbage under Mg deficiency (Sung et al. 2018).

6.2.2 Tolerance Mechanisms to Mg Stress

Alteration of photosynthetic carbon metabolism, a restriction in carbon-di-oxide fixation are resulted due to an accumulation of carbohydrates in source tissue during Mg deficient conditions.

(Hermans et al. 2004). In cabbage, *B. napa* L. ssp. *pekinensis*, Mg stress resulted in a quantitative increase in the levels of amino acids due to the restriction of various pathways involved in protein production. An increase in the levels of quinate, a precursor of the shikimate pathway was recorded under conditions of Mg stress (Sung et al. 2018). Magnesium has a protective effect on Al toxicity through mechanisms involving regulating the pH in cytoplasm, organic acid anions exudation, shoot to roots carbon partitioning, and activity of H^+-ATPase on plasma membranes in a better way. Toxicity of Mg is rare and in soils with higher Mg indicating its toxicity, competition with Ca and K uptake is found resulting in their deficiencies in plant tissue.

6.3 Sulphur

Sulphur has structural, regulatory, and catalytic functions in the plant system. It is an integral part of chlorophyll and amino acids. These amino acids are involved in metabolic activities of vitamins, biotins, thiamine, and coenzyme-A. It is required for the formation of vitamins and enzymes. Among field crops, *Brassica* species require high amount of S. Rapeseed and mustard are the most common species of *Brassica* grown as oilseed crops which are generally referred as Oleiferous brassicas. These require more S than that of other crops and hence more sensitive towards S deficiency (Walker and Booth 2003) and it ranges from 0.1 to 1.5% of dry weight depending on species and developmental stage (Anjum et al. 2012). For production of 1 tonne of rapeseed 16 kg of S is required (Blake-Kalff et al. 2001; McGrath and Zhao 1996). All oilseed crops especially belonging to Brassicaceae family have comparatively more requirement of sulphur due to their high contents of essential oils and amino acids containing S. It is a known fact that brassicas and mustards need adequate nitrogen and sulphur for their growth, production, and quality.

6.3.1 Significance of N: S Ratio in *Brassica* Crops

N: S ratio is very important in *Brassica* crops. In general for *Brassica* species, N: S ratios of 4:1–8:1 are found to be ideal. For rapeseed crop, optimum N: S ratio in soil is 7:1, while in cabbage, a N:S ratio of 6–25 is found ideal for S nutrition. To recognize deficiency of S in soils, ratio of N: S in plants is often used. In oleiferous brassicas, the requirement of S is closely related to N nutrition (Fazili et al. 2008). The assimilation pathways of N and S are very similar (Abdin et al. 2003). Deficiency of N represses the S pathway and vice versa (Scherer 2008). In oilseed brassicas, the positive interaction between N and S is found to be beneficial (Anjum et al. 2012). These two nutrients are essential in oilseed brassicas as N deficiency reduces the efficiency of S uptake and S deficiency reduces the efficiency of N uptake (Fismes et al. 2000).

6.4 Sulphur Deficiency, Response, and Stress Tolerance

Sulphur is not readily mobile within plants; hence there should be a continuous availability of sulphur from emergence to crop maturity. Sulphur and nitrogen deficiency symptoms are similar except that the symptoms of S deficiency appear in younger leaves initially as S is immobile in plants. Deficiency of S results in stunted growth and thin stem. In broccoli, S deficiency leads to chlorosis of young leaves and as the deficiency progresses the symptoms appear in older leaves also. The deficiency in cabbage appears as purple coloured leaves with marginal parts curling inward like a cup and the roots remain long but thin. Symptoms begin as chlorosis in the leaves in the intercostal fields, which turn uniform light green or yellow in the later stages. Purplish colouration in the stem, petiole, and leaf also occur. The leaves are stiff and curl inward. Stem of leaves become long and malformed leaves appear narrow in the shape of a spoon in *Brassica* plants. In *B. napus* L., sulphur stress decreases the accumulation of S-rich seed storage protein and as a result it decreases the protein quality of seeds. A decrease in the accumulation of C18 derivatives results in a decrease in lipid quality as a result of S starvation (D'Hooghe et al. 2014).

6.4.1 Tolerance Mechanisms to S Stress

Brassica plants have higher requirement for S as it is essential in the production of oil and glucosinates. Sulphur uptake capacity of *Brassica* is also exceedingly higher than majority of other plant species. Many secondary plant products and coenzymes have amino acids that contain S, which act as precursors to these products. Pest and disease resistance in *Brassica* are influenced by typical secondary S-containing glucosinolates. These perform as resistance barriers to plants which provide mechanisms for plant defense. The quality of oil gets reduced if the levels of glucosinolates remain high.

The fertilizers, Ammonium Sulphate and Single Super Phosphate are recommended for use in S deficit soils. Sulphur is taken up as sulphates and once it reaches the leaves through xylem transport, it gets reduced to cysteine which may get converted to methionine or incorporated to proteins and cysteine-containing peptides such as glutathione. This proves the inevitable role played by S in protein formation, especially in rapeseeds for its high protein content. Chloroplasts have most protein located, especially in case of green leaves. A decrease in the chlorophyll content and its breakdown are resulted due to S stress, which decreases the green colour of leaves turning them yellow (Bergmann, 1992). The expected outcome of S fertilization is resulted in 3–7 days if the duration of S deficiency is not too lengthy. This will result into normal dark green leaves. The levels of applied fertilizer nitrogen, status of S in soil, yield potential are required to know the optimal dosage of S to be applied. Nitrogen and sulphur are essential for protein synthesis; hence the dosages of N and S should be related.

7 Micronutrients

There are eight micronutrients essential for plant growth and development. These are iron (Fe), manganese (Mn), zinc (Zn), boron (B), copper (Cu), molybdenum (Mo), nickel (Ni), and chloride (Cl). Micronutrients are those nutrients required in micro quantities and are as important to plant nutrition as primary and secondary macronutrients. However, a deficiency in one of the micronutrients may limit the growth of plants, with sufficient quantities of all the remaining nutrients.

7.1 Iron (Fe)

Iron (Fe) is an important component of many enzymes associated with nitrogen reduction and fixation, energy transfer. It performs many processes in plant physiology by involving in chlorophyll synthesis and lignin formation. Iron is a constituent of iron-containing (heme) proteins in plants, i.e. cytochromes. Cytochromes are involved in the electron transfer systems in chloroplasts and mitochondria. Iron is also a component of certain non-heme proteins such as ferredoxin.

7.1.1 Iron Deficiency, Response, and Stress Tolerance

Iron is fairly immobile after it is incorporated in compounds in the upper parts of the plant tissues. The re-translocation of Fe is negligible from one plant part to another, due to which Fe deficiency symptoms first appear in the younger leaves. The Fe deficient young leaves exhibit yellowing, referred as "iron chlorosis". Calcareous soils with a $pH > 7.2$ frequently show Fe deficiency because under alkaline pH conditions Fe is mainly in the oxide and unavailable form. In acidic soils also Fe deficiency occurs mainly due to competition between Mn and Fe. Over-liming and resulted high calcium carbonate in soil can also lead to Fe deficiencies ("lime-induced Fe chlorosis") mainly because of the high pH and competition of Ca with Fe uptake by the root. Also long time use irrigation water with high concentrations of carbonates can lead to chlorosis.

Fe deficiencies are rarely seen in *Brassica* plants. Chlorosis (yellowing) of younger leaves is usually seen associated with Fe deficiency. In cabbage, due to Fe deficiency leaf blades are smaller and petioles are longer than usual. The leaves are not cupped inward and head formation is inhibited. Young leaves show a faint yellowish green tinge. In case of cauliflower, young leaves show typical chlorosis. In the initial stage of the symptom the leaf laminas become chlorotic with greenish leaf veins. As the deficiency progresses, the growth of roots and overhead parts are depressed without any leaf deformation. In Chinese cabbage (*B. chinensis* L.) iron deficiency resulted in stunted plant growth, chlorosis of leaves, reduction in photosynthetic pigments, net photosynthetic rate, Fv/Fm, Φ PS II and activities of antioxidant enzymes, and increase in lipid peroxidation (Ding et al. 2008). Iron deficiency resulted in decreased contents of chlorophyll or carotenoid and caused damages to photosynthetic apparatus in oilseed rape (*B. napus* L.) (Muneer et al. 2014).

7.1.2 Tolerance Mechanisms to Fe Stress

Different strategies such as either enhancement or attenuation of root growth or an alternation between these two, are developed by plants to cope up with lower levels of iron or phosphate in soils. Primary root growth reduced with an increase in secondary root growth along with an increase in root hair length and density during phosphorus deficient conditions in *Arabidopsis thaliana* (a member of Brassicaceae family). Root growth accelerated during iron deficiency as a result of activating brassinosteroid signaling and root growth inhibited due to P deficiency induced high accumulation of iron in *Arabidopsis thaliana.* This alteration in hormone signaling intensity resulted in controlling iron accumulation in the root zones responsible for elongation and differentiation. This constitutes a feedback response between brassinosteroid and iron. This feedback loop consists of the transcription factors BES1/BZR, the brassinosteroid receptor inhibitor BKI1, and the ferroxidase LPR1. Iron regulatory components and brassinosteroid drive the adaptive response to soil Fe status by linking that with root morphology (Singh et al. 2018). Deprivation of Fe decreased the ferric chelate reducing activity. Thylakoid lamellae underwent disorganization with the loss of grana under conditions of Fe deprivation in oilseed rape (Muneer et al. 2014).

There are two mechanisms of iron uptake in plants. They are the strategy I mechanism by non-graminaceous monocots and dicots and strategy II mechanism by grasses (Marschner and Roemheld, 1994). In strategy I mechanism, acidification, reduction, and transport are the major activities involved. Iron arrives into soil solution through acidification of the rhizosphere. During iron-limited conditions in response to Fe stress, a proton ATPase pumps protons across the plasma membrane. The acidification helps to reduce Fe^{3+} to Fe^{2+} and make it possible for the Fe uptake easier. In grasses, the strategy II mechanism is involved in which synthesis of phytosiderophores by the plant roots will release into the rhizosphere and these will complex with the Fe present in the soil solution. This phytosiderophore-Fe complex can move across the membrane of the roots.

Iron toxicity is equally harmful and in *Arabidopsis*, under conditions of excess iron, iron is taken up into vacuoles by AtVIT1 transporter (Kim et al. 2006). In rapeseed, mechanisms to iron tolerance is complex and to enhance Fe tolerance in oilseed rape, the vacuolar iron transporter BnMEB2 is localized on the vacuolar membrane and part in vacuolar sequestration if iron storage (Zhu et al. 2016).

7.2 Manganese (Mn)

Manganese (Mn), plays a key role by assisting in chlorophyll synthesis. It has an essential role in maintenance and activating enzyme systems in plants: examples are Mn-catalase, pyruvate carboxylase, Mn-superoxide dismutase, and phosphoenolpyruvatecarboxykinase. It activates several important metabolic reactions in the plant system. It regulates carbohydrate metabolism and affects energy budgeting.

Manganese improves germination and maturity along with an increase in the availability of phosphorus (P) and calcium (Ca). It is important for immobilization of free radicals in the system. Manganese assists in photosynthesis by providing electrons as a result of involving in the water-splitting system of photosystem II. Manganese acts as a redox shuttle involved in mediating the process of adding monolignols to phenolic groups that exist in the cell wall, hence plays a role in lignin polymer biosynthesis. It is also involved in RuBP carboxylase reactions, ATP synthesis, and the biosynthesis of proteins, lipids, and fatty acids. It also takes part in the biosynthetic pathway of isoprenoids and assimilation of nitrate. Manganese is also involved in several metabolic processes such as photosynthesis, respiration, synthesis of amino acids, and hormone activation (IAA) through the IAA-oxidases.

7.2.1 Manganese Deficiency, Response, and Stress Tolerance

Manganese being immobile in the plant system, its deficiency symptoms occur in younger leaves first and include mottling and interveinal chlorosis. In cauliflower, the deficiency of manganese results in brightening of the interveinal areas of young leaves leaving the veins green, developing yellow mottling. Young and medium leaves become crinkled. In addition, the leaves roll or curl upward. As the deficiency progresses, the young leaves completely turn yellow. Usually leaves are smaller than normal. In broccoli, manganese deficiency symptoms include yellow-green mottling of young leaves due to chlorosis. Almost all leaves will turn pale green than normal. Leaf blades may enlarge. In cabbage, deficiency of Mn results in yellow-green mottling of young leaves due to interveinal chlorosis. Leaf blades are enlarged, no inward cupping occurs, and head formation is inhibited. Petioles are longer than normal giving the plant a loose appearance.

7.2.2 Tolerance Mechanisms to Mn Stress

Tolerance to environmental stresses such as salt stress can be enhanced by the activities of enzyme manganese-super oxide dismutase. Manganese in excess damages the photosynthetic apparatus. The Mn toxicity symptoms vary greatly between plant species. Typical symptoms of Mn toxicity occur as brown spots surrounded by chlorotic zones on older leaves. Accumulation of Mn^{4+} in pockets is reported to be the reason behind the development of such necrotic brown spots. Manganese toxicity is common in acid soils. At low pH, Mn^{2+} increases in the soil solution.

7.3 Zinc (Zn)

Zinc plays an important role in the metabolic activities of plants. Zinc influences stabilization of ribosomal fractions, enzymatic activities of carbonic anhydrase (CA) and dehydrogenase, and synthesis of cytochrome. Zinc activates many plant enzymes which are involved in protein synthesis, carbohydrate metabolism, regulation of auxin synthesis, maintenance of the integrity of cellular membranes, and pollen formation. Gene expression during environmental stress tolerance is regulated and maintained by Zn. Zinc is helpful in negating the effects of salt and heat stress of shorter duration. Uptake of water and its transport within plants are influenced by Zn. It is also involved in the formation of chlorophyll and carbohydrate.

7.4 Zinc Deficiency, Response, and Stress Tolerance

In plant system, the immobile nature of the nutrient results in developing the initial deficiency symptoms in the newer leaves. Symptoms of zinc deficiency vary with crop. Zinc deficiency symptoms are fairly typical and relatively easy to identify. Severe deficiency of zinc in different plants is characterized by different symptoms in different plant parts such as interveinal chlorosis called as mottle leaf, development of reddish brown or bronze tinge called as bronzing, inward curling of leaf lamina called as goblet leaf, reduction in leaf size as little leaf (in leaves), shortening of internodes termed as rosetting, and necrosis in the root apex called dieback.

Product quality is influenced by Zn, hence its deficiency deteriorates the same. Severe light and temperature intensity causes injury to the plants that are deficient in Zn. Such Zn deficient plants are also more prone to the incidence of fungal infections. Since zinc is needed for the biosynthesis of tryptophan, a precursor of indol acetic acid, it has a prominent role in auxin synthesis. In broccoli zinc deficiency results in stunted growth and shortened internodes. Zinc deficient cabbage crop shows stunted growth and faint yellow-green leaf colour.

Zinc is one of the micronutrients, found to be deficient in Indian soils. An application of 10 kg Zn ha^{-1} is the recommended dosage to alleviate this deficiency. However, application of Zn is not a necessary practice if the previous crop was applied with Zn fertilizer or 10–15 tonnes of farmyard manure per hectare. Critical concentration of Zn required for optimum growth in crops is 15–20 mg Zn kg^{-1}dry weight. Zinc toxicity may develop when the values go beyond 200 mg Zn kg^{-1}dry weight. Reduction in leaf size, leaf tip necrosis, chlorosis, etc., is the major symptoms of Zn toxicity along with decreased root growth and a reduction in the total growth of the plant. Zinc is proved to have competitive relationships with other elements like P, Fe, Mn, Cu indicating that Zn toxicity in soils can lead to deficiency of these nutrients in plants.

7.5 Copper (Cu)

Copper plays important roles in the process of photosynthesis and respiratory electron transport chains, ethylene sensing, cell wall metabolism, oxidative stress protection, and biosynthesis of molybdenum cofactor. Also it activates many enzymes in the plant which are involved in lignin synthesis and it assists in the metabolism of carbohydrates and proteins. It has a role in colour and flavour of vegetables and flowers. The optimum concentration of copper in soils is 0.05–0.5 mg kg^{-1}, but it varies from 3 to 10 mg kg^{-1} in plants.

7.5.1 Copper Deficiency, Response, and Stress Tolerance

Copper is immobile in plants, and hence its deficiency symptoms occur in the newer leaves. Symptoms are different in different crops. Newer leaves begin to exhibit slight yellowing of leaves, either as whole or in between the veins of them. Necrotic leaf spots develop on leaf margins and the cupping of leaves is also seen. As the deficiency worsens, the leaves become smaller and wilting results. Growth of lateral branches is hindered due to the necrosis in apical meristems. Due to the shortened stem, plants assume a compact appearance. Flowers also appear lighter in colour due to Cu deficiency. In cabbage copper deficiency leads to elongated leaf which may bend down. Colour of the leaf turn into pale with broad white leaf vein. In cauliflower chlorosis and heart necrosis occurs due to copper deficiency. Copper deficiencies are largely reported on sandy soils with low organic matter. As soil pH increases, uptake of copper by plants reduces. Copper uptake by plants is also declined due to an increased phosphorus and iron availability in soils. In oilseed rape, Cu deficiency resulted in decrease in the plant growth by 19% and content in older leaves by 61.4% (Billard et al. 2014).

7.5.2 Tolerance Mechanisms to Cu Stress

Even though deficiencies and toxicities of copper are uncommon, either extremes need to be avoided as both can negatively influence on crop growth and quality. During Cu deficiency in *B. napus*, the transcription factor SPL7 (SQUAMOSA-promoter binding link protein 7) was found to be differentially expressed, which otherwise actually targets the Cu response element (CuRE) in the promoter of genes and hence coordinates the early response to Cu deficiency (Yamasaki et al. 2009; Andre´s-Cola´s et al. 2013). In *B. napus,* conditions of Cu deficiency induced Cu mobilization and also increased the Cu transporter expression in roots (COPT2) and leaves (HMA1) and increased the Mo uptake by induction of MOT1 gene encoding Mo transporter (Billard et al. 2014).

7.6 Boron

Boron (B) plays important roles in plant system such as in the formation of cell wall, maintenance of structural, functional integrity and stability of cell membrane, carbohydrate metabolism and transport of sugar, cell division, amino acid production, pollination and seed set, phenol and auxin metabolism, regulation of water intake by cell, reproduction and germination of seeds, disease resistance, improves photosynthetic activity, enhances activity of enzymes, protein, and nucleic acid metabolism, etc. Cruciferous plants have a high requirement of boron and the critical boron concentration in *Brassica* crops and root crops is 25–50 ppm indicating <15 ppm to be deficient, 25–100 ppm adequate, and >200 ppm toxic for growth and production. Boron requirement of oilseed rape is around five times greater than that of cereals. At vegetative growth stages up to stem elongation, critical B concentration in oilseed rape (*B. napus* cv Eureka) was found to be 10–14 mg kg^{-1} dry matter in the youngest open leaf (Huang et al. 1996). Even though boron is an essential micronutrient for plant growth and development, its range between deficiency and toxicity is narrower than that of any other element (Goldberg 1997).

7.6.1 Boron Deficiency, Response, and Stress Tolerance

Boron is mobile in soil but it is immobile in plant system, hence, the deficiency is commonly observed in growing tips, i.e. deficiency first develop in young leaves and plant reproductive structures. Deficiency of boron is more common in some soil types. Boron deficiency affects many biochemical and physiological processes. Since it is important in cell wall maintenance, a deficiency of B affects the plasticity and extensibility of cell walls, destroying the structure and function of cell walls.

In broccoli, deficiency may cause hollow stem. Initially watery tissues arise within the stem, which turn brown or black as the deficiency progresses. Then the affected tissue gets dehumidified and hollow area develops and brown patches are visible in the stem. In cauliflower, deficiency symptoms start on young leaves. The leaves get deformed in a curled down pattern with a reduction in size. Chlorotic mottling appears on the edges and between the veins in the leaves. Leaf margins and intercostal areas turn necrotic. Plant growth is stunted and light heads are produced. Leaf colour may be brighter than usual. In cauliflower, discolouration of stem and hollow stem are other symptoms of B deprivation. The curds exhibit browning symptoms due to B stress. The symptom starts in younger leaves and the leaves remain small, stiff and the margins turn reddish brown. Development of curd is retarded and only light curd is produced. Older leaves are curled down and may show chlorosis or reddish discolouration. In oilseed rape, B deficiency results in curled, rolled leaves, with chlorotic margins and dwarfing of plants (Huang et al. 1996).

7.6.2 Tolerance Mechanisms to B Stress

Transpiration is responsible for the transport of B in plants as it is mainly carried out by the xylem. Meristematic growing points die resulting in stunted growth and development during severe B deficiency in soils. Along with this, a decreased root elongation and seed set are seen. However, leaves seldom exhibit B deficiency symptoms. Reduced pollination and fruit abortion are also resulted due to low B supply.

Boron deficiency adversely affects the seed yield and oil quality of *B. napus*, which is highly sensitive to B stress due to its high demand for the nutrient. To develop new cultivars of better B efficiency, the responses of crops to B stress need to be understood in detail. In *B. napus* different varieties have significant genotypic differences under B deficient conditions in soil. The genotypes which are B-efficient possess larger root system than B inefficient genotypes and also have normal growth and yield. Absorption of B from soils, its transportation within the plant system, and utilization are the processes contributing to B efficiency in *B. napus*. Cell walls possess lesser B-binding sites in the B-efficient line indicating its role in tolerance of *B. napus* to boron deficient conditions in soil. Thus it is evident that the B stress response of *B. napus* is dependent on genotype which is also related to the B uptake, translocation and requirement in the cell walls, biomass allocation, and plant physiological development. Carbohydrate metabolism in plants, activities of antioxidants, development of cell wall, cellar detoxification of reactive oxygen species, and signaling networks are some of the important processes in plant metabolism which are affected by the deficiency of B, and this is revealed by proteomics analyses. Genetic analyses have shown that B efficiency of *B. napus* is polygenic (Zhang et al. 2014).

7.7 *Molybdenum (Mo)*

Molybdenum is a very important micronutrient for Brassicaceae plants. It is involved in enzyme systems for nitrogen-fixing bacteria, formation of pollen fruit and grain, nitrogen and sulphur metabolism, and protein synthesis. Since it is required in minor quantities its plants do not commonly exhibit Mo deficiency symptoms. It is also important for potassium uptake. Since it is important for the functions of nitrogenase and nitrate reductase the nitrogen fixation and nitrogen reduction is affected by its deficiency.

7.7.1 Molybdenum Deficiency and Response

In *Brassica* plants such as cauliflower, broccoli, and Brussels sprouts, the deficiency of Mo is known as whiptail. Deficiency of Mo in rapeseed produces a brown substance on the leaves which is sticky and syrupy. This is the basic reason for the deficiency symptoms exhibited mainly by the younger leaves. Leaves exhibit local chlorosis and necrotic lesions along the main veins and leaf blades become smaller. The midrib twists at the centre and the margins curl upward. This is known as whip-tail in cauliflower (Marschner 1995). There will not be any formation of leaf lamina except that the midrib continues to grow in case of severe deficiency of Mo. Necrotic lesions develop along the active growing points in plants which hinders the growth and the plant dies eventually. Older leaves with increased nitrate content exhibit marginal chlorosis and necrosis. The leaf margins turn gray or white and the leaf blades remain green. Growth and yield parameters decreased during Mo deficiency as indicated by decrease in number of branches per plant, pods per plant, seed set, and seed weight (Sinha et al 1990). In cauliflower, in addition to whiptail small curds and often the small leaves and bracts penetrate the curd surface. Sandy soils in humid regions with acidic pH are found to be low in Mo. Molybdenum deficiency reduces the activities of enzymes like cytochrome-c-oxidase and nitrate reductase, whereas increases acid phosphatase and ribonuclease in mustard (*B. campestris*). The deficiency results in a lowered concentration of DNA due to the decreased number of cells in the interphase. A rise in soil pH increases uptake of molybdenum by plants whereas the uptake of other micronutrients decreases with an increase in soil pH.

7.7.2 Molybdenum Toxicity

Molybdenum toxicity in plants is rare. Some reports in cauliflower are available. When plants are grown in high concentrations of Mo, leaves accumulate anthocyanin and turn purple. The main threat associated with high Mo levels in plants is with crops used for grazing or silage production. Consumption of plant tissues high in molybdenum content by ruminant animals leads to a condition called molybdenosis, which is a disorder that induces copper deficiencies. In the rumen diet, molybdenum to copper ratio needs to be maintained to control molybdenosis. This can also be achieved by adjusting the soil pH so as to alter the availability of molybdenum.

7.8 Chlorine (Cl)

Chlorine is that micronutrient which is taken up by plants in its anionic form Cl^-. Chlorine is distributed mainly in the vegetative organs like shoots and leaves. Around 80% of the total $Cl^-$is present in the nutritional organs and it accumulates more in the older leaves than the upper or the younger leaves. The most important function of Cl is in the maintenance of electrical charge balance and osmotic and stomatal regulation. Chlorine is essential in imparting resistance to diseases and it also plays an inevitable role in evolution of oxygen in photosynthesis.

7.8.1 Chlorine Deficiency, Response, and Stress Tolerance

The critical chlorine concentration is 2 mg kg^{-1} by dry weight below which it becomes deficient. Soils are rarely deficient in Cl, however, soils developed from parent materials low in chlorides and sandy soils in heavy rainfall areas may develop chloride deficiencies. The main chloride-deficiency symptoms include wilting and restricted highly branched root systems. Chloride, than its importance in plant nutrition, is essential in imparting resistance to various diseases, from a practical viewpoint.

7.8.2 Tolerance Mechanisms to Cl Stress

Salinity stress deals with excess Cl as one of its major component. When excess, like every element Cl is also toxic to plants. In broccoli inflorescences, total glucosinolates content increased as a result of increased salinity (40 mM) (Lopez-Berenguer et al. 2008). Florets recorded higher glucosinolates than the young fully expanded leaves, probably due to increased transport to this physiological sink via the phloem or a higher de novo synthesis (Lopez-Berenguer et al. 2009). Low water potential resulted in increased accumulation of glucosinolates as a result of restricted primary metabolism and growth due to salinity stress. The increase in glucosinolates was related to the synthesis of osmoprotective compounds (Ballesta et al 2013). Salinity alters defense allocation patterns in *B. rapa* exhibited by the decreased concentration of gluconapin (Steinbrenner et al. 2012).

7.9 Nickel (Ni)

Nickel has its critical concentration in plant tissue varying between 0.05 and 5 mg kg^{-1} indicating its low requirement for crop nutrition. Due to this, it is found in abundance as a contaminant in soil, fertilizer, water, etc. The major function of Ni is as a constituent in several metallo-enzymes and these enzymes include urease, superoxide dismutase, acetyl coenzyme-A synthase, carbon monoxide dehydrogenase, hydrogenases, NiFehydrogenases, methyl coenzyme M reductase, and RNase-A. In leguminous crops, enzymes that assist in fixing nitrogen uses Ni as a catalyst. Nickel, like Cl is said to impart tolerance to plants, however, the mechanisms are still not clear.

7.9.1 Nickel (Ni) Deficiency, Response, and Stress Tolerance

A stress due to a deficiency in Ni can result in increased levels of urea in the plant tissue. This is exhibited as necrotic regions on leaf tips. Thus, indicating urea toxicity due to Ni deficiency. Urease in plants is seen in two forms, ubiquitous form occurring in the vegetative tissues and another form found in seeds. The former is less active than the latter, despite its significance in N recycling in plants. Processes such as urease activity, N assimilation, scavenging of superoxide free radical are disturbed due to Ni deficiency. Nickel deficiency rarely exhibits any visible symptoms due to which it often goes unnoticed; however it results in chlorosis of leaf and meristem, changes in metabolism of nitrogen, reduction in the uptake of Fe, induction of senescence, stunted growth, and reduced yield in plants.

8 Hyperaccumulators

Brassicaceae are among the hyperaccumulators of Ni and around 82 species in Brassicaceae accumulate Ni in the shoots to the level >1000 mg kg^{-1} of dry weight (Seregin and Kozhevnikova 2006).

9 Conclusion

The plant family Brassicaceae is diverse with many economically important species including vegetables, oil yielding crops, weeds, and ornamentals that are grown worldwide. Mustard and rapeseed (oleiferous brassicas) are grown as oil yielding seeds and are major providers of edible oil for a large population. Brassicas are not to be nutrient responsive crops especially to Nitrogen, Sulphur, Phosphorus, and Molybdenum as some of the nutrients are directly involved production and metabolic pathways of essential amino acids, glucosinolates, erucic acid, etc. In depth understanding of the physiological role of nutrients, their requirement and removal could be a significant strategy for improvement of growth and productivity of Brassicaceae crops.

References

Abdallah M, Dubousset L, Meuriot F, Etienne P, Avice JC, Ourry A (2010) Effect of mineral sulphur availability on nitrogen and sulphur uptake and remobilization during the vegetative growth of Brassica napus L. J Exp Bot 61(10):2635–2646

Abdin MZ, Ahmad A, Khan N, Khan I, Jamal A, Iqbal M (2003) Sulphur interaction with other nutrients. In: Abrol YP, Ahmad A (eds) Sulphur in plants. Kluwer, Dordrecht, pp 359–374

Andre´s-Cola´s N, Perea-Garcı´a A, de Andre´s SM, Dorcey E, Garcia-Molina A et al (2013) Comparison of global responses to mild deficiency and excess copper levels in Arabidopsis seedlings. Metallomics 5:1234–1246

Anjum AN, Gill SS, Omar S, Ahmad I, Duarte AC, Pereira E (2012) Improving growth and productivity of oleiferous brassicas under changing environment: significance of Nitrogen and sulphur nutrition and underlying mechanisms. Sci World J. https://doi.org/10.1100/2012/657808

Arnon DI, Stout PR (1939) An essentiality of certain elements in minute quantity for plants with special reference to copper. Plant Physiol 14:371–375

Bailey LD, Grant CA (1990) Fertilizer placement studies on calcareous and non-calcareous chernozemic soils: growth, P-uptake, oil content and yield of Canadian rape. Comm Soil Sci Plant Anal 21(17–18):2089–2104

Ballesta MMC, Moreno DA, Carvajal M (2013) The physiological importance of glucosinolates on plant response to abiotic stress in *Brassica*. Int J Mol Sci 14:11607–11625. https://doi.org/10.3390/ijms140611607

Bergmann W (1992) Nutritional disorders of plants. Gustav Fischer Verlag, Jena

Blake-Kalff MMA, Zhao FJ, Hawkesford MJ, McGrath SP (2001) Using plant analysis to predict yield losses caused by sulphur deficiency. Ann Appl Biol 138(1):123–127

Billard V, Ourry A, Maillard A, Garnica M, Coquet L et al (2014) Copper-deficiency in brassica napus induces copper remobilization, molybdenum accumulation and modification of the expression of chloroplastic proteins. PLoS ONE 9(10):e109889. https://doi.org/10.1371/journal.pone.0109889

Collett MG, Stegelmeier BL, Tapper BA (2014) Could nitrile derivatives of turnip (Brassica rapa) glucosinolates be hepato-or cholangiotoxic in cattle? J Agric Food Chem 62:7370–7375

Collins RP, Harris PJC, Bateman MJ, Henderson J (2008) Effect of calcium and potassium nutrition on yield, ion content and salt tolerance of *Brassica campestris* (rapa). J Plant Nutr 31(8):1461–1481. https://doi.org/10.1080/01904160802208444

Coulombe J, Villeneuve S, Lamy P, Yelle S, Bélec C, Tremblay N (1998) Evaluation of soil and petiole sap nitrate quick tests for broccoli in Québec. Acta Hortic 506:147–152

D'Hooghe P, Dubousset L, Gallardo K, Kopriva S, Avice J-C, Trouverie J (2014) Evidence for proteomic and metabolic adaptations associated to alterations of seed yield and quality in sulphur-limited *Brassica napus* L. Mol Cell Proteomics 13:1165–1183

Ding F, Wang X, Shi Q, Wang M, Yang F, Gao, Q (2008) Exogenous nitric oxide alleviated the inhibition of photosynthesis and antioxidant enzyme activities in iron-deficient Chinese cabbage (*Brassica chinensis* L.). Agri Sci China (2):168–179

Ekbladh G, Writter E (2010) Determination of the critical nitrogen concentration of white cabbage. Eur J Agron 33(4):276–284

Fazili IS, Jamal A, Ahmad S, Masoodi M, Khan JS, Abdin MZ (2008) Interactive effect of sulfur and nitrogen on nitrogen accumulation and harvest in oilseed crops differing in nitrogen assimilation potential. J Plant Nutr 31(7):1203–1220

Fismes J, Vong PC, Guckert A, Frossard E (2000) Influence of sulfur on apparent N-use efficiency, yield and quality of oilseed rape (*Brassica napus* L.) grown on a calcareous soil. European J Agron. 12(2):127–141

Fohse D, Claassen H, Jungk A (1991) Phosphorus efficiency of plants. II. Significance of root radius, root hairs and cation-anion balance for phosphorus influx in seven plant species. Plant Soil 132:261–272

Goldberg S (1997) Reactions of boron in soil. Plant Soil 193:35–48

Grant CA, Bailey LD (1993) Fertility management in canola production. Can J Plant Sci 73:651–670

Greenwood DJ, Stone DA (1998) Prediction and measurement of the secline in the critical potassium, the maximum potassium and total cation plant concentration during the growth of field/vegetable crops. Ann Bot 82:871–881

Gupta RK, Abrol IP (1990) Salt-affected soils—their reclamation and management for crop production. Adv Soil Sci 12:233–275

Hermans C, Johnson CN, Strasser RJ, Verbruggen N (2004) Physiological characterization of magnesium deficiency in sugar beet: acclimation to low magnesium differentially affects photosystems I and II. Planta 220:344–355

Hirschi KD (2004) The calcium conundrum. Both versatile nutrient and specific signal. Plant Physiol 136:2438–2442

Huang L, Ye Z, Bell WR (1996) The importance of sampling immature leaves for the diagnosis of boron deficiency in oilseed rape (*Brassica napus* cv Eureka). Plant Soil 183:187–198. https://doi.org/10.1007/BF00011434

Jung JY, Shin R, Schachtman DP (2009) Ethylene mediates response and tolerance to potassium deprivation in Arabidopsis. Plant Cell 21:607–621

Karthika KS, Rashmi I, Parvathi MS (2018) Biological functions, uptake and transport of essential nutrients in relation to plant growth. In: Hasanuzzaman M, Fujita M, Oku H, Nahar K, Hawrylak-Nowak B (eds) Plant nutrients and abiotic stress tolerance. Springer Nature Singapore Pte. Ltd., pp 1–49

Kim SA, Punshon T, Lanzirotti A, Li L, Alonso JM, Ecker JR (2006) Localization of iron in Arabidopsis seed requires the vacuolar membrane transporter VIT1. Science 314:1295–1298. https://doi.org/10.1126/science.1132563e

Kochian VL (1995) Cellular mechanisms of aluminium toxicity and resistance in plants. Ann Rev Plant Physiol Plant Mol Biol 46:237–260

Koprivova A, Suter M, Op den Camp R, Brunold C, Kopriva S (2000) Regulation of sulfate assimilation by nitrogen in *Arabidopsis*. Plant Physiol 122:737–746

Lemaire and Gastal (1997) N uptake and distribution in plant canopies. In: Lemaire G (ed) Diagnosis of the nitrogen status in crops. Springer, Berlin Heidelberg, pp 3–43

Leustek T, Martin MN, Bick JA, Davies JP (2000) Pathways and regulation of sulfur metabolism revealed through molecular and genetic studies. Ann Rev Plant Physiol Plant Mol Biol 51:141–165

Lopez-Berenguer C, Martínez-Ballesta MC, García-Viguera C, Carvajal M (2008) Leaf water balance mediated by aquaporins under salt stress and associated glucosinolate synthesis in broccoli. Plant Sci 174:321–328

Lopez-Berenguer C, Martínez-Ballesta MC, Moreno DA, Carvajal M, García-Viguera C (2009) Growing hardier crops for better health: salinity tolerance and the nutritional value of broccoli. J Agric Food Chem 57:572–578

Lynch J (1995) Root architecture and plant productivity. Plant Physiol 109(1):7–13

Marschner H (1995) Mineral nutrition of higher plants, 2nd. Academic Press Ltd., London

Marschner H, Roemheld V (1994) Strategies of plants for acquisition of iron. Pl Soil 165:261–274

McGrath SP, Zhao FJ (1996) Sulphur uptake, yield responses and the interactions between nitrogen and sulphur in winter oilseed rape (*Brassica napus*). J Agric Sci 126(1):53–62

Muneer S, Lee B, Kim KY, Park SH, Zhang Q (2014) Involvement of sulphur nutrition in modulating iron deficiency responses in photosynthetic organelles of oilseed rape (*Brassica napus* L.). Photosynt Res 119:319–329. https://doi.org/10.1007/s11120-01-9953-8

Omirou MD, Papadopoulou KK, Papastylianou I, Constantinou M, Karpouzas DG, Asimakopoulos I, Ehaliotis C (2009) Impact of nitrogen and sulfur fertilization on the composition of glucosinolates in relation to sulfur assimilation in different plant organs of broccoli. J Agric Food Chem 57:9408–9417

Pittman JK, Hirsch KD (2001) Regulation of CAX1, an Arabidopsis Ca2 + /H + antiporter. Identification of an N-terminal autoinhibitory domain. Plant Physiol 127(3):1020–1029

Scherer HW (2008) Impact of sulfur on N2 fixation of legumes. In: Khan NA, Singh S, Umar S (eds) Sulfur assimilation and abiotic stress in plants. Springer, The Netherlands, pp 43–54

Schonhof I, Blankenburg D, Müller S, Krumbein A (2007) Sulfur and nitrogen supply influence growth, product appearance, and glucosinolate concentration of broccoli. J Plant Nutr Soil Sci 170:65–72

Seregin IV, Kozhevnikova AD (2006) Physiological role of nickel and its toxic effects on higher plants. Russian J. Plant Physiol 53(2):257–277

Singh AP, Fridman Y, Holland, N, AckermanLavert M, Zananiri R, Jaillais Y, Henn A, Savaldi-Goldstein S (2018) Interdependent nutrient availability and steroid hormone signals facilitate root growth plasticity. Dev Cell 46, 59–72. e4. https://doi.org/10.1016/j.devcel.2018.06.002

Sinha AC, Jana PK, Mandal BB (1990) Effect of micronutrients on rapeseed grown on acid soils of Eastern India. Indian J Agron 35:126–130

Spragg J (2016) Australian feed grain supply and demand report. JCS Solutions Pty Ltd., North Victoria, Australia, pp 1–42

Steinbrenner AD, Agerbirk N, Orians CM, Chew FS (2012) Transient abiotic stresses lead to latent defense and reproductive responses over the *Brassica rapa* life cycle. Chemoecology 22:239–250

Su T, Yu S, Zhang F, Yu Y, Zhang D, Zhao X, Wang W (2015) Loss of function of the carotenoid isomerase gene BrCRTISO confers orange color to the inner leaves of Chinese cabbage (*Brassica rapa* L. ssp. pekinensis). Plant Mol Biol Rep 33:648–659

Sung J, Yun H, Back S, Fernie AR, Kim XY, Lee Y, Lee S, Lee D, Kim J (2018) Changes in mineral nutrient concentrations and CN metabolism in cabbage shoots and roots following macronutrient deficiency. J Plant Nutr Soil Sci 181:777–786. https://doi.org/10.1002/jpln.201800001

Tandon HLS (2004) Fertilisers in Indian agriculture—from 20th to 21st century. New Delhi, Fertiliser Development and Consultation Organisation, p 239

Troufflard S, Mullen W, Larson TR, Graham IA, Crozier A, Amtmann A, Armengaud P (2010) Potassium deficiency induces the biosynthesis of oxylipins and glucosinolates in *Arabidopsis thaliana*. BMC Plant Biol 10:172–185

Tyagi NK (2000) Management of salt-affected soils. In: Singh GB, Sharma BR (eds) Fifty years of natural resource management research. Indian Council of Agricultural Research, New Delhi, pp 363–401

Walker KC, Booth EJ (2003) Sulphur nutrition and oilseed quality. In: Abrol YP, Ahmad A (eds) Sulphur in plants. Kluwer, Dordrecht, pp 323–339

Yamasaki H, Hayashi M, Fukazawa M, Kobayashi Y, Shikanai T (2009) SQUAMOSA promoter binding protein–like7 is a central regulator for copper homeostasis in Arabidopsis. Plant Cell Online 21:347–361

Zhang D, Zhao H, Shi L, Xu F (2014) Physiological and genetic responses to boron deficiency in *Brassica napus*: a review. Soil Sci Plant Nutr 60(3):304–313. https://doi.org/10.1080/00380768.2014.893537

Zhu W, Zuo R, Zhou R, Huang J, Tang M, Cheng X, Liu Y, Tong C, Xiang Y, Dong C, Liu S (2016) Vacuolar iron transporter BnMEB2 is involved in enhancing iron tolerance of *Brassica napus*. Front Plant Sci. https://doi.org/10.3389/fpls2016.01353

Brassicaceae Plants Response and Tolerance to Metal/Metalloid Toxicity

Shyamashree Roy and Sanchita Mondal

Abstract Brassicaceae is an important Family having numerous numbers of genera and species within it. Some of these species are important food crops specifically oilseed and vegetables, some are weeds also. They have significant role in accumulation of heavy metals like cadmium, arsenic, lead, caesium, uranium, zinc, etc. in their tissues and have the potential genes within them which attribute tolerance/resistance against the ill effect of these metals. These plants naturally clean and ameliorate the soil contaminated with metal/metalloids. Under Brassicaceae family, members like *Brassica juncea, Brassica oleracea, Brassica napus, Brassica carinata* and many more are known for their phytoremediation property. Like, the seedlings of *B. oleracea, B. juncea* and *Raphanus sativus* are found to uptake arsenic by their root and shoot both. Also, some of the species are known for huge accumulation of heavy metals in tissues which may be a source of potential new genes widening the future research work on phytoremediation. Therefore keeping the worlds' increasing population in mind many of them can be well adapted for growing commercially in heavy metal contaminated soil to contribute to the food basket of the world. The goal of this article is to review the use of species under Brassicaceae family in phytoremediation.

Keywords Heavy metals · Brassica crops · Oilseeds · Phytoextraction · Phytoremediation

Abbreviations

Cd	Cadmium
Cr	Chromium

S. Roy (✉)
Regional Research Station, Old Alluvial Zone, Uttar Banga Krishi Viswavidyalaya, Majhian, Dakshin Dinajpur, West Bengal, India
e-mail: shree.agr@gmail.com

S. Mondal
Department of Agronomy, Bidhan Chandra Krishi Viswavidyalaya, Mohanpur, Nadia, West Bengal, India

M. Hasanuzzaman (ed.), *The Plant Family Brassicaceae*,
https://doi.org/10.1007/978-981-15-6345-4_12

Mo	Molybdenum
Zn	Zinc
Mn	Manganese
Pb	Lead
As	Arsenic
Ni	Nickel
DW	Dry weight
EDTA	Ethylene Diamine Tetra Acetic Acid
ROS	Reactive Oxygen Species

1 Introduction

Brassicaceae is an important plant family contains genus Brassica. This genus is also important from agriculture point of view because it contains 39 species. It ranks among world's first ten most important plant families with high economic value of the world (Warwick et al. 2000). There are 375 genera and 3200 species under the umbrella Brassicaceae. It is also called as mustard family. Also there are a number of cultivated and hybrids of cultivated species. The species with their high dietary fibre has possessed a valuable place in the list of food crops with their large variation of economic parts starting from oil bearing seeds, stems, buds, flowers, roots and leaves which are used as vegetables, leafy greens, salad, spices, oils, etc. (Hasanuzzaman 2008). Within genus Brassica, *Brassica juncea, Brassica carinata, Brassica rapa, Brassica campestris* and *Brassica napus*—these six species are altogether called as Rapeseed-mustard group. Rapeseed-Mustard being the main oilseed crop is planted on more than 80% area covered under oilseeds. Their seed bear oil and contributes 35% of total oilseeds crops in India (Table 1) (Nmoop 2018). According to FAOSTAT (2013) Brassica oilseeds occupies more than 34 million ha land annually over the world. In 1935, UN has given the relationship between these six species in a form of a triangle, designated as U's triangle (Fig. 1).

Also some very common vegetables like cabbage, cauliflower, broccoli, Brussels sprout, kohlrabi, collards and kale come under *Brassica oleraceae* species. They are known for high nutritional value. The other vegetable Brassicaceae includes *B. napus, B. rapa, Raphanus sativus, Lepidium sativum* and *Nasturtium officinale.* The condiment crops include *B. juncea, Sinapis alba, Brassica nigra, Armoracia rusticana* and *Eutrena japonica.* Still there are numbers of crops which are being used as potherbs and salads. Among a number of weedy species, *Sinapis arvensis, Raphanus raphanistrum, B. rapa* and *Hirschfeldia incana* are of greatest interest with regard to cross-pollination with *B. napus.* Among wild relatives of Brassica, species like *Begonia. adpressa, Brassica fruticulosa, Brassica pinescens, Brassica oxyrrhina, Brassica barrelieri and Brassica tournefortii* comes under the group *Brassica coenospecies.* They are known for their good agronomic characteristics and used

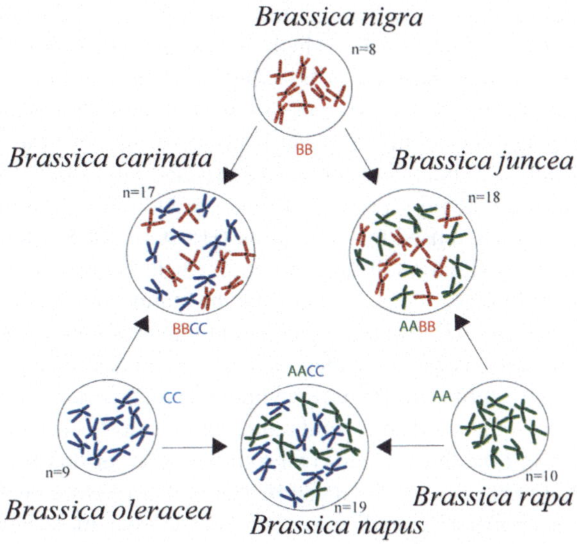

Fig. 1 U's triangle (*Source* UN 1935)

widely in hybridization programme (Warwick 1993). The genus *Brassica* is classified as follows:

Order Brassicales (= Cruciales)

Family Brassicaceae (= Cruciferae)

Tribe Brassiceae

Subtribe Brassicinae

Genus *Brassica* L.

Under Brassicaceae, there are 25 tribes with a supplementary 5 currently under study (Al-Shehbaz et al. 2006). Brassiceae contains the genus *Brassica* and wild relatives has 48 genera and near about 240 species (Table 2) (Warwick and Hall 2009). As the population is increasing, it is providing pace to the industrialization also; and the whole, leads to the exposure of the toxic matters as waste materials directly or indirectly to the environment. All the major parts of the environment—air, soil and water are contaminated highly by these toxics to such extent that it has become a life threat to humankind (Zhuang et al. 2007), as well as for animal and plant kingdom (Singh et al. 2011; Hasanuzzaman et al. 2012; Hasanuzzaman and Fujita 2012). These include metal, metalloids, radio nucleotides and different kinds of environmental contaminants. Human activities related to industries and agricultural practices expose the environment to various heavy metals and metalloids like Molybdenum, Cadmium, Arsenic, Nickel, Cobalt, etc. (Pilon-Smits 2005). There is very less or narrow chance to find out a remedy (Cunningham et al. 1997; Prasad and Freitas 2003). Though there are policies and routine precautions and a big list of Do's and DON'Ts, which is not enough to control the heavy metal pollution of the earth (Zayad and Tery 2003). There is a huge pressure of providing enough food to supply a square meal to this burgeoning population of the world. Horizontal expansion of cultivable land is not possible until the degraded, polluted and waste fallows are being converted to cultivated one. But these contaminated and marginal lands are always a threat to food production with a huge chance of food contamination when used for cultivation (He et al. 2005). But nature has always an answer to every situation. Thangahu et al. (2011) reported about development of different techniques to judge and observe the status and movement of metals inside the soil, water and wastewater. Again some plants/adopted a technique to extract, sequester and detoxify the toxic elements. The plant is known as accumulator plant. The technique is termed as Phytoremediation (Fig. 2) where the Greek originated prefix *Phyto* means plant and latin word remedium means to correct (Erakhrumen and Agbontalor 2007; Tangahu et al. 2011). From the observation of (USEPA 2000; Sarma 2011 and Tangahu et al. 2011), it can be stated that phytoremediation is comparatively cheap and ecofriendly method to get rid of the toxic materials from soil, water or air. Some specific plants are known to successfully engross contaminants include metals and metalloids responsible for environmental contamination. Phytoextraction being the significant process, used to eliminate heavy metals from soil utilizing its ability to take up the essential nutrients which are contaminants too under specific conditions (Fe, Mn, Zn, Cu, Mg, Mo, and Ni). Besides, other metals with unidentified function (Cd, Cr, Pb, Co, Ag, Se, Hg) can also be stored (Cho-Ruk et al. 2006). Among angiosperms, Brassicaceae is the most important family having the huge number

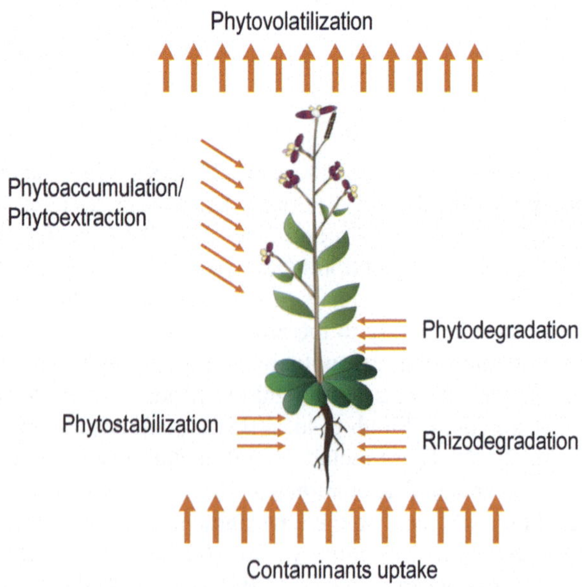

Fig. 2 Phytoremediation-the mechanism for heavy metal uptake by plants

of (25%) metal accumulating species (90) (Prasad and Freitas 2003; Palmer et al. 2001a, b; Krämer 2010; Sarma 2011). Some member of the family Brassicaceae can accumulate metals even the toxic ones in relatively high amount (Kumar et al. 1995 and Anjumet al. 2012). They are potential phytoextraction plants (Van Ginneken et al. 2007; Gall and Rajkaruna 2013) who show visible symptoms of metal accumulation and being the food crops they contaminate the food chain (Gall et al. 2015). Brassicas accumulate heavy metals in their stem due to the inherent tolerance to metals present in excess. Tangahu et al. (2011) observed that some of the member of Brassica genus (*B. campestris* L., *B. carinata* A. Br., *B. juncea* (L.) *Czern*) can accumulate lead in their tissue more than 50 mg g^{-1} dry weight of plant whereas *B. nigra* Koch. can gather more than 100 mg g^{-1} dry weight. From the same study we came to know that in case of mercury, *B. juncea* accumulated more than 1 mg g^{-1} dry weight compared to other species (<0.2 mg/g dry weight). Therefore they may be used as tool for phytoextraction (Neison and Rajkaruna 2012). Indian mustard or *B. juncea*, an important oil contributor in India has been reported to have a reduced oil percentage in seed due to accumulation of Cadmium (Ahmad et al. 2015; Mourato et al.2015). Not only Cadmium, according to Alkorta et al. (2004) and Szczyglowska et al. (2011), *B. juncea* has a superior capacity of different toxic metal accumulation. *Brassica napus* also known as rapeseed is the chief source of eatable oil in the world can accumulate Zn in its plant biomass (Grispenet al. 2006; Carrier et al. 2003). *Brassica campestris* also known as turnip rape, also an oilseed can also accumulate metals in its below ground parts (Gleba et al. 1999). It is considered for the phytoremediation of Chromium which can be accumulated in huge quantity by this species (Dheri et al. 2007). Kabata-Pendias (2001), from a study confirmed that *B. oleracea* accumulated lead (Pb) in good quantity. Kumar et al. (1995) observed that roots are more efficient in Pb accumulation than above ground parts. Gisbert found that not only Pb, it also accumulate Zn and Cu in its shoot. Addae (2010) observed in a study that *B. oleracea* species can extract significant amount of Cd and Pb. All these research work has been conducted with a focus in the matter to find out a plant which will be pretty good to be used for phytoremediation with the characters like, more accumulation efficiency in shoot, tolerance capacity in soils with high amount of metals, fast growing and easy to grow and harvest (Table 3) (Marchiol et al. 2004). Some of the best studies citing the role of Brassicaceae in tolerance to metals and metalloids are presented hereunder for better understanding.

2 Amelioration of Cadmium (Cd) Toxicity by *Coronopus didymus*

Coronopus didymus, a native of South America is an annual herb of family Brassicaceae and widely distributed across the world (Yannitsaros 1986). In Northern India, it is being grown along the road sides and gardens specifically during October to February. Sidhu et al. (2017) worked with *C. didymus* to assess the development and survival of *C. didymus* under varied Cd treatments and also on the tolerance and extraction efficiency of C. *didymus* under various levels of Cd concentrations. They reported that under 400 mg kg^{-1} Cd, there was an increase in Cd content in roots and shoots (867.2 and 864.5 mg kg^{-1} DW), reflecting the potential of this herb *C. didymus* in ameliorating the toxicity effect of Cd in soils. The possible explanation for this tolerance is the compartmentalization of metals in vacuoles that restricted the excess Cd transport in the plants. Besides, Phytochelatins has a significant role in accumulating Cd within roots (Zeng et al. 2009). Toxicity of Cd induces stress stimulates the generation and build-up of Reactive Oxygen Species (Devi et al. 2007). Enzymes like super oxide dismutase, catalase and peroxidase play a crucial role in quenching Reactive Oxygen Species (ROS) produced as because of heavy metal stress (Zhang et al. 2013). Due to the antioxidative responses, *C. didymus* could be one of the possible options to be used in Cd contaminated soils with no risk to human or animals as it is unpalatable and cannot enter the food chain.

3 Response of *Hirschfeldia incana* to Lead (Pb)

Lead (Pb) is one of the toxic elements having ill effect on the quality of environment as well as human health (Lee et al. 2005). Although it is not essential for plants, but still imposes a toxicity threat to plants and animals. Auguy et al. (2013) identified *Hirschfeldia incana*, a member of the Brassicaceae family, with a high potential to accumulate lead in their biomass. *Hirschfeldia incana* was formerly known as *Brassica geniculata* is a species of flowering plant in the mustard family with common names as shortpod mustard, buchan weed, hoary mustard and Mediterranean mustard. In their study, chosen area was heavily affected by lead mining contamination with a concentration range of 26–9479 mg kg^{-1} (Smouni et al. 2010). The Brassicaceae *H. incana* was found to be important for its ability to accumulate toxic heavy metals in leaves. *H. incana* was able to accumulate 106 mg Pb g^{-1} DW in roots at their vegetative growth and 77 mg Pb g^{-1} DW in roots at the floral stage after 60 days of lead exposure. In one of the study, Kaur (2018) identified the ability of Indian mustard with and without chelants like EDTA and salicylic acid. Pb concentrations increased in root and shoot organs of the plant with EDTA treatment, but it decreased with salicylic acid treatment. The results also revealed that *B. juncea var. arawali* accumulated high amount of Pb in roots with low translocation in the parts of shoots; nevertheless it could be used for phytoremediation of Pb (Tables 1 and 2).

4 Arsenic Tolerance by Brassicaceae Seedlings

Arsenic is one of the potential toxic elements with a high risk of entering food chains and one report by Bhattacharya et al. (2010) emphasize that 93% of the total arsenic are being consumed by humans. This consumption is always accompanied

Table 1 Year-wise production of major oilseeds in India (in Million tonnes)

Year	Groundnut	Rapeseed-mustard	Soybean	Sunflower
1950–51	3.48	0.76	–	–
1960–61	4.81	1.35	–	–
1970–71	6.11	1.98	0.01	0.08
1980–81	5.01	2.30	0.44	0.07
1990–91	7.51	5.23	2.60	0.87
2000–01	6.41	4.19	5.28	0.65
2010–11	8.26	8.18	12.74	0.65
2011–12	6.96	6.60	12.21	0.52
2012–13	4.70	8.03	14.67	0.54
2013–14	9.71	7.88	11.86	0.43
2014–15	7.40	6.28	10.37	0.43
2015–16	6.73	6.80	8.57	0.30
2016–17	7.46	7.92	13.16	0.25
2017–18	9.25	8.43	10.98	0.22
2018–19	6.52	8.78	13.74	0.20

Source Modified from Directorate of Economics and Statistics
Third Advance Estimates of Production of Commercial Crops for 2018–19

with a health hazard to human population as it is considered one of the most unsafe ingredients to humans (Khan et al. 2009). Numerous species of Brassicaceae are designated as arsenic tolerant to metals (Srivastava et al. 2009). Freitas-Silva et al. (2016) showed that even Brassicaceae seedlings at initial growing conditions can accumulate arsenic in their biomass (Table 4). Roots accumulated high arsenic when compared to shoots suggesting little movement of these metalloids to the shoots.

5 Hyperaccumulation of Selenium by Brassicaceae Crops

Seleniferous soils are those with high levels of selenium and there are some Brassicaceae species that can accumulate high levels of Se within their biomass (Brown and Shrift 1981). Plants can normally accumulate around 0.05–1 mg kg^{-1} Se dry weight, but there are hyperaccumulators too which can engross higher concentrations and occasionally hundreds of times more than the usual range of Se accumulated by normal plants. Hladun et al. (2011) studied the accumulation of Se in the plant parts of *Stanleya pinnata* and *B. Juncea. Stanleya pinnata was* found to be the efficient accumulator of selenium which has a potential to accumulate 22% (Nectar) and 85% (pollen) excess over *B. juncea. Stanleya pinnata* is a species under mustard family and also known as desert prince's plume and is a native to North America. The mechanism behind the accumulation of Se in *S. pinnata* is the mobilization of

Table 2 Cultivated *Brassica* species and related genera

Species name	Common name
Brassica rapa L.	
subsp. *Campestris* (L.) A.R. Clapham	Summer turnip rape, canola
subsp. *oleifera* (DC.) Metzg	Winter turnip rape
subsp. *campestris* (L.) A.R. Clapham	Bird or wild turnip rape
subsp. *trilocularis* (Roxb.) Hanelt	Yellow and brown Sarson
subsp. *dichotoma* (Roxb.) Hanelt	Toria
subsp. *chinensis* (L.) Hanelt	Pak-choi or bokchoy, Chinese mustard, chinese broccoli, GaiLan
subsp. *pekinensis* (Lour.) Hanelt	Pe-tsai, Chinese cabbage
subsp. *nipposinica* (L.H. Bailey) Hanelt	Curled mustard
subsp. *rapa*	Turnip
B.tournefortii Gouan	Wild turnip
B.nigra (L.) W.D.J. Koch	Black mustard
B. oleracea L.	
var. *viridis* L.	Kale, collard
var. *botrytis* L.	Cauliflower and broccoli
var. *capitata* L.	Cabbage
var. *Gongylodes*	Kohlrabi
var. *gemmifera* (DC.) Zenker	Brussels sprout
var. *italic* Plenck	Broccoli
var. *Oleracea*	Wild cabbage
subsp. *alboglabra* L.H. Bailey	Chinese kale, Kailan
B. juncea (L.) Czern.	Brown and oriental mustard, rai
B. napus L.	
var. *napus*	Summer oilseed rape, canola
var. *napus*	Winter oilseed rape, winter Canola
var. *pabularia* (DC.) Rchb.	Rape-kale
var. *napobrassica* (L.) Rchb.	Rutabaga, swede
B. carinata A. Braun.	Abyssinian mustard
Hirschfeldia incana (L.) Lagr.-Foss.	Hoary mustard

(continued)

Table 2 (continued)

Species name	Common name
Sinapis arvensis L.	Wild mustard, charlock
Sinapis alba L.	Yellow or white mustard
Raphanus sativus L.	Radish
Raphanus raphanistrum L.	Wild radish
Diplotaxis muralis (L.) DC.	Annual wall-rocket
Eruca strumgallicum (Willd.) O.E. Schulz	Dog mustard
Eruca vesicaria (L.) Cav. subsp. *sativa* (Mill.) Thell.	Rocket salad

Modified from OECD (2016)

Table 3 Heavy metal accumulation by *Brassica* and related genera

Heavy metal accumulated	Species name
Nickel (Ni)	*Alyssum lesbiacum*
	Astronidium inflatum
	Noccaea goesingensis
	B. juncea
Zinc (Zn)	*Arabidopsis halleri*
	Noccaea caerulescens
	B. napus
	B. rapa L.
Cadmium (Cd)	*B. juncea*
Chromium (Cr)	*B. juncea*
Copper (Cu)	*B. juncea*
Selenium (Se)	*B. oleracea* L.
	B. napus L.
Uranium (U)	*B. juncea*
	B. nigra (L.) Koch.
	Brassica chinensis L.
	Brassica narinosa L.
Lead (Pb)	*B. juncea*
	Brassica campetris L.
	B. carinata A. Br.
	B. napus L.
	B. nigra (L.) Koch.
Caesium (Cs)	*B. juncea*
	B. oleracea L.

Modified from Dar et al. (2015) and Pantola and Alam (2014)

Table 4 Arsenic content ($\mu g\ g^{-1}$) in root and shoot dry matter of Brassicaceae 12-days seedling stage (Freitas-Silva et al. 2016)

Species	As (μM)	Root ($\mu g\ g^{-1}$ DM)	Shoot ($\mu g\ g^{-1}$ DM)
B. oleracea	0	6.43	3.05
	250	17.66	3.64
	350	16.40	4.24
	450	24.86	7.20
B. juncea	0	7.45	1.58
	250	12.96	2.75
	350	20.53	1.82
	450	25.35	2.69
R. sativus	0	3.85	1.49
	250	4.68	3.51
	350	6.73	3.85
	450	15.28	4.45

the element in the leaves into the reproductive organs of the plants; some reports are also there relating the volatilization of Se from leaves directly into the atmosphere. Accumulation of selenium by *S. pinnata* in the reproductive parts (flowers and seeds) is also being confirmed in a number of field studies with corresponding decrease in the leaf system (Galeas et al. 2007). Although *B. juncea* can also accumulate selenium but they basically prefer soils with moderate levels of Se; on the other hand, *Stanleya pinnata* can take up high amount of selenium even under soils with low amount of selenium (Terry et al. 2000).

6 Zinc (Zinc) Tolerance of *Arabidopsis halleri*

Arabidopsis halleri (syn: *Cardaminopsis halleri*) can tolerate higher concentrations of heavy metals (Brooks 1998). Zinc tolerance was investigated in five populations of *A. halleri* (syn.: *C. halleri*). All these were raised from the seeds collected from the even sites, i.e. both contaminated and uncontaminated. Irrespective of the origin, *A. halleri* seedlings showed a continuous growth of roots at 100 μM Zn, with no toxic symptoms (dry or chlorotic leaves) in the plants. In contrast to *A. halleri*, roots of other species like *Arabidopsis thaliana* and *Arabidopsis lyrata* subsp. *petraea* seedlings were completely repressed at 100 μMZn concentration (Bert et al. 2000). This study was the first to demonstrate the ability of *A. halleri* to tolerate high level of zinc in the soil. Even, uncontaminated site seedlings showed a tolerance suggested a constitutive trait in *A. halleri*.

Table 5 Nickel concentration in the roots of various species of Brassicaceae

Species	Treatments		
	Control	1 mM Ni	2 mM Ni
A. alyssoides	13.5	132	197
A. campestre	3.21	75.2	123
A. murale	128	816	1014
E. ibericum	7.15	146	386

Adopted from Drozdova et al. (2017)

7 Ni Tolerance of Brassicaceae Species

With an idea to identify potential accumulators, Drozdova et al. (2017) conducted a comparative study and reported the outcome of higher doses of Nickel on few wild growing plant species of Brassicaceae. Results of the experiments revealed distinct differences in the species in terms of tolerance to Ni. There were clear symptoms like chlorosis and necrosis on the leaves of species which are non-hyperaccumulators. After 11 days of Ni exposure, about 80% of plants *Alyssum alyssoides* and *Alyssum campestre* and 50% *Erysimum ibericum* started showing the toxicity symptoms. Surprisingly, *Alyssum murale* developed normally even in the same concentrations of Ni applied for *A. alyssoides, A. campestre*a nd *E. ibericum.* A comparative study of the accumulation of Ni is shown in Table 5. The accumulation of Ni was highest in the roots of *A. murale* suggesting its high efficiency of extracting Ni ions from the soil solution, even in areas with low Ni content. In terms of tolerance to high concentrations of Ni, the species could be arranged in the order: *A. murale* > *E. ibericum* > *A. alyssoides* > *A. campestre.*

8 Conclusion

Across many studies, it has been confirmed that the family Brassicaceae is remarkably vital for remediation of heavy metals and metalloids across the world. Brassicaceae comprises a number of species with better accumulation efficiency signifying a great tolerance level to metals. Although a number of species are identified with proper scientific theories on the mechanisms leading to the tolerance or hyperaccumulation; but their genetic level study linking with their metal tolerance ability needs to be explored for better understanding.

References

Addae C, Piva M, Bednar AJ, Zaman MS (2010) Cadmium and lead bioaccumulation in cabbage plants grown in metal contaminated soils. AdvSciTechnol 4:79–82

Ahmad P, Sarwat M, Bhat NA, Wani MR, Kazi AG, Tran LSP (2015) Alleviation of cadmium toxicity in *Brassica juncea*L. (czern. & coss.) by calcium application involves various physiological and biochemical strategies. PLoS ONE10e0114571

Alkorta I, Hernandez-Allica J, Becerril JM, Amezaga I, Albizu I, Garbisu C (2004) Recent findings on the phytoremediation of soils contaminated with environmentally toxic heavy metals and metalloids such as zinc, cadmium, lead, and arsenic. Rev Environ Sci Biotechnol 3:71–90

Anjum NA, Ahmad I, Pereira ME, Duarte AC, Umar S, Khan NA (2012) The plant family Brassicaceae: an introduction. In: Anjum NA, Ahmad I, Pereira ME, Duarte AC, Umar S, Khan NA (eds) The plant family Brassicaceae: contribution towards phytoremediation, Environmental pollution series no. 21. Springer, Dordrecht

Auguy F, Fahr M, Moulin P, Brugel A, Laplaze L, Mzibri ME, Filali-Maltouf A, Doumas P, Smouni A (2013) Lead tolerance and accumulation in Hirschfeldia incana, a Mediterranean Brassicaceae from metalliferous mine spoils. PLoS ONE 8(5):e61932

Al-Shehbaz LA, Beilstein, MA, Kellogg EA (2006) Systematics and phylogeny of the Brassicaceae (Cruciferae): an overview. Plant Syst Evol 259(2):89–120

Bert V, Macnair MR, De-Laguerie P, Saumitou-Laprade P, Petit D (2000) Zinc tolerance and accumulation in metallicolous and nonmetallicolous populations of *Arabidopsis helleri* (Brassicaceae). New Phytologists 146:225–233

Bhattacharya P, Samal AC, Majumdar J (2010) Arsenic contamination in rice, wheat, pulses, and vegetables: a study in an arsenic affected area of West Bengal, India. Water Air Soil Pollut 21:3–13

Brooks RR (1998) Plants that hyperaccumulate heavy metals. Wallingford, UK: CAB International

Brown TA, Shrift A (1981) Exclusion of selenium from proteins of selenium tolerant *Astragalus* species. Plant Physiol 67:1051–1053

Carrier P, Baryla A, Havaux M (2003) Cadmium distribution and microlocalization in oilseed rape (*Brassica napus*) after long-term growth on cadmium contaminated soil. Planta 216:939–950

Cho-Ruk K, Kurukote J, Supprung P, Vetayasuporn S (2006) Perennial plants in the phytoremediation of lead contaminated soils. Biotechnology 5(1):1–4

Cunningham SD, Shann JR, Crowley DE, Anderson TA (1997) Phytoremediation of contaminated water and soil. In: Kruger EL, Anderson TA, Coats JR (eds) Phytoremediation of soil and water contaminants, ACS symposium series 664. American Chemical Society, Washington, DC

Dar MI, Khan FA, Rehman F, Masoodi A, Ansari AA, Varshney D, Naushin F, Naikoo MI (2015) Roles of Brassicaceae in phytoremediation of metals and metalloids. In: Ansari A, Gill S, Gill R, Lanza G, Newman L (eds) Phytoremediation. management of environmental contaminants, vol 1. https://doi.org/10.1007/978-3-319-10395-2_14, © Springer International Publishing Switzerland 2015. pp 201–215

Devi R, Munjral N, Gupta AK, Kaur N (2007) Cadmium induced changes in carbohydrate status and enzymes of carbohydrate metabolism, glycolysis and pentose phosphate pathway in pea. Environ Exp Bot 61:167–174

Dheri GS, Brar MS, Malhi SS (2007) Comparative phytoremediation of chromium-contaminated soils by Fenugreek, Spinach, and Raya. Commun Soil Sci Plant Anal 38:1655–1672

Drozdova IV, Alekseeva-Popova NV, Kalimova IB, Belyaeva AI, Smirnova NA (2017) The accumulating ability and nickel tolerance of Brassicaceae species of the North Caucasus in connection with the problem of phytoremediation. J Geo chem. Expln 182:235–241

Erakhrumen A, Agbontalor A (2007) Phytoremediation: an environmentally sound technology for pollution prevention, control and remediation in developing countries. Educ Res Rev 2:151–156

FAOSTAT (2013) FAO Statistics online database, Production/Crops,– rapeseed, Year 2013. Food and Agriculture Organization of the United Nations. http://faostat3.fao.org/home/E

Freitas-Silva LD, Araujo TOD, Silva LCD, Oliveira JAD, Araujo JMD (2016) Arsenic accumulation in Brassicaceae seedlings and its effect on plant growth and plant anatomy. Eco Toxicol Environ Saf 124:1–9

Galeas ML, Zhang LH, Freeman JL, Wegner M, Pilon-Smits EAH (2007) Seasonal fluctuations of selenium and sulfur accumulation in selenium hyperaccumulators and related nonaccumulators. New Phytol 173:517–525

Gall JE, Rajakaruna N (2013) The physiology, functional genomics, and applied ecology of heavy metal-tolerant brassicaceae. In: Lang M (ed) Brassicaceae: characterization, functional genomics and health benefits. Nova Science Publishers: Hauppauge, NY, USA, pp 121–148

Gall JE, Boyd RS, Rajakaruna N (2015) Transfer of heavy metals through terrestrial food webs: a review. Environ Monit Assess 187:201

Gleba D, Borisjuk MV, Borisjuk LG, Kneer R, Poulev A, Skarzhinskaya M (1999) Use of plant roots for phytoremediation and molecular farming. Proc Natl Acad Sci USA 96:5973–5977

Grispen VMJ, Nelissen HJM, Verkleij JAC (2006) Phytoextraction with *Brassica napus* L.: a tool for sustainable management of heavy metal contaminated soils. Environ Pollut 144:77–83

Hasanuzzaman M (2008) Siliqua and seed development in rapeseed (*Brassica campestris* L.) as affected by different irrigation levels and row spacings. Agric Conspectus Scientificus 73(4):221–226

Hasanuzzaman M, Fujita M (2012) Heavy metals in the environment: current status, toxic effects on plants and possible phytoremediation. In: Anjum NA, Pereira MA, Ahmad I, Duarte AC, Umar S, Khan NA (eds) Phytotechnologies: remediation of environmental contaminants. CRC Press, Boca Raton, pp 7–73

Hasanuzzaman M, Hossain MA, Teixeira da Silva JA, Fujita M (2012) Plant Responses and tolerance to abiotic oxidative stress: antioxidant defense is a key factor. In: Bandi V, Shanker AK, Shanker C, Mandapaka M (eds) Crop stress and its management: perspectives and strategies. Springer, New York, pp 261–316

He ZL, Yang XE, Stoffella PJ (2005) Trace elements in agroecosystems and impacts on the environment. J Trace Elem Med Biol 19:125–140

Hladun KR, Parker DR, Trumble JT (2011) Selenium, accumulation in floral tissues of two Brassicaceae species and its impacts on floral traits and plant performance. Environ Exp Bot 74:90–97

Kabata-Pendias A (2001) Trace elements in soils and plants, 3rd edn.CRC, Boca Raton, FL

Kaur L (2018) Accumulation potential of Indian mustard (*Brassica juncea* var. *arawali*) and fenugreek (*Trigonellafoenum-graecum* L.) planted on Lead and Nickel contaminated soil. Trop Plant Res 5(2):217–223

Khan I, Ahmad A, Iqbal M (2009) Modulation of antioxidant defense system for arsenic detoxification in Indian mustard. Ecotoxicol Environ Saf 72:626–634

Krämer U (2010) Metal hyperaccumulation in plants. Annu Rev Plant Biol 61:517–534

Kumar PBAN, Dushenkov V, Motto H, Raskin I (1995) Phytoextraction: the use of plants to remove heavy metals from soils. Environ Sci Technol 29:1232–1238

Lee M, Lee K, Lee J, Noh EW, Lee Y (2005) AtPDR12 contributes to lead resistance in *Arabidopsis*. Plant Physiol 138:827–836

Marchiol L, Assolari S, Sacco P, Zerbi G (2004) Phytoextraction of heavy metals by canola (*Brassica napus*) and radish (*Raphanus sativus*) grown on multi contaminated soil. Environ Pollut 132:21–27

Mourato MP, Moreira IN, Leitão I, Pinto FR, Sales JR, Martins LL (2015) Effect of heavy metals in plants of the *Genus Brassica*. Int J Mol Sci 16:17975–17998

Neilson S, Rajakaruna N (2012) Roles of rhizospheric processes and plant physiology in applied phytoremediation of contaminated soils using brassica oilseeds. In: Anjum NA, Ahmad I, Pereira ME, Duarte AC, Umar S, Khan NA (eds) The Plant Family Brassicaceae, vol. 21. Springer Netherlands: Dordrecht, The Netherland, pp 313–330

NMOOP (2018) Present status of oilseed crops and vegetable oils in India
OECD (2016) Brassica crops (*Brassica species*). In: Safety assessment of transgenic organisms in the environment, vol 5, OECD Consensus Documents. OECD Publishing, Paris. https://doi.org/10.1787/9789264253018-6-en
Palmer CE, Warwick S, Keller W (2001a) Brassicaceae (Cruciferae) family, plant biotechnology, and phytoremediation. Int J Phytorem 3(3):245–287
Palmer CE, Warwick S, Keller W (2001b) Brassicaceae (Cruciferae) family, plant biotechnology, and phytoremediation. Int J Phytoremediation 3:245–287
Pantola RC, Alam A (2014) Potential of Brassicaceae burnett (Mustard family; Angiosperms) in phytoremediation of heavy metals. Int J Sci Res Environ Sci 2(4):120–138
Pilon-Smits, EAH (2005) Phytoremediation. Ann. Rev. Plant Biol. 56:15–39
Prasad MNV, Freitas HMO (2003) Metal hyperaccumulation in plants—biodiversity prospecting for phytoremediation technology. Electron J Biotechnol 6:285–321
Sarma H (2011) Metal hyperaccumulation in plants: a review focusing on phytoremediation technology. Environ Sci Technol 4:118–138
Sidhu GPS, Singh HP, Batish DR, Kohli RK (2017) Tolerance and hyperaccumulation of cadmium by a wild, unpalatable herb *Coronopus didymus* (L.) Sm. (Brassicaceae). Eco toxicol Environ Saf 135:209–215
Singh J, Kalamdhad Ajay S (2011) Effects of heavy metals on soil, plants, human health and aquatic life. Int J Res Chem Environ 1(2):15–21
Smouni A, Ater M, Auguy F, Laplaze L, El Mzibri M (2010) Assessment of contamination by metallic trace elements in a mining area of eastern Morocco. Cah Agric 19:26–76
Srivastava S, Srivastava AK, Suprasanna P, D´Souza SF (2009) Comparative biochemical and transcriptional profiling of two contrasting varieties of *Brassica juncea* L. in response to arsenic exposure reveals mechanisms of stress perception and tolerance. J Exp Bot 60:3419–3431
Szczyglowska M, Piekarska A, Konieczka P, Namiesnik J (2011) Use of Brassica plants in the phytoremediation and biofumigation processes. Int J Mol Sci 12:7760–7771
Tangahu BV, Abdullah SRS, Basri H, Idris M, Anuar N, Mukhlisin M (2011) A review on heavy metals (As, Pb, and Hg) uptake by plants through phytoremediation. Int J Chem Eng. https://doi.org/10.1155/2011/939161
Terry N, Zayed AM, de Souza MP, Tarun AS (2000) Selenium in higher plants. Annu Rev Plant Physiol Plant MolBiol 51:401–432
UN (1935) Genomic analysis in Brassica with special reference to the experimental formation of Brassica napus and peculiar mode of fertilization. Jpn J Bot 7:389–452
USEPA (2000) Introduction to phytoremediation. EPA, Washington, DC
Van Ginneken L, Meers E, Guisson R, Ruttens A, Elst K, Tack FMG, Vangronsveld J, Diels L, Dejonghe W (2007) Phytoremediation for heavy metal-contaminated soils combined with bioenergy production. J Environ Eng Land Sc Manag 15:227–236
Warwick SI (1993) Guide to wild germplasm of Brassica and allied crops Part IV: wild species in the tribe Brassicaceae (cruciferae) as sources of agronomic trait. Technical Bulletin 17E-1993. Centre for Land and Biological Resources Research, Agriculture Canada, Ottawa, Ontaria
Warwick SI, Hall JC (2009) Phylogeny of *Brassica* and wild relatives. In: Gupta SK (ed) Biology and breeding of Crucifers. CRC Press, Boca Raton, Florida, pp 19–36
Warwick SI, Francis A, Lafleche J (2000) Guide to Wild Germplasm of Brassica and allied crops (*Tribe Brassiceae*, Brassicaceae), 2nd edn. Agriculture and Agri-food Canada. http://www.brassica.info/resources/crucifer_genetics/guidewild/htm
Yannitsaros A (1986) New data on the naturalization and distribution of *Coronopusdidymus* (Cruciferae) in Greece. Willdenowia 61–64
Zayed AM, Terry N (2003) Chromium in the environment: factors affecting biological remediation. Plant Soil 249:139–156

Zeng X, Ma LQ, Qiu R, Tang Y (2009) Responses of non-protein thiols to Cd exposure in Cd hyperaccumulator *Arabispaniculata* Franch. Environ Exp Bot 66:242–248
Zhang S, Lin H, Deng L, Gong G, Jia Y, Xu X, Li T, Li Y, Chen H (2013) Cadmium tolerance and accumulation characteristics of *Siegesbeckiaorientalis* L. Ecol Eng 51:133–139
Zhuang X, Chen J, Shim H, Bai Z (2007) New advances in plant growth-promoting rhizobacteria for bioremediation. Environ Int 33:406–413

Toxic Metals/Metalloids Accumulation, Tolerance, and Homeostasis in *Brassica* Oilseed Species

Muhammad Mudassir Nazir, Zaid Ulhassan, Muhammad Zeeshan, Sharafat Ali, and Muhammad Bilal Gill

Abstract *Brassica* oilseed are economically important crops with greater heavy metals/metalloids tolerance ability and produce higher biomass (due to their genetically inherited traits). These species have different adaptive mechanisms to handle the toxic metal ions which attract plant scientists to fully understand the accumulation and tolerance mechanisms in order to minimize/remove the heavy metals/metalloids phytotoxicity. At physiological level, toxic heavy metal/metalloids impair the carbon assimilation (photosynthesis, etc.) and nitrate assimilation process that lead to yield reduction. A very common response of higher plants against heavy metals/metalloids toxicity is the overproduction of reactive oxygen species (ROS) and methylglyoxal (MG) which can cause proteins oxidation, enzymes inactivation/dysfunction, lipid peroxidation as MDA, DNA damages and interaction with cellular components. At the cellular level, toxic metal ions restrict the mechanisms of nutrients transport, disturbance in polynucleotide functions, disruption in cellular integrity which may be due to their interaction with biomolecules. These higher plants evolved antioxidant defense and glyoxalase enzymatic systems to scavenge the ROS-MG components. Also, the entry of toxic metal ions into the cell is restricted/sequestered (via efflux mechanisms) through the involvement of amino acids, glutathione, particularly heavy metals/metalloids binding ligands. GSH, being a central molecule of the defense system, directly/indirectly control (via metabolic enzymes, etc.) the reaction products of ROS-MG in plant cells, hence protection against heavy metals/metalloids induced oxidative damages. *Brassica* species synthesized chelating ligands (phytochelatins, metallothionein's and organic acids etc.) ensured the detoxification, complexation,

Muhammad Mudassir Nazir, Zaid Ulhassan have equal contribution to this chapter.

M. M. Nazir · Z. Ulhassan · M. Zeeshan · S. Ali
Institute of Crop Science and Zhejiang Key Laboratory of Crop Germplasm, Zhejiang University, Hangzhou 310058, China

M. B. Gill (✉)
International Centre for Environmental Membrane Biology, Foshan University, Foshan 528000, China
e-mail: muhammad.bilal@utas.edu.au

Tasmanian Institute of Agriculture, College of Science and Engineering, University of Tasmania, Hobart, TAS 7005, Australia

M. Hasanuzzaman (ed.), *The Plant Family Brassicaceae*,
https://doi.org/10.1007/978-981-15-6345-4_13

and compartmentation of toxic heavy metals/metalloids. These plants also possess transporter genes for their resistance against toxic metal ions and homeostasis. Thus, *Brassica* oilseed species utilize accumulation and transport mechanisms for improved tolerance. That's, why *Brassica* species are ideal candidates to clean up the heavy metals/metalloids contaminated soils and their adaptive mechanisms against harsh environmental condition enhanced the suitability for phytoremediation technology. The adaptive potential of these plant species can assist to boost our understandings concerning the heavy metals/metalloids tolerance mechanisms and crops cultivation in contaminated soils.

Keywords *Brassica* oilseed · Metals toxicity · Phytoremediation · Tolerance · Chelation · Sequestration

Abbreviations

Al	Aluminum
APX	Ascorbate peroxidase
As	Arsenic
B	Boron
Be	Beryllium
CAT	Catalase
CDF	Cation diffusion facilitator
Co	Cobalt
Cu	Copper
DHAR	Dehydroascorbate reductase
ERF	Ethylene-responsive factor
Fe	Iron
GPX	Glutathione peroxidase
GSH	Reduced glutathione
GSH/GSSG	Reduced/oxidized glutathione
Hg	Mercury
HSPs	Heat-shock proteins
IRT	Iron-regulated transporter
MDHAR	Monodehydroascorbate reductase
MG	Methylglyoxal
Mn	Manganese
MRP	Multidrug resistance-associated proteins
MTs	Metallothionein's
Ni	Nickel
NRAMP	Natural resistance-associated macrophage protein
PAs	Polyamines
Pb	Lead
PCS	Phytochelatins synthase

PCs	Phytochelatins
PDR	Pleiotropic drug resistance
POD	Peroxidase
RNS	Reactive nitrogen species
ROS	Reactive oxygen species
Sb	Antimony
Se	Selenium
SOD	Superoxide dismutase
TE	Trace elements
MG	Methylglyoxal
TF	Transcription factor
YSL	Yellow-stripe-like
Zn	Zinc
ZRT	Zinc-regulated transporter

1 Introduction

Oilseed crops are largely cultivated and potential producer of edible oils worldwide since ancient times. Nowadays there are numerous alternatives in the form of diesel fuels which are eco-friendly and renewable. Among cultivated oilseed plants, considerable efforts have been made toward rapeseed, sunflower, soybean, mustard, and cotton. Top countries in oilseed production are China, India, USA, and Brazil. Ideally, oilseed crops are potential candidates to cope with the rising demands of consumers. Due to the greater interest in vegetable oils and higher market value, the production of oilseed crop has been greatly increased (Hassan et al. 2017a). Various environmental (abiotic and biotic) factors negatively affect the production of oilseed crops. Toxic metals/metalloids contamination in soil not only limit the growth and yield of oilseed crops but also impose health hazards. The excessive accumulation of toxic metals/metalloids in oilseed crop (*Brassica napus* L.) disturb the nutrients uptake, carbon and nitrogen assimilation, ROS and methylglyoxal induced oxidative stress, lipid peroxidation, desynchronized the antioxidant defense system, impair tissues (leaf and root) ultrastructure (Hasanuzzaman and Fujita 2012; Nahar et al. 2016, 2017; Ali et al. 2018a, b; Ulhassan et al. 2018, 2019a, c), impairs amino acids metabolism (Ulhassan et al. 2019c), induced he thiol biosynthesis (mainly phytochelatins and glutathione) and altered the osmotic metabolites (Ulhassan et al. 2019b). Regardless of metals/metalloids toxicity, some crops such as oilseed crops are tolerant to their higher concentration. Therefore, significant attention has been made regarding the use of these crops for the remediation of toxic metals/metalloids govern contaminated soils. Based on genetic shreds of evidence, various tolerance strategies have been made for these oilseed crops such as phytoremediation. The potential use of these crops is important for the higher plant biomass through appropriate agricultural management. To regulate the redox homeostasis (disturbed by toxic metals/metalloids) in oilseed crops, the application of enzymatic and non-enzymatic antioxidant machinery plays a central role. Besides, conjugators such as phenolic compounds, thiols, carboxylic and organic acids as well as metals/metalloids binding protein, peptides, and sequestration or compartmentation of toxic metals/metalloids are the main contributing factors to maintain cellular homeostasis in oilseed crops.

2 Fate of Toxic Elements (Metals/Metalloids) in the Environment

The origin of trace elements (TEs) including metals/metalloids in soil environment may be either anthropogenic (human-induced) or non-anthropogenic (lithogenic) activities. Various natural activities introduce TEs to the soil through mineral weathering, volcanic eruption and erosion. The total contents of these TEs in topsoil soil depend on their lithogenic derivatives (Toth et al. 2016). Anthropogenic inputs of these TEs are greatly related to industrial activities such as mining, electroplating, waste water discharge, smelting, and aerial deposition, in addition to, agriculture-based wastes and fertilizers application (Rinklebe and Shaheen 2014). Overall, major contributors of TEs in the soil are anthropogenic activities especially geographical vicinity of a polluted industrial area. Although small amounts of these TEs can be beneficial due to higher bioavailability in soils and normal plant growth. These TEs act as a micronutrient for plant metabolic functions such as energy metabolism, signal transduction, gene regulation, signal, and hormonal perception (Hebbern et al. 2009). However, the elevated levels of TEs are harmful to living organisms (microorganisms, animals, plants and humans) (Antoniadis et al. 2017a; Ulhassan et al. 2018, 2019a) and their higher accumulation in soil may enhance the contamination of underground water bodies and sediments after leaching (Shahen et al. 2017). Ultimately, the excessive accumulation of TEs enhances the risk of their transfer into human food (Antoniadis et al. 2017b). In outside environment, the non-degradable nature of these toxic TEs (at excessive concentrations) impose serious threats to living organisms. Thus, it is prerequisite to minimize the damaging effects induced by the toxic TEs both soil-plant interface and their possible capability to enter the food chains. The source and sink of toxic metals/metalloids in the environment are illustrated in Fig. 1.

3 Bioavailability of Toxic Metals/Metalloids in Soil-Plant Interface

Plants assimilate toxic metal/metalloids via foliar and adsorption process from natural and harsh environments. However, the potential of plants to absorb toxic metals/metalloids depends upon their bioavailability (Hassan et al. 2017b). The extent and rate of bioavailability of these metals/metalloids in soil environment are dependent on various key factors such as soil pH, redox potential, organic matter, clay

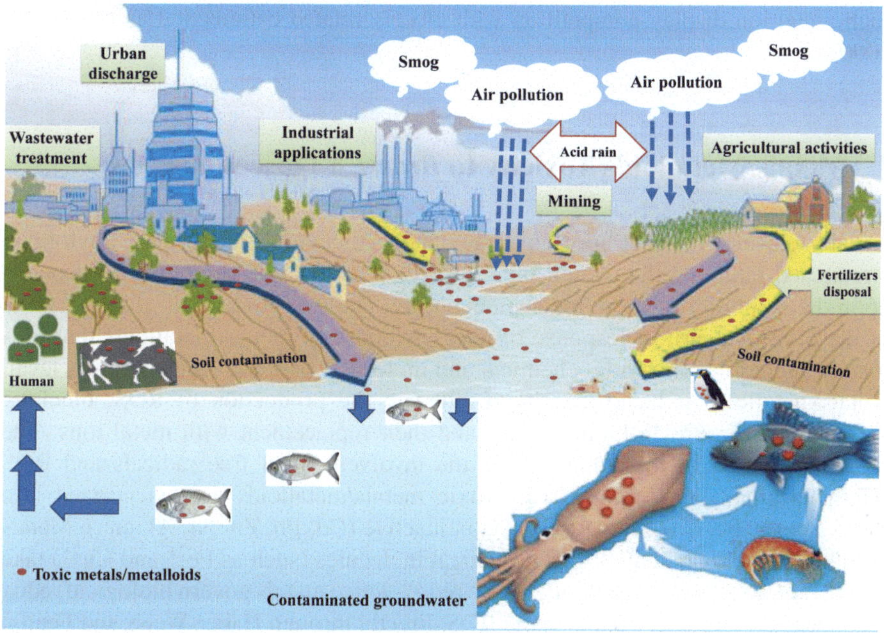

Fig. 1 The source and sink of toxic metals and metalloids in the environment

content, cation exchange capacity, ageing of added elements, nature and quantity of added metals/metalloids (Antoniadis et al. 2017a, b). And plant factors such as kinetic parameters, ions replacement, regulation of metals absorption and metabolic behavior of toxic metals/metalloids (Lehto et al. 2006). These soil and plant characteristics govern the availability of metals/metalloids that affect the potential mechanisms directed by both root exudates (outside) and selective uptake and translocation processes inside plant tissues (Antoniadis et al. 2017a). To make sure the successful uptake of metal ions in the soil-plant interface, water-soluble species closely interact with the root membrane for a defined time period. This interaction between metal ions and root membrane directly affect the longevity and transport/mobility of complex formation in soil. While elements mobility depends upon the nature of target metal ions. Solitary small fractions of intrinsic elements are bioavailable for plant absorption. Therefore, the bioavailability of these elements is mentioned as a concentration ratio. Only that particular higher concentration ratio provides the bioavailability of metal ions for plant absorption. Primarily metal absorption is accumulated in root epidermis and cortex. Overall uptake mechanisms are regulated by target plants. The efficiency of the transport mechanism is governed by kinetic constants, Km and Vmax. Depending upon the relevant Km values, every metal ion denotes distinct kinetic phase. The concentration of metal ions also plays a key role in their uptake systems. Usually, a lower concentration of metal ions is more efficient in comparison with higher concentrations in soil. Nevertheless, the uptake mechanisms for each metal ion display competition with closely related cations and anions (Lasat 2000).

4 Metals/Metalloids Toxicity to *Brassica* Oilseed Species

Usually, elevated concentrations of heavy metals/metalloids induced the phytotoxic effects and inhibited the overall plant growth. The level of phytotoxicity depends upon the solubility of these metals/metalloids. The overproduction of ROS is one of the primitive earliest response of excessive dose of metals/metalloids that ultimately altered the physio-biochemical and molecular process. At the cellular level, metals/metalloids induced toxicity resulted in the production of ROS, blocking of functional groups in biomolecules and their replacement with metal ions. The possible mechanism of metal ions is the involvement of free radicals and ROS (Hasanuzzaman and Fujita 2012). Toxic metals/metalloids are categorized into redox-active (Fe, Cu, Cr, etc.) and redox inactive (Cd, Pb, Zn, Ni, Al, etc.). Metals with higher redox potential than biological molecules (such as Fe^{2+} and Cu^{2+}) take part in biological redox reactions. These redox-active metals govern biological redox reactions and ultimately induced the ROS directly through Haber-Weiss and Fenton reactions. Remaining metals/metalloids such as Cd, Pb, Al, Ni, Zn, etc. improved the ROS generation without redox potential by impairing the antioxidant defense system (mainly glutathione pool), inducing NADPH oxidase, displacement of obligatory cations from binding sites of enzymes, activation of Ca-dependent system and Fe-mediating process (Fig. 2; Al Mahmud et al. 2019). The formulation of general concept about the toxic effects of metals/metalloids on oilseed crops is difficult due to the multiple variants regarding parameters under diverse environmental conditions such as type of metal, exposure time, target plant tissue, plant age, etc. Here we explain the possible toxic impacts of heavy metals/metalloids on the morphological, physio-biochemical, and molecular attributes of oilseed crops.

Cadmium (Cd) has been supposed as serious environmental contaminant regarding plant growth and human health issues. Its excessive accumulation in soil-plant interface resulted in its entry into the food chain. The increased Cd levels disturb the plant metabolisms and decline the key growth traits of *Brassica juncea* L. (Al Mahmud et al. 2018). The exposure of increasing Cd concentrations reported a decline in biomass production, light harvesting pigments, leaf water levels, whereas induced the H_2O_2, MDA, proline, lipoxygenase activity, and MG contents in the tissues of *Brassica* species (*B. napus*, *B. campestris,* and *B. juncea*) in dose dependent manner (Al Mahmud et al. 2017) Previous findings also observed that excessive Cd levels stopped the root elongation, declined the antioxidant defense system, enhanced the oxidative stress induced by ROS and impaired the ultrastructure in root tip cells of *B. napus* L. (Ali et al. 2013). Due to the non-redox-active nature of Cd, it did not induce ROS directly through Fenton or Haber process but induced ROS by blocking the photosynthetic electron system.

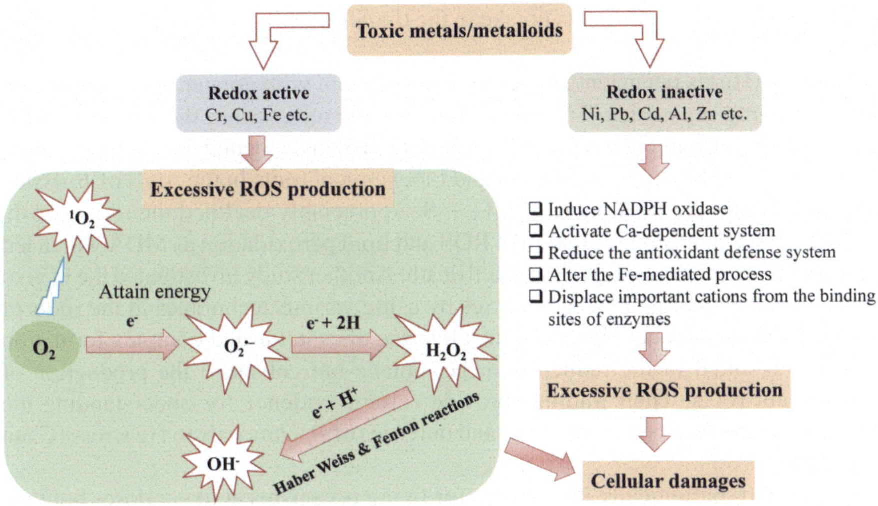

Fig. 2 Mechanisms of heavy metals or metalloids induced reactive oxygen species (ROS) and related oxidative damages in oilseed crops

Chromium (Cr) is one of the most widespread and toxic contaminants which rarely occur in nature. Its availability in soil-plant interface resulted in human health problems due to its entry in food chain (Shahid et al. 2017). The higher levels of chromium (Cr) in the soil can disturb the biological activities and fertility of soils that lead to a reduction in crop yield (Hassan et al. 2016). As roots are the primary targets of Cr and more sensitive parts than leaves (Ulhassan et al. 2019c). However, the application of Cr also impairs the other growth traits of *Brassica* species as revealed by the reduction in plant biomass, light harvesting pigments, oxidative stress due to the overproduction of ROS, lipid peroxidation (Gill et al. 2015a, b; Ahmad et al. 2017; Ulhassan et al. 2019c). In addition, the greater accumulation of Cr in the tissues of *B. napus* L. caused the ultrastructural damages in leaf mesophyll and root tip cells (Gill et al. 2015a; Ulhassan et al. 2019c), reduction in nutrients uptake, excessive production of methylglyoxal (MG) levels, impairment in the synthesis of essential amino acids, antioxidant enzymes of AsA-GSH cycle and glyoxalase system (Gly I and Gly II) (Ulhassan et al. 2019c).

Lead (Pb) is well-known for its ecotoxicology due to persistent occurrence and non-biodegradable nature (Ashraf and Tang 2017). In Pb toxic conditions, a decline in chlorophyll contents, photosynthetic activity, reduction in the uptake of maco or micronutrients, and enhanced production of ROS-MDA contents and ultrastructural damages in plant tissues (leaf and root) was observed that ultimately led to the reduction in plant growth and biomass production. *Brassica* plants strengthen their defense system against the ROS-overproduction by activating the enzymatic (SOD, POD, CAT, APX) and non-enzymatic (total glutathione and ascorbic acid) defense systems (Ali et al. 2014a, b). The exposure of Pb also altered the metabolic functions of cells and inhibit the nutrients uptake in the tissues of *Brassica* plants (Ashraf et al. 2011).

Mercury (Hg) is persistent environmental pollutant with greater bioaccumulation ability in living organisms. The main source of Hg-release into the environment is human activity (Chang et al. 2009). Bian et al. (2016) investigated the combined toxic effects of MG + Se on physiological and biochemical traits in the roots of *Brassica rapa*. The combined treatments of MG + Se appreciably declined the root growth, root biomass, enhanced production of ROS and lipid peroxidation as MDA which led to the loss in membrane integrity and cell death. Another study investigated the In vivo and Hg-phytochelatins (PCs) complexes by using various techniques in the roots of *Brassica chinesis*. Under Hg stress, the presence of cysteine and complex formation (Hg-PCs) resulted in the reduction in Hg uptake but enhanced the production of PCs in plant roots. Their findings provide certain evidence for understanding the potential mechanisms (accumulation and defense) of *B. chinensis* to Hg stress (Chen et al. 2009).

Cobalt (Co) biologically importance for living organisms at lower doses but they can be phytotoxic at excessive levels. It has been reported that excessive concentrations of Co (0, 100, 200, 400 μM) drastically damage the *B. napus* plants by altering the transcript levels of antioxidant enzyme activities, induction of ROS, lipid peroxidation as MDA, disruption in leaf and root cellular structures, reduction in nutrients uptake, total soluble protein contents, plant growth and biomass production (Ali 2018b, c). In addition, Karuppanapandian and Kim (2013) reported that excess Co (100 μM) significantly enhanced the proline levels, carbonylated protein, malondialdehyde, superoxide anion, and hydrogen peroxide which resulted in the greater accumulation of Co.

Beryllium (Be) has gained worldwide attention due to its potential use in nuclear power, electronic industry, and aircraft structures. However, its presence in the environment is becoming hazardous for the ecosystem (Shah et al. 2016). In plants, few studies have been documented on Be-toxicity. A study reported the toxic effects of Be (0, 100, 200, 400 μM) on the physio-biochemical and ultrastructural alterations in the six-days grown seedlings of *B. napus*. Reports suggested that excessive Be doses decline the plant growth, biomass, chlorophyll contents, and water-soluble proteins. While, Be-induced the lipid peroxidation as MDA, ROS, and cellular damages in the plant tissues (leaf and root) (Ali et al. 2018a). First and foremost, Be disturbs the energy production which requires the metabolic process. Be inhibits phosphoglucomutase activity and limits the conversion of glucose-1-phosphate to glucose-6-phosphate and ultimately growth reduction (Shah et al. 2016). However, the exact mechanisms (regarding Be-caused plant growth reduction) at molecular levels are missing and require further investigation in oilseed crops. As grain parts (potential food chain) are major sources of Be-phytotoxicity (Tanveer and Wang 2019). Thus, it is imperative to enhance crops tolerance by developing strategies against Be-toxicity.

Antimony (Sb) is an ubiquitous toxic element and require further studies on its pollution traits. Previous investigations observed that antimony Sb potentiated its toxic effects on the germination rate, germination index, vitality index, and root elongation of rapeseed (*B. napus* L.). Reports suggested that root elongation was more sensitive to Sb than germination rate (Liang et al. 2018). Another study reported that toxic effects of Sb (1, 2.5 and 5 g/l) on the metabolic pathways of *B. rapa*. Findings suggested that increasing Pb levels significantly declined the germination rate, G-index value, and overall plant growth. Also, Sb stress reduced the antioxidant 2, 2-diphenyl-1-picrylhydrazyl (DPPH)-scavenging activity (Siddiqui et al. 2014).

Aluminum (Al) is the abundantly found metal in the earth crust and limiting factor for crops productivity mainly in acidic soils. The Al-induced toxic effects on the physiological attributes of *B. napus* plants were reported. Results indicated that Al (0 and 100 μM) applications drastically hampered the root elongation and root activities, in addition to, the induction in malondialdehyde through the increase in Al accumulation in plant tissues (Yan et al. 2018). Another study documented that Al imposed toxic effects on the growth, the capacity of the antioxidant system and ultrastructural alterations in the leaf and root tissues of *B. napus*. Findings suggested that increasing doses of Al (0, 0.1, and 0.3 mM) stress declined the growth, biomass production and disturbed the antioxidant system by inducing the production of lipid peroxidation as MDA, H_2O_2 and higher accumulation of Al in the tissues of *B. napus*. Furthermore, Al impaired the leaf mesophyll and root tip cells. The observed chloroplasts were in spongy shape with dissolved thylakoid membrane and a greater number of starch grains. A visible toxic symptom in root tip cells was noticed in the form of greater Al deposition in vacuole and impairment in whole-cell organelle (Qian et al. 2014).

Metalloids, although the displayed promotive role and essential nutrient for the plant growth. But their excessive concentrations can be phytotoxic. Recent studies showed that selenium (Se) exhibited dual effects (positive/negative) on the overall growth of *B. napus* plants. It was noticed that a lower dose of Se promoted the plant growth and biomass by enhancing the nutrients level, photosynthetic efficiency (light harvesting pigments and gas exchange parameters), water-soluble protein and sugar contents. In addition, lower doses of Se supply minimized the ROS and MG generation by activating the activities of enzymes taking part in AsA-GSH cycle and thus detoxication of ROS and MG components. Excessive levels of Se impaired oxidative metabolism as revealed in the desynchronization of antioxidant enzyme activities. Also, elevated doses of Se induce thiol biosynthesis and their metabolic enzymes, deplete nutrients uptake, impairs photosynthesis, induce osmotic stress (by proline accumulation), and ultrastructural damages in the leaf mesophyll and root tip cells (Ulhassan et al. 2018, 2019a, b). Also, lower concentrations of Se improved the nutritional quality, biomass, and Se-metabolism related gene family of *B. rapa* varieties (Li et al. 2018).

Copper (Cu) is not poisonous in metallic state but some of its salts displayed damaging effects. Feigl et al. (2015) noticed the effects of excess Cu in the leaves of two *Brassica* species (*B. napus* and *B. juncea*). Results delineated that *B. juncea* is more sensitive to Cu stress, as compared to *B. napus*. Elevated Cu levels declined the microelement contents such as Fe and Mn, decline the light harvesting components and induction of ROS. Mwamba et al. (2016) reported the interactive effects of Cd and Cu metals to drive the synergetic toxicity in *B. napus*. Findings suggested that combine (Cd + Cu) applications further induce the oxidative damages and growth retardation as compared to individuals. Higher Cd accumulation in *B. napus* tissues was observed than Cu uptake and Cu-induced more damaging effects on micronutrient contents.

Arsenic (As) is the most toxic metalloids which has a non-essential role in the plant. It was observed that elevated levels of As imposed the phytotoxic effects which resulted in the morphological and physio-biochemical impairments in the leaves and roots of *B. napus* plants (Farooq et al. 2016). Pandey et al. (2016) reported the impacts of exogenous As (150 μM) on the roots of 15 days grown seedlings of *B. juncea.* Findings depicted that As induced the production of ROS, nitric oxide and intracellular production of superoxide radical (O_2^-). The activation of proline, cysteine, and antioxidant enzymes (SOD, POD, and APX) under As stress indicated the greater tolerance ability of *B. juncea.*

Zinc (Zn) is one of the most essential micronutrients for plants (ul Hassan et al. 2017a) and animals but it can display phytotoxic effects at elevated doses (Hambidge and Krebs 2007). Feigl et al. (2014) reported that Zn alters the metabolism of reactive oxygen species (ROS) and reactive nitrogen species (RNS) in the roots of two *Brassica* species (*B. napus* and *B. juncea*). Results indicated that *B. juncea* was more sensitive to Zn as compared to *B. napus.* Overall, ROS components were found more prominent than RNS in the roots of both *Brassica* species and ultimately decline the growth and yield of the crop. Wang et al. (2009) observed that elevated Zn levels (0.07–1.12 mM) for seven days induced toxic effects on the seedlings of rapeseed (*B. napus*). Inhibition of plant growth, leaf chlorosis, decreased chlorophyll contents, decline nutrients uptake, and root damages. Meanwhile, Zn induced the lipid peroxidation, NADH oxidase, and POD enzyme, but decreased the other antioxidant enzymes (SOD, CAT, and APX) activities. In conclusion, Zn potentiated its toxicity by disturbing the nutrients uptake/balance and inducing oxidative stress via ROS.

Nickel (Ni) is an essential micronutrient but can be phytotoxic at higher concentrations. It has been reported that higher concentrations of Ni impaired the growth, photosynthetic parameters, as well as, induced the ROS, MDA production, protein, and DNA damages by altering the antioxidant defense system in *B. napus* plants (Farid et al. 2017). The increasing levels of Ni triggered the lipid peroxidation, hydrogen peroxide, methylglyoxal, electrolyte leakage, enzymes of AsA-GSH cycle, osmolyte biosynthesis while inhibited the mineral nutrients and photosynthetic attributes in the tissues of *B. juncea* (Zaid et al. 2019). Higher concentrations of Ni significantly declined the leaf water, osmotic potentials, and free amino acids. In addition, higher accumulation of histidine, cysteine, and serine was positively correlated with the Ni tolerance in canola (*B. napus*) (Ali et al. 2009).

Manganese (Mn) is an important micronutrient for plants which play a critical role in the enzymatic activation of metabolic pathways. The excessive levels of Mn have been found to induce the phytotoxicity in concentration dependent manner and inhibited the plant growth, photosynthetic attributes, and disturbed water relations. Also, Mn induced the ROS, lipid peroxidation as MDA, electrolyte leakage and carbonic anhydrase in the tissues of *B. juncea* (Parashar et al. 2014; Fariduddin et al. 2015). Another study noted that excess Mn declined the growth traits (length and weights of tissues), chlorophyll contents, photochemical efficiency (Fv/Fm), total free amino acids, microelements (Ca, Mg, Fe, Zn, Cu, and K) and CO_2 assimilation in a concentration-dependent manner (Lee et al. 2011).

Boron (B) is an essential micronutrient for the growth and development of plants, but its deficiency and toxicity threshold is narrow (Singh et al. 2010). Savić et al. (2013) analyzed the response of 16 genotypes of *B. napus* against B toxicity. Results indicated that the reduction in root length was the better indicator of B toxicity in *B. napus* as compared to other growth traits. Another study reported the toxic effects of B on the photosynthesis and antioxidant defense machinery of *B. napus*. Findings from the field experiment proposed that excess B levels impaired the photosynthetic attributes and accumulation of lipid peroxidation as MDA contents. The escalation in antioxidant enzyme activities indicated the tolerant ability of *B. napus* plants against B toxicity (Hossain et al. 2015).

Iron (Fe) is a well-known plant nutrient which is beneficial at lower concentrations (ul Hassan et al. 2017b) but can be phytotoxic at higher concentrations. The excessive concentrations of Fe were found toxic as noticed by the induction in ROS components in many plant species. In addition, Fe caused protein oxidation, lipid peroxidation as MDA induction, carbohydrates and DNA damages which lead to cell death (Baruah and Bharali 2015). Most common toxic effects of metals/metalloids in oilseed crops are displayed in Fig. 3.

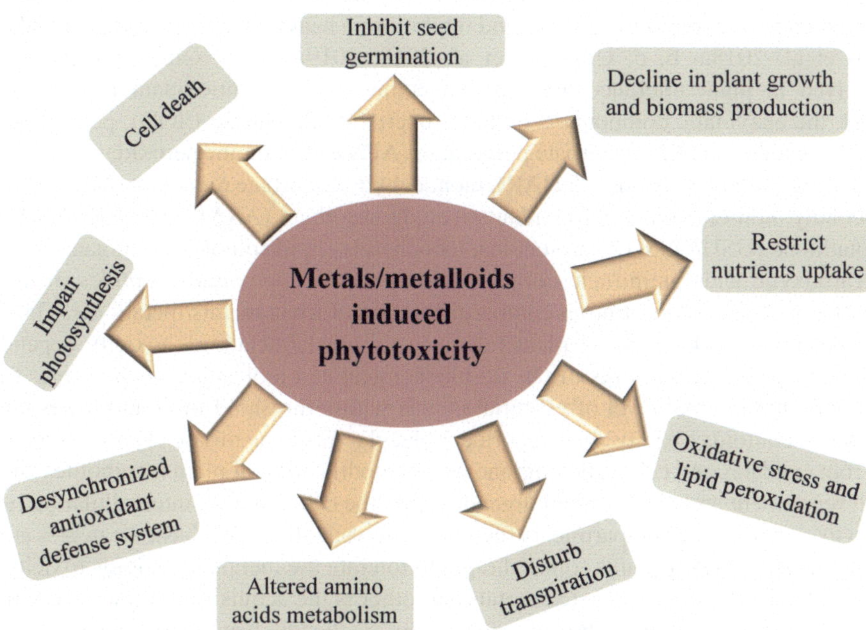

Fig. 3 Toxic heavy metals/metalloids triggered possible impacts on the morphological, physio-biochemical, and molecular attributes of oilseed crops

5 Tolerance Mechanisms Adapted by Oilseed Rape Species Against Metals/Metalloids Toxicity

Plant tolerance to toxic metals/metalloids can be defined as their ability to withstand or survive in a toxic environment. When plants are exposed to different metals or metalloids toxicity, they may limit crops potency and yield. Mostly, the above-mentioned stresses prompt changes in cellular redox homeostasis as the result of excessive ROS generation. Either lower or higher production of cellular ROS plays a key role as a signal molecule in plants and control the plant physiological functions under stress response. Therefore, the balance between ROS production and its elimination is crucial for the normal cellular functioning of plants. At the cellular level, plants utilize a wide range of adaptive mechanisms against toxic metals/metalloids stress conditions. By doing this, plants maintain redox homeostasis by providing tolerance against stresses such as sequestration of toxic metals, chelates formation with lower molecular organic compounds, complex formation of toxic metals in plant tissues and transport to specified cellular compartments. Against metals/metalloids stress, plants mechanisms include chelating agents such as cysteine-rich peptides and small proteins such as phytochelatins and metallothionein (Chen et al. 2006). Another general response of oilseed crops against heavy metals/metalloids stress is the activation of antioxidant or non-antioxidant defense system, biosynthesis of thiolic compounds, amino acids metabolism, sequestration of complexed or uncomplexed metal/metalloid in vacuole, and inhibition in heavy metals/metalloids uptake (Ali et al. 2018a, b, c; Ulhassan et al. 2018, 2019a, b, c). Other key players involved in plant tolerance are the two components of antioxidant machinery. First, the enzymatic components include superoxide dismutase (SOD), peroxidase (POD), catalase (CAT), ascorbate peroxidase (APX), glutathione peroxidase (GPX), dehydroascorbate reductase (DHAR), monodehydroascorbate reductase (MDHAR). Secondly, non-enzymatic antioxidants include ascorbate (ASA), reduced/oxidized glutathione (GSH/GSSG), carotenoids, tocopherols (lipid soluble), conjugates such as carboxylic acids, amino acids, organic acids, phenolic compounds, peptides, metal-binding proteins, sequestration/compartmentation of toxic metals/metalloids. Thus, the potential mechanisms of plant tolerance to metals/metalloids toxicity depend on the interrelated networks of physio-biochemical and molecular mechanisms. To maintain the internal level of essential metals within threshold limit and lower the metals/metalloids toxicity, well-reported molecular mechanisms are (1) involvement of root exudates and cell wall components, (2) modification of plant metabolites and cell repairment and (3) Chelating ligands inside the cell and their intercellular partition to extra nuclear compartments (cytoplasm to vacuoles). Two basic strategies are being used by higher plants (oilseed rape) to tolerate the metals/metalloids toxicity. One is the excluding of targeted metals and other is the acquisition of the metal. In excluding strategy, plants adapted mechanisms are the involvement of root exudates and cell wall constituents in order to exclude the entry of toxic metal/metalloid inside the cell. In metals acquisition strategy, plants minimized/alleviate the toxic impacts of the existing metals inside the cell through the involvement of molecular players such as alteration in transcript levels, repair of protein damages, and intracellular compartmentation of toxic metals/metalloids.

6 Molecular Aspects of Metals/Metalloids Tolerance Mechanisms and Homeostasis in Oilseed Crops

Plants tolerance to metals/metalloids stress are governed by interlinked networks of physio-biochemical and molecular mechanisms to fully understand these mechanisms and develop tolerant plants as ideal candidates for phytoremediation. These approaches (physio-biochemical and molecular mechanisms) are employed to recognize the accumulation, transport, and tolerant/adaptive mechanisms. The adaptive mechanisms proceeded by tolerant plants such as oilseed crops are the immobilization, exclusion from the plasma membrane, restriction in the uptake and transport, particular transporter proteins, chelation via conjugation, sequestration, biosynthesis of osmolytes and induction of stress proteins (Hossain et al. 2012).

6.1 Barriers/Restrictions of Metal/Metalloids Uptake and Transport

6.1.1 Exclusion Strategies of Metals/Metalloids from Plants

The entry of toxic metals/metalloids inside plant cells can be potentially restricted by the extracellular exudates that actively or passively secreted in plant roots (Bertrand and Poirier 2005). Root exudates containing organic molecules including amino acids, phenolics, and other organic acids. These molecules govern the detoxification of toxic metals/metalloids internally and externally. Usually, mucilage uronic acid binds with toxic metals/metalloids at the root surface. The binding potential of uronic acid depends upon the type of target metal/metalloid. For toxic metals, these compounds act as chelating agents and declined the soil pH by converting them into reduced form (Weis and Weis 2004). Ultimately, these compounds directly influence nutrients availability. Previously, it was noticed that plant exudates such as organic acids, phosphate, and mucilage in response to excessive Al concentration and prevented its uptake in Al-tolerant species (Salt et al. 2000). Soil pH, redox potential, type of soil, cation exchange capacity, metal accumulation, and competing ions directly/indirectly influence the bioavailability of toxic metals/metalloids. The regulation of metals uptake by plants depends upon the bioavailability of toxic metals/metalloids (Lasat 2000). Usually, toxic metals enter in a plant cell in the form of cationic elements such as Zn^{2+}, some enters as anionic groups such as As^{2+} and organic compounds such as CH_3Hg. In addition, the root interception of metal ions governs the uptake, entry, and translocation of toxic metals/metalloids in plant parts through mass flow/diffusion (mainly). The uptake of metal ions from soil to plant is facilitated by the root exudates such as proton extrusion and metal reductase. These metal ions become bioavailable inside plant parts through extracellular (apoplast) and intercellular (symplast) transportation. Mainly, cation exchange capacity hinders and upregulate the transport system (Jabeen et al. 2009). Extracellular transport includes root cortex and epidermis. Also, membrane facilitated a strong electrochemical gradient governed by the intercellular transportation for the influx of toxic metal ions (Ghosh and Singh 2005).

6.1.2 Cell Wall Constituents-Metal Ions Interaction

Cell wall constituents also play a key role in inhibiting the uptake of toxic metals/metalloids. The interaction between the cell wall and toxic metal ions resulted in the exclusion of specific transporter proteins and protoplast governed exocytosis mechanisms. Against toxic metals/metalloids, cell wall behaves as defensive, sink, and modifier of toxic metal ions stress. The altered cell wall was found to accumulate more toxic metal ions and thus decline their permeability in the cytosol. Another constituent of cell wall such as polysaccharides significantly binds with metal cations (divalent and trivalent). Therefore, the action of polysaccharide in the cell wall determines the binding capacity of the cell wall with metallic cations. In higher plants, major constituents of the cell wall are Polysaccharide pectins. Four different domains reside in pectins. Among them, the HGA domain is well reported regarding its binding capacity with toxic metal cations due to the presence of carboxyl group. The interaction between metal cations and the carboxyl group of HGA domain depends upon the charge density of pectins and ionic strength of toxic metal ions. Usually, Pb^{2+} and Cu^{2+} cations make strong binding with pectins whereas Zn^{2+}, Ca^{2+}, and Ni^{2+} bound less strongly with pectins (Krzeslowska 2011). In addition, calcium ions are replaced by strong cations such as Zn^{2+}, Cd^{2+}, Cu^{2+}, Al^{3+}, and Pb^{2+}. By doing this, cell wall constituents take part in the make-up of inactive cations form and excessive accumulation of toxic metal cations hinder the uptake of toxic metal cations inside the cell. Under toxic metals/metalloid stress, cell wall modifications enhance the polysaccharides especially pectin constituents and induce the callos (which act as a physical barrier against penetration of toxic metal ions) synthesis (Krzeslowska et al. 2009; Mwamba et al. 2020). Many of the hyperaccumulator (oilseed crops) adopt this strategy to grow plants in toxic metal/metalloid contaminated soils.

6.1.3 Complexation of Metal Ions at Cell Wall Plasma Membrane

One of the primary targets of metal ions is the plasma membrane. Toxic metal ions disturbed the functions of the plasma membrane via ions leakage from the cells. The oxidation of protein thiols and restriction of other membrane protein resulted in the modulation of plasma membrane composition (Hall 2002) and collectively disturbed the ions permeability. One of the particular feature of metals/metalloids tolerant

Brassica oilseed crops is the maintenance of plasma membrane integrity. Among other *Brassica* oilseed species, *B. juncea* contain more stable cellular membrane against cadmium stress in comparison with *B. napus*. In addition to having improved repair mechanism and resistant plasma membrane against toxic metal ions, these metal/metalloids tolerant species sustain ionic homeostasis via efflux mechanisms and restrict metals/metalloids entry into the cell. The efflux of toxic metal/metalloids outside in plants is governed by a number of transporters. These tolerant plant species release chemicals in the soil that restricts the availability of metal ions for their transport and lessen their concentrations inside the cell via transporter proteins (Nauairi et al. 2006).

6.2 Complexation and Partition of Metals/Metalloids in Plant Cell

6.2.1 Intracellular Sequestration or Compartmentation Within Vacuoles (Involvement of Carrier Proteins/Transporters)

Toxic metal ions are partitioned between the root-shoot system after their absorption. Various factors are involved in the transportation of metal ions from root to shoot such as target plants, type of metal and applied concentration. Oilseed crops are well-known for their ability to transfer toxic metal ions to shoot in a greater amount which is perquisites for phytoremediation. The transportation of these metal ions from root to shoot is dependent on the transporters for sequestration, absorption, and storage of both essential and toxic metals/metalloids in vacuoles. A number of metal ions carriers/channels take part in the transportation of metal ions in plant cells (Jabeen et al. 2009). These include vacuole proton pumps, co or anti transporters and channel proteins. A number of metals/metalloids transporter proteins are well-considered in various plants such as copper transporter (COPT) family proteins, cation diffusion facilitator (CDF) family of proteins, ATP-binding cassette (ABC) transporters, iron-regulated transporter (IRT) like protein ZIP family, ABC transporters of the mitochondria (ATM), zinc-regulated transporter (ZRT), Ca^{2+}: cation antiporter (CAX), natural resistance-associated macrophage protein (NRAMP) family, pleiotropic drug resistance (PDR) transporters, the P-type metal ATPases, yellow-stripe-like (YSL) transporter and multidrug resistance-associated proteins (MRP) (Kramer et al. 2007). Bhuiyan et al. (2011) reported that a yeast cadmium factor 1 (YCF1), member of ATP-binding cassette (ABC) transporter family was introduced in *B. juncea* via genetic transformation. Overexpression of YCF1 in *B. juncea* enhanced tolerance for Cd and Pb stress, in addition to, their higher accumulation in transgenic plants. Wu et al. (2015) investigated that Cd uptake in *B. napus* is done by the roots (*IRT1*) and transported by xylem (*HMA2* and *HMA4*) were consistent with absorption and transport of Cd in *B. napus* tissues. In another study, cation-efflux transporter *BjCET2* enhanced the tolerance for Zn and Cd through mediating their accumulation in the leaves of *B. juncea* (Xu et al. 2009). Recently, Wu et al. (2019) revealed the pivotal transporters (*BcIRT1* and *BcZIP2*) involved in the reduction of Cd accumulation in pak choi (*B. chinensis*) and probably involved in the transportation of Fe, Zn, Mn and Cd. Gill et al. (2017) reported that Cr induced the novel transcripts (*BnaC01g29930D* and *BnaA07g14320D*) which were responsible for secondary active transmembrane transporter and protein transporter activities. These findings suggested the improved Cr-tolerance in *B. napus* plants. Meng et al. (2017) investigated that *BnNRAMP1b* (NRAMP) transporter was responsible for Zn, Mn and Cd uptake, transport and homeostasis in *B. napus*. Zhu et al. (2016) also reported that vacuolar Fe transporter (*BnMEB2*) is actively involved in Fe-tolerance of *B. napus*. This *VIT* gene in *B. napus* provides harmless storage and vacuole detoxification system of excessive Fe which lead to improving the Fe-tolerance. Zhang et al. (2018) annotated and identified ATP-binding cassette (ABC) transporters for the uptake of Cd in *B. napus*. A recent study suggested that phosphate transporter (*OsPT2*) was actively involved in the uptake of Se (selenite) in *B. napus* plants (Ulhassan et al. 2019b). There is less information available in the literature regarding the transport functions of metals/metalloids and their molecular location in *Brassica* oilseed rape species remain unknown which require further investigations.

6.2.2 Metals/Metalloids Complex Formation by Phytochelatins (PCs)

Chelation of metals/metalloids in the cytosol via the involvement of high-affinity ligands is the key strategy for the detoxification and tolerance mechanisms. Higher plants make peptide metal-binding ligands which are phytochelatins (PCs) and metallothioneins (MTs). PCs are synthesized from GSH (acts as substrate for PCs biosynthesis). Metals bind to highly expressed γ-glutamylcysteinyl dipeptidyl transpeptidase (PC synthase) and activate it to catalyze the GSH-PCs conversion (Oven et al. 2002). Metals/metalloids usually induced the biosynthesis of PCs such as Cd, Zn, Ni, Ag, Pb, As, Hg, Cu, and Se, while Cd is the strongest PCs-inducer. In this way, plants make Cd-PCs complexes through thiolic group of cysteine (amino acids residue) and gathered in vacuole via ABC transporters, hence limited the Cd ions transport inside cytosol (Cobbett and Goldsbrough 2002). Ulhassan et al. (2019b) reported that excessive levels of metalloid (Se) make complexes with thiolic compounds (PCs, etc.) with induction in linked metabolic enzymes such as phytochelatins synthase (PCS) which sequester Se there in roots and inhibits further translocation to shoot in *B. napus*. As GSH is the precursor of phytochelatins (PCs), which bind Se, Be, Co, and Cr by minimizing their cellular free concentrations in the tissues of *B. napus* (Ulhassan et al. 2018, 2019a; Ali et al. 2018a, c; Gill et al. 2015a). Iglesia-Turiño et al. (2006) reported that Hg-accumulation in *B. napus* plants induced the PCs production. The induction of PCS under metals/metalloids stress suggested that PC production and metals/metalloids accumulation was positively correlated in both above and below plant parts. PCs are also involved in the activation of the antioxidant defense system and a reduction in oxidative stress. In addition to the detoxification process via chelation of metal ions and activation of PC synthase, this metal ions complex formation is transported to the vacuole and stabilized thereby forming complexes with organic acids. PCs also play a key role in this transportation system to facilitate their detoxication capability via complex mechanisms. However, further research is needed regarding the thorough estimation of PCs detoxication potential at cellular levels.

6.2.3 Metals/Metalloids Complex Formation by Metallothionein's (MTs)

Metallothionein's (MTs) are low molecular weight, cysteine-rich, gene-encoded polypeptides that can bind metals/metalloids via thiolic compounds. The exact physiological functions of MTs are still needed further verifications. However, possible actions performed by MTs are; sequestration of toxic metal ions, maintain homeostasis for essential metal ions and protection against metals/metalloids oxidative damages. MTs induction is done by various factors such as plant hormones, cytotoxic agents, and metals/metalloids stress (Hossain et al. 2012). Gene expression studies were carried out to quantify the qRNA levels in plant tissues especially development stages under metals/metalloids stress. MT genes are regulated in tissue-specific manner and linked with the plant development stages. Under different metals stress conditions, MT genes (*BrMT1*, *BrMT2*, and *BrMT3*) belong to *B. rapa* were differentially regulated. While, *BrMT1* and *BrMT2* genes in Fe-applied seedlings were not induced, except the induction of *BrMT3* transcripts. The expression of *BrMT1/BrMT2* was upregulated/downregulated and *BrMT3* remains unchanged in both Cu and Zn stressed seedlings. An increase in the expression of *BrMT1* and *BrMT3* up to 12 h and then downregulated, while *BrMT2* did not change) (Ahn et al. 2012). Most of the previous findings proved the involvement of MTs in metals/metalloids and their homeostasis, but lack of MTs induction by metals/metalloids which require further study. Also, further investigation is required regarding possible mechanisms of MTs functions and its structural properties. Possible function/involvement of cysteine (Cys) rich phytochelatins (PCs), metallothionein's (MTs), and glutathione (GSH) in heavy metals/metalloids chelation via complex formation and then sequestration in the vacuole is illustrated in Fig. 4.

6.2.4 Metals/Metalloids Chelation by Glutathione

Glutathione, a precursor of PCs, is non-protein sulphur comprising peptide which is directly/indirectly interrelated with the sequestration of metals/metalloids and participate in the endogenous defense system by the cellular components protection (Anjum et al. 2012). It exists in oxidized (GSSG) and reduced (GSH) (90% abundance) states. The interaction of reduced glutathione with unstable molecules such as reactive oxygen species get it oxidize (Tausz et al. 2004). Usually, plants conjugate with toxic organic compounds or their metabolites to GSH for additional

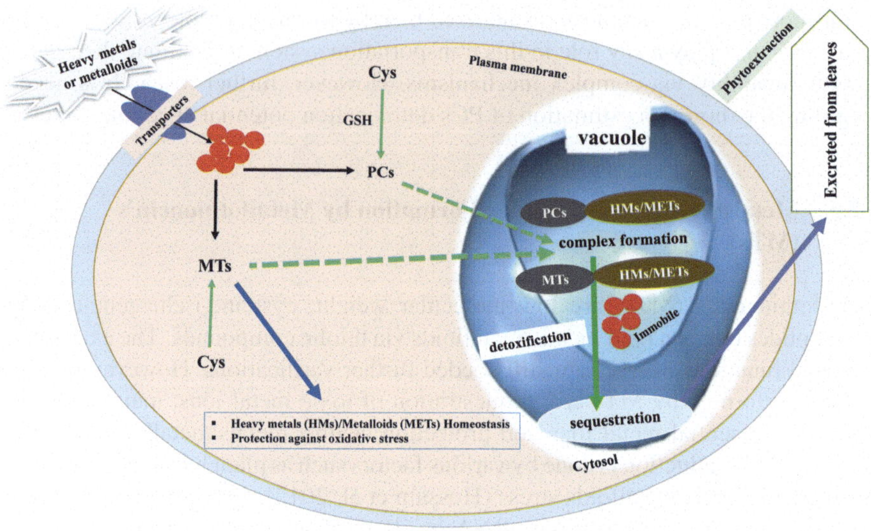

Fig. 4 Detoxification/homeostasis of toxic heavy metals and metalloids in oilseed crops; involvement of glutathione (GSH), phytochelatins (PCs), and metallothioneins (MTs)

metabolism that is catalyzed by glutathione S-transferase (GSTs). For cellular detoxification, the conjugation of glutathione is the crucial mechanism which (glutathione-conjugation) can increase the activity of glutathione S-transferase enzyme and occur spontaneously (Nagalakshmi and Prasad 2001). With nucleophilic quality, GSH form mercaptide bonds with heavy metals/metalloids and facilitate the mitigation/sequestration of toxic metal ions (Zagorchev et al. 2013). The chemical reactivity and relative stability of GSH make it novel biochemical component that reduced the heavy metals/metalloids stress in *Brassica* oilseed species (Gill et al. 2017; Zlobin et al. 2017).

6.2.5 Metals/Metalloids Chelation by Organic Acids, Amino Acids, and Phosphate Derivatives

The tolerance and detoxification mechanisms of metals/metalloids are categorized into external removal or internal tolerance. In external exclusion process, root exudates such as organic acids form ligand complexes with toxic metals/metalloids which resulted in the restriction of their translocation in target plant tissues or avoid storage of toxic metal ions in the sensitive sites of roots. Internal detoxification includes the chelate formation between organic acids and metal ions in the cytosol, where toxic ions may alter into less toxic ions or their complete removal. Various plant including *B. napus* produce a wide range of ligands for metals/metalloids (Cd, Cu, Ni, Zn, Al, Se), carboxylic and amino acids (histidine, citrate, oxalate, malate, nicotianamine), and phosphate derivatives (phytate) which are potential ligands for the tolerance and detoxification of metals/metalloids (Hossain et al. 2012; Ulhassan et al. 2019c). Citrate has high capacity to chelate with metal ions especially Fe. Metals/metalloids such as Cd, Ni, and Zn have strong affinity for citrate. Previous studies documented that citric acid enhanced the uptake of Cu (Zaheer et al. 2015), Cr (Afshan et al. 2015), Ni (Farid et al. 2017), and may be beneficial in accelerating the phytoremediation potential of *B. napus* plants. Citric acids play a dual role in *B. juncea* as it boosted its phytoremediation capacity and improved the Cd-tolerance through the involvement of chelating agents, and activation of antioxidant defense-glyoxalase systems. Oxalate (root-secreted) has great potential to detoxify (externally/internally) the toxic metal ions. The combined effects of oxalic acid and ethylene diamine tetraacetic acid (EDTA) enhanced the bioavailability of Cd and Zn in the soil through complexation and assisted the phytoextraction capacity of *B. juncea* to remove toxic metals (Guo et al. 2019). Histidine has also been involved in the tolerance and detoxification of toxic metal ions. The exogenously applied histidine has been found to induce the Ni tolerance in *B. juncea* (Kerkeb and Kramer 2003). The histidine-linked root to shoot transport of Ni governed the Ni-hyperaccumulation. Nicotianamine also has the ability to detoxify toxic metals. It has been mostly studied in context with Fe-nutrition. The carboxyl group of nicotianamine and N-donor center form complexes with toxic metals unlike phytochelatins-metals complexation. This complex maintains metals/metalloids homeostasis in plant cells and other tissues. The root to shoot transportation of toxic metal ions is also regulated by nicotianamine (Sharma and Dietz 2006). These studies exposed that organic acids are fundamental chunks for the detoxification mechanisms against toxic metals/metalloids stress. Phosphate derivates such as phytate are also known for their complex formation with toxic metals. It has been reported that phytate tends to be interconnected with the metals (Al) deposit in AL-tolerant plant vacuoles. The globules (Zn-phytate) was found to be abundantly more common in the tolerant plant species as compared to sensitive species. Yet, still, it is ambiguous that either phytate complexes induced tolerance or response to stress. Above-mentioned findings suggested that organic acids, amino acids and phosphate are the byproducts of heavy metals/metalloids detoxification. Further investigations are required to figure out the production of organic acids as a result of toxicity-tolerance response in different plant species by the use of various heavy metals/metalloids.

6.3 Involvement of Polyamines

Polyamines (PAs) are low molecular organic cations distributed and ubiquitous in living organisms. Most commonly existed PAs such as spermidine (Spd), spermine (Spm), and putrescine (Put) have been reported in plants especially in response to environmental stresses. These PAs have the potential to protect cellular membrane by acting as an inhibitor of lipid peroxidation and ROS production in response to toxic metal ions, thus induced the antioxidant defense system. They also facilitated the compartmentation of toxic metal ions by decreasing the ion conductance at the vacuolar membrane (Sharma and Dietz 2006). Overexpression of PAs biosynthetic genes enhanced the tolerance of *Brassica* oilseed crops under multiple environmental stresses. In addition, exogenous applications of PAs have been documented to enhance the plant tolerance in response to the applications of heavy metals/metalloids such as Cd and Mn, possibly via regulation of redox homeostasis (Soudek et al. 2016; Hussain et al. 2019).

6.4 Antioxidant and Glyoxalase Systems

First response of plants against heavy metals/metalloids stress is the induction in ROS production which causes oxidative stress. To scavenge the ROS-induced oxidative stress, plants evolved their antioxidant defense system such as superoxide dismutase (SOD), peroxidase (POD), catalase (CAT), ascorbate peroxidase (APX), glutathione peroxidase (GPX), dehydroascorbate reductase (DHAR), monodehydroascorbate reductase (MDHAR. Free radical scavenging molecules such as AsA and GSH and, their related metabolic enzymes play a vital role in enhancing plant tolerance by limiting the production of ROS and MG components. In addition, MG is a cytotoxic compound, which induced in response to heavy metals/metalloids stress. Detoxification of heavy metals/metalloid induced MG and its indulger oxidative stress is carried out by the upsurge in glyoxalase system (Gly I and Gly II) enzymes. *Brassica* oilseed species induced different components of antioxidant and glyoxalase systems to enhance plant tolerance via limiting the ROS or MG induced oxidative stress (Ulhassan et al. 2019a, c; Hasanuzzaman et al. 2017a, b).

6.4.1 Heat-Shock Proteins (HSPs) and Phenolic Compounds

Heat shock proteins (*HSPs*) are molecular chaperones which repair or protect proteins from stressful environmental conditions. Previous investigations reported the induction of *HSPs* in response to different toxic metal ions. It has been reported that Cr and Co stress-induced the transcripts of *HSP90-1*, *GR-1*, and *MT-1* stress-responsive HSPs in the leaves and roots of *B. napus* (Ali et al. 2018c; Gill et al. 2015b). Still the suspected roles of *HSPs* in metals/metalloids tolerance are largely unknown and need further investigations. Also, there is a need for advanced technologies (microarray and omics) to fully understand the potential role of *HSPs* and their signal pathways under stress conditions in different plant (including *Brassica* oilseed) species.

Phenolic compounds such as flavonoids, phenylpropanoids, and anthocyanin are reported to induce/accumulate in response to different heavy metals/metalloids applications. The biosynthesis of phenolic compounds was found to induce in response to Pb, Cr in *B. juncea* (Kohli et al. 2018; Handa et al. 2018) and Zn, Se in *Brassica oleracea* (Barrameda-Medina et al. 2017; Bachiega et al. 2016). Above studies revealed that these phenolic compounds act as a scavenger of free radicals in antioxidant systems and facilitate detoxification of toxic metal ions via conjugation with metals/metalloids and then sequestration in the vacuole.

6.4.2 Regulation of Transcription Factor (TF) via Signal Transduction

Toxic metal ions transmit information to each cell through signal transduction pathways. Toxic metals induced signals startup pathways containing receptor coupled phosphorylation, phosphor-inositol coupled Ca^{+2} fluctuations, protein kinases cascades and transcription activation of stress-responsive genes. The signal generates by these pathways lead to the expression of a wide range of genes that cause metabolic changes. Toxic metal ions induced the signal by calcium and calmodulin messengers, in addition to, the involvement of protein posttranslational modification and protein-protein interaction. Extracellular signals and perception are governed by the receptor coupled phosphorylation like kinases (Quelhadj et al. 2007). Intracellular receptor coupled kinases interact with downstream regulatory factors are interconnected in signaling pathways. During toxic metals stress, intracellular signal transduction is carried out by mitogen-activated protein kinases. Toxic metals have been reported to activate the MAPKs activities in different plants (especially tolerant species). In addition, toxic metal disturbed the reduced/oxidized glutathione ratio which activated other defense system involved in redox signaling. Decreased GSH/GSSG ratio directly modulate the cell cycle, growth, gene expression, and protein activity due to the influence on redox status of cells (Anjum et al. 2012). The elevated levels of toxic metals induced the ROS which triggered signal transduction mechanisms by antioxidant defense system and metabolic modifications to control ROS (act as a signal molecule). Toxic metals induced signal pathways can be used to enhance plant tolerance against other stress conditions. These toxic metal ions induced the jasmonate pathways and enhanced ethylene production. Nitric oxide (NO) and melatonin (MT) are widespread cellular messenger/signal molecule which regulated the different physiological process. Toxic metals induced the NO and MT signals inside the cell and also act as an antioxidant by ROS-scavenging mechanisms. Brassinosteroids are phytohormones and ameliorating role in plants against stress conditions. Various studies documented the involvement of these phytohormones/signal molecules to induce the tolerance of *Brassica* oilseed species against toxic metals/metalloids stress (Ulhassan et al. 2019b; Masood et al. 2017; Farooq et al. 2016; Khan et al. 2019; Per et al. 2017; Kanwar and Bhardwaj 2015). The signal transduction by toxic metal ions is transmitted by various pathways and crosstalk. The activation of ROS signaling, Ca-calmodulin system, phosphorylation cascades and stress-responsive phytohormones ultimately regulated the transcription factors (TFs). These TFs further activated the set of genes response against toxic metals. In this way, toxic metal ions modified the plant metabolisms via regulation/modulation of TFs or activation of transcript levels of stress associated genes responsible for metals transport and detoxification. TFs-linked various families such as WRKY, basicleucine zipper (bZIP), ethylene-responsive factor (ERF) and myeloblastosis protein (MYB) play key role in controlling the transcription of stress-responsive genes (Hossain et al. 2012; Yang et al. 2005). Gill et al. (2017) reported that Cr induced the transcripts encoded nucleic acid, metal ions binding proteins, protein kinases and phosphotransferase activities, signal transduction and molecular transporter proteins in *B. napus*. The induction of TFs in response to Cr stress revealed the enhanced plant tolerance. The induction in the expression of *OsMYB4* TF in transgenic canola plants indicated the improved tolerance in response to Zn and Cu stress. Higher tolerance of transgenic plants against Zn and Cu stress-induced the phenolic compounds which indicated the increase in proline and phenylpropanoids biosynthesis. Thus, canola *OsMYB4* transgenic plants could be employed in the phytoremediation of heavy metals/metalloids contaminated soils (Raldugina et al. 2018).

6.5 Control of DNA Damages

Toxic metal ions disturbed the growth patterns and plant metabolic profiles via direct interaction with the nucleic acid. Toxic metals induced ROS production and DNA damages. The cell cycle is temporarily halted during DNA repairment, genes induction and cellular protection against metals toxicity. Repairment of DNA damages is important for maintaining genomic integrity and plant tolerance against the exposure of toxic metal ions. The first group of gene products is the cellular protection from dehydration, biosynthesis enzymes of osmoprotectants, chaperones, and detoxification systems. The second group includes transcription factors, protein kinases, and enzymes of phosphoinositide metabolism (Hamid et al. 2010). The above discussion indicated that heavy metals/metalloids tolerance and homeostasis is complex and interconnection among physio-biochemical, metabolic and molecular process is prerequisite for the better understanding of tolerance-homeostasis mechanisms.

7 Possible Strategies to Improve Phytoremediation Potential of *Brassica* Oilseed Crops

Brassica oilseed crops are well-known for metals extraction ability from the soil and higher plant biomass which makes them suitable candidates for phytoremediation purpose, unlike other hyperaccumulators with slow growth and low biomass. These tolerant plants either exclude the toxic metal ions either by restricting their uptake from roots to shoot, retained them in roots or detoxification by chelation in cytoplasm and storage in leaf vacuoles. Among other *Brassica* oilseed crops, *Brassica* genus plants such as *B. juncea* are identified as ideal plants for phytoremediation due to their fast growth rate and higher biomass production. Both *B. napus* and *B. juncea* are famous worldwide for phytoremediation research due to their higher tolerance abilities in response to heavy metals/metalloids such as Cd, Cr, Pb, Co, Cu, As and

Se, strong ability to accumulate and transport in plant tissues (Ali et al. 2013, 2014a, 2018c; Farooq et al. 2016; Gill et al. 2015a, b; Handa et al. 2018; Mwamba et al. 2016; Ulhassan et al. 2019a). These plants have a well-developed root system and greater above-ground biomass, which make them useful for phytoextraction. They have adaptive mechanisms to grow in cold/warm conditions which attract researchers to understand the accumulation and tolerance potential to detoxify toxic metal ions. Recently, *B. juncea* has gained considerable attention worldwide by plant genetics and breeder researchers for their phytoextraction ability. Therefore, their plant species play a key role in phytoremediation technology and cleaning/remediation of contaminated soils. The removal of toxic metal ions from contaminated soils can be carried out by the utilization of *Brassica* oilseed species. For this, different physiochemical and biotechnology approaches can be employed such as lowering pH (by sulphuric acids and organic fertilizers, etc.), redox potential, chelators (to enhance metals bioavailability), organic acids (for effective metals accumulation), lime/organic amendments (for high biomass), use of microorganisms, plant growth-promoting rhizobacteria (to depollute toxic metal ions) and plant breeding (by selecting mutants/variant of desire traits, etc.) and classical genetics (generate metal tolerant species) for improving phytoremediation. Different biotechnological approaches could be effective to enhance the detoxification of *Brassica* oilseed species such as selection and generation of transgenic with characteristics; fast growth under stressful environment, capacity to store toxic metals into harvestable biomass, avoid weed growth and safe consumption for humans. To improve phytoremediation ability, various strategies could be adaptive for the generation of transgenics *Brassica* oilseed species such as; induction of gene-encoded transporter molecules, increase proteins/peptides/molecules in plant cells having greater affinity for toxic metals, change redox state by introduce genes encoding enzymes, enhance the defense capacity (enzymatic and non-enzymatic) by introducing genes of antioxidant/detoxification systems and metabolic engineering (Varkey et al. 2012). These approaches/strategies could be employed to enhance the phytoremediation capacity of *Brassica* oilseed species in order to detoxify toxic metal ions and improve plant tolerance against toxic metals/metalloids.

8 Conclusions and Future Prospective

The excessive concentrations of toxic heavy metals/metalloids in soil-plant interface cause detrimental effects on living organisms. The molecular and cellular adaptation of *Brassica* oilseed species to heavy metals/metalloids stress facilitates interlinked plant tolerance and homeostasis against these stresses that eventually restricts the entry of metal ions into the food chain. Plants boosted their antioxidant and glyoxalase defense systems by the activation of enzymatic and non-enzymatic components. Different heavy metals/metalloids employ diverse mechanisms to induce phytotoxicity and also various resistant reactions as tolerant mechanisms. Tolerance mechanisms and reaction response are also varied for different growth conditions (hydroponic/field conditions), type and dose of applied heavy metal/metalloid.

These involved factors make it difficult for the researchers to produce a particular/common tolerance mechanism in response to all heavy metals/metalloids. *Brassica* oilseed species make use of accumulation and transport (uptake and translocation) mechanisms of toxic metal ions for effective tolerance. A number of biochemical approaches further induced the adaptive ability of these plants to tolerate heavy metals/metalloids toxicity. These plant species utilize the source and expression of transporter genes, which provide genetic evidence for the phytoremediation purpose that could be helpful for other plants. In addition, the physio-biochemical and molecular approaches enhanced the accumulation and tolerance mechanisms which can be effectively employed to resolve phytoremediation issues.

References

Afshan S, Ali S, Bharwana SA, Rizwan M, Farid M, Abbas F, Ibrahim M, Mehmood MA, Abbasi GH (2015) Citric acid enhances the phytoextraction of chromium, plant growth, and photosynthesis by alleviating the oxidative damages in Brassica napus L. Environ Sci Pollut Res 22:11679–11689

Ahmad R, Ali S, Hannan F, Rizwan M, Iqbal M, Hassan Z, Akram NA, Maqbool S, Abbas F (2017) Promotive role of 5-aminolevulinic acid on chromium-induced morphological, photosynthetic, and oxidative changes in cauliflower (Brassica oleracea botrytis L.). Environ Sci Pollut Res 24:8814–8824

Ahn YO, Kim SH, Lee J, Kim H, Lee HS, Kwak SS (2012) Three Brassica rapa metallothionein genes are differentially regulated under various stress conditions. Mol Biol Rep 39:2059–2067

Al Mahmud JA, Hasanuzzaman M, Nahar K, Rahman A, Fujita M (2017) Relative tolerance of different species of Brassica to cadmium toxicity: coordinated role of antioxidant defense and glyoxalase systems. Plant Omics 10:107–117

Al Mahmud J, Hasanuzzaman M, Nahar K, Bhuyan MB, Fujita M (2018) Insights into citric acid-induced cadmium tolerance and phytoremediation in Brassica juncea L.: coordinated functions of metal chelation, antioxidant defense and glyoxalase systems. Ecotoxicol Environ Saf 147:990–1001

Al Mahmud J, Bhuyan MB, Anee TI, Nahar K, Fujita M, Hasanuzzaman M (2019) Reactive oxygen species metabolism and antioxidant defense in plants under metal/metalloid stress. In: Plant abiotic stress tolerance. Springer, pp 221–257

Ali MA, Ashraf M, Athar HR (2009) Influence of nickel stress on growth and some important physiological/biochemical attributes in some diverse canola (Brassica napus L.) cultivars. J Hazard Mater 172:964–969

Ali B, Tao Q, Zhou Y, Gill RA, Ali S, Rafiq MT, Xu L, Zhou W (2013) 5-Aminolevolinic acid mitigates the cadmium-induced changes in Brassica napus as revealed by the biochemical and ultra-structural evaluation of roots. Ecotoxicol Environ Saf 92:271–280

Ali B, Xu X, Gill RA, Yang S, Ali S, Tahir M, Zhou W (2014a) Promotive role of 5-aminolevulinic acid on mineral nutrients and antioxidative defense system under lead toxicity in Brassica napus. Ind Crop Prod 52:617–626

Ali B, Song WJ, Hu WZ, Luo XN, Gill RA, Wang J, Zhou W (2014b) Hydrogen sulfide alleviates lead-induced photosynthetic and ultrastructural changes in oilseed rape. Ecotoxicol Environ Saf 102:25–33

Ali S, Jin R, Gill RA, Mwamba TM, Zhang N, Ulhassan Z, Islam F, Ali S, Zhou WJ (2018a) Beryllium stress-induced modifications in antioxidant machinery and plant ultrastructure in the seedlings of black and yellow seeded oilseed rape. Biomed Res Int 1–14

Ali S, Gill RA, Mwamba TM, Zhang N, Lv MT, Ulhassan Z, Islam F, Zhou WJ (2018b) Differential cobalt-induced effects on plant growth, ultrastructural modifications, and antioxidative response among four Brassica napus L. cultivars. Int J Environ Sci Technol 15:1–16

Ali S, Gill RA, Ulhassan Z, Najeeb U, Kanwar MK, Abid M, Mwamba TM, Huang Q, Zhou W (2018c) Insights on the responses of Brassica napus cultivars against the cobalt-stress as revealed by carbon assimilation, anatomical changes and secondary metabolites. Environ Exp Bot 156:183–196

Anjum NA, Ahamd I, Mohmood I, Pacheco M, Duarte AC, Pereira E, Umar S, Ahmad A, Khan NA, Iqbal M, Prasad MNV (2012) Modulation of glutathione and its related enzymes in plants' responses to toxic metals and metalloids—a review. Environ Exp Bot 75:307–324

Antoniadis V, Levizou E, Shaheen SM, Ok YS, Sebastian A, Baum C, Prasad MN, Wenzel WW, Rinklebe J (2017a) Trace elements in the soil-plant interface: phytoavailability, translocation, and phytoremediation—a review. Earth-Sci Rev 171:621–645

Antoniadis V, Shaheen SM, Boersch J, Frohne T, Du Laing G, Rinklebe J (2017b) Bioavailability and risk assessment of potentially toxic elements in garden edible vegetables and soils around a highly contaminated former mining area in Germany. J Environ Manag 186:192–200

Ashraf U, Tang X (2017) Yield and quality responses, plant metabolism and metal distribution pattern in aromatic rice under lead (Pb) toxicity. Chemosphere 176:141–155

Ashraf MY, Azhar N, Ashraf M, Hussain M (2011) Influence of lead on growth and nutrient accumulation in canola (Brassica napus L.) cultivars. J Environ Biol 32:659–666

Bachiega P, Salgado JM, de Carvalho JE, Ruiz ALT, Schwarz K, Tezotto T, Morzelle MC (2016) Antioxidant and antiproliferative activities in different maturation stages of broccoli (Brassica oleracea Italica) biofortified with selenium. Food Chem 190:771–776

Barrameda-Medina Y, Blasco B, Lentini M, Esposito S, Baenas N, Moreno DA, Ruiz JM (2017) Zinc biofortification improves phytochemicals and amino-acidic profile in Brassica oleracea cv. Bronco. Plant Sci 258:45–51

Baruah KK, Bharali A (2015) Physiological basis of iron toxicity and its management in crops: recent advances in crop physiology. Daya Publishing House, New Delhi, India pp 203–224

Bertrand M, Poirier I (2005) Photosynthetic organisms and excess of metals. Photosynthetica 43:345–353

Bhuiyan MSU, Min SR, Jeong WJ, Sultana S, Choi KS, Song WY, Lee Y, Lim YP, Liu JR (2011) Overexpression of a yeast cadmium factor 1 (YCF1) enhances heavy metal tolerance and accumulation in Brassica juncea. Plant Cell Tiss Org 105:85–91

Bian ZW, Chen J, Li H, Liu DD, Yang LF, Zhu YL, Zhu WL, Liu W, Ying ZZ (2016) The phytotoxic effects of selenium–mercury interactions on root growth in Brassica rapa (LvLing). Hortic Environ Biotechnol 57:232–240

Chang TC, You SJ, Yu BS, Chen CM, Chiu YC (2009) Treating high-mercury-containing lamps using full-scale thermal desorption technology. J Hazard Mater 162:967–972

Chen A, Kornives EA, Schroeder JI (2006) An improved grafting technique for mature Arabidopsis plants demonstrates long-distance shoot-to-shoot transport of phytochelatins in Arabidopsis. Plant Physiol 141:108–120

Chen L, Yang L, Wang Q (2009) In vivo phytochelatins and Hg–phytochelatin complexes in Hg-stressed Brassica chinensis L. Metallomics 1:101–106

Cobbett C, Goldsbrough P (2002) Phytochelatins and metallothioneins: roles in heavy metal detoxification and homeostasis. Annu Rev Plant Biol 53:159–182

Farid M, Ali S, Rizwan M, Saeed R, Tauqeer HM, Sallah-Ud-Din R, Azam A, Raza N (2017) Microwave irradiation and citric acid assisted seed germination and phytoextraction of nickel (Ni) by Brassica napus L.: morpho-physiological and biochemical alterations under Ni stress. Environ Sci Pollut Res 24:21050–21064

Fariduddin Q, Ahmed M, Mir BA, Yusuf M, Khan TA (2015) 24-Epibrassinolide mitigates the adverse effects of manganese induced toxicity through improved antioxidant system and photosynthetic attributes in Brassica juncea. Environ Sci Pollut Res 22:11349–11359

Farooq MA, Gill RA, Islam F, Ali B, Liu H, Xu J, He S, Zhou W (2016) Methyl jasmonate regulates antioxidant defense and suppresses arsenic uptake in Brassica napus L. Front Plant Sci 7:468

Feigl G, Lehotai N, Molnár Á, Ördög A, Rodríguez-Ruiz M, Palma JM, Corpas FJ, Erdei L, Kolbert Z (2014) Zinc induces distinct changes in the metabolism of reactive oxygen and nitrogen species (ROS and RNS) in the roots of two Brassica species with different sensitivity to zinc stress. Ann Bot 116:613–625

Feigl G, Kumar D, Lehotai N, Pető A, Molnár Á, Rácz É, Ördög A, Erdei L, Kolbert Z, Laskay G (2015) Comparing the effects of excess copper in the leaves of Brassica juncea (L. Czern) and Brassica napus (L.) seedlings: growth inhibition, oxidative stress and photosynthetic damage. Acta Biol Hung 66:205–221

Ghosh M, Singh SP (2005) A review on phytoremediation of heavy metals and utilization of its by products. Appl Ecol Environ Res 3:1–18

Gill RA, Zang L, Ali B, Farooq MA, Cui P, Yang S, Ali S, Zhou W (2015a) Chromium-induced physio-chemical and ultrastructural changes in four cultivars of Brassica napus L. Chemosphere 120:154–164

Gill RA, Ali B, Islam F, Farooq MA, Gill MB, Mwamba TM, Zhou W (2015b) Physiological and molecular analyses of black and yellow seeded Brassica napus regulated by 5-aminolivulinic acid under chromium stress. Plant Physiol Biochem 94:130–143

Gill RA, Ali B, Yang S, Tong C, Islam F, Gill MB, Mwamba TM, Ali S, Mao B, Liu S, Zhou W (2017) Reduced glutathione mediates pheno-ultrastructure, kinome and transportome in chromium-induced Brassica napus L. Front Plant Sci 8:2037

Guo D, Ali A, Ren C, Du J, Li R, Lahori AH, Xiao R, Zhang Z, Zhang Z (2019) EDTA and organic acids assisted phytoextraction of Cd and Zn from a smelter contaminated soil by potherb mustard (Brassica juncea, Coss) and evaluation of its bioindicators. Ecotoxicol Environ Saf 167:396–403

Hall JL (2002) Cellular mechanisms for heavy metal detoxification and tolerance. J Exp Bot 52:1–11

Hambidge KM, Krebs NF (2007) Zinc deficiency: a special challenge. J Nutr 137:1101–1105

Hamid N, Bukhari N, Jawaid F (2010) Physiological responses of Phaseolus vulgaris to different lead concentrations. Pak J Bot 42:239–246

Handa N, Kohli SK, Thukral AK, Bhardwaj R, Alyemeni MN, Wijaya L, Ahmad P (2018) Protective role of selenium against chromium stress involving metabolites and essential elements in Brassica juncea L. seedlings. Biotech 8:66

Hasanuzzaman M, Fujita M (2012) Heavy metals in the environment: current status, toxic effects on plants and phytoremediation. In: Phytotechnologies. CRC Press, pp 30–97

Hasanuzzaman M, Nahar K, Gill SS, Alharby HF, Razafindrabe BH, Fujita M, (2017a) Hydrogen peroxide pretreatment mitigates cadmium-induced oxidative stress in Brassica napus L.: an intrinsic study on antioxidant defense and glyoxalase systems. Front Plant Sci 8:115

Hasanuzzaman M, Nahar K, Anee TI, Fujita M (2017b) Exogenous silicon attenuates cadmium-induced oxidative stress in Brassica napus L. by modulating AsA-GSH pathway and glyoxalase system. Front Plant Sci 8:1061

Hassan Z, Ali S, Ahmad R, Rizwan M, Ab-bas F, Yasmeen T, Iqbal M (2017a) Biochemical and molecular responses of oilseed crops to heavy metal stress. In: Oilseed crops: yield and adaptations under environmental stress. Oilseed Crops 15:236–248

Hassan Z, Ali S, Rizwan M, Ibrahim M, Nafees M, Waseem M (2017b) Role of bioremediation agents (bacteria, fungi, and algae) in alleviating heavy metal toxicity. In: Kumar V, Kumar M, Sharma S, Prasad R (eds) Probiotics in agroecosystem. Springer, Singapore, pp 517–537

Hossain MA, Piyatida P, da Silv JAT, Fujita M (2012) Molecular mechanism of heavy metal toxicity and tolerance in plants: central role of glutathione in detoxification of reactive oxygen species and methylglyoxal and in heavy metal chelation. J Bot 2012

Hossain MF, Shenggang P, Meiyang D, Zhaowen M, Karbo MB, Bano A, Xiangru T (2015) Photosynthesis and antioxidant response to winter rapeseed (Brassica napus L.) as affected by boron. Pak J Bot 47:675–684

Hussain A, Nazir F, Fariduddin Q (2019) Polyamines (spermidine and putrescine) mitigate the adverse effects of manganese induced toxicity through improved antioxidant system and photosynthetic attributes in Brassica juncea. Chemosphere 236:124830

Iglesia-Turiño S, Febrero A, Jauregui O, Caldelas C, Araus JL, Bort J (2006) Detection and quantification of unbound phytochelatin 2 in plant extracts of Brassica napus grown with different levels of mercury. Plant Physiol 142:742–749

Jabeen R, Ahmad A, Iqbal M (2009) Phytoremediation of heavy metals: physiological and molecular mechanisms. Bot Rev 75:339–364

Kanwar MK, Bhardwaj R (2015) Arsenic induced modulation of antioxidative defense system and brassinosteroids in Brassica juncea L. Ecotoxicol Environ Saf 115:119–125

Karuppanapandian T, Kim W (2013) Cobalt-induced oxidative stress causes growth inhibition associated with enhanced lipid peroxidation and activates antioxidant responses in Indian mustard (Brassica juncea L.) leaves. Acta Physiol Plant 35:2429–2443

Kerkeb L, Kramer U (2003) The role of free histidine in xylem loading of nickel in Alyssum lesbiacum and Brassica juncea. Plant Physiol 131:716–724

Khan MIR, Jahan B, Alajmi MF, Rehman MT, Khan NA (2019) Exogenously-sourced ethylene modulates defense mechanisms and promotes tolerance to zinc stress in mustard (Brassica juncea L.). Plants 8:540

Kohli SK, Handa N, Sharma A, Gautam V, Arora S, Bhardwaj R, Alyemeni MN, Wijaya L, Ahmad P (2018) Combined effect of 24-epibrassinolide and salicylic acid mitigates lead (Pb) toxicity by modulating various metabolites in Brassica juncea L. seedlings. Protoplasma 255:11–24

Kramer U, Talke IN, Hanikenne M (2007) Transition metal transport. FEBS Lett 581:2263–2272

Krzeslowska M (2011) The cell wall in plant cell response to trace metals: polysaccharide remodeling and its role in defense strategy. Acta Physiol Plant 33:35–51

Krzesłowska M, Lenartowska M, Mellerowicz EJ, Samardakiewicz S, Wozny A (2009) Pectinous cell wall thickenings formation—a response of moss protonemata cells to Pb. Environ Exp Bot 65:119–131

Lasat MM (2000) Phytoextraction of metals from contaminated soil: a review of plant/soil/metal interaction and assessment of pertinent agronomic issues. J Hazard Subst Res 2:1–25

Lee TJ, Luitel BP, Kang WH (2011) Growth and physiological response to manganese toxicity in Chinese cabbage (Brassica rapa L. ssp. campestris). Hortic Environ Biotechnol 52:252

Lehto NJ, Davison W, Zhang H, Tych W (2006) Theoretical comparison of how soil processes affect uptake of metals by diffusive gradient in thinfilms and plants. J Environ Qual 35:1903–1913

Li N, Xie W, Zhou X, Chai Y, Xu W (2018) Comparative effects on nutritional quality and selenium metabolism in two ecotypes of Brassica rapa exposed to selenite stress. Environ Exp Bot 150:222–231

Liang SX, Gao N, Li X, Xi X (2018) Toxic effects of antimony on the seed germination and seedlings accumulation in Raphanus sativus L. radish and Brassica napus L. Mol Biol Rep 45:2609–2614

Meng JG, Zhang XD, Tan SK, Zhao KX, Yang ZM (2017) Genome-wide identification of Cd-responsive NRAMP transporter genes and analyzing expression of NRAMP 1 mediated by miR167 in Brassica napus. Biometals 30:917–931

Mwamba TM, Li L, Gill RA, Islam F, Nawaz A, Ali B, Farooq MA, Lwalaba JL, Zhou W (2016) Differential subcellular distribution and chemical forms of cadmium and copper in Brassica napus. Ecotoxicol Environ Saf 134:239–249

Mwamba TM, Islam F, Ali B, Lwalaba JLW, Gill RA, Zhang F, Farooq MA, Ali S, Ulhassan Z, Huang Q, Zhou W, Wang J (2020) Comparative metabolomic responses of low- and high-cadmium accumulating genotypes reveal the cadmium adaptive mechanism in Brassica napus. Chemosphere 250:126308

Nagalakshmi N, Prasad MNV (2001) Responses of glutathione cycle enzymes and glutathione metabolism to copper stress in Scenedesmus bijugatus. Plant Sci 160:291–299

Nahar K, Hasanuzzaman M, Alam MM, Rahman A, Suzuki T, Fujita M (2016) Polyamine and nitric oxide crosstalk: antagonistic effects on cadmium toxicity in mung bean plants through upregulating the metal detoxification, antioxidant defense, and methylglyoxal detoxification systems. Ecotoxicol Environ Saf 126:245–255

Nahar K, Hasanuzzaman M, Suzuki T, Fujita M (2017) Polyamine-induced aluminium tolerance in mung bean: a study on antioxidant defense and methylglyoxal detoxification systems. Ecotoxicology 26:58–73

Nauairi I, Ammar WB, Youssef NB, Daoud DBM, Ghorbal MH, Zarrouk M (2006) Comparative study of cadmium effects on membrane lipid composition of Brassica juncea and Brassica napus leaves. Plant Sci 170:511–519

Oven M, Page JE, Zenk MH, Kutchan TM (2002) Molecular characterization of the homophytochelatins synthase of soybean Glycine max. J Biol Chem 277:4747–4754

Pandey C, Khan E, Panthri M, Tripathi RD, Gupta M (2016) Impact of silicon on Indian mustard (Brassica juncea L.) root traits by regulating growth parameters, cellular antioxidants and stress modulators under arsenic stress. Plant Physiol Biochem 104:216–225

Parashar A, Yusuf M, Fariduddin Q, Ahmad A (2014) Salicylic acid enhances antioxidant system in Brassica juncea grown under different levels of manganese. Int J Biol Macromol 70:551–558

Per TS, Masood A, Khan NA (2017) Nitric oxide improves S-assimilation and GSH production to prevent inhibitory effects of cadmium stress on photosynthesis in mustard (Brassica juncea L.). Nitric Oxide 68:111–124

Qian P, Sun R, Ali B, Gill RA, Xu L, Zhou W (2014) Effects of hydrogen sulfide on growth, antioxidative capacity, and ultrastructural changes in oilseed rape seedlings under aluminum toxicity. J Plant Growth Regul 33:526–538

Quelhadj A, Kaminski M, Mittag M, Humbeck K (2007) Receptor-like protein kinase HvLysMR1 of barley (Hordeum vulgare L.) is induced during leaf senescence and heavy metal stress. J Exp Bot 58:1381–1396

Raldugina GN, Maree M, Mattana M, Shumkova G, Mapelli S, Kholodova VP, Karpichev IV, Kuznetsov VV (2018) Expression of rice OsMyb4 transcription factor improves tolerance to copper or zinc in canola plants. Biol Plantarum 62:511–520

Rinklebe J, Shaheen SM (2014) Assessing the mobilization of cadmium, lead, and nickel using a seven-step sequential extraction technique in contaminated floodplain soil profiles along the Central Elbe River, Germany. Water Air Soil Poll 225:2039

Salt DE, Kato N, Kramer U, Smith RD, Raskin I (2000) The role of root exudates in nickel hyperaccumulation and tolerance in accumulator and nonaccumulator species of Thlaspi. In: Terry N, Banuelos GS (eds) Phytoremediation of contaminated soil and water. CRC Press LLC, Boca Raton, FL, USA, pp 191–202

Savić J, Marjanović-Jeromela A, Glamočlija Đ, Prodanović S, Savić J, Jeromela AM, Glamočlija Đ, Prodanović S (2013) Oilseed rape genotypes response to boron toxicity. Genetika 45:565–574

Shah AN, Tanveer M, Hussain S, Yang G (2016) Beryllium in the environment: Whether fatal for plant growth? Rev Environ Sci Biotechnol 15:549–561

Shahen SM, Kwon EE, Biswas JK, Tack FMG, Ok YS, Rinklebe J (2017) Arsenic, chromium, molybdenum, and selenium: geochemical fractions and potential mobilization in riverine soil profiles originating from Germany and Egypt. Chemosphere 180:453–463

Shahid M, Shamshad S, Rafiq M, Khalid S, Bibi I, Niazi NK, Dumat C, Rashid MI (2017) Chromium speciation, bioavailability, uptake, toxicity and detoxification in soil-plant system: a review. Chemosphere 178:513–533

Sharma SS, Dietz KJ (2006) The significance of amino acids and amino acid-derived molecules in plant responses and adaptation to heavy metal stress. J Exp Bot 57:711–726

Siddiqui MM, Abbasi BH, Ahmad N, Ali M, Mahmood T (2014) Toxic effects of heavy metals (Cd, Cr and Pb) on seed germination and growth and DPPH-scavenging activity in Brassica rapa var. turnip. Toxicol Ind Health 30:238–249

Singh AL, Jat RS, Chaudhari V, Bariya H, Sharma SJ (2010) Toxicities and tolerance of mineral elements boron, cobalt, molybdenum and nickel in crop plants. Plant Stress 4:31–56

Soudek P, Ursu M, Petrová Š, Vaněk T (2016) Improving crop tolerance to heavy metal stress by polyamine application. Food Chem 213:223–229

Tanveer M, Wang L (2019) Potential targets to reduce beryllium toxicity in plants: a review. Plant Physiol Bioch 139:691–696

Tausz M, Sircelj H, Grill D (2004) The glutathione system as a stress marker in plant ecophysiology: is a stress-response concept valid? J Exp Bot 55:1955–1962

Toth G, Hermann T, Szatmari G, Pasztor L (2016) Maps of heavy metals in the soils of the European Union and proposed priority areas for detailed assessment. Sci Total Environ 565:1054–1062

ul Hassan Z, Ali S, Rizwan M, Hussain A, Akbar Z, Rasool N, Abbas F (2017a) Role of zinc in alleviating heavy metal stress. In: Essential plant nutrients. Springer, Cham, pp 351–366

ul Hassan Z, Ali S, Rizwan M, Ali Q, Haider MZ, Adrees M, Hussain A (2017b) Role of Iron in Alleviating Heavy Metal Stress. In: Essential Plant Nutrients. Springer, Cham, pp 335–350

Ulhassan Z, Ali S, Gill RA, Mwamba TM, Abid M, Li L, Zhang N, Zhou W (2018) Comparative orchestrating response of four oilseed rape (*Brassica napus*) cultivars against the selenium stress as revealed by physio-chemical, ultrastructural and molecular profiling. Ecotoxicol Environ Saf 161:634–647

Ulhassan Z, Gill RA, Ali S, Mwamba TM, Ali B, Wang J, Huang Q, Aziz R, Zhou W (2019a) Dual behaviour of selenium: insights into physio-biochemical, anatomical and molecular analyses of four Brassica napus cultivars. Chemosphere 225:329–341

Ulhassan Z, Huang Q, Gill RA, Ali S, Mwamba TM, Ali B, Hina F, Zhou W (2019b) Protective mechanisms of melatonin against selenium toxicity in Brassica napus: insights into physiological traits, thiol biosynthesis, and antioxidant machinery. BMC Plant Biol 19:507

Ulhassan Z, Gill RA, Huang H, Ali S, Mwamba TM, Ali B, Huang Q, Hamid Y, Khan AR, Wang J, Zhou W (2019c) Selenium mitigates the chromium toxicity in Brassica napus L. by ameliorating nutrients uptake, amino acids metabolism and antioxidant defense system. Plant Physiol Biochem 145:142–152

Varkey M, Lal N, Khan ZH (2012) Phytoremediation: strategies to enhance the potential for toxic metal remediation of Brassica oilseed species. In: Anjum NA, Pereira ME, Ahmad I, Duarte AC, Umar S, Khan NA (eds) Phytotechnologies: remediation of environmental contaminants. Taylor and Francis, New York, pp 293–308

Wang C, Zhang SH, Wang PF, Hou J, Zhang WJ, Li W, Lin ZP (2009) The effect of excess Zn on mineral nutrition and antioxidative response in rapeseed seedlings. Chemosphere 75:1468–1476

Weis JS, Weis P (2004) Metal uptake, transport and release by wetland plants: implications for phytoremediation and restoration. Environ Int 30:685–700

Wu Z, Zhao X, Sun X, Tan Q, Tang Y, Nie Z, Hu C (2015) Xylem transport and gene expression play decisive roles in cadmium accumulation in shoots of two oilseed rape cultivars (Brassica napus). Chemosphere 119:1217–1223

Wu X, Zhu ZB, Chen JH, Huang YF, Liu ZL, Zou JW, Chen YH, Su NN, Cui J (2019) Transcriptome analysis revealed pivotal transporters involved in the reduction of cadmium accumulation in pak choi (Brassica chinensis L.) by exogenous hydrogen-rich water. Chemosphere 216:684–697

Xu J, Chai T, Zhang Y, Lang M, Han L (2009) The cation-efflux transporter BjCET2 mediates zinc and cadmium accumulation in Brassica juncea L. leaves. Plant Cell Rep 28:1235–1242

Yan L, Riaz M, Wu X, Wang Y, Du C, Jiang C (2018) Interaction of boron and aluminum on the physiological characteristics of rape (Brassica napus L.) seedlings. Acta Physiol Plant 40:33

Yang XE, Jin XF, Feng Y, Islam E (2005) Molecular mechanisms and genetic basis of heavy metal tolerance/hyperaccumulation in plants. J Integr Plant Biol 47:1025–1035

Zagorchev L, Seal CE, Kranner I, Odjakova M (2013) A central role for thiols in plant tolerance to abiotic stress. Int J Mol Sci 14:7405–7432

Zaheer IE, Ali S, Rizwan M, Farid M, Shakoor MB, Gill RA, Najeeb U, Iqbal N, Ahmad R (2015) Citric acid assisted phytoremediation of copper by Brassica napus L. Ecotoxicol Environ Saf 120:310–317

Zaid A, Mohammad F, Wani SH, Siddique KM (2019) Salicylic acid enhances nickel stress tolerance by up-regulating antioxidant defense and glyoxalase systems in mustard plants. Ecotoxicol Environ Saf 180:575–587

Zhang XD, Zhao KX, Yang ZM (2018) Identification of genomic ATP binding cassette (ABC) transporter genes and Cd-responsive ABCs in Brassica napus. Gene 664:139–151

Zhu W, Zuo R, Zhou R, Huang J, Tang M, Cheng X, Liu Y, Tong C, Xiang Y, Dong C, Liu S (2016) Vacuolar iron transporter BnMEB2 is involved in enhancing iron tolerance of Brassica napus. Front Plant Sci 7:1353

Zlobin IE, Kartashov AV, Shpakovski GV (2017) Different roles of glutathione in copper and zinc chelation in Brassica napus roots. Plant Physiol Bioch 118:333–341

Phytoremediation of Toxic Metals/Metalloids and Pollutants by Brassicaceae Plants

Neerja Srivastava

Abstract Contamination through heavy metals/metalloids as well as other pollutants is a major concern to the environment. There is a gradual increase in the concentration of these toxic metal/metalloids and pollutants in the soil because of human as well as natural actions. To remove these contaminants from soil, phytoremediation is a very efficient, cheap and environmental friendly approach. Recently, for phytoremediation of these metals/metalloids contaminated soil, a novel method of phytoextraction and phytomining with hyperaccumulator plants is gaining popularity. Of these hyperaccumulator plant species, Brassicaceae has shown potential for removing these pollutants completely and permanently. Members of this family like *T. caerulescens*, *A. murale*, *A. halleri*, *B. napus*, *B. nigra*, *R. sativus* are very efficient phytoremediators. Indian mustard (*B. juncea*), belonging to this family is known to have a very high capacity for phytoremediation.

Keywords Phytoremediations · Hyperaccumulators · Metals · Metalloids · Pollutants · Brassicaceae

1 Introduction

High amounts of toxic wastes are added in the environment due to very fast growth of human population as well as industrialization, which influences human health around the world (Zhuang et al. 2007; Hasanuzzaman et al. 2012; Hasanuzzaman and Fujita 2012; Nahar et al. 2016, 2017). Major pollutants in the environment including metals/metalloids and radionuclides with some additional inorganic elements cannot be remediated easily in soils and sediments (Cunningham et al. 1997; Prasad and Freitas 2003). The toxicity of these pollutants is a major issue because of which various useful methods to evaluate the existence and movement of metals in soil with water as well as wastewater are developed (Tangahu et al. 2011). Among them, phytoremediation is a cost-effective and ecofriendly technique which can remove or

N. Srivastava (✉)
Department of Biochemistry, IBSBT, CSJM University, Kanpur, UP, India
e-mail: neerjas70@gmail.com

M. Hasanuzzaman (ed.), *The Plant Family Brassicaceae*,
https://doi.org/10.1007/978-981-15-6345-4_14

decrease toxic contaminants from various parts of environment (USEPA 2000; Sarma 2011; Tangahu et al. 2011). Hyperaccumulator plants are utilized for extracting, compartmentalizing and/or detoxifying pollutants in phytoremediation. Phytoremediation utilizes the capacity of plant root systems for being exclusive and choosy for intake as well as the capacity of hyperaccumulators for effectively translocating, bioaccumulating and degrading pollutants (Hinchman et al. 1996). According to Palmer et al. (2001), those plants which have widespread rooting absorbs and transports huge amount of pollutants preferably to the shoots which is ultimately removed by harvesting the biomass. This process is also called as phytoextraction and utilized for removing heavy metals present in the soil. Some of the heavy metals which have no biological function and are harmful like Cd, Cr, Pb, Co, Ag, Se and Hg are transferred along with necessary elements and stored in the plants (Cho-Ruk et al. 2006; Tangahu et al. 2011). These plants which are now having high concentration of the above said metals can be taken out from the fields and should be discarded. These stored metal elements then should be retrieved and possibly be recycled (Salt et al. 1998; Palmer et al. 2001). Various such phytoremediating plants species have been recognized, of which several are hyperaccumulators and competently store these harmful metals even 50–500 times more than normal plants (Baker et al. 1994a; Reeves and Baker 2000; Palmer et al. 2001; Sarma 2011; Tangahu et al. 2011; Mellem et al. 2012). These hyperaccumulators which have >1 ratio of metal concentration between root and shoot, moves metals very effectively from below ground to upper parts of the ground (Baker 1981; Gabbrielli et al. 1990; Homer et al. 1991; Baker et al. 1994a, b; Krämer et al. 1996; Shen et al. 1997; Zhao et al. 2000; McGrath et al. 2002; Alkorta et al. 2004). High bioconcentration factor is the plant's capacity for extracting metals present in the soil which together with effective root-to-shoot transfer as well as greater metal resistance increases the necessary detoxification capability of hyperaccumulator plants (Krämer 2010). The limit of particular metal which a plant can store without adversely affecting it decides the standard for hyperaccumulators. These standards of hyperaccumulation of a plant differ according to the metal. A plant is categorized as hyperaccumulator if it can store 1,000 μg Ni per gram of leaf dry mass, 10,000 μg Zn or Mn per gram of shoot dry mass, 100 μg Cd per gram of shoot dry matter and 1,000 μg Co/Cu/Pb or Se per gram of shoot dry matter (Brooks et al. 1977; Brooks 1998; Baker et al. 2000). Around 500 plant species have special capacity to store heavy metals at high concentrations (Krämer 2010) which are mainly angiosperms (Palmer et al. 2001; Baker and Whiting 2002; Krämer 2010; Sarma 2011). The largest group of hyperaccumulators belongs to families of Asteraceae, Brassicaceae, Caryophyllaceae, Cyperaceae, Cunoniaceae, Fabaceae, Flacourtiaceae, Lamiaceae, Poaceae, Violaceae as well as Euphorbiaceae (Prasad and Freitas 2003). About one fourth of all hyperaccumulators of angiosperms belongs to family Brassicaceae (Palmer et al. 2001; Krämer 2010; Sarma 2011) including 90 species of 11 genera (Prasad and Freitas 2003; Krämer 2010). The members of Brassicaceae have special capacity for storing and resisting large amount of variety of harmful metals (Kumar et al. 1995; Anjum et al. 2012).

Most of the metal hyperaccumulators of Brassicaceae known till now are hyperaccumulating Ni after Zn. While, *A. gemmifera* (Kubota and Takenaka 2003), *A. paniculata* (Küpper et al. 2000; Tang et al. 2009), *A. halleri* (Zhao et al. 2000; Yang et al. 2004), *N. caerulescens* (Basic et al. 2006; Küpper and Kochian 2010), *N. praecox* (Küpper et al. 2000; Yang et al. 2004; Vogel-Mikuŝ et al. 2005; Hassan and Aarts 2010; Maestri et al. 2010) and *N. goesingensis* (Lombi et al. 2000) of Brassicaceae are Cd hyperaccumulators, Nickle was hyperaccumulated in 72 species belonging to seven genera of family Brassicaceae (Reeves et al. 1996, 1999) and Zinc was accumulated in 20 species belonging to 3 genera (Delorme et al. 2001; Prasad 2005). Mostly, they store only specific metals as they are either stores of Nickle or Zinc, but few also store multiple heavy metals. These plants are suitable indicators of Nickle as well as of Zinc because of their limited allocation in site and indigenous to serpentine soils as well as particular kind of rocks (Brooks and Radford 1978; Baker and Brooks 1989; Raskin et al. 1997; Palmer et al. 2001; Alkorta et al. 2004). Higher amount of essential as well as non-essential heavy metals are present in Serpentine soils and species thriving in these soils have either established processes for removing the metals or storing them in higher amounts (Reeves et al. 1983; Palmer et al. 2001) which does not impact negatively on their growth In the family Brassicaceae, species like *Alyssum*, *Noccaea* and *Arabidopsis* are known as classical hyperaccumulators because of having the capacity for storing various heavy metals as well as metalloids in extra amount (Anjum et al. 2012). This heavy metal storage is reported in various members of Brassicaceae like *Alyssum* (Krämer et al. 1996; Sarma 2011), *Arabidopsis* sp. (Zhao et al. 2000; Chiang et al. 2006), *Brassica* sp. (Blaylock et al. 1997; Huang et al. 1997) and *Noccaea* sp. (Baker et al. 1994a, b) which are regarded as suitable phytoremediators (Dar et al. 2015).

2 Environmental Pollution and Sources of Contamination

Pollution in the environment is caused by the presence of elements in toxic concentration in terrestrial, water as well as air. When environments is contaminated by the discharge produced through intentional or unintentional anthropogenic activities, it is regarded as pollution. The environment is polluted by organic as well as inorganic contaminants which are discharged in the environment through various means. Uncontrollable population of the world together with industries pollutes land, waterbodies as well as air. Environment is very severely polluted through different metals like Cadmium, Copper, Cromium, Nickle, Zinc and lead (Ahmadpour et al. 2012).

3 Risks of Metals and Metalloids in Soils

Metals/metalloids like As and Se are very dangerous when they accumulate in soils because of various anthropogenic activities. Some of them like Zn, Cu, Mn, Ni, Se, Co, Cr, and Mo are necessary for living beings and their deficiency is either due to very low level of these metals in soil or due to low bioavailability because of soil chemical properties. In these situations, when metals are increased, they give positive biological effects represented through growth as well as health of organisms. However, those metals which are essential as well as non-essential one such as lead, arsenic, mercury and cadmium accumulate in soils which cause higher bioavailability of them and ultimately cause toxicity. They can have negative impact on soil microbes, soil fauna, higher animals, plants and even humans. Additionally, radionuclides like U, ^{137}Cs, ^{90}Sr and ^{3}H in soil and water are also dangerous (Negri and Hinchman 2000). In fact, these elements are naturally present at higher levels in soils and waters which affects biodiversity as well as animal and human health like evolution of metal resistant vegetation from metal-mineralized soils (Baker and Proctor 1990), the effect of high Se on human health (Yang et al. 1983), and Cd storage in the tissues of white-tailed ptarmigan (Lagopus leucurus) of Colorado Rocky Mountains which are harmful (Larison et al. 2000). In these situations, cleaning of soil is not always possible or required but plants can be used to decrease the exposure of biota for these elements either through reduction in uptake as well as removal from tissues, or removing Se from geogenically laden water (Ohlendorf et al. 1986; Wu et al. 1995). In fact, particular fauna and flora are present in these natural hot spots such as metallophyte vegetation which requires preservation (Reeves and Baker 2000). Metals and metalloids get entry in soils and waters through various mechanisms like atmospheric deposition by industries or by generation of power; removal of by-products like sludge of sewage, manures of animals, ash, by-products of household and industries, by irrigating and flooding or seeping water as well as through the use of fertilizers, lime or agrochemicals. Certain regions may accumulate radionuclides because of known or unintentional discharges associated with the utilization for energy production or for military reasons. Contrary to nitrate or chloride, various elements are comparatively powerfully bound to the surface of soils and do not easily percolate which results in the storage and cause danger to humans and biota. Although in certain situations some portion of these elements also percolate which is problematic for irrigation or drinking water. Main examples are radionuclides, As, Se, and Cr (Chiou et al. 1995; Kimbrough et al. 1999; Negri and Hinchman 2000; Ohlendorf et al. 1986). In these situations, phytoremediation is emerging as a tool for removing these elements from soil or water. It is cheap and may be practiced in those regions where other techniques prove costly as well as pollutants level is so low that other processes become ineffective or not economically viable (McGrath et al. 2002).

4 Remediation of Contamination

The remediation of soils polluted through toxic chemicals is very costly and technically complicated process. Traditional amendments at the site or the outside of the site depend on different techniques like

(1) **Leaching**—contaminant are leached through draining along with water or by chelate. This leachate is retrieved and remediated on the spot or outside of site.
(2) **Solidification** or stabilization—It is done either through physical insertion or chemical relations of stabilizer with contaminants.
(3) **Vitrification**—It utilizes thermal energy in soil fusion permitting physical or chemical stabilization.
(4) **Electrokinetical treatment**—In this contaminant ions move towards electrodes implanted in the soil.
(5) **Chemical oxidation** or **reduction**—contaminants are chemically oxidized or reduced to get less harmful chemicals having more stability and low mobility.
(6) "**dig and dump**" or **Excavation**—treatment out of site or accumulation at more suitable place (Saxena et al. 1999).

These methods are costly as well as confined in comparatively small places in majority of the cases. These procedural problems along with better information of processes like intake, movement, resistance and elimination of heavy metals as well as some other probable harmful pollutants in microbes as well as plants, in recent times encouraged the development of novel technique known as bioremediation. Bioremediation is dependent on living organisms' capacity particularly microorganisms as well as plants for remediation of environment (Anderson and Coats 1994). People are aware about the capability of plants to remediate the environment from the eighteenth century when Joseph Priestley, Antoine Lavoissier, Karl Scheele and Jan Ingenhousz have shown through experiments that atmosphere is detoxified through plants under light. The value of green areas for conservation of quality of air is commonly acknowledged but all times are not appreciated. Recently, plants are utilized for remediating the polluted soils as well as water. During 1970s, remediation proposals of mining sites produced methods for layering soil with vegetation for stability as well as for visual appeal (Williamson and Johnson 1981). Before 1990s people never thought about phytoremediation as a novel technique which implements plants for remediation or reducing the damage to soil, surface as well as wastewaters polluted through metals, organic xenobiotics and explosives/radionuclides (Baker et al. 1994a; Chaney 1983; Chaney et al. 1997; Cunningham et al. 1995; Ernst 1995; Lasat et al. 1998; Macek et al. 2000; McGrath 1998; Raskin et al. 1997; Salt et al. 1995; Barceló and Poschenrieder 2003).

5 Phytoremediation and Its Mechanism

Phytoremediation word has been taken from Greek word phyto meaning plant and Latin word remedium meaning correcting or eliminating an evil (Cunningham et al. 1996). Phytoremediation is the process where remediation is done on the site itself and it exploits the characteristic potential of plants which are alive. This clean-up method is ecofriendly and driven by solar energy and depends upon the idea of exploiting nature for purification of environment. Plants may assist in remediating various types of contaminants such as metals, pesticides, explosives as well as oil. The plants assist in averting wind, rain, as well as groundwater from taking contaminants away from polluted region to other places. The procedure and competence of phytoremediation is based on the pollutant species, bioavailability as well as soil characteristics (Cunningham and Ow 1996). Plants adapt various methods for detoxification or remediation of polluted region. The intake of pollutants in plants takes place mainly by root system which has a major process for foiling toxicity. The root system gives big surface area for absorbing and storing water along with nutrients needed for growth, as well as those pollutants which are unrequired (Raskin et al. 1997; Mahajan et al. 2016).

6 Phytoremediation Strategies

Phytoremediation involves various modalities which are based on the chemical identity as well as characteristics of the pollutants like being inert, unstable or degradable inside the plant or soil along with plant species. Therefore phytoremediation basically has six diverse approaches and may be more than one is used by the plant at the same time.

6.1 Phytodegradation or Phytotransformation

Metabolization or mineralization of the organic pollutants in the plant cells is done through particular enzymes. While nitroreductases degrade nitroaromatic compounds, dehalogenases metabolize chlorinated solvents with pesticides, laccases degrade anilines. These enzymes are present in *Populus* species as well as *M. spicatum* (Schnoor et al. 1995; Rylott and Bruce 2008).

6.2 Phytostabilization or Phytoimmobilization

It is observed that organic/inorganic pollutants bind to the humus and lignin of the root cell wall. Root exudates directly function on metals and precipitate them in insoluble forms which ultimately get captured into soil matrix. The prime focus here is stoppage of pollutants movement as well as restriction of their dispersal in soil (Prasad 2004; Berti and Cunningham 2000; Domínguez et al. 2009; Ali et al. 2013). Some of the genera whose species are grown for this purpose are *Haumaniastrum*, *Eragrostis*, *Ascolepis*, *Gladiolus* and *Alyssum*.

6.3 Phytovolatilization

This process depends upon the capacity of some plants for absorbing and volatilizing particular metals and metalloids. Roots absorb certain element ions present in groups IIB, VA and VIA of the periodic table such as Hg, Se and As to change them into harmless forms and release them in the environment. *Astragalus bisulcatus* and *Stanleya pinnata* absorb Se while transgenic plants having bacterial genes of *Arabidopsis thaliana, Nicotiana tabacum, Liriodendron tulipifera* or *B. napus* absorb Hg (Ali et al. 2013; Brooks 1998; Pilon-Smits and Pilon 2000; Poschenrieder and Barceló 2004; Ruiz and Daniell 2009; Pilon-Smits and Leduc 2009). This process can also be utilized for organic pollutants.

6.4 Phytoextraction or Phytoaccumulation or Phytoabsorption or Phytosequestration

Aerial organs translocate and store the pollutants which are absorbed through roots in the process of phytoextraction. This is basically used for metals like cadmium, nickle, copper, zinc and lead, but may also be extended to additional elements like selenium and arsenic as well as to organic compounds. This process has preference for those hyperaccumulators which stores higher levels of some metals in the aerial regions which can be 0.01–1% of dry weight on the basis of particular metal. Certain hyperaccumulator plants like *E. splendens, A. bertolonii, T. caerulescens* and *P. vittata* accumulate copper, nickle, zinc/cadmium and arsenic, respectively (Prasad 2004; McGrath 1998; Blaylock and Huang 2000; Ma et al. 2001; McGrath and Zhao 2003; Hernández-Allica et al. 2008; Pedron et al. 2009; Xie et al. 2009; Van der Ent et al. 2013).

6.5 Phytofiltration

This process utilizes plants for absorbing, concentrating and/or precipitating pollutants, especially heavy metals as well as radioactive substances from water by its roots or other underwater parts. The plants are grown hydroponically through which effluents are passed and filtered by roots which is called Rhizofiltration, or through those organs which absorb as well as concentrate contaminants (Ali et al. 2013; Frers 2009; Dhote and Dixit 2009). Aquatic hyperaccumulators plants having more root biomass, or larger absorption surface and greater storage ability as well as more resistance power provide best results like *H. annus, B. juncea, P. australis, F. antipyretica* and various other species of *Salix, Populus, Lemna* and *Callitriche* (Prasad 2004; Poschenrieder and Barceló 2004; Dushenkov and Kapulnik 2000; Pratas et al. 2012; Favas et al. 2012).

6.6 Rhizodegradation or Phytostimulation

Growth of the roots stimulates the propagation of those microorganisms which degrade rhizosphere and use exudates as well as metabolites of plants as carbon and energy source. Additionally, these plants also secrete enzymes for biodegradation. The use of phytostimulation is restricted for organic pollutants only (Prasad 2004; Frers 2009). The microbial population present on rhizosphere is mixed because nutrients have inconsistent spatial distribution, although species of the genus *Pseudomonas* are main organisms linked with roots (Ali et al. 2013; Crowley et al. 1997; Khan et al. 2009). Some authors also categorized additional approaches under phytoremediation, although they are combinations of these different approaches or deviated from them and are mentioned below.

6.7 Hydraulic Barriers

Certain big plants especially with profound roots like *Populus* sp., eliminate groundwater in huge amounts through transpiration. Plant enzymes process the pollutants present in the groundwater which are vaporized along with water or compartmentalized in simple manner in tissues of the plants (Prasad 2004; Schnoor 2000).

6.8 Vegetation Covers

To reduce the rain water penetration as well as for restricting contaminants extension, generally landfills/tailings grasses like herbs as well as shrubs or trees are employed. Biodegradation, evaporation as well as transpiration rises with aeration in the soil which is increasing because of roots (Mendez and Maier 2008; Williamson et al. 1982; Brooks et al. 1998; Jorba and Vallejo 2008). The problem here is that tailings usually are inappropriate in developing roots. But several studies focussed on developing the procedure for cultivating tailings like Hungarian agronomists used organic soil made from sawdust, plant remnants and certain NPK fertilizers which were placed on the surface (González 1990). It was observed that solitary biological cycle can produce 76 diverse species of plants comprising cereals, shrubs, fruit trees and also big plants such as oaks as well as pines.

6.9 *Constructed Wetlands*

Those ecosystems where water level exists at or near the surface at least for certain time in the year, possess organic soils, microbes, algae as well as those aquatic plants which are vascular. Different parts are joined together for effluent treatment by the process of filtration, ion exchange, adsorption and precipitation (Frers 2009; Vymazal 2009; Fonder and Headley 2013). It is the foremost process for treating wastewater which is even not exactly phytoremediation as it is depending upon the participation of the complete scheme (Prasad 2004; Horne 2000). The major advantages are better cleaning proficiency, cost-effective construction with simple operation and maintenance. It is extensively used for treating domestic, agricultural as well as industrial wastewater, but is also fit to treat acid mine drainages (Sobolewski 1999; López Pamo et al. 2002; Nyquist and Greger 2009; Adams et al. 2013).

6.10 *Phytodesalination*

It is a newly discovered (Ali et al. 2013; Zorrig et al. 2012) developing procedure to treat saline soils which exploits halophytes for eliminating extra salts. *S. maritima* and *S. portulacastrum* are capable of eliminating as well as accumulating NaCl from the extremely saline soils (Ravindran et al. 2007). While it has its own distinctiveness, this process is actually a type of phytoextraction (Favas et al. 2014).

7 Plants Selections Criteria for Phytoremediation

For phytoremediation, selection of plants is a very important matter. One of the remarkable candidates is grasses as they possess fibrous roots which provide stability to soil as well as big surface area for interaction of root with soil (Kulakow et al. 2000). Probably, very prominent issue for successful phytoremediation approach is the selection of plants. The utilization of local plant species is usually more favourable for phytoremediation because they need less effort and adapts effectively in local environmental conditions as well as seasonal cycle. Although, certain rare plant species act better in remediating particular metals and can be utilized securely when their possible offensive behaviour has been (USEPA 2000). Certain major factors for choosing plant species for phytoremediation are: (1) The resistance limit for the metal present on the site. (2) The limit for acceptable storage, translocation and metal intake capacity (3) better growth with production of biomass (4) resistance to flooding and severe water stress environments, (5) Accessibility and choice for living place like land, water and semi-aquatic, etc., (6) resistance to basicity as well as salinity (Sarma 2011).

7.1 Metal Hyperaccumulator Plants

About four hundred species of forty families hyperaccumulate metals (Brooks 1998). Hyperaccumulators are defined as those herbaceous or woody plants which store and resist about hundred times more amount of metals in shoots in comparison to those which are not hyperaccumulators, with no noticeable signs. Baker and Brooks (1989) reported that the lowest threshold limit in the tissues for cobalt, copper, chromium, lead or nickle hyperaccumulators plants is 0.1%, but is 1% for zinc and manganese (Baker et al. 2000). Hyperaccumulators are metallophytes and native of metal rich soils (Ernst 2000a, b; Pollard et al. 2000). They have developed inner processes which permit these hyperaccumulators for taking as well as resisting high amount of metals which is highly dangerous for other organisms (Clemens 2001; Lasat 2002). These plants very efficiently adjusted to the specific climatic conditions of their surroundings where storage of high levels of metal may defend them against herbivores as well as infections caused by fungus (Boyd and Martens 1998; Martens and Boyd. 2002; Tolrà et al. 2001). But generally, they are unable to fight effectively on unadulterated soil along with non-metallophytes because of metabolic as well as energetic expenses of adjustment strategies. Development of metal hyperaccumulation is global phenomena and places like New Caledonia, Australia, Central and South Europe, the Mediterranean Area, South-East Asia, Cuba, Dominican Republic, California, Zimbabwe, Transvaal in South Africa, Goiàs in Brazil, Hokkaido in Japan and Newfoundland in Canada are prominent areas for accumulating germplasm (Baker and Brooks 1989; Ernst 1974). Several theories are given to elucidate the process of metal hyperaccumulation as well as the evolutionary benefit of this approach which are as follows.

7.2 Complex Formation and Compartmentation

Chelators are produced by hyperaccumulators to degrade metal ions through complex formation. The complex of an organic compound with metal is solvable, not much dangerous, and is transferred in cell compartments, has reduced metabolic activity and deposited in cell wall and vacuole as an unchanging organic/inorganic substance (Barceló et al. 2003; Barceló and Poschenrieder 1999, 2002; Ernst et al. 1992; Hall 2002; Tolrà et al. 1996; Vázquez et al. 1992, 1994).

7.3 Deposition Hypothesis

Hyperaccumulators withdraw metals through the root and store in old abscised leaves, epidermis, hairs which percolated through rain or burnt.

7.4 *Inadverted Uptake*

Metal hyperaccumulation is considered as the result of an adjustment process for other unfavourable soil properties like serpentinophytes hyperaccumulates Ni.

8 Hyperaccumulation as a Protection Mechanism Against Abiotic or Biotic Stress Conditions

Metals have great effects on some pathogens like fungi and bacteria as well as leaf-eating herbivores (Boyd 1998). But, parasites of phloem (Ernst et al. 1990) are not affected, possibly due to less movement of the metals in the phloem. Greater amount of metal inside leaves may curb feeding or later on absorption is going to decrease breeding rate in herbivores or may prove toxic for them. Metal hyperaccumulation replaces organic resistances and is more beneficial (Robinson et al. 1997). It was proposed that greater leaf metal levels are for osmotic adaptation in drought stress (Baker and Walker 1989; Poschenrieder and Barceló 1999). This hypothesis was proved by different studies done in *A. murale* which hyperaccumulates Ni and *T. caerulescens* which hyperaccumulates Zn (Whitting et al. 2003; Barceló and Poschenrieder 2003).

9 Role of Brassicaceae in Phytoremediation

Brassicaceae is grouped under order Brassicales with other fifteen families. Other significant families which are placed under this order are Capparaceae (sometimes grouped in subfamily Capparoideae of Brassicaceae), Tropaeolaceae with Caricaceae (recognized for papaya) which possess fascinating characters of pharmacology (Breithaupt et al. 2003; Nayak et al. 2007; Okeniyi et al. 2007; Amazu et al. 2009; Abdullah et al. 2011). Mustard family Brassicaceae generally have herbs which are hermaphrodite, occasionally shrubs which consist of simple lobbed/divided, spiral as well as exstipulate leaves. Flowers are bisexual, cross-shaped, generally actinomorphic with racemic inflorescence and dichlamydeous perianth. Androecium has two small stamens outside, while four are inside. It has syncarpous gynoecium and possesses 2 carpels with silique/silicle fruit (Ehrendorfer and Neuffer 2006). Brassicaceae family has been classified into 338 genera and 3,709 species which is very complex (Crespo et al. 2000; Marhold et al. 2004; Johnston et al. 2005; Bremer et al. 2009; Haston et al. 2009; Huang et al. 2010; Qiu et al. 2010; Al-Shehbaz et al. 2006; Bailey et al. 2006; Koch and Mummenhoff 2006; German et al. 2009; Endress 2011; Ge et al. 2011). This family is found to be distributed globally. There are various species which are economically valuable like vegetable plants, are utilized for production dyes and oil, etc. Many ornamental plants are also placed in this family. The best family member is Arabidopsis thaliana which is like a model plant in molecular biology. The Brassicaceae have several species which hyperaccumulates lNi in Thlaspi and Alyssum genera, Cd as well as Zn in *T. caerulescens, T. praecox, T. goesingense* and *A. halleri*. One-fourth of the hyperaccumulators are Brassicaceae members (Rascio and Navari-Izzo 2011; Marques et al. 2004; Doran and Nedelkoska 1999; Babula et al. 2012).

Various spill regions as well as industrial effluents dumping areas are polluted jointly by several metals. They pose very big threat to the atmosphere and health of human beings for which maintainable reclamation approaches are immediately required. Phytoremediation seems to be cheap and ecofriendly choice for remediating these dangerous metals/metalloids. It is the most significant task to enhance phytoremediation effectively by increasing metals storage in plants, or through improvement in crucial plant biological characters that should increase the intake of metal (Wu and Tang 2009). Storage of these metals in various plants species is targeted such as *Thlaspi* sp. (Baker et al. 1994b), *Brassica* sp. (Blaylock et al. 1997; Huang et al. 1997) and Alyssum (Krämer et al. 1996). At present more than 500 species of metal hyperaccumulators are known and their number is rising gradually. Some of them are very particular for metal, have less biomass, slow development and need careful breeding management (Gleba et al. 1999) and appropriate to use economically. Therefore, recognition of newer species of plant which has greater biomass production linked to capacity for resisting and storing numerous metals and is major part of phytoremediation studies. Scented geranium (*Pelargonium* sp.), a metal hyperaccumulator with various metals, has been discovered which resists and stores various types of metals like cadmium, nickle and lead as well as preserve usual mechanism of metabolism (Dan et al. 2000; KrishnaRaj et al. 2000). Evidence for the model plants came through the discovery of perennial shrub *S. drummondii* present in southern coastal region of USA which accumulates lead and produces more biomass (Sahi et al. 2002). The metals bioaccumulation capacity of different species has been recognized globally by various scientists during the past years. By studying the physiology of *V. zizanioides*, it has been proved that it can be used in different phytoremediation applications (Danh et al. 2009). This plant has the capacity to store numerous heavy metals in roots and shoots which can be measured experimentally. Gupta et al. (2008) investigated numerous metal storage as well as metal transfer factors (TFS) in the soil of wild macrophyte species *Ipomea* sp., *Eclipta* sp. and *Marsilea* sp. It was found that *Ipomea* sp. have transfer factors (TFS) for cadmium, Copper, Manganese as well as Zinc, while *Eclipta* sp. as well as *Marsilea* sp. have transfer factors $\geq$1 for iron, copper as well as Cadmium. Metals concentration in soil as well as in plant portions (TFS) is a very prominent factor for choosing model plant species for phytoremediation, while proportion of more than one reflects that plant parts store more metals than soil (Barman et al. 2000). Among the plants which are aquatic, *E. densa* exhibited higher storage of Hg inside leaves, shoots as well as roots taken up through natural soil rich in CH_3HgCl (Ochiai 1987) that provides it the capacity to remediate mercury (Sarma 2011).

9.1 Metal Tolerance in the Brassicaceae

Physiological and genetic processes for metal resistance and hyperaccumulation and their ecological consequences can be investigated through well documented 93 metal-hyperaccumulating species of Brassicaceae. In this family, certain most investigated genera for hyperaccumulation are *Arabidopsis*, *Brassica*, *Alyssum*, *Noccaea* (earlier known as Thlaspi), Stanleya, with Streptanthus (Bhargava et al. 2012; Boyd et al. 2009; Freeman et al. 2010; Vamerali et al. 2010; Verbruggen et al. 2009). *Noccaea caerulescens* from Brassicaceae hyperaccumulates metals and possibly the best studied plant which hyperaccumulates (Milner and Kochian 2008). It stores about 36,900 μg per gram of zinc with 1800 μg per gram of cadmium with no symptoms of toxicity (Bhargava et al. 2012). *N. caerulescens* is easily developed in the lab like several other Brassicaceae model taxa and investigated very extensively demonstrating various processes of metal intake, transportation, as well as localization (Cosio et al. 2004). But, because of low biomass its application is restricted in phytoremediation purpose (Bhargava et al. 2012). While metals are not stored by *A. thaliana* in nature, this is a common model plant in the studies which are dependent on plants (Bevan and Walsh 2005). Mapping the genome of *A. thaliana* (Weigel and Mott 2009) revealed similarity in sequence with metal-storing analogue *A. halleri* (Becher et al. 2004; Meyer and Verbruggen 2012; Weber et al. 2004). Therefore *A. thaliana* and *A. halleri* are generally utilized for investigating the role of genetics in metal tolerance as well as hyperaccumulation (Bevan and Walsh 2005; Cho et al. 2003; Chaffai and Koyama 2011; Courbot et al. 2007; Hanikenne et al. 2008). *B. juncea* i.e. Indian mustard is used very commonly for phytoremediation which is a spice crop of North America as well as Europe, (Lim et al. 2004; Neilson and Rajakaruna 2012). Though, it is non-hyperaccumulator but has capacity for storing Cadmium, Zinc, Selenium and lead and possesses a biomass which is almost 10 times higher in comparison to *N. caerulescens*, due to which it is successfully implemented in many phytoremediation investigations as well as tests (Bhargava et al. 2012; Szczyglowska et al. 2011; Warwick 2011). The molecular basis behind selenium resistance as well as hyperaccumulation was studied in *S. pinnata* by comparing to Se-resistant analogue, *S. albescens*, through the joint physiological, structural, genomic, as well as biochemical strategies (Freeman et al. 2010). Furthermore, the environmental roles of Stanleya and certain other plants (El Mehdawi and Pilon-Smits 2012) with its uses in hyperaccumulation of selenium as well as its capacity for phytoremediation of selenium were also studied (Banuelos 2001).

Brassicaceae family has several hyperaccumulators. Previously, people thought these hyperaccumulators were unfit to phytoextraction and investigators searched for more members of the Brassicaceae with greater biomass which can store higher amounts of harmful metals (Dushenkov et al. 1995; Kumar et al. 1995). Kumar et al. (1995) verified several rapidly developing Brassicas for their potential to resist and store metals such as *B. juncea*, *B. nigra* Koch, *B. campestris* L., *B. napus* L. as well as *B. oleracea* L. Though every Brassicas store various metals, *B. juncea* has high capacity for storage as well as translocation of Copper, Chromium VI, Cadmium,

Nickle, Lead and Zinc in the stems. Kumar et al. (1995) have similarly studied prospects of genetic variability in various *B. juncea* accessions in expectation to get a few varieties with greater phytoextraction capacity in comparison to others. *B. juncea* is an oilseed crop cultivated in India to western Egypt, from central Asia to Europe (Nishi 1980). Prakash (1980) reported that the *B. juncea* was first described in Sanskrit as "Rajika". Its carbonized seeds of which were recovered from the prehistoric locations of the Harappa civilization (2300–1750 B.C.). Regardless of historians and researchers hard work, the exact source for this crop is a mystery. There may be the highest possibility that it is originated from the areas where *B. nigra* as well as *B. campestris* its parents are jointly distributed. Most probably, it has originated from African (Zeven and Zhukovsky 1975), Chinese (Chen et al. 1995), the Middle Eastern, Southwest Asian and Indian region (Sauer 1993). Chinese use it like leaf vegetable while Indians cultivate it mainly to produce oil from its seeds (Prakash 1980) as to use like cooking oil as well as spice (Nishi 1980; Krzymaski 1997). In one hectare Indian mustard generates eighteen tons of biomass per crop (Kumar et al. 1995). Plants grown on nutrient solution culture give better performance as they produce full plants from the four-leaf phase that produce approximately fifty gram of fresh mass of shoots after twenty one days. Though in short-day environments (less than 12 h light) there is higher compact growth, less height as well as restricted production of leaf (Bhaskar and Vora 1994). While, storage of biomass is more in comparison to long-day environments i.e. in 9–10 h of optimum light (Neelam et al. 1994). Long-day environment increase premature formation of flower (Bhaskar and Vora 1994) and unrequired in flower growth. The plants have uncertain development and have continuous branching through the nodes for storing biomass following the development of flower as well as seed pod. The prescribed rate of fertility to get maximal growth in *B. juncea* under non-polluted environment is 75–120 kg N/hectare and 30–50 kg phosphorus pentaoxide per hectare (Gurjar and Chauhan 1997; Thakral et al. 1995; Tomar et al. 1997). According to Zaurov et al. (1999) *B. juncea* stored highest biomass in those plants which were grown on soil containing two hundred kilogram Nitrogen, hundred kilogram of phosphate pentaoxide and sixty six kilogram of potassium oxide per hectare. Although, tissues have highest amount of cadmium in the absence of Nitrogen. Indian mustard has got substantial consideration by recent workers due to its special genome which is polyploid. *B. nigra* and *B. campestris* are diploids which produce *B. juncea* an allotetraploid. In recent propagation plans, selection of *B. juncea* is depended upon various properties. Some plant breeders have specific purpose to enhance the quality of oil as well as metal through removal of nutritionally unwanted erucic acid or through alteration in fatty acid component of oil (Banga 1997). Further choices depend upon insect resistance like diamondback moth *Plutella xylostella L.* (Andrahennadi and Gillott 1998), disease tolerance like blackleg *Leptosphaeria maculans* (Desm.) Ces. (Pang and Halloran 1996) or different temperature adjustments (Banga 1997). In recent times, interest in choosing Indian mustard lines has been increased which depends upon its capacity for resisting and storing heavy metals. Many accessions of *B. juncea's* varieties accumulate metals modestly which have been preserved by the USDA-ARS, Plant Introduction Station in Iowa State University, Ames, Iowa. The advantage in utilizing seeds of *B. juncea* from plant introduction station is in maintaining genetic integrity of the varieties by proper breeding methods. Experiments which use these seeds are more accurate in comparison to those which are performed through commercially available seeds. It has more accuracy as future scientists can get the identical accessions to utilize in tests. USDA-ARS Plant Introduction Station preserves the global assembly of accessions of identified metal accumulators of *B. juncea* as well as distribute seeds free of cost for public as well as private research organization (Pantola and Alam 2014).

9.2 Phytoremediation by **B. juneca**

Phytoremediation technique depends upon implementing metal hyperaccumulators to eliminate harmful elements together with radioactive substances in soil as well as aquatic system. In recent times, Phytoremediation is the topic of strong public and scientific concern and a subject of various current studies. This has resulted in the search of plants for phytoremediation purpose with characteristics given below:

(i) resistance to the higher amount of metals in soils; (ii) rapid development and greater biomass accumulation (iii) potential to store heavy metals in the aerial portions (iv) simple to develop agricultural crop and easy to harvest (Ghosh and Singh 2005). Therefore, on the basis of level of heavy metal ion concentration of the polluted soil as well as the target points needed to treat soil, phytoextraction may consists of repeated cycles of crops till the amount of metal becomes tolerable (Vijayarengan 2005). Despite the low cost as well as being environment friendly, phytoremediation has been applied in fields only in developed countries. It is still not a commercially accessible procedure in most of the developing countries. Though, there are more than 400 taxa of heavy metal hyperaccumulators known, mostly are unusual species and produces less biomass. Therefore, list of plants available for phytoremediation should be enhanced. Edible crops like Mustard, maize, sorghum and alfalfa have greater capacity for heavy metal storage (Hemingway 1976). In fact, the need of irrigation water in fields is fulfilled through the utilization of wastewaters produced by industries but it has several pollutants like heavy metals. Recently, certain reports are published with the risk of utilizing remedied discarded water from industries to irrigate because the crops like *B. juneca* are consumed by humans (Axelsson et al. 2000). The investigation was done for assessing the capacity of *Brassica* for phytoextracting heavy metals from moderately polluted soils.

Brassica is placed under Cruciferae i.e. Brassicaceae family, generally known by the name of mustard family. Term crucifer was derived by the shape of flower which possess 4 petals arranged diagonally opposite like cross. *B. juncea* consists of light green leaves having first leaves as well as leaf blades with little hairs which ends at the petiole. Mature *B. juncea* plants develop up to a length of one to two meters. The lower leaves have deep lobes, however possess thin as well as complete upper leaves. *B. juncea* does not have similarity with the close relatives *B. napus* as well as *B. rapa* because upper leaves in *B. juncea* does not embrace. It has elongated racemic inflorescence with light yellow flowers which gradually up from raceme base. The seed pods are somewhat appressed as well as 2.5–5 cm long without counting beak. The beak is 0.5–1 cm in length with yellow and brown round seeds. *B. juncea's* vegetable as well as oilseed varieties probably have varied origins (Miguel et al. 2015). Both forms are regarded as natural amphidiploids having AABB genome, $2n = 36$ from *B. rapa* with AA genome, $2n = 20$ and through *B. nigra* with BB genome, $2n = 16$, crosses. Axelsson et al. (2000) have demonstrated conserved genomes of the ancestor species in *B. juncea* through molecular studies. Investigations in *Brassica* species for phytoremediation of heavy metals have mainly targeted Cadmium, Copper, Zinc and lead. *B. juncea* is a hyperaccumulator which is used in phytoremediation because of which most investigations targeted this particular species. *Brassica* generally performs well in pots under controlled environment in most of the studies, but few also reported its performance in field conditions. Some of them have performed well hydroponically while few in heavy metals polluted soil while others in normal polluted soils (Singh and Prasad 2014). Mhalappa et al. (2013) reported an increase in storage of metal in plants as the starting concentration of metal rises. According to them *Brassica* posses maximum TF value for Cr next to Ni and Cd. According to Singh et al. (2014) Arsenic can be effectively remediated at all levels through more biomass producing *B. juncea* in hydroponic solution. This also resulted in higher storage of Arsenic in biomass with the rise in each selected level and time concept of Arsenic in the roots and shoots of *B. juncea* (Ariyakanon and Winaipanich 2006). Comparable results were observed in the studies where *B. juncea* was utilized in removing copper from soil (Mellem et al. 2009). Almost similar outcomes were also observed in the storage design of heavy metals (Sharma 2016). Studies were done for eliminating heavy metals from the solution containing 5, 10, 20 and 50 ppm, respectively. Roots of *B. juncea* stored arsenic about 3045 Âμg per gram and 26,650 Âμg per gram, respectively, while shoots stored 1075 Âμg per gm and 5462 Âμg per gram of the stem dry biomass at 5–50 ppm. 4165 and 20,850 μg per gram of dry weight Arsenic are stored in roots at 10 and 20 ppm, respectively. Kathal et al. (2016) have reported effective uptake of Lead and Nickle through *B. juncea* in contaminated soil of Delhi University nursery confirming it to be a hyperaccumulator plant. Their investigation demonstrated that by rotating *B. juncea* and food crops toxic metals biomagnification can be regulated inside the food chain.

10 Conclusion

Industrial development and human activities are responsible for producing huge amount of non-biodegradable compounds. These harmful compounds are the cause for public concern. They are divided in organic as well as inorganic compounds. Highly inorganic compounds consisting of heavy metals, like Pb, Zn, Cd, Cu, Ni, Cr with radionuclides like U have widespread occurrence. Mostly are discharged in the surroundings which result in serious threat to the ecosystem and cause a negative impact on human health because of non-biodegradable nature as well as due to inevitable bioaccumulation. In majority of situations, the single method to remediate the environment is to eliminate the contaminants. Therefore, earlier procedures utilized for remediating the surroundings depend upon the physico-chemical processes such as volatilization, verification, excavation, soil washing, soil incineration, chemical extraction, solidification and landfills. Though all of them are effective for various situations, they are mostly costly, aggressive and are inappropriate to implement over big areas. Therefore, proposal of phytoextraction was put forward which ultimately created rapid interest in workers for making it as a common working mode. Following the current progress in scientific as well as technical information, people should be made techno-commercial and more aware about recent methodologies for increasing remediation capacity of the plants. The field application on big scale can be attained through development of pre-defined methodologies with benefits and limits of process as well as plant species utilized for treating heavy metal polluted locations. Scientists have tried for perfecting this remediation technique during the past. Most of the phytoextraction work has targeted in getting the perfect hyperaccumulators for storing metal as well as the method through which discharge of metals for the uptake of roots from the soil can take place. Currently, *B. juncea* with some other possible members of family Brassicaceae the most practical candidate to phytoextract different metals like Cadmium, Chromium (IV), Cesium, Copper, Nickle, lead, Uranium as well as Zn. Suitable plant nutrition is an effective solution for less costly agronomic exercise to increase heavy metals phytoextracion through members of this specific family of angiospermic flowering plants.

References

Abdullah M, Chai PS, Loh CY, Chong MY, Quay HW, Vidyadaran S, Seman Z, Kandiah M, Seow HF (2011) Carica papaya increases regulatory T cells and reduces IFN-gamma(+)CD4(+) T cells in healthy human subjects. Mol Nutr Food Res 55:803–806

Adams A, Raman A, Hodgkins D (2013) How do the plants used in phytoremediation in constructed wetlands, a sustainable remediation strategy, perform in heavy-metalcontaminated mine sites? Water Environ J 27(3):373–386

Ahmadpour P, Ahmadpour F, Mahmud TMM, Abdu A, Soleimani M, Tayefeh FH (2012) Afri J Biotechnol 11(76):14036–14043. http://www.academicjournals.org/AJB, https://doi.org/10.5897/ajb12.459. ISSN 1684-5315 © 2012 Academic Journals

Amazu LU, Ebong OO, Azikiwe CCA, Unekwe PC, Siminialayi MI, Nowsu PJC, Ezeani MC, Obidiya OS, Ajugwo AO (2009) Effects of the methanolic seeds extract of Carica Papaya on plasmodium Berghei infected mice. Asian Pac J Trop Med 2:1–6

Ali H, Khan E, Sajad MA (2013) Phytoremediation of heavy metals—Concepts and applications. Chemosphere 91:869–881

Alkorta I, Hernandez-Allica J, Becerril JM, Amezaga I, Albizu I, Garbisu C (2004) Recent findings on the phytoremediation of soils contaminated with environmentally toxic heavy metals and metalloids such as zinc, cadmium, lead, and arsenic. Rev Environ Sci Biotechnol 3:71–90

Al-Shehbaz IA, Beilstein MA, Kellogg EA (2006) Systematics and phylogeny of the Brassicaceae (Cruciferae): an overview. Plant Syst Evol 259:89–120

Anderson TA, Coats JR (1994) Bioremediation through rhizosphere technology. ACS symposium series, vol 563. American Chemical Society, Washington, DC, 249 pp

Andrahennadi R, Gillott C (1998) Resistance of *Brassica*, especially B. juncea (L.) Czern, genotypes to the diamondback moth, *Plutella xylostella (L.)*. Crop Prot 17:85–94

Anjum NA, Ahmad I, Pereira ME, Duarte AC, Umar S, Khan NA (2012) The plant family Brassicaceae: an introduction. In: Anjum NA, Ahmad I, Pereira ME, Duarte AC, Umar S, Khan NA (eds) The plant family Brassicaceae: contribution towards phytoremediation, environmental pollution series no. 21. Springer, Dordrecht

Ariyakanon N, Winaipanich B (2006) Phytoremediation of copper contaminated soil by *Brassica juncea* (L.) Czern and Bidens alba (L.) DC. var. *radiata*. J Sci Res Chula Univ 31(1):49–56

Axelsson T, Bowman CM, Sharoe AG, Lydiate DJ, Lagercrantz U (2000) Genome 43:679–688

Babula P, Adam V, Havel L, Kizek R (2012) Cadmium accumulation by plants of Brassicaceae family and its connection with their primary and secondary metabolism. In: Anjum NA et al (eds) The plant family Brassicaceae: contribution towards phytoremediation, environmental pollution, vol 21. https://doi.org/10.1007/978-94-007-3913-0_3, # Springer Science + Business Media Dordrecht 2012

Bailey CD, Koch MA, Mayer M, Mummenhoff K, O'Kane SL, Warwick SI, Windham MD, Al-Shehbaz IA (2006) Toward a global phylogeny of the Brassicaceae. Mol Biol Evol 23:2142–2160

Baker AJM (1981) Accumulators and excluders—strategies in the response of plants to heavy metals. J Plant Nutr 3:643–654

Baker AJM, Brooks RR (1989) Terrestrial higher plants which hyperaccumulate metallic elements—a review of their distribution, ecology and phytochemistry. Biorecovery 1:81–126

Baker AJM, McGrath SP, Sidoli CMD, Reeves RD (1994a) The possibility of in situ heavy metal decontamination of polluted soils using crops of metal-accumulating plants. Resour Conserv Recycl 11:41–49

Baker AJM, Reeves RD, Hajar ASM (1994b) Heavy metal hyperaccumulation and tolerance in British populations of the metallophyte Thlaspi caerulescens J. & C. Presl (Brassicaceae). New Phytol 127:61–68

Baker AJM, McGrath SP, Reeves DR, Smith JAC (2000) Metal hyperaccumulators plants: a review of the ecology and physiology of a biological resource for phytoremediation of metal-polluted soils. In: Terry N, Banuelos G (eds) Phytoremediation of contaminated soil and water. CRC, Boca Raton, FL

Baker AJM, Proctor J (1990) The influence of cadmium, copper, lead, and zinc on the distribution and evolution of metallophytes in the British Isles. Plant Syst Evol 173:91–108

Baker AJM, Walker PL (1989) Ecophysiology of metal uptake by tolerant plants. In: Shaw AJ (ed) Heavy metal tolerance in plants. Evolutionary aspects. CRC Press, Boca Raton FL, USA, pp 155–177

Baker AJM, Whiting SN (2002) In search of the holy grail—a further step in understanding metal hyperaccumulation? New Phytol 155:1–7

Banga SS (1997) Genetics and breeding in *Brassica* oilseed crops. In: Thomas G, Monteiro AA (eds) Proceedings of the inter national symposium on Brassicas. Acta Hortic. No. 459. ISHS, Rennes, France, pp 389–395

Banuelos GS (2001) The green technology of selenium phytoremedian. Biofactor 14:255–260

Barceló J, Poschenrieder C (1999) Structural and ultrastructural changes in heavy metal exposed plants. In: Prasad MNV, Hagemeyer J (eds) Heavy metal stress in plants. From molecules to ecosystems. Springer, Berlin, pp 183–205

Barceló J, Poschenrieder C (2002) Fast root growth responses, root exudates, and internal detoxification as clues to the mechanisms of aluminium toxicity and resistance: a review. Environ Exp Bot 48:75–92

Barceló J, Poschenrieder C (2003) Phytoremediation: principles and perspectives contributions to science 2(3):333–344

Barceló J, Poschenrieder C, Tolrà RP (2003) Importance of phenolics in rhizosphere and roots for plantmetal relationships. In: Gobran G (ed) Extended abstracts 7th ICOBTE Upsala 15–19 June, pp 162–163

Barman SC, Sahu RK, Bhargava SK, Chaterjee C (2000) Distribution of heavy metals in wheat, mustard and weed grown in fields irrigated with industrial effluents. Bull Environ Contam Toxicol 64:489–496

Basic N, Keller C, Fontanillas P, Besnard G, Galland N (2006) Cadmium hyperaccumulation and reproductive traits in natural Thlaspi caerulescens populations. Plant Biol 8:64–72

Becher M, Talke IN, Krall L, Kreamer U (2004) Cross-species microarray transcript profiling reveals high constitutive expression of metal homeostasis genes in shoots of the zinc hyperaccumulator Arabidopsis halleri. Plant J 37:251–268

Berti WR, Cunningham SD (2000) Phytostabilization of metals. In: Raskin I, Ensley BD (eds) Phytoremediation of toxic metals. Using plants to clean up the environment. Wiley, New York, pp 71–88

Bevan M, Walsh S (2005) The Arabidopsis genome: a foundation for plant research. Genome Res 15:1632–1642

Bhargava A, Carmona FF, Bhargava M, Srivastava S (2012) Approaches for enhanced phytoextraction of heavy metals. J Env Manag 105:103–120

Bhaskar VV, Vora AV (1994) Growth and yield analysis of *Brassica juncea* L. under three photoperiods. J Environ Biol 15:55–61

Blaylock MJ, Huang JW (2000) Phytoextraction of metals. In: Raskin I, Ensley BD (eds) Phyto remediation of toxic metals. Using plants to clean up the environment. Wiley, New York, pp 53–70

Blaylock MJ, Salt DE, Dushenkov S, Zakharova O, Gussman C, Kapulnik Y (1997) Enhanced accumulation of Pb in Indian mustard by soil-applied chelating agents. Environ Sci Technol 31:860–865

Boyd RS (1998) Hyperaccumulation as a plant defensive strategy. In: Brooks RR (ed) Plants that hyperaccumulate heavy metals. Their role in phytoremediation, microbiology, archaeology, mineral exploration and phytomining. CAB International, Wallingford, UK, pp 181–201

Boyd RS, Martens SN (1998) The significance of metal hyperaccumulation for biotic interactions. Chemoecology 8:1–7

Boyd RS, Wall MA, Santos SR, Davis MA (2009) Variation of morphology and elemental concentrations in the California nickel hyperaccumulator *Streptanthus polygaloides* (*Brassicaceae*). Northeast Nat 16:21–38

Breithaupt DE, Weller P, Wolters M, Hahn A (2003) Plasma response to a single dose of dietary beta-cryptoxanthin esters from papaya (Carica papaya L.) or non-esterified beta-cryptoxanthin in adult human subjects: a comparative study. Braz J Nutr 90:795–801

Bremer B, Bremer K, Chase MW, Fay MF, Reveal JL, Soltis DE, Soltis PS, Stevens PF et al (2009) An update of the Angiosperm Phylogeny Group classification for the orders and families of flowering plants: APG III. Bot J Linn Soc 161:105–121

Brooks RR (1998) Phytoremediation by volatilisation. In: Brooks RR (ed) Plants that hy peraccumulate heavy metals: their role in phytoremediation, microbiology, archaeology, mineral exploration and phytomining. CAB International, New York, pp 289–312

Brooks RR, Chiarucci A, Jaffré T (1998) Revegetation and stabilization of mine dumps and other degraded terrain. In: Brooks RR (ed) Plants that hyperaccumulate heavy metals: their role in phytoremediation, microbiology, archaeology, mineral exploration and phytomining. CAB International, New York, pp 227–247

Brooks RR, Lee J, Reeves RD, Jaffre T (1977) Detection of nickeliferous rocks by analysis of herbarium specimens of indicator plants. J Geochem Explor 7:49–57

Brooks RR, Radford CC (1978) Nickel accumulation by European species of the genus Alyssum. Proc R Soc Lond B 200:217–224

Chaffai R, Koyama H (2011) Heavy metal tolerance in Arabidopsis thaliana. Adv Bot Res 60:149

Chaney RL (1983) Plant uptake of inorganic waste constitutes. In: Parr JF, Marsh PB, Kla JM (eds) Land treatment of hazardous wastes. Noyes Data Corp, Park Ridge, USA, pp 50–76
Chaney RL, Malik M, LI YM, Brown SL, Brewer EP, Angle JS, Baker AJM (1997) Phytoremediation of soil metals. Curr Opin Biotechnol 8:279–284
Chen CL, Zhou GF, Fan YH, Zhou Y, Chen XQ (1995) Discussion on the origin of mustard (*Brassica juncea*) in China. Acta Hortic 402:317–320
Chiang HC, Lo J-C, Yeh K-C (2006) Genes associated with heavy metal tolerance and accumulation in Zn/Cd hyperaccumulator Arabidopsis halleri: a genomic survey with cDNA microarray. Environ Sci Technol 40:6792–6798
Chiou HY, Hueh YM, Liaw KF, Horng SF, Chiang MH, Pu YS, Lin JSN, Huang H Chen CJ (1995) Incidence of internal cancers and ingested organic arsenic: as even years follow up study in Taiwan. Cancer Res 55:1296–1300
Cho M, Chardonnens AN, Dietz K-J (2003) Differential heavy metal tolerance of Arabidopsis halleri and Arabidopsis thaliana: a leaf slice test. New Phytol 158:287–293
Cho-Ruk K, Kurukote J, Supprung P, Vetayasuporn S (2006) Perennial plants in the phytoremediation of lead contaminated soils. Biotechnology 5:1–4
Clemens S (2001) Molecular mechanisms of plant metal homeostasis and tolerance. Planta 212:475–486
Courbot M, Willems G, Motte P, Arvidsson S, Roosens N, Saumitou-Laprade P, Verbruggen N (2007) A major quantitative trait locus for cadmium tolerance in *Arabidopsis halleri* colocalizes with HMA4, a gene encoding a heavy metal ATPase. Plant Physiol 144:1052–1065
Cosio C, Martinoia E, Keller C (2004) Hyperaccumulation of cadmium and zinc in *Thlaspi caerulescens* and *Arabidopsis halleri* at the leaf and cellular level. Plant Physiol 134:716–725
Crespo MB, Lledo MD, Fay MF, Chase MW (2000) Subtribe *Vellinae (Brassiceae, brassicaceae)*: a combined analysis of ITS nrDNA sequences and morphological data. Ann Bot 86:53–62
Crowley DE, Alvey S, Gilbert ES (1997) Rhizosphere ecology of xenobiotic-degrading microorganisms. In: Kruger EL, Anderson TA, Coats JR (eds) Phytoremediation of soil and water contaminants. ACS Symposium Series, Washington, pp 20–36
Cunningham SD, Anderson TA, Schwab AP, Hsu FC (1996) Phytoremediation of soils contaminated with organic pollutants. Adv Agron 56(1):55–114
Cunningham SD, Ow DW (1996) Promises and prospects of phytoremediation. Plant Physiol 110(3):715
Cunningham SD, Berti WR, Huang JW (1995) Remediation of contaminated soils and sludges by green plants. In: Hinchee RE, Means JL, Burris DR (eds) Bioremediation of Inorganics. Battelle Press, Columbus, OH, USA, pp 33–54
Cunningham SD, Shann JR, Crowley DE, Anderson TA (1997) Phytoremediation of contaminated water and soil. In: Kruger EL, Anderson TA, Coats JR (eds) Phytoremediation of soil and water contaminants, ACS symposium series 664. American Chemical Society, Washington, DC
Dan TV, Krishna Raj S, Saxena PK (2000) Metal tolerance of *Scented geranium (Pelargonium sp. Frensham)*: effects of cadmium and nickel on chlorophyll fluorescence kinetics. Int J Phytorem 2:91–104
Danh LT, Truong P, Mammucari R, Tran T, Foster N (2009) Vetiver grass, *Vetiveria zizanioides*: a choice plant for phytoremediation of heavy metals and organic wastes. Int J Phytorem 11:664–691
Dar MI, Khan FA, Rehman F, Masoodi A, Ansari AA, Varshney D, Naushin F. Naikoo MI (2015) Roles of Brassicaceae in phytoremediation of metals and metalloids. In: Ansari AA et al (eds) Phytoremediation: management of environmental contaminants, vol 1. https://doi.org/10.1007/978-3-319-10395-2_14, © Springer International Publishing, Switzerland
Delorme TA, Gagliardi JV, Angle JS, Chaney RL (2001) Influence of the zinc hyperaccumulator *Thlaspi caerulescens* J. & C. Presl. and the non-metal accumulator *Trifolium pratense L.* on soil microbial populations. Can J Microbiol 47:773–776
Dhote S, Dixit S (2009) Water quality improvement through macrophytes—a review. Envi Monitor Assess 152:149–153

Domínguez MT, Madrid F, Marañón T, Murillo JM (2009) Cadmium availability in soil and retention in oak roots: potential for phytostabilization. Chemosphere 76:480–486
Doran PM, Nedelkoska TV (1999) Uptake of heavy metals by hairy roots of hyperaccumulating plant species. Abstr Pap Am Chem Soc 217:U168
Dushenkov S, Kapulnik Y (2000) Phytofiltration of metals. In: Raskin I, Ensley BD (eds) Phytoremediation of toxic metals. Using plants to clean up the environment. Wiley, New York, pp 89–106
Dushenkov V, Kumar PBAN, Motto H, Raskin I (1995) Rhizofilteration: the use of plants to remove heavy metal from aqueous streams. Environ Sci Technol 29:1239–1245
Ehrendorfer F, Neuffer B (2006) Herbert Hurka: research in botany, particularly in the evolutionary systematics of *Brassicaceae*. Plant Syst Evol 259:85–87
El Mehdawi AF, Pilon-Smits EA (2012) Ecological aspects of plan selenium hyperaccumulation. Plant Biol (Stuttg) 14:1–10
Endress PK (2011) Evolutionary diversification of the flowers in angiosperms. Am J Bot 98:370–396
Ernst WHO (1974) Schwermetallvegetation der Erde. Gustav Fischer Verlag, Stuttgart
Ernst WHO, Schat H, Verkleij JAC (1990) Evolutionary biology of metal resistance in *Silene vulgaris*. Evolutionary Trends in Plants 4:45–51
Ernst WHO, Verkleij JAC, Schat H (1992) Metal tolerance in plants. Acta Bot Neerl 41:229–248
Ernst WHO (1995) Decontamination or consolidation of metal contaminated soils by biological means. In: Förstner U, Salomons W, Meder P (eds) Heavy metals: problems and solutions. Springer, Berlin, pp 141–149
Ernst WHO (2000 a) Evolution and ecophysiology of metallophytes in Africa and Europe. In: Breckle SW, Schweizen B, Arndt U (eds) Results of worldwide ecological studies. 1st symposium AFW Schimper foundation (H & E Walter) Stuttgart 1998, Heimbach Verlag, Stutgart, pp 23–35
Ernst WHO (2000b) Evolution of metal hyperaccumulation and phytoremediation hype. New Phyto 146:357358
Favas PJC, Pratas J, Prasad MNV (2012) Accumulation of arsenic by aquatic plants in largescale field conditions: opportunities for phytoremediation and bioindication. Sci Total Environ 433:390–397
Favas PJ, Pratas J, Varun M, D'Souza R, Paul MS (2014) Phytoremediation of soils contaminated with metals and metalloids at mining areas: potential of native flora. Environ Risk Asses Soil Contam 3:485–516
Fonder N, Headley T (2013) The taxonomy of treatment wetlands: a proposed classification and nomenclature system. Ecol Eng 51:203–211
Freeman JL, Tamaoki M, Stushnoff C, Quinn CF, Cappa JJ, Devonshire J, Fakra SC, Marcus MA, McGrath SP, Hoewyk DV, Pilon-Smits EAH (2010) Molecular mechanisms of selenium tolerance and hyperaccumulation in Stanleya pinnata. Plant Physiol 153:1630–1652
Frers C (2009) El uso de plantas acuáticas en el tratamiento de aguas residuales. El Planeta Azul, Carmen de Areco, Argentina
Gabbrielli R, Pandolfini T, Vergnano O, Palandri MR (1990) Comparison of two serpentine species with different nickel tolerance strategies. Plant Soil 122:271–277
Ge Y, Ramchiary N, Wang T, Liang C, Wang N, Wang Z, Choi SR, Lim YP, Piao ZY (2011) Development and linkage mapping of unigene-derived microsatellite markers in *Brassica rapa* L. Breed Sci 61:160–167
German DA, Friesen N, Neuffer B, Al-Shehbaz IA, Hurka H (2009) Contribution to ITS phylogeny of the *Brassicaceae*, with special reference to some Asian taxa. Plant Syst Evol 283:33–56
Ghosh M, Singh S (2005) A review on phytoremediation of heavy metals and utilization of its by-products. Appl Ecol Environ Res 3. https://doi.org/10.15666/aeer/0301_001018
Gleba D, Borisjuk NV, Borisjuk L, Kneer R, Poulev A et al (1999) Use of plant roots for phytoremediation and molecular farming. Proc Natl Acad Sci 96:5973–5977
González V (1990) A indústria extractiva e o ambiente. Boletim de Minas 27(3):311–323
Gupta S, Nayek S, Saha RN, Satpati S (2008) Assessment of heavy metal accumulation in macrophyte, agricultural soil and crop plants adjacent to discharge zone of sponge iron factory. Environ Geol 55:731–739

Gurjar BS, Chauhan DVS (1997) Yield attributes and seed yield of Indian mustard (*Brassica juncea*) as influenced by varieties, fertility levels and spacing in Harsi Command area. Indian J Agron 42:142–144

Hall JL (2002) Cellular mechanisms for heavy metal detoxification and tolerance. J Exp Bot 53:1–11

Hanikenne M, Talke IN, Haydon MJ, Lanz C, Nolte A, Motte P, Kroymann J, Weigel D, Kramer U (2008) Evolution of metal hyperaccumulation required cis-regulatory changes and triplication of HMA4. Nature 453:391–395

Hasanuzzaman M, Hossain MA, Silva JAT, Fujita M (2012) Plant responses and tolerance to abiotic oxidative stress: antioxidant defense is a key factor. In: Bandi V, Shanker AK, Shanker C, Mandapaka M (eds) Crop stress and its management: perspectives and strategies. Springer, New York, pp 261–316

Hasanuzzaman M, Fujita M (2012) Heavy metals in the environment: current status, toxic effects on plants and possible phytoremediation. In: Anjum NA, Pereira MA, Ahmad I, Duarte AC, Umar S, Khan NA (eds) Phytotechnologies: remediation of environmental contaminants. CRC Press, Boca Raton, pp 7–73

Hassan Z, Aarts MGM (2010) Opportunities and feasibilities for biotechnological improvement of Zn, Cd or Ni tolerance and accumulation in plants. Environ Exp Bot 72:53–63

Haston E, Richardson JE, Stevens PF, Chase MW, Harris DJ (2009) The linear angiosperm phylogeny group (LAPG) III: a linear sequence of the families in APG III. Bot J Linn Soc 161:128–131

Hemingway JS (1976) Evolution of crop plants. Longman New York 1:19–21

Hernández-Allica J, Becerril JM, Garbisu C (2008) Assessment of the phytoextraction potential of high biomass crop plants. Environ Pollut 152:32–40

Hinchman RR, Negri CM, Gatliff EG (1996) Phytoremediation: using green plants to clean up contaminated soil, groundwater and wastewater. In: Proceedings of the international topical meeting on nuclear and hazardous waste management, Spectrum 96, Seattle, WA

Homer FA, Morrison RS, Brooks RR, Clement J, Reeves RD (1991) Comparative studies of nickel, cobalt, and copper uptake by some nickel hyperaccumulators of the genus Alyssum. Plant Soil 138:195–205

Horne AJ (2000) Phytoremediation by constructed wetlands. In: Terry N, Bañuelos G. (eds.) Phytoremediation of contaminated soil and water. New York: Lewis Publishers, pp 13–39

Huang JW, Chen J, Berti WR, Cunningham SD (1997) Phytoremediation of lead contaminated soil: role of synthetic chelates in lead phytoextraction. Environ Sci Technol 31:800–805

Huang JL, Sun GL, Zhang DM (2010) Molecular evolution and phylogeny of the angiosperm ycf2 gene. J Syst Evol 48:240–248

Johnston JS, Pepper AE, Hall AE, Chen ZJ, Hodnett G, Drabek J, Lopez R, Price HJ (2005) Evolution of genome size in Brassicaceae. Ann Bot 95:229–235

Jorba M, Vallejo R (2008) La restauración ecológica de canteras: un caso con aplicación de enmiendas orgánicas y riegos. Ecosistemas 17(3):119–132

Kathal R, Malhotra P, Kumar L, Uniyal PL (2016) Phytoextraction of Pb and Ni from the polluted soil by *Brassica juncea* L. J Environ Anal Toxicol 6:394. https://doi.org/10.4172/2161-0525.1000394

Khan MS, Zaidi A, Wani PA, Oves M (2009) Role of plant growth promoting rhizobacteria in the remediation of metal contaminated soils. Environ Chem Lett 7:1–19

Kimbrough DE, CohenY Winer AM, Creelman L, Mabuni C (1999) A critical assessment of chromium in the environment. Crit Rev Env Sci Technol 29:1–46

Koch MA, Mummenhoff K (2006) Evolution and phylogeny of the Brassicaceae. Plant Syst Evol 259:81–83

Krämer U (2010) Metal hyperaccumulation in plants. Annu Rev Plant Biol 61:517–534

Krämer U, Cotter-Howells JD, Charnock JM, Baker AJM, Smith JAC (1996) Free histidine as a metal chelator in plants that accumulate nickel. Nature 379:635–638

KrishnaRaj S, Dan T, Saxena P (2000) A fragrant solution to soil remediation. Int J Phytorem 2:117–132

Krzymaski J (1997). Agronomy of oilseed Brassicas. In: Thomas G, Monteiro AA (eds) Proceedings of the international symposium on Brassicas. Acta Horticultureae No 459 ISHS Rennes France, pp 55–60

Kubota H, Takenaka C (2003) *Arabis gemmifera* is a hyperaccumulator of Cd and Zn. Int J Phytoremediation 5:197–201

Kulakow PA, Schwab AP, Banks AP (2000) Screening plant species for growth on weathered, petroleum hydrocarbon-contaminated sediments. Int J Phytorem 2:297–317

Kumar PBAN, Dushenkov V, Motto H, Raskin I (1995) Phytoextraction: the use of plants to remove heavy metals from soils. Environ Sci Technol 29:1232–1238

Küpper H, Kochian LV (2010) Transcriptional regulation of metal transport genes and mineral nutrition during acclimatization to cadmium and zinc in the Cd/Zn hyperaccumulator, *Thlaspi caerulescens* (Ganges population). New Phytol 185:114–129

Küpper H, Lombi E, Zhao FJ, McGrath SP (2000) Cellular compartmentation of cadmium and zinc in relation to other elements in the hyperaccumulator Arabidopsis halleri. Planta 212:75–84

Larison JR, Likens GE, Fitzpatrick JW, Crock JG (2000) Cadmium toxicity among wildlife in the Colorado Rocky mountains. Nature 406:181–183

Lasat MM, Fuhrman M, Ebbs SD, Cornish JE, Kochian LV (1998) Phytoextraction of radiocesium-contaminated soil: evaluation of cesium-137 bioaccumulation in the shoots of three plant species. J Environ Qual 27:165–169

Lasat MM (2002) Phytoextraction of toxic metals: a review of biological mechanisms. J Environ Qual 31:109–120

Lim JM, Salido AL, Butcher DJ (2004) Phytoremediation of lead using Indian mustard (*Brassica juncea*) with EDTA and electrodics. Microchem J 76:3–9

Lombi E, Zhao FJ, Dunham SJ, McGrath SP (2000) Cadmium accumulation in populations of *Thlaspi caerulescens* and *Thlaspi goesingense*. New Phytol 145:11–20

López Pamo E, Aduvire O, Barettino D (2002) Tratamientos pasivos de drenajes ácidos de mina: estado actual y perspectivas de futuro. Boletín Geológico y Minero 113(1):3–21

Ma LQ, Komar KM, Tu C, Zhang W, Cai Y, Kennelley ED (2001) A fern that hyperaccumulates arsenic. Nature 409579

Macek T, Mackova M, Kas J (2000) Exploitation of plants for the removal of organics in environmental remediation. Biotechnol Adv 18:23–34

Maestri E, Marmiroli M, Visioli G, Marmiroli N (2010) Metal tolerance and hyperaccumulation: cost and trade-offs between traits and environment. Environ Exp Bot 68:1–13

Mahajan P, Singla S, Kaushal J (2016) Phytoremediation of heavy metals using *Brassica juneca*—a Review. J Chem Environ Sci Appl 2(2):157–173

Mhalappa, NJ, Mohan VK, Puranik, PR (2013) Phytoremediation of metal contaminated soils with special reference to *Brassica juncea* (l.) czern., *Macrotyloma uniflorum lam verdc. (Dolichos biflorus) and Medicago sativa.* Trends Bio Res 2(2):1–19

Marhold K, Lihova J, Perny M, Bleeker W (2004) Comparative ITS and AFLP analysis of diploid Cardamine (Brassicaceae) taxa from closely related polyploid complexes. Ann Bot 93:507–520

Marques L, Cossegal M, Bodin S, Czernic P, Lebrun M (2004) Heavy metal specificity of cellular tolerance in two hyperaccumulating plants, *Arabidopsis halleri* and *Thlaspi caerulescens*. New Phytol 164:289–295

Martens SN, Boyd RS (2002) The defensive role of Ni hyperaccumulation by plants: a field experiment. Am J Bot 89:998–1003

McGrath SP (1998) Phytoextraction for soil remediation. In: Brooks RR (ed) Plants that hyper-accumulate heavy metals: their role in phytoremediation, microbiology, archaeology, mineral exploration and phytomining. CAB International, New York, pp 261–287

McGrath SP, Zhao FJ (2003) Phytoextraction of metals and metalloids from contaminated soils. Curr Opin Biotechnol 14:277–282

McGrath SP, Zhao FJ, Lombi E (2002) Phytoremediation of metals, metalloids, and radionuclides. Adv Agron 75:1–56

Mellem JJ, Baijnath H, Odhav B (2009) Translocation and accumulation of Cr, Hg, As, Pb, Cu and Ni by Amaranthus dubius (Amaranthaceae) from contaminated sites. J Environ Sci Health Part A 44(6):568–575

Mellem JJ, Baijnath H, Odhav B (2012) Bioaccumulation of Cr, Hg, As, Pb, Cu and Ni with the ability for hyperaccumulation by Amaranthus dubius. Afr J Agric Res 7:591–596

Mendez MO, Maier RM (2008) Phytoremediation of mine tailings in temperate and arid environments. Rev Environ Sci Biotechnol 7:47–59

Meyer CL, Verbruggen N (2012) Use of the model species Arabidopsis halleri towards phytoextraction of cadmium polluted soils. N Biotechnol 30:9–14

Mhalappa NJ, Kulkarni M, Puranik P (2013) Trends Bio Res 2:1–19

Miguel PM, Inês N, Pinto FR, Joana RS, Martins LL (2015) Int J Mol Sci 16:17975–17998. https://doi.org/10.3390/ijms160817975

Milner MJ, Kochian LV (2008) Investigating heavy metal hyperaccumulation using *Thlaspi caerulescens* as a model system. Ann Bot 102:3–13

Nahar K, Hasanuzzaman M, Alam MM, Rahman A, Suzuki T, Fujita M (2016) Polyamine and nitric oxide crosstalk: antagonistic effects on cadmium toxicity in mung bean plants through upregulating the metal detoxification, antioxidant defense, and methylglyoxal detoxification systems. Ecotoxicol Environ Saf 126:245–255

Nahar K, Hasanuzzaman M, Suzuki T, Fujita M (2017) Polyamine-induced aluminium tolerance in mung bean: a study on antioxidant defense and methylglyoxal detoxification systems. Ecotoxicology 26(1):58–73

Nayak BS, Pereira LP, Maharaj D (2007) Wound healing activity of *Carica papaya L.* in experimentally induced diabetic rats. Ind J Exp Biol 45:739–743

Neelam K, Bhargava SC, Bhardwaj SN, Kumari N (1994) Physiological basis of yield and its components in rapeseed-mustard with reference to photoperiod. Indian J Plant Physiol 37:142146

Negri MC, Hinchman RR (2000) The use of plant for the treatment of radionuclides. In: Raskin I, Ensley BD (eds) Phytoremediation of toxic metals. Wiley, New York, pp 107–132

Neilson S, Rajakaruna N (2012) Roles of rhizospheric processes and plant physiology in phytoremediation of contaminated sites using oilseed Brassicas. In: Anjum NA, Ahmad I, Pereira ME, Duarte AC, Umar S, Khan NA

Nishi S (1980) Differentiation of *Brassica* crops in Asia and the breeding of "Hakuran", a newly synthesized leafy vegetable. In: Tsunoda S et al (eds) *Brassica* crops and wild allies—biology and breeding. Jpn Sci Soc Press, Tokyo, pp 133–150

Nyquist J, Greger MA (2009) Field study of constructed wetlands for preventing and treating acid mine drainage. Ecol Eng 35:630–642

Ochiai EI (1987) General principles of biochemistry of the elements. Plenum Press, New York

Ohlendorf HM, HoffmanDJ Saiki MK, Aldrich TW (1986) Embryonic mortality and abnormalities of birds: apparent impacts of selenium from irrigation drain water. Sci Total Environ 52:49–63

Okeniyi JAO, Ogunlesi TA, Oyelami OA, Adeyemi LA (2007) Effectiveness of dried *Carica papaya* seeds against human intestinal parasitosis: a pilot study. J Med Food 10:194–196

Palmer CE, Warwick S, Keller W (2001) Brassicaceae (Cruciferae) family, plant biotechnology, and phytoremediation. Int J Phytoremediation 3:245–287

Pang ECK, Halloran GM (1996) The genetics of adult plant blackleg (*Leptosphaeria maculans)* resistance from *B. juncea* in *B. napus*. Theor Appl Genet 92:382–387

Pantola RC, Alam A (2014) Potential of *Brassicaceae Burnett* (Mustard family; Angiosperms) in phytoremediation of heavy metals. Int J Sci Res Environ Sci 2(4):120–138

Pedron F, Petruzzelli G, Barbafieri M, Tassi E (2009) Strategies to use phytoextraction in very acidic soil contaminated by heavy metals. Chemosphere 75:808–814

Pilon-Smits EAH, LeDuc DL (2009) Phytoremediation of selenium using transgenic plants. Curr Opin Biotechnol 20:207–212

Pilon-Smits E, Pilon M (2000) Breeding mercury-breathing plants for environmental clean-up. Trends Plant Sci 5(6):235–236

Pollard AJ, Dandridge KL, Shee EM (2000) Ecological genetics and the evolution of trace element hyperaccumulation in plants. In: Terry N, Bañuelos G (eds) Phytoremediation of contaminated soils and waters. CRC Press LLC, Boca Raton, FL, USA, pp 251–264

Poschenrieder C, Barceló J (1999) Water relations in heavy metal stressed plants. In: Prasad MNV, Hagemeyer J (eds) Heavy metal stress in plants. From molecules to ecosystems. Springer, Berlin, pp 207–229

Poschenrieder CH, Barceló J (2004) Estrés por metales pesados. In: Reigosa MJ, Pedrol N, Sánchez A (eds) La ecofisiologia vegetal: Una ciencia de síntesis. Thomson, Madrid, pp 413–442

Prakash S (1980) Cruciferous oilseeds in India. In: Tsunoda S et al (eds) *Brassica* crops and wild allies—biology and breeding. Jpn Sci Soc Press, Tokyo, pp 151–166

Prasad MNV (2004) Phytoremediation of metals and radionuclides in the environment: the case for natural hyperaccumulators, metal transporters, soil-amending chelators and transgenic plants. In: Prasad MNV (ed) Heavy metal stress in plants: from biomolecules to ecosystems, 2nd edn. Springer, Berlin, pp 345–391

Prasad MNV (2005) Nickelophilous plants and their significance in phytotechnologies. Braz J Plant Physiol 17:113–128

Prasad MNV, Freitas HMO (2003) Metal hyperaccumulation in plants—biodiversity prospecting for phytoremediation technology. Electron J Biotechnol 6:285–321

Pratas J, Favas PJC, Paulo C, Rodrigues N, Prasad MNV (2012) Uranium accumulation by aquatic plants from uranium-contaminated water in Central Portugal. Int J Phytorem 14:221–234

Qiu YL, Li LB, Wang B, Xue JY, Hendry TA, Li RQ, Brown JW, Liu Y, Hudson GT, Chen ZD (2010) Angiosperm phylogeny inferred from sequences of four mitochondrial genes. J Syst Evol 48:391–425

Rascio N, Navari-Izzo F (2011) Heavy metal hyperaccumulating plants: how and why do they do it? And what makes them so interesting? Plant Sci 180:169–181

Raskin I, Smith RD, Salt DE (1997) Phytoremediation of metals: using plants to remove pollutants from the environment. Curr Opin Biotechnol 8:221–226

Ravindran KC, Venkatesan K, Balakrishnan V, Chellappan KP, Balasubramanian T (2007) Restoration of saline land by halophytes for Indian soils. Soil Biol Biochem 39:2661–2664

Reeves RD, Baker AJM, Borhidi A, Berazain R (1996) Nickel-accumulating plants from the ancient serpentine soils of Cuba. New Phytol 133:217–224

Reeves RD, Baker AJM, Borhidi A, Berazain R (1999) Nickel hyperaccumulation in the serpentine flora of Cuba. Ann Bot 83:29–38

Reeves RD, Baker AJM (2000) Metal-accumulating plants. In: Raskin I, Ensley BD (eds) Phytoremediation of toxic metals—using plants to clean up the environment. Wiley, New York

Reeves RD, Brooks RR, Dudley TR (1983) Uptake of nickel by species of Alyssum, Bornmuellera and other genera of old world Tribus Alysseae. Taxon 32:184–192

Robinson BH, Brooks RR, Howes AW, Kirkman JH, Gregg PEH (1997) The potential of the high-biomass nickel hyperaccumulator *Berkheya coddiifor* phytoremediation and phytomining. J Geochem Explor 60:115–126

Ruiz ON, Daniell H (2009) Genetic engineering to enhance mercury phytoremediation. Curr Opin Biotechnol 20:213–219

Rylott EL, Bruce NC (2008) Plants disarm soil: engineering plants for the phytoremediation of explosives. Trends Biotechnol 27(2):73–81

Sahi SV, Bryant NL, Sharma NC, Singh SR (2002) Characterization of a lead hyperaccumulator shrub, *Sesbania drummondii*. Environ Sci Technol 36:4676–4680

Salt DE, Smith RD, Raskin I (1998) Phytoremediation. Annu Rev Plant Physiol Plant Mol Biol 49:643–668

Salt DE, Blaylock M, Nanda Kumar PBA, Dushenkov V, Ensley BD, Chet I, Raskin I (1995) Phytoremediation: a novel strategy for the removal of toxic metals from the environment using plants. Biotechnol 13:468–474

Sarma H (2011) Metal hyperaccumulation in plants: a review focussing on phytoremediation technology. Environ Sci Technol 4:118–138

Sharma H (2016) Phytoremediation of lead using *Brassica juncea* and Vetiveria zizanioides. Int J Life Sci Res 4(1):91–96

Sauer JD (1993) Historical geography of crop plants—a select roster. CRC Press, Boca Raton, FL

Saxena PK, Raj SK, Dan T, Perras MR, Vettakkorumakankav NN (1999) Phytoremediation of heavy metal contaminated and polluted soils. In: Prasad MNV, Hagemayr J (eds) Heavy metal stress in plants. From molecules to ecosystems. Springer, Berlin, pp 305–329

Schnoor JL, Licht LA, McCutcheon SC, Wolfe NL, Carreira LH (1995) Phytoremediation of organic and nutrient contaminants. Environ Sci Technol 29:318A–323A

Schnoor JL (2000) Phytostabilization of metals using hybrid poplar trees. In: Raskin I, Ensley BD (eds) Phytoremediation of toxic metals. Using plants to clean up the environment. Wiley, New York, pp 133–150

Shen ZG, Zhao FJ, McGrath SP (1997) Uptake and transport of zinc in the hyperaccumulator *Thlaspi caerulescens* and the non-hyperaccumulator *Thlaspi ochroleucum*. Plant Cell Environ 20:898–906

Singh A, Kumar D, Sahu A (2014) J Environ Biol 283:59–365

Singh A, Prasad P (2014) Evaluation of potential of Brassica juncea for removal of arsenic from hydroponic solution. Int J Curr Microbiol App Sci 3(12):246–252

Sobolewski A (1999) A review of processes responsible for metal removal in wetlands treating contaminated mine drainage. Int J Phytorem 1(1):19–51

Szczyglowska M, Piekarska A, Konieczka P, Namiesnik J (2011) Use of *Brassica* plants in the phytoremediation and biofumigation processes. Int J Mol Sci 12:7760–7771

Tang Y-T, Qiu R-L, Zeng X-W, Ying R-R, Yu F-M, Zhou X-Y (2009) Lead, zinc, cadmium accumulation and growth simulation in Arabis paniculata Franch. Environ Exp Bot 66:126–134

Tangahu BV, Abdullah SRS, Basri H, Idris M, Anuar N, Mukhlisin M (2011) A review on heavy metals (As, Pb, and Hg) uptake by plants through phytoremediation. Int J Chem Eng 939161:1–31

Thakral SK, Singh BP, Faroda AS, Gupta SK, Kumar S (1995) Fatty acids composition in seeds of *Brassica* species as affected by moisture and fertility levels. Crop Res 10:137–140

Tolrà RP, Poschenrieder C, Barceló J (1996) Zinc hyperaccumulation in *Thlaspi caerulescens*. II Influence on organic acids. J Plant Nutr 19:1541–1550

Tolrà RP, Poschenrieder C, Alonso R, Barceló D, Barceló J (2001) Influence of zinc hyperaccumulation on glucosinolates in *Thlaspi caerulescens*. New Phytol 151:621–626

Tomar TS, Singh S, Kumar S, Tomar S (1997) Response of Indian mustard (*Brassica juncea*) to nitrogen, phosphorus and sulfur fertilization. Indian J Agron 42:148–151

USEPA (2000) Introduction to phytoremediation. EPA, Washington, DC

Vamerali T, Bandiera M, Mosca D (2010) Field crops for phytoremediation of metal contaminated land. A review. Environ Chem Lett 8:1–17

Van der Ent A, Baker AJM, Reeves RD, Pollard AJ, Schat H (2013) Hyperaccumulators of metal and metalloid trace elements: facts and fiction. Plant Soil 362:319–334

Vázquez MD, Barceló J, Poschenrieder C, Mádico J, Hatton P, Baker AJM, Cope GH (1992) Localization of zinc and cadmium in *Thlaspi caerulescens* (*Brassicaceae*), a metallophyte that can hyperaccumulate both metals. J Plant Physiol 140:350–355

Vázquez MD, Poschenrieder C, Barceló J, Baker AJM, Hatton P, Cope GH (1994) Compartmentation of zinc in roots and leaves of the zinc hyperaccumulator *Thlaspi caerulescens*. J & C Presl. Bot Acta 107:243–250

Verbruggen N, Hermans C, Schat H (2009) Molecular mechanisms of metal hyperaccumulation in plants. New Phytol 181:759–776

Vijayarengan P (2005) Nitrogen and potassium status of greengram (*Vigna radiata*) cultivars under nickel stress. Nat Environ Pollut Technol 4(1):65–69

Vogel-Mikuŝ K, Drobne D, Regvar M (2005) Zn, Cd and Pb accumulation and arbuscular mycorrhizal colonisation of pennycress *Thlaspi praecox* Wulf. (*Brassicaceae*) from the vicinity of a lead mine. Environ Pollut 133:233–242

Vymazal J (2009) The use constructed wetlands with horizontal sub-surface flow for various types of wastewater. Ecol Eng 35:1–17

Warwick SI (2011) Brassicaceae in agriculture. In: Schmidt R, Bancroft I (eds) Genetics and genomics of the Brassicaceae. Plant Gene. Geno. Crop Models 9:33–65

Weber M, Harada E, Vess C, Roepenack-Lahaye EV, Clemens S (2004) Comparative microarray analysis of *Arabidopsis thaliana* and *Arabidopsis halleri* roots identifies nicotianamine synthase, a ZIP transporter and other genes as potential metal hyperaccumulation factors. Plant J 37:269281

Weigel D, Mott R (2009) The 1001 genomes project for *Arabidopsis thaliana*. Gen Bio 10:07

Whitting SN, Neumann PM, Baker AJM (2003) Nickel and zinc hyperaccumulation by *Alyssum murale* and *Thlaspi caerulescens* (*Brassicaceae*) do not enhance survival and whole—plant growth under drought stress. Plant Cell Environ 26:351–360

Williamson A, Johnson MS (1981) Reclamation of metalliferous mine wastes. In: Lepp NW (ed) Effect of heavy metal pollution on plants, vol. 2. Metals in the environment. Applied Science Publishers, London, pp 185–212

Williamson NA, Johnson MS, Bradshaw AD (1982) Mine wastes reclamation. The establishment of vegetation on metal mine wastes. Mining Journal Books, London

Wu L, Chen J, Tanji KK, Bañuelos GS (1995) Distribution and biomagnification of selenium in a restored upland grassland contaminated by selenium from agricultural drain water. Environ Tox Chem 14:733–742

Wu HB, Tang SR (2009) Using elevated CO_2 to increase the biomass of a *Sorghum vulgare* x *Sorghum vulgare* var. *sudanense* hybrid and *Trifolium pratense* L. and to trigger hyperaccumulation of cesium. J Hazard Mater 170:861–870

Xie QE, Yan XL, Liao XY, Li X (2009) The arsenic hyperaccumulator fern Pteris vittata L. Environmen Sci Technol 43(22):8488–8495

Yang XE, Long XX, Ye HB, He ZL, Calvert DV, Stoffella PJ (2004) Cadmium tolerance and hyperaccumulation in a new Zn hyperaccumulating plant species (*Sedum alfredii* Hance). Plant Soil 259:181–189

Yang G, Wang S, Zhou R, Sun S (1983) Endemic selenium in toxication of humans in China. Am J Clin Nutr 37:872–881

Zaurov DE, Perdomo P, Raskin I (1999) Optimizing soil fertility and pH to maximize cadmium removed by Indian mustard from contaminated soils. J Plant Nutr 22(6):977–986

Zhao FJ, Lombi E, Breedon T, McGrath SP (2000) Zinc hyperaccumulation and cellular distribution in *Arabidopsis halleri*. Plant Cell Environ 23:507–514

Zhuang X, Chen J, Shim H, Bai Z (2007) New advances in plant growth-promoting rhizobacteria for bioremediation. Environ Int 33:406–413

Zeven AC, Zhukovsky PM (1975) Dictionary of cultivated plants and their centres of diversity—excluding ornamentals, forest trees and lower plants. Pudoc, Wageningen

Zorrig W, Rabhi M, Ferchichi S, Smaoui A, Abdelly C (2012) Phytodesalination: a solution for salt-affected soils in arid and semi-arid regions. J Arid Land Stud 22:299–302

Molecular and Biotechnological Interventions for Improving Brassicaceae Crops for Abiotic Stress Tolerance

Pankaj Kumar and Dinesh Kumar Srivastava

Abstract *Brassica* genus recognized worldwide as economic important genus, nutritionally rich (vegetables, oilseed crops, forage crops) and having medicinally important phytochemical compounds. Furthermore, *Brassica* genus has diverse crop species considered as model species for plants scientific community due to remarkable metabolite potential. However, different biotic and abiotic stresses limit the productivity as well as quality of produce worldwide. Among different productivity constraints, abiotic stresses such as salinity, drought, heavy metals, submergence, elevated temperature and nutrient deficiencies are the major limiting factors resulted in approximately 80% of crop yield reduction. Global climatic disruption and unsustainable growth of world population mounted huge pressure on agricultural land to increase agricultural productivity per unit area. Under such scenario, scientific interventions based on modern biotechnological approaches to breed abiotic stress tolerant crops with increased productivity and improved qualities are urgently needed. Advancement in modern molecular biology and biotechnological techniques during past two decades such as molecular breeding, crop genetic engineering, identification of genes (gene discovery) that serve as valuable candidates or targets for crop improvement and "omics" based approaches (functional genomics, proteomics and metabolomics) have revealed innumerable mechanisms which confer improved productivity and adaptation to abiotic stresses. Significant advancement has been accomplished in the cellular and molecular biology of *Brassica* species by utilizing biotechnological interventions. This chapter provides insight towards the cited literature on abiotic stress tolerance in *Brassica* crops that possibly will be valuable for researchers working on this economically important *Brassica* group for its genetic advancement.

P. Kumar (✉)
Science and Engineering Research Board, Department of Science & Technology, Government of India, New Delhi 110070, India
e-mail: pksharmabiotech@gmail.com

CSIR-IHBT, Palampur, India

D. K. Srivastava
Department of Biotechnology, Dr. Y. S. Parmar University of Horticulture and Forestry, Solan 173230, Himachal Pradesh, India

M. Hasanuzzaman (ed.), *The Plant Family Brassicaceae*,
https://doi.org/10.1007/978-981-15-6345-4_15

Keywords Abiotic stresses · Breeding · Genetic engineering · Genomics · Transcription factor

1 Introduction

In today's world; agriculture biotechnology practices are especially meaningful, when our global population is growing and our demand for food is increasing, especially in developing countries. Feeding a world population of 9.1 billion in 2050 will require raising of overall food production by 70% (nearly 100% in developing countries), which could only possible through molecular and biotechnological interventions in conjunction with conventional practices. Advances in understanding plant biology, novel genetic resources, genome modification, and omics technologies generate new solutions for food security and strengthen the farming community to grow more food on less land using cultivation practices that are environmentally sustainable through biotechnology (Parmar et al. 2017; Kumar and Srivastava 2015, 2016; Kumar et al. 2018a, Pathak et al. 2018).

The genus *Brassica* is worldwide known for its economically and nutritionally enriched different species, commercially cultivated as agricultural and horticultural crops such as oil seed, vegetable, condiment and forage crops rich in anti-oxidants, minerals, vitamins, and having bioactive molecules possessing anticancerous activities (Ravanfar et al. 2017; Hanschen et al. 2015; Odongo et al. 2017; Kumar et al. 2015a, b, 2018a, b, c). *Brassica* species and varieties have been the subject of great scientific interest owing to their agricultural significance and in particular six species {*B. carinata* (BBCC, 2n = 34; Ethiopian mustard), *B. juncea* (AABB; 2n-36; Indian mustard), *B. napus* (AACC, 2n-38; canola), *B. oleracea* (CC; 2n-18; cabbage, cauliflower, broccoli, brussel sprouts), *B. nigra* (BB; 2n-16; black mustard), and *B. rapa* (AA; 2n-20; turnip, chinese cabbage)} have been evolved through chromosomes combining from three earlier species (named as A, B, and C genomes) as reported in the triangle of U hypothesis depicted in Fig. 1 (Nagaharu 1935). The *Brassica* species originally emerges from Western Europe, Asia's Mediterranean and temperate region. Some wild species grow as weeds, particularly in North and South America and in Australia and different *Brassica* species grown worldwide under diverse agro-climatic environments.

In family Brassicaceae, the major economically important species worldwide emphasized are *B. balearica* (Mallorca cabbage); *B. carinata* (Abyssinian mustard or Abyssinian cabbage—used to produce biodiesel); *B. elongata* (elongated mustard); *B. fruticulosa* (Mediterranean cabbage); *B. hilarionis* (St. Hilarion cabbage); *B. juncea* (Indian mustard/brown and leaf mustards/Sarepta mustard); *B. napus* (rapeseed/canola/rutabaga); *B. narinosa* (broad beaked mustard); *B. nigra* (black mustard); *B. oleracea* (kale, cabbage, collard greens, broccoli, cauliflower, kai-lan, Brussels sprouts, kohlrabi); *B. perviridis* (tender green/mustard spinach); *B. rapa* syn. *B. campestris* (Chinese cabbage/turnip/rapini/komatsuna); *B. rupestris* (brown mustard); *B. tournefortii* (Asian mustard). Entire genome of rapeseed (canola,

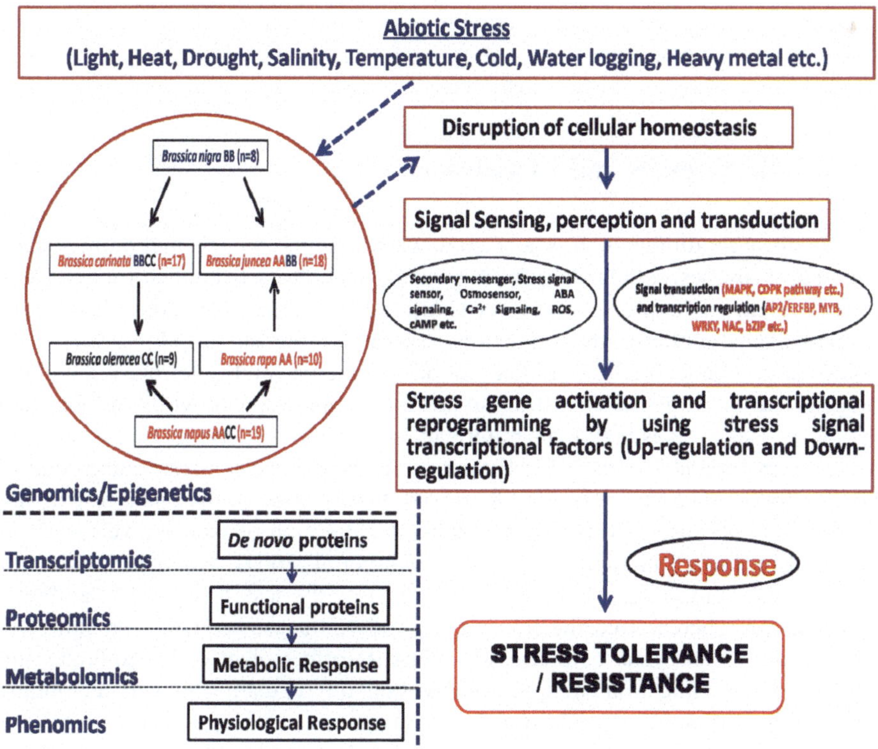

Fig. 1 Biotechnological interventions for abiotic stress tolerance in Brassicaceae

B. napus) along with its constituent genomes of *B. rapa* and *B. oleracea* were sequenced and announced via multi-consortium genome sequencing project i.e. Bayer Crop Science, BGI-Shenzhen, China; Keygene N.V., the Netherlands and University of Queensland, Australia (Wang et al. 2011). In family Brassicaceae, vegetables and oilseeds *Brassica* species are of highest economic value worldwide. However; various environmental stresses both biotic and abiotic highly affect the quality and quantity of produce. Abiotic stresses mainly drought, salinity cold and heat are the foremost factors for productivity and quality losses in the family Brassicaceae.

Recent molecular intervention and genomics era led to the identification of large number of stress responsive/inducible genes, transcription factors, micro RNA etc., therefore plant genetic transformation procedure and genomic assisted breeding gained a impetus shift towards crop improvement programme via incorporation of various abiotic stress tolerance factors. However, Atkinson and Urwin (2012); Kayum et al. (2016) discussed that different genotypes responds to different abiotic stresses via highly coordinated complex signal transduction pathway that undergoes changes at cellular, physiological, transcriptome and proteome levels. But to understand the physiological effect of transgene at complete plant level and assessment of transgenic plants to different stress conditions in an open field still remains as major challenge to overcome.

2 Abiotic Stresses and Brassicaceae

Abiotic stress conditions such as heat, cold, water logging, drought and salinity leads to oxidative stresses, cellular dehydration and triggered the activation of various identified differentially expressed gene(s) (Akula and Gill 2018). However; plant has to respond against multiple stress situations at a time to prevent crop loss. Therefore, molecular and biotechnological interventions offer a possibility to reveal the physiological mechanism whereby different stress tolerant crop can withstand multiple abiotic stresses such as high temperature and water logging (Mittler and Blumwald 2010; Atkinson and Urwin 2012; Kumar et al. 2018b). Heat stress caused due to increase in the temperature beyond threshold, highly affected the plant growth and development because of functional proteins denaturation of their native form and increased in the cell membranes fluidity (Wang et al. 2004). Whereas, cold stress resulted from chilling injury ($\geq$0 °C) and form gelly, less fluidy cell membrane conformation whereas freezing injury (at sub 0 °C) causes dehydration cell state (Thomashow 1994). The cold stresses primarily affect plant cell membrane and involved the modification of lipid composition for regulation of cell membrane permeability. In response to cold stress, different genes and transcription factors have been characterized in modal plant *A. thaliana* such as cold binding factors/proteins (*CBF*); Dehydration responsive element binding proteins (*DREB1*); early dehydration inducible genes (*ERD*); cold regulated genes (*COR*) and cold inducible genes (*KIN*). Ishitani et al. (1997) used genetic screening strategies by changing the activity of reporter gene, which was under the control of *RD29a* promoter and reported complex networking and interactions in expression to *COR*/*ERD*/*RD* genes.

Brassica genus is also highly affected by drought and salt stress. Plant growth, development and productivity is highly affected resulted from ion toxicity, poor water uptake, disruption of intracellular homeostasis Na^+/K^+ (Na^+ concentration high and K^+ uptake inhibition) (Serrano et al. 1999; Jamil et al. 2011). In such situations; plants cope up salinity stress via stress avoiding and stress tolerating mechanism. The salt overly sensitive (*SOS*) pathway genes such as (*SOS1*; *SOS2*, *SOS3*, *SOS4*, *SOS5* etc.) along with calcium regulated signaling cascade and protein kinases activated to overcome salinity stress (Shi et al. 2000, 2003). The *SOS* pathway functions to maintain cellular homeostasis via Na^+/H^+ antiporter protein transport and tonoplast (vacuolar compartmentation) antiporter mechanism. Enhanced salt tolerance were achieved using overexpression of *SOS* pathway genes that involved in cell surface adhesion (*SOS5*) and kinase (*SOS4*) actively participated in pyridoxal-5-phosphate biosynthesis (Zhu 2001; Shi et al. 2003).

In salinity, drought and freezing stress plants undergoes dehydration conditions due to water loss and loss in turgor pressure. In such conditions, plants perceive its signal to osmotic stress via two signaling pathway i.e. ABA dependent pathway and

ABA independent pathway. Both theses pathways involve certain cross-talks point with transcriptions factors such as *NAC*, *MYB* etc., protein kinases and signaling molecule secondary messenger (Ca^{2+}) to counteract stress conditions (Chinnusamy et al. 2004; Yoshida et al. 2014). Calcium ion ($Ca^{2+)}$ plays an important role as a ubiquitous secondary messenger molecule to overcome the abiotic stresses via its increased concentration which otherwise highly regulated in eukaryotic cell at concentration level of 100–200 nM (Knight 1999; Cheng et al. 2002). Water deficit/drought conditions leads to plant wilting and loss of turgor pressure. To overcome the effect of drought stress, plant express certain proteins i.e. late embryogenesis proteins (*LEA*), molecular chaperones; alteration in drought responsive genes and also increased the level of certain metabolites such as abscisic acid (*ABA*), osmolytes, osmo-protectants (mannitol and sorbitol) and antioxidants (Hasanuzzaman et al. 2019).

3 Brassicaceae—Plant Genetic Engineering for Abiotic Stress Tolerance

Transgenic approaches of plant genetic engineering include various methods of gene transfer for *Brassica* species improvement. Differential expression of abiotic stress resistant/tolerant genes and transcription factors can be engineered into new combinations of desired traits, ectopically expressed and transferred to other crops/genotypes, where they do not exist or vary naturally (Kumar and Srivastava 2016; Kumar et al. 2018b). The ability to transform different *Brassica* species with transgene from any biological source (plant, insect, microbial) is therefore a highly prevalent method for the molecular crop breeding. Transgenic *Brassica* with improved traits such as abiotic stress tolerance and resistance serve as novel source of variations and development of new cultivar to improve socioeconomic status of farmers and to improve ecological status.

Plant genetic engineering/transgenic approach involve the introduction of desirable agronomically important traits such as abiotic stress responsive genes/transcription factors etc. However, genetic engineering is only promising option for crop improvement; when desirable gene for particular trait which we want to improve is not present in it germplasm and depend upon cross-barrier species. With the recent molecular and biotechnological intervention such as high throughput genomics approaches, advance next generation sequencing technologies resulted in identification of abiotic stress induced genes, which serve as promising targets for plant genetic engineering and molecular breeding purpose (Deshmukh et al. 2014; Chunqing et al. 2015).

Genetic improvement of different species of family Brassicaceae via genetic transformation procedures proved an effective and time saving approach for abiotic stress tolerance than conventional means (Kumar and Srivastava 2016; Srivastava et al. 2016; Ravanfar et al. 2017). The most scientifically investigated and applied method for genetic transformation of *Brassica* spp. is *Agrobacterium tumefaciens* mediated gene transfer method (Kumar and Srivastava 2015; Ravanfar et al. 2017; Kumar et al. 2018b; Shah et al. 2018; Ahmar et al. 2019). But to carryout successful plant genetic transformation, optimization of large number of factors such as explant type, high frequency in vitro regeneration protocol *Agrobacterium* strain, culture parameter optimization, growth medium, plant growth regulators, sugar source; pre-incubation, co-cultivation time, bacterial inoculum concentration, antibiotics, selectable marker genes etc. are critically required. From literature reviewed; lots of studies have been carried out to optimize all the above mentioned factors in Brassicaceae family (Kumar et al. 2015a, b, 2017, 2018a, b, c; Kumar and Srivastava 2015; Srivastava et al. 2016; Kumar and Srivastava 2016; Ravanfar et al. 2017; Parmar et al. 2017).

Therefore, plant genetic engineering proved as an important, most promising and sustainable approach for crop improvement. Transgenic approaches now a day's become as consumer acceptable approach with its higher adoption rate worldwide i.e. 191.7 million hectares in the year 2018 as per recent report published by International Service for the Acquisition of Agri-biotech Applications. Largest coverage under transgenic crops as follows USA (soybeans, maize, cotton, canola), Brazil (soybeans, maize, cotton), Argentina (soybeans, maize, cotton), Canada (canola, soybeans, maize, canola) and India (cotton). Till date; large numbers of abiotic stress tolerant genes has been transferred in *Brassica* species to cope up with abiotic stresses using *Agrobacterium*—mediated gene transfer method such as genes that control salt uptake and transport (Na^+/H^+ antiporter); osmotic/protective role (Glycine betaine); late embryogenesis abundant (*LEA*) protein; antioxidants {superoxide dismutase (*SOD*) and ascorbate peroxidase (*APX*) genes}; lectin; dehydration-responsive element-binding (*DREB*) transcription factors (DRE/CRT); Glyoxalase (*Gly I and Gly II*); abscisic acid (*ABA*); pathogenesis related protein (PR10); 5-Aminolevulinic acid (*5-ALA*) etc. (Kumar et al. 2018b; Shah et al. 2018; Ahmar et al. 2019). Transgenic *B. napus* and *B. juncea* expressing *CodA* transgene (Choline oxidase) provide salinity tolerance (Huang et al. 2000; Prasad et al. 2000). Transgene *CodA* also provided increased tolerance to salinity in *B. compestris* L. spp. chinensis (Wang et al. 2010). Similarly; Transgenic *B. oleracea* var. *capitata* expressing bacterial *betA* gene resulted in osmoprotection and salinity tolerance. Metwali et al. (2012) transferred the *SOD* and *APX* genes via *Agrobacterium tumefaciens*-mediated gene transfer method, which resulted in tolerance to salinity. Kumar et al. (2015a, b) over-expressed the chickpea lectin gene in *B. juncea* to provide fungal resistance against *Alternaria Brassicae* but transgenic *B. juncea* also resulted in tolerance to drought and salinity. Similarly; Saxena et al. (2011) and Rajwanshi et al. (2016) developed transgenic *B. juncea* using transgene (*Gly I and Gly II*) for detoxification of methylglyoxal, that significantly provide stress tolerance to salinity. Transgenic *B. napus* expressing *AtNHX1*, *BnNHX1* (Vacuolar Na^+/H^+ antiporter) (Zhang et al. 2001); *PR10* (Srivastava et al. 2004); *AtCBF1* (*LEA4-1* (Dalal et al. 2009); *OsNASI* (Kong et al. 2011); *YHem1* that increase the endogenous 5-ALA metabolism (Sun et al. 2015); *DREB* (Qamarunnisa et al. 2015); *BnSIP1-1*, *BnABI5*, *BnNAC485* (Luo et al. 2017) significantly contributed to provide abiotic stress tolerance/resistance i.e. drought, salinity, frost etc. To overcome heat stress; heat shock proteins as molecular chaperones plays an important role. Ravanfar and Aziz (2014) developed heat tolerant transgenic broccoli (*B. oleracea* L. var. *italica* cv. Green Marval) using *Agrobacterium*-mediated gene transformation method containing transgene *AtHSP101* from *Arabidopsis thaliana.* Molecular confirmation using gene specific primers and RT-qPCR expression confirms the transgene integration and confirmation. *fad7* is an important regulatory enzyme involved in the biosynthesis of polyunsaturated fatty acids and also engineered using genetic transformation procedure which confirm heat stress tolerance (Murakami et al. 2000). Transgenic *B. juncea* developed using transgenes *PgNHX1* (Rajagopal et al. 2007); *AtLEA4-1* (Saha et al. 2016) and *AnnBj2* that resulted in upregulation of gene expression i.e. *RAB18* (ABA-dependent) and *DREB2B* (ABA independent pathway genes (Ahmed et al. 2017) provides tolerance to salinity stress.

4 Molecular Advent and Omics Approaches for Stress Tolerance

To cope up with the environmental abiotic stresses i.e. stress avoidance/tolerance responses; large numbers of genes/TFs have been identified in different plant genotypes including modal plants such as *Arabidopsis* and rice (Yamaguchi-Shinozaki and Shinozaki 2006; Perez-Clemente et al. 2013). Through signaling cascade pathways, stress inducible/responsive genes have been classified broadly into two categories i.e. functional genes and regulatory genes. Functional genes directly involved in cell protection from stresses via detoxification enzymes, water channel and functional proteins such as late embryogenesis abundant (LEA) proteins. However, regulatory genes involved the numbers of important regulatory proteins, transcription factors and protein kinases in response to abiotic stresses. Transcription factors (TFs), naturally acts as key regulators in plant cellular process and also known to have pivotal role in abiotic stress tolerance by perceiving stress signal and conversion to stress responsive gene expression. In the promoter region of various stress responsive genes, different transcription factors with its cis acting elements serve as molecular switches for gene expression to provide stress tolerance (Golldack et al. 2011). Biotechnological interventions such as plant genetic engineering and transgenic technologies resulted in targeted modification of plant regulatory pathways to provide stress tolerance by use of transcription factors such as *MYB, bZIP, HSF, CBF/DREB1, NAC, TGA6, BOS1, AP2/EREBP, WRKY, ERF/AP2* (Ambawat et al. 2013; Shao et al. 2015). In *Brassica* species; Supriya et al. (2006) and Sharma et al. (2010) reported *MYB* transcription factors and *AtMYB2* gene provide tolerance to drought stress. Similarly; Wang et al. (2015) and Kayum et al. (2016) reported *MYB* transcription factors such as *BrMYB88, BrMYB137, BrMYB154, BrMYB210* and *BrMYB222* in response to osmotic and cold stress in *Brassica* species. Ahmed et al. (2012) and Saha et al. (2015) reported function of different *BZR* transcription factor in *B. rapa* against drought, cold, and ABA stresses. Liu et al. (2014) reported 204 NAC transcription factors in Chinese cabbage (*B. rapa*) involved in stress tolerance and cellular processes. Ahmed et al. (2015) also revealed differential expression of cold stress responsive transcription factors (*BoCRG54*, *BoCRG56*, *BoCRG59*, *BoCRG62*, *BoCRG70*, *BoCRG72* and *BoCRG99*) in cabbage (*B. oleracea*) genotypes.

In last decade, high throughput approaches of omics (functional genomics, transcriptomics, proteomics and metabolomics) revolutionized the biological research. Model plant i.e. *A. thaliana* of family Brassicaceae shared a close phylogenetic relationship with *Brassica* species. Therefore; significant understanding of *A. thaliana* stress physiology directly resulted in *Brassica* crop improvement. In *Arabidopsis* and *Brassica*, genomics approaches have been extensively studied to understand the complex mechanism of abiotic stress tolerance with their multi-faceted plant response (Stephen et al. 2012; Kumar et al. 2018b). Creation of novel mutants by means of physical, chemical and genetic perturbations through utilization of genetics approaches such as forward genetics and reverse genetics not only resulted in identification of the improved phenotypes but also led to the gene discovery via targeting loci that are responsible for stress tolerance (Thorlby et al. 2004; Xin et al. 2007). This approach successfully applied for identification of *Arabidopsis* mutant lines i.e. *Eskimo1* to tolerate cold/freezing stress. Also; Ishitani et al. (1997) used genetic screening to reveled complex networking interactions in expression to *COR/ERD/RD* genes under the control of *RD29a* promoter. *RD29a* expression is induced in exposure to low temperature, salinity, cold and dehydration. Zhuang et al. (2009) reported expressed sequence tags, gene discovery and gene expression pattern of *AP2/ERF* family in *B. rapa*. High throughput genetic screening of resulted mutants helpful in pin up the underlying mechanism of stress signal transduction (Shinozaki and Yamaguchi-Shinozaki 2007; Kumar and Wigge 2010). With the availability of genomic resources in modal organisms like *Arabidopsis thaliana* and sequenced *Brassica* genome will continue to supplement additional information and targets to improve *Brassica* species also allow refinement in the ongoing *Brassica* improvement strategies. Recently, Verma et al. (2019) carried out genome wide identification and characterization in *B. juncea* and *B. rapa* for abiotic stress responsive superoxide dismutase (*SOD*) gene family. SOD family proved as a metallo-enzyme gene family to provide tolerance against abiotic stress tolerance. Genome wide analysis resulted in 29 *SOD* genes in *B. juncea* and 18 *SOD* genes in *B. rapa*. 14 *SOD* genes in *B. rapa* and 10 *SOD* genes in *B. juncea* reported to provide drought, heat and salt stress. Identified abiotic-stress responsive genes could be further exploit for abiotic stress tolerance in different *Brassica* species.

Proteomic analysis is an important high throughput approach to study plant stress response in stress tolerant/adopted genotypes. Different crop/cultivars/genotypic difference to various abiotic stress conditions resulted in alteration of cellular homeostasis and localization, protein isoforms, post-translation modification, protein-protein interactions etc. Protein isoforms resulted in primary sequences differences whereas, protein post-translation modification variations revealed distinctive alteration in amino acid residues. Simultaneously, interactomics studies also provide evidence and proof of concept to identify and validate protein biological functions within given molecular context. Comparative and differential proteomic studies underlying the mechanism to understand the stress physiology i.e. stress adaptation of different genotypes in response to various abiotic stress. Genotypic stress adaptability revealed increased level of constitutive expression of stress responsive proteins i.e. molecular chaperones, heat and cold shock proteins, reactive oxygen species scavenging) and also adjust metabolic flux/energy metabolism in the tolerant genotype than the susceptible ones due to photosynthetic and aerobic respiration activation (Kosova et al. 2018). Abat and Deswal (2009) reported the nitric oxide (*NO*) effect as a stress signaling molecule by differential changes in the S-nitrosoproteome of *B. juncea* due to low temperature. Exposure to cold stress leads to the decrease in RubisCO carboxylation activity. Tolerance and adaptability to stress condition are two important terms of stress-coping strategies to the particular stress condition. Abiotic stress tolerant versus susceptible plants/genotypes reported to have significant differences in the development of stress related proteins. In stress susceptible genotypes; programmed cell death process activation takes place due to cellular damage. However, in stress tolerant plants, growth and development process restored due to activation of signal transduction pathway with respect to different environmental abiotic stresses. Lin et al. (2015) reported proteomics analysis in *B. oleracea* L. var. *italica* commonly known as broccoli as highly affected under high temperature and waterlogging stresses. Through proteome profiling, thirty one differentially expressed proteins were identified in heat tolerant and susceptible cultivar. Furthermore, cloning of stress-responsive Rubisco genes resulted in better H_2O_2 scavenging and improved carbon fixation efficiency to give ample energy to facilitate stress tolerance in waterlogging stress condition and high temperature 40 °C. For comprehensive understanding of plant proteome analysis to abiotic stress; plants responds in dynamic way via different phases such as alarm, acclimation, resistance, exhaustion and recovery. As per the available literature, it has been suggested that further detail scientific investigation is needed to understand plant stress complexity because multiple stress treatments are not equal to the sum of the effects of single stress treatment. Kumar et al. (2018b) reviewed recent biotechnological advances made in abiotic stress tolerance in broccoli—an important vegetable crop *Brassica* genus. Similarly; Hasanuzzaman et al. (2019) described in details about different approaches in enhancing abiotic stress tolerance in plants in perspective to plant physiology and chemistry, metabolic and molecular changes and genomics and proteomics. Recently in family Brassicaceae; Luo and Zhang (2019) studied proteomic changes in response to Cd (heavy metal) stress in *B. napus*. It has been reported that different xylem sap proteins were differentially induced and expressed under cadmium stress and particularly, plant defensin protein (*BnPDFL*) that belongs to stress/oxidoreductase category have a critical role in Cd tolerance. Finally molecular advent and omics approaches provide an efficient platform to counteract environmental stress conditions for genotypic improvement via better understanding of metabolic pathways and their cross-talk under abiotic stresses.

5 Conclusion

Brassica being an economically important genus grown worldwide in different agro-climatic conditions are highly affected by salinity, drought, heat and frost stresses. These abiotic stresses not only affect the plant morph-physiological performance but also crop productivity. To tackle this abiotic stresses problem, molecular and biotechnology interventions such as plant genetic engineering, RNAi technology, QTL (Quantitative trait loci) mapping, Genome wide association studiers, association mapping, next generation sequencing and the recent omics and targeted gene editing (CRISPR/Cas9) approaches plays significant role for stress resistance and tolerance in family Brassicaceae. In, *Brassica* crop improvement program, transgenic technology plays a major role worldwide by incorporation of abiotic stress responsive genes/transcription factors such as (antiporters, osmolytes, manitol, glycine betain, proline, *ABA*, *DREBI*, *HSP*, *MAPK* etc.) into the transgenic *Brassica* genotypes/cultivars that resulted in enhanced abiotic stress tolerance. Modern approaches like omics, next generation sequencing led to the screening, identification and characterization of abiotic stress tolerant/resistant traits and resulted in development of stress tolerant cultivars and supplementing the conventional breeding in time saving and eco-friendly manner.

Acknowledgements The senior author (PK) thankfully acknowledges the award of National Post Doctoral Fellowship, Science and Engineering Research Board, Department of Science & Technology, Government of India, New Delhi, India.

References

Abat JK, Deswal R (2009) Differential modulation of S-nitrosoproteome of *Brassica juncea* by low temperature: change in S-nitrosylation of Rubisco is responsible for the inactivation of its carboxylase activity. Proteomics 9:4368–4380

Ahmar S, Liaqat N, Hussain M, Salim MA, Shabbir MA, Ali MY et al (2019) Effect of abiotic stresses on *Brassica* species and role of transgenic breeding for adaptation. Asian J Res Crop Sci 3(1):1–10

Ahmed NU, Park JI, Jung HJ, Lee IH, Song IJ, Yang SY et al (2012) Identification and characterization of longevity assurance gene related to stress resistance in *Brassica.* Afr J Biotech 11:12721–12727

Ahmed NU, Jung HJ, Park JI, Cho YG, Hur Y, Nou IS (2015) Identification and expression analysis of cold and freezing stress responsive genes of *Brassica oleracea.* Gene 554:215–223

Ahmed I, Yadav D, Shukla P, Vineeth TV, Sharma PC, Kirti PB (2017) Constitutive expression of *Brassica juncea* annexin, *AnnBj2* confers salt tolerance and glucose and ABA insensitivity in mustard transgenic plants. Plant Sci 265:12–28

Akula R, Gill SS (2018) Metabolic adaptations in plants during abiotic stress. Taylor & Francis (CRC Press), USA

Ambawat S, Sharma P, Yadav NR, Yadav RC (2013) *MYB* transcription factor genes as regulators for plant responses: an overview. Physiol Mol Biol Plants 19:307–321

Atkinson NJ, Urwin PE (2012) The interaction of plant biotic and abiotic stresses: from genes to the field. J Exp Bot 63(10):3523–3544

Cheng S, Willmann MR, Chen HC, Sheen J (2002) Calcium signaling through protein kinases. The *Arabidopsis* calcium-dependent protein kinase gene family. Plant Physiol 129:469–485

Chinnusamy V, Schumaker K, Zhu JK (2004) Molecular genetic perspectives on cross-talk and specificity in abiotic stress signaling in plants. J Exp Bot 55:225–236

Chunqing L, Xuekun Z, Zhang K, Hong A, Hu K, Wen J, Shen JX et al (2015) Comparative analysis of the *Brassica napus* root and leaf transcript profiling in response to drought stress. Int J Mol Sci 16:18752–18777

Dalal M, Tayal D, Chinnusamy V, Bansal KC (2009) Abiotic stress and ABA-inducible Group 4 LEA from *Brassica napus* plays a key role in salt and drought tolerance. J Biotechnol 139(2):137–145

Deshmukh R, Sonah H, Patil G, Chen W, Prince S, Mutava R, Vuong T, Valliyodan B, Nguyen HT (2014) Integrating omic approaches for abiotic stress tolerance in soybean. Front Plant Sci 5:244

Golldack D, Luking I, Yang O (2011) Plant tolerance to drought and salinity: stress regulating transcription factors and their functional significance in the cellular transcriptional network. Plant Cell Rep 30:1383–1391

Hanschen FS, Herz C, Schlotz N, Kupke F, BartoloméRodríguez MM, Schreiner M, Rohn S, Lamy E (2015) The *Brassica* epithionitrile 1-cyano-2, 3-epithiopropane triggers cell death in human liver cancer cells *in vitro*. Mol Nutr Food Res 59:2178–2189

Hasanuzzaman M, Nahar K, Fujita M, Oku H, Islam MT (2019) Approaches for enhancing abiotic stress tolerance in plants. CRC Press, Taylor & Francis, London

Huang J, Hirji R, Adam L, Rozwadowski KL, Hammerlind JK, Keller WA, Selvaraj G (2000) Genetic engineering of glycine betaine production toward enhancing stress tolerance in plants: metabolic limitations. Plant Physiol 122(3):747–756

Ishitani M, Xiong L, Stevenson B, Zhu JK (1997) Genetic analysis of osmotic and cold stress signal transduction in *Arabidopsis*: interactions and convergence of abscisic acid dependent and abscisic acid independent pathways. Plant Cell 9:1935–1949

Jamil A, Riaz S, Ashraf M, Foolad MR (2011) Gene expression profiling of plants under salt stress. Crit Rev Plant Sci 30:435–458

Kayum MA, Park JI, Ahmed NU, Saha G, Chung MY, Kang JG et al (2016) Alfin-like transcription factor family: characterization and expression profiling against stresses in *Brassica oleracea*. Acta Physiol Plant 38:127

Knight H (1999) Calcium signaling during abiotic stress in plants. Int Rev Cytol 195:269–324

Kong F, Mao S, Du K, Wu M, Zhou X, Chu C, Wang Y (2011) Comparative proteomics analysis of *OsNAS1* transgenic *Brassica napus* under salt stress. Chin Sci Bull 56(22):2343–2350

Kosova K, Vitamvas P, Urban MO, Prasil IT, Renaut J (2018) Plant abiotic stress proteomics: the major factors determining alterations in cellular proteome. Front Plant Sci 9:122

Kumar P, Srivastava DK (2015) High frequency organogenesis in hypocotyl, cotyledon, leaf and petiole explants of broccoli (*Brassica oleracea* L. var. *italica*), an important vegetable crop. Physiol Mol Biol Plants 21(2):279–285

Kumar P, Srivastava DK (2016) Biotechnological advancement in genetic improvement of broccoli (*Brassica oleracea* L. var. *italica*), an important vegetable crop. Biotechnol Lett 38(7):1049–1063

Kumar SV, Wigge PA (2010) *H2A*. Z-containing nucleosomes mediate the thermosensory response in *Arabidopsis*. Cell 140:136–147

Kumar D, Shekhar S, Bisht S, Kumar V, Varma A, Kumar M (2015a) Ectopic overexpression of lectin in transgenic Brassica juncea plants exhibit resistance to fungal phytopathogen and showed alleviation to salt and drought stress. J Bioeng Biomed Sci 5:147

Kumar P, Gambhir G, Gaur A, Srivastava DK (2015b) Molecular analysis of genetic stability in in vitro regenerated plants of broccoli (*Brassica oleracea* L. var. *italica*). Curr Sci 109(8):1470–1475

Kumar P, Gaur A, Srivastava DK (2017) *Agrobacterium*—Mediated insect resistance gene (cry1Aa) transfer studies pertaining to antibiotic sensitivity on cultured tissues of broccoli (*Brassica oleracea* L. var. *italica*): an important vegetable crop. Int J Veg Sci 23:523–535

Kumar P, Gambhir G, Gaur A, Thakur AK, Sharma KC, Srivastava DK (2018a) Development of transgenic broccoli with *cryIAa* gene for resistance against diamondback moth (*Plutella xylostella*) 3 Biotech 8(7):299

Kumar P, Thakur AK, Srivastava DK (2018b) Genetic engineering approaches for abiotic stress tolerance in broccoli: recent progress. In: Akula R, Gill SS (eds) Metabolic adaptations in plants during abiotic stress. Taylor & Francis (CRC Press), USA, pp 363–367

Kumar P, Dhiman K, Srivastava DK (2018c) Morphogenic potential of different explants of broccoli (*Brassica oleracea* L. var. *italica*): important "nutrient rich" vegetable, using Thidiazuron. In: Ahmad N, Faisal M (eds) Thidiazuron: from urea derivative to plant growth regulator. Springer publishing house, pp 373–392

Lin HH, Lin KH, Chen SC, Shen YH, Lo HF (2015) Proteomic analysis of broccoli (*Brassica oleracea*) under high temperature and waterlogging stresses. Bot Stud 56:18

Liu TK, Song XM, Duan WK, Huang ZN, Liu GF, Li Y, et al (2014) Genome-wide analysis and expression patterns of NAC transcription factor family under different developmental stages and abiotic stresses in Chinese cabbage. Plant Mol Biol Rep 32:1041–1056

Luo J, Zhang Z (2019) Proteomic changes in response to Cd under cadmium stress and functional validation. BMC Plant Biol 19:280

Luo J, Tang S, Mei F, Peng X, Li J, Li X, Wu G (2017) *BnSIP1-1*, a trihelix family gene, mediates abiotic stress tolerance and ABA signaling in *Brassica napus*. Front Plant Sci 8:1–14

Metwali EMR, Fuller MP, Jellings AJ (2012) *Agrobacterium* mediated transformation of anti-stress genes into cauliflower (*Brassica oleracea* var. *botrytis* L.). 2. Transformation and confirmation of stress tolerance. Aust J Basic Appl Sci 6(5):31–39

Mittler R, Blumwald E (2010) Genetic engineering for modern agriculture: challenges and perspectives. Annu Rev Plant Biol 61:443–462

Murakami Y, Tsuyama M, Kobayashi Y, Kodama H, Iba K (2000) Trienoic fatty acids and plant tolerance of high temperature. Science 287:476–479

Nagaharu U (1935) Genome analysis in *Brassica* with special reference to the experimental formation of *B. napus* and peculiar mode of fertilisation. Jpn J Bot 7:389–452

Odongo GA, Schlotz N, Herz C, Hanschen FS, Baldermann S, Neugart S et al (2017) The role of plant processing for the cancer preventive potential of Ethiopian kale (*Brassica carinata*). Food Nutr Res 61(1):1–11

Parmar N, Singh KH, Sharma P, Singh L, Kumar P, Nanjundan J, Khan YJ, Chauhan DK, Thakur AK (2017) Genetic engineering strategies for biotic and abiotic stress tolerance and quality enhancement in horticultural crops: a comprehensive review. 3 Biotech 7:239

Pathak RK, Baunthiyal M, Pandey D, Kumar A (2018) Augmentation of crop productivity through interventions of omics technologies in India: challenges and opportunities. 3 Biotech 8:454

Perez-Clemente RM, Vives V, Zandalinas SI, Lopez-Climent MF, Munoz Gomez-Cadenas A (2013) Biotechnological approaches to study plant responses to stress. Biomed Res Int. https://doi.org/10.1155/2013/654120

Prasad KVSK, Sharmila P, Kumar PA, Saradhi PP (2000) Transformation of *Brassica juncea* (L.) Czern with bacterial *codA* gene enhances its tolerance to salt stress. Mol Breed 6(5):489–499

Qamarunnisa S, Jamil I, Raza S, Azhar A, Naqvi SHM (2015) Genetic improvement of canola against abiotic stress through incorporation of *DREB* gene. Asian J Agric Biol 3(3):77–104

Rajagopal D, Agarwal P, Tyagi W, Singla-Pareek SL, Reddy MK, Sopory SK (2007) Pennisetum glaucum Na^+/H^+ antiporter confers high level of salinity tolerance in transgenic *Brassica juncea*. Mol Breed 19:137–151

Rajwanshi R, Kumar D, Yusuf MA, DebRoy S, Sarin NB (2016) Stress-inducible overexpression of *glyoxalase I* is preferable to its constitutive overexpression for abiotic stress tolerance in transgenic *Brassica juncea*. Mol Breed 36(6):1–15

Ravanfar SA, Aziz MA (2014) Shoot tip regeneration and optimization of Agrobacterium tumefaciens-mediated transformation of Broccoli (*Brassica oleracea* var. *italica*) cv. Green Marvel. Plant Biotechnol Rep. https://doi.org/10.1007/s11816-014-0340-5

Ravanfar SA, Orbovic V, Moradpour M, Abdul M, Aziz MA, Karan R, Wallace S, Parajuli S (2017) Improvement of tissue culture, genetic transformation, and applications of biotechnology to *Brassica*. Biotechnol Genet Eng Rev. https://doi.org/10.1080/02648725.2017.1309821

Saha G, Park JI, Jung HJ, Ahmed NU, Kayum MA, Kang JG et al (2015) Molecular characterization of *BZR* transcription factor family and abiotic stress induced expression profiling in *Brassica rapa*. Plant Physiol Biochem 92:92–104

Saha B, Mishra S, Awasthi JP, Sahoo L, Panda SK (2016) Enhanced drought and salinity tolerance in transgenic mustard [*Brassica juncea* (L.) Czern & Coss.] overexpressing *Arabidopsis* group 4 late embryogenesis abundant gene (*AtLEA4-1*). Environ Exp Bot 128:99–111

Saxena M, Roy SD, Singla-Pareek SL, Sopory SK, Bhalla-Sarin N (2011) Overexpression of the *glyoxalase II* gene leads to enhanced salinity tolerance in *Brassica juncea*. Open Plant Sci J 5:23–28

Serrano R, Mulet JM, Rios G, Marquez JA, De-Larrinoa IF, Leube MP, Mendizabal I et al (1999) A glimpse of the mechanisms of ion homeostasis during salt stress. J Exp Bot 50:1023–1036

Shah N, Anwar S, Xu J, Hou Z, Salah A et al (2018) The response of transgenic *Brassica species* to salt stress: a review. Biotech Lett 40:1159–1165

Shao H, Wang H, Tang X (2015) *NAC* transcription factors in plant multiple abiotic stress responses: progress and prospects. Front Plant Sci 6:902

Sharma P, Kumar H, Supriya, Kumar S, Yadav NR, Pradeep, Yadav RC, Singh D (2010) Comparative myb gene expression in Brassica species under drought stress through semi-quantitative RT-PCR. Cruciferae Newslett 29:23–25

Shi H, Ishitani M, Kim C, Zhu JK (2000) The *Arabidopsis thaliana* salt tolerance gene *SOS1* encodes a putative Na^+/H^+ antiporter. Proc Natl Acad Sci 97:6896–6901

Shi H, Lee BH, Wu SJ, Hu JK Z (2003) Overexpression of a plasma membrane Na^+/H^+ antiporter gene improves salt tolerance in *Arabidopsis thaliana*. Nat Biotechnol 21:81–85

Shinozaki K, Yamaguchi-Shinozaki K (2007) Gene networks involved in drought stress response and tolerance. J Exp Bot 58:221–227

Srivastava S, Fristensky B, Kav NNV (2004) Constitutive expression of a *PR10* protein enhances the germination of *Brassica napus* under saline conditions. Plant Cell Physiol 45:1320–1324

Srivastava DK, Kumar P, Sharma S, Gaur A, Gambhir G (2016) Genetic engineering for insect resistance in economically important vegetable crops. In: Ahmad N, Anis M (eds) Plant tissue culture: propagation, conservation and crop. Springer Publishing House, pp 343–378

Stephen J, Robinson Isobel, Parkin AP (2012) *Brassica* crops: improving abiotic stress tolerance—current status and prospects. In: Tuteja N, Gill SS, Tiburcio AF, Tuteja R (eds) Improving crop resistance to abiotic stress. Wiley-VCH Verlag GmbH & Co. KGaA, USA, pp 1315–1349

Sun XE, Feng XX, Li C, Zhang ZP, Wang LJ (2015) Study on salt tolerance with *YHem1* transgenic canola (*Brassica napus*). Physiol Plant 154(2):223–242

Supriya YNR, Yadav RC, Singh D (2006) Isolation of biologically active RNA for generating *MYB* transcription factor cDNAs in *Brassica species*. Brassica 8:95–98

Thomashow MF (1994) *Arabidopsis thaliana* as a model for studying mechanisms of plant cold tolerance. Arabidopsis 807–834

Thorlby G, Fourrier N, Warren G (2004) The SENSITIVE TO FREEZING2 GENE, required for freezing tolerance in Arabidopsis thaliana, encodes a beta—glucosidase. Plant Cell 16:2192–2203

Verma D, Lakhanpal N, Singh K (2019) Genome-wide identification and characterization of abiotic-stress responsive *SOD* (superoxide dismutase) gene family in *Brassica juncea* and *B. rapa*. BMC Genomics 20:227

Wang W, Vinocur B, Shoseyov O, Altman A (2004) Role of plant heat-shock proteins and molecular chaperones in the abiotic stress response. Trends Plant Sci 5:244–252

Wang QB, Xu W, Xue QZ, Su WA (2010) Transgenic *Brassica chinensis* plants expressing a bacterial *codA* gene exhibit enhanced tolerance to extreme temperature and high salinity. J Zhejiang Univ Sci B 11(11):851–861

Wang X, Wang H, Wang J, Sun R, Wu J, Liu S, Bai Y, Mun J et al (2011) The genome of the mesopolyploid crop species *Brassica rapa*. Nat Genet 43(10):1035–1039

Wang Z, Tang J, Hu R, Wu P, Hou XL, Song XM et al (2015) Genome-wide analysis of the *R2R3-MYB* transcription factor genes in Chinese cabbage (*Brassica rapa* ssp. *pekinensis*) reveals their stress and hormone responsive patterns. BMC Genomics 16:17

Xin Z, Mandaokar A, Chen J, Last RL, Browse J (2007) *Arabidopsis ESK1* encodes a novel regulator of freezing tolerance. Plant J 49:786–799

Yamaguchi-Shinozaki K, Shinozaki K (2006) Transcriptional regulatory networks in cellular responses and tolerance to dehydration and cold stresses. Annu Rev Plant Biol 57:781–803

Yoshida T, Mogami J, Yamaguchi-Shinozaki K (2014) ABA-dependent and ABA-independent signaling in response to osmotic stress in plants. Curr Opin Plant Biol 21:133–139

Zhang HX, Hodson JN, Williams JP, Blumwald E (2001) Engineering salt-tolerant *Brassica* plants: characterization of yield and seed oil quality in transgenic plants with increased vacuolar sodium accumulation. Proc Natl Acad Sci 98(22):12832–12836

Zhu JK (2001) Plant salt tolerance. Trends Plant Sci 6:66–71

Zhuang J, Xiong AS, Peng RH, Gao F, Zhu B, Zhang J, Fu XY, Jin XF, Chen JM, Zhang Z, Qiao YS, Yao QH (2009) Analysis of *Brassica rapa* ESTs: gene discovery and expression patterns of family genes. Mol Biol Rep 37(5):2485–2492

Biotechnological Approach for Enhancing Capability of *Brassica oleracea* var. italica Against Stresses Under Changing Climate

Mohammad Mafakheri and Mojtaba Kordrostami

Abstract By drastic increment in frequency fluctuation of climatic phenomena worldwide which the consequences majorly negatively have been multifacetedly affecting world's economy with agriculture section at the pinnacle, is threatening the food security of a large portion of growing population on the Earth to a dangerous level in addition to the percentage that already dealing with food inadequacy. The impacts on this section mainly targeting the productivity and quality of temperate vegetable crops (mainly negative and positive is some cases) and among them Brassicas unquestionably key crops that are an essential source of plant-based protein which increases the incidence of environmental stressors, high temperature, in particular, leads to yield losses to a large quantity since they are vernalization required, a low-temperature period to reach the reproduction phase. In case of the point, Broccoli (*Brassica oleracea* L. var. italica) a nutritionally and medicinally well-known staple vegetable that in the recent two or three decades has greatly mirrored the devastating effects of global warming owing to its supersensitive crud initiation period and degradation in the salability of heads as a result of heat stress which hopefully with the aim of biotechnology toolbox initial steps toward untwining the complex polygenetically regulated characteristics of heat tolerance and crud quality have been taken and lines of broccoli with non-vernalization requirements have been identified that could expand the broccoli cultivation to tropics. Putative QTLs associated with high-temperature resistance and crud initiation have been detected through gene pyramiding and molecular assistant selection aims to develop commercial cultivars. Gene transformation in broccoli to improve the capability of the current cultivars in addition to elevated CO_2 scenario with its influences on broccoli have been studied extensively. The aim of this chapter is to reflect the result of studies concerning reaction and identifying the underlying genetic or protein bases involved in resistance to abiotic or biotic stresses.

M. Mafakheri
Department of Horticultural Sciences, Faculty of Agricultural Sciences, University of Guilan, P.O. Box 41635-1314, Rasht, Iran

M. Kordrostami (✉)
Nuclear Agriculture Research School, Nuclear Science and Technology Research Institute (NSTRI), Karaj, Iran
e-mail: kordrostami009@gmail.com

M. Hasanuzzaman (ed.), *The Plant Family Brassicaceae*,
https://doi.org/10.1007/978-981-15-6345-4_16

Keywords High-temperature stress · Elevated CO_2 · Putative QTL · Plant biotechnology · Vernalization · Climate

1 Introduction

Vegetables are unquestionably critical owing to their nutritional values since they are a major source to supply a large number of mineral elements, micronutrients, vitamins, antioxidants, phytochemicals, and dietary fibers to the public daily diet (Béliveau and Gingras 2017; Dias 2019; Siddiq and Uebersax 2018; Davis 2017). While almost every vegetable is known for its own nutritional content, tastes, and etc., but some are significantly more known and important, as a case in point, Broccoli (*Brassica oleracea* L. var. italica) is a worldwide important staple cold-season (i.e., preferred temperature below 18 °C) vegetable crop, in particular in developed countries belong to family Brassicaceae ($2n = 18$)(Kumar and Srivastava 2016; Lin et al. 2018; Raiola et al. 2018; Šamec and Salopek-Sondi 2019), a family that harbors top ten economically important food crops with broccoli at topmost (Francisco et al. 2017). Being a rich source of health-enhancing primary and secondary constituents that confer broccoli anti-carcinogenic as well as anti-cardiovascular diseases and anti-aging potentials (Baenas et al. 2017; Suresh et al. 2018; Li et al. 2018; Nawaz et al. 2018; dos Santos et al. 2018) have gained the attention for broccoli in which based on recent public surveys now it's the most popular vegetable in the US (Wida 2019). The annual production of broccoli on the worldscale reached over 50 million tones which has been in constant increase since 1997 with an annual export value of 1.49 billion US dollar and having China, Spain, and the US as the biggest producer, exporter, and importer, respectively, and the available data of yearly production value of broccoli (i.e., fresh and processed frozen) in 2018 in the US was 739 million US dollar (Shahbandeh 2019). Further, Spain and China which hold 44 and 2.4% share of the world's production averagely had 382 and 123 million US dollars worth export of broccoli as of 2018, respectively ("Broccoli suppliers" 2018).

A combination of enormous health benefits from having broccoli in diet and a growing demand with high profitability has led to a significant increase in the cultivation of broccoli where climatic condition is in favor, a requirement that broccoli is highly selective for and its production confined to specific areas or low-temperature periods throughout the year (Mark and Thomas 2011; Branham and Farnham 2019), in view of the fact that its flowering initiation is sparked by facultative vernalization and it's cool temperature that facilitates the formation of not misshapen normal curds whereas exposure to the temperature range higher than its restricted temperature window halts the differentiation of floral meristems which has deformed irregular floral buds (Lin et al. 2018; Björkman and Pearson 1998; Akter et al. 2018; Irwin et al. 2012). To be more precise, increasing temperature to 30 °C and above during the generative phases when the buds are forming has negative impacts or even could be detrimental, that's why hybrid commercial broccoli cultivars mainly have a cold and dry climatic condition in interest (Kałużewicz et al. 2009; Mark and Thomas 2011).

Considering such a limited temperature range to grow properly and with saleable products on one hand and a constant increase in global temperature, on the other hand, have left scholars in this industry with serious challenges. Even a small increment in mean temperatures could generate numerous positive or negative impacts on vegetable products, with dependency on the climatic condition of where they are farmed (McDonald and Warland 2018; McKeown et al. 2004). The estimated increase in global temperature by 2050 is 1.5 °C (Rumjaun et al. 2018). The consequences of rising global temperature predicted to onset "a knock on-chain effect" which every aspect on climate will be influenced by such as an abnormal fluctuation in extreme weather events, including "heatwaves" (Tigchelaar et al. 2018; Yadav et al. 2019), which such phenomenon can harm the productivity of broccoli at a large scale.

Radical changes in climate are accompanied by a population explosion that reaches 10 billion by the middle of twenty-first century which expands the gap between supply versus demand of food products, jeopardizes food security of a large portion of the world's population, and limits the accessibility of nutritious food for public particularly in developing countries (Motesharrei et al. 2016; Béné et al. 2015; Mann 2018). To make the matter more complex, increasing in temperature which has been accompanied by rising sea levels will add salt to soils as well as the underground water table, which all ends up degrading arable lands to harming the productivity of lands worldwide. Although broccoli is moderately salinity tolerant (Tian et al. 2014), world-scale surveys indicated that of the 20% cultivated arable lands and 33% of agricultural lands that receive regular irrigation experienced a high concentration of salt (Pimentel et al. 2004; Machado and Serralheiro 2017). Moreover, the lands that lose to salinity is as much as 10% yearly and turn to unproductive lands, which in addition to application of damaging management strategies, saltwater, and etc., climate change has been the true original cause of this disastrous growing event that until 2050, over half of the world's arable lands would be unproductive containing high level of salt (Jamil et al. 2011; Shrivastava and Kumar 2015). Additionally, of the effects of climate change that considered positive for a large number of agricultural crops is the elevated CO_2 in the atmosphere which many vegetables have reacted positively. For example, the elevated CO_2 significantly improved the growth rate and tolerance of broccoli to salt stress (Zaghdoud et al. 2016a). The effects of climate change are complicated and intertwined, that is, the elevated CO_2 might be advantageous for growth rate of crops, however, degradation in protein content of food products under elevated CO_2 climate have frequently investigated, which the reduction in vegetables expected to be as high as −17.3% (Dong et al. 2018; Huang and Xu 2015; Schmid et al. 2016; Medek et al. 2017; Myers et al. 2019), besides, considering the devastating impacts of such phenomena in nutritional prospective and percentage of people in the world majorly rely on plant source protein are another concerning part of the climate change scenario (Toreti et al. 2019; Medek et al. 2017). Moreover, elevation of CO_2 accompanied by high temperature which early maturation and significant decrease in dry matter of plants in a linear manner observed to the consequences of their combination, this is in contrast to the common perception that increasing in CO_2 under the climate change would be in favor of many plant species (Hasanuzzaman et al. 2013a).

The prevalence level of the impacts of climate change can be seen at the disruption of very fundamental rather infinitesimal ecological relationships and tranquilities between crops, pathogens, pest, and weeds to an extensive level. Climate change-driven outbreaks of pests and pathogens enhanced in frequency since increasing temperature as well as CO_2 is advantageous for some fungi and insects. Thus, it's of prime importance to gain knowledge of climate change impacts in all aspects which enables us to address the challenges in an efficient way (Medek et al. 2017; Fanzo et al. 2018; El-Sayed et al. 2018; Rehman and Kumar 2018).

Overall, changing climate generates multifactor stresses (Gray and Brady 2016; Räsänen et al. 2016; Deryng et al. 2014), there is strong evidence of the fact that plants have been receiving the environmental upheaval by increasing the occurrence of environmental stress as a result of climate change since the last decades, considering the point that these impacts of climate change add up eventually which makes the need for alleviation and adaptation strategies are a matter of urgency (Fanzo et al. 2018) since developing stress-tolerant cultivars is of the best possible options to minimize the toll of climate change. In this chapter main concentrations are on how climate change induces abiotic and biotic stresses (i.e., some particular stress of both types, depending on the quantity of research that has been devoted) for broccoli and so far to what extent biologists have taken the prerequisite steps (such as in vitro tissue culture) and utilized biotechnological toolbox to fight the consequences of climate change for broccoli by gathering genetical background knowledge and paving the road to produce superior cultivars.

2 High-Temperature Stress

If the projected increase in average global temperature comes to reality (its effects have been already experienced by plants to a large degree), acceleration of breeding programs and establishing real-time strategies to develop crop varieties resistance to high temperature are the matters of urgency for plant breeders (Hasanuzzaman et al. 2013a, b; Nahar et al. 2015a, b). Major temperate vegetables, cole crops belong to *B. oleracea* species are cool-weather vegetables that involve several groups of popular vegetable crops such as cauliflower (*B. oleracea* var. Botrytis), cabbage (*B. oleracea* var. Capitata), and most importantly in our case broccoli. With having an optimum temperature threshold of 15–18 °C in which these crops produce products with appropriate market quality (Maynard and Hochmuth 2007; Wurr et al. 1995; Romo and Tylianakis 2013; McKeown et al. 2004; Hassini et al. 2017). Hence, these crops are in favor of temperate regions with mild winters. Naturally, close crops and broccoli in particular limited to a short-list of temperate areas around the world which by increasing the global temperature at an alarming rate the quality of curds in broccoli is threatened even though many farmers of broccoli have attempted to grow this crop at regions with considerably high-temperature regime (e.g., 25–30 °C). Given that transmission of phases from vegetative to reproduction depends on temperature and not on daylight, hence developing tolerant cultivars to high temperature using modern approaches is essential for broccoli's ever-expanding market (Miller et al. 1985; Mourão and Hadley 1998; Tan et al. 2000; Akter et al. 2018; Witzel et al. 2015).

2.1 QTL Identification

In general, broccoli cultivars are chiefly cool-season, hence they have their optimal productivity in regions with specific average temperature status. As described above, by having specific temperature requirements, the global warming which increases the average annual temperature, as well as incidences of heat shocks, can be indeed a serious threat owing to the very fact that broccoli is highly sensitive and to complete the transition from vegetative to reproduction phase needs vernalization that involves a duration of low temperature, which makes the establishment of extensive breeding programs essential to develop non-vernalization required cultivars of broccoli resistance to heat stress that if such promising cultivars would be achieved, cultivation of broccoli can be continued in a temperate region and also expand to warm regions such as tropics and its production as a part climate change resilient agriculture will be possible. To this end, conventional breeding procedures are not efficient and time-saving as biotechnological methodologies, hopefully, during the last two decades, comprehensive studies have been conducted to explore putative and determinative underlying QTLs responsible for regulating flowering time and heat tolerance in broccoli to lower or even eliminate season or area dependency (Table 1), which the first breeding program initiated by USAD in 1990 to develop broccoli cultivars with that capabilities is of the superb instances, the main approach in that program has been selected and accumulate related putative alleles in a specific number of lines for high-temperature resistance and consequently produce a cultivar(s) with the ability to provide florets with appropriate quality grow at the temperature of 30 °C and above. The perspective of that group toward heat tolerance in broccoli was the quantitative nature of this feature which foreseeably controls by a number of alleles located on various loci.

Of the progresses that group have made to date is an extensive work reported by Branham et al. (2017) on mapping double haploid populations of broccoli transgressive segregation that was results of the cross of a tolerant line (USVL138) with an intolerant line (USVL108) detected five QTLs and an epistatic interaction responsive to high-temperature stress. Unexpectedly, from the USVL108 non-vernalization side, a major QTL identified, nonetheless, the market-related qualities of the head were not fulfilled when subjected to high temperature, which reflects the multidimensionality of high-temperature resistance in broccoli.

Most recently, from that program, another breakthrough was reported by Branham and Farnham (2019) involving the development of another double haploid derived from an F_1 that developed by crossing two nonrelated relatively heat-tolerant parents, which one of them (USVL123) was the sister of USVL138 line while another line, USVL118, had a separated pedigree. Considering the genetic distance of USVL118

Table 1 Studies involving Identification of quantitative trait loci (QTLs) for high-temperature tolerance and early flowering in broccoli

Population	Plant material	Locus	Traits governed	Reference
Double haploid	Cross of contrasting genotypes: A12DHd × GDDH33	BolC.FRI.a and BolC.FRI.b	Regulating the vernalization requirements	Irwin (2012)
Double haploid	Cross of contrasting genotypes: USVL138 × USVL108	five QTLs and an epistatic interaction responsive to high-temperature stress	Heat tolerance	Branham et al. (2017)
Inbred lines	Cross of contrasting genotypes: 195 (cabbage) and 93219 (broccoli)	Ef2.1	Early flowering induction	Shu et al. (2018)
Subtropical elite lines	Cross of contrasting genotypes: BLM29 × BLM25	Two major effect QTLs: qDCI-6 and qCQ-6	The former involves in days to crud induction while the latter influences the quality of cruds under high temperature of the subtropical zone	Lin et al. (2015)
Double haploid	Cross of contrasting genotypes: USVL118 × USVL123	Two major effect QTLs: QHT_C06 and QHT_C09.2	Heat tolerance	Branham and Farnham (2019)
Double haploid	BolTBDH	Four QTLs (OQ_C03@57.0, OQ_C04@33.3, OQ_CC08@25.5, and OQ_C09@49.7) and two FLC homologs (Bo9g173400 and Bo9g173370)	The QTLs associated with crud qualitative features and two homologs related to early flowering	Stansell et al. (2019)

from other two lines, the authors speculated that this line possibly holds putative QTLs relevant to high-temperature stress, which they further conducted a two-year period field experiment at summer to investigate the quality of heads beside another two summer field experiment for phenotypic validation and probing high-temperature tolerance-related QTLs by resequencing the whole genome of the bulked segregants which they found two QTLs (QHT_C06 and QHT_C09.2) completely different from formerly introduced ones. Additionally, nonsynonymous SNPs located at a genetic block of flowering time in QHT_C09.2 which provides a possible explanation for significantly negatively correlated two characteristics of time requirement for head maturation and high-temperature resistance. The end goal of such programs is to identify reliable QTLs as backbones of developing novel cultivars, in this case, heat tolerance, which now through a combination of the putative QTLs in those two studies under gene pyramiding bring the development of broccoli cultivars resistance to high-temperature stress while producing saleable heads will be in within reach.

Similarly, Lin et al. (2018) developed 112 lines under an environmental condition with high average temperature and humid in the subtropical zone (Taiwan) investigate and locate potential responsive QTLs for features including the time of flowering, induction, and formation of curd as cornerstone traits of breeding in broccoli which the knowledge can be further applied to other species of *Brassica*. A line with early maturation (BLM29) originated from kale was opted from those lines and crossed with a late flowering line of broccoli (BLM25) under an average temperature of 28 °C, from the progenies, only three witnessed to have lower number of days to curd induction (DCI) compared to BLM29, the early-maturing line, however, similar to Branham et al. (2017) the individuals with favorable heat tolerance failed to produce curds with proper quality and leaves appeared in the curds, only one produced crud quality (CQ) of one (i.e., absence of leaf in the crud) but with average days of crud induction higher than BLM29. Further genetic evaluations, four associated QTLs with DCI and CQ which two major ones (DCI: qDCI-6; CQ: qCQ-6) found to be contributed from BLM29 decreased DCI by 3.81 days and leafiness intensity of crud by 0.44°. Through introgression of those two alleles above into BLM25, validation of their effectiveness was performed. In searching candidate genes where those two potential QTLs located, between two parents only variation in Bol008758, defined as FLOWERING LOCUS C (FLC) observed, the linkage map showed correspondence of Bol008758 and BoFLC3 which had exactly the same sequence pattern between parents and identical to *B. oleracea* var. alboglabra and var. capitate. The CI in temperate broccoli needs vernalization, a physiological process which is controlled by BoFLC2 while BoFLC3 and PAN (PERIANTHIA, Bol024000, a transcription factor previously identified in Arabidopsis) are regulating CI in subtropical broccolis which enabled them to develop a novel pathway for flowering under a high-temperature environment. This is an examiner groundbreaking study that sheds light on developmental regulation of flowering in *Brassica* and the possibility to address the challenges in changing the climate by introducing temperate vegetable crops to tropics.

A renown model plant species for biotechnological studies in *Arabidopsis thaliana* that is similar to *B. oleracea* has vernalization requirement for flowering induction which is regulated by FRIGIDA (FRI), therefore the imposition of yearly winter habitation through enhancing the expression intensity of FLC. Given the fact that FRI variability is a key factor influencing the strategies of *A. thaliana* including autonomous loss-of-function of alleles which its effects can be traced back in short-cycling pattern in various accessions that seems to have been numerously evolved, considering the systematical closeness of *A. thaliana* to *B. oleracea* and the importance of FRI, Irwin et al. (2012) conducted a study to detect possible orthologues of FRIs in *B. oleracea* which they successfully identified two orthologues of FRIs in *B. oleracea* (BolC.FRI.a and BolC.FRI.b). Mapping of the genes revealed the high similarity of their regions on *B. oleracea* chromosomes a section of *A. thaliana* chromosome 5 which evokes the loss of a copy of the FRI long since the formation the genome of *B. oleracea* in the triplication phenomena. The absence of this genomic location with FRI in *A. thaliana* which a recombination occurrence inside the promotor of FRI identified when both comparatively assessed. Following the analyzing of allelic diversity at BolC.FRI.a in domesticated germplasm of *B. oleracea*, two alleles with major effects detected, that seem operational either to each other or to *A. thaliana* FRI, that the latter part acquired in a complementary study. While this experiment surely provided first-hand important information but added to the complexity of former FRI association assessments in the genus of *Brassica*. Moreover, such information of *B. oleracea* FRI gene number, functions and physical location on chromosome provides this unique opportunity to conduct large-scale experiments to unravel the influence of diversity in alleles at FRI on flowering induction, vernalization requirements, and life-history strategies. Finally, analyzing the closeness of BoFRI to BoFLC, BoFY, and BoCO generate novel challenging questions on the possibilities that they might affect the flowering paradigm. It's specifically essential when opting for particular alleles in breeding projects that supposed to improve the capabilities of vegetable crops of this species under changing the climate.

Simultaneous subjecting of two varieties (TSS-AVRDC-2 and B-75) of broccoli to high temperature (40 °C) and waterlogging stresses (for a three-day period) to unveil the physiological and protein expression paradigms Lin et al. (2015) evaluated the expression intensity of rubisco subunits (small: RubS or large: RubL) since these photosynthetically involved proteins are accounted for carbon fixation and their regulation under stress condition has been reported. Uniquely detected proteins in reaction to simultaneous stress in TSS-AVRDC-2 including RubL, RubS, and ribosome-binding factor for the first time offer information that could be utilized to unwind the complicated metabolical mechanisms and selection parameter of potential lines of broccoli resistance to heat stress as well as waterlogging which the occurrence of their combination in subtropical zone as Taiwan is threatening the broccoli cultivation.

The importance of flowering time in broccoli and the fact of the matter that earliness or lateness of curd initiation and flowering can make an enormous difference by the occurrence of those developmental process early before the elevation of temperature in a particular region or late that prevents the completion of the transition or damages the quality of heads in broccoli and cabbage that's why figuring out the genetic regulators of these agnomical features are pivotal in which Shu et al. (2018) using population of BC_1P_1, BC_1P_2, F_2, and $F_{2:3}$ originated from crossing of inbred lines, one late flowering habit (195; cabbage), and another early flowering (93219; broccoli) that this indicated the polygene control of this trait. Afterward, in an attempt to identify responsible QTLs, the whole genome sequenced, as a result of mapping an F_2 population a considerably large genomic region conserving a potent QTL located on chromosome number 2 of cabbage, Ef2.1, associated with early flowering. The physical position of a related gene of cabbage, Bol024659, was determined, which is the correspondence of GRF6 a large effect gene controlling flowering in Arabidopsis known as BolGRF6 which its significantly high expression level in early flowering genotypes observed. Such accomplishment suggests BolGRF6 as a potential gene conditioning early flowering in the broccoli line 93219 which may greatly contribute to breeding programs having an improvement in a flowering habit in the target.

Of the most recent and updated study on polygenic controlled morphological, developmental, quality, and yield-related characteristics on broccoli is a comprehensive investigation conducted by Stansell et al. (2019) involved bi-parental mapping of a double haploid BolTBDH derived from a cross of Chinese kale × broccoli, for 25 horticulturally important features classified in 6 classes (architecture, biomass, phenology, leaf morphology, floral morphology, and head quality) using QTL mapping of 1,881 genotype-by-sequencing SNPs. The position of 56 single and 41 epistatic QTLs determined. Four QTLs (OQ_C03@57.0, OQ_C04@33.3, OQ_CC08@25.5, and OQ_C09@49.7) associated with crud qualitative features explained a cumulative 81.9% of the phenotypical variation in the heading phenotypic of broccoli, composed of the FLC homologs Bo9g173400 and Bo9g173370, which showed epistatic impacts. Interestingly enough, 3 pivotal hotspots related to pleiotropic regulation of the broccoli heading phenotype were detected. A phenological hotspot identified that lessens the number of days to flowering by 7.0 days and consists of an extra FLC homolog Bo3g024250 which has no epistatic influence with the 3 qualitative horticultural hotspots. Potent genes related to other horticultural morphological features were located: BoLMI1 (Bo3g002560) tied with serrated leaf margins and leaf apex shape, BoCCD4 (Bo3g158650) implicated in flower color, and BoAP2 (Bo1g004960) implicated in the hooked sepal horticultural trait.

The outcomes and findings of the entire studies summarized here, in particular, the last studies on BolTBDH population, on probing for potential QTLs all culminate in preparing a framework for broccoli and even other varieties of *B. oleracea* improvement by having essential genomic loci influencing on horticultural associated qualitative traits of broccoli and developing novel cultivars of broccoli resilient to climate change era through multi-trait marker-assisted selection.

2.2 Genetic Transformation

Attempts to identify favorable genes and develop transgenic plants of broccoli have been numerous. First study date back to the early 90s which Wagoner et al. (1992)

introduced a novel methodology to produce transgenic plants and regenerate as well as infection tissues of broccoli by *Agrobacterium.*

Considering the preparation that has made to effectively transfer specific genes to broccoli, sensitivity of this crop to high temperature to enable broccoli cultivars to resist high temperature, and the cardinality of heat stress-associated proteins known as heat-shock proteins (HSPs) that are operationally involved in provoking a set of other proteins, many studies have paid attention to this area. The radical increase in the expression level of HSPs in the cell at the time high-temperature stress exposure (or other stresses) have extensively been published (Narberhaus 2010; Omar et al. 2011; Su and Li 2008). Of the mostly noted HSPs that often witnessed to confer a significant level of high-temperature tolerance is HSP101 (Wahid et al. 2007; Su and Li 2008). Multistress resistance in genetically modified broccoli plants observed, in particular, dehydration as a result of heat stress (Vinocur and Altman 2005). By transferring the famous HSP101 to broccoli through transformation mediated by *Agrobacterium* from Arabidopsis, Ravanfar and Aziz (2015) successfully developed transgenic broccoli with high-temperature tolerance. Using a highly optimized PCR technique, the efficiency of transformation determined to be 5%. To further validate the transformation and transcription of AtHSP101, a reporter gene of luciferase in the targeted cells and RT-PCR method utilized. The peak of sensitivity of broccoli to high temperature is when transition from vegetative to generative stage occurs which requires a period of low temperature to complete this process, to this end, two genes (FCA and CONSTANS (CO) derived from *A. thaliana* which their overexpression is responsible for advanced induction of flowering by overriding the prerequisite of vernalization which Irwin et al. (2002) inserted those two genes in target into chosen lines of *B. oleracea*, afterward subjected them to two different conditions of vernalization and non-vernalization, further evaluation showed the absence of any tangible variation in flowering response. Of course, significant variability in days to curd induction observed, however, FCA or CO failed to nullify the essentiality of vernalization for flowering initiation, probably due to modulation in some other passes. It can be inferred that vernalization a requirement that controls the production of broccoli is much more complicated and involves numerous still not known mechanisms and genes.

3 Salinity and Elevated CO_2

As an important part of the projected climate change, increasing in atmospheric CO_2 concentration is a "two blades knife" since the elevation of CO_2 might actually benefit many plant species by enhancing the growth rate and improving water use efficiency while degradation of the protein content of food crops under high concentrations of CO_2 is the concerning part of this story in addition to agro-ecological turmoil that this phenomenon is responsible for owing to increase in temperature which exacerbates the negative effects of elevation in CO_2. Hitherto, searches on databases for studies involving QTLs or genes identified associated with elevated CO_2 in broccoli

(*B. oleracea* var. italica) yielded no results. However, here we tried to summarize investigations on the advantageous or problematic influence of elevated CO_2 on broccoli from a physiological perspective. From the effects of salinity stress on broccoli, some are associated with restriction of proteins involved in cell wall metabolism and boosting up a set of defense mechanism-relevance proteins (Fernandez-garcia et al. 2011) (Fig. 1).

The comparative responses of two broccoli cultivars (*B. oleracea* var. italica, cv. Parthenon, and cv. Naxos) to a 15-day exposure to different NaCl levels were investigated. Salinity led to increased concentrations of Na^+ and Cl^- ions in both cultivars, a disruption of the endogenous minerals levels in the shoots and roots that varied with the cultivar and salt concentration and decreases the osmotic potential ($\Psi\pi$), root hydraulic conductance (L0), and stomatal conductance (Gs). The reduced biomass of Naxos at moderate NaCl indicates its greater sensitivity to salinity, compared with Parthenon. Parthenon accumulated more soluble sugars, for osmotic adjustment, whereas Naxos accumulated proline, which gave the two cultivars differing nutritional characteristics. The total glucosinolates (GSLs) content was not affected by salinity in Parthenon while it decreased significantly in Naxos as a consequence of

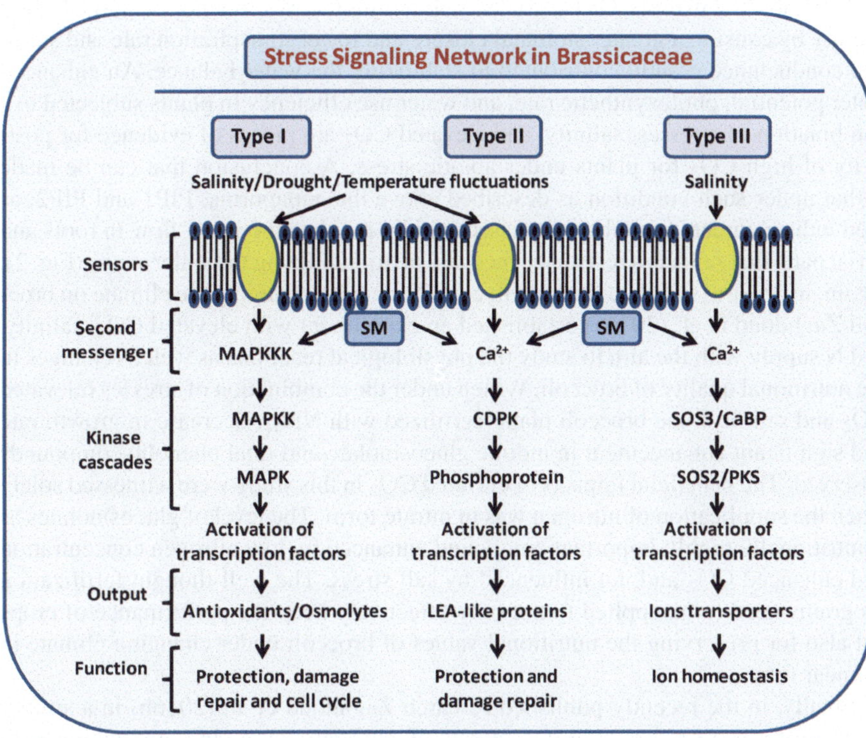

Fig. 1 Proposed schematic representing the processes from sensing the abiotic stresses (salinity, extreme temperatures and drought) to generating compounds assisting the exposed plant by enhancing ion hemostasis and repairing the damaged organs

the decrease in the indole GSL. However, Naxos accumulated more aliphatic GSLs under salt stress than Parthenon, which confers on this cultivar a greater nutritional value when cultivated under salinity. These results suggest that, at distinct salinity levels, each broccoli cultivar adopts a specific strategy, indicating the crucial role of the genetic background information on the organoleptic and nutritional properties that each cultivar acquires. In abiotic stress imposed by elevation in CO_2 or salinity, nutrient supplication can indeed be determinative. The most notable studies have been dedicated to exploring the effect of CO_2 in combination with other stresses on various aspects of physiological responses of broccoli is a series of studies conducted by Zaghdoud with different co-workers from 2013 to 2016, from which Zaghdoud et al. (2013) by considering the probably advantageous impact of elevated CO_2 under climate change condition in the decades ahead which has salinity of soil in company, on water use efficiency they conducted an investigation on effects of simultaneous application of elevated CO_2 and salinity on water status related to PIP aquaporins in underground and aboveground parts of broccoli. A sharp increment in stomatal conductance and transpiration rate by enhancing CO_2 observed while water potential remained unaffected and static which had a reduction in hydraulic conductance of roots as an aftermath. The elevation in CO_2 also negatively influenced the frequency of PIP1 and PIP2 in roots. On the contrary, the combinatory influence of high CO_2 and salinity by causing a greater stomatal closure and lower transpiration rate and stomatal conductance greatly contributed to stabilizing the water balance. An enhanced water potential, photosynthetic rate, and water use efficiency in plants subjected to a combination of stresses, salinity, and elevated CO_2 are pieces of evidence for positivity of high CO_2 for plants under abiotic stress. A conclusion that can be made is that under such condition as described above the aquaporins, PIP1 and PIP2, as their enhancement in number witnessed, rigidly regulate the water flow in roots and aerial parts that reflects a resistance mechanism to stabilizing the water status (Fig. 2). Again, in order to simulate the growth condition under changing the climate on broccoli Zaghdoud et al. (2016a) established an experiment with elevated CO_2, salinity, and N supply with the aim to study the physiological reactions as well as changes in the nutritional quality of broccoli. Which under the combination of stresses (elevated CO_2 and salinity), the broccoli plants fertilized with NH_{4+}, decrease in growth rate and significant enhancement in indolic glucosinolate and total phenolic compounds observed. The beneficial impacts of elevated CO_2 in this study were witnessed solely when the supplication of nitrogen was in nitrate form. The level of glucosinolates as a nutritionally highly important compound enhanced by low nitrogen concentration and enhanced CO_2 and not influenced by salt stress. The well-thought fertilization programs need to be applied to not only effectively keep the performance of crops but also for preserving the nutritional values of broccoli under changing climate in the near future.

Finally, in the recently published research Zaghdoud et al. (2016b) in a similar study to the last one in 2016, the effects of nitrogen source in combination with elevated atmospheric CO_2 on plants subjected to salt stress evaluated. The outcomes indicated the suitability of medium level of both nitrate and ammonium nitrogen for broccoli and physiological responses in plants treated with salinity under elevated

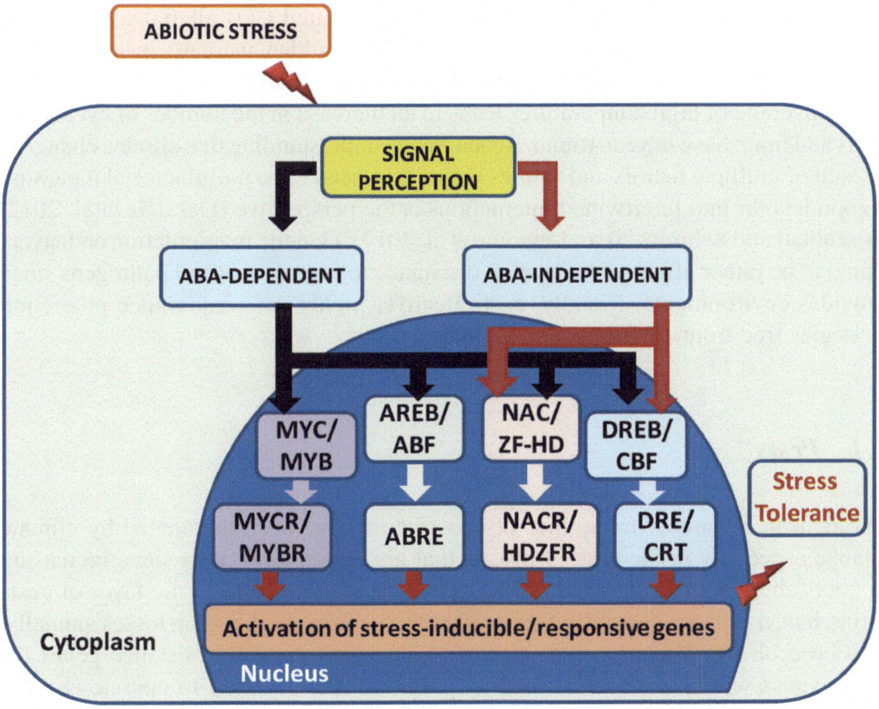

Fig. 2 Overview of the representative transcriptional networks mediated by ABA-dependent and independent transcription factors in brassicaceae under abiotic stresses

CO_2 were mainly alleviated even in low nitrogen in consistence with Zaghdoud et al. (2016a). These studies resonated the positive influence of elevated CO_2 on plants' tolerant under drought stress and the important role that nitrogen forms can have in increasing their resistance which all contribute to an optimistically perspective toward broccoli production under changing climate although the matters are far more complicated and a safe conclusion is tedious to make.

4 Biotic Stress

Majorly anthropogenic-driven radical increment in global temperature and CO_2 level is an inevitable reality (Schuur et al. 2015; Hansen et al. 2016) which by the middle and end of the current century the concentration of this greenhouse gas and average global temperature rise by 710 ppm and 1.5 °C, respectively (Cubasch et al. 2013). Fluctuations resulted from changes in climate especially elevated CO_2 and increasing temperature have significantly disturbed the life cycle of insects at an alarming rate which can cause an outbreak and damage to the cultivated lands. The quality of prey species as the nutritional value of plants under elevated CO_2 altered which might cause the reduction in predators' population and sudden increase in the level of damage to an economic threshold and above (Xu et al. 2019; Yifei et al. 2018). A higher average of high temperatures leads to an increase in the number of cycles that pests and fungi have in year-round. Reaching the understanding that climate change is a result of multiple factors and studies in this area need to be multifactorial if gaining a good insight into intertwined interactions is the perspective (DeLucia et al. 2012; Rosenblatt and Schmitz 2016; Lemoine et al. 2017). Genetic manipulation endeavors found to be rather effective to develop resistance to insect pests and pathogens since provides environmental-friendly, cost-effective, highly safe, endurance protection strategies free from environmental variables.

4.1 Pests

The delicate nature of tranquility of ecosystems have been threatened by climate change especially in the case of insects that are agricultural pests since increasing in atmospheric CO_2 and temperature has been suggested to be in the favor of pests through having more generations in a year which means higher crop losses annually, also broccoli has a narrow gene pool with no much capable resistance genes for biotic stress which makes it's breeding program more difficult to operate (Kumar and Srivastava 2016). In the last decades the traditional breeding method a granted significant developments in pest resistance broccolis, however, the constraints of traditional breeding have been problems which can be effectively addressed with biotechnological approaches, for instance, genetic engineering and skillfully manipulating genes is now a critical tool for the improvement of resistance in broccoli to biotic stresses as well as decreasing the farming-related concerns. These novel methodologies through making possible the transformation of favorable genes from unrelated species to confer resistance to broccoli cultivars and cole crops on the whole therefore their significant vulnerability to many biotic stressors decreased. Of the important studies in this field are a handful of investigations from the late 80s to recent years. Of the effective renown proteins with insecticidal impacts, *Bacillus thuringiensis* has enabled the scholars to generate transgenic plants with considerable resistance to a wide variety of pests belong to lepidopteran and coleopteran in various important crops (Fischhoff et al. 1987; Hofte and Whiteley 1989; Perlak et al. 1990; Bottrell et al. 1992; Fujimoto et al. 1993; Pardo-López et al. 2013). By successful transferring of resistance gene of cry1A(c) (Metz 2001; Metz et al. 1995), a significant quantity of transgenic plants of broccoli produced, the mortality rate for most of them reached 100% for a notorious pest of broccoli, diamondback moth(DBM). However, the opted plants with a 100% mortality rate of vulnerable larvae enabled a strain of the pest to acquire tolerance to Bt. In another similar study, Cao et al. (1999) regenerated high Bt(cry1C protein) contented transgenic broccoli to tackle cry1A or cry1C tolerance diamondback moth larvae resistant to DBM. They claimed that biosynthesis of a high concentration of cry1C protein makes transgenic broccoli safe from vulnerable larvae of DBM and additionally from moderately cry1C resistance DBM. The resistance of these resistance broccolis to two other lepidopteran pests of crucifers, a cabbage worm and looper.

Investigating the expression of cry1Ac and cry1C genes from Bt in regenerated transgenic clones of broccoli under in vitro condition employing related markers of kanamycin and hygromycin, Cao and Earle (2003) showed the PCR-proven stability of the two aforementioned genes. The similarity of the production level of the two targeted proteins determined in all the clones using the ELISA test, but the concentration level of cry1C in clones was significantly lower in comparison to the original one. In a comprehensive study Viswakarma et al. (2004) established a procedure for successful transfer of cryIA(b) gen regulating Bt proteins with insecticidal effects to a popular cultivar of broccoli through co-cultivation of hypocotyl segments with *A. tumefaciens.*

4.2 Fungal Resistance

Attempts to develop broccoli cultivars have been limited from which Mora and Earle (2001) generated broccoli transgenic plant containing the endochitinase gene of *Trichoderma harzianum.* Introduction of the genes from *T. harzianum* carried out using Agrobacterium and selective marker as neomycin phosphotransferase-II which as result of one transformation two dozen of transgenic plants achieved, in 19 of them the endochitinase and antifungal activities were assessed, the outcomes indicated 15–20-fold higher activity of endochitinase in transgenic line than control, by comparison.

In an interesting experiment, Jiang et al. (2016) considering the importance of antioxidant based mechanisms that plants use to fight back stresses, using Agrobacterium-aimed transformation for ascorbate peroxidase, generated seven overexpressing lines of BoAPX gene, these lines showed a notably higher *Hyaloperonospora parasitica* infection as well as high-temperature stress tolerance in comparison to the wild broccoli. In an outstanding manner, four of the overexpression lines including oe-apx07, oe-apx15, oe-apx32, and oe-apx33 manifested a much higher resistance to downy mildew. This study was a reliable confirmation for the significant contributions of overexpression of BoAPX genes in improving the resistance of transgenic broccoli lines against biotic and abiotic stresses.

By evaluation of the influence of the combined effect of the elevated atmospheric CO_2 and high temperature in subtropical areas on the interactions of at a tri-trophic system (plant-insect–parasitoid), Teawkul and Hwang (2019) larva of *Spodoptera litura* (Fabricius) (Lepidoptera: Noctuidae) fed on broccoli until reaching pupal stage afterward at the second instar they exposed to *Snellenius manilae* (Ashmead) (Hymenoptera: Braconidae), the pest's parasitoid, in the exact same condition. Increase in temperature in combination with enhanced CO_2 accelerated developmental changes as well as the level of activity in both the pest and the parasitoid in addition to increasing the carbohydrate content and C/N ratio. Taking these outcomes into account, it seems the climate change ahead won't be favorable for broccoli.

In the case of viral diseases, several studies developed transgenic plants having genes conferring a considerable level of virus resistance (Lomonossoff 1995), such as coat protein genes (Beachy et al. 1990) and genes controlling viral motions (Cooper et al. 1995; Tacke et al. 1996). So far, no studies related to the development of lines in broccoli with bacterial resistance has been reported.

5 Conclusion and Future Prospects

A great deal of advancements has been achieved through employing biotechnology tools and many more come onstream during the next decade, which identification of putative QTLs as a crucial necessity for breeding programs will be a significant part of this ongoing attempt to improve the physiological reactions of broccoli to environmental stress on a large scale, in particular, high temperature which has significant negative effects on curd induction and its quality. Considering the fact that the vernalization requirement in broccoli is a polygene controlled characteristic, pinpointing its underlying genes is a tedious task. However, several investigations so far have provided breeding programs of broccoli with substantial information. Further, genetic engineering as another part of this arms race of developing cultivars to resist abiotic and biotic stressors have found to significant since the transformation of genes in target to broccoli in an efficient manner beyond the species limitation is possible. Overall, expanding the genetic knowledge on broccoli aims to create a reliable insight into actual regulating genes controlling the tolerance to heat stress and carefully establishment of strategies to that end. Of the most economically efficient ways to address the climate change-driven stresses imposed on important vegetable crops, in this case, broccoli is to develop resistant cultivars. Developing potential cultivars of broccoli to cope with challenging growth conditions under the climate change era additionally can have a considerable contribution in improving other characteristics especially those related to quality and marketability. Overall, comprehensive uses of biotechnology can illuminate the complex principles regulating the responses to environmental factors that will aim to enhance plant reactions in broccoli and cole crops on the whole.

Utilizing the biotechnological toolbox to develop crops in order to ameliorate the negative impacts of climate change is contemporary. This means that constant research is required to make this strategy applicable in every important crop. The sustainability of this strategy is notably high which makes it highly superior to others. Developing cultivars of broccoli with no vernalization requirement that consequently can help to expand the cultivation of broccoli far beyond the original origins of broccoli. Additionally, the characteristics in broccoli, in particular, those associated with resistance to abiotic and biotic stress are polygenic and to untwine the complexity required until a cultivar obtained with the ability to efficiently grow under the high-temperature condition and provide salable products. More importantly, since broccoli is an exceptionally nutritious, its expansion to warm regions of the world can indeed aim to end malnutrition and hanger in tropics.

References

Akter A, Nishida N, Takada S, Itabashi E, Osabe K, Shea DJ, Fujimoto R (2018) Genetic and epigenetic regulation of vernalization in Brassicaceae. *Brassica* Germplasm: Characterization, Breeding and Utilization: 75

Baenas N, Gómez-Jodar I, Moreno DA, García-Viguera C, Periago PM (2017) Broccoli and radish sprouts are safe and rich in bioactive phytochemicals. Postharvest Biol Technol 127:60–67

Béliveau R, Gingras D (2017) Foods to fight cancer: what to eat to reduce your risk. Dorling Kindersley Ltd

Béné C, Barange M, Subasinghe R, Pinstrup-Andersen P, Merino G, Hemre G-I, Williams M (2015) Feeding 9 billion by 2050–Putting fish back on the menu. Food Security 7(2):261–274

Björkman T, Pearson KJ (1998) High temperature arrest of inflorescence development in broccoli (*Brassica oleracea* var. italica L.). J Exp Bot 49(318):101–106

Bottrell DG, Aguda RM, Gould F, Theunis W, Demayo C, Magalit VF (1992) Potential strategies for prolonging the usefulness of *Bacillus thuringiensis* in engineered rice. Korean J Appl Entomol 31

Branham SE, Farnham MW (2019) Identification of heat tolerance loci in broccoli through bulked segregant analysis using whole genome resequencing. Euphytica 215(2):34. https://doi.org/10.1007/s10681-018-2334-9

Branham SE, Stansell ZJ, Couillard DM, Farnham MW (2017) Quantitative trait loci mapping of heat tolerance in broccoli (*Brassica oleracea* var. italica) using genotyping-by-sequencing. Theor Appl Genet 130(3):529–538. https://doi.org/10.1007/s00122-016-2832-x

Broccoli suppliers wp, and global market information (2018) Broccoli suppliers, wholesale prices, and global market information. https://www.tridge.com/intelligences/brocolli

Cao J, Earle ED (2003) Transgene expression in broccoli (*Brassica oleracea* var. italica) clones propagated in vitro via leaf explants. Plant Cell Rep 21(8):789–796. https://doi.org/10.1007/s00299-003-0589-6

Cao J, Tang JD, Strizhov N, Shelton AM, Earle ED (1999) Transgenic broccoli with high levels of *Bacillus thuringiensis* Cry1C protein control diamondback moth larvae resistant to Cry1A or Cry1C. Mol Breeding 5(2):131–141. https://doi.org/10.1023/A:1009619924620

Cooper B, Lapidot M, Heick JA, Dodds JA, Beachy RN (1995) A defective movement protein of TMV in transgenic plants confers resistance to multiple viruses whereas the functional analog increases susceptibility. Virology 206(1):307–313

Cubasch U, Wuebbles D, Chen D, Facchini MC, Frame D, Mahowald N, Winther J-G (2013) Introduction. In: Plattner K, Tignor M, Allen SK, Boschung J, Nauels A, Xia Y, Bex V, Midgley PM (eds) Climate change 2013: the physical science basis. Contribution of working group I to the fifth assessment report of the intergovernmental panel on climate change. Cambridge University Press, Cambridge, UK, and New York. http://www.climatechange2013.org/images/report/WG1AR5_Chapter01_FINAL.pdf

Davis M (2017) Are all fruits and vegetables created equal? The importance of plant genetics in cancer prevention

DeLucia EH, Nabity PD, Zavala JA, Berenbaum MR (2012) Climate change: resetting plant-insect interactions. Plant Physiol 160(4):1677–1685

Deryng D, Conway D, Ramankutty N, Price J, Warren R (2014) Global crop yield response to extreme heat stress under multiple climate change futures. Environ Res Lett 9(3):034011. https://doi.org/10.1088/1748-9326/9/3/034011

Dias JS (2019) Nutritional quality and effect on disease prevention of vegetables. Food and Nutrition Sciences 10(04):369

Dong J, Gruda N, Lam S, Li X, Duan Z (2018) Effects of elevated CO_2 on nutritional quality of vegetables: a review. Front Plant Sci. https://doi.org/10.3389/fpls.2018.00924

dos Santos TC, Gomes TM, Pinto BAS, Camara AL, de Andrade Paes AM (2018) Naturally occurring acetylcholinesterase inhibitors and their potential use for Alzheimer's disease therapy. Front Pharmacol 9

El-Sayed AA, Nada MA, El-Fattah SMA (2018) Impacts of climate change on insect pests of main crops in Egypt. In: Sustainability of agricultural environment in Egypt: part II. Springer, pp 189–214

Fanzo J, Davis C, McLaren R, Choufani J (2018) The effect of climate change across food systems: Implications for nutrition outcomes. Glob Food Secur 18:12–19

Fernandez-garcia N, Hernandez M, Casado-Vela J, Bru R, Elortza F, Hedden P, Olmos E (2011) Changes to the proteome and targeted metabolites of xylem sap in *Brassica oleracea* in response to salt stress. Plant, Cell Environ 34(5):821–836. https://doi.org/10.1111/j.1365-3040.2011.02285.x

Fischhoff DA, Bowdish KS, Perlak FJ, Marrone PG, McCormick SM, Niedermeyer JG, Dean DA, Kusano-Kretzmer K, Mayer EJ, Rochester DE (1987) Insect tolerant transgenic tomato plants. Biotechnology 5(8):807

Francisco M, Tortosa M, Martínez-Ballesta MdC, Velasco P, García-Viguera C, Moreno DA (2017) Nutritional and phytochemical value of *Brassica* crops from the agrifood perspective. Ann Appl Biol 170(2):273–285. https://doi.org/10.1111/aab.12318

Fujimoto H, Itoh K, Yamamoto M, Kyozuka J, Shimamoto K (1993) Insect resistant rice generated by introduction of a modified δ-endotoxin gene of *Bacillus thuringiensis*. Biotechnology 11(10):1151–1155. https://doi.org/10.1038/nbt1093-1151

Gray SB, Brady SM (2016) Plant developmental responses to climate change. Dev Biol 419(1):64–77. https://doi.org/10.1016/j.ydbio.2016.07.023

Hansen J, Sato M, Hearty P, Ruedy R, Kelley M, Masson-Delmotte V, Russell G, Tselioudis G, Cao J, Rignot E (2016) Ice melt, sea level rise and superstorms: evidence from paleoclimate data, climate modeling, and modern observations that 2 C global warming could be dangerous. Atmos Chem Phys 16(6):3761–3812

Hasanuzzaman M, Nahar K, Alam MM, Roychowdhury R, Fujita M (2013a) Physiological, biochemical, and molecular mechanisms of heat stress tolerance in plants. Int J Mol Sci 14(5):9643–9684

Hasanuzzaman M, Nahar K, Fujita M (2013b) Extreme temperatures, oxidative stress and antioxidant defense in plants. In: Vahdati K, Leslie C (eds) Abiotic stress: plant responses and applications in agriculture. InTech, Rijeka, pp 169–205

Hassini I, Martinez-Ballesta MC, Boughanmi N, Moreno DA, Carvajal M (2017) Improvement of broccoli sprouts (*Brassica oleracea* L. var. italica) growth and quality by KCl seed priming and methyl jasmonate under salinity stress. Sci Hortic 226:141–151

Hofte H, Whiteley HR (1989) Insecticidal crystal proteins of *Bacillus thuringiensis*. Microbiol Rev 53(2):242–255

Huang B, Xu Y (2015) Cellular and molecular mechanisms for elevated CO_2–regulation of plant growth and stress adaptation. Crop Sci 55(4):1405–1424

Irwin J, Bird N, Richardson A, Sparrow P, Townsend T, Dean C, Coupland G, Dale P (2002) Control of the transition from vegetative to flowering mode in horticultural Brassicas. In: Abstracts of the 13th crucifer genetics workshop, California

Irwin JA, Lister C, Soumpourou E, Zhang Y, Howell EC, Teakle G, Dean C (2012) Functional alleles of the flowering time regulator FRIGIDA in the *Brassica oleracea* genome. BMC Plant Biol 12(1):21. https://doi.org/10.1186/1471-2229-12-21

Jamil A, Riaz S, Ashraf M, Foolad MR (2011) Gene expression profiling of plants under salt stress. Crit Rev Plant Sci 30(5):435–458. https://doi.org/10.1080/07352689.2011.605739

Jiang M, Jiang JJ, He CM, Guan M (2016) Broccoli plants over-expressing a cytosolic ascorbate peroxidase gene increase resistance to downy mildew and heat stress. Journal of Plant Pathology 98(3):413–420

Kałużewicz A, Krzesiński W, Knaflewski M (2009) Effect of temperature on the yield and quality of broccoli heads. Veg Crops Res Bullet 71:51–58

Kumar P, Srivastava DK (2016) Biotechnological advancement in genetic improvement of broccoli (*Brassica oleracea* L. var. italica), an important vegetable crop. Biotechnol Lett 38(7):1049–1063. https://doi.org/10.1007/s10529-016-2080-9

Lemoine NP, Doublet D, Salminen JP, Burkepile DE, Parker JD (2017) Responses of plant phenology, growth, defense, and reproduction to interactive effects of warming and insect herbivory. Ecology 98(7):1817–1828

Li L, Song S, Nirasawa S, Hung Y-C, Jiang Z, Liu H (2018) Slightly acidic electrolyzed water treatment enhances the main bioactive phytochemicals content in broccoli sprouts via changing metabolism. J Agric Food Chem 67(2):606–614

Lin H-H, Lin K-H, Chen S-C, Shen Y-H, Lo H-F (2015) Proteomic analysis of broccoli (*Brassica oleracea*) under high temperature and waterlogging stresses. Bot Stud 56(1):18. https://doi.org/10.1186/s40529-015-0098-2

Lin Y, Lee J, Tseng M, Lee C, Shen C, Wang C, Liou C, Shuang L, Paterson AH, Hwu K (2018) Subtropical adaptation of a temperate plant (*Brassica oleracea* var. italica) utilizes non-vernalization-responsive QTLs. Sci Rep 8(1):13609. https://doi.org/10.1038/s41598-018-31987-1

Lomonossoff GP (1995) Pathogen-derived resistance to plant viruses. Annu Rev Phytopathol 33(1):323–343

Machado RMA, Serralheiro RP (2017) Soil salinity: effect on vegetable crop growth. Management practices to prevent and mitigate soil salinization. Horticulturae 3(2):30

Mann CC (2018) Can planet earth feed 10 billion people? The Atlantic

Mark WF, Thomas B (2011) Breeding vegetables adapted to high temperatures: a case study with broccoli. HortScience Horts 46(8):1093–1097. https://doi.org/10.21273/HORTSCI.46.8.1093

Maynard D, Hochmuth G (2007) Vegetables and the vegetable industry. Knott's Handbook for Vegetable Growers:1–53

McDonald M, Warland J (2018) Vegetables and climate change. Agroclimatology: Linking Agriculture to Climate (agronmonogr60)

McKeown A, Warland J, Mcdonald M, Hutchinson C (2004) Cool season crop production trends: a possible signal for global warming. Acta Horticulturae:241–248

Medek DE, Schwartz J, Myers SS (2017) Estimated effects of future atmospheric CO_2 concentrations on protein intake and the risk of protein deficiency by country and region. Environ Health Perspect 125(8):087002

Metz TD (2001) Transgenic broccoli (*Brassica oleracea* var. italica) and cabbage (var. capitata). In: Bajaj YPS (ed) Transgenic crops II. Springer Berlin Heidelberg, Berlin, Heidelberg, pp 69–86. https://doi.org/10.1007/978-3-642-56901-2_6

Metz TD, Roush RT, Tang JD, Shelton AM, Earle ED (1995) Transgenic broccoli expressing a *Bacillus thuringiensis* insecticidal crystal protein: implications for pest resistance management strategies. Mol Breeding 1(4):309–317. https://doi.org/10.1007/BF01248408

Miller C, Konsler T, Lamont W (1985) Cold stress influence on premature flowering of broccoli. HortScience

Mora AA, Earle ED (2001) Resistance to *Alternaria brassicicola* in transgenic broccoli expressing a *Trichoderma harzianum* endochitinase gene. Mol Breed 8(1):1–9. https://doi.org/10.1023/A:1011913100783

Motesharrei S, Rivas J, Kalnay E, Asrar GR, Busalacchi AJ, Cahalan RF, Cane MA, Colwell RR, Feng K, Franklin RS (2016) Modeling sustainability: population, inequality, consumption, and bidirectional coupling of the earth and human systems. Natl Sci Rev 3(4):470–494

Mourão I, Hadley P (1998) Environmental control of plant growth development and yield in broccoli (*Brassica oleracea* L. var. italica Plenk): crop responses to light regime. Acta Hort 459:71–78. https://doi.org/10.17660/ActaHortic.1998.459.6

Myers MB, Roberts JJ, White C, Stalker L (2019) An experimental investigation into quantifying CO_2 leakage in aqueous environments using chemical tracers. Chem Geol 511:91–99

Nahar K, Hasanuzzaman M, Alam MM, Fujita M (2015a) Exogenous glutathione confers high temperature stress tolerance in mung bean (*Vigna radiata* L.) by modulating antioxidant defense and methylglyoxal detoxification system. Environ Exp Bot 112:44–54

Nahar K, Hasanuzzaman M, Fujita M (2015b) Heat stress responses and thermotolerance in soybean. In: Miransari M (ed) Abiotic and biotic stresses in soybean production: soybean production, vol 1, 1st edn. Academic Press/Elsevier Inc, New York, USA, pp 261–284

Narberhaus F (2010) Translational control of bacterial heat shock and virulence genes by temperature-sensing mRNAs. RNA Biol 7(1):84–89. https://doi.org/10.4161/rna.7.1.10501

Nawaz H, Shad MA, Muzaffar S (2018) Phytochemical composition and antioxidant potential of *Brassica*. *Brassica* Germplasm: Characterization, Breeding and Utilization:7

Omar SA, Fu QT, Chen MS, Wang GJ, Song SQ, Elsheery NI, Xu ZF (2011) Identification and expression analysis of two small heat shock protein cDNAs from developing seeds of biodiesel feedstock plant *Jatropha curcas*. Plant Sci 181(6):632–637. https://doi.org/10.1016/j.plantsci.2011.03.004

Pardo-López L, Soberón M, Bravo A (2013) *Bacillus thuringiensis* insecticidal three-domain Cry toxins: mode of action, insect resistance and consequences for crop protection. FEMS Microbiol Rev 37(1):3–22

Perlak FJ, Deaton RW, Armstrong TA, Fuchs RL, Sims SR, Greenplate JT, Fischhoff DA (1990) Insect resistant cotton plants. Biotechnology (N Y) 8(10):939–943. https://doi.org/10.1038/nbt1090-939

Pimentel D, Berger B, Filiberto D, Newton M, Wolfe B, Karabinakis E, Clark S, Poon E, Abbett E, Nandagopal S (2004) Water resources: agricultural and environmental issues. Bioscience 54(10):909–918

Raiola A, Errico A, Petruk G, Monti DM, Barone A, Rigano MM (2018) Bioactive compounds in brassicaceae vegetables with a role in the prevention of chronic diseases. Molecules 23(1):15

Räsänen A, Juhola S, Nygren A, Käkönen M, Kallio M, Monge A, Kanninen M (2016) Climate change, multiple stressors and human vulnerability: a systematic review. Reg Environ Change 16:2291–2302. https://doi.org/10.1007/s10113-016-0974-7

Ravanfar SA, Aziz MA (2015) Shoot tip regeneration and optimization of Agrobacterium tumefaciens-mediated transformation of Broccoli (*Brassica oleracea* var. italica) cv. Green Marvel. Plant Biotechnol Rep 9(1):27–36. https://doi.org/10.1007/s11816-014-0340-5

Rehman SA, Kumar R (2018) Scenario of insect pests under changing climatic situations. Int J Chem Stud 6(3):77–81

Romo CM, Tylianakis JM (2013) Elevated temperature and drought interact to reduce parasitoid effectiveness in suppressing hosts. PLoS ONE 8(3):e58136. https://doi.org/10.1371/journal.pone.0058136

Rosenblatt AE, Schmitz OJ (2016) Climate change, nutrition, and bottom-up and top-down food web processes. Trends Ecol Evol 31(12):965–975

Rumjaun AB, Borde B, Guilyardi E, Lescarmontier L, Matthews R, Niewöhner C, Point SP (2018) IPCC special report "Global Warming of 1.5 C": summary for teachers. Office for Climate Education, Paris. https://www.ipcc.ch/site/assets

Šamec D, Salopek-Sondi B (2019) Cruciferous (Brassicaceae) vegetables. In: Nonvitamin and nonmineral nutritional supplements. Elsevier, pp 195–202

Schmid I, Franzaring J, Müller M, Brohon N, Calvo O, Högy P, Fangmeier A (2016) Effects of CO_2 enrichment and drought on photosynthesis, growth and yield of an old and a modern barley cultivar. J Agron Crop Sci 202(2):81–95

Schuur EA, McGuire AD, Schädel C, Grosse G, Harden J, Hayes DJ, Hugelius G, Koven CD, Kuhry P, Lawrence DM (2015) Climate change and the permafrost carbon feedback. Nature 520(7546):171–179

Shahbandeh M (2019) Broccoli: fresh and processing market production value 2018. https://www.statista.com/statistics/1030612/us-market-broccoli-production-value/. Accessed 14 Nov 2019

Shrivastava P, Kumar R (2015) Soil salinity: a serious environmental issue and plant growth promoting bacteria as one of the tools for its alleviation. Saudi J Biol Sci 22(2):123–131. https://doi.org/10.1016/j.sjbs.2014.12.001

Shu J, Liu Y, Zhang L, Li Z, Fang Z, Yang L, Zhuang M, Zhang Y, Lv H (2018) QTL-seq for rapid identification of candidate genes for flowering time in broccoli × cabbage. Theor Appl Genet 131(4):917–928. https://doi.org/10.1007/s00122-017-3047-5

Siddiq M, Uebersax MA (2018) Handbook of vegetables and vegetable processing. Wiley Online Library

Stansell Z, Farnham M, Björkman T (2019) Complex horticultural quality traits in broccoli are illuminated by evaluation of the immortal BolTBDH mapping population. Front Plant Sci 10(1104). https://doi.org/10.3389/fpls.2019.01104

Su P-H, Li H-m (2008) Arabidopsis stromal 70-kD heat shock proteins are essential for plant development and important for thermotolerance of germinating seeds. Plant Physiol 146(3):1231. https://doi.org/10.1104/pp.107.114496

Suresh S, Waly MI, Rahman MS (2018) Broccoli (*Brassica oleracea*) as a preventive biomaterial for cancer. In: Bioactive components, diet and medical treatment in cancer prevention. Springer, pp 75–87

Tacke E, Salamini F, Rohde W (1996) Genetic engineering of potato for broad-spectrum protection against virus infection. Nat Biotechnol 14(11):1597

Tan DKY, Birch CJ, Wearing AH, Rickert KG (2000) Predicting broccoli development: II. Comparison and validation of thermal time models. Scientia Horticulturae 86(2):89–101. doi:https://doi.org/10.1016/S0304-4238(00)00203-X

Teawkul P, Hwang S-Y (2019) Carbon dioxide—and temperature-mediated changes in plant defensive compounds alter food utilization of herbivores. J Appl Entomol 143(3):289–298. https://doi.org/10.1111/jen.12593

Tian Y, Tian Y, Luo X, Zhou T, Huang Z, Liu Y, Qiu Y, Hou B, Sun D, Deng H, Qian S, Yao K (2014) Identification and characterization of microRNAs related to salt stress in broccoli, using high-throughput sequencing and bioinformatics analysis. BMC Plant Biol 14:226. https://doi.org/10.1186/s12870-014-0226-2

Tigchelaar M, Battisti DS, Naylor RL, Ray DK (2018) Future warming increases probability of globally synchronized maize production shocks. Proc Natl Acad Sci 115(26):6644–6649. https://doi.org/10.1073/pnas.1718031115

Toreti A, Cronie O, Zampieri M (2019) Concurrent climate extremes in the key wheat producing regions of the world. Sci Rep 9(1):5493. https://doi.org/10.1038/s41598-019-41932-5

Vinocur B, Altman A (2005) Recent advances in engineering plant tolerance to abiotic stress: achievements and limitations. Curr Opin Biotechnol 16(2):123–132. https://doi.org/10.1016/j.copbio.2005.02.001

Viswakarma N, Bhattacharya R, Chakrabarty R, Dargan S, Bhat S, Kirti PB, Shastri NV, Chopra VL (2004) Insect resistance of transgenic broccoli ('Pusa Broccoli KTS-1') expressing a synthetic cryIA(b) gene. J Horticult Sci Biotechnol 79:182–188. https://doi.org/10.1080/14620316.2004.11511746

Wagoner WJ, Kellogg JA, Bestwick RK, Stamp JA (1992) Superior regeneration and Agrobacterium infectability of broccoli and cauliflower tissues and the identification of a procedure for the generation of transgenic plants. HortSci 27:620–621

Wahid A, Gelani S, Ashraf M, Foolad MR (2007) Heat tolerance in plants: an overview. Environ Exp Bot 61(3):199–223. https://doi.org/10.1016/j.envexpbot.2007.05.011

Wida E (2019) This is America's favorite vegetable. https://www.today.com/food/what-america-s-favorite-vegetable-green-giant-says-it-s-t156398. Accessed 14 Nov 2019

Witzel K, Neugart S, Ruppel S, Schreiner M, Wiesner M, Baldermann S (2015) Recent progress in the use of 'omics technologies in brassicaceous vegetables. Front Plant Sci 6(244). https://doi.org/10.3389/fpls.2015.00244

Wurr DCE, Fellows JR, Phelps K, Reader RJ (1995) Vernalization in calabrese (*Brassica oleracea* var. italica): a model for apex development. J Exp Bot 46(10):1487–1496. https://doi.org/10.1093/jxb/46.10.1487

Xu H, Xie H, Wu S, Wang Z, He K (2019) Effects of elevated CO_2 and increased N fertilization on plant secondary metabolites and chewing insect fitness. Front Plant Sci 10:739

Yadav RK, Datta A, Dagar JC (2019) Future research needs: way forward for combating salinity in climate change scenario. In: Dagar JC, Yadav RK, Sharma PC (eds) Research developments in saline agriculture. Springer Singapore, Singapore, pp 883–899. https://doi.org/10.1007/978-981-13-5832-6_31

Yifei Z, Yang D, Guijun W, Bin L, Guangnan X, Fajun C (2018) Effects of elevated CO2 on plant chemistry, growth, yield of resistant soybean, and feeding of a target lepidoptera pest, *Spodoptera litura* (Lepidoptera: Noctuidae). Environ Entomol 47(4):848–856

Zaghdoud C, Carvajal M, Ferchichi A, del Carmen Martínez-Ballesta M (2016a) Water balance and N-metabolism in broccoli (*Brassica oleracea* L. var. Italica) plants depending on nitrogen source under salt stress and elevated CO_2. Sci Total Environ 571:763–771

Zaghdoud C, Carvajal M, Moreno DA, Ferchichi A, del Carmen Martínez-Ballesta M (2016b) Health-promoting compounds of broccoli (*Brassica oleracea* L. var. italica) plants as affected by nitrogen fertilisation in projected future climatic change environments. J Sci Food Agric 96(2):392–403. https://doi.org/10.1002/jsfa.7102

Zaghdoud C, Mota-Cadenas C, Carvajal M, Muries B, Ferchichi A, Martínez-Ballesta MdC (2013) Elevated CO_2 alleviates negative effects of salinity on broccoli (*Brassica oleracea* L. var Italica) plants by modulating water balance through aquaporins abundance. Environ Exp Bot 95:15–24. https://doi.org/10.1016/j.envexpbot.2013.07.003

Genome Editing for the Improvement of Brassicaceae for Abiotic Stress Tolerance

Syed Uzma Jalil and Mohammad Israil Ansari

Abstract Abiotic stresses are the major prevailing forms of environmental contaminants that result in harmful effects in plants and cause foremost environmental problems globally. A general effect of abiotic stress is the extreme accumulation of reactive oxygen species that can cause lipid peroxidation, oxidation of protein, inactivation of enzymes, DNA damage and interact with other imperative constituents of plant cells. Higher plants have evolved an intricate antioxidant defence system to scavenge reactive oxygen species during abiotic stress conditions. Genome editing approaches propose numerous applications in the improvement of crops towards abiotic stress tolerance, and produce quality improvement. Utilizing transgenic approaches, functional validation of several target genes engage in different processes, viz signalling, transcription, homeostasis, antioxidant defence for enhanced abiotic stress resistance has been employed in different plants including Brassicaceae crop plants. This chapter provides an inclusive outline to illustrate the interest of researchers for a better understanding of genome editing advancements in relation to abiotic stress tolerance in different Brassicaceae crops.

Keywords Abiotic stress · Reactive oxygen species · Brassicaceae crops · Genome editing

1 Introduction

Abiotic stresses viz. drought, excessive temperature and salinity drastically decrease crop production globally through effecting plant development (Pandey et al. 2017). The adverse environment conditions are the outcome from global climate change that deteriorates these stresses, conceivably leading to tremendous reduction in worldwide

S. U. Jalil
Amity Institute of Biotechnology, Amity University Uttar Pradesh, Lucknow Campus, Lucknow, India

M. I. Ansari (✉)
Department of Botany, University of Lucknow, Lucknow, India
e-mail: ansari_mi@lkouniv.ac.in; ansari_mi@hotmail.com

M. Hasanuzzaman (ed.), *The Plant Family Brassicaceae*,
https://doi.org/10.1007/978-981-15-6345-4_17

agricultural profitability. Plants endure different abiotic stresses by various mechanisms, which usually involved the stress-related gene expression (Kuzuoglu-Ozturk et al. 2012; Golldack et al. 2011). Specifically, transcription factors (TFs) are the keystone in genes regulation network, which controlled the expression of numerous genes associated with abiotic stress regulation (Singh et al. 2002). Progression in genetics and genomics has enhanced the perception of the intricate mechanism of environmental stresses that are associated with signalling, gene regulation and metabolic pathway mechanisms (Mickelbart et al. 2015). Several important genes were recognized and their transformation was performed with conventional genetic engineering approaches for enhancing the resistance in Brassicaceae family species against adverse environmental conditions (Bidhan et al. 2011). Because of intricate adverse environmental conditions, a very limited research on genome editing has been performed in this field (Gong and Liu 2013).

Genome editing is a group of cutting edge molecular study tools, which facilitates accurate, efficient and targeted modification at genomic loci (Gao et al. 2017). Genome editing for targeted genes expression is extensively utilized in the area of agricultural science to improve the required qualities in economically important crops. Generally four approaches are utilized for focused genome editing until now. There are four kinds of engineered nucleases that are used in genome editing technology (i) Zinc Finger Nucleases (ZFNs), (ii) Meganucleases (iii) Transcription Activator-Like Effectors Nucleases (TALENs) and (iv) Clustered Regularly Interspaced Short Palindromic Repeats (CRISPR). These are increasingly an accumulation of various firmly associated approaches that adjusted to utilize in genome editing technology (Puchta 2016). Presently, various researches in plant system have been performed by utilizing genome editing approaches such as TALENs or CRISPR systems for the improvement of plant traits. In most of the studies plant transformation involved the genome editing approaches into the plant genome. Consequently, the plants are self pollinated or crossed pollinated to get rid of incorporated DNA, only the intended mutation remains . In several studies, transient expression of the genome editing technique by using plasmids instigated these mutations but these strategies utilize recombinant DNA at least in an intermediate stage. Recently, the genome editing tools have been designed utilizing total RNA, preassembled Cas9 protein-guide RNA ribonucleoproteins (RNPs) or TALEN proteins to create mutations. Those are totally free of DNA so the possibility of DNA integration into the genome can be avoided.

The Brassicaceae family includes numerous species and subspecies, which are devoured as shoots, leaves, roots, or in the form of seeds. Vegetative parts of plants are provided primarily as raw products, while generative parts are utilized mostly in processed form as oils, meals, powders, proteins, or condiments (Möellers 2017). They are extensively utilized as foodstuff and silage and are highly valuable in nutrition and pharmaceutical areas (Kumar and Andy 2012). The family has diversified into various genus and species that are agriculturally important due to domestication and further breeding. In spite of their valuable economic importance, advanced biotechnological tools for research in Brassicaceae plants are still deficient. In this chapter, we have discussed the genome editing approaches for the improvement of stress tolerance in Brassicaceae plants.

2 Genome Editing Approaches for Abiotic Stress Tolerance

Genome editing technologies arbitrated via different sequence-specific nucleases have appeared as powerful tools for plant genes functions studies and crop improvement. Accessibility of the advanced genome editing tools proposes tremendous opportunity for quality improvement of economically important crops. This can extend the variety of improved qualities, thus resulting in highly proficient and targeted development of crop traits, particularly resistance against the abiotic stress (Zhang et al. 2017). The innovation of engineered nucleases, which creates double-stranded break, has been transforming the molecular biology paving the new way for genome editing of targeted genes. Genome editing using ZFNs (Kim et al. 1996) and TALENs (Christian et al. 2010) have been approx. for two decades, but in recent times it becomes highlighted by the introduction of clustered regularly interspaced short palindromic repeats (CRISPR)/Cas systems (Jinek et al. 2012) that provide effortlessness and easiness of targeted gene editing. These gene editing tools utilized sequence-specific nucleases, which induced recognizing targeted DNA sequences and to create double-strand breaks. An endogenous repair system of plants repairs the double-stranded breaks in two ways either via joining of non-homologous ends that can results in the insertion or deletion of nucleotides thus producing knockout mutants, or through homologous recombination that could result in gene insertions and substitution (Symington and Gautier 2011). Various knockout mutants and the number of gene substitution and insertion mutant has been developed by utilizing genome editing tools in different plants, and a number of these mutants have been exposed to be functional for the development of improved plant traits. Here, we converse the genome editing tools that can potentially be employed for the improvement of abiotic stress tolerance of Brassicaceae.

2.1 Zinc Finger Nucleases (ZFNs)

These nucleases are the gene editing approach that utilization illusorily designed nucleases that were created following the disclosure of functioning of the useful Cys2-His2 zinc finger (ZF) area (Kim et al. 1996; Gaj et al. 2013; Palpant and Dudzinski 2013; Pabo et al. 2001). Every ZFs space consists of thirty amino acids that are collapsed up to arrangement (Pabo et al. 2001; Cathomen and Keith Joung 2008). Crystallographic examination demonstrated that these nucleases are bounded to DNA by embedding an -helix of the protein in significant section of DNA-twofold helix (Pavletich and Pabo 1991). Every ZF protein can perceive 3 couple nucleotides in the DNA. Together this nuclease monomer consists of 2 diverse spaces: ZF Cys2-His2 nuclease at the N-terminal and a Fok I DNA cleavage space at the C terminal. Fok I area dimerization is basic for this nucleases enzymatic movement (Kim et al. 1996). The perception, which isolated the recognition of ZF areas exhibit as an arrangement to the relating, continuous 3 bp targets empowered an acknowledgement that every ZF spaces can be compatible and that the control of the request for the areas would prompt one of a kind restricting specific to the proteins harbouring them in this manner empowering focusing of explicit, interesting successions in the genome. This nucleases dimer, comprising of two, three, or four ZF spaces, perceives an objective arrangement of 18 or 24 base sets, which factually structure interesting destinations in the genomes of most life forms. The structure and use of these nucleases include secluded plan, get together, and streamlining of ZF against explicit objective DNA successions pursued by connecting of individual ZFs towards focusing on bigger arrangements. Throughout the years, ZF areas have been created to perceive countless triplet nucleotides.

While the principal reports on ZF in 1996, they effectively utilized in a few living beings including plants (Gaj et al. 2013). Models incorporate focused on inactivation of endogenous qualities in Brassicaceae (Osakabe et al. 2010; Zhang et al. 2010), and other plants (Townsend et al. 2009; Shukla et al. 2009). ZFNs have disturbed the genome editing by exhibiting the capacity to control genomic destinations of intrigue and cleared a path of fundamental and applied researches. ZFNs furnish preferences over different apparatuses regarding effectiveness, high particularity and insignificant non-target impacts and present endeavours are highlighted on recuperating structure and conveyance just as growing their applications in assorted harvests of intrigue. A specially design zinc finger nuclease along with heat-shock promotor have been utilized to prompt changes in an AP2/ERF family transcription factor gene, ABA-INSENSITIVE 4, engaged with abiotic stress reactions (Osakabe et al. 2010). A high recurrence of quality transformations bringing about the ideal phenotypes was observed.

2.2 *Transcription Activator-like Effector Nucleases (TALENs)*

The journey of productive and particular control of targeted DNA prompted the one of a kind transcriptional activator-like effectors (TALE) protein, which perceive and initiate explicit plants through a lot of pair rehashes that framed the reason for the making of another genome editing approach including effective nucleases called TALENs (Jankele and Svoboda 2014). Protein consists of the focal space liable to bind DNA, atomic limitation signals, and an area which fills as activator of interpretation of objective quality (Schornack et al. 2006). Just because, the DNA-restricting capacity of these proteins (Römer et al. 2007), and after years, researchers decoded the recognition code of target DNA arrangement with TALE proteins (Boch et al.

2009). It is demonstrated that the DNA-restricting area in TALE monomers thus consist of a central recurrent space (CRD) which presents DNA. The CRD consist of couple rehashes of 34 amino acids and each amino acid long rehash in the central recurrent space bind to one nucleotide in the objective nucleotide arrangement. Two of the amino acids of the rehash, situated at positions 12 and 13, are exceptionally factor (rehash variable diresidue (RVD) and are answerable for the acknowledgement of explicit nucleotides with decadence of restricting a few nucleotides with divergence productivity. The final pair continued official to nucleotides at the 3′ end of the acknowledgement site consist of 20 amino acids in particular and, subsequently, it is named as half-rehash. While TALE transcriptional proteins, can be intended to bind any DNA grouping of intrigue have shown that the 5′ most nucleotide base of the DNA arrangement bind by a TALE proteins ought to consistently be a Thymidine (Lamb et al. 2013).

Subsequent to the DNA code recognition necessities by TALE proteins have been broken, the absolute first exertion attempted was the production of deceptive TALENs (Weeks et al. 2016). The researchers encoding the DNA-restricting TALE spaces were embedded into a plasmid vector recently utilized to make ZFNs (Christian et al. 2010). These brought about the making of an engineered, fanciful arrangement explicit nuclease genetically developed the DNA-restricting area of TALEs and the reactant space of FokI limitation endonuclease. These structures made false nucleases with DNA-restricting area and distinctive RVDs, which focused on any nucleotide succession of intrigue (Nemudryi et al. 2014; Gaj et al. 2013). A few investigations have likewise indicated that the use of RVDs for explicit official of guanine decreases hazards of non-specific impacts (Cong et al. 2012; Christian et al. 2012).

It demonstrated that in RVDs the principal amino acids are liable for the adjustment of spatial compliance in spite of the fact that it doesn't straightforwardly tie to a nucleotide, though the 2nd amino acids ties to a nucleotide either by hydrogen holding by nitrogenous bases or by van der Waals powers (Streubel et al. 2012). In view of the method of activity and particularity of these nucleases, it must be conceivable to present double-stranded breaks in a few areas of the genome till then that area harbours the acknowledgement grouping relating to the DNA-restricting spaces of this nucleases. In different circumstances, the prerequisite of the nearness of Thymidine before the 5′ finish of the planned objective succession since it shown that W232 structure in the N-terminal part of the DNA-restricting space interfaces along with Thymidine and impacts the coupling productivity (Mak et al. 2012). It has been additionally conceivable to beat this 5′ Thymidine limitation by creating freak variations of this nuclease N- terminal space that may bind to different nucleotides (Lamb et al. 2013). Considering the ease of site-coordinated control utilizing these nucleases, inside a brief timeframe after the disentangling of this genome editing method of activity, the gene modified by this approach have been utilized effectively in plants of Brassicaceae such as *Arabidopsis* and the other plant models incorporate rice, wheat, potato and tomato (Xiong et al. 2015).

2.3 Clustered Regularly Interspaced Short Palindromic Repeat (CRISPR)/CRISPR-Related Protein-9 (CRISPR/Cas9)

Abiotic stresses represent the significant challenge for crop production and cause considerable yield decrease around the world. Rearing resistant varieties against various adverse environmental conditions is the highly supportable and eco-accommodating way to deal by adapting these approaches. Progression in genome editing innovations gives new chances to edit improvement by utilizing the targeted genome for specific qualities of crops. Be that as it may, the determination of the up-and-comer qualities is basic for the achievement of accomplishing the ideal characteristics. In any case, the disclosure of 'Clustered Regularly Interspaced Short Palindromic Repeat (CRISPR)/CRISPR-related protein-9, the Cas9)' got extraordinary consideration of researchers because of its evident advantages over ZFNs and TALENs (Mao et al. 2013). Dissimilar to these nucleases that used proteins for target ID, CRISPR-Cas9 relies upon RNA–DNA acknowledgement to make the twofold strand break. Different favourable circumstances of CRISPR-Cas9 over these nucleases are (i) straightforwardness of the objective plan, (ii) effectiveness of presenting transformations by legitimately infusing the RNAs encoding Cas9 protein and guide RNA and (iii) the simplicity of multiplexing, focused on changes in various qualities in a solitary occasion (Zafar et al. 2019). It is simpler, adaptable and productive on the grounds that structuring of CRISPR-Cas9 vector is moderately less dubious in comparison to past procedures, for example, ZFNs or TALENs because of the accessibility and simple access to the enhanced bioinformatics approaches that can be utilized to distinguish the highly suitable arrangements to designed the guide RNAs without the screening libraries to perspective out the highly effective objective. As of late, a few alterations and upgrade has been made in this genome editing approach which empowered specialists to make exact changes in several life forms of enthusiasm with 'nucleotide-level' exactness in an incredibly quick way. This progression has fundamentally added to the more extensive flexibility CRISPR/CAS9 amongst the eukaryotic organism.

Furthermore in mutagenesis, this approach can be utilized to actuate (CRISPR initiation or CRISPRa) or quell (CRISPR impedance or CRISPRi) quality articulation by combining Cas9 along transcriptional activator or repressor (Bortesi and Fischer 2015). In this way, it can possibly supplant standard genetic modification-based overexpression and gene editing techniques. As a result of the exact change or modification of gene of interest, genome editing system have included advantage over ordinary genetic modification techniques where coordination of transgene is random (if there should be an occurrence of overexpression) which can curb or initiate different genes and undesirable editing of different can occur by siRNA-mediated RNA interference. In this approach, suitable structuring of guide RNA defeated this confinement of off-focusing on. Despite the fact that there are fruitful instances of CRISPRa/I functions in animal cells (Gilbert et al. 2013; Zhang et al. 2018), models from plants system is now starting to be focused. In plant system, this approach intervened genome altering and has been effectively utilized in various yields.

In any case, reports in regard to focusing on abiotic stress resilience qualities are meagre as the vast majority of such examinations so far have concentrated on biotic stresses like wounding and creepy crawly bugs. As of late, this approach intervened genome altering for heat resistance has been accomplished by focusing on a stress gene, SlAGAMOUS-LIKE 6 (SlAGL6) in tomato that enhanced the plant tolerance during high-temperature condition (Klap et al. 2017). Furthermore, focusing on numerous genes utilizing CRISPR-Cas9 has additionally been presented and effectively enhanced for various yields including Brassicaceae and others crops (Miao et al. 2013; Wang et al. 2016; Char et al. 2017; Gao et al. 2017). In this way, this genome editing approach holds an incredible prospective to create crops tolerant to numerous stresses by focusing on a few stress genes at the same time in a world-class high yielding yet delicate harvest cultivar. Additionally, tolerant genes, as recently examined, can be overexpressed utilizing CRISPR mediated gene modification (CRISPRa) which is generally being utilized for gene knockouts. Additionally, the utilization of this approach in genome editing further expanded the utilization of this approach to genome-wide screening for focused quality enhancement (Rodríguez-Leal et al. 2017; Mahas and Mahfouz 2018). It might be an alternative traditional plant breeding methodologies that depend on developing plant varities with adequate genetic variation to bring desirable qualities into crop cultivars. This approach can produce new allelic variations in plants, and therefore the new alleles relating to specific desirable phenotype would then be able to be recognized by sequencing of guide RNA (Eid et al. 2018). In this manner, this approach holds prominent prospective for the development of novel varieties for agricultural benefits.

3 Conclusion

Over the recent years the genome editing technique has developed into a progressive improvised tool that can be effectively used to reform the plants genetically in order to get better yield varieties. It can be concluded that the genome editing approaches can be considered as a ubiquitous molecular tool for crop improvement, particularly exhibiting developed nucleases, which has resulted into better yield outputs. The yield improvement advances largely depend upon the modified endonuclease that creates double-stranded breaks at the focussed loci. The research concludes that CRISPR/Cas9 genome methodology can be considered as an innovative technique because of its straightforwardness, convenience, flexibility, exactness and encouraging impactful effects. The genome editing tools hold extraordinary guarantee in producing crop assortments with upgraded diseases obstruction, enhanced oil content, biotic and abiotic stress resistant, improved productivity, quality and novel agronomic attributes that may be resourceful for farmers' and other agricultural stakeholders.

References

Bidhan R, Noren SK, Asit BM, Asit KB (2011) Genetic engineering for abiotic stress tolerance in agricultural crops. Biotechnology 10:1–22

Boch J, Scholze H, Schornack S et al (2009) Breaking the code of DNA binding specificity of TAL-type III effectors. Science 326:1509–1512

Bortesi L, Fischer R (2015) The CRISPR/Cas9 system for plant genome editing and beyond. Biotechnol Adv 33:41–52

Cathomen T, Keith Joung J (2008) Zinc-finger nucleases: the next generation emerges. Mol Ther 16:1200–1207

Char SN, Neelakandan AK, Nahampun H, Frame B, Main M, Spalding MH, Becraft PW, Meyers BC, Walbot V, Wang K et al (2017) An agrobacterium-delivered CRISPR/Cas9 system for high-frequency targeted mutagenesis in maize. Plant Biotechnol J 15:257–268

Christian M, Cermak T, Doyle EL et al (2010) Targeting DNA double-strand breaks with TAL effector nucleases. Genetics 186:757–761

Christian ML, Demorest ZL, Starker CG et al (2012) Targeting G with TAL effectors: a comparison of activities of TALENs constructed with NN and NK repeat variable Di-residues. PLoS ONE 7:e45383

Cong L, Zhou RH, Kuo Y-C, Cunniff M, Zhang F (2012) Comprehensive interrogation of natural TALE DNA-binding modules and transcriptional repressor domains. Nat Commun 3:968

Eid A, Alshareef S, Mahfouz MM (2018) CRISPR base editors: genome editing without double-stranded breaks. Biochem J 475:1955–1964

Gaj T, Gersbach CA, Barbas CF III (2013) ZFN, TALEN, and CRISPR/Cas-based methods for genome engineering. Trends Biotechnol 31:397–405

Gao W, Long L, Tian X, Xu F, Liu J, Singh PK, Botella JR, Song C (2017) Genome editing in cotton with the CRISPR/Cas9 system. Front Plant Sc 8:1364

Gilbert LA, Larson MH, Morsut L, Liu Z, Brar GA, Torres SE, Stern-Ginossar N, Brandman O, Whitehead EH, Doudna JA, Lim WA, Weissman JS, Qi LS (2013) CRISPR-mediated modular RNA-guided regulation of transcription in eukaryotes. Cell 154:442–451

Golldack D, Lüking I, Yang O (2011) Plant tolerance to drought and salinity: stress regulating transcription factors and their functional significance in the cellular transcriptional network. Plant Cell Rep 30:1383–139110

Gong X-Q, Liu J-H (2013) Genetic transformation and genes for resistance to abiotic and biotic stresses in *Citrus* and its related genera. Plant Cell Tiss Org 113:137–147

Jankele R, Svoboda P (2014) TAL effectors: tools for DNA targeting. Briefings Funct Genomics 13(5):409–419

Jinek M, Chylinski K, Fonfara I, Hauer M, Doudna JA, Charpentier E (2012) A programmable dual-RNA-guided DNA endonuclease in adaptive bacterial immunity. Sci 337(6096):816–821

Kim Y-G, Cha J, Chandrasegaran S (1996) Hybrid restriction enzymes: zinc finger fusions to Fok I cleavage domain. Proc Natl Acad Sci 93:1156–1160

Klap C, Yeshayahou E, Bolger AM, Arazi T, Gupta SK, Shabtai S, Usadel B, Salts Y, Barg R (2017) Tomato facultative parthenocarpy results from SlAGAMOUS-LIKE 6 loss of function. Plant Biotechnol J 15:634–647

Kumar S, Andy A (2012) Health promoting bioactive phytochemicals from Brassica. Int Food Res J 19:141–152.

Kuzuoglu-Ozturk D, Cebeci Yalcinkaya O, Akpinar BA et al (2012) Autophagy-related gene, *TdAtg8*, in wild emmer wheat plays a role in drought and osmotic stress response. Planta 236:1081–1092

Lamb BM, Mercer AC, Barbas CF III (2013) Directed evolution of the TALE N-terminal domain for recognition of all 50 bases. Nucleic Acids Res 41:9779–9785

Lorraine S, Symington, Gautier J (2011) Double-Strand break end resection and repair pathway choice. Annu Rev Genet 45(1):247–271

Mahas A, Mahfouz M (2018) Engineering virus resistance via CRISPR–Cas systems. CurrOpin Virol 32:1–8

Mak AN-S, Bradley P, Cernadas RA, Bogdanove AJ, Stoddard BL (2012) The crystal structure of TAL effector PthXo1 bound to its DNA target. Science 335:716–719

Mao Y, Zhang H, Xu N, Zhang B, Gou F, Zhu J-K (2013) Application of the CRISPR–Cas system for efficient genome engineering in plants. Mol Plant 6:2008–2011

Miao J, Guo D, Zhang J, Huang Q, Qin G, Zhang X, Wan J, Gu H, Qu LJ (2013) Targeted mutagenesis in rice using CRISPR-Cas system. Cell Res 23:1233–1236

Möellers C (2017) Quality aspects in breeding Brassica species, in Brassica 2017: VII International Symposium on Brassicas, Pontevedra

Mickelbart MV, Hasegawa PM, Bailey-Serres J (2015) Genetic mechanisms of abiotic stress tolerance that translate to crop yield stability. Nat Rev Genet 16:237

Nemudryi AA, Valetdinova KR, Medvedev SP, Zakian SM (2014) TALEN and CRISPR/Cas genome editing systems: tools of discovery. Acta Naturae 6(3):19–40

Osakabe K, Osakabe Y, Toki S (2010) Site-directed mutagenesis in Arabidopsis using custom-designed zinc finger nucleases. Proc Natl Acad Sci 107:12034–12039

Pabo CO, Peisach E, Grant RA (2001) Design and selection of novel Cys2His2 zinc finger proteins. Annu Rev Biochem 70:313–340

Palpant NJ, Dudzinski D (2013) Zinc finger nucleases: looking toward translation. Gene Ther 20:121–127

Pandey P, Irulappan V, Bagavathiannan MV, Senthil-Kumar M (2017) Impact of combined abiotic and biotic stresses on plant growth and avenues for crop improvement by exploiting physio-morphological traits. Front Plant Sci 8:537

Pavletich NP, Pabo CO (1991) Zinc finger-DNA recognition: crystal structure of a Zif268-DNAcomplex at 2.1 A. Science 252:809–817

Puchta H (2016) Using CRISPR/Cas in three dimensions: towards synthetic plant genomes, transcriptomes and epigenomes. Plant J 87:5–15

Römer P, Hahn S, Jordan T, Strauß T, Bonas U, Lahaye T (2007) Plant pathogen recognition mediated by promoter activation of the pepper Bs3 resistance gene. Science 318: 645–648

Rodríguez-Leal D, Lemmon ZH, Man J, Bartlett ME, Lippman ZB (2017) Engineering quantitative trait variation for crop improvement by genome editing. Cell 171:470–480

Schornack S, Meyer A, Römer P, Jordan T, Lahaye T (2006) Gene-for-gene-mediated recognition of nuclear-targeted AvrBs3-like bacterial effector proteins. J Plant Physiol 163: 256–272

Shukla VK, Doyon Y, Miller JC et al (2009) Precise genome modification in the crop species *Zea mays* using zinc-finger nucleases. Nature 459:437–441

Singh KB, Foley RC, Oñate-Sánchez L (2002) Transcription factors in plant defense and stress responses. Curr Opin Plant Biol 5:430–436

Streubel J, Blücher CA, Landgraf, Boch J (2012) TAL effector RVD specificities and efficiencies. Nat Biotechnol 30: 593–595

Townsend JA, Wright DA, Winfrey RJ et al (2009) High frequency modification of plant genes using engineered zinc finger nucleases. Nature 459(442–445):2009

Wang W, Akhunova A, Chao S, Akhunov E (2016) Optimizing multiplex CRISPR/Cas9-based genome editing for wheat. BioRxiv 051342

Weeks DP, Spalding MH, Yang B (2016) Use of designer nucleases for targeted gene and genome editing in plants. Plant Biotechnol J 14:483–495

Xiong J, Ding J, Li Y (2015) Genome-editing technologies and their potential application in horticultural crop breeding. Hortic Res 2:15019

Zafar SA, Zaidi SS, Gaba Y, Pareek SLS, Dhanhker OP, Li X, Mansoor S, Pareek A (2019) Engineering abiotic stress tolerance via CRISPR/Cas-mediated genome editing. J Exp Bot. https://doi.org/10.1093/jxb/erz476

Zhang F, Maeder ML, Unger-Wallaced E et al (2010) High frequency targeted mutagenesis in *Arabidopsis thaliana* using zinc finger nucleases. Proc Nat Acad Sci USA 107:12028–12033

Zhang M, Zhao J, Li L, Gao Y, Zhao L, Patil SB, Fang J, Zhang W, Yang Y, Li M et al (2017) The Arabidopsis U-box E3 ubiquitin ligase PUB30 negatively regulates salt tolerance by facilitating BRI1 kinase inhibitor 1 (BKI1) degradation. Plant, Cell Environ 40:2831–2843

Zhang, Y, Li S, Xue S, Yang S, Huang J, Wang, L (2018). Phylogenetic and CRISPR/Cas9 studies in deciphering the evolutionary trajectory and phenotypic impacts of rice ERECTA genes. Front Plant Sci 9:473.https://doi.org/10.3389/fpls.2018.00473

Bioinformatics Studies on the Identification of New Players and Candidate Genes to Improve *Brassica* Response to Abiotic Stress

Heba T. Ebeed

Abstract *Brassica* genus includes many important species that represent important oil and vegetable crops. *Brassica* production is greatly affected by various environmental conditions and abiotic stresses such as drought, salt, and high temperature. Comprehensively studying the molecular responses of *Brassica* species to these abiotic stresses would help in developing tolerant and more adaptive species. However, this requires information at the omics levels. Using bioinformatics and computational tools, analysis and identification of new and novel genes became easier and very fast. Bioinformatics have made obvious aids to the up-to-date improvements in our understanding of *Brassica* stress tolerance. This chapter will explore the recent studies on *Brassica* species that used integrative tools and high-throughput technologies depending on Bioinformatics to investigate the *Brassica* responses to abiotic stresses and identify key players under these stresses.

Keywords Abiotic stress · Bioinformatics · *Brassica* · Genes · Omics

1 Introduction

A diverse important variety of crops have emerged from the wild ancestral of *Brassica* species. *Brassica* species are important for the food and nonfood industries. The genus *Brassica* contains 37 different species belonging to the crucifer family (Gómez-Campo 1980). Climatic and environmental conditions are major factors that restrict many crops to be planted and thereby lessen its production. Therefore, we are now growing crops in the conditions that maximize the production and minimize adversities (Atkinson and Urwin 2012).

Various environmental conditions contribute to the overall stress effect (Ebeed 2019; Hassan et al. 2020). The responses to the environmental stresses are very complicated. Engineering genes that play key functional and structural roles in cellular components could enhance adaptation to stresses. Thus, identifying the

H. T. Ebeed (✉)
Botany and Microbiology Department, Faculty of Science, Damietta University, Damietta, Egypt
e-mail: hebeed@du.edu.eg

M. Hasanuzzaman (ed.), *The Plant Family Brassicaceae*,
https://doi.org/10.1007/978-981-15-6345-4_18

stress-responsive genes and characterizing their regulation under stress conditions is essential (Liu and Baird 2003). Given the range of abiotic stresses to which *Brassica* is exposed, it might be thought that a wide range of different tools and integrate studies would be required to understand and hence enhance stress responses. This chapter will concentrate on the bioinformatic studies that had been done to identify stress-responsive genes in *Brassica* species. The best way to improve crop productivity through enhancing their tolerance or resistance to biotic and abiotic stresses is by using biotechnology. In biotechnology different approaches are involved for the improvement of crop yield and quality (Kumar et al. 2018).

Examination of genomes through sequence information isn't just significant for giving a profound comprehensive considering of the molecular and cellular functions of gene in addition to giving knowledge into evolutionary relationships among organisms or genes. Assessment of different species phylogenetically assists in understanding the evolutionary relationships between plant genomes. Sequence information is likewise significant for the identification of genes that could be economically important and the development of breeding programs. As further sequences become accessible for whole genomes, it will speed up the studies to examine epigenomes, transcriptomes, and metabolomes and bring down the expenses (Okay 2016). Omics technologies including studies at genomic, transcriptomic, proteomic, metabolomic, epigenomic, and metagenomic levels are mainly dependent on bioinformatics for data analysis. The molecular data acquired from the high-throughput technologies are very powerful, comprehensive, and might be multifaceted on other occasions. This could make the incorporation of omics data fairly a challenging task if the test investigations were not intended to add to downstream data analyses. The integration of omics information would realize a far-reaching outline of data on different biological variables, consequently permitting analysts to have a comprehensive way in which they could investigate relationships among biological variables within a biological system. Thus, these advancements would empower researchers to figure out which gene region could be manipulated to advance the levels of a required metabolite without disturbing other biological systems. This development in *Brassica* sequence information has led to the prerequisite for tools for the structuring and cross-examination of this information. There are already several functional *Brassica* sequence databases developed by the *Brassica* community.

For *Brassica*, there are public resources that provide data on *Brassica* species supporting the worldwide *Brassica* Genome Sequencing Projects depending on five different modules, ESTDB, Microarray, QTL Markers, CMap, and EnsEMBL. ESTDB hosts expressed genes and corresponding annotation derived from comparison with genomic databases. QTL Markers integrates information on the genetic markers, maps, genotypes, and traits. (Erwin et al. 2007).

This chapter presents the latest research on *Brassica* to study abiotic stress using bioinformatics and pinpoint the progress of using high-throughput technologies to uncover the molecular responses and new genes in attempts to improve abiotic stress tolerance in *Brassica*. It also lists the available resources and databases for *Brassica* that developed and progressed recently to provide a powerful tool to study *Brassica* genome and evolutionary relationships.

2 Importance and History of *Brassica* Species

Most *Brassica* species are of great economic importance to vegetable or oil crops. *Brassica* vegetables include *B. oleracea*, *B. rapa,* and Leaf mustards (*B. juncea*). The major *Brassica* vegetable species is *B. oleracea*, which includes cabbage types, cauliflower, broccoli, sprouts, and others.

The basic species *B. nigra* (L.) (n = 8) is found as a weed, the unplanted species of the *B. oleracea* (n = 9). *B. oleracea* used as a leaf vegetable and can be classified into six groups; kales, cabbages (var. capitata, var. sabauda, var. bullata), kohlrabi, inflorescence kales including broccoli, cauliflower, and others, branching bush and Chinese kale (Snogerup 1980). *B. rapa* (n = 10) grows at very low temperatures in mountainous regions and it is thought that *B. rapa* was introduced through western Asia into China. *B. rapa* is grown for oilseed and vegetable. Chinese cabbage has some relationship to the oilseed type that is grown there.

Brassica napus (n = 19) is an amphidiploid species resulting from interspecific crosses between *B. oleracea* (n = 9) and *B. rapa* (n = 10). It is attainable that *B. napus* could bang settled at divergent places from crosses between diverse forms of *B. oleracea* and *B. rapa. B. napus* are cultivated as oilseeds in many regions in the world. *B. carinata* (n = 17) is an amphidiploid species resulting from interspecific crosses between *B. nigra* (n = 8) and *B. oleracea* (n = 9). No wild forms of *B. carinata* have been reported. Its cultivation is limited to Ethiopia, where it might have originated from hybridization between kale and *B. nigra. B. carinata* grows slowly, a trait which it might have hereditary from its *B. oleracea* parent, and its seed contains mustard oil as *B. nigra.* Farmers in Ethiopia grow *B. carinata* for its seed oil and the leafy vegetable in their gardens.

Brassica oilseed production is an important source of oil. India has the maximal *Brassica* cultivated expanse, followed by China and Canada. Nonetheless, the yield in China and Canada is higher than India and this might refer to the low yield and oil of seeds in India. England, France, and Germany have the highest seed yields because of the advantageous growth conditions of Europe. More than 50% of the globally marketed rapeseed oil is purified in Europe. The *Brassica* oil is one of the healthiest vegetative oils because of its fatty acid composition. *Brassica* oil uses for frying purposes in the fast food because of its oxidative firmness at high temperatures. Canola oil with zero erucic acid is obtained from *B. napus* varieties which have about 60% oleic acid content and around 20% linoleic and 10% alpha-linolenic and about 7% saturated fats.

Brassica juncea (n = 18) is an amphidiploid species derivative from interspecific crosses between *B. nigra* (n = 9) and *B. rapa* (n = 10). Wild types of *B. juncea* have been reported in the East and cultivated types as oilseed and as vegetable in India and China. Likewise, *B. juncea* is also grown to produce seasoning mustard, oilseed in India and Canada. The oil has a content of about 40% erucic acid. Efforts are made for producing zero erucic acid *B. juncea* by breeding from Australian zero erucic acid lines to improve oil quality.

3 Responses of *Brassica* Species to Abiotic Stresses

Plants are continuously subjected to many environmental stresses as high or low temperatures, low water availability, mineral excess or insufficiency, and high salinity. Stress can be defined as any unfavorable condition(s) or material(s) that limits the plant growth or development (Hasanuzzaman et al. 2012, 2013a, 2013b; 2014; 2018a, 2018b, 2019). All these conditions restrict the growth and production of plant crops. The continuous change in the climatic conditions limits efforts made for sustainable harvest efficiency under abiotic stresses.

Each plant needs an ideal environment for development that helps it to arrive at the creation stage just as finishing its life cycle. Stresses alter the biological, cellular, and physiological processes (Hassan et al. 2015; Ebeed et al. 2017, 2019). Hence, they include various types of responses at the cellular level to the whole organism's level that let plants to diminish the costs of stress adversities (Ebeed et al. 2018; Hassan et al. 2020). At the cellular levels in response to water shortage or high level of salts, for example, high salt concentration results in low water potential. This low availability of water reasons a reduction in turgor in the cell and adversely affects the cell polarity and macromolecules.

When a plant is exposed to a stress factor, its response divided into stages. These stages differ according to other factors such as stress length and strength (Kranner et al. 2010). At the point when at least one factor influences the plant's ideal environment, a special machinery is used to perceive this change called "Stress sensing." This triggers responses at the whole plant level to stress (Kumar et al. 2018). Responses to stress conditions happen in time-course stages. Thus, the genes response to these conditions can be divided into early-responsive genes and late-responsive genes which represent majority of the stress-responsive genes. The early-responsive genes encode transcription factors that regulate the expression of the late-responsive genes (Ebeed 2019).

4 Integrating the Bioinformatic Tools and Platforms for *Brassica* Genome Research

Recent high-throughput technologies provide valuable results. These results are largely complex, the issue that makes it hard for comprehensive understanding. The need to access and study these data make researchers develop numerous bioinformatic tools and databases. Nevertheless, these tools and databases host data with different format as it developed independently the something that limit search between these databases. The integrative databases that host different types of data from gene structure to gene functions have resolved these limitations. These integrative databases developed in silico analysis of the genes and gene families and prediction of gene function for the newly discovered genes. Table 1 lists several major web-based genome databases. The browsers host sequence information for different organisms

Table 1 List of major multiple-species integrate databases

Name	URL
Phytozome	http://www.phytozome.net/
NCBI	http://www.ncbi.nlm.nih.gov/
Ensembl Plants	http://plants.ensembl.org/index.html
PlantGDB	http://www.plantgdb.org/prj/GenomeBrowser/

as well as facilitate comparative genomic research between species. For *Brassica*, specific databases that host information about *Brassica* species only are listed in Table 2.

Computational tools have developed quickly as a result of the advancement of the complete sequence of the Brassica genome. Some efforts had been made to

Table 2 List of *Brassica*-specific web-based databases

Name	URL	Description
Brassica napus GBrowse	http://appliedbioinformatics.com.au/gb2/gbrowse/BnapusPan/	Including SNPs and functional annotation of genes
BRAD	http://brassicadb.org/brad/toolsOverview.php	Provides several datasets as genome sequence of (*B. rapa*), predicted genes, annotations, and genetic markers
B. rapa Genome	http://www.plantgdb.org/BrGDB/	Displays high-quality alignments.
Brassica Information Portal	https://bip.earlham.ac.uk/	A web repository for population and trait scoring information related to the *Brassica* breeding
The European *Brassica* Database	https://www.actahort.org/books/459/459_28.htm	Information about the *Brassica* germplasm
B. napus Genome Browser	https://www.genoscope.cns.fr/brassicanapus/	Shows orthologous pairs between *B. napus*, *B. oleracea*, and *B. rapa*
B. oleracea GBrowse	http://appliedbioinformatics.com.au/gb2/gbrowse/BolePan/	Including snps and functional annotation of genes
B. rapa GBrowse	http://appliedbioinformatics.com.au/gb2/gbrowse/XA/	Including snps and functional annotation of genes
Brassibase	https://brassibase.cos.uni-heidelberg.de/	Online-accessible knowledge and database system related to taxonomy, systematics, and evolution
*Brassica*DB	http://brassica.nbi.ac.uk/BrassicaDB/	Includes blast search and mapping data
Brassica IGF project	http://brassica.nbi.ac.uk/IGF/?page=body/database.htm	Including BAC libraries and physical maps of *Brassica*

acquire the bioinformatic tools for *Brassica*. The beginning of the sequencing of *B. rapa* genome as well as the increase in the volume of sequences entries to databases provide development for the *Brassica* bioinformatic research. The bioinformatics developed too many tools for Brassica to discover molecular markers and to aid in Brassica gene mapping. Molecular markers as the single nucleotide polymorphisms (SNPs) and simple sequence repeats (SSRs) are important for applied studies in agriculture and plant breeding programs (Love et al. 2004).

5 Relevance of Bioinformatics and Omics Research

The modern technologies with high-throughput data facilitate genome-wide analysis of different plant species and aid researchers to compare different genomes (Flagel and Blackman 2012). Bioinformatics or computational biology is a multidisciplinary branch that is widely used by researchers to analyze and interpret results of different omics approaches. Analyzing genome or proteome sequences by using bioinformatics helps scientists to figure out different biological processes. Different bioinformatic resources that depend on web-based tools can be used by researchers to process data, interpret data, and analyze data in a comprehensive way. Using algorithms, bioinformatics also creates standard formats by using mathematical and statistical models which help bioinformatists to store, retrieve, and share data. The knowledge of bioinformatics along with understating of the biological processes are important and useful for comparisons throughout species.

Omics approaches led to the appearance of innovative terms like metabolomics, metagenomics, cytomics, proteomics, and transcriptomics (Ebeed 2019). Another new approach is system biology which integrates and relates the data to pathways or networks. The wide-ranging omics data is being successfully accomplished by bioinformatics specialists through the biological databases. Management and mining of the data are essential stages for high-throughput data analysis which require involving bioinformatics. The RNA-seq and microarray technologies and quantitative polymerase chain reaction (RT-PCR) are the main technologies to study gene expression and require the use of statistical analysis and bioinformatics. RT-PCR is the most sensitive quantitative technique for studying gene expression through quantifying gene expression or determining gene copy number, and validation of RNA-seq and microarray results (Liu et al. 2012).

6 Bioinformatic Studies for Identification of Genes and Gene Families in *Brassica* Species

Bioinformatics have been used widely for identifying genes and genomic structure as well as predicting protein and transcripts characteristics. SMART, Pfam, and NCBI databases were used to analyze the conserved domains. TAIR, Phytozome, and PGDD databases were also helpful in retrieving proteins from different plants. Comparative genomics tools are helpful and important to understand gene function and the genome evolution. Our understanding of genome complexity, evolution, and organization changed greatly after the accessibility of complete genome sequences. However, as a result of the biological variation between species complicates the evolutionary analyses. Therefore, using comparative genomics databases may overwhelm these tasks. Table 3 lists main web-based tools that can be used to compare and analyze data related to *Brassica* species.

Chen et al. (2010) investigated salinity- and drought-responsive genes in *B. napus* by microarray. They identified 536 clones, 172 induced and 141 suppressed by high salinity and 288 induced and 189 suppressed by drought. These genes were predicted to belong to regulatory factors, metabolic enzymes, hormone responses, stresses-related proteins, components of signal transduction, and other growth- and development-related processes in *B. napus*. Zhang et al. (2015) selected 66 lines from 140 accessions of *B. napus* varying in their tolerance to drought conditions using the Illumina Infinium *Brassica* 60KSNP-array. The SNP array sequences were used to perform a BLAST search against the public *B. napus* genome sequence, and the 4794 identified SNPs were used for Genome-Wide Association Studies (GWAS). Further analysis using RNA-seq for two drought-sensitive inbred lines. Xin et al. (2019) aimed to investigate the molecular mechanisms to chilling and freezing conditions in *B. napus*. By RNA-seq, they identified 47,328 differentially expressed genes (DEGs) to chilling and freezing stress. Via KEGG analysis, they showed that two conserved and two novel signaling pathways were significantly enriched with differentially expressed transcripts. Luo et al. (2019) investigated the transcriptional regulation of germinated *B. napus* under low temperature using RNA-seq analysis. About 10233 and 9111 DEGs up- and down-regulated, respectively. Genes that up-regulated were involved in the regulation of phytohormones, signals transduction, the pentose phosphate pathway, and lipid metabolism whereas, genes that down-regulated were associated with ubiquitin-mediated proteolysis.

Cheng et al. (2019) used bioinformatics to find homologous genes of *AtPYLs* in *B. juncea* using TAIR and BRAD to download sequences. Then, the phylogenetic trees were constructed by MEGA5, the gene structure analysis was done by GSDS, the protein sequences were aligned by ESPript 3.0 and WebLogo software. The promoter cis-element analysis was done using PlantCARE and PLACE. Then the *BjuPYLs* gene expression in response to NaCl, low temp, and ABA was used and investigated by qRT-PCR.

WRKY, the supergene family was studied in *B. oleracea* by Yao et al. (2015). The genome-wide analysis was done using different tools and databases as Bolbase

Table 3 Researches for studying the responses to abiotic stress in *Brassica* species using different bioinformatics analyses

Species	Gene family	Genes identified/studied by Bioinformatics	Technology	Abiotic stress investigated	Reference
B. rapa	Heterotrimeric G-Protein Subunit Gene Family	9 G-Protein Subunit Genes	qRT-PCR	Abscisic acid, salinity, low and high temperature	Arya et al. (2014)
B. rapa	*NAC* transcription factor family	188 putative NAC factors	qRT-PCR	Temperature	Ma et al. (2014)
B. rapa	Growth regulating factors (*GRFs*) multigene family	17 *BrGRFs*	qRT-PCR	Gibberellic acid	Wang et al. (2014)
B. rapa	*MADS*-box gene family	160 MADS-box transcription factors	RNA-seq	ABA, cold and heat stresses	Duan et al. (2015)
B. rapa	*WRKY* family genes	41 *BrWRKY* genes	microarray	Cold, salt, and drought	Kayum et al. (2015)
B. oleracea	nucleotide-binding site (*NBS*) encoding resistance genes	80 up-regulated *NBS* genes	RNA-seq	Heat	Kim et al. (Kim et al. 2014)
B. rapa	*R2R3-MYB* transcription factor	256 *R2R3-MYB* genes	RNA-seq	Cold, ABA, osmotic stress	Wang et al. (2015)
B. oleracea	*WRKY* gene family	148 *WRKY* genes	RNA-seq	–	Yao et al. (2015)
B. napus	*LAC* Gene Family	45 *LAC* gene and miR397a, miR397b and miR6034	RNA-seq	Drought, wounding, Cd^{2+}, and NH_4^+	Ping et al. (2019)
B. rapa	*LAC* Gene Family	25 *LAC* genes	–	–	
B. oleracea	*LAC* Gene Family	8 *LAC* genes	–	–	
B. rapa	glyoxalase I (*GLYI*) and glyoxalase II (*GLYII*)	16 *BrGLYI* and 15 *BrGLYII*	qRT-PCR	Heavy metal	Yan et al. (2018)
B. juncea	ABA Receptor Family	25 *PYL* family genes	qRT-PCR	NaCl, low temp, and ABA	Cheng et al. (2019)
B. napus		4794 SNPs	RNA-seq	Drought	Zhang et al. (2015)

(continued)

Table 3 (continued)

Species	Gene family	Genes identified/studied by Bioinformatics	Technology	Abiotic stress investigated	Reference
B. rapa	*PDI* family gene	32 *PDI* genes	Microarray	Cold, salt, drought, and ABA	Kayum et al. (2017)
B. napus		2167 TFs	RNA-seq	Drought, salt, and ABA	Wang et al. (2018)
B. rapa	*E2* Gene Family	83 *E2* genes	RNA-seq	ABA, PEG, NaCl, and heat and cold stress	Khan et al. (2018)
B. rapa	*PP2C* Gene Family	131 *PP2C* genes	RNA-seq	Heat, cold, ABA, and drought	Khan et al. (2019)
B. rapa	*ALDH* Superfamily	23 *ALDH* genes	qRT-PCR	NaCl, and H2O2	Gautam et al. (2019)
B. napus		536 clones as salinity or drought-responsive genes	Microarray	High salinity and drought	Chen et al. (2010)
B. napus		47,328 differentially expressed genes (DEGs)	RNA-seq	Cold and freezing	Xin et al. (2019)
B. napus		19344 DEGs	RNA-seq	Low temperature	Luo et al. (2019)

database, Pfam, HMMER, MUSCLE, and SMART. They retrieved RNA-seq data from SRA, then mapped the sequence reads using Bowtie and TopHat, and analyzed the conserved motifs by MEME, and constructed the phylogenetic tree using MEGA. Kim et al. (2014) studied the NBS-LRR gene family in *B. oleracea.* RNA-seq reads were assembled, validated, and mapped to the assembled genes by Bowtie. Then the mapped reads were normalized. The NBS-LRR proteins and the conserved motifs were analyzed and the genes were localized to chromosomes using MapChart.

Arya et al. (2014) studied the family of subunit genes of the heterotrimeric G-protein in *B. rapa.* They retrieved protein sequences via Phytozome database and aligned the sequences using ClustalW and investigated the phylogenetic relationships by MEGA software. The gene duplications were investigated by DnaSP. MADS-box gene family in *B. rapa* was identified by Duan et al. (2015) by using the *Brassica* database (BRAD). They aligned MADS protein sequences, constructed the phylogenetic trees, and identified the orthologous genes. They analyzed the conserved motifs by MEME and SMART and used GSDS to determine the gene structures. The gene duplications were determined using Ka/Ks calculator. Finally, they build the cellular network using Cytoscape software. Using BRAD database, family of NAC transcription factor in *B. rapa* was studied by Ma et al. (2014). They used ClustalW to align the sequences, MEME to predict the conserved motifs and MEGA to construct the phylogenetic tree. Wang et al. (2014) studied the GRF gene family in *B. rapa* using BRAD. The DNAMAN (http://www.lynnon.com) was used to align the sequences and the phylogenetic trees were constructed using MEGA software. The GSDS was used to predict the gene structure, the DNASTAR to calculate the content of GC, MEME to predict the conserved motifs and the ProtParam to predict the protein properties. The Simple Sequence Repeat markers were detected using SSR-Identification Tool.

Using BRAD, Duan et al. (2015) analyzed the sequences of the ascorbic acid-related genes in *B. rapa* using FGENESH, the homologous genes were predicted using Phytozome and Amborella Genome Database, aligned the sequences by ClustalW, constructed the phylogenetic tree using MEGA, and calculated the gene duplication values by Ka/Ks tool. They analyzed the conserved motifs by MEME and used InterProScan for annotation. Kayum et al. (2015) analyzed the WRKY transcription factors. They used multiple tools for analyses as they used BRAD, SMART, and ExPASy. Locations on the chromosomes were identified by MapChart. Wang et al. (2015) studied *B. rapa* R2R3-MYB transcription factors. They analyzed proteins and the conserved motifs, then aligned the sequences, constructed the phylogenetic tree, and analyzed the orthologous genes and gene duplication.

Kayum et al. (2017) identified 32 genes BrPDI (disulfide isomerase; PDI) using SWISS-PROT of the *B. rapa* genomic database BRAD, NCBI, and Br135K microarray dataset. Using bioinformatics, they characterized their sequences in silico. Genes were clustered by constructing a phylogenetic tree, the presence of a thioredoxin domain in the protein sequences were analyzed using SMART and NCBI. Protein length, molecular weight, and isoelectric point were analyzed using ExPasy, sequence alignment of protein sequences was carried out using CLUSTAL Omega, and the number of introns and exons were analyzed by GSDS. The evolutionary constraint (Ka/Ks) was calculated using Mega 6.0 software. Furthermore, the expression profiles of these genes were analyzed in response to abiotic stresses; cold, salt, drought, and ABA.

Using the publicly available sequences, (Wang et al. 2018) identified 2167 TFs belonging to five families AP2/EREBPs, bZIPs, MYBs, NACs, and WRKYs in *A. thaliana*, *B. rapa*, *B. oleracea* and *B. napus*. Genomic and protein sequences were downloaded from TAIR, BRAD, Ensembl, and Genoscope. Genome-wide study identified 518 *BnAP2/EREBPs*, 252 *BnbZIPs*, 721 *BnMYBs*, 398 *BnNACs,* and 278 *BnWRKYs* in *B. napus* which contained novel members. Khan et al. (2018) studied ubiquitin conjugating enzymes *E2*. They retrieved protein sequences from BRAD and verified sequences using NCBI database and SMART, analyzed protein characters by ExPASy protparam and used WoLF PSORT server for prediction of the subcellular localization. Protein sequences were aligned using CLUSTALW and constructed the phylogenetic tree and calculated Ks/Ka using MEGA 7.0. Motif analysis was analyzed by MEME and the exon–intron structure was examined by GSDS. Chromosomal localization of *BraE2* was drawn by Mapchart. Yan et al. (2018) studied glyoxalase I and glyoxalase II (*GLYI* and *GLYII*). They identified 16 *BrGLYI* and 15 *BrGLYII* genes from *B. rapa* using bioinformatics and also linked genes to chromosomes. Further analyses were also done as exon–intron structures, identification of the conserved motifs and cis-elements of the promoter. They also studied the expression of these genes in different tissues and in response to biotic and abiotic stresses using publicly available data and quantitative real-time polymerase chain reaction (qRT-PCR) qRT-PCR.

Using bioinformatics, Ping et al. (2019) identified 45 *LAC* genes in *B. napus,* 25 *LAC* genes in *B. rapa,* and 8 *LAC* genes in *B. oleracea.* Then, they used RNA sequencing (RNA-seq) and qRT-PCR to study the evolution and functions of LAC gene family in *B. napus* and results in identifications of three miRNAs (miR397a, miR397b, and miR6034) target 11 *BnLACs* differentially expressed under different series of stresses. Khan et al. (2019) studied one of the protein phosphatases (PP2C) group, the type 2C protein. They identified 131 *PP2C* genes in *B. rapa* by retrieving sequences from BRAD and used *A. thaliana* sequences as queries and the domains of BraPP2C proteins were verified using NCBI-CD and SMART databases. The phylogenetic tree was constructed using MEGA 7 software and Ka/Ks ratios were calculated for *BraPP2C* genes by MEGA 7 and the exon–intron structure by GSDS server. Gautam et al. (2019) studied aldehyde dehydrogenase (*ALDH*) genes and identified family members in *B. rapa.* Protein and CDS were retrieved using the BRAD, the conserved domains were confirmed using SMART, Pfam, and InterPro. Phylogenetic tree was done by MEGA v.7 and the putative promoter elements of *BrALDH* genes were predicted using Plant Care database.

7 Conclusion

Too many bioinformatics tools have been used to study how *Brassica* species respond to abiotic stress conditions. In *B. rapa* and *B. napus*, transcriptional and translational studies have progressed as expected but the other species still need to be further studied. These bioinformatics studies are very useful and helpful in predicting key players in response to abiotic stresses in *Brassica* and to understand the *Brassica* biological systems under abiotic stresses.

References

Arya GC, Kumar R, Bisht NC (2014) Evolution, expression differentiation and interaction specificity of heterotrimeric G-protein subunit gene family in the mesohexaploid *Brassica rapa*. PLoS ONE 9:e105771. https://doi.org/10.1371/journal.pone.0105771

Atkinson NJ, Urwin PE (2012) The interaction of plant biotic and abiotic stresses: from genes to the field. J Exp Bot 63:3523–3543

Chen L, Ren F, Zhong H et al (2010) Identification and expression analysis of genes in response to high-salinity and drought stresses in *Brassica napus*. Acta Biochim Biophys Sin 42:154–164. https://doi.org/10.1093/abbs/gmp113.Advance

Cheng C, Zhong Y, Cai Z, et al (2019) Genome-wide identification and gene expression analysis of ABA receptor family genes in *Brassica juncea* var. tumida. Genes 10:460

Duan W, Song X, Liu T et al (2015) Genome-wide analysis of the MADS-box gene family in *Brassica rapa* (Chinese cabbage). Mol Genet Genomics 290:239–255. https://doi.org/10.1007/s00438-014-0912-7

Ebeed HT (2019) Omics approaches for developing abiotic stress tolerance in wheat. In: Wheat production in changing environments. Springer, pp 443–463

Ebeed HT, Hassan NM, Aljarani AM (2017) Exogenous applications of Polyamines modulate drought responses in wheat through osmolytes accumulation, increasing free polyamine levels and regulation of polyamine biosynthetic genes. Plant Physiol Biochem 118:438–448. https://doi.org/10.1016/j.plaphy.2017.07.014

Ebeed HT, Hassan NM, Keshta MM, Hassanin OS (2019) Comparative analysis of seed yield and biochemical attributes in different sunflower genotypes under different levels of irrigation and salinity 59:339–355. https://doi.org/10.21608/ejbo.2019.5043.1205

Ebeed HT, Stevenson SR, Cuming AC, Baker A (2018) Conserved and differential transcriptional responses of peroxisome associated pathways to drought, dehydration and ABA. J Exp Bot 69:4971–4985. https://doi.org/10.1093/jxb/ery266

Erwin TA, Jewell EG, Love CG et al (2007) BASC: an integrated bioinformatics system for Brassica research. Nucleic Acids Res 35:870–873. https://doi.org/10.1093/nar/gkl998

Flagel LE, Blackman BK (2012) The first ten years of plant genome sequencing and prospects for the next decade. In: Plant Genome Diversity, vol 1. Springer, pp 1–15

Gautam R, Ahmed I, Shukla P et al (2019) Genome-wide characterization of ALDH superfamily in *Brassica rapa* and enhancement of stress tolerance in heterologous hosts by BrALDH7B2 expression. Sci Rep 9:7012. https://doi.org/10.1038/s41598-019-43332-1

Gómez-Campo C (1980) Morphology and morpho-taxonomy of the tribe Brassiceae. Morphol morpho-taxonomy tribe Brassiceae 3–31

Hasanuzzaman M, Hossain MA, Silva JAT, Fujita M (2012) Plant responses and tolerance to abiotic oxidative stress: antioxidant defense is a key factor. In: Bandi V, Shanker AK, Shanker C, Mandapaka M (eds) Crop stress and its management: perspectives and strategies. Springer, New York, pp 261–316

Hasanuzzaman M, Nahar K, Alam MM, Roychowdhury R, Fujita M (2013a) Physiological, biochemical, and molecular mechanisms of heat stress tolerance in plants. Int J Mol Sci 14(5):9643–9684

Hasanuzzaman M, Nahar K, Fujita M (2013b) Extreme temperatures, oxidative stress and antioxidant defense in plants. In: Vahdati K, Leslie C (eds) Abiotic stress—plant responses and applications in agriculture. InTech, Rijeka, pp 169–205

Hasanuzzaman M, Nahar K, Fujita M (2014) Role of tocopherol (vitamin E) in plants: abiotic stress tolerance and beyond. In: Parvaiz A, Rasool S (eds) Emerging technologies and management of crop stress tolerance, vol 2. A sustainable approach. Academic Press and Elsevier, New York, pp 267–289

Hasanuzzaman M, Nahar K, Rohman MM, Anee TI, Huang Y, Fujita M (2018a) Exogenous silicon protects Brassica napus plants from salinity-induced oxidative stress through the modulation of AsA-GSH pathway, thiol-dependent antioxidant enzymes and glyoxalase systems. Gesunde Pflanzen 70:185–194

Hasanuzzaman M, Nahar K, Rahman A, Mahmud JA, Alharby HF, Fujita M (2018b) Exogenous glutathione attenuates lead-induced oxidative stress in wheat by improving antioxidant defense and physiological mechanisms. J Plant Interact 13(1):203–212

Hasanuzzaman M, Bhuyan MHMB, Anee TI, Parvin K, Nahar K, Mahmud JA, Fujita M (2019) Regulation of ascorbate-glutathione pathway in mitigating oxidative damage in plants under abiotic stress. Antioxidants 8(9):384. https://doi.org/10.3390/antiox8090384

Hassan N, El-bastawisy Z, Ebeed H, Nemat Alla M (2015) Role of defense enzymes, proteins, solutes and Δ1-pyrroline-5-carboxylate synthase in wheat tolerance to drought. Rend Lincei 26. https://doi.org/10.1007/s12210-015-0429-y

Hassan N, Ebeed H, Aljaarany A (2020). Exogenous application of spermine and putrescine mitigate adversities of drought stress in wheat by protecting membranes and chloroplast ultra-structure. Physiol Mol Biol Plants 26(2):233–245

Kayum A, Jeong H, Jong J, Park I (2015) Identification and expression analysis of WRKY family genes under biotic and abiotic stresses in *Brassica rapa*. Mol Genet Genomics 209:79–95. https://doi.org/10.1007/s00438-014-0898-1

Kayum A, Park J, Nath UK, et al (2017) Genome-wide characterization and expression profiling of PDI family gene reveals function as abiotic and biotic stress tolerance in Chinese cabbage (*Brassica rapa* ssp . pekinensis). BMC Genomics 18:885. https://doi.org/10.1186/s12864-017-4277-2

Khan N, Hu C, Huijie D, et al (2018) Evolution and expression divergence of E2 gene family under multiple abiotic and phytohormones stresses in *Brassica rapa*. Biomed Res Int

Khan N, Ke H, Hu C, et al (2019) Genome-wide identification, evolution, and transcriptional profiling of PP2C gene family in *Brassica rapa*. Biomed Res Int

Kim Y, Jung H, Park J et al (2014) Response of NBS encoding resistance genes linked to both heat and fungal stress in *Brassica oleracea*. Plant Physiol Biochem 86:130–136. https://doi.org/10.1016/j.plaphy.2014.11.009

Kranner I, Minibayeva FV, Beckett RP, Seal CE (2010) What is stress? Concepts, definitions and applications in seed science. New Phytolog 188(3):655–673

Kumar A, Sengar RS, Singh A, et al (2018) Biotechnological tools for enhancing abiotic stress tolerance in plant. In: Eco-friendly agro-biological techniques for enhancing crop productivity. Springer, pp 147–172

Liu L, Li Y, Li S, et al (2012) Comparison of next-generation sequencing systems. Biomed Res Int

Liu X, Baird WM (2003) Differential expression of genes regulated in response to drought or salinity stress in sunflower. Crop Sci 43:678–687

Love CG, Batley J, Lim G et al (2004) New computational tools for Brassica genome research. Comp Funct Genomics 5:276–280. https://doi.org/10.1002/cfg.394

Luo T, Xian M, Zhang C, et al (2019) Associating transcriptional regulation for rapid germination of rapeseed (*Brassica napus* L.) under low temperature stress through weighted gene co-expression network analysis. Sci Rep 9:55. https://doi.org/10.1038/s41598-018-37099-0

Ma J, Wang F, Li M et al (2014) Genome wide analysis of the NAC transcription factor family in Chinese cabbage to elucidate responses to temperature stress. Sci Hortic 165:82–90. https://doi.org/10.1016/j.scienta.2013.11.005

Okay S (2016) Identification of gene families using genomics and/or transcriptomics data. In: Plant omics: trends and applications. Springer, pp 217–254

Ping X, Wang T, Lin N et al (2019) Genome-wide identification of the LAC gene family. Molecules 24:1985

Snogerup S (1980) The wild forms of the Brassica oleracea group (2n= 18) and their possible relations to the cultivated ones. Brassica crops and wild allies.[I]. 121–132

Wang F, Qiu N, Ding Q, et al (2014) Genome-wide identification and analysis of the growth-regulating factor family in Chinese cabbage (*Brassica rapa* L . ssp. pekinensis). BMC Genomics 15:807

Wang P, Yang C, Chen H et al (2018) Exploring transcription factors reveals crucial members and regulatory networks involved in different abiotic stresses in *Brassica napus* L. BMC Plant Biol 18:202

Wang Z, Tang J, Hu R, et al (2015) Genome-wide analysis of the R2R3-MYB transcription factor genes in Chinese cabbage (*Brassica rapa* ssp . pekinensis) reveals their stress and hormone responsive patterns. BMC Genomics 16:17. https://doi.org/10.1186/s12864-015-1216-y

Xin H, Xianchao N, Pan X, et al (2019) Comparative transcriptome analyses revealed conserved and novel responses to cold and freezing stress in *Brassica napus* L . G3 Genes, Genomes, Genet g3-400229. https://doi.org/10.1534/g3.119.400229

Yan G, Xiao X, Wang N, et al (2018) Genome-wide analysis and expression profiles of glyoxalase gene families in Chinese cabbage (*Brassica rapa* L.). PLoS One 13:e0191159

Yao Q, Xia E, Liu F, Gao L (2015) Genome-wide identification and comparative expression analysis reveal a rapid expansion and functional divergence of duplicated genes in the WRKY gene family of cabbage. Brassica Oleracea Var Capitata Gene 557:35–42. https://doi.org/10.1016/j.gene.2014.12.005

Zhang J, Mason AS, Wu J et al (2015) Identification of putative candidate genes for water stress tolerance in canola (*Brassica napus*). Front Plant Sci 6:1058. https://doi.org/10.3389/fpls.2015.01058

Use of Biostimulants for Improving Abiotic Stress Tolerance in Brassicaceae Plants

M. H. M. Borhannuddin Bhuyan, Sayed Mohammad Mohsin, Jubayer Al Mahmud, and Mirza Hasanuzzaman

Abstract The present climatic change is depleting natural resources and exerting a negative impact on the crop production. Moreover, modern agriculture is shifting toward organic environment amicable and everlasting systems to improve yield as well as crop quality without increasing the inputs. To achieve sustainability, species-specific breeding program is going on, which is time-consuming. Contrary an improved cultivar from breeding program may not be tolerant of the abiotic stresses; if so, may tolerate one or two specific abiotic stress conditions. In this context, exogenous application of phytoprotectants to enhance abiotic stress tolerance is popular among the scientific community for the last two decades. But, modern agriculture requires not only abiotic stress protection, but also need a low-cost improvement in plant performance within a shorter duration with high-quality yield. Accordingly, biostimulants could be an excellent and viable alternative in this condition, which are capable to enhance the growth of plants, improve the nutrient uptake, increase tolerance to biotic and abiotic stresses, and expand crop quality traits along with a good yield. In leafy vegetables, biostimulants increased root growth, photosynthetic pigments, and antioxidant potential, which further upregulated plant growth. A large number of research articles already reported about the biostimulation of crops using various substances and microorganisms. But still, there are controversies regarding

M. H. M. Borhannuddin Bhuyan
Citrus Research Station, Bangladesh Agricultural Research Institute, Jaintapur, Sylhet 3556, Bangladesh

S. M. Mohsin
Laboratory of Plant Stress Responses, Department of Applied Biological Sciences, Faculty of Agriculture, Kagawa University, Miki-Cho, Kita-Gun, Kagawa 761-0795, Japan

Department of Plant Pathology, Sher-e-Bangla Agricultural University, Sher-e-Bangla Nagar, Dhaka 1207, Bangladesh

J. A. Mahmud
Department of Agroforestry and Environmental Science, Sher-e-Bangla Agricultural University, Sher-e-Bangla Nagar, Dhaka 1207, Bangladesh

M. Hasanuzzaman (✉)
Department of Agronomy, Sher-e-Bangla Agricultural University, Sher-e-Bangla Nagar, Dhaka 1207, Bangladesh
e-mail: mhzsauag@yahoo.com

M. Hasanuzzaman (ed.), *The Plant Family Brassicaceae*,
https://doi.org/10.1007/978-981-15-6345-4_19

the definition of biostimulants. Moreover, for developing a science-based biostimulants industry, the functional and biological basis of the biostimulants should be elucidated. In addition, proper regulations for controlling these compounds are also a prerequisite. For example, the European Union (EU) already has drawn a line between biostimulants and other plant growth-related chemicals such as pesticides or biocontrol agents and fertilizers. In this chapter, we overviewed the definition and main categories of biostimulants. Moreover, the future prospects, opportunities, and challenges of biostimulants are also discussed highlighting Brassicaceae crops.

Keywords Abiotic stress · Bioeffectors · Food security · Plant growth regulating microorganism · Nutrient use efficiency · Sustainable agriculture

1 Introduction

Climatic change and rapid population growth are increasing food demand worldwide. The extreme environment created a harmful impact on the production of food by reducing agricultural land and crop yield (Hawrylak-Nowak et al. 2019). Abiotic stresses are linked to climatic, edaphic, and environmental components, and plants are always affected by different abiotic stresses in the whole life cycle from seed germination to harvest. Occurrences of stresses in susceptible phonological phases of plants hamper plant growth and development, which ultimately reduces crop productivity (Bulgari et al. 2019). To consider future food demand, environmental conditions need to be managed by improving agricultural systems; for example, the development of tolerant crop varieties or the invention of new agronomic techniques (Mariani and Ferrante 2017). However, to alleviate the effect of abiotic stress on the crop, plants need to develop effective methods because tolerance mechanisms in plants are so much complicated under stress conditions (Mou 2011). Therefore, to improve quickly plant's tolerance, exogenous phytoprotectants application is the possible way (Hasanuzzaman et al. 2018a). A promising and effective agronomic tool to reduce abiotic stress would be the application of "biostimulants" because biostimulants contain different bioactive molecules to improve plants' capability to tolerate unfavorable environmental stresses (Bulgari et al. 2019).

Biostimulants are a group of substances or microorganisms, whose have a positive impact to enhance plant growth, availability of nutrient, and improvement of plant stress tolerance (Calvo et al. 2014; Shubha et al. 2017). Diverse natures and activities of biostimulants have been discovered. The nature of biostimulants can be described based on substances or microorganisms. Substances may be a solitary compound (for example, glycine betaine) or a compounds mixture (such as seaweed extracts), as well as they can be organic compounds or inorganic compounds (Przybysz et al. 2014). Similarly, for microorganisms it can be single (for example, *Trichoderma*) or a combination of microorganisms (commercial products). Biostimulants activity can be explained by its physiological function on plant mechanisms. For example, biostimulants protected photosynthetic damage (Shabala et al. 2012), scavenged reactive oxygen species (ROS) by antioxidants (Chen and Murata 2011), improved auxin synthesis (Dobbelaere et al. 1999), increased nutrient uptake (Billard et al. 2014), and finally enhanced the abiotic stress tolerance (Huang et al. 2010).

Abiotic stresses are the vital limiting factors for the growth and productivity of plants, which can reduce worldwide crop yields above 50% (Hinojosa et al. 2018). Most common abiotic stresses are salinity, drought, extreme temperature, metal/metalloid(s) toxicity, waterlogging, nutrient insufficiency, UV-radiation, and ozone; whose are responsible for crop quality reduction and yield losses (Hasanuzzaman et al. 2017a, 2018b; Hinojosa et al. 2018). The abiotic stresses hampered the plant's physiological and molecular mechanisms such as reduced uptake of nutrient and photosynthetic pigment, disordered mitochondrial activity, and created oxidative stress thus plant development and productivity are affected (Hasanuzzaman et al. (2012a, b; 2017a). The main symptom of abiotic stresses is oxidative damage, which produced by excess amounts of ROS (Hasanuzzaman et al. 2018b). Non-radical singlet oxygen ($^1O_2$) and hydrogen peroxide (H_2O_2), and free radical superoxide ($O_2^{\bullet -}$) and hydroxyl ($OH^{\bullet}$) are the common ROS and can destroy plant cell by interfering with cellular molecules and ultimately cause plant death (Hasanuzzaman et al. 2017a). Another important cytotoxic compound is methylglyoxal (MG) generated by abiotic stresses and both ROS and MG cause serious damage to crop plants especially in Brassicaceae plants (Hasanuzzaman et al. 2018a).

Brassicaceae is one of the major families of the plant kingdom, which contain 340–360 genera and more than 3,700 species; among them, most of the species have enormous economic and scientific value, including the worldwide recognized model plant *Arabidopsis* (Warwick et al. 2006; Sharma et al. 2014). The common plants under Brassicaceae family are mustard (*Brassica napus* and *B. juncea*), oilseed crops; turnip (*B. rapa*), leafy vegetables; cabbage, cauliflower, kale, and broccoli (*B. olercaea*), vegetables; radish (*Raphanus sativus*), vegetables; and *Arabidopsis thaliana*, model plant (Sharma et al. 2014). However, growth as well as yield of Brassicaceace plants are destructively lost by various abiotic stresses, among them most common are drought stress (Hasanuzzaman et al. 2017b, c; Khan et al. 2019), salinity stress (Wani et al. 2016; Zhao et al. 2018; Hasanuzzaman et al. 2018c), metal/metalloid(s) stress (Hasanuzzaman et al. 2017d; Mahmud et al. 2017a, b, c, 2018), extreme temperature stress (Hasanuzzaman et al. 2014b; Jia et al. 2019), and waterlogging stress (Habibzadeh et al. 2013; Zou et al. 2014).

Besides ROS production under abiotic stress, plants also developed different protective systems (enzymatic and non-enzymatic antioxidants) to reduce and stabilize the damaging effect of ROS as well as developing stress tolerance (Hasanuzzaman et al. 2014b). Moreover, plants are immobile organisms, and they have the ability for activating different mechanisms toward surviving in unfavorable situations, but extreme environmental changes might irreversibly affect plant growth (Bechtold and Field 2018). Some agronomic strategies such as the selection of variety, growing period, crop density, etc., could reduce the harmful effect of abiotic stresses (Mariani and Ferrante 2017), but this is not enough for all situations. In addition, genetic transformation is an effective way to make stress tolerance, but it is difficult due to genetic variability of the species (Singhal et al. 2016; Wang et al. 2016). In this situation, it has been examined that the application of biostimulants is an effective technique to develop stress-tolerant crops (Van Oosten et al. 2017; Yamauchi 2018; Bulgari et al. 2019). Therefore, in this chapter, we tried to summarize all the possible mechanisms of different biostimulants for developing abiotic stress tolerance and to concise the interactive role with signaling molecules and others in Brassicaceae plants. Finally, we attempted to represent an overview of recent findings on the functions of biostimulants in response to abiotic stress tolerance of Brassicaceae plants with special focus on opportunities and challenges of biostimulants on crops yield and quality.

2 Plant Biostimulants

2.1 Definitions and Common Features

The idea of biostimulation goes back to the 1930s, when Prof. V. P. Filatov first gave the proposal about the biogenic stimulant. He proposed that various organisms, when exposed to stressors, could synthesize some biological materials, which could affect metabolic and energetic processes in humans, animals, and plants (Filatov 1944). Later other researchers illustrated some experimental evidence to prove the efficacy of these biological materials especially the organic acids under practical conditions. But the recent advancement in biostimulants science was initiated by Hervé (1994), who first provided a practical concept for biostimulants. He suggested that novel "bio-rational products" should be synthesized with a systemic approach from chemical synthesis to plant physiological, agricultural, and ecological prospects. He also suggested that these products should be functioned at low doses, be ecologically safe, and have reproducible benefits in agricultural plant cultivation (Hervé 1994).

Afterward du Jardin (2012) first provided the profound analysis of phyto-biostimulants with an emphasis on their systemic categorization based on their cellular physiological and biochemical functions, their origin and method of action in cells, which further influence the European Union (EU) to develop legislation and regulation for biostimulants industry. At present, a number of approaches have been implicated for developing novel biostimulants utilizing several methodology including physicochemical composition characterization of stimulating substances (Aremu et al. 2015), growth, development, and yield analysis of plants (Tian et al. 2015). Moreover, *omic* strategies, for instance, physiological and microarray analysis (Jannin et al. 2012), genomic analysis (Santaniello et al. 2013), phenomic and molecular analysis (Petrozza et al. 2014), metabolomic analysis (Ertani et al. 2014), transcriptomic analysis (Goñi et al. 2016), and proteomic analysis (Martínez-Esteso et al. 2016) are also implicated. A number of important scientific conferences on biostimulants held over the last decades. Among them "World Congresses on the Use of Biostimulants in Agriculture' 2012" took place in France, and the second "World Congresses on the Use of Biostimulants in Agriculture' 2015" held in Italy were significant for conceptualization and developing a novel methodology for biostimulants. Nevertheless, many researchers have coined definitions from their experimental point of view, which are also of significant importance. Some of the recently derived definitions are illustrated (Table 1).

Therefore, considering all the above-mentioned definitions, the biostimulants can be termed as "Natural and/or formulated substances derived from diverse inorganic and organic (plant/animal) sources and/or microorganisms, having potentiality to increase morphophysiological, qualitative and yield attributes along with abiotic stress tolerance when applied at a lower dose." For more clarification, biostimulants should neither be a fertilizer nor be a pesticide, but trace elements having bioactivity are included in biostimulants. Among other vitamins, amino acids, polysaccharides, oligosaccharides, phenolic compounds, low molecular weight organic acids, trace phytohormones, etc., are the well known constituents in biostimulants. Nevertheless, scientists came in a common decision that biostimulants must not contain major plant nutrients and natural plant hormones; or the mechanism of biostimulants must not depend on the nutrient and hormonal contents of the substance.

The mechanisms activated by biostimulants are still under investigation. They act directly on the physiology of plants as well as improve soil quality. They can also alter some cellular molecular mechanisms to perk up nutrient along with water utilizing efficiency, enhancing plant growth and development, regulate primary as well as secondary metabolism as well as stand against abiotic stresses.

2.2 Main Categories

However, there is no definite or legal definition of biostimulants; despite this, many scientists, stakeholders, and regulators categorized the biostimulants in some key groups, which recognized widely (Calvo et al. 2014; Halpern et al. 2015). The source of biostimulants could be from synthetic or natural origins having several inorganic and organic components (Calvo et al. 2014).

Worldwide recognized common biostimulants are organic acids having low molecular weight, phenolic compounds, humic substances (humic and fulvic acids), hormone-containing substances, like extract of seaweeds (Battacharyya et al. 2015), vegetable oils (Canellas et al. 2015), protein hydrolysates; mixture of amino acids, oligopeptides, and polypeptides (Colla et al. 2015), N-containing compounds; glycine betaines, polyamines, and non-protein amino acids (du Jardin 2015), chitosan (Pichyangkura and Chadchawan 2015), inorganic compounds; essential micro elements (silicon, Si; sodium, Na; selenium, Se; cobalt, Co; and aluminum, Al) and their salts (chlorides, phosphates, phosphites, silicates, and carbonates) (Hasanuzzaman et al. 2014a; du Jardin 2015; Gómez-Merino and Trejo-Téllez 2015; Savvas and Ntatsi 2015), beneficial fungi; mycorrhiza (Rouphael et al. 2015) and *Trichoderma* (Fiorentino et al. 2018), and beneficial bacteria; (plant growth-promoting rhizobacteria [PGPR]) (Ruzzi and Aroca 2015).

Table 1 The definitions of biostimulants coined by different workers over the years (adopted from Yakhin et al. 2017)

Terms used	Definitions/hypotheses	References
Biostimulant	Biostimulants are exogenous compound(s) or microorganisms (neither fertilizer nor pesticide), which applied in plant or root zone (rhizosphere) to improve nutrient uptake, plant growth, crop quality, and abiotic stress tolerance	EBIC (2012)
Biostimulant	Biostimulants enhance plant growth and development, are from organic origin, and showed their efficacy in a small quantity, but their responses are not attributed to the essential plant nutrients	Sharma et al. (2014)
Biostimulant Microorganisms	Both plant growth-promoting microorganisms (PGPMs) and biocontrol agent microorganisms (BCMs) can be grouped as "biostimulant microorganisms" are able to promote growth of plants and can defend them against pathogens all through the crop cycle, from seed to seed	Sofo et al. (2014)
Phytostimulators/Biostimulators	Phytostimulators or biostimulators are the group of microorganisms that could promote growth of plants by hormone-like actions	Aguado-Santacruz et al. (2014)
Plant Biostimulant	Plant Biostimulants are substances or microorganisms and/or any combinations, in a formulated form, applied to seeds or plants or the rooting media intended for stimulating plant biological processes advantaging nutrient utilizing efficiency and/or abiotic stress tolerance, not considering the nutrients content	Traon et al. (2014)
Biostimulants	Biostimulants are bioactive compounds obtained from the extracts of organic raw materials	Bulgari et al. (2015)

(continued)

Table 1 (continued)

Terms used	Definitions/hypotheses	References
Biostimulant	Biostimulants are substances rather than fertilizers that could enhance the growth and development of plants in a small quantity. These substances are eco-friendly, from natural origins, foster morphological growth, nutrient efficiency as well as abiotic stress tolerance	Chojnacka et al. (2015)
Biostimulants	Biostimulants are materials or microorganisms applied exogenously to plants intended to improve nutrient efficiency, crop quality as well as tolerance to abiotic stress, despite nutrients composition. Moreover, a commercial formulation of these materials and/or microbes is also designated as plant biostimulants	du Jardin (2015)
Biostimulants	Biostimulants are any substances, except pesticides and nutrients, having the capacity for modifying plant physiology and potentially enhance growth and development, along with abiotic stress tolerance when applied in a specific formulation to growing media, seeds or growing plants	Halpern et al. (2015)
Biostimulants	Biostimulants referring to the compounds or a combination of compounds enhance plant growth and health, from neither a pesticide nor fertilizer	Lovatt (2015)
Biostimulants	Biostimulants are products based on organic raw materials, applied in a very small dose for regulating physicobiochemical processes within plants aiming at the stimulation of growth, development by realizing of genetic productivity potential by changing hormonal profile, increase nutrition efficiency and metabolic processes as well as strengthening capability to tolerate negative abiotic stresses	Yakhin et al. (2017)

3 The Potential Use of Biostimulants from Lab to Field

Over the last few decades, the science—plant physiology emerged as a potential solution for the problems related to crop production and invented many technical breakthroughs to cope up with the changing climatic conditions as well as taking the tough challenges to feed the world population. Generally, an experiment is always based on one or more than one model organism in controlled conditions. On the other hand, a classical course of actions may be followed to the development of a successful biostimulant, where the active ingredients or microorganisms are tested in a laboratory or strongly regulated conditions to select committing candidates. But moving this promising active substance obtained from the laboratory study to a more practical situation (field condition) is a challenging job. In these aspects some of the potential biostimulants are used under field conditions on the outstretched cultivated plants, therefore those initiatives are overcoming the cavity in between lab and field data of a single or mixtures of biostimulants (most often combined with fertilizers), which is also important.

Up to date, many fertilizers are formulated or fortified with biostimulants. For example, 5′-aminolevulinic acid (ALA)—the key precursor for tetrapyrroles biosynthesis (chlorophylls and hemes), for example, chlorophyll (C) *a*, C *b*, protoheme, heme, and siroheme, contains Mg or Fe at the active center. Many of the enzyme antioxidants (catalase, CAT; ascorbate peroxidase, APX; nitrate reductase, NR; nitrite reductase, NiR; sulfite reductase, SiR) are related with ALA metabolites and plays important roles in plants. Japanese researchers formulated liquid ALA fertilizer composing 5-ALA-HCl (0.3%), $MgNO_3$, urea Fe(III)-diethylene triamine pentaacetic acid (Fe-DTPA urea), and some other micronutrients like Mo, Mn, Cu, Zn, and B. The formulation is reported to exhibit increased growth, fruit quality and yield, along with abiotic stress (salt and low temperature) tolerance of crops (Iwai et al. 2004). This fertilizer increased paprika yield (Iwai et al. 2005). As the fertilizer contained ALA, Fe-DTPA as well as some trace plant nutrient elements, it showed positive interacting effects for promoting growth in several crops (Iwai et al. 2004). Moreover, combined Fe-DTPA urea and ALA also exhibited increased number of node and fresh biomass in paprika. Exogenous ALA stimulated the activity of NiR that stimulated the NO_3-N assimilation and increased the yield. The ALA fertilizer also increased fruit yield, which is attributed to ALA induced increased stomatal opening hours as well as improved photosynthesis (Iwai et al. 2004). Additionally ALA fertilizer also increased water uptake in hydroponically grown plants. In their study, Shahrekizad et al. (2015) examined the efficacy of EDTA-engrafted nanoparticles of Fe_3O_4 (Nano-Fe_3O_4-EDTA) on plants both through spray or soil amendment. The results demonstrated a dramatic increase of Fe content in plants (137% higher compared to the respective control) after applying Nano-Fe_3O_4-EDTA, which creates this novel compound and fertilizer, a committing candidate to prevent Fe deficiency crisis to plants.

Silicon (13 and 26 mM K_2SiO_3) enhanced the growth and synthesis of Chl in canola (*B. napus* L. cv. Okapi) plant by preventing the accumulation of Na^+

and MDA content under salinity in field condition (Bybordi 2016). Further, this research group also observed that Se (9 and 18 mM K_2SeO_4) mitigated the salinity effect under field conditions in canola (*B. napus* L. cv. Okapi) plant. By Se application plant height, siliqua per plant, seeds per siliqua, oil percentage, yield, and harvest index were increased along with decreasing respiration and the content of malondialdehyde (MDA), proline, and Na^+ concentration.

Another example of biostimulants is the root growth-promoting soil bacteria. These bacterial stimulating effects are over and over again demonstrated at laboratory but in the practical field conditions a little beneficial effects were found. For example, in annual maize crop soil scientists have observed the variability in interacting effects of plant cultivars and rhizobacteria and found that the bacterial populations changed over cultivars as well as in growing seasons (Aira et al. 2010; Philippot et al. 2013). They concluded that crop whether genotype-dependent alteration of soil microbes have a positive impact on plant health, is thus an unwrapped question. These strategies therefore motivate the scientists to a new path for developing new fertilizers and managing crops sustainably, by breeding advantageous endospheric and rhizospheric microbes (Rastall and Gibson 2015).

4 Biostimulants and Abiotic Stress Tolerance

The exogenous application of biostimulants depends on time and environmental conditions. For example, sometimes biostimulants need to apply in crop plant before the stress or during/after the stress. Occasionally, biostimulants are applied in the early stages of plant growth or during the developmental period (Kunicki et al. 2010). Based on the compound, biostimulants, which contain anti-stress compounds (e.g., proline, glutamic acid), should be applied during the stress condition, but which helps the synthesis of bioactive compounds must be applied before occurring stress. The application of biostimulants also can differ within plant species (Bulgari et al. 2019). Therefore, identification of proper time and dose of biostimulants is most important to know the stress tolerance mechanisms in Brassicaceae plants under different abiotic stress conditions.

4.1 Salinity

Salinity is a common and vital factor for restraining plant growth and biochemical metabolisms due to the creation of osmotic and oxidative stress as well as deficiency of plant nutrients (Kumari and Parida 2018; Mohsin et al. 2019). To cultivate the crops in the coastal area, the main restriction is salinity, where saline water often used for irrigation purpose (Lucini et al. 2015). Plant nutrient balance hampered by salt stress limits the nutrient accumulation from the soil and transfer into plant growing parts (Kumari and Parida 2018). The availability of Na^+ concentration is increased by salinity, which limited the uptake of other nutrients such as K^+, Ca^{2+}, and P and created problems to transfer within plant parts. Micronutrients (Fe, Cu, Zn Mo, and Mn) solubility is also affected by salinity, therefore, reduced their availability (Xu and Mou 2016; Hossain et al. 2019). Under saline conditions, the activity of CAT, superoxide dismutase (SOD), and peroxidase (POD) significantly reduced, as well as decreased the amount of total soluble proteins and phenolic compounds (Bano et al. 2012). Several plant metabolic processes also altered by salinity, for instance, respiration (Moud and Maghsoudi 2008), photosynthesis (Sayyad-Amin et al. 2016), nitrogen assimilation and biosynthesis of protein (Flores et al. 2004), and oxidative stress production by overproducing ROS (Hossain et al. 2019). Therefore, crop quality and production gradually reduced due to leaf chlorosis and growth retardation (Colla et al. 2010). The exogenous uses of biostimulants could be an effective way to reduce the salt toxicity of Brassicaceae plants (Table 2).

The inorganic compound Si is an important biostimulant to improve salt tolerance in Brassicaceae plant. Hashemi et al. (2010) declared that when grew canola (*B. napus* L. cv. Hayola) plant hydroponically to investigate the effect of Si (2 mM Na_2SiO_3) under salinity (150 mM NaCl), they observed that salt stress reduced the growth parameters and Chl content of canola by improving the accumulation of Na^+ ion, lignin content and peroxidation of lipid. On the other hand, Si treatment enhanced the plant growth and the content of Chl by inhibiting the higher Na^+ accumulation and production of higher lignin content and lipid peroxidation. In addition, Si also increased CAT and POD activity to reduce the overproduction of ROS, thus helped to maintain membrane integrity by removing negative effects of salinity. Another group of scientists also found similar types of results. They conducted an experiment under controlled conditions in growth chamber on hydroponically grown canola plants (*B. napus* L. cv. Hayola) by using 100 mM NaCl and with or without Si (1.7 mM Na_2SiO_3). They observed salt stress significantly reduced fresh weight of both shoot and root, leaf area, rate of transpiration, and stomatal density of plants by increasing higher concentration of Na^+ and Cl^-. Therefore, reduced the concentration of K^+, P, Fe, and B by increasing the content of H_2O_2 and lipid peroxidation. Consequently, the application of Si mitigated the oxidative stress by increasing antioxidant enzyme activity and decreasing the content of H_2O_2 and lipid peroxidation (Farshidi et al. 2012).

In 2016, Bybordi exhibited that Si (13 and 26 mM K_2SiO_3) enhanced the growth and synthesis of Chl in canola (*B. napus* L. cv. Okapi) plant by preventing the accumulation of Na^+ and MDA content under salinity in field condition. The potential role of Si to mitigate salinity damage also observed by Hasanuzzaman et al. (2018c), they studied on semi-hydroponically grown rapeseed (*B. napus* L. cv. BinaSarisha-3) seedlings by exogenous application of Si (1 mM SiO_2) under salinity (100 and 200 mM NaCl) condition. Salt stress created a negative effect by enhancing H_2O_2 and MDA content, as well as the toxicity of MG. Thus, reduced the activity of enzymes and disrupted the antioxidant defense and MG detoxification systems. While Si application upregulated the activities of enzymatic antioxidants (CAT; APX; monodehydroascorbate reductase, MDHAR; dehydroascorbate reductase, DHAR; glutathione reductase, GR; glutathione peroxidase, GPX; and glutathione *S*-transferase, GST)

Table 2 Beneficial effects of exogenously applied biostimulants in Brassicaceae plant under salinity and drought stresses

Biostimulants	Plant species	Stressors	Protective effects	References
Si; 2 mM Na_2SiO_3	*B. napus* L. cv. Hayola	Salinity; 150 mM NaCl	Increased growth parameters and chlorophyll (Chl) content Improved enzyme activity (CAT, POD) Reduced Na^+ accumulation, lignin production, and lipid peroxidation	Hashemi et al. (2010)
Si; 1.7 mM Na_2SiO_3	*B. napus* L. cv. Hayola	Salinity; 100 mM NaCl	Mitigated H_2O_2 content and lipid peroxidation Enhanced antioxidant enzyme activity	Farshidi et al. (2012)
Si; 13 and 26 mM K_2SiO_3	*B. napus* L. cv. Okapi	Salinity; Field condition	Improved plant growth and Chl synthesis Decreased Na^+ accumulation and MDA content	Bybordi (2016)
Si; 1 mM SiO_2	*B. napus* L. cv. BinaSarisha-3	Salinity; 100 and 200 mM NaCl	Upregulated non-enzymatic (GSH and AsA) and enzymatic (APX, DHAR, MDHAR, GR, GST, GPX, and CAT) antioxidant activity Improved Gly I and Gly II activities and mitigated MG toxicity	Hasanuzzaman et al. (2018c)
Se; 25 μM Na_2SeO_4	*B. napus* L. cv. BinaSarisha-3	Salinity; 100 and 200 mM NaCl	Improved activity of enzymatic antioxidants Mitigated peroxidation of lipids	Hasanuzzaman et al. (2011)

(continued)

Table 2 (continued)

Biostimulants	Plant species	Stressors	Protective effects	References
Se; 9 and 18 mM K_2SeO_4	*B. napus* L. cv. Okapi	Salinity; Field condition	Increased growth parameters Decreased respiration, MDA, proline, and Na^+ content	Bybordi (2016)
ALA; 0.228 mM	*B. napus* L. cv. ZS 758	Salinity; 100 and 200 mM NaCl	Improved plant growth, water content, and enzymatic and non-enzymatic antioxidant activity Inhibited overproduction of ROS and MDA content	Naeem et al. (2011)
Proline; 20 mM	*B. juncea* L. cv. Varuna and RH-30	Salinity; 28, 42, and 56 mM NaCl	Improved growth parameters, photosynthesis, and yield Enhanced antioxidant defense system	Wani et al. (2016)
Plant growth-promoting bacteria (PGPB)	*R. sativus* L. cv Cherry Belle	Salinity; 100 mM NaCl	Increased growth of plants, Chl and relative water content (RWC) Lowered EL and Na^+ uptake	Yildrim et al. (2008)
Arbuscular mycorrhizal fungi (AMF)	*B. juncea* L.	Salinity; 100 and 200 mM NaCl	Reduced peroxidation of lipids and decline proline content Increased enzymatic antioxidants (SOD, APX, GR) activity Improved growth, yield, and biomass as well as RWC and Chl content	Sarwat et al. (2016)

(continued)

Table 2 (continued)

Biostimulants	Plant species	Stressors	Protective effects	References
A. nodosum extract	*A. thaliana*	Salinity; 100 and 150 mM NaCl	Augmented plant biomass Regulated the expression of salinity stress responsible genes	Jithesh et al. (2019)
Si; 2.73 mM Na_2SiO_3	*B. napus* L. cv. Okapi	Drought; no irrigation	Increased dry biomass, RWC, and photosynthesis Increased SOD and POD activity and reduce peroxidation of lipids	Habibi (2014)
Si; 1 mM SiO_2	*B. napus* cv. BinaSarisha-3	Drought; 10 and 20% PEG	Increased enzymatic (CAT, MDHAR, APX, GST, GR, and DHAR) and non-enzymatic (GSH and AsA) antioxidants Improved RWC and Chl contents Reduced MDA and H_2O_2 content	Hasanuzzaman et al. (2018a)
Se; 25 μM Na_2SeO_4	*B. napus* L. cv. BinaSarisha-3	Drought; 10 and 20% PEG	Increased antioxidant and glyoxalase enzyme activities Reduced lipid peroxidation and ROS production	Hasanuzzaman and Fujita (2011)
ALA; 0.75, 7.5, and 75 μM	*B. napus* L. cv. ZS 758	Drought; 15% PEG	Improved plant biomass, photosynthetic pigment, and water potential, and antioxidant system Regulated higher MDA and ROS production	Liu et al. (2011)

(continued)

Table 2 (continued)

Biostimulants	Plant species	Stressors	Protective effects	References
Putrescine; 0.01 mM	*B. napus* L.	Drought; no irrigation	Increased the content of seed glucosinolates, oil erucic acid and proline Improved plant growth and oil quality	Ullah et al. (2012)
Proline, Glycinebetain, Trehalose; 0.5 mM	*B. rapa* L. cv. BARI Sharisha-15	Drought; 20% PEG	Improved plant growth, biomass, Chl content, and water balance Regulated antioxidant and glyoxalase defense system Reduced MDA, H_2O_2, and MG content	Bhuiyan et al. (2019)
Amino acid + *A. nodosum* extract	*B. oleracea* var. *italica*	Drought; no irrigation	Increased stomatal conductance, photosynthetic rate, and internal CO_2 concentration	Kałużewicz et al. (2017)
Amino acid + *A. nodosum* extract	*B. oleracea* var. *italica*	Drought; no irrigation	Increased photochemical fluorescence quenching and quantum yield Decreased non-photochemical quenching	Kałużewicz et al. (2018)

and non-enzymatic antioxidants (ascorbate, AsA and glutathione, GSH) to improve antioxidant defense system. Furthermore, Si increased the glyoxalase I (Gly I) and glyoxalase II (Gly II) activities to mitigate MG toxicity.

Selenium is another biostimulant, which also play a vital role to reduce salt toxicity. Hasanuzzaman et al. (2011) studied the phytoprotective role of Se (25 μM Na_2SeO_4) under salt stress (100 and 200 mM NaCl) in rapeseed (*B. napus* L. cv. BinaSarisha-3) seedlings. They found that salinity enhanced the MDA and H_2O_2 contents by reducing antioxidant enzyme activity, while Se treatment improved the enzyme activity and mitigated the lipid peroxidation damage. Bybordi (2016) also observed that Se (9 and 18 mM K_2SeO_4) mitigated the salinity effect under field condition in canola (*B. napus* L. cv. Okapi) plant. Selenium application enhanced plant height, number of siliqua, number of seed per siliqua, oil percentage, yield, and harvest index by decreasing respiration and the content of MDA, proline, and Na^+ concentration.

Nitrogen-containing compound ALA (0.228 mM) alleviated the harmful effect of salinity (100 and 200 mM NaCl) in rapeseed (*B. napus* L. cv. ZS 758) plant. Researchers focused that salt stress limited the shoot and root growth of plant and reduced the leaf water content, whereas increased the content of MDA and electrolytic conductivity, as well as ROS production. But the exogenous application of ALA increased plant growth and water content, as well as enzymatic and non-enzymatic antioxidant activity by inhibiting overproduction of ROS and MDA content (Naeem et al. 2011). Foliar application of proline (20 mM) also mitigated the salinity-induced (28, 42, and 56 mM NaCl) negative impact in *B. juncea* L. (cv. Varuna and RH-30) plant. Researchers found that growth parameters, photosynthesis, and yield of both varieties are significantly affected by salinity. However, the exogenous application of proline improved plant growth by alleviating salt toxicity (Wani et al. 2016).

Biostimulant, PGPB showed the protective role in mineral uptake and growth physiology of radish (*R. sativus* L. cv Cherry Belle) plant under salt stress. Yildrim et al. (2008) reported that the growth of radish plant negatively affected by salinity, but seeds treated with PGPB suspension, improved plant growth parameters such as fresh and dry weight of shoot and root, relative water content (RWC), and Chl content by reducing Na^+ uptake. Furthermore, PGPB reduced the electrolyte leakage (EL), whereas enhanced the uptake of mineral content except Na^+ in the plants. Sarwat et al. (2016) studied on *B. juncea* L. plant by using arbuscular mycorrhizal fungi (AMF) under salt stress (100 and 200 mM NaCl). They observed that salt stress, altered the plant morphology (growth, water content, and Chl content), biomolecules, and antioxidant defense by increasing lipid peroxidation and proline content. However, AMF treated plant increased the antioxidant defense system by regulating the function of common antioxidant enzymes (such as SOD, APX, and GR) and ultimately improved plant growth parameters. Alga (*Ascophyllum nodosum*) is an effective biostimulant to mitigate the salinity effect. The extract of *A. nodosum* was used in *A. thaliana* plant under saline condition (100 and 150 mM NaCl) to study the salt tolerance mechanisms. *A. nodosum* extract increased plant biomass about 50% and induced salt tolerance by controlling the salt stress-related genes expression (Jithesh et al. 2019). So, biostimulants have protective mechanisms against oxidative stress, in order to reduce the osmotic and ionic toxicity.

4.2 Drought

Drought is the major abiotic stress because of the diversified damaging character of plants (Cao et al. 2017). The limitation of both ground and surface water is the main cause to create drought stress (Bhuiyan et al. 2019). Drought stress significantly influenced the plant physiological functions such as changing the transpiration and photosynthetic rates resulting in reduced growth, biomass, and yield (Hasanuzzaman et al. 2016). Insufficient water disturbed the normal function of stomata; consequently hampered energy balance, this leads to higher ROS production (Hasanuzzaman et al. 2012a, b). Reactive oxygen species can destroy the plant biomolecules, thus leads to metabolic disorders in the plant cell (Nahar et al. 2015). However, plants also possess various organized mechanisms for ROS detoxification and antioxidant defense is the main way to counter the ROS successfully (Hasanuzzaman et al. 2019a). Exogenous application of biostimulants could be the most effective strategy to reduce water deficiency in crop plants (Table 2), because these can improve the biosynthesis of osmolytes and plant hormones to reduce drought injury (Calvo et al. 2014).

Inorganic biostimulant, Si can improve the plant growth and mechanisms under drought stress. A result showed that the supplementation of Si (2.73 mM, Na_2SiO_3) increased the shoot and root DW, RWC, and net assimilation rate in canola plant under drought stress. Silicon application enhanced photosynthetic rate by increasing the absorbance of CO_2. Lower lipid peroxidation also found after applying Si in canola plant under drought stress, because of the enhancement of SOD and POD activities significantly (Habibi 2014). The role of Si in mitigating drought toxicity was also studied by Hasanuzzaman et al. (2018a). They studied the function of Si (1 mM SiO_2) on drought-induced (10 and 20%; polyethylene glycol, PEG) tolerance in rapeseed (*B. napus* cv. BinaSarisha-3) plant. Silicon protected photosynthetic component and reduced lipid peroxidation (MDA) and ROS (H_2O_2) by increasing enzymatic (CAT, APX, GST, MDHAR, DHAR, and GR), and non-enzymatic (AsA and GSH) antioxidants under drought stress. In addition, Si also significantly improved the glyoxalase system and RWC in drought-induced plant. Another inorganic biostimulants, selenium (Se) has also a regulatory role to improve tolerance mechanisms under drought stress. Hasanuzzaman and Fujita (2011) observed that Se (25 μM, Na_2SeO_4) application improved the growth and morphology of drought-induced (10% and 20% PEG) rapeseed (*B. napus* cv. BinaSarisha-3) plant. The antioxidant (enzymatic and non-enzymatic) activities effectively increased by Se application, whereas the MDA and H_2O_2 content significantly reduced. Selenium also regulated glyoxalase system by improving the activity of Gly I and Gly II enzymes under drought stress.

The negative effect of water deficit can be reduced by the exogenous application ALA at low doses. A result was found that drought stress (15% PEG) developed negative effects on rapeseed (*B. napus* L. cv. ZS758) seedlings by reducing plant biomass, photosynthetic pigment, and water potential due to higher lipid peroxidation and ROS production. However, the application of ALA (0.75, 7.5, and 75 μM) improved enzymatic antioxidant activities and the content of non-enzymatic antioxidant, consequently reduced the MDA and H_2O_2 contents, as a result, enhanced growth and biomass of plants (Liu et al. 2011). The application of polyamine also acts as biostimulants to make tolerance in plants under water deficit condition. Ullah et al. (2012) found that the exogenous application of putrescine (Put; 0.01 mM) recovered the growth and quality of oil in drought-induced canola (*B. napus* L.) plant. They observed that water deficiency reduced soluble protein, Chl and RWC while augmented the seed glucosinolates, oil erucic acid and proline level. But the application of Put mitigated the undesirable effect of drought, thus improved growth and oil quality of the canola plant. Osmolytes are also effective biostimulants, which has an adverse effect on growth and physiology under drought stress. Bhuiyan et al.

(2019) investigated the role of three common osmolytes; proline (0.5 mM), glycine-betaine (GB, 0.5 mM) and trehalose (Tre, 0.5 mM) under drought stress (20% PEG) on the seedlings of rapeseed (*B. rapa* L. cv. BARI Sharisha-15). They reported that drought stress reduced the plant growth parameters, biomass, content of Chl, and RWC by increasing ROS production, lipid peroxidation, and MG content. However, the application of these three osmolytes separately in drought-induced plants significantly mitigated the harmful effect by regulating the antioxidant defense and MG systems.

Amino acids and *A. nodosum* both are widely used biostimulants to enhance drought tolerance in plants. An experiment was carried out by Kałużewicz et al. (2017), to explore the effect of amino acids and *A. nodosum* in two broccoli (*B. oleracea* var. *italica*) cultivars under drought stress. They observed that the biostimulants application increased the stomatal conductance, photosynthetic rate, and internal CO_2 concentration in drought-stressed plants. The next year, Kałużewicz et al. (2018) again observed that the application of amino acids and *A. nodosum* improved drought tolerance in broccoli plants by increasing photochemical fluorescence quenching and quantum yield of electron transport while decreasing the non-photochemical fluorescence quenching.

4.3 Extreme Temperature

The most detrimental environmental factor to reduce plant growth, metabolism and physiological activities is considered the extreme temperature. Low temperature, decreases plant metabolism and leads to inhibition of the photosystem activity, as well as cell membrane and phospholipid layers were also damaged (Bulgari et al. 2019). Similarly, high temperatures could alter the plant biochemical reactions, which are involved in plant growth and productivity (Hasanuzzaman et al. 2013a). For enzyme activity and cell structural integrity the optimum temperature is 30–40 °C, but above this range temperature affected the photosynthesis and respiration mechanisms (Bulgari et al. 2019), and produced toxic compounds like ROS that leads to oxidative damage (Hasanuzzaman et al. 2013b). High temperature inhibits the germination of seed and crop yield by interfering with various growth and reproductive phases of plants. To cope with the negative impact of extreme temperature, plants possess many tolerance strategies such as synthesized protective solutes to regulate cell homeostasis and structure, regulated stomatal activity to reduce water loss, and expressed various genes to activate enzyme activity and maintained antioxidant defense system (Hasanuzzaman et al. 2013a). The application of biostimulants could be one of the effective protective mechanisms to develop heat tolerance in Brassicaceae plants (Table 3).

The protective role of Se was investigated by Hasanuzzaman et al. (2014b) in rapeseed (*B. napus* L. cv. BinaSarisha-3) under high temperature (38 °C) stress. The results showed that Se (25 μM Na_2SeO_4) reduced the high temperature-induced damages. At high temperature, photosynthetic pigment and RWC were decreased,

Table 3 Beneficial effects of exogenously applied biostimulants in Brassicaceae plants under extreme temperature and waterlogging stresses

Biostimulants	Plant species	Stressors	Protective effects	References
Se; 25 μM Na_2SeO_4	*B. napus* L. cv. BinaSarisha-3	High temperature; 38 °C	Improved Chl and RWC Reduced MDA, H_2O_2, and MG content Enhanced antioxidant and glyoxalase related enzyme activities	Hasanuzzaman et al. (2014b)
L-proline; 5, 10, 15, and 20 mM	*B. napus* L.	Low temperature; 4 °C	Improved cell membrane stability, endogenous proline, and soluble sugar content	Jonytiene et al. (2012)
Proline, Glycinebetain; 2 mM	*B. rapa* L. cv. BARI Sharisha-13	High temperature; 38 °C	Reduced H_2O_2, LOX, and MDA level Increased AsA-GSH redox pool and related enzyme activities Reduced MG content and increased glyoxalase-related enzyme activities	Hasanuzzaman et al. (2019b)
A. nodosum extracts	*A. thaliana*	Freezing temperature; −7.5 °C (in vitro) and −5.5 °C (in vivo)	Reduced the expression of chlorophyllase genes *AtCHL1* and *AtCHL2* Cold related gene expression reregulation (*COR15A*, *RD29A*, and *CBF3*)	Rayirath et al. (2009)
PGPB	*B. napus* L. cv. Westar	Flooding	Increased tolerance mechanisms and shoot biomass	Farwell et al. (2007)
PGPB	*B.napus* L. cv. Hayola 401	Waterlogging	Increased dry weight, shoot and root length Enhanced enzyme activities (SOD, CAT, and POX) Reduced lipid peroxidation and ethylene production	Habibzadeh et al. (2012)

whereas increased lipid peroxidation, amount of ROS, MG, and proline content. However, the supplementation of Se increased Chl content and water status and alleviated oxidative damage and glyoxalase toxicity by decreasing MDA, H_2O_2, and MG content. Selenium enhanced the antioxidant and glyoxalase systems involved enzyme activities, as well as improved the content of non-enzymatic antioxidant (AsA and GSH).

An experiment was shown that the application of proline (5, 10, 15, and 20 mM) improved cold stress (4 °C) tolerance in winter rapeseed shoots. Cold stress damaged cell membranes and reduced endogenous proline and soluble sugar content. But exogenous proline improved cell membrane stability with increasing endogenous proline under cold stress (Jonytiene et al. 2012). Exogenous proline also can mitigate the harmful effect of high temperature stress. Hasanuzzaman et al. (2019b) studied the protective role of proline (2 mM) and GB (2 mM) in rapeseed (*B. napus* L.) seedlings under high temperature (38 °C). High temperature increased the H_2O_2, lipoxygenase (LOX), and MDA level. However, exogenous proline and GB increased the AsA-GSH redox pool and related enzyme activities, thus improved antioxidant defense system and reduced lipid peroxidation and ROS production. Proline and GB application also regulated the glyoxalase system by reducing MG content and increasing glyoxalase enzyme activities.

The extract of brown seaweed (*A. nodosum*) has an effective role to enhance plant tolerance under extreme temperatures. Rayirath et al. (2009) investigated the role of *A. nodosum* extracts in *A. thaliana* plant for tolerance mechanisms under freezing temperature (–7.5 °C, in vitro and –5.5 °C, in vivo). They observed that freezing temperature severely damaged the plant tissue and caused leaf chlorosis. However, *A. nodosum* treated plant reduced CHL damage by inhibiting the expression of *AtCHL1* and *AtCHL2* chlorphyllase genes. *A. nodosum* also enhanced freezing tolerance by regulating the expression of cold-related genes (such as *COR15A*, *RD29A*, and *CBF3*) in *A. thaliana* plant.

4.4 Waterlogging

Among the abiotic stresses, waterlogging is a significantly detrimental factor for plant growth because it can reduce not only plant development and biochemical reactions, but also total yield of crop plants (Yu and Chen 2013; Anee et al. 2019). The waterlogging condition depleted the oxygen (O_2) level, which causes a multifarious effect on the physiology and anatomy of plants (Capon et al. 2009; Zou et al. 2014). The deficiency of O_2 created hypoxia, thus caused leaf chlorosis, stunted plant growth, delayed maturity and finally yield loss occurred (Chang et al. 2000). Waterlogging inhibited the respiration rate in plant root, therefore, reduced the energy level in root cells (Anee et al. 2019). Furthermore, waterlogging stress also destroyed the root permeability, thus hydraulic conductivity decreased in plant roots (Else et al. 2001). In addition, waterlogging destructed Chl synthesis by inhibiting the transport of electron in photosystem II (Ashraf 2012), which leads to the generation of a higher amount of ROS (Ashraf 2009). To minimize these harmful effects plants showed some protective mechanisms itself, among them the formation of aerenchyma is the most important one (Thomas et al. 2005). The exogenous biostimulants also can improve the tolerance mechanisms under waterlogging stress (Table 3).

The application of PGPB (*Pseudomonas putida* UW4) enhanced tolerance mechanisms in canola (*B. napus* L. cv. Westar) plant under flooding conditions. Flood stress reduced the shoot biomass of canola plant, while *P. putida* treated plants improved shoot biomass by increasing tolerance mechanisms (Farwell et al. 2007). Another experiment also reported that PGPB can improve antioxidant defense mechanisms to reduce oxidative stress. Habibzadeh et al. (2012) observed that waterlogging stress reduced dry weight, shoot and root length, and enzyme activities (SOD, CAT, and POX) in canola (*B. napus* L. cv. Hayola 401) plants, whereas increased the production of lipid peroxidation and ethylene. However, the application PGPB altered the plant mechanisms and enhanced plant growth by alleviating waterlogging damage.

4.5 *Metal/Metalloid(s)*

The rapid growth of the world population is accelerating industrial insurgency and urbanization, which discharge a substantial quantity of toxic metal/metalloid(s) to the environment including soil, water, and air. Therefore, toxic metal/metalloid(s) are one of the worst yield-reducing causes of world agriculture. They create oxidative stress in plants through the over generation of ROS and MG (Hasanuzzaman et al. 2017a; Mahmud et al. 2019). Recently biostimulants are gaining the attention of plant scientists and farmers as it plays a significant role to enhance tolerance in different crops against abiotic stress, including toxic metal/metalloid(s) stress (Van Oosten et al. 2017; Bulgari et al. 2019). Different biostimulants confirmed their positive function against a range of metal/metalloid(s) stress by improving the antioxidative status of plant. Choudhary et al. (2012) carried out an experiment using epibrassinolide and spermidine in *R. sativus* L. plant against Cr stress. They reported that both epibrassinolide and spermidine lessened Cr-induced oxidative damage caused by ROS generation. The activity of SOD also increased due to their action that helped *R. sativus* L. to confer stress tolerance against Cr toxicity. Mahmud et al. (2017b) checked the performance of γ-aminobutyric acid (GABA) in *B. juncea* L. under Cr toxicity and observed almost similar results. They confirmed that GABA increased the content of non-enzymatic antioxidants (AsA and GSH) and enzymatic antioxidants (SOD, CAT, APX, MDHAR, DHAR, GR, GPX, Gly I, and Gly II) activities, which help plant to decrease ROS generation, MG content and LOX activity. Selenium, another prominent biostimulant, also works against metal-induced ROS mitigation and upregulation of antioxidants. Hasanuzzaman et al. (2012a, b) exposed Se pretreated *B. napus* L. seedlings to Cd stress and found that Se is able to counteract Cd toxicity by improving AsA and GSH contents and GSH/GSSG ratio with increasing antioxidant enzyme activities. Later they also got an almost similar result from another study by using Si as a protectant/biostimulant (Hasanuzzaman et al. 2017b, c). Ehsan et al. (2014) observed that citric acid plays a significant role against Cd stress in *B. napus* L. seedlings. Citric acid increased antioxidant enzyme activities and decreased production of ROS as well as El and MDA. Afterward, citric acid proved its positive efficiency under different study with Cd and Pd in Brassicaceae plants (Shakoor et al. 2014; Zaheer et al. 2015; Mahmud et al. 2018). Ali et al. (2014) reported that ALA can alleviate Pb-induced toxicity in *B. napus.* Maleic acid and GABA are also helpful for mitigation of Cr-induced toxicity in *B. juncea* L. (Mahmud et al. 2017a, b). Arsenic, a toxic metalloid, damaged *B. napus* L. plant but supplementation of methyl jasmonate recovers plants from damages (Farooq et al. 2016). They recorded that methyl jasmonate increased enzyme activities and gene expression of key antioxidants, including SOD, APX, CAT, etc., and reduced the generation of toxic ROS. Recently Vishwakarma et al. (2020) suggested that indole acetic acid and plant growth-promoting rhizobacterium are effective to mitigate toxic metal stress (silver nanoparticles, AgNPs) by improving the status of antioxidant defense system. In Table 4, we have included detail information about the abovementioned reports.

4.6 Nutrient Deficiencies and Toxicity

Nutrient deficiency or toxicity disturbs the normal functioning of plants and reduces growth and productivity by creating oxidative stress. Biostimulants have the capability to augment nutrient uptake in the deficient condition through diverse techniques (du Jardin 2015). For example, they are capable of modifying soil property, solubility of nutrients, roots morphology, which helps to increase stress tolerance of plants (Halpern et al. 2015; Toscano et al. 2019). Moreover they have the ability to lessen excessive nutrient-induced cellular damages (Van Oosten et al. 2017; Bulgari et al. 2019). A number of experiments have been carried out to explore the function of biostimulants against nutrient deficiencies and toxicity. Alam et al. (2007) conducted an experiment with *B. juncea* L. under Ni toxicity. They supplemented 28-homobrassinolide (HBL) as a foliar spray and observed that it neutralized the lethal effect of Ni by improving antioxidant enzyme activities. Similarly, Fariduddin et al. (2009) reported that HBL increased CAT and SOD activity and reduced H_2O_2 content in excess Cu affected plants. Recently, Ren et al. (2019) observed that PGPB can alleviate nutrient toxicity (excessive Cu) by decreasing thiobarbituric acid-reactive substances (TBARS) and improving the status of the antioxidant defense system. On the other hand, Navarro-León et al. (2016) found that ethylene precursors such as aminocyclopropane-1-carboxylic acid, salicylic acid, abscisic acid, and jasmonic acid, and CKs might be involved in Zn uptake of the *B. oleracea* plants in Zn deficient condition. The information related to findings are listed down in Table 4.

5 Biostimulants and Yield Improvement

Due to climate change, severe abiotic stresses like salinity, drought, temperature extreme, waterlogging, metal/metalloid(s) toxicity, etc., are becoming major yield-limiting factors for the world agriculture (Drobek et al. 2019). Thus progresses in crop production are struggling at present. The main target of contemporary agricultural practices is to trim down inputs without dropping crop growth, quality as well as productivity (Kocira 2019). Supplementation of different biostimulants to the plants is a great technique, which helps to enhance crop yield and quality particularly during diverse stress situations (Bulgari et al. 2019). Many studies confirmed the positive performance of biostimulants in a different group of plants in aspect to yield and

Table 4 Beneficial effects of exogenous application of biostimulants in Brassicaceae plant under different toxic metal/metalloid(s) and nutrient deficiencies or toxicities stresses

Biostimulants	Plant species	Stressors	Protective effects	References
Epibrassinolide, 10^{-9} M and spermidine, 1 mM	*R. sativus* L.	Toxic metal: Cr (1.2 mM K_2CrO_4,VI)	Reduced MDA content and ROS generation Increased activity of SOD	Choudhary et al. (2012)
Selenium, 50 and 100 μM Na_2SeO_4	*B. napus* L. cv. BinaSarisha-3	Toxic metal: Cd (0.5 and 1.0 mM $CdCl_2$)	Increased AsA and GSH and their redox ratio Enhanced the activities antioxidant and glyoxalase Reduced ROS and lipid peroxidation	Hasanuzzaman et al. (2012a, b)
5-aminolevulinic acid (ALA), 25 mg L^{-1})	*B. napus* L.	Toxic metal: Pb (100 and 400 μM)	Decreased ROS generation Increased the performance of antioxidant	Ali et al. (2014)
Citric acid, 2.5 mM	*B. napus* L.	Toxic metal: Cd (10 and 50 μM $CdCl_2$)	lessened MDA and H_2O_2 production Decreased electrolyte leakage (EL) Enhanced the activities of antioxidant enzymes	Ehsan et al. (2014)
Citric acid, 2.5 mM	*B. napus* L.	Toxic metal: Pb (50 and 100 μM)	Decreased MDA and H_2O_2 production Decreased membrane damage Increased APX, CAT and SOD activity	Shakoor et al. (2014)

(continued)

Table 4 (continued)

Biostimulants	Plant species	Stressors	Protective effects	References
Citric acid, 2.5 mM	*B. napus* L.	Toxic metal: Pb (50 and 100 μM)	Reduced MDA and H_2O_2 production Enhanced protein concentration Increased APX, CAT and SOD activity	Zaheer et al. (2015)
Methyl jasmonate, 1 μM	*B. napus* L.	Toxic metalloids: As (200 μM $NaAsO_2$)	Decreased ROS (H_2O_2 and $OH^•$) generation Increased enzyme activities and *SOD*, *APX*, and *CAT* gene expression	Farooq et al. (2016)
Silicon, 1.0 mM SiO_2	*B. napus* L. cv. BinaSarisha-3	Toxic metal: Cd (0.5 and 1.0 mM $CdCl_2$)	Enhanced CAT DHAR,MDHAR, GR, and APX activities Increased glyoxalase activity Decreased MDA and H_2O_2 levels	Hasanuzzaman et al. (2017b, c)
γ-aminobutyric acid, 0.125 mM	*B. juncea* L. cv. BARI Sharisha-11	Toxic metal: Cr (0.15 and 0.3 mM K_2CrO_4)	Increased the content of AsA and GSH Enhanced antioxidant enzyme activities Increased Gly I and Gly II activities Decreased ROS generation, MG content and LOX activity	Mahmud et al. (2017b)

(continued)

Table 4 (continued)

Biostimulants	Plant species	Stressors	Protective effects	References
Maleic acid, 0.25 mM	*B. juncea* L. cv. BARI Sharisha-11	Toxic metal: Cr (0.15 and 0.3 mM K_2CrO_4)	Augmented AsA-GSH cycle Improved performance of the glyoxalase system Reduced oxidative stress trough lessening the production of ROS and MG Increased phytochelatin (PC) content	Mahmud et al. (2017c)
Citric acid, 0.5 and 1.0 mM	*B. juncea* L. cv. BARI Sharisha-11	Toxic metal: Cd (0.5 and 1.0 mM $CdCl_2$)	Increased content of AsA and GSH Enhanced activities of antioxidant enzymes Enhanced activities of GPX, SOD, and CAT Improve performance of the glyoxalase system Decreased ROS and MG production Enhanced PC content	Mahmud et al. (2018)
Indole acetic acid and plant growth-promoting rhizobacterium	*B. juncea* L.	Toxic metal: Ag (Ag nanoparticle)	Decreased MDA and H_2O_2 production Alleviated AgNPs-induced stress by improving the status of antioxidants	Vishwakarma et al. (2020)
28-Homobrassinolide (HBR), 10^{-8} M	*B. juncea* L. cv. T-59	Nutrient toxicity: Ni, (50 or 100 μM $NiCl_2$)	Reduced the lethal effect of Ni Enhanced CAT activity	Alam et al. (2007)
HBR, 10^{-4} M	*B. juncea* L.	Nutrient toxicity: Cu, 50, 100 and 150 mg $CuSO_4 \cdot 5H_2O$ kg^{-1} soil	Decreased ROS (H_2O_2) content Enhanced SOD and CAT activities	Fariduddin et al. (2009)

(continued)

Table 4 (continued)

Biostimulants	Plant species	Stressors	Protective effects	References
PGPB: *Burkholderia cepacia* J62, *Pseudomonas thivervalensis* Y1-3-9, and *Microbacterium oxydans* JYC17	*B. napus* L.	Nutrient toxicity: Cu, 1068.25 mg kg^{-1}	Decreased thiobarbituric acid-reactive substances (TBARS) Lessened POD activity Increased AsA and GSH content	Ren et al. (2019)

quality but research on the member of the Brassicaceae family on yield is little. Gajc-Wolska et al. (2013) performed a comparative study among three cultivars of broccoli: Monaco, Parthenon, and Ironman to observe the growth and yield performance among them. They also used seaweed extracts (Goëmar BM 86 2 L ha^{-1}) as biostimulators and reported that the yield and quality had been increased. Gugala et al. (2018) applied three types of biostimulators e.g., Tytanit®, Asahi®SL, Silvit® to the *B. napus* L. and found higher production over control plants.

6 Legislation of Plant Biostimulants

The regulation of biostimulants in the European Union and the United States is still very complicated. In EU, two routes are used to place any biostimulant in the market—the European pesticides law and the national fertilizer regulations. In European countries, any types of biostimulants could be submitted as plant protection products ('PPPs') according to EC regulation No 1107/2009, as biostimulants could influence the physiological processes not likely to plant nutrients, therefore biostimulants might be categorized as plant protection agents from a regulatory point of view. According to this regulation, natural and formulated products as well as microorganisms—all could be covered. Point to be noted that, substances interacting with basic plant physiology, for example, herbicides and growth regulators are also registered under PPP regulation.

The other route—fertilizer route regulation [(EC) No 2003/2003] amended by another article [(EC) No 1107/2008] is very restricted and biostimulants cannot be included. This article defines fertilizer mainly as nutrient providing substances. But later some consideration on fertilizer additives came out; as a result, urease or nitrification inhibitors like substances achieved access to the market by this regulation. Furthermore, some biostimulants could act as enhancers of fertilizer efficiency, thus this special character offers regulatory advances and biostimulants' inclusion under EU fertilizer law. However, this option cannot be followed for all biostimulants.

The US situation is almost similar to the EU: where still biostimulants have no recognized definition. There is no harmonization among 50 states. At present, the fertilizer laws are used for placing biostimulants in the market at the state levels. But they are hopeful and working hard with all the stakeholders and federal agencies for improving the laws for biostimulants. Already the functions for officials of the American Association of Plant Food Control are underlined, especially regarding the formal definitions and recognition of fertilizer categories (AAPFCO 2012). Yet we have to wait to see the regulation policies of biostimulants by the Environmental Protection Agency (EPA) and the United States Department of Agriculture (USDA). Some scenario seems that some biostimulants will be covered by EPA jurisdiction, whereas others will be placed as soil amendments at specific states. Furthermore, USDA could list down and acknowledge some biostimulants as soil health-promoting products via the Natural Resources Conservation Service (NRCS) agency.

7 Opportunities and Challenges

Due to climate change, crop production is decreasing significantly. One of many important tasks of plant science experts is the exploration of suitable gateways in improving plant performance under adverse conditions. The use of biostimulants in the crop field is becoming popular because they not only improve crop growth and yield but also confer stress tolerance. Some of the biostimulants act as plant nutrients, growth regulators, and phytoprotectors. Therefore, there is a great scope of using these stimulants as regular inputs for plants. Some of the biostimulants are of organic origins and these can stimulate plant growth without any negative effect. However, some inorganic stimulants warrant further research on their synthesis, application dose, residual effects, etc. Finding a crop-specific suitable dose for biostimulants is a challenging task because every biostimulant has a specific behavior and mode of action. Manipulating genes that are responsible for the biosynthesis of biostimulants or their precursor is another good option in improving plant health. However, this approach is still unsuccessful due to the lack of proper research and the legislative process. A coordinated action for the potential use of biostimulants is needed to get the maximum benefit.

8 Conclusion and Ways Ahead

This chapter discusses the recent progress on biostimulants, for improving tolerance against abiotic stresses in crops especially of the Brassicaceae family. In their life tenure, *Brassica* crops may get exposed to several abiotic stresses, singly or in combination, which reduce yield as well as crop quality. Therefore, biostimulants might be a suitable option to augment crop production. Biostimulant application might improve water use efficiency, essential nutrients metabolism, subsequent yield, and quality enhancement as well as abiotic stress (salinity, drought, extreme temperature, metal/metalloids, etc.) tolerance. A considerable number of substances are included in the biostimulants; therefore, it is difficult to justify the efficiency of a single biostimulating substance or a heterogeneous mixture as well as the rhizobacteria. Therefore, a product may produce different effects depending on the genetic variability, variety of cultivars, as well as the stress severity. Therefore, the actual mode of action for any specific biostimulant should be evaluated based on morphophysiological and biochemical attributes of plants.

Moreover, agricultural use of biostimulants requires locally adapted solutions, as well as field evaluation considering ecological and environmental aspects. Companies developing biostimulants may take initiative to test the efficacy of their product and can integrate them into new agrosystem as one element. In addition, stakeholders like farmers, researchers, and other regulatory authorities should come in a decision that biostimulants can be a tool for profitable and sustainable plant productions.

Acknowledgements We thank Abdul Awal Chowdhury Masud, Lecturer, Department of Agronomy and Tasnim Farha Bhuiyan, Lecturer, Department of Agricultural Botany for assisting in collecting secondary data for writing this manuscript.

References

AAPFCO (Association of American Plant Food Control Officials Inc.) (2012) Stabilized fertilizer. Official Publication No. 65. West lafayette, Ind.: AAPFCO

Aguado-Santacruz GA, Moreno-Gómez B, Rascón-Cruz Q, Aguirre-Mancilla C, Espinosa-Solís JA, González-Barriga CD (2014) Biofertilizers as complements to synthetic and organic fertilization. In: López-Valdes F, Fernández-Luqueno F (eds) Components, uses in agriculture and environmental impacts. Nova Science Publishers Inc, New York, US pp 155–180

Aira M, Gómez-Brandón M, Lazcano C, Bååth E, Domïnguez J (2010) Plant genotype strongly modifies the structure and growth of maize rhizosphere microbial communities. Soil Biol Biochem 42:2276–2281

Alam MM, Hayat S, Ali B, Ahmad A (2007) Effect of 28-homobrassinolide treatment on nickel toxicity in *Brassica juncea*. Photosynthetica 45(1):139–142

Ali B, Xu X, Gill RA, Yang S, Ali S, Tahir M, Zhou W (2014) Promotive role of 5-aminolevulinic acid on mineral nutrients and antioxidative defense system under lead toxicity in *Brassica napus*. Ind Crops Prod 52:617–626

Anee TI, Nahar K, Rahman A, Mahmud JA, Bhuiyan TF, Alam MU, Fujita M, Hasanuzzaman M (2019) Oxidative damage and antioxidant defense in *Sesamum indicum* after different waterlogging durations. Plants 8:196. https://doi.org/10.3390/plants8070196

Aremu AO, Masondo NA, Rengasamy KR, Amoo SO, Gruz J, Bíba O, Šubrtová M, Pěnčík A, Novák O, Doležal K, Van Staden J (2015) Physiological role of phenolic biostimulants isolated from brown seaweed *Ecklonia maxima* on plant growth and development. Planta 241:1313–1324

Ashraf M (2009) Biotechnological approach of improving plant salt tolerance using antioxidants as markers. Biotech Adv 27:84–93

Ashraf MA (2012) Waterlogging stress in plants: A review. Afr J Agril Res 7:1976–1981

Bano S, Ashraf M, Akram NA, Al-Qurainy F (2012) Regulation in some vital physiological attributes and antioxidative defense system in carrot (*Daucus carota* L.) under saline stress. J Appl Bot Food Qual Angew Bot 85:105–115

Battacharyya D, Babgohari MZ, Rathor P, Prithiviraj B (2015) Seaweed extracts as biostimulants in horticulture. Sci Hortic 196:39–48

Bechtold U, Field B (2018) Molecular mechanisms controlling plant growth during abiotic stress. J Exp Bot 69:2753–2758

Bhuiyan TF, Ahamed KU, Nahar K, Mahmud JA, Bhuyan MHMB, Anee TI, Fujita M, Hasanuzzaman M (2019) Mitigation of PEG-induced drought stress in rapeseed (*Brassica rapa* L.) by exogenous application of osmolytes. Biocatal Agric Biotechnol 20:101197. https://doi.org/10.1016/j.bcab.2019.101197

Billard V, Etienne P, Jannin L, Garnica M, Cruz F, Garcia-Mina JM, Yvin JC, Ourry A (2014) Two biostimulants derived from algae or humic acid induce similar responses in the mineral content and gene expression of winter oilseed rape (*Brassica napus* L.). J Plant Growth Regul 33:305–316

Bulgari R, Cocetta G, Trivellini A, Vernieri P, Ferrante A (2015) Biostimulants and crop responses: a review. Biol Agric Hort 31:1–17

Bulgari R, Franzoni G, Ferrante A (2019) Biostimulants application in horticultural crops under abiotic stress conditions. Agronomy 9:306. https://doi.org/10.3390/agronomy9060306

Bybordi A (2016) Influence of zeolite, selenium and silicon upon some agronomic and physiologic characteristics of canola grown under salinity. Commun Soil Sci Plant Anal 47:832–850

Calvo P, Nelson L, Kloepper JW (2014) Agricultural uses of plant biostimulants. Plant Soil 383:3–41

Canellas LP, Olivares FL, Aguiar NO, Jones DL, Nebbioso A, Mazzei P, Piccolo A (2015) Humic and fulvic acids as biostimulants in horticulture. Sci Hortic 196:15–27

Cao Y, Luo Q, Tian Y, Meng F (2017) Physiological and proteomic analyses of the drought stress response in *Amygdalus mira* (Koehne) *Yü et Lu* roots. BMC Plant Biol 17:53–69

Capon SJ, James CS, Williams L, Quinn GP (2009) Responses to flooding and drying in seedlings of a common Australian desert floodplain shrub: *Muehlenbeckia florulenta* Meisn. Environ Exp Bot 66:178–185

Chang WP, Huang L, Shen M, Webster C, Burlingame AL, Roberts JK (2000) Protein synthesis and tolerance of anoxia in root tips of maize seedlings acclimated to a low oxygen environment and identification of protein by mass spectrometry. Plant Physiol 122:295–318

Chen THH, Murata N (2011) Glycinebetaine protects plants against abiotic stress: mechanisms and biotechnological applications. Plant, Cell Environ 34:1–20

Chojnacka K, Michalak I, Dmytryk A, Wilk R, Gorecki H (2015) Innovative natural plant growth biostimulants. In: Sinha S, Pant KK, Bajpai S, Govil JN (eds) Fertilizer technology: II Biofertilizer, Studium Press LLC, Houston, TX, US, pp 451–489

Choudhary SP, Kanwar M, Bhardwaj R, Yu J-Q, Tran L-SP (2012) Chromium stress mitigation by polyamine-brassinosteroid application involves phytohormonal and physiological strategies in *Raphanus sativus* L. PLoS ONE 7(3):e33210. https://doi.org/10.1371/journal.pone.0033210

Colla G, Rouphael Y, Leonardi C, Bie Z (2010) Role of grafting in vegetable crops grown under saline conditions. Sci Hortic 127:147–155

Colla G, Nardi S, Cardarelli M, Ertani A, Lucini L, Canaguier R, Rouphael Y (2015) Protein hydrolysates as biostimulants in horticulture. Sci Hortic 196:28–38

Dobbelaere S, Croonenborghs A, Thys A, Broek AV, Vanderleyden J (1999) Phytostimulatory effect of *Azospirillum brasilense* wild type and mutant strains altered in IAA production on wheat. Plant Soil 212:155–164

Drobek M, Frac M, Cybulska J (2019) Plant biostimulants: Importance of the quality and yield of horticultural crops and the improvement of plant tolerance to abiotic stress—a review. Agronomy 9:335. https://doi.org/10.3390/agronomy9060335

du Jardin P (2012) The science of plant biostimulants—a bibliographic analysis. Ad hoc Study Report to the European Commission DG ENTR. http://ec.europa.eu/enterprise/sectors/chemicals/files/fertilizers/finaLreport_bio_2012_en.pdf.

du Jardin P (2015) Plant biostimulants: definition, concept, main categories and regulation. Sci Hortic 196:3–14

EBIC (2012) European biostimulants industry council and biostimulants in brief. http://www.biostimulants.eu/.

Ehsan S, Ali S, Noureen S, Mehmood K, Farid M, Ishaque W, Shakoo MB, Rizwan M (2014) Citric acid assisted phytoremediation of Cd by *Brassica napus* L. Ecotoxicol Environ Saf 106:164–172

Else MA, Coupland D, Dutton L, Jackson MB (2001) Decreased root hydraulic conductivity reduces leaf water potential, initiates stomatal closure and slows leaf expansion in flooded plants of castor oil (*Riccinus communis*) despite diminished delivery of ABA from the roots to shoots in the xylem sap. Physiol Plant 111:46–54

Ertani A, Pizzeghello D, Francioso O, Sambo P, Sanchez-Cortes S, Nardi S (2014) *Capsicum chinensis* L. growth and nutraceutical properties are enhanced by biostimulants in a long-term period: chemical and metabolomic approaches. Front Plant Sci 5:375. https://doi.org/10.3389/fpls.2014.00375

Fariduddin Q, Yusuf M, Hayat S, Ahmad A (2009) Effect of 28-homobrassinolide on antioxidant capacity and photosynthesis in *Brassica juncea* plants exposed to different levels of copper. Environ Exp Bot 66:418–424

Farooq MA, Gill RA, Islam F, Ali B, Liu H, Xu J, He S, Zhou W (2016) Methyl jasmonate regulates antioxidant defense and suppresses arsenic uptake in *Brassica napus* L. Front Plant Sci 7:468. https://doi.org/10.3389/fpls.2016.00468

Farshidi M, Abdolzadeh A, Sadeghipour HR (2012) Silicon nutrition alleviates physiological disorders imposed by salinity in hydroponically grown canola (*Brassica napus* L.) plants. Acta Physiol Plant 34:1779–1788

Farwell AJ, Vesely S, Nero V, Rodriguez H, McCormack K, Shah S, Dixon DG, Glick BR (2007) Tolerance of transgenic canola plants (*Brassica napus*) amended with plant growth-promoting bacteria to flooding stress at a metal-contaminated field site. Environ Pollut 147:540–545

Filatov VP (1944) Tissue therapy in ophthalmology. Am Rev Sov Med 2:53–66

Fiorentino N, Ventorino V, Woo SL, Pepe O, De Rosa A, Gioia L, Romano I, Lombardi N, Napolitano M, Rouphael Y (2018) *Trichoderma*-based biostimulants modulate rhizosphere microbial populations and improve N uptake efficiency, yield and nutritional quality of leafy vegetables. Front Plant Sci 9:743. https://doi.org/10.3389/fpls.2018.00743

Flores P, Botella MÁ, Cerdá A, Martínez V (2004) Influence of nitrate level on nitrate assimilation in tomato (*Lycopersicon esculentum*) plants under saline stress. Can J Bot 82:207–213

Gajc-Wolska J, Spiżewski T, Grabowska A (2013) The effect of seaweed extracts on the yield and quality parameters of broccoli (*Brassica oleracea* var. *cymosa* l.) in open field production. Acta Hortic 1009:83–89

Gómez-Merino FC, Trejo-Téllez LI (2015) Biostimulant activity of phosphite in horticulture. Sci Hortic 196:82–90

Goñi O, Fort A, Quille P, McKeown PC, Spillane C, O'Connell S (2016) Comparative transcriptome analysis of two *Ascophyllum nodosum* extract biostimulants: same seaweed but different. J Agric Food Chem 64:2980–2989

Gugala M, Sikorska A, Findura P, Kapela K, Malaga-Tobola U, Zarzecka K, Domanski L (2018) Effect of selected plant preparations containing biologically active compounds on winter rape (*Brassica napus* L.) yielding. Appl Ecol Env Res 17:2779–2789

Habibi G (2014) Silicon supplementation improves drought tolerance in canola plants. Russ J Plant Physiol 61:784–791

Habibzadeh F, Sorooshzadeh A, Pirdashti H, Modarres-Sanavy SAM (2012) A comparison between foliar application and seed inoculation of biofertilizers on canola (*Brassica napus* L.) grown under waterlogged conditions. Aust J Crop Sci 6:1435–1440

Habibzadeh F, Sorooshzadeh A, Pirdashti H, Modarres-Sanavy SAM (2013) Alleviation of waterlogging damage by foliar application of nitrogen compounds and tricyclazole in canola. Aust J Crop Sci 7:401–406

Halpern M, Bar-Tal A, Ofek M, Minz D, Muller T, Yermiyahu U (2015) The use of biostimulants for enhancing nutrient uptake. In: Sparks DL (ed) Advances in Agronomy. Academic press, Cambridge, pp 141–174

Hasanuzzaman M, Fujita M (2011) Selenium pretreatment upregulates the antioxidant defense and methylglyoxal detoxification system and confers enhanced tolerance to drought stress in rapeseed seedlings. Biol Trace Elem Res 143:1758–1776

Hasanuzzaman M, Hossain MA, Fujita M (2011) Selenium-induced up-regulation of the antioxidant defense and methylglyoxal detoxification system reduces salinity-induced damage in rapeseed seedlings. Biol Trace Elem Res 143:1704–1721

Hasanuzzaman M, Hossain MA, da Silva JAT, Fujita M (2012a) Plant responses and tolerance to abiotic oxidative stress: antioxidant defense is a key factor. In: Bandi V, Shanker AK, Shanker C, Mandapaka M (eds) Crop Stress and its management: perspectives and strategies. Springer, Berlin, pp 261–316

Hasanuzzaman M, Hossain MA, Fujita M (2012b) Exogenous selenium pretreatment protects rapeseed from cadmium-induced oxidative stress by upregulating antioxidant defense and methylglyoxal detoxification systems. Biol Trace Elem Res 149:248–261

Hasanuzzaman M, Nahar K, Alam M, Roychowdhury R, Fujita M (2013a) Physiological, biochemical, and molecular mechanisms of heat stress tolerance in plants. Int J Mol Sci 14:9643–9684

Hasanuzzaman M, Nahar K, Fujit M (2013b) Extreme temperature responses, oxidative stress and antioxidant defense in plants. In: Vahdati K, Leslie C (eds) Abiotic stress—plant responses and applications in agriculture. InTech Open, Rijeka, pp 169–205

Hasanuzzaman M, Nahar K, Alam MM, Fujita M (2014a) Modulation of antioxidant machinery and the methylglyoxal detoxification system in selenium-supplemented *Brassica napus* seedlings confers tolerance to high temperature stress. Biol Trace Elem Res 161:297–307

Hasanuzzaman M, Nahar K, Fujita M (2014b) Silicon and selenium: two vital trace elements that confer abiotic stress tolerance to plants. In: Ahmad P, Rasool S (eds) Emerging technologies and management of crop stress tolerance. Academic Press, Cambridge, pp 377–422

Hasanuzzaman M, Nahar K, Mahmud JA, Ahmad P, Fujita M (2016) Nitric oxide: a Jack of all trades for drought stress tolerance in plants. In: Ahmad P (ed) Water stress and crop plants: a sustainable approach. Wiley, West Sussex, pp 628–648

Hasanuzzaman M, Nahar K, Hossain MS, Anee TI, Parvin K, Fujita M (2017a) Nitric oxide pretreatment enhances antioxidant defense and glyoxalase systems to confer PEG-induced oxidative stress in rapeseed. J Plant Interact 12:323–331

Hasanuzzaman M, Nahar K, Hossain MS, Mahmud JA, Rahman A, Inafuku M, Oku H, Fujita M (2017b) Coordinated actions of glyoxalase and antioxidant defense systems in conferring abiotic stress tolerance in plants. Int J Mol Sci 18:200. https://doi.org/10.3390/ijms18010200

Hasanuzzaman M, Nahar K, Anee TI, Fujita M (2017c) Exogenous silicon attenuates cadmium-induced oxidative stress in *Brassica napus* L. by modulating AsA-GSH pathway and glyoxalase system. Front Plant Sci 8:1061. https://doi.org/10.3389/fpls.2017.01061

Hasanuzzaman M, Nahar K, Anee TI, Fujita M (2017d) Exogenous silicon attenuates cadmium-induced oxidative stress in *Brassica napus* L. by modulating AsA-GSH pathway and glyoxalase system. Front Plant Sci 8:1061. https://doi.org/10.3389/fpls.2017.01061

Hasanuzzaman M, Hossain MS, Bhuyan MB, Mahmud JA, Nahar K, Fujita M (2018a) The role of sulfur in plant abiotic stress tolerance: molecular interactions and defense mechanisms. In: Hasanuzzaman M, Fujita M, Oku H, Nahar K, Hawrylak-Nowak B (eds) Plant nutrients and abiotic stress tolerance. Springer, Singapore, pp 221–252

Hasanuzzaman M, Nahar K, Anee TI, Khan MIR, Fujita M (2018b) Silicon-mediated regulation of antioxidant defense and glyoxalase systems confers drought stress tolerance in *Brassica napus* L. S Afr J Bot 115:50–57

Hasanuzzaman M, Nahar K, Rohman MM, Anee TI, Huang Y, Fujita M (2018c) Exogenous silicon protects *Brassica napus* plants from salinity-induced oxidative stress through the modulation of AsA-GSH pathway, thiol-dependent antioxidant enzymes and glyoxalase systems. Gesunde Pflanzen 70:185–194

Hasanuzzaman M, Bhuyan MHMB, Anee TI, Parvin K, Nahar K, Mahmud JA, Fujita M (2019a) Regulation of ascorbate-glutathione pathway in mitigating oxidative damage in plants under abiotic stress. Antioxidants 8:384. https://doi.org/10.3390/antiox8090384

Hasanuzzaman M, Nahar K, Khan MIR, Mahmud JA, Alam MM, Fujita M (2019b) Regulation of reactive oxygen species metabolism and glyoxalase systems by exogenous osmolytes confers thermotolerance in *Brassica napus*. Gesunde Pflanzen. https://doi.org/10.1007/s10343-019-00476-4

Hashemi A, Abdolzadeh A, Sadeghipour HR (2010) Beneficial effects of silicon nutrition in alleviating salinity stress in hydroponically grown canola, *Brassica napus* L., plants. Soil Sci Plant Nutr 56:244–253

Hawrylak-Nowak B, Hasanuzzaman M, Wójcik M (2019) Biostimulation and biofortification of crop plants–new challenges for modern agriculture. Acta Agrobot 72:1777. https://doi.org/10.5586/aa.1777

Hervé JJ (1994) Biostimulant, a new concept for the future and prospects offered by chemical synthesis and biotechnologies. Comptes-Rendus del'Académie d'Agriculture de France 80:91–102

Hinojosa L, González JA, Barrios-Masias FH, Fuentes F, Murphy KM (2018) Quinoa abiotic stress responses: a review. Plants 7:106. https://doi.org/10.3390/plants7040106

Hossain MS, Hasanuzzaman M, Sohag MMH, Bhuyan MHMB, Fujita M (2019) Acetate-induced modulation of ascorbate: glutathione cycle and restriction of sodium accumulation in shoot confer salt tolerance in *Lens culinaris* Medik. Physiol Mol Biol Plants 25:443–455

Huang J, Gu M, Lai Z, Fan B, Shi K, Zhou Y-H, Yu J-Q, Chen Z (2010) Functional analysis of the Arabidopsis *PAL* gene family in plant growth, development, and response to environmental stress. Plant Physiol 153:1526–1538

Iwai K, Kuramochi H, Takeuchi Y (2004) Effects of 5-aminolevulinic acid, iron, and urea on growth of Japanese lawngrass seedlings. The 33rd annual meeting of the JSTS, pp. 68–69

Iwai K, Saito A, van Leeuwen J, Tanaka T, Takeuchi Y (2005) A new functional fertilizer containing 5-aminolevulinic acid promoted hydroponically-grown vegetables in the Netherlands. Acta Hort 697:351–355

Jannin L, Arkoun M, Etienne P, Laine P, Goux D, Garnica M, Fuentes M, San Francisco M, Baigorri R, Cruz F, Houdusse F, Gracia-Mina J, Yvin J, Ourry A (2012) *Brassica napus* growth is promoted by *Ascophyllum nodosum* (L.) Le Jol. seaweed extract: microarray analysis and physiological characterization of N, C, and S metabolisms. J Plant Growth Regul 32:31–52

Jia C, Yu X, Zhang M, Liu Z, Zou P, Ma J, Xu Y (2019) Application of melatonin-enhanced tolerance to high-temperature stress in cherry radish (*Raphanus sativus* L. var. *radculus* pers). J Plant Growth Regul. https://doi.org/10.1007/s00344-019-10006-1

Jithesh MN, Shukla PS, Kant P, Joshi J, Critchley AT, Prithiviraj B (2019) Physiological and transcriptomics analyses reveal that *Ascophyllum nodosum* extracts induce salinity tolerance in *Arabidopsis* by regulating the expression of stress responsive genes. J Plant Growth Regul 38:463–478

Jonytiene V, Burbulis N, Kupriene R, Blinstrubiene A (2012) Effect of exogenous proline and de-acclimation treatment on cold tolerance in *Brassica napus* shoots cultured in vitro. J Food Agric Environ 10:327–330

Kałużewicz A, Krzesiński W, Spiżewski T, Zaworska A (2017) Effect of biostimulants on several physiological characteristics and chlorophyll content under drought stress and re-watering. Not Bot Hortic Agrobot 45:197–202

Kałużewicz A, Bączek-Kwinta R, Krzesiński W, Spiżewski T, Zaworska A (2018) Effect of biostimulants on chlorophyll fluorescence parameters of broccoli (*Brassica oleracea* var. *italica*) under drought stress and rewatering. Acta Sci Pol Hortorum Cultus 17:97–106

Khan MN, Zhang J, Luo T, Liu J, Rizwan M, Fahad S, Xu Z, Hu L (2019) Seed priming with melatonin coping drought stress in rapeseed by regulating reactive oxygen species detoxification: antioxidant defense system, osmotic adjustment, stomatal traits and chloroplast ultrastructure perseveration. Ind Crops Prod 140:111597. https://doi.org/10.1016/j.indcrop.2019.111597

Kocira S (2019) Effect of amino acid biostimulant on the yield and nutraceutical potential of soybean. Chil J Agr Res 79:17–25

Kumari A, Parida AK (2018) Metabolomics and network analysis reveal the potential metabolites and biological pathways involved in salinity tolerance of the halophyte *Salvadora persica*. Environ Exp Bot 148:85–99

Kunicki E, Grabowska A, Sękara A, Wojciechowska R (2010) The effect of cultivar type, time of cultivation, and biostimulant treatment on the yield of spinach (*Spinacia oleracea* L.). Folia Hortic 22:9–13

Liu D, Pei ZF, Naeem MS, Ming DF, Liu HB, Khan F, Zhou WJ (2011) 5–Aminolevulinic acid activates antioxidative defence system and seedling growth in *Brassica napus* L. under water–deficit stress. J Agron Crop Sci 197:284–295

Lovatt CJ (2015) Use of a natural metabolite to increase crop production. US Patent No. US 9,603,366 B2. https://patentimages.storage.googleapis.com/66/b4/72/78ab7e35f3a8da/US9603366.pdf

Lucini L, Rouphael Y, Cardarelli M, Canaguier R, Kumar P, Colla G (2015) The effect of a plant-derived biostimulant on metabolic profiling and crop performance of lettuce grown under saline conditions. Sci Hortic 182:124–133

Mahmud JA, Hasanuzzaman M, Nahar K, Rahman A, Fujita M (2017a) Relative tolerance of different species of *Brassica* to cadmium toxicity: coordinated role of antioxidant defense and glyoxalase systems. Plant Omics 10:107–117

Mahmud JA, Hasanuzzaman M, Nahar K, Rahman A, Hossain MS, Fujita M (2017b) γ-aminobutyric acid (GABA) confers chromium stress tolerance in *Brassica juncea* L. by modulating the antioxidant defense and glyoxalase systems. Ecotoxicology 26:675–690

Mahmud JA, Hasanuzzaman M, Nahar K, Rahman A, Hossain SM, Fujita M (2017c) Maleic acid assisted improvement of metal chelation and antioxidant metabolism confers chromium tolerance in *Brassica juncea* L. Ecotoxicol Environ Saf 144:216–226

Mahmud JA, Hasanuzzaman M, Nahar K, Bhuyan MHMB, Fujita M (2018) Insights into citric acid-induced cadmium tolerance and phytoremediation in *Brassica juncea* L.: coordinated functions of metal chelation, antioxidant defense and glyoxalase systems. Ecotoxicol Environ Saf 147:990–1001

Mahmud JA, Bhuyan MHMB, Anee TI, Nahar K, Fujita M, Hasanuzzaman M (2019) Reactive oxygen species metabolism and antioxidant defense in plants under metal/metalloid stress. In: Hasanuzzaman M, Hakeem KR, Nahar K, Alharby HF (eds) Plant abiotic stress tolerance. Switzerland, Springer, Cham., pp 221–257

Mariani L, Ferrante A (2017) Agronomic management for enhancing plant tolerance to abiotic stresses—drought, salinity, hypoxia, and lodging. Horticulturae 3:52. https://doi.org/10.3390/horticulturae3040052

Martinez-Esteso MJ, Vilella-Antón M, Sellés-Marchart S, Martínez-Márquez A, Botta-Català A, Piñol-Dastis R, Bru-Martínez R (2016) A DIGE proteomic analysis of wheat flag leaf treated with TERRA-SORB® foliar, a free amino acid high content biostimulant. J Integr OMICS 6:9–17

Mohsin SM, Hasanuzzaman M, Bhuyan MHMB, Parvin K, Fujita M (2019) Exogenous tebuconazole and trifloxystrobin regulates reactive oxygen species metabolism toward mitigating salt-induced damages in cucumber seedling. Plants 8:428. https://doi.org/10.3390/plants8100428

Mou B (2011) Improvement of horticultural crops for abiotic stress tolerance: an introduction. Hort Science 46:1068–1069

Moud AM, Maghsoudi K (2008) Salt stress effects on respiration and growth of germinated seeds of different wheat (*Triticum aestivum* L.) cultivars. World J Agric Sci 4:351–358

Naeem MS, Rasheed M, Liu D, Jin ZL, Ming DF, Yoneyama K, Takeuchi Y, Zhou WJ (2011) 5-Aminolevulinic acid ameliorates salinity-induced metabolic, water-related and biochemical changes in *Brassica napus* L. Acta Physiol Plant 33:517–528

Nahar K, Hasanuzzaman M, Alam MM, Fujita M (2015) Roles of exogenous glutathione in antioxidant defense system and methylglyoxal detoxification during salt stress in mung bean. Biol Plant 59:745–756

Navarro-León E, Albacete A, Torre-Gonzáleza A, Ruiz JM, Blascoa B (2016) Phytohormone profile in Lactuca sativa and Brassica oleracea plants grown under Zn deficiency 130:85–89

Petrozza A, Santaniello A, Summerer S, Di Tommaso G, Di Tommaso D, Paparelli E, Piaggesi A, Perata P, Cellini F (2014) Physiological responses to Megafol® treatments in tomato plants under drought stress: a phenomic and molecular approach. Sci Hortic 174:185–192

Philippot L, Raaijmakers JM, Lemanceau P, Van Der Putten WH (2013) Going back to the roots: the microbial ecology of the rhizosphere. Nat Rev Microbiol 11:789–799

Pichyangkura R, Chadchawan S (2015) Biostimulant activity of chitosan in horticulture. Sci Hortic 196:49–65

Przybysz A, Gawronska H, Gajc-wolska J (2014) Biological mode of action of a nitrophenolates-based biostimulant: case study. Front Plant Sci 5:713. https://doi.org/10.3389/fpls.2014.00713

Rastall RA, Gibson GR (2015) Recent developments in prebiotics to selectively impact beneficial microbes and promote intestinal health. Curr Opin Biotechnol 32:42–46

Rayirath P, Benkel B, Hodges DM, Allan-Wojtas P, MacKinnon S, Critchley AT, Prithiviraj B (2009) Lipophilic components of the brown seaweed, *Ascophyllum nodosum*, enhance freezing tolerance in *Arabidopsis thaliana*. Planta 230:135–147

Ren X-M, Guo S-J, Tian W, Chen Y, Han H, Chen E, Li B-L, Li Y-Y, Chen Z-J (2019) Effects of Plant Growth-Promoting Bacteria (PGPB) inoculation on the growth, antioxidant activity, Cu uptake, and bacterial community structure of rape (*Brassica napus* L.) grown in cu-contaminated agricultural soil. Front Microbiol 10:1455. https://doi.org/10.3389/fmicb.2019.01455

Rouphael Y, Franken P, Schneider C, Schwarz D, Giovannetti M, Agnolucci M, De Pascale S, Bonini P, Colla G (2015) Arbuscular mycorrhizal fungi act as biostimulants in horticultural crops. Sci Hortic 196:91–108

Ruzzi M, Aroca R (2015) Plant growth-promoting rhizobacteria act as biostimulants in horticulture. Sci Hortic 196:124–134

Santaniello A, Giorgi FM, Di Tommaso D, Di Tommaso G, Piaggesi A, Perata P (2013) Genomic approaches to unveil the physiological pathways activated in *Arabidopsis* treated with plant-derived raw extracts. Acta Hortic 1009:161–174

Sarwat M, Hashem A, Ahanger M.A, Abd_Allah EF, Alqarawi AA, Alyemeni MN, Ahmad P, Gucel S (2016) Mitigation of NaCl stress by arbuscular mycorrhizal fungi through the modulation of osmolytes, antioxidants and secondary metabolites in mustard (*Brassica juncea* L.) plants. Front Plant Sci 7: 869. https://doi.org/10.3389/fpls.2016.00869

Savvas D, Ntatsi G (2015) Biostimulant activity of silicon in horticulture. Sci Hortic 196:66–81

Sayyad-Amin P, Jahansooz MR, Borzouei A, Ajili F (2016) Changes in photosynthetic pigments and chlorophyll-a fluorescence attributes of sweet-forage and grain sorghum cultivars under salt stress. J Biol Phys 42:601–620

Shabala L, Mackay A, Tian Y, Jacobsen S-E, Zhou D, Shabala S (2012) Oxidative stress protection and stomatal patterning as components of salinity tolerance mechanism in quinoa (*Chenopodium quinoa*). Physiol Plant 146:26–38

Shahrekizad M, Gholamalizadeh AA, Mir N (2015) EDTA-coated Fe_3O_4 nanoparticles: a novel biocompatible fertilizer for improving agronomic traits of sunflower (*Helianthus annuus*). J Nanostruct 5:117–127

Shakoor MB, Ali S, Hameed A, Farid M, Hussain S, Yasmeen T et al (2014) Citric acid improves lead (Pb) phytoextraction in *Brassica napus* L. by mitigating Pb-induced morphological and biochemical damages. Ecotoxicol Environ Saf 109:38–47

Sharma A, Li X, Lim YP (2014) Comparative genomics of Brassicaceae crops. Breed Sci 64:3–13

Shubha K, Mukherjee A, Kumari M, Tiwari K, Meena VS (2017) Bio-stimulants: an approach towards the sustainable vegetable production. In: Meena V, Mishra P, Bisht J, Pattanayak A (eds) Agriculturally important microbes for sustainable agriculture. Springer, Singapore, pp 259–277

Singhal P, Jan AT, Azam M, Haq QMR (2016) Plant abiotic stress: a prospective strategy of exploiting promoters as alternative to overcome the escalating burden. Front Life Sci 9:52–63

Sofo A, Nuzzaci M, Vitti A, Tataranni G, Scopa A (2014) Control of biotic and abiotic stresses in cultivated plants by the use of biostimulant microorganisms. In: Ahmad P, Wani MR, Azooz MM, Phan Tran LS (eds) Improvement of crops in the era of climatic changes. Springer, New York, US, pp 107–11

Thomas AL, Guerreiro SM, Sodek L (2005) Aerenchyma formation and recovery from hypoxia of the flooded root system of nodulated soybean. Ann Bot 96:1191–1198

Tian S, Lu L, Xie R, Zhang M, Jernstedt J, Hou D, Ramsier C, Brown P (2015) Supplemental macronutrients and microbial fermentation products improve the uptake and transport of foliar applied zinc in sunflower (*Helianthus annuus* L.) plants. Studies utilizing micro X-ray florescence. Front Plant Sci 5:808. https://doi.org/10.3389/fpls.2014.00808

Toscano S, Romano D, Massa D, Bulgari R, Franzoni G, Ferrante A (2019) Biostimulant applications in low input horticultural cultivation systems. Italus Hortus 25:27–36

Traon D, Amat L, Zotz F, du Jardin P (2014) A legal framework for plant biostimulants and agronomic fertiliser additives in the EU. Report to the European Commission, DG Enterprise & Industry, Arcadia International, p 115

Ullah F, Bano A, Nosheen A (2012) Effects of plant growth regulators on growth and oil quality of canola (*Brassica napus* L.) under drought stress. Pak J Bot 44:1873–1880

Van Oosten MJ, Pepe O, De Pascale S, Silletti S, Maggio A (2017) The role of biostimulants and bioeffectors as alleviators of abiotic stress in crop plants. Chem Biol Technol Agric 4:5. https://doi.org/10.1186/s40538-017-0089-5

Vishwakarma K, Singh VP, Prasad SM, Chauhan DK, Tripathi DK, Sharma S (2020) Silicon and plant growth promoting rhizobacteria differentially regulate AgNP-induced toxicity in *Brassica juncea*: implication of nitric oxide. J Hazard Mater 390:121806.

Wang H, Wang H, Shao H, Tang X (2016) Recent advances in utilizing transcription factors to improve plant abiotic stress tolerance by transgenic technology. Front Plant Sci 7:67. https://doi.org/10.3389/fpls.2016.00067

Wani AS, Ahmad A, Hayat S, Tahir I (2016) Is foliar spray of proline sufficient for mitigation of salt stress in *Brassica juncea* cultivars? Environ Sci Pollut Res 23:13413–13423

Warwick SI, Francis A, Al-Shehbaz IA (2006) Brassicaceae: species checklist and database on CD-ROM. Plant Syst Evol 259:249–258

Xu C, Mou B (2016) Responses of spinach to salinity and nutrient deficiency in growth, physiology, and nutritional value. J Am Soc Hortic Sci 141:12–21

Yakhin OI, Lubyanov AA, Yakhin IA, Brown PH (2017) Biostimulants in plant science: a global perspective. Front Plant Sci 7:2049. https://doi.org/10.3389/fpls.2016.02049

Yamauchi Y (2018) Integrated chemical control of abiotic stress tolerance using biostimulants. In: Andjelkovic V (ed) Plant, abiotic stress and responses to climate change. InTech Open, Rijeka, pp 133–143

Yildrim E, Donmez MF, Turan M (2008) Use of bioinoculants in ameliorative effects on radish plants under salinity stress. J Plant Nutr 31:2059–2074

Yu M, Chen GY (2013) Conditional QTL mapping for waterlogging tolerance in two RILs populations of wheat. Springerplus 2:245. https://doi.org/10.1186/2193-1801-2-245

Zaheer IE, Ali S, Rizwan M, Farid M, Shakoor MB, Gill RA et al (2015) Citric acid assisted phytoremediation of copper by *Brassica napus* L. Ecotoxicol Environ Saf 120:310–317

Zhao G, Zhao Y, Yu X, Kiprotich F, Han H, Guan R, Wang R, Shen W (2018) Nitric oxide is required for melatonin-enhanced tolerance against salinity stress in rapeseed (*Brassica napus* L.) seedlings. Int J Mol Sci 19:1912. https://doi.org/10.3390/ijms19071912

Zou X, Hu C, Zeng L, Cheng Y, Xu M, Zhang X (2014) A comparison of screening methods to identify waterlogging tolerance in the field in *Brassica napus* L. during Plant Ontogeny. PLoS ONE 9: e89731. https://doi.org/10.1371/journal.pone.0089731

The manufacturer's authorised representative in the EU is Springer Nature Customer Service Centre GmbH, Europaplatz 3, 69115 Heidelberg, Germany. If you have any concerns regarding our products, please contact ProductSafety@springernature.com

Printed and bound by CPI Group (UK) Ltd, Croydon, CR0 4YY
30/09/2026
02225559-0019